Lecture Notes in Computer Science 9235

Commenced Publication in 1973
Founding and Former Series Editors:
Gerhard Goos, Juris Hartmanis, and Jan van Leeuwen

Advanced Research in Computing and Software Science

Subline of Lecture Notes in Computer Science

More information about this series at http://www.springer.com/series/7407

Giuseppe F. Italiano · Giovanni Pighizzini
Donald T. Sannella (Eds.)

Mathematical Foundations of Computer Science 2015

40th International Symposium, MFCS 2015
Milan, Italy, August 24–28, 2015
Proceedings, Part II

 Springer

Editors

Giuseppe F. Italiano
Università di Roma "Tor Vergata"
Rome
Italy

Donald T. Sannella
University of Edinburgh
Edinburgh
UK

Giovanni Pighizzini
Università degli Studi di Milano
Milan
Italy

ISSN 0302-9743   ISSN 1611-3349   (electronic)
Lecture Notes in Computer Science
ISBN 978-3-662-48053-3   ISBN 978-3-662-48054-0   (eBook)
DOI 10.1007/978-3-662-48054-0

Library of Congress Control Number: 2015945159

LNCS Sublibrary: SL1 – Theoretical Computer Science and General Issues

Springer Heidelberg New York Dordrecht London

Printed on acid-free paper

Springer-Verlag GmbH Berlin Heidelberg is part of Springer Science+Business Media
(www.springer.com)

Preface

The series of MFCS symposia has a long and well-established tradition of encouraging high-quality research into all branches of theoretical computer science. Its broad scope provides an opportunity to bring together researchers who do not usually meet at specialized conferences. The first symposium was held in 1972. Until 2012 MFCS symposia were organized on a rotating basis in Poland, the Czech Republic, and Slovakia. The 2013 edition took place in Austria, the 2014 edition in Hungary, while in 2015 MFCS was organized for the first time in Italy.

The 40th International Symposium on Mathematical Foundations of Computer Science (MFCS 2015) was held in Milan during August 24–28, 2015. The scientific program of the symposium consisted of five invited talks and 81 contributed papers.

To celebrate the 40th edition of the conference, a special invited talk was given by:

– Zoltán Ésik (University of Szeged, Hungary)

This talk was sponsored by the European Association of Theoretical Computer Science (EATCS). The other invited talks were given by:

– Anindya Banerjee (IMDEA Software Institute, Spain)
– Paolo Boldi (University of Milan, Italy)
– Martin Kutrib (University of Giessen, Germany)
– Yishay Mansour (Microsoft Research, Hertzelia and Tel Aviv University)

We are grateful to all invited speakers for accepting our invitation and for their excellent presentations at the symposium.

The 81 contributed papers were selected by the Program Committee (PC) out of a total of 201 submissions. All submitted papers were peer reviewed and evaluated on the basis of originality, quality, significance, and presentation. To support the selection process, approximatively 600 reviews were written by PC members with the help of external experts.

As is the MFCS tradition, a Best Paper Award and a Best Student Paper Award sponsored by EATCS were assigned. The PC decided to assign these awards to the following papers:

– "Strong Inapproximability of the Shortest Reset Word" by Paweł Gawrychowski and Damian Straszak (Best Paper Award)
– "Maximum Minimal Vertex Cover Parameterized by Vertex Cover" by Meirav Zehavi (Best Student Paper Award)

We thank all authors who submitted their work for consideration to MFCS 2015. We wish to thank all PC members and external reviewers for their competent and timely handling of the submissions. The success of the scientific program is due to their hard work. During the selection process and for preparing these proceedings, we used the EasyChair conference management system, which provided excellent support.

Owing to the large number of accepted papers, the proceedings of the conference were divided into two volumes on a thematical basis: *Logic, Semantics, Automata and Theory of Programming* (Vol. I) and *Algorithms, Complexity and Games* (Vol. II).

We gratefully acknowledge the support of the University of Milan (Università degli Studi di Milano, Dipartimento di Informatica) and EATCS. Special thanks for the local organization are due to Violetta Lonati (University of Milan). We also thank Bruno Guillon (University Paris-Diderot, France) for the website design and maintenance.

June 2015
Giuseppe F. Italiano
Giovanni Pighizzini
Don Sannella

Conference Organization

Program Committee Chairs

Giuseppe F. Italiano, co-chair	University of Rome "Tor Vergata", Italy
Giovanni Pighizzini, chair	University of Milan, Italy
Donald Sannella, co-chair	University of Edinburgh, UK

Program Committee

Hee-Kap Ahn	POSTECH, Korea
Andris Ambainis	University of Latvia
Marie-Pierre Béal	University of Paris-Est Marne-la-Vallée, France
Lars Birkedal	Aarhus University, Denmark
Jarosław Byrka	University of Wrocław, Poland
Luis Caires	University of Lisbon "Nova", Portugal
Bruno Codenotti	CNR Pisa, Italy
Adriana Compagnoni	Stevens Institute of Technology, USA
Erzsébet Csuhaj-Varjú	Eötvös Loránd University, Budapest, Hungary
Artur Czumaj	University of Warwick, UK
Rocco de Nicola	IMT Lucca, Italy
Martin Dietzfelbinger	Technical University of Ilmenau, Germany
Devdatt Dubashi	Chalmers, Sweden
Amos Fiat	Tel Aviv University, Israel
Enrico Formenti	Nice Sophia Antipolis University, France
Pierre Fraigniaud	CNRS and University Paris Diderot, France
Matt Franklin	UC Davis, USA
Loukas Georgiadis	University of Ioannina, Greece
Jan Holub	Czech Technical University in Prague, Czech Republic
Markus Holzer	University of Giessen, Germany
Martin Lange	University of Kassel, Germany
Massimo Lauria	KTH Royal Institute of Technology, Sweden
Inge Li Gørtz	Technical University of Denmark
Alberto Marchetti-Spaccamela	University of Rome "La Sapienza", Italy
Elvira Mayordomo	University of Zaragoza, Spain

Pierre McKenzie	University of Montréal, Canada
Friedhelm Meyer auf der Heide	University of Paderborn, Germany
Prakash Panangaden	McGill University, Canada
Dana Pardubská	Comenius University, Bratislava, Slovakia
Kunsoo Park	Seoul National University, Korea
Alexander Rabinovich	Tel Aviv University, Israel
Rajeev Raman	University of Leicester, UK
Jean-Francois Raskin	University of Brussels "Libre", Belgium
Liam Roditty	Bar-Ilan University, Israel
Marie-France Sagot	Inria and University of Lyon 1, France
Piotr Sankowski	University of Warsaw, Poland
Philippe Schnoebelen	LSV, CNRS and ENS Cachan, France
Marinella Sciortino	University of Palermo, Italy
Jiří Sgall	Charles University, Prague, Czech Republic
Arseny Shur	Ural Federal University, Russia
Mariya Soskova	Sofia University, Bulgaria
Tarmo Uustalu	Tallinn University of Technology, Estonia
Peter van Emde Boas	University of Amsterdam, The Netherlands
Jan van Leeuwen	Utrecht University, The Netherlands
Dorothea Wagner	Karlsruhe Institute of Technology, Germany
Peter Widmayer	ETH Zürich, Switzerland
Jiří Wiedermann	Academy of Sciences, Czech Republic
Christos Zaroliagis	University of Patras, Greece
Norbert Zeh	Dalhousie University, Halifax, Canada

Steering Committee

Juraj Hromkovič	ETH Zürich, Switzerland
Antonín Kučera, chair	Masaryk University, Czech Republic
Jerzy Marcinkowski	University of Wrocław, Poland
Damian Niwiński	University of Warsaw, Poland
Branislav Rovan	Comenius University, Bratislava, Slovakia
Jiří Sgall	Charles University, Prague, Czech Republic

Additional Reviewers

Akutsu, Tatsuya	Barto, Libor	Bevern, René van
Allender, Eric	Bärtschi, Andreas	Beyersdorff, Olaf
Almeida, Jorge	Basset, Nicolas	Bi, Jingguo
Amir, Amihood	Baum, Moritz	Bianchi, Maria Paola
Ananichev, Dmitry	Becchetti, Luca	Bienvenu, Laurent
Asarin, Eugene	Berkholz, Christoph	Bille, Philip
Azar, Yossi	Bernasconi, Anna	Bioglio, Livio
Bampas, Evangelos	Bernstein, Aaron	Bläsius, Thomas

Blondin, Michael
Blumensath, Achim
Boella, Guido
Böhm, Martin
Bohmova, Katerina
Boker, Udi
Bollig, Benedikt
Bonacina, Ilario
Bonelli, Eduardo
Borassi, Michele
Bosek, Bartłomiej
Bozianu, Rodica
Bradfield, Julian
Bradley, Jeremy
Bremer, Joachim
Bresolin, Davide
Breveglieri, Luca
Bruse, Florian
Bulteau, Laurent
Cadilhac, Michaël
Canonne, Clément
Carayol, Arnaud
Cardinal, Jean
Carpi, Arturo
Cassez, Franck
Caucal, Didier
Cave, Andrew
Cerone, Andrea
Čevorová, Kristína
Chailloux, André
Chechik, Shiri
Cho, Dae-Hyung
Cicalese, Ferdinando
Cleophas, Loek
Colcombet, Thomas
Cording, Patrick Hagge
Dal Lago, Ugo
Damaschke, Peter
D'Angelo, Gianlorenzo
Davies, Peter
Dell, Holger
Della Monica, Dario
Dennunzio, Alberto
Dibbelt, Julian
Diestel, Reinhard
Dobrev, Stefan

Doyen, Laurent
Drees, Maximilian
Droste, Manfred
Drucker, Andrew
Ďuriš, Pavol
Eden, Alon
Englert, Matthias
Eppstein, David
Epstein, Leah
Erde, Joshua
Fasoulakis, Michail
Feldotto, Matthias
Fenner, Stephen
Fici, Gabriele
Fijalkow, Nathanaël
Filiot, Emmanuel
Flammini, Michele
Forejt, Vojtech
Forišek, Michal
Franciosa, Paolo
Frid, Anna
Frigioni, Daniele
Fuchs, Fabian
Fukuda, Komei
Gajardo, Anahi
Galesi, Nicola
Gavinsky, Dmitry
Gazdag, Zsolt
Giannopoulos, Panos
Giannopoulou, Georgia
Girard, Vincent
Gogacz, Tomasz
Goldenberg, Elazar
Goldwurm, Massimiliano
Göller, Stefan
Gordon, Colin S.
Goubault-Larrecq, Jean
Green, Fred
Grigorieff, Serge
Grosshans, Nathan
Grossi, Giuliano
Guillon, Pierre
Guo, Heng
Gusev, Vladimir
Habib, Michel
Halldorsson, Magnus M.

Hamann, Michael
Haviv, Ishay
Hoogeboom, Hendrik Jan
Hoyrup, Mathieu
Hrubes, Pavel
Huang, Chien-Chung
Huang, Sangxia
Hundeshagen, Norbert
Iliev, Petar
Itsykson, Dmitry
Jansen, Klaus
Jeandel, Emmanuel
Jecker, Ismaël
Jeż, Łukasz
Johannsen, Jan
Johnson, Matthew
Jones, Mark
Jung, Daniel
Kari, Jarkko
Kavitha, Telikepalli
Kempa, Dominik
Kernberger, Daniel
Kikot, Stanislav
Kim, Min-Gyu
Kim, Sang-Sub
Kis, Tamas
Klasing, Ralf
Klein, Kim-Manuel
Komusiewicz, Christian
Kontogiannis, Spyros
Kopczynski, Eryk
Korman, Matias
Koucký, Michal
Koutris, Paraschos
Krajíček, Jan
Královič, Rastislav
Krebs, Andreas
Kuich, Werner
Kulkarni, Janardhan
Kumar, Mrinal
La Torre, Salvatore
Laura, Luigi
Lázár, Katalin A.
Lazic, Ranko
Li, Shouwei
Limouzy, Vincent

Lin, Jianyi
Löding, Christof
Loff, Bruno
Lombardy, Sylvain
López-Ortiz, Alejandro
Loreti, Michele
MacKenzie, Kenneth
Mäcker, Alexander
Mahajan, Meena
Malatyali, Manuel
Mamageishvili, Akaki
Mandrioli, Dino
Marathe, Madhav
Markarian, Christine
Martens, Wim
Martin, Russell
Mary, Arnaud
Massazza, Paolo
Mazoit, Frédéric
Mederly, Pavol
Meduna, Alexander
Mehrabi, Ali D.
Mendes de Oliveira,
    Rafael
Mertzios, George
Mezzina, Claudio Antares
Miksa, Mladen
Miller, Joseph S.
Montanari, Angelo
Moscardelli, Luca
Müller, Moritz
Mundhenk, Martin
Nakagawa, Kotaro
Obraztsova, Svetlana
Oh, Eunjin
Okhotin, Alexander
Otachi, Yota
Ott, Sebastian
Otto, Martin
Oum, Sang-Il
Paluch, Katarzyna
Panagiotou, Konstantinos
Pantziou, Grammati

Papadopoulos, Charis
Parotsidis, Nikos
Paryen, Haim
Paul, Christophe
Pavlogiannis, Andreas
Pich, Ján
Pilipczuk, Marcin
Ponse, Alban
Pouly, Amaury
Praveen, M.
Pribavkina, Elena
Protti, Fabio
Provillard, Julien
Prutkin, Roman
Rabinovitch, Alex
Ramanujan, M.S.
Ramyaa, Ramyaa
Randour, Mickael
Rao, Michaël
Rasin, Oleg
Regan, Kenneth
Restivo, Antonio
Riveros, Cristian
Romashchenko, Andrei
Rutter, Ignaz
Rybicki, Bartosz
Sabharwal, Yogish
Salo, Ville
Salvail, Louis
Saurabh, Saket
Schaudt, Oliver
Schewe, Klaus-Dieter
Schmid, Markus L.
Schmidt, Jens M.
Schmitz, Sylvain
Seki, Shinnosuke
Serre, Olivier
Seto, Kazuhisa
Shen, Alexander
Shpilka, Amir
Siggers, Mark
Slaman, Theodore
Sloth, Christoffer

Smith, Adam
Son, Wanbin
Sornat, Krzysztof
Spoerhase, Joachim
Srinivasan, Srikanth
Stougie, Leen
Suchy, Ondřej
Taati, Siamak
Tagliaferri, Roberto
Tan, Tony
Thapen, Neil
Tichler, Krisztián
Tiezzi, Francesco
Todinca, Ioan
Torres Vieira, Hugo
Tribastone, Mirco
Trystram, Denis
Ucar, Bora
Uznański, Przemysław
Vaananen, Jouko
van Leeuwen, Erik Jan
Vandin, Andrea
Vanier, Pascal
Velner, Yaron
Veselý, Pavel
Vinyals, Marc
Vollmer, Heribert
Wacker, Arno
Ward, Justin
Watrigant, Rémi
Watrous, John
Weihrauch, Klaus
Williams, Ryan
Wollan, Paul
Yakaryilmaz, Abuzer
Yoon, Sang-Duk
Zeitoun, Marc
Ziadi, Tewfik
Zielinski, Pawel
Živný, Stanislav
Zündorf, Tobias

Contents – Part II

Contents – Part I

Near-Optimal Asymmetric Binary Matrix Partitions

Fidaa Abed[1], Ioannis Caragiannis[2], and Alexandros A. Voudouris[2]([✉])

[1] Technische Universität Berlin, Berlin, Germany
[2] Computer Technology Institute and Press "Diophantus" and
Department of Computer Engineering and Informatics,
University of Patras, 26504 Rion, Greece
voudouris@ceid.upatras.gr

Abstract. We study the asymmetric binary matrix partition problem that was recently introduced by Alon et al. (WINE 2013) to model the impact of asymmetric information on the revenue of the seller in take-it-or-leave-it sales. Instances of the problem consist of an $n \times m$ binary matrix A and a probability distribution over its columns. A *partition scheme* $B = (B_1, ..., B_n)$ consists of a partition B_i for each row i of A. The partition B_i acts as a smoothing operator on row i that distributes the expected value of each partition subset proportionally to all its entries. Given a scheme B that induces a smooth matrix A^B, the partition value is the expected maximum column entry of A^B. The objective is to find a partition scheme such that the resulting partition value is maximized. We present a 9/10-approximation algorithm for the case where the probability distribution is uniform and a $(1 - 1/e)$-approximation algorithm for non-uniform distributions, significantly improving results of Alon et al. Although our first algorithm is combinatorial (and very simple), the analysis is based on linear programming and duality arguments. In our second result we exploit a nice relation of the problem to submodular welfare maximization.

1 Introduction

We study the *asymmetric binary matrix partition problem* (ABMP) proposed by Alon et al. [2]. Consider a matrix $A \in \{0, 1\}^{n \times m}$ and a probability distribution p over its columns; p_j denotes the probability associated with column j. We distinguish between two cases for the probability distribution over the columns of the given matrix, depending on whether it is uniform or non-uniform. A partition scheme $B = (B_1, ..., B_n)$ for matrix A consists of a partition B_i of $[m]$ for each row i of A. More specifically, B_i is a collection of k_i pairwise disjoint subsets $B_{ik} \subseteq [m]$ (with $1 \le k \le k_i$) such that $\bigcup_{k=1}^{k_i} B_{ik} = [m]$. We can think of each partition B_i as a smoothing operator, which acts on the entries of row i and

---

This work was partially supported by the European Social Fund and Greek national funds through the research funding program Thales on "Algorithmic Game Theory" and by the Caratheodory research grant E.114 from the University of Patras.

G.F. Italiano et al. (Eds.): MFCS 2015, Part II, LNCS 9235, pp. 1–13, 2015.
DOI: 10.1007/978-3-662-48054-0_1

changes their value to the expected value of the partition subset they belong to. Formally, the *smooth value* of an entry (i, j) such that $j \in B_{ik}$ is defined as

$$A_{ij}^B = \frac{\sum_{\ell \in B_{ik}} p_\ell \cdot A_{i\ell}}{\sum_{\ell \in B_{ik}} p_\ell}. \tag{1}$$

Given a partition scheme B that induces the *smooth matrix* A^B, the resulting *partition value* is the expected maximum column entry of A^B, namely,

$$v^B(A, p) = \sum_{j \in [m]} p_j \cdot \max_i A_{ij}^B. \tag{2}$$

The objective of ABMP is to find a partition scheme B such that the resulting partition value $v^B(A, p)$ is maximized.

Alon et al. [2] were the first to consider the asymmetric matrix partition problem. They prove that the problem is APX-hard and provide a 0.563- and a 1/13-approximation for uniform and non-uniform probability distributions, respectively. They also consider input matrices with non-negative non-binary entries and present a 1/2- and an $\Omega(1/\log m)$-approximation algorithm for uniform and non-uniform distributions, respectively. This interesting combinatorial optimization problem has apparent relations to revenue maximization in *take-it-or-leave-it sales*. For example, consider a setting with m items and n potential buyers. Each buyer has a value for each item. Nature selects at random (according to some probability distribution) an item for sale and, then, the seller approaches the highest valuation buyer and offers the item to her at a price equal to her valuation. Can the seller exploit the fact that she has much more accurate information about the items for sale compared to the potential buyers? In particular, information asymmetry arises since the seller knows the realization of the randomly selected item whereas the buyers do not. The approach that is discussed in [2] is to let the seller define a buyer-specific signalling scheme. That is, for each buyer, the seller can partition the set of items into disjoint subsets (bundles) and report this partition to the buyer. After nature's random choice, the seller can reveal to each buyer the bundle that contains the realization, thus enabling her to update her valuation beliefs. The relation of this problem to asymmetric matrix partition should be apparent. Interestingly, the seller can achieve revenue from items for which no buyer has any value.

This scenario falls within the line of research that studies the impact of information asymmetry to the quality of markets. Akerlof [1] was the first to introduce a formal analysis of "markets for lemons", where the seller has more information than the buyers regarding the quality of the products. Crawford and Sobel [7] study how, in such markets, the seller can exploit her advantage in order to maximize revenue. In [17], Milgrom and Weber provide the "linkage principle" which states that the expected revenue is enhanced when bidders are provided with more information. This principle seems to suggest full transparency but, in [15] and [16] the authors suggest that careful bundling of the items is the best way to exploit information asymmetry. Many different frameworks that reveal information to the bidders have been proposed in the literature.

More recently, Ghosh et al. [12] consider full information and propose a clustering scheme according to which, the items are partitioned into bundles and, then, for each such bundle, a separate second-price auction is performed. In this way, the potential buyers cannot bid only for the items that they actually want; they also have to compete for items that they do not have any value for. Hence, the demand for each item is increased and the revenue of the seller is higher. Emek et al. [10] present complexity results in similar settings and Miltersen and Sheffet [19] consider fractional bundling schemes for signaling.

In this work, we focus on the simplest binary case of asymmetric matrix partition which has been proved to be APX-hard. We present a 9/10-approximation algorithm for the uniform case and a $(1-1/e)$-approximation algorithm for non-uniform distributions. Both results significantly improve previous bounds of Alon et al. [2]. The analysis of our first algorithm is quite interesting because, despite its purely combinatorial nature, it exploits linear programming techniques. Similar techniques have been used in a series of papers on variants of set cover (e.g. [3–6]) by the second author; however, the application of the technique in the current context requires a quite involved reasoning about the structure of the solutions computed by the algorithm.

In our second result, we exploit a nice relation of the problem to submodular welfare maximization and use well-known algorithms from the literature. First, we discuss the application of a simple greedy 1/2-approximation algorithm that has been studied by Lehmann et al. [14] and then apply Vondrák's smooth greedy algorithm [20] to achieve a $(1 - 1/e)$-approximation. Vondrák's algorithm is optimal in the value query model as Khot et al. [13] have proved. In a more powerful model where it is assumed that demand queries can be answered efficiently, Feige and Vondrák [11] have proved that $(1 - 1/e + \epsilon)$-approximation algorithms — where ϵ is a small positive constant — are possible. We briefly discuss the possibility/difficulty of applying such algorithms to asymmetric binary matrix partition and observe that the corresponding demand query problems are, in general, NP-hard.

The rest of the paper is structured as follows. We begin with preliminary definitions and examples in Sect. 2. Then, we present our 9/10-approximation algorithm for the uniform case in Sect. 3 and our $(1 - 1/e)$-approximation algorithm for the non-uniform case in Sect. 4. Due to lack of space, most proofs are omitted.

2 Preliminaries

Let $A^+ = \{j \in [m] :$ there exists a row i such that $A_{ij} = 1\}$ denote the set of columns of A that contain at least one 1-value entry, and $A^0 = [m]\backslash A^+$ denote the set of columns of A that contain only 0-value entries. In the next sections, we usually refer to the sets A^+ and A^0 as the sets of *one-columns* and *zero-columns*, respectively. Furthermore, let $A_i^+ = \{j \in [m] : A_{ij} = 1\}$ and $A_i^0 = \{j \in [m] : A_{ij} = 0\}$ denote the sets of columns that intersect with row i at a 1- and 0-value entry, respectively. All columns in A_i^+ are one-columns and, furthermore, $A^+ = \cup_{i=1}^{n} A_i^+$. The columns of A_i^0 can be either one- or zero-columns and, thus, $A^0 \subseteq \cup_{i=1}^{n} A_i^0$. Also, denote by $r = \sum_{j \in A^+} p_j$ the total probability of the one-columns. As an example, consider the 3×6 matrix

$$A = \begin{pmatrix} 0\ 1\ 1\ 0\ 1\ 0 \\ 0\ 1\ 1\ 0\ 1\ 0 \\ 0\ 1\ 1\ 0\ 0\ 0 \end{pmatrix}$$

and a uniform probability distribution over its columns. We have $A^+ = \{2, 3, 5\}$ and $A^0 = \{1, 4, 6\}$. In the first two rows, the sets A_i^+ and A_i^0 are identical to A^+ and A^0, respectively. In the third row, the sets A_3^+ and A_3^0 are $\{2, 3\}$ and $\{1, 4, 5, 6\}$. Finally, the total probability of the one-columns r is $1/2$.

A partition scheme B can be thought of as consisting of n partitions B_1, B_2, ..., B_n of the set of columns $[m]$. We use the term *bundle* to refer to the elements of a partition B_i; a bundle is a non-empty set of columns. For a bundle b of partition B_i corresponding to row i, we say that b is an *all-zero* bundle if $b \subseteq A_i^0$ and an *all-one* bundle if $b \subseteq A_i^+$. A singleton all-one bundle of partition B_i is called *column-covering bundle in row i*. A bundle that is neither all-zero nor all-one is called *mixed* and corresponds to a set of columns that intersect with row i at both 1- and 0-value entries.

Let us examine the following partition scheme B and its corresponding smooth matrix A^B defined by Eq. (1).

B_1	$\{1,2,3,4\}, \{5,6\}$
B_2	$\{1,2\}, \{3\}, \{4,6\}, \{5\}$
B_3	$\{1,4,6\}, \{2,3,5\}$

A^B	1/2	1/2	1/2	1/2	1/2	1/2
	1/2	1/2	1	0	1	0
	0	2/3	2/3	0	2/3	0
$\max_i A_{ij}^B$	1/2	2/3	1	1/2	1	1/2

The bundle $\{1, 2, 3, 4\}$ of the partition B_1 in row 1 is mixed. The bundle $\{3\}$ of B_2 is all-one and column-covering in row 2. The bundle $\{1, 4, 6\}$ of B_3 is all-zero.

By Eq. (2), the partition value is $25/36$ and it can be further improved. Observe that the leftmost zero-column is included in two mixed bundles (in rows 1 and 2), and the mixed bundle in row 3 contains a one-column that is covered by a column-covering bundle in row 2 and intersects with row 3 at a 0-value entry. Let us modify these bundles.

B_1'	$\{1\}, \{2,3,4\}, \{5,6\}$
B_2'	$\{1,2\}, \{3\}, \{4,6\}, \{5\}$
B_3'	$\{1,4,5,6\}, \{2,3\}$

$A^{B'}$	0	2/3	2/3	2/3	1/2	1/2
	1/2	1/2	1	0	1	0
	0	1	1	0	0	0
$\max_i A_{ij}^{B'}$	1/2	1	1	2/3	1	1/2

The partition value is $7/9 > 25/36$. By merging the mixed bundles $\{2, 3, 4\}$ and $\{5, 6\}$ in row 1, we obtain a partition value $47/60 > 7/9$. Observe that the contribution of column 4 to the partition value decreases but, overall, the partition value increases due to the increase in the contribution of column 6. Actually, such merges never decrease the partition value (see Lemma 1).

B_1''	{1}, {2,3,4,5,6}
B_2''	{1,2}, {3}, {4,6}, {5}
B_3''	{1,4,5,6}, {2,3}

$A^{B''}$	0	3/5	3/5	3/5	3/5	3/5
	1/2	1/2	1	0	1	0
	0	1	1	0	0	0
$\max_i A_{ij}^{B''}$	1/2	1	1	3/5	1	3/5

Finally, by merging the bundles {1,2} and {3} in the second row and decomposing the bundle {2,3} in the last row into two singletons, the partition value becomes $73/90 > 47/60$ which can be verified to be optimal.

B_1'''	{1}, {2,3,4,5,6}
B_2'''	{1,2,3}, {4,6}, {5}
B_3'''	{1,4,5,6}, {2}, {3}

$A^{B'''}$	0	3/5	3/5	3/5	3/5	3/5
	2/3	2/3	2/3	0	1	0
	0	1	1	0	0	0
$\max_i A_{ij}^{B'''}$	2/3	1	1	3/5	1	3/5

We will now give some more definitions that will be useful in the following. We say that a one-column j is *covered* by a partition scheme B if there is at least one row i in which $\{j\}$ is column-covering. For example, in B''', the singleton $\{5\}$ is a column-covering bundle in the second row and the singletons $\{2\}$ and $\{3\}$ are column-covering in the third row. We say that a partition scheme *fully covers* the set A^+ of one-columns if all of them are covered. In this case, we use the term *full cover* to refer to the pairs of indices (i,j) of the 1-value entries A_{ij} such that $\{j\}$ is a column-covering bundle in row i. For example, the partition scheme B''' has the full cover $(2,5), (3,2), (3,3)$.

It turns out that optimal partition schemes always have a special structure like the one of B'''. Alon et al. [2] have formalized such observations into the next statement.

Lemma 1 (Alon et al. [2]). *Given a uniform instance of ABMP with a matrix A, there is an optimal partition scheme B with the following properties:*

P1. B fully covers the set A^+ of one-columns.

P2. For each row i, B_i has at most one bundle containing all columns of A_i^+ that are not included in column-covering bundles in row i (if any). This bundle can be either all-one (if it does not contain zero-columns) or the unique mixed bundle of row i.

P3. For each $j \in A^0$, there exists at most one row i such that j is contained in the mixed bundle of B_i (and j is contained in the all-zero bundles of the remaining rows).

P4. For each row i, the zero-columns that are not contained in the mixed bundle of B_i form an all-zero bundle.

Properties P1 and P3 imply that we can think of the partition value as the sum of the contributions of the column-covering bundles and the contributions of the zero-columns in mixed bundles. Property P2 should be apparent; the columns of A_i^+ that do not form column-covering bundles in row i are bundled together with zero-columns (if possible) in order to increase the contribution of the latter to the partition value. Property P4 makes B consistent to the definition of a partition scheme where the disjoint union of all the partition subsets in a row should be $[m]$. Clearly, the contribution of the all-zero bundles to the partition value is 0. Also, the non-column-covering all-one bundles do not contribute to the partition value either.

The proof of Lemma 1 in [2] extensively uses the fact that the instance is uniform. Unfortunately, as we will see later in Sect. 4, the lemma (in particular, property P1) does not hold for non-uniform instances. Luckily, it turns out that non-uniform instances also exhibit some structure which allows us to consider the problem of computing an optimal partition scheme as a *welfare maximization problem*. In welfare maximization, there are m items and n agents; agent i has a valuation function $v_i : 2^{[m]} \to \mathbb{R}^+$ that specifies her value for each subset of the items. I.e., for a set S of items, $v_i(S)$ represents the value of agent i for S. Given a disjoint partition (or allocation) $S = (S_1, S_2, ..., S_n)$ of the items to the agents, where S_i denotes the set of items allocated to agent i, the social welfare is the sum of values of the agents for the sets of items allocated to them, i.e., $\text{SW}(S) = \sum_i v_i(S_i)$. The term welfare maximization refers to the problem of computing an allocation of maximum social welfare. We discuss only the variant of the problem where the valuations are monotone and submodular; following the literature, we use the term *submodular welfare maximization* to refer to it.

Definition 1. *A valuation function v is monotone if $v(S) \le v(T)$ for any pair of sets S, T such that $S \subseteq T$. A valuation function v is submodular if $v(S \cup \{x\}) - v(S) \ge v(T \cup \{x\}) - v(T)$ for any pair of sets S, T such that $S \subseteq T$ and for any item $x \notin T$.*

An important issue in (submodular) welfare maximization arises with the representation of valuation functions. A valuation function can be described in detail by listing explicitly the values for each of the possible subsets of items. Unfortunately, this is clearly inefficient due to the necessity for exponential input size. A solution that has been proposed in the literature is to assume access to these functions by queries of a particular form. The simplest such form of queries reads as "what is the value of agent i for the set of items S?" These are known as *value queries*. Another type of queries, known as *demand queries*, are phrased as follows: "Given a non-negative price for each item, compute a set S of items for which the difference of the valuation of agent i minus the sum of prices for the items in S is maximized." Approximation algorithms that use a polynomial number of valuation or demand queries and obtain solutions to submodular welfare maximization with a constant approximation ratio are well-known in the literature. Our improved approximation algorithm for the non-uniform case of asymmetric binary matrix partition exploits such algorithms.

3   The Uniform Case

In this section, we present the analysis of a greedy approximation algorithm when the probability distribution p over the columns of the given matrix is uniform. Our algorithm uses a *greedy completion procedure* that was also considered by

Alon et al. [2]. This procedure starts from a full cover of the matrix, i.e., from column-covering bundles in some rows so that all one-columns are covered (by exactly one column-covering bundle). Once this initial full cover is given, the set of columns from A_i^+ that are not included in column-covering bundles in row i can form a mixed bundle together with some zero-columns in order to increase the contribution of the latter to the partition value. Greedy completion proceeds as follows. It goes over the zero-columns, one by one, and adds a zero-column to the mixed bundle of the row that maximizes the marginal contribution of the zero-column. The marginal contribution of a zero-column to the partition value when it is added to a mixed bundle that consists of x zero-columns and y one-columns is given by the quantity

$$\Delta(x,y) = (x+1)\frac{y}{x+y+1} - x\frac{y}{x+y} = \frac{y^2}{(x+y)(x+y+1)}.$$

The right-hand side of the first equality is simply the difference between the contribution of $x+1$ and x zero-columns to the partition value when they form a mixed bundle with y one-columns. Alon et al. [2] proved that, in the uniform case, this greedy completion procedure yields the maximum contribution from the zero-columns to the partition value among all partition schemes that include a given full cover. We extensively use this property as well as the fact that $\Delta(x,y)$ is non-decreasing with respect to y.

So, our algorithm consists of two phases. In the first phase, called the *cover phase*, the algorithm computes an arbitrary full cover for set A^+. In the second phase, called the *greedy phase*, it simply runs the greedy completion procedure mentioned above. In the rest of this section, we will show that this simple algorithm obtains an approximation ratio of $9/10$; we will also show that our analysis is tight. Even though our algorithm is purely combinatorial, our analysis exploits linear programming duality.

Overall, the partition value obtained by the algorithm can be thought of as the sum of contributions from column-covering bundles (this is exactly r) plus the contribution from the mixed bundles created during the greedy phase (i.e., the contribution from the zero-columns). Denote by ρ the ratio between the total number of appearances of one-columns in the mixed bundles of the optimal partition scheme (so, the number each one-column is counted equals the number of mixed bundles that contain it) and the number of zero-columns. For example, in the partition scheme B''' in the example of the previous section, the two mixed bundles are $\{2,3,4,5,6\}$ in the first row and $\{1,2,3\}$ in the second row. So, the one-columns 2 and 3 appear twice while the one-column 5 appears once in these mixed bundles. Since we have three zero-columns, the value of ρ is $5/3$. We can use the quantity ρ to upper-bound the optimal partition value as follows.

Lemma 2. *The optimal partition value is at most $r + (1-r)\frac{\rho}{\rho+1}$.*

In our analysis, we distinguish between two main cases depending on the value of ρ. The first case is when $\rho < 1$; in this case, the additional partition value obtained during the greedy phase of the algorithm is lower-bounded by the partition value we would have by creating bundles containing exactly one one-column and either $\lceil 1/\rho \rceil$ or $\lfloor 1/\rho \rfloor$ zero-columns.

Lemma 3. *If $\rho < 1$, then the partition value obtained by the algorithm is at least 0.97 times the optimal one.*

Proof. We will lower-bound the partition value returned by the algorithm by considering the following formation of mixed bundles as an alternative to the greedy completion procedure used in the greedy phase. For each appearance of a one-column in a mixed bundle in the partition scheme computed by the algorithm (observe that the total number of such appearances is exactly $\rho m(1 - r)$), we include this one-column in a mixed bundle together with either $\lceil 1/\rho \rceil$ or $\lfloor 1/\rho \rfloor$ distinct zero-columns. By Lemma 2, this process yields an optimal partition value if $1/\rho$ is an integer. Otherwise, denote by $x = m(1 - r)(1 - \rho \lfloor 1/\rho \rfloor)$ the number of mixed bundles containing $\lceil 1/\rho \rceil$ zero-columns. Then, the number of mixed bundles containing $\lfloor 1/\rho \rfloor$ zero-columns will be $\rho m(1 - r) - x = m(1 - r)(\rho \lceil 1/\rho \rceil - 1)$. Observe that the smooth value of a zero-column is $\frac{1}{1+\lceil 1/\rho \rceil}$ in the first case and $\frac{1}{1+\lfloor 1/\rho \rfloor}$ in the second case. Hence, we can bound the partition value obtained by the algorithm as follows:

$$\text{ALG} \geq r + (1 - r)(1 - \rho \lfloor 1/\rho \rfloor)\frac{\lceil 1/\rho \rceil}{1 + \lceil 1/\rho \rceil} + (1 - r)(\rho \lceil 1/\rho \rceil - 1)\frac{\lfloor 1/\rho \rfloor}{1 + \lfloor 1/\rho \rfloor}.$$

Now, assuming that $\rho \in (\frac{1}{k+1}, \frac{1}{k})$ for some integer $k \geq 1$, we have that $\lfloor 1/\rho \rfloor = k$ and $\lceil 1/\rho \rceil = k + 1$ and, hence,

$$\text{ALG} \geq r + (1 - r)\frac{1 + \rho k(k + 1)}{(k + 1)(k + 2)}.$$

Using Lemma 2, we have

$$\frac{\text{ALG}}{\text{OPT}} \geq \frac{r + (1 - r)\frac{1+\rho k(k+1)}{(k+1)(k+2)}}{r + (1 - r)\frac{\rho}{\rho+1}} \geq \frac{\frac{1+\rho k(k+1)}{(k+1)(k+2)}}{\frac{\rho}{\rho+1}} = \frac{(1 + 1/\rho)(1 + \rho k(k + 1))}{(k + 1)(k + 2)}.$$

This last expression is minimized (with respect to ρ) for $1/\rho = \sqrt{k(k + 1)}$. Hence,

$$\frac{\text{ALG}}{\text{OPT}} \geq \frac{\left(1 + \sqrt{k(k + 1)}\right)^2}{(k + 1)(k + 2)},$$

which is minimized for $k = 1$ to approximately 0.97.   □

For the case $\rho \geq 1$, we use completely different arguments. We will reason about the solution produced by the algorithm by considering a particular decomposition of the set of mixed bundles computed in the greedy phase. Then, using again the observation of Alon et al. [2, Lemma 1], the contribution of the zero-columns to the partition value in the solution computed by the algorithm is lower-bounded by their contribution to the partition value when they are part of the mixed bundles obtained after the decomposition.

The decomposition is defined as follows. It takes as input a bundle with y zero-columns and x one-columns and decomposes it into y bundles containing exactly one zero-column and either $\lfloor x/y \rfloor$ or $\lceil x/y \rceil$ one-columns. Note that if x/y is not an integer, there will be $x - y\lfloor x/y \rfloor$ bundles with $\lceil x/y \rceil$ one-columns. The solution obtained after the decomposition of the solution returned by the algorithm has a very special structure as our next lemma suggests.

Lemma 4. *There exists an integer $s \geq 1$ such that each bundle in the decomposition has at least s and at most $3s$ one-columns.*

Now, our analysis proceeds as follows. For every triplet $r \in [0,1], \rho \geq 1$ and integer $s \geq 1$, we will prove that any solution consisting of an arbitrary cover of the rm one-columns and the decomposed set of bundles containing at least s and at most $3s$ one-columns yields a 9/10-approximation of the optimal partition value. By the discussion above, this will also be the case for the solution returned by the algorithm. In order to account for the worst-case contribution of zero-columns to the partition value for a given triplet of parameters, we will use the following linear program $\mathrm{LP}(r, \rho, s)$:

$$\text{minimize} \quad \sum_{k=s}^{3s} \frac{k}{k+1} \theta_k$$

$$\text{subject to:} \quad \sum_{k=s}^{3s} \theta_k = 1 - r; \quad \sum_{k=s}^{3s} k\theta_k \geq \rho(1-r) - r; \quad \theta_k \geq 0, k = s, ..., 3s$$

The variable θ_k denotes the total probability of the zero-columns that participate in decomposed mixed bundles with k one-columns. The objective is to minimize the contribution of the zero-columns to the partition value. The equality constraint means that all zero-columns have to participate in bundles. The inequality constraint requires that the total number of appearances of one-columns in bundles used by the algorithm is at least the total number of appearances of one-columns in mixed bundles of the optimal partition scheme minus one appearance for each one-column, since for every selection of the cover, the algorithm will have the same number of (appearances of) one-columns available to form mixed bundles. Informally, the linear program answers (rather pessimistically) to the question of how inefficient the algorithm can be. In particular, given an instance with parameters r and ρ, the quantity $\min_{\text{int } s \geq 1} \mathrm{LP}(r, \rho, s)$ yields a lower bound on the contribution of the zero-columns to the partition value and $r + \min_{\text{int } s \geq 1} \mathrm{LP}(r, \rho, s)$ is a lower bound on the partition value. The next lemma completes the analysis of the greedy algorithm for the case $\rho \geq 1$.

Lemma 5. *For every $r \in [0,1]$ and $\rho \geq 1$, $r + \min_{\text{int } s \geq 1} \mathrm{LP}(r, \rho, s) \geq \frac{9}{10}\mathrm{OPT}$.*

The next statement summarizes the discussion above.

Theorem 1. *The greedy algorithm always yields a 9/10-approximation of the optimal partition value in the uniform case.*

Our analysis is tight as our next counter-example suggests.

Theorem 2. *There exists an instance of the uniform ABMP for which the greedy algorithm computes a partition scheme with value (at most) 9/10 of the optimal one.*

Proof. Consider the ABMP instance that consists of the matrix

$$A = \begin{pmatrix} 1\,0\,0\,0 \\ 0\,1\,0\,0 \\ 1\,1\,0\,0 \\ 1\,1\,0\,0 \end{pmatrix}$$

with $p_i = 1/4$ for $i = 1, 2, 3, 4$. The optimal partition value is obtained by covering the one-columns in the first two rows and then bundling each of the two zero-columns with a pair of one-columns in the third and fourth row, respectively. This yields a partition value of 5/6. The greedy algorithm may select to cover the one-columns using the 1-value entries A_{31} and A_{42}. This is possible since the greedy algorithm has no particular criterion for breaking ties when selecting the full cover. Given this full cover, the greedy completion procedure will assign each of the two zero-columns with only one one-column. The partition value is then 3/4, i.e., 9/10 times the optimal partition value. □

4   Asymmetric Binary Matrix Partition as Welfare Maximization

We now consider the more general non-uniform case. Interestingly, property P1 of Lemma 1 does not hold any more as the following statement shows.

Lemma 6. *For every $\epsilon > 0$, there exists an instance of ABMP in which any partition scheme containing a full cover of the columns in A^+ yields a partition value that is at most $8/9 + \epsilon$ times the optimal one.*

Proof. Consider the ABMP instance consisting of the matrix

$$A = \begin{pmatrix} 1\,0\,0\,0 \\ 0\,1\,0\,0 \\ 0\,1\,0\,0 \\ 1\,0\,1\,0 \end{pmatrix}$$

with column probabilities $p_j = \frac{1}{\beta+3}$ for $j = 1, 2, 3$ and $p_4 = \frac{\beta}{\beta+3}$ for $\beta > 2$. There are four partition schemes containing a full cover (depending on the rows that contain the column-covering bundle of the first two columns) and, in each of them, the zero-column is bundled together with a 1-value entry. By making calculations, we obtain that the partition value in these cases is $\frac{4\beta+3}{(\beta+1)(\beta+3)}$. Here is one of these partition schemes:

B_1	$\{1\}, \{2,3,4\}$
B_2	$\{2\}, \{1,3,4\}$
B_3	$\{1,3\}, \{2,4\}$
B_4	$\{1\}, \{3\}, \{2,4\}$

	1	0	0	0
A^B	0	1	0	0
	0	$\frac{1}{\beta+1}$	0	$\frac{1}{\beta+1}$
	1	0	1	0
$p_j \cdot \max_i A^B_{ij}$	$\frac{1}{\beta+3}$	$\frac{1}{\beta+3}$	$\frac{1}{\beta+3}$	$\frac{\beta}{(\beta+1)(\beta+3)}$

Now, consider the partition scheme B' in which the 1-value entries A_{11} and A_{22} form column-covering bundles in rows 1 and 2, A_{32} and A_{33} are bundled together in row 3 and A_{41}, A_{43}, and A_{44} are bundled together in row 4. As it can be seen from the tables below (recall that $\beta > 2$), the partition value becomes $\frac{4.5\beta+5}{(\beta+2)(\beta+3)}$.

B'_1	$\{1\}, \{2,3,4\}$
B'_2	$\{2\}, \{1,3,4\}$
B'_3	$\{1,4\}, \{2,3\}$
B'_4	$\{2\}, \{1,3,4\}$

	1	0	0	0
$A^{B'}$	0	1	0	0
	0	$1/2$	$1/2$	0
	$\frac{2}{\beta+2}$	0	$\frac{2}{\beta+2}$	$\frac{2}{\beta+2}$
$p_j \cdot \max_i A^{B'}_{ij}$	$\frac{1}{\beta+3}$	$\frac{1}{\beta+3}$	$\frac{1}{2(\beta+3)}$	$\frac{2\beta}{(\beta+2)(\beta+3)}$

Clearly, the ratio of the two partition values approaches $8/9$ from above as β tends to infinity. The theorem follows by selecting β sufficiently large for any $\epsilon > 0$. $\qquad\square$

Still, as the next statement indicates, the optimal partition scheme has some structure which we will exploit later.

Lemma 7. *Consider an ABMP instance consisting of a matrix A and a probability distribution p over its columns. There is an optimal partition scheme B that satisfies properties P2, P3, P4 and the following one:*

P5. *Given any column j, denote by $H_j = \arg\max_i A^B_{ij}$ the set of rows through which column j contributes to the partition value $v^B(A,p)$. For every $i \in H_j$ such that $A_{ij} = 1$, the bundle of partition B_i that contains column j is not mixed.*

What Lemma 7 says is that the contribution of column $j \in A^+$ to the partition value comes from a row i such that either $j \in A^+_i$ and $\{j\}$ forms a column-covering bundle or $j \in A^0_i$ and j belongs to the mixed bundle of row i. The contribution of a column $j \in A^0$ to the partition value always comes from a row i where j belongs to the mixed bundle. Hence, the problem of computing the optimal partition scheme is equivalent to deciding the row from which each column contributes to the partition value.

Let B be a partition scheme and S be a set of columns whose contribution to the partition value of B comes from row i (i.e., i is the row that maximizes the smooth value A^B_{ij} for each column j in S). Denoting the sum of these contributions by $R_i(S) = \sum_{j \in S} p_j \cdot A^B_{ij}$, we can equivalently express $R_i(S)$ as

$$R_i(S) = \sum_{j \in S \cap A_i^+} p_j + \frac{\sum_{j \in S \cap A_i^0} p_j \sum_{j \in A_i^+ \setminus S} p_j}{\sum_{j \in S \cap A_i^0} p_j + \sum_{j \in A_i^+ \setminus S} p_j}.$$

The first sum represents the contribution of columns of $S \cap A_i^+$ to the partition value (through column-covering bundles) while the second sum represents the contribution of the columns in $S \cap A_i^0$ which are bundled together with all 1-value entries in $A_i^+ \setminus S$ in the mixed bundle of row i. Then, the partition scheme B can be thought of as a collection of disjoint sets S_i (with one set per row) such that S_i contains those columns whose entries achieve their maximum smooth value in row i. Hence, the partition value of B is $v^B(A, p) = \sum_{i \in [n]} R_i(S_i)$ and the problem is essentially equivalent to welfare maximization where the rows act as the agents who will be allocated bundles of items (corresponding to columns).

Lemma 8. *For every row i, the function R_i is non-decreasing and submodular.*

Lehmann et al. [14] studied the submodular welfare maximization problem and provided a simple algorithm that yields a $1/2$-approximation of the optimal welfare. Their algorithm considers the items one by one and assigns item j to the agent that maximizes the marginal valuation (the additional value from the allocation of item j). In our setting, this algorithm can be implemented as follows. It considers the one-columns first and the zero-columns afterwards. Whenever considering a one-column j, a column-covering bundle $\{j\}$ is formed at an arbitrary row i with $j \in A_i^+$ (such a decision definitely maximizes the increase in the partition value). Whenever considering a zero-column, the algorithm includes it to a mixed bundle so that the increase in the partition value is maximized. Using the terminology of [2], the algorithm essentially starts with an arbitrary cover of the one-columns and then it runs the greedy completion procedure. Again, we will use the term greedy for this algorithm.

Theorem 3. *The greedy algorithm for ABMP has a tight approximation ratio of $1/2$.*

We can use the more sophisticated smooth greedy algorithm of Vondrák [20], which uses value queries to obtain the following.

Corollary 1. *There exists a $(1 - 1/e)$-approximation algorithm for ABMP.*

One might hope that better approximation guarantees might be possible using the $(1 - 1/e + \epsilon)$-approximation algorithm of Feige and Vondrák [11] which requires that demand queries of the form "given a price q_j for every item $j \in [m]$, select the bundle S that maximizes the difference $R_i(S) - \sum_{j \in S} q_j$" can be answered in polynomial time. Unfortunately, in our setting, this is not the case in spite of the very specific form of the function R_i.

Lemma 9. *Answering demand queries associated with ABMP are NP-hard.*

References

1. Akerlof, G.A.: The market for lemons: quality uncertainty and the market mechanism. Quaterly J. Econ. **84**, 488–500 (1970)
2. Alon, N., Feldman, M., Gamzu, I., Tennenholtz, M.: The asymmetric matrix partition problem. In: Chen, Y., Immorlica, N. (eds.) WINE 2013. LNCS, vol. 8289, pp. 1–14. Springer, Heidelberg (2013)
3. Athanassopoulos, S., Caragiannis, I., Kaklamanis, C.: Analysis of approximation algorithms for k-set cover using factor-revealing linear programs. Theor. Comput. Sys. **45**(3), 555–576 (2009)
4. Athanassopoulos, S., Caragiannis, I., Kaklamanis, C., Kyropoulou, M.: An improved approximation bound for spanning star forest and color saving. In: Královič, R., Niwiński, D. (eds.) MFCS 2009. LNCS, vol. 5734, pp. 90–101. Springer, Heidelberg (2009)
5. Caragiannis, I.: Wavelength management in WDM rings to maximize the number of connections. SIAM J. Discrete Math. **23**(2), 959–978 (2009)
6. Caragiannis, I., Kaklamanis, C., Kyropoulou, M.: Tight approximation bounds for combinatorial frugal coverage algorithms. J. Comb. Optim. **26**(2), 292–309 (2013)
7. Crawford, V., Sobel, J.: Strategic information transmission. Econometrica **50**, 1431–1451 (1982)
8. Cremer, J., McLean, R.P.: Optimal selling strategies under uncertainty for a discriminating monopolist when demands are interdependent. Econometrica **53**, 345–361 (1985)
9. Cremer, J., McLean, R.P.: Full extraction of the surplus in bayesian and dominant strategy auctions. Econometrica **56**, 1247–1257 (1988)
10. Emek, Y., Feldman, M., Gamzu, I., Paes Leme, R., Tennenholtz, M.: Signaling schemes for partition value maximization. In: 13th ACM Conference on Electronic Commerce (EC), pp. 514–531 (2012)
11. Feige, U., Vondrák, J.: The submodular welfare problem with demand queries. Theor. Comput. **6**, 247–290 (2010)
12. Ghosh, A., Nazerzadeh, H., Sundararajan, M.: Computing optimal bundles for sponsored search. In: Deng, X., Graham, F.C. (eds.) WINE 2007. LNCS, vol. 4858, pp. 576–583. Springer, Heidelberg (2007)
13. Khot, S., Lipton, R., Markakis, E., Mehta, A.: Inapproximability results for combinatorial auctions with submodular utility functions. Algorithmica **52**, 3–18 (2008)
14. Lehmann, B., Lehmann, D.J., Nisan, N.: Combinatorial auctions with decreasing marginal utilities. Games Econ. Behav. **55**, 270–296 (2006)
15. Levin, J., Milgrom, P.: Online advertising: Heterogeneity and conflation in market design. Am. Econ. Rev. **100**, 603–607 (2010)
16. Milgrom, P.: Simplified mechanisms with an application to sponsored-search auctions. Games Econ. Behav. **70**, 62–70 (2010)
17. Milgrom, P.R., Weber, R.J.: A theory of auctions and competitive bidding. Econometrica **50**, 1089–1122 (1982)
18. Milgrom, P.R., Weber, R.J.: The value of information in a sealed-bid auction. J. Math. Econ. **10**, 105–114 (1982)
19. Miltersen, P. B., Sheffet, O.: Send mixed signals - Earn more, work less. In: 13th ACM Conference on Electronic Commerce (EC), pp. 234–247 (2012)
20. Vondrák, J.: Optimal approximation for the submodular welfare problem in the value oracle model. In: 40th ACM Symposium on Theory of Computing (STOC), pp. 67–74 (2008)

Dual VP Classes

Eric Allender[1]([⊠]), Anna Gál[2], and Ian Mertz[1]

[1] Department of Computer Science, Rutgers University,
Piscataway, NJ, USA
`allender@cs.rutgers.edu, iwmertz@gmail.com`
[2] Department of Computer Science, University of Texas,
Austin, TX, USA
`panni@cs.utexas.edu`

Abstract. We consider the complexity class ACC^1 and related families of arithmetic circuits. We prove a variety of collapse results, showing several settings in which no loss of computational power results if fan-in of gates is severely restricted, as well as presenting a natural class of arithmetic circuits in which no expressive power is lost by severely restricting the algebraic degree of the circuits. These results tend to support a conjecture regarding the computational power of the complexity class VP over finite algebras, and they also highlight the significance of a class of arithmetic circuits that is in some sense dual to VP.

1 Introduction

Most of the well-studied subclasses of P are defined in terms of Boolean or arithmetic circuits. The question of the relative power of $\mathsf{NC}^1, \mathsf{LogCFL},$ and AC^1, or of $\#\mathsf{NC}^1$ and $\#\mathsf{LogCFL}$ boils down to the question of how the computational power of a (log-depth, polynomial-size) circuit model depends on the *fan-in* of gates in the model.

Our main contribution is to present several settings where fan-in can be severely restricted for log-depth, polynomial-size circuits, with *no* loss of computational power.

1.1 ACC^1 and TC^1

There is a large literature exploring the connections between Boolean and arithmetic circuit complexity; see [18]. For instance, the Boolean class TC^1 (log-depth MAJORITY circuits) corresponds to $\#\mathsf{AC}^1(\mathbb{F}_{p_n})$ (log-depth *unbounded* fan-in arithmetic circuits where the circuits for inputs of size n operate over the field $\mathbb{F}_{p_n}$, where p_n is the n-th prime [12]). We show here that $\mathsf{ACC}^1 = \bigcup_p \#\mathsf{AC}^1(\mathbb{F}_p)$. Is unbounded fan-in necessary for these characterizations?

No, it is not! The *semiunbounded* fan-in model, where the $+$ gates have fan-in two, also yields ACC^1 (Corollary 1), and the same is true (modulo logspace-Turing reductions) for TC^1 (Theorem 6).

The usual definition of ACC^1 is in terms of *unbounded* fan-in AND and OR gates, along with MOD_m gates for different m, and we observe here that TC^1 has

© Springer-Verlag Berlin Heidelberg 2015
G.F. Italiano et al. (Eds.): MFCS 2015, Part II, LNCS 9235, pp. 14–25, 2015.
DOI: 10.1007/978-3-662-48054-0_2

an analogous characterization with MOD_{p_n} gates. Here, too, the fan-in of the AND and OR gates can be restricted, to *constant* fan-in for ACC^1 (Theorem 9) (while AND and OR gates are not needed at all for the TC^1 characterization (Theorem 6)).

1.2 Algebraic Degree

In the previous section's discussion, the arithmetic circuit families that character-ize ACC^1 and TC^1 have algebraic degree $n^{O(\log n)}$. Much more has been written about poly-size arithmetic circuits with degree $n^{O(1)}$: VP. Similar to our charac-terization of ACC^1 in terms of semiunbounded circuits, VP also corresponds to semiunbounded circuits, but with the (more common) restriction of having the $\times$ gates have fan-in two [2,16]. VP is usually studied as a class of *polynomials*, but it is also common to study the *Boolean part* of VP over a given semiring R, where (fol-lowing [5]), the Boolean part of an arithmetic circuit class is the class of languages whose characteristic functions are computed by circuits in the class. Especially over finite fields, there is little to distinguish VP from its Boolean part.

Immerman and Landau conjectured that computing the determinant of inte-ger matrices is complete for TC^1 [11]. This would have several consequences, including providing a characterization of TC^1 in terms of $\mathsf{VP}(\mathbb{Q})$. Buhrman et al. [7] have argued that the Immerman-Landau conjecture is unlikely, in that this would imply that arbitrary polynomials having degree $n^{O(\log n)}$ and polynomial-size arithmetic circuits mod p_n could be simulated by arithmetic circuits of *much lower degree* over $\mathbb{Q}$. This raises the question: When can high-degree polynomials over one algebra be simulated by low-degree polynomials over another?

Our degree-reduction theorem (Corollary 7) gives one natural class of polyno-mials of degree $n^{O(\log n)}$ over one algebra ($\mathbb{F}_2$) that *can* be simulated by polyno-mials having much smaller degree. We show that restricting the fan-in of $\times$ gates in $\#\mathsf{AC}^1(\mathbb{F}_2)$ circuits to be logarithmic results in *no loss of expressive power*; the restricted class (whose polynomials have algebraic degree only $n^{O(\log \log n)}$) repre-sents the same class of functions as the unrestricted class (with degree $n^{O(\log n)}$). We believe that this weakens the arguments against the Immerman-Landau con-jecture that were raised in [7], and we suspect that there are other such examples, where restricting the fan-in of $\times$ gates causes no loss of power. We also see no reason why degree $n^{O(\log \log n)}$ should be optimal. Lowering the degree to $n^{O(1)}$ would imply $\#\mathsf{AC}^1(\mathbb{F}_2) = \mathsf{AC}^1[2] = \mathsf{VP}(\mathbb{F}_2)$. (We omit "Boolean part" if it causes no confusion.)

1.3 Duality

We have mentioned that VP corresponds to semiunbounded arithmetic circuits with bounded-fan-in $\times$ gates. Over the Boolean semiring, logarithmic depth polynomial-size semiunbounded fan-in circuits (with bounded fan-in AND gates and unbounded fan-in OR gates, with NOT gates only at the input level) char-acterize the complexity class LogCFL, also known as SAC^1, which has been the subject of numerous investigations [9,13].

Because LogCFL is closed under complement [6], it can be characterized in terms of semiunbounded fan-in circuits by restricting either the AND gates or the OR gates to have bounded fan-in. It is unknown if there is any other algebraic structure for which a similar phenomenon occurs. In particular, it is not known how the complexity of functions in $\mathsf{VP}(\mathbb{F}_p)$ compares to that of the functions in the classes defined by logarithmic depth polynomial-size semiunbounded fan-in circuits with bounded fan-in $+$ gates and unbounded fan-in $\times$ gates.

A large part of the motivation for this paper is to understand the computational power of these semiunbounded fan-in circuit classes, which are in some sense dual to Valiant's classes $\mathsf{VP}(\mathbb{F}_p)$. We use the notation $\mathit{\Lambda}\mathsf{P}(\mathbb{F}_p)$ to refer to the class of problems characterized by logarithmic depth polynomial-size semiunbounded fan-in circuits with bounded fan-in addition gates and unbounded fan-in multiplication gates. Formal definitions appear in Sect. 2. We show that each class $\mathit{\Lambda}\mathsf{P}(\mathbb{F}_p)$ corresponds exactly to a particular subclass of ACC^1, and that the union over all p of $\mathit{\Lambda}\mathsf{P}(\mathbb{F}_p)$ is exactly equal to ACC^1 (Corollary 1).

We conjecture that ACC^1 is precisely the class of languages logspace-Turing reducible to $\bigcup_m \mathsf{VP}(\mathbb{Z}_m)$. If the conjecture is true, then ACC^1 can be defined using either kind of semiunbounded circuits, with bounded fan-in $+$ or bounded fan-in $\times$.

2   Preliminaries, and Definitions of $\mathit{\Lambda}$-classes

We assume that the reader is familiar with Boolean circuit complexity classes such as AC^0 and ACC^0; a good source for this background material is the excellent text by Vollmer [18]. The following notation is used by Vollmer, and we follow those conventions here:

Definition 1.   – AC^i is the class of languages accepted by Dlogtime-uniform circuit families of size $n^{O(1)}$ and depth $O(\log^i n)$, with NOT gates, and unbounded fan-in (AND, OR).

- $\mathsf{AC}^i[m]$ is defined as AC^i, but in addition unbounded fan-in MOD_m gates are allowed, which output 1 iff the number of input wires carrying a value of 1 is a multiple of m.

- For any finite set $S \subset \mathbb{N}$, $\mathsf{AC}^i[S]$ is defined analogously to $\mathsf{AC}^i[m]$, but now the circuit families are allowed to use MOD_r gates for any $r \in S$. It is known that, for any $m \in \mathbb{N}, \mathsf{AC}^i[m] = \mathsf{AC}^i[\mathsf{Supp}(m)]$, where – following the notation of [8] – $\mathsf{Supp}(m) = \{p : p$ is prime and p divides $m\}$ [14]. Thus, in particular $\mathsf{AC}^i[6] = \mathsf{AC}^i[2,3]$ and $\mathsf{AC}^i = \mathsf{AC}^i[\emptyset]$. (We omit unnecessary brackets, writing for instance $\mathsf{AC}^i[2,3]$ instead of $\mathsf{AC}^i[\{2,3\}]$.)

- $\mathsf{ACC}^i = \bigcup_m \mathsf{AC}^i[m]$.

- TC^i is the class of languages accepted by Dlogtime-uniform circuit families of size $n^{O(1)}$ and depth $O(\log^i n)$, consisting of unbounded fan-in MAJORITY gates, and NOT gates.

- SAC^i is the class of languages accepted by Dlogtime-uniform circuit families of polynomial size and depth $O(\log^i n)$, consisting of unbounded fan-in OR gates and bounded fan-in AND gates, along with NOT gates at (some of) the leaves.

Note that the restriction that NOT gates appear only at the leaves in SAC^i circuits is essential; if NOT gates could appear everywhere, then these classes would coincide with AC^i. Similarly, note that we do not bother to define $\mathsf{SAC}^i[m]$, since a MOD_m gate with a single input is equivalent to a NOT gate, and thus $\mathsf{SAC}^i[m]$ would be the same as $\mathsf{AC}^i[m]$.

The algebraic complexity classes $\mathsf{VP}(R)$ for various algebraic structures R were originally defined [15] as classes of families of n-variate polynomials of degree $n^{O(1)}$ that can be represented by polynomial-size (*nonuniform*) arithmetic circuits over R. Here, we let $\mathsf{VP}(R)$ denote the corresponding *uniform* class, and recall that the $\log^2 n$ depth bound of [16] can be made logarithmic:

Theorem 1. *[2] For any commutative semiring R, $\mathsf{VP}(R)$ coincides with the class of families of polynomials over R represented by logspace-uniform circuit families of polynomial size and logarithmic depth with unbounded fan-in $+$ gates, and fan-in two $\times$ gates.*

Note that over $\mathbb{F}_p$, many different polynomials yield the same function. For example, since $x^3 = x$ in $\mathbb{F}_3$, every function on n variables has a polynomial of degree at most $2n$. Very likely there are functions represented by polynomials in $\mathsf{VP}(\mathbb{F}_3)$ of degree, say, n^5, but not by any VP polynomial of degree $2n$. On the other hand, there is a case to be made for focusing on the *functions* in these classes, rather than focusing on the *polynomials* that represent those functions. For instance, if the Immerman-Landau conjecture is true, and TC^1 is reducible to problems in $\mathsf{VP}(\mathbb{Q})$, it would suffice for every *function* in $\mathsf{TC}^1 = \#\mathsf{AC}^1(\mathbb{F}_{p_n})$ to have a representation in $\mathsf{VP}(\mathbb{Q})$, even though the *polynomials* represented by $\#\mathsf{AC}^1(\mathbb{F}_{p_n})$ circuits have large degree, and thus cannot be in any VP class.

In the literature on VP classes, one standard way to focus on the *functions* represented by polynomials in VP is to consider what is called the *Boolean Part* of $\mathsf{VP}(R)$, which is the set of *languages* $A \subseteq \{0,1\}^*$ such that, for some sequence of polynomials (Q_n), for $x \in A$ we have $Q_{|x|}(x) = 1$, and for $x \in \{0,1\}^*$ such that $x \notin A$ we have $Q_{|x|}(x) = 0$.

When the algebra R is a finite field, considering the Boolean part of $\mathsf{VP}(R)$ captures the relevant complexity aspects, since the computation of any function represented by a polynomial in $\mathsf{VP}(R)$ (with inputs and outputs coming from R) is logspace-Turing reducible to some language in the Boolean Part of $\mathsf{VP}(R)$.

In this paper, we are concerned exclusively with the "Boolean Part" of arithmetic classes. For notational convenience, we refer to these classes using the "VP" *notation, rather than constantly repeating the phrase "Boolean Part".*

Following the naming conventions of [18], for any Boolean circuit complexity class $\mathcal{C}$ defined in terms of circuits with AND and OR gates, we define $\#\mathcal{C}(R)$ to be the class of functions represented by arithmetic circuits defined over the algebra R, where AND is replaced by $\times$, and OR is replaced by $+$ (and NOT gates at the leaves are applied to the $\{0,1\}$ inputs). (The classes $\#\mathsf{P}, \#\mathsf{L}$, and $\#\mathsf{LogCFL}$ also fit into this naming scheme, using established connections between Turing machines and circuits.) In particular, we will be concerned with the following two classes:

Definition 2. *Let R be any suitable semiring[1]. Then*

- *#AC^1(R) is the class of functions $f : \{0,1\}^* \to R$ given by families of logspace-uniform circuits of unbounded fan-in $+$ and $\times$ gates having depth $O(\log n)$ and size $n^{O(1)}$.*
- *#SAC^1(R) is the class of functions $f : \{0,1\}^* \to R$ represented by families of logspace-uniform circuits of unbounded fan-in $+$ gates and $\times$ gates of fan-in two, having depth $O(\log n)$ and polynomial size.*

Input variables may be negated. Where no confusion will result, the notation #$\mathcal{C}$(R) will also be used to refer to the related class of languages.

Hence from Theorem 1 we see that $\mathsf{VP}(\mathbb{F}_p) = \#\mathsf{SAC}^1(\mathbb{F}_p)$ for any prime p.

Now we introduce classes that are dual to the #SAC^1(R) classes. Define #SAC^{1,*}(R) to be the class of functions $f : \{0,1\}^* \to R$ represented by families of logspace-uniform circuits of unbounded fan-in $\times$ gates and $+$ gates of fan-in two, having depth $O(\log n)$ and size $n^{O(1)}$. Because of the connection between VP and #SAC^1, we use the convenient notation $\Lambda\mathsf{P}(R)$ to denote the dual notation, rather than the more cumbersome #SAC^{1,*}(R).

Of course, the set of formal polynomials represented by $\Lambda\mathsf{P}$ circuits is not contained in any VP class, because $\Lambda\mathsf{P}$ contains polynomials of degree $n^{O(\log n)}$. However, as discussed in the previous section, we are considering the "Boolean Part" of these classes. More formally:

Definition 3. *Let p be a prime power. $\Lambda\mathsf{P}(\mathbb{F}_p)$ is the class of all languages $A \subseteq \{0,1\}^*$ with the property that there is a logspace-uniform family of circuits $\{C_n : n \in \mathbb{N}\}$, each of depth $O(\log n)$ consisting of input gates, $+$ gates of fan-in two, and $\times$ gates of unbounded fan-in, such that for each string x of length n, x is in A if and only if $C_n(x)$ evaluates to 1, when the $+$ and $\times$ gates are evaluated over $\mathbb{F}_p$. Furthermore, if $x \notin A$, then $C_n(x)$ evaluates to 0.*

Another way of relating arithmetic classes (such as VP and $\Lambda\mathsf{P}$) to complexity classes of languages would be to consider the languages that are logspace-Turing reducible to the polynomials in $\mathsf{VP}(R)$ or $\Lambda\mathsf{P}(R)$, via a machine M with a polynomial p as an oracle, which obtains the value of $p(x_1, \ldots, x_n)$ when M writes $x_1, \ldots x_n$ on a query tape. It is worth mentioning that (the Boolean parts of) both $\mathsf{VP}(\mathbb{F}_p)$ and $\Lambda\mathsf{P}(\mathbb{F}_p)$ are closed under logspace-Turing reductions, although this is still open for classes over $\mathbb{Z}_m$ when m is not prime.

Proposition 1. $\Lambda\mathsf{P}(\mathbb{F}_p) = \mathsf{L}^{\Lambda\mathsf{P}(\mathbb{F}_p)}$ *and* $\mathsf{VP}(\mathbb{F}_p) = \mathsf{L}^{\mathsf{VP}(\mathbb{F}_p)}$

(Proofs are omitted; see [3].) VP over fields of the same characteristic yield the same class of languages.

---

[1] Our primary focus in this paper is on *finite* semirings, as well as countable semirings such as $\mathbb{Q}$, where we use the standard binary representation of constants (say, as a numerator and denominator) when a logspace uniformity machine makes use of constants in the description of a circuit. It is not clear to us which definition would be most useful in describing a class such as #AC^1($\mathbb{R}$), and so for now we consider such semirings to be "unsuitable".

Proposition 2. *Let p be a prime, and let $k \geq 1$. Then* $\mathsf{VP}(\mathbb{F}_p) = \mathsf{VP}(\mathbb{F}_{p^k})$.

It is also appropriate to use the VP and ΛP notation when referring to the classes defined by Boolean semiunbounded fan-in circuits with negation gates allowed at the inputs. With this notation, $\mathsf{VP}(B_2)$ corresponds to the Boolean class SAC^1, and $\Lambda\mathsf{P}(B_2)$ corresponds to the complement of SAC^1 (with bounded fan-in OR gates, and unbounded fan-in AND gates). It has been shown by [6] that SAC^1 is closed under complement. Thus we close this section with the equality that serves as a springboard for investigating the ΛP classes.

Theorem 2. *[6]* $\mathsf{VP}(B_2) = \Lambda\mathsf{P}(B_2)(= \mathsf{SAC}^1 = \mathsf{LogCFL})$.

We believe $\mathsf{VP}(\mathbb{F}_p) \neq \Lambda\mathsf{P}(\mathbb{F}_p)$ for every prime p; see Sect. 5.

3 Subclasses of ACC^1

In this section, we present our characterizations of ACC^1 in terms of $\Lambda\mathsf{P}(\mathbb{F}_p)$.

Theorem 3. *For any prime p and any $k \geq 1$,* $\Lambda\mathsf{P}(\mathbb{F}_{p^k}) = \mathsf{AC}^1[\mathsf{Supp}(p^k - 1)]$. *(Recall that $\mathsf{Supp}(m)$ is defined in Definition 1.)*

Corollary 1. $\mathsf{ACC}^1 = \bigcup_p \Lambda\mathsf{P}(\mathbb{F}_p)$.

Note also that several of the $\Lambda\mathsf{P}(\mathbb{F}_p)$ classes coincide. This is neither known nor believed to happen with the $\mathsf{VP}(\mathbb{F}_p)$ classes. Augmenting the $\Lambda\mathsf{P}(\mathbb{F}_p)$ classes with unbounded fan-in addition gates increases their computation power only by adding MOD_p gates, as the following theorem demonstrates.

Theorem 4. *For each prime p and each $k \geq 1$,* $\#\mathsf{AC}^1(\mathbb{F}_{p^k}) = \mathsf{AC}^1[\{p\} \cup \mathsf{Supp}(p^k - 1)]$.

Corollary 2. $\mathsf{ACC}^1 = \bigcup_p \Lambda\mathsf{P}(\mathbb{F}_p) = \bigcup_p \#\mathsf{AC}^1(\mathbb{F}_p) = \bigcup_m \#\mathsf{AC}^1(\mathbb{Z}_m)$.

Corollary 3. *For any prime p there is a prime q such that* $\#\mathsf{AC}^1(\mathbb{F}_p) \subseteq \Lambda\mathsf{P}(\mathbb{F}_q)$.

$\mathsf{VP}(\mathbb{F}_p)$ also has a simple characterization in terms of Boolean circuits. For this, we need a more general definition:

Definition 4. *Let $m \in \mathbb{N}$, and let g be any function on $\mathbb{N}$. Define $g\text{-}\mathsf{AC}^1[m]$ to be the class of languages with logspace-uniform circuits of polynomial size and depth $O(\log n)$, consisting of unbounded-fan-in MOD_m gates, along with AND gates of fan-in $O(g(n))$. Clearly $g\text{-}\mathsf{AC}^1[m] \subseteq \mathsf{AC}^1[m]$.*

Observe that, since a MOD_m gate can simulate a NOT gate, $g\text{-}\mathsf{AC}^1[m]$ remains the same if OR gates of fan-in $O(g)$ are also allowed.

Corollary 4. *For every prime p,* $\mathsf{VP}(\mathbb{F}_p) = 2\text{-}\mathsf{AC}^1[p] \subseteq \mathsf{AC}^1[p]$.

We remark that the same proof shows that, for any $m \in \mathbb{N}$, $\mathsf{VP}(\mathbb{Z}_m) \subseteq 2\text{-}\mathsf{AC}^1[m]$. However, the converse inclusion is not known, unless m is prime. We remark that the proofs of Theorems 3 and 4 carry over also for depths other than $\log n$. (Related results for constant-depth unbounded-fan-in circuits can be found already in [1,14].)

Corollary 5. *For any prime p, $\#\mathsf{SAC}^{i,*}(\mathbb{F}_p) = \mathsf{AC}^i[\mathsf{Supp}(p-1)]$ and $\#\mathsf{AC}^i(\mathbb{F}_p)$* *$= \mathsf{AC}^i[p \cup \mathsf{Supp}(p-1)]$.*

3.1   Comparing $\mathit{\Lambda}\mathsf{P}$ and VP.

How do the $\mathit{\Lambda}\mathsf{P}$ and VP classes compare to each other? As a consequence of Corollary 4 and Theorem 3, $\mathsf{VP}(\mathbb{F}_p) \subseteq \mathit{\Lambda}\mathsf{P}(\mathbb{F}_q)$ whenever p divides $q - 1$. In particular, $\mathsf{VP}(\mathbb{F}_2) \subseteq \mathit{\Lambda}\mathsf{P}(\mathbb{F}_q)$ for any prime $q > 2$. No inclusion of any $\mathit{\Lambda}\mathsf{P}$ class in any VP class is known unconditionally, although $\mathit{\Lambda}\mathsf{P}(B_2)(= \mathsf{SAC}^1)$ is contained in every $\mathsf{VP}(\mathbb{F}_p)$ class in the nonuniform setting [9,13], and this holds also in the uniform setting under a plausible derandomization hypothesis [4].

No $\mathit{\Lambda}\mathsf{P}(\mathbb{F}_q)$ class can be contained in $\mathsf{VP}(\mathbb{F}_p)$ unless $\mathsf{AC}^1 \subseteq \mathsf{VP}(\mathbb{F}_p)$, since $\mathsf{AC}^1 = \mathit{\Lambda}\mathsf{P}(\mathbb{F}_2) \subseteq \mathit{\Lambda}\mathsf{P}(\mathbb{F}_3) \subseteq \mathit{\Lambda}\mathsf{P}(\mathbb{F}_q)$ for every prime $q \geq 3$. AC^1 is not known to be contained in any VP class.

4   Threshold Circuits and Small Degree

The inspiration for the results in this section comes from the following theorem of Reif and Tate [12] (as re-stated by Buhrman et al. [7]):

Theorem 5. $\mathsf{TC}^1 = \#\mathsf{AC}^1(\mathbb{F}_{p_n})$.

Here, the class $\#\mathsf{AC}^1(\mathbb{F}_{p_n})$ consists of the languages whose (Boolean) characteristic functions are computed by logspace-uniform families of arithmetic circuits of logarithmic depth with unbounded fan-in $+$ and $\times$ gates, where the arithmetic operations of the circuit C_n are interpreted over $\mathbb{F}_{p_n}$, where $p_1, p_2, p_3, \ldots$ is the sequence of all primes $2, 3, 5, \ldots$ That is, circuits for inputs of length n use the n-th prime to define the algebraic structure.

This class is closed under logspace-Turing reductions – but when we consider *other* circuit complexity classes defined using $\mathbb{F}_{p_n}$, it is *not* clear that these other classes are closed.

As an important example, we mention $\mathsf{VP}(\mathbb{F}_{p_n})$. As we show below, this class has an important connection to $\mathsf{VP}(\mathbb{Q})$, which is perhaps the canonical example of a VP class. Vinay [17] proved that $\mathsf{VP}(\mathbb{Q})$ has essentially the same computational power as $\#\mathsf{LogCFL}$ (which counts among its complete problems the problem of determining how many distinct parse trees a string x has in a certain context-free language). Here, we mention one more alternative characterization of the computational power of $\mathsf{VP}(\mathbb{Q})$.

Proposition 3. $\mathsf{L}^{\mathsf{VP}(\mathbb{F}_{p_n})} = \mathsf{L}^{\mathsf{VP}(\mathbb{Q})} = \mathsf{L}^{\#\mathsf{LogCFL}}$.

With arithmetic circuits of superpolynomial algebraic degree (such as $\Lambda\mathsf{P}$), evaluating the circuits over $\mathbb{Z}$ produces output that needs a superpolynomial number of bits to express in binary. Thus, when we consider such classes, it will always be in the context of structures (such as $\mathbb{F}_{p_n}$) where the output can be represented in a polynomial number of bits.

Our first new result in this section, is to improve Theorem 5. (Recall the definition of $g\text{-}\mathsf{AC}^1[m]$ from Definition 4.)

Theorem 6. $\mathsf{TC}^1 = \#\mathsf{AC}^1(\mathbb{F}_{p_n}) = \mathsf{L}^{\Lambda\mathsf{P}(\mathbb{F}_{p_n})} = \mathsf{AC}^1[p_n] = 0\text{-}\mathsf{AC}^1[p_n]$.

We also mention that Theorem 6 generalizes to other depths, in a way analogous to Corollary 5:

Corollary 6. $\mathsf{TC}^i = \#\mathsf{AC}^i(\mathbb{F}_{p_n}) = \mathsf{AC}^i[p_n] = 0\text{-}\mathsf{AC}^i[p_n]$.

For $i \geq 1$ the equality $\mathsf{TC}^i = \mathsf{L}^{\#\mathsf{SAC}^{i,*}(\mathbb{F}_{p_n})}$ also holds, but for $i = 0$ a more careful argument is needed, using AC^0-Turing reducibility in place of logspace-Turing reducibility.

In order to set the context for the results of the next section, it is necessary to consider an extension of Theorem 6, involving arithmetic circuits over certain *rings*. Thus we require the following definition.

Definition 5. *Let (m_n) be any sequence of natural numbers (where each $m_n > 1$) such that the mapping $1^n \mapsto m_n$ is computable in logspace. We use the notation $\#\mathsf{AC}^1(\mathbb{Z}_{m_n})$ to denote the class of functions f with domain $\{0,1\}^*$ such that there is a logspace-uniform family of arithmetic circuits $\{C_n\}$ of logarithmic depth with unbounded fan-in $+$ and $\times$ gates, where the arithmetic operations of the circuit C_n are interpreted over $\mathbb{Z}_{m_n}$, and for any input x of length n, $f(x) = C_n(x)$. We use the notation $\#\mathsf{AC}^1(\mathbb{Z}_\mathsf{L})$ to denote the union, over all logspace-computable sequences of moduli (m_n), of $\#\mathsf{AC}^1(\mathbb{Z}_{m_n})$.*

Since the sequence of primes (p_n) is logspace-computable, we have that TC^1 $(= \#\mathsf{AC}^1(\mathbb{F}_{p_n}))$ is clearly contained in $\#\mathsf{AC}^1(\mathbb{Z}_\mathsf{L})$. Conversely, each function in $\#\mathsf{AC}^1(\mathbb{Z}_\mathsf{L})$ is in TC^1. Thus, arithmetic circuits over the integers mod m_n for reasonable sequences of moduli m_n give yet another arithmetic characterization of TC^1.

4.1 Degree Reduction

In this subsection, we introduce a class of circuits that is intermediate between the unbounded fan-in circuit model and the semiunbounded fan-in model, for the purposes of investigating when arithmetic circuits of superpolynomial algebraic degree can be simulated by arithmetic circuits (possibly over a different algebra) with much smaller algebraic degree.

The starting point for this subsection is Theorem 4.3 in [2], which states that every problem in AC^1 is reducible to a function computable by polynomial-size arithmetic circuits of degree $n^{O(\log \log n)}$. In this section, we refine that result and put it in context with the theorems about TC^1 that were presented in the previous subsection. Those results show that TC^1 reduces to semiunbounded fan-in arithmetic circuits in the $\mathit{\Lambda}\mathsf{P}(\mathbb{F}_{p_n})$ model, but leave open the question of whether TC^1 also reduces to semiunbounded fan-in arithmetic circuits in the $\mathsf{VP}(\mathbb{F}_{p_n})$ model (which coincides with $\mathsf{VP}(\mathbb{Q})$). We are unable to answer this question, but we show some interesting inclusions occur if we relax the VP model, by imposing a less-stringent restriction on the fan-in of the $\times$ gates.

Definition 6. *Let (m_n) be any sequence of natural numbers (where each $m_n > 1$) such that the mapping $1^n \mapsto m_n$ is computable in L. $\#\mathsf{WSAC}^1(\mathbb{Z}_{m_n})$ is the class of functions represented by logspace-uniform arithmetic circuits $\{C_n\}$, where C_n is interpreted over $\mathbb{Z}_{m_n}$, where each C_n has size polynomial in n, and depth $O(\log n)$, and where the $+$ gates have unbounded fan-in, and the $\times$ gates have fan-in $O(\log n)$. Thus these circuits are not semiunbounded, but have a "weak" form of the semiunbounded fan-in restriction. We use the notation $\#\mathsf{WSAC}^1(\mathbb{Z}_\mathsf{L})$ to denote the union, over all logspace-computable sequences of moduli (m_n), of $\#\mathsf{WSAC}^1(\mathbb{Z}_{m_n})$. In the special case when $m_n = p$ for all n, we obtain the class $\#\mathsf{WSAC}^1(\mathbb{F}_p)$.*

We refrain from defining a weakly semiunbounded analog of $\mathit{\Lambda}\mathsf{P}$, because it would coincide with $\mathit{\Lambda}\mathsf{P}$, since AC^0 circuits can add $O(\log n)$ numbers.

We improve on [2, Theorem 4.3] by showing that AC^1 is contained in the class $\#\mathsf{WSAC}^1(\mathbb{F}_2)$; note that all polynomials in $\#\mathsf{WSAC}^1(\mathbb{F}_p)$ have degree $n^{O(\log\log n)}$, and note also that the class of functions considered in [2] is not obviously even in TC^1. In addition, we improve on [2] by reducing not merely AC^1, but also $\mathsf{AC}^1[p]$ for any prime p. Also, we obtain an equality.

Theorem 7. *Let p be any prime. Then $\mathsf{AC}^1[p] = \#\mathsf{WSAC}^1(\mathbb{F}_p)$.*

We especially call attention to the following corollary, which shows that, over $\mathbb{F}_2$, polynomial size logarithmic depth arithmetic circuits of degree $n^{O(\log n)}$ and of degree $n^{O(\log\log n)}$ represent precisely the same functions!

Corollary 7. $\#\mathsf{WSAC}^1(\mathbb{F}_2) = \#\mathsf{AC}^1(\mathbb{F}_2) = \mathsf{AC}^1[2] = \mathit{\Lambda}\mathsf{P}(\mathbb{F}_3)$.

If we focus on the Boolean classes, rather than on the arithmetic classes, then we obtain a remarkable collapse.

Theorem 8. *Let $m \in \mathbb{N}$. Then $\mathsf{AC}^1[m] = \log\text{-}\mathsf{AC}^1[m]$.*

It follows that arithmetic AC^1 circuits over *any* finite field $\mathbb{F}_p$ can be simulated by Boolean circuits with MOD gates and small fan-in AND gates. It remains open whether this in turn leads to small-degree arithmetic circuits over $\mathbb{F}_p$ when $p > 2$, and also whether the fan-in of the AND gates can be sublogarithmic, without loss of power.

When m is composite, Theorem 8 can be improved to obtain an even more striking collapse, by invoking the work of Hansen and Koucký [10].

Theorem 9. *Let m not be a prime power. Then $\mathsf{AC}^1[m] = 2\text{-}\mathsf{AC}^1[m]$.*

Corollary 8. $\mathsf{ACC}^1 = \bigcup_p \varLambda\mathsf{P}(\mathbb{F}_p) = \bigcup_p \#\mathsf{AC}^1(\mathbb{F}_p) = \bigcup_m \#\mathsf{AC}^1(\mathbb{Z}_m) = \bigcup_m 2\text{-}\mathsf{AC}^1[m]$.

It might be useful to have additional examples of algebras, where some degree reduction can be accomplished. Thus we also offer the following theorem:

Theorem 10. *Let p be any prime. Then* $\mathsf{AC}^1[p] \subseteq \mathsf{L}^{\#\mathsf{WSAC}^1(\mathbb{Z}_\mathsf{L})}$.

Using Theorems 3 and 4 we obtain the following.

Corollary 9. *If p is a Fermat prime, then* $\varLambda\mathsf{P}(\mathbb{F}_p) \subseteq \mathsf{L}^{\#\mathsf{WSAC}^1(\mathbb{Z}_\mathsf{L})}$.

5   Conclusions, Discussion, and Open Problems

We have introduced the complexity classes $\varLambda\mathsf{P}(R)$ for various algebraic structures R, and have shown that they provide alternative characterizations of well-known complexity classes. Furthermore, we have shown that arithmetic circuit complexity classes corresponding to polynomials of degree $n^{O(\log \log n)}$ also yield new characterizations of complexity classes, such as the equality $\mathsf{AC}^1[p] = \log\text{-}\mathsf{AC}^1[p] = \#\mathsf{WSAC}^1(\mathbb{F}_p)$. In the case when $p = 2$, we additionally obtain $\#\mathsf{AC}^1(\mathbb{F}_2) = \mathsf{AC}^1[2] = \log\text{-}\mathsf{AC}^1[2] = \#\mathsf{WSAC}^1(\mathbb{F}_2)$, showing that algebraic degree $n^{O(\log n)}$ and $n^{O(\log \log n)}$ have equivalent expressive power, in this setting.

We have obtained new characterizations of ACC^1 in terms of restricted fan-in: $\mathsf{ACC}^1 = \bigcup_p \#\mathsf{AC}^1(\mathbb{F}_p) = \bigcup_p \varLambda\mathsf{P}(\mathbb{F}_p) = \bigcup_m 2\text{-}\mathsf{AC}^1[m]$. That is, although ACC^1 corresponds to unbounded fan-in arithmetic circuits of logarithmic depth, and to unbounded fan-in Boolean circuits with modular counting gates, no power is lost if the addition gates have bounded fan-in (in the arithmetic case) or if only the modular counting gates have unbounded fan-in (in the Boolean case). It remains unknown if every problem in ACC^1 is reducible to a problem in $\bigcup_m \mathsf{VP}(\mathbb{Z}_m)$, although we believe that our theorems suggest that this is likely. It would be highly interesting to see such a connection between ACC^1 and VP.

We believe that it is fairly likely that several of our theorems can be improved. For instance:

* Perhaps Theorems 8 and 9 can be improved, to show that for all m, $\mathsf{AC}^1[m] = 2\text{-}\mathsf{AC}^1[m]$. Note that this is known to hold if m is not a prime power. By Corollary 4 this would show that $\mathsf{VP}(\mathbb{F}_p) = \mathsf{AC}^1[p]$ for all primes p. It would also show that $\#\mathsf{AC}^1(\mathbb{F}_2) = \mathsf{VP}(\mathbb{F}_2) = \varLambda\mathsf{P}(\mathbb{F}_p)$ for every Fermat prime p. (We should point out that this would imply that $\mathsf{AC}^1 \subseteq \mathsf{VP}(\mathbb{F}_p)$ for every prime p, whereas even the weaker inclusion $\mathsf{SAC}^1 \subseteq \mathsf{VP}(\mathbb{F}_p)$ is only known to hold non-uniformly [9].)

* Can Corollary 9 be improved to hold for all primes p, or even for $\varLambda\mathsf{P}(\mathbb{F}_{p_n})$? The latter improvement would show that $\mathsf{TC}^1 \subseteq \mathsf{L}^{\#\mathsf{WSAC}^1(\mathbb{Z}_\mathsf{L})}$.

* Perhaps one can improve Theorem 10, to achieve a simulation of degree $n^{O(1)}$. Why should $n^{O(\log \log n)}$ be optimal? Perhaps this could also be improved to hold for composite moduli?

Note that if some combinations of the preceding improvements are possible, TC^1 would reduce to $\mathsf{VP}(\mathbb{Q})$, which would be a significant step toward the Immerman-Landau conjecture.

It appears as if $\mathsf{VP}(\mathbb{F}_p)$ and $\mathit{\Lambda}\mathsf{P}(\mathbb{F}_p)$ are incomparable for every non-Fermat prime $p > 2$, since $\mathsf{VP}(\mathbb{F}_p) = 2\text{-}\mathsf{AC}^1[p]$ and $\mathit{\Lambda}\mathsf{P}(\mathbb{F}_p) = 2\text{-}\mathsf{AC}^1[\mathsf{Supp}(p-1)]$, involving completely different sets of primes. For Fermat primes we have $\mathit{\Lambda}\mathsf{P}(\mathbb{F}_p) = \log\text{-}\mathsf{AC}^1[2]$ and again the VP and $\mathit{\Lambda}\mathsf{P}$ classes seem incomparable. When $p = 2$, we have $\mathsf{VP}(\mathbb{F}_2) = 2\text{-}\mathsf{AC}^1[2]$ and $\mathit{\Lambda}\mathsf{P}(\mathbb{F}_2) = \mathsf{AC}^1$; if $\mathsf{VP}(\mathbb{F}_2) = \mathsf{AC}^1[2]$ (which may be possible), then it would appear that the VP class could be *more* powerful than the $\mathit{\Lambda}\mathsf{P}$ class. But based on current knowledge it also appears possible that the VP and $\mathit{\Lambda}\mathsf{P}$ classes are incomparable even for $p = 2$.

Some of our theorems overcome various hurdles that would appear to stand in the way of a proof of our conjecture that $\mathsf{ACC}^1 = \bigcup_m \mathsf{L}^{\mathsf{VP}(\mathbb{F}_{\mathbb{Z}_m})}$. First, recall that $\mathsf{VP}(\mathbb{Z}_m) \subseteq 2\text{-}\mathsf{AC}^1[m]$ (Corollary 4). Thus, if the conjecture is correct, then *unbounded* fan-in AND and OR gates would have to be simulated efficiently with *bounded* fan-in gates. But this is true in this context: $\mathsf{AC}^1[m] = 2\text{-}\mathsf{AC}^1[m]$, if m is not a prime power (Theorem 9). If m is a prime power, then the fan-in can be reduced to $\log n$ (Theorem 8). If the fan-in can be reduced to $O(1)$ also in the case of prime power moduli, or if ACC^1 circuits with *bounded* fan-in AND and OR (which have the full power of ACC^1, by Corollary 8) can be simulated by $\mathsf{VP}(\mathbb{Z}_m)$ circuits, then the conjecture holds. (The latter simulation is possible if the MOD gates in the ACC^1 circuits are for a prime modulus; see Corollary 4.)

A second objection that might be raised against the conjecture deals with algebraic degree. ACC^1 corresponds precisely to polynomial-size logarithmic depth unbounded fan-in arithmetic circuits over finite fields (Corollary 2). Such circuits represent polynomials of degree $n^{O(\log n)}$, whereas VP circuits represent polynomials of degree only $n^{O(1)}$. One might assume that there are languages represented by polynomial-size log-depth arithmetic circuits of degree $n^{O(\log n)}$ that actually *require* such large degree in order to be represented by arithmetic circuits of small size and depth.

Our degree-reduction theorem (Corollary 7) shows that this assumption is incorrect. Every Boolean function that can be represented by an arithmetic AC^1 circuit over $\mathbb{F}_2$ (with algebraic degree $n^{O(\log n)}$) can be represented by an arithmetic AC^1 circuit over $\mathbb{F}_2$ where the multiplication gates have fan-in $O(\log n)$ (and thus the arithmetic circuit has algebraic degree $n^{O(\log \log n)}$).

Acknowledgments. The first and third authors acknowledge the support of NSF grants CCF-0832787 and CCF-1064785. The second author was supported in part by NSF grant CCF-1018060. We also acknowledge stimulating conversations with Meena Mahajan, which occurred at the 2014 Dagstuhl Workshop on the Complexity of Discrete Problems (Dagstuhl Seminar 14121), and illuminating conversations with Stephen Fenner and Michal Koucký, which occurred at the 2014 Dagstuhl Workshop on Algebra in Computational Complexity (Dagstuhl Seminar 14391). We also thank Igor Shparlinski and our Rutgers colleagues Richard Bumby, John Miller and Steve Miller, for helpful pointers to the literature, as well as helpful feedback from Pascal Koiran and Russell Impagliazzo.

References

1. Agrawal, M., Allender, E., Datta, S.: On TC^0, AC^0, and arithmetic circuits. J. Comput. Syst. Sci. **60**, 395–421 (2000)
2. Allender, J., Jiao, J., Mahajan, M., Vinay, V.: Non-commutative arithmetic circuits. Theoret. Comp. Sci. **209**, 47–86 (1998)
3. Allender, E., Gál, A., Mertz, I.: Dual VP classes. In: ECCC (2014). TR14-122
4. Allender, E., Reinhardt, K., Zhou, S.: Isolation, matching, and counting: Uniform and nonuniform upper bounds. J. Comput. Syst. Sci. **59**(2), 164–181 (1999)
5. Blum, L., Cucker, F., Shub, M., Smale, S.: Complexity and Real Computation. Springer, Heidelberg (1998)
6. Borodin, A., Cook, S.A., Dymond, P.W., Ruzzo, W.L., Tompa, M.: Two applications of inductive counting for complementation problems. SIAM J. Comput. **18**, 559–578 (1989). See Erratum in SIAM. J. Comput. 18, 1283 (1989)
7. Buhrman, H., Cleve, R., Koucký, M., Loff, B., Speelman, F.: Computing with a full memory: catalytic space. In: STOC, pp. 857–866 (2014)
8. Corrales-Rodrigáñez, C., Schoof, R.: The support problem and its elliptic analogue. J. Num. Theor. **64**(2), 276–290 (1997)
9. Gál, A., Wigderson, A.: Boolean complexity classes vs. their arithmetic analogs. Random Struct. Algorithms **9**(1–2), 99–111 (1996)
10. Hansen, K.A., Koucký, M.: A new characterization of ACC^0 and probabilistic CC^0. Comput. Complex. **19**(2), 211–234 (2010)
11. Immerman, N., Landau, S.: The complexity of iterated multiplication. Inf. Comput. **116**, 103–116 (1995)
12. Reif, J., Tate, S.: On threshold circuits and polynomial computation. SIAM J. Comput. **21**, 896–908 (1992)
13. Reinhardt, K., Allender, E.: Making nondeterminism unambiguous. SIAM J. Comput. **29**, 1118–1131 (2000)
14. Smolensky, R.: Algebraic methods in the theory of lower bounds for Boolean circuit complexity. In: STOC, pp. 77–82 (1987)
15. Valiant, L.G.: Completeness classes in algebra. In: Proceedings of the 11th ACM STOC, pp. 249–261 (1979)
16. Valiant, L.G., Skyum, S., Berkowitz, S., Rackoff, C.: Fast parallel computation of polynomials using few processors. SIAM J. Comput. **12**(4), 641–644 (1983)
17. Vinay, V.: Counting auxiliary pushdown automata and semi-unbounded arithmetic circuits. In: Proceedings of 6th Structure in Complexity Theory Conference, pp. 270–284 (1991)
18. Vollmer, H.: Introduction to Circuit Complexity: A Uniform Approach. Springer, Heidelberg (1999)

On Tinhofer's Linear Programming Approach to Isomorphism Testing

V. Arvind[1], Johannes Köbler[2]($\boxtimes$), Gaurav Rattan[1], and Oleg Verbitsky[2,3]

[1] The Institute of Mathematical Sciences, Chennai 600 113, India
{arvind,grattan}@imsc.res.in
[2] Institut für Informatik, Humboldt Universität zu Berlin, Berlin, Germany
{koebler,verbitsk}@informatik.hu-berlin.de
[3] On leave from the Institute for Applied Problems of Mechanics
and Mathematics, Lviv, Ukraine

Abstract. Exploring a linear programming approach to Graph Isomorphism, Tinhofer (1991) defined the notion of *compact graphs*: A graph is *compact* if the polytope of its fractional automorphisms is integral. Tinhofer noted that isomorphism testing for compact graphs can be done quite efficiently by linear programming. However, the problem of characterizing and recognizing compact graphs in polynomial time remains an open question. In this paper we make new progress in our understanding of compact graphs. Our results are summarized below:

- We show that all graphs G which are distinguishable from any non-isomorphic graph by the classical color-refinement procedure are compact. In other words, the applicability range for Tinhofer's linear programming approach to isomorphism testing is at least as large as for the combinatorial approach based on color refinement.
- Exploring the relationship between color refinement and compactness further, we study related combinatorial and algebraic graph properties introduced by Tinhofer and Godsil. We show that the corresponding classes of graphs form a hierarchy and we prove that recognizing each of these graph classes is P-hard. In particular, this gives a first complexity lower bound for recognizing compact graphs.

1 Introduction

Consider the following natural linear programming formulation [15, 17] of Graph Isomorphism. Let G and H be two graphs on n vertices with adjacency matrices A and B, respectively. Then G and H are isomorphic if and only if there is an $n \times n$ permutation matrix X such that $AX = XB$. A linear programming relaxation of this system of equations is to allow X to be a doubly stochastic matrix. If such an X exists, it is called a *fractional isomorphism* from G to H, and these graphs are said to be *fractionally isomorphic*.

This work was supported by the Alexander von Humboldt Foundation in its research group linkage program. The second author and the fourth author were supported by DFG grants KO 1053/7-2 and VE 652/1-2, respectively.

© Springer-Verlag Berlin Heidelberg 2015
G.F. Italiano et al. (Eds.): MFCS 2015, Part II, LNCS 9235, pp. 26–37, 2015.
DOI: 10.1007/978-3-662-48054-0_3

It turns out remarkably, as shown by [15], that two graphs G and H are fractionally isomorphic if and only if they are indistinguishable by the classical color-refinement procedure (to be abbreviated as CR; we outline this approach to isomorphism testing in Sect. 3).

For which class of graphs is testing fractional isomorphism equivalent to isomorphism testing? We call a graph G *amenable* (to this approach to isomorphism testing) if for any other graph H it holds that G and H are fractionally isomorphic exactly when G and H are isomorphic.

The characterization of pairs of fractionally isomorphic graphs in terms of the CR algorithm implies that amenable graphs include some well-known graph classes with easy isomorphism testing algorithms, e.g. unigraphs (i.e. graphs characterizable by their degree sequences), trees, and graphs for which CR terminates with singleton color classes. We call graphs with the last property *discrete*. Babai, Erdős, and Selkow [4] have shown that a random graph $G_{n,1/2}$ is discrete with high probability. Thus, almost all graphs are amenable, which makes graph isomorphism efficiently solvable in the average case. Recently, we showed that the class of *all* amenable graphs can be recognized in nearly linear time [2]; a similar result was obtained independently in [13]. This reduces testing of isomorphism for G and H to computing a fractional isomorphism between these graphs, once G passes the amenability tests of [2,13].

The concept of a fractional isomorphism was used by Tinhofer in [19] as a basis for yet another linear-programming approach to isomorphism testing. Tinhofer calls a graph G *compact* if the polytope of all its fractional automorphisms is integral; more precisely, if A is the adjacency matrix of G, then the polytope in $\mathbb{R}^{n^2}$ consisting of the doubly stochastic matrices X such that $AX = XA$ has only integral extreme points (i.e. all coordinates of these points are integers).

If a compact graph G is isomorphic to another graph H, then the polytope of fractional isomorphisms from G to H is also integral. If G is not isomorphic to H, then this polytope has *no* integral extreme point (and in fact no integral point at all). Thus, isomorphism testing for a compact graph G and an arbitrary graph H can be done in polynomial time by using linear programming to compute an extreme point of the polytope and testing if it is integral. Before testing isomorphism in this way, we need to know that G is compact. Unfortunately, no efficient characterization of compact graphs is currently known.

As our main result, in Sect. 3 we show that all amenable graphs are compact. This implies that Tinhofer's approach to Graph Isomorphism [19] has at least as large an applicability range as color refinement. More precisely, whenever the restriction of Graph Isomorphism to input graphs G and H such that G belongs to a class C is solvable by the latter approach, then it is also solvable by the former approach. Consider, for example, the class C of unigraphs. The restricted graph isomorphism problem can obviously be solved by the CR algorithm in this case. As a particular consequence of our general result, it can also be solved by computing an extreme point of the polytope of fractional isomorphisms for the input graphs. In general, Tinhofer's approach is even more powerful than color refinement because it is known that the class of compact graphs contains many regular graphs (for example, all cycles [17]), for which CR cannot refine even the initial coloring.

In Sect. 4, we look at the relationship between the concepts of compactness and color refinement also from the other side. Let us call a graph G *refinable* if the color partition produced by CR coincides with the orbit partition of the automorphism group of G. It is interesting to note that the CR procedure gives an efficient algorithm to check if a given refinable graph has a nontrivial automorphism. It follows from the results in [19] that all compact graphs are refinable. The inclusion Amenable $\subset$ Compact, therefore, implies that all amenable graphs are refinable as well. The last result is independently obtained in [13] by a different argument. In the particular case of trees, this fact was observed long ago by several authors; see a survey in [20].

Taking a finer look at the inclusion Compact $\subset$ Refinable, we discuss algorithmic and algebraic graph properties that were introduced by Tinhofer [19] and Godsil [9]. We note that, along with the other graph classes under consideration, the corresponding classes Tinhofer and Godsil form a hierarchy under inclusion:

$$\text{Discrete} \subset \text{Amenable} \subset \text{Compact} \subset \text{Godsil} \subset \text{Tinhofer} \subset \text{Refinable}. \qquad (1)$$

We show the following results on these graph classes:

- The hierarchy (1) is strict.
- Testing membership in any of these graph classes is P-hard.

We prove the last fact by giving a suitable uniform AC^0 many-one reduction from the P-complete monotone boolean circuit-value problem MCVP. More precisely, for a given MCVP instance (C, x) our reduction outputs a graph $G_{C,x}$ such that if $C(x) = 1$ then $G_{C,x}$ is discrete and if $C(x) = 0$ then $G_{C,x}$ is not refinable. In particular, the graph classes Discrete and Amenable are P-complete. We note that Grohe [10] established, for each $k \geq 2$, the P-completeness of the equivalence problem for first-order k-variable logic with counting quantifiers; according to [12], this implies the P-completeness of indistinguishability of two input graphs by color refinement. We adapt the gadget constructions in [10] to show our P-hardness results.

Related Work. Particular families of compact graphs were identified in [5,9,16,21]; see also Chap. 9.10 in the monograph [6]. The concept of compactness is generalized to *weak compactness* in [7,8].

The linear programming approach of [15,17] to isomorphism testing is extended in [3,11,14], where it is shown that this extension corresponds to the k-dimensional Weisfeiler-Leman algorithm (which is just color refinement if $k = 1$).

Notation. The vertex set of a graph G is denoted by $V(G)$. The vertices adjacent to a vertex $u \in V(G)$ form its neighborhood $N(u)$. A set of vertices $X \subseteq V(G)$ induces a subgraph of G, that is denoted by $G[X]$. For two disjoint sets X and Y, $G[X, Y]$ is the bipartite graph with vertex classes X and Y formed by all edges of G connecting a vertex in X with a vertex in Y. The vertex-disjoint union of graphs G and H will be denoted by $G + H$. Furthermore, we write mG for the disjoint union of m copies of G. The *bipartite complement* of a bipartite graph

G with vertex classes X and Y is the bipartite graph G' with the same vertex classes such that $\{x, y\}$ with $x \in X$ and $y \in Y$ is an edge in G' if and only if it is not an edge in G. We use the standard notation K_n for the complete graph on n vertices, $K_{s,t}$ for the complete bipartite graph whose vertex classes have s and t vertices, and C_n for the cycle on n vertices.

All the proofs omitted in this extended abstract can be found in [1].

2   Amenable Graphs

For convenience, we will consider graphs to be vertex-colored in the paper. A *vertex-colored graph* is an undirected simple graph G endowed with a vertex coloring $c : V(G) \rightarrow \{1, \ldots, k\}$. Automorphisms of a vertex-colored graph and isomorphisms between vertex-colored graphs are required to preserve vertex colors. We get usual graphs when c is constant.

Given a graph G, the *color-refinement* algorithm (to be abbreviated as *CR*) iteratively computes a sequence of colorings C^i of $V(G)$. The initial coloring C^0 is the vertex coloring of G, i.e., $C^0(u) = c(u)$. Then,

$$C^{i+1}(u) = \left(C^i(u), \{\!\{ C^i(a) : a \in N(u) \}\!\} \right),$$

where $\{\!\{ \ldots \}\!\}$ denotes a multiset.

The partition $\mathcal{P}^{i+1}$ of $V(G)$ into the color classes of C^{i+1} is a refinement of the partition $\mathcal{P}^i$ corresponding to C^i. It follows that, eventually, $\mathcal{P}^{s+1} = \mathcal{P}^s$ for some s; hence, $\mathcal{P}^i = \mathcal{P}^s$ for all $i \geq s$. The partition $\mathcal{P}^s$ is called the *stable partition* of G and denoted by $\mathcal{P}_G$.

Since the colorings C^i are preserved under isomorphisms, for isomorphic G and H we always have the equality

$$\{\!\{ C^i(u) : u \in V(G) \}\!\} = \{\!\{ C^i(v) : v \in V(H) \}\!\} \tag{2}$$

for all $i \geq 0$ or, equivalently, for $i = 2n$. The CR algorithm accepts two graphs G and H as isomorphic exactly under this condition. To avoid an exponential growth of the lengths of color names, CR renames the colors after each refinement step.

We call a graph G *amenable* if CR works correctly on the input G, H for every H, that is, Equality (2) is false for $i = 2n$ whenever $H \not\cong G$.

The elements of the stable partition $\mathcal{P}_G$ of a graph G will be called *cells*. We define the auxiliary *cell graph* $C(G)$ of G to be the complete graph on the vertex set $\mathcal{P}_G$. A vertex X of the cell graph is called *homogeneous* if the graph $G[X]$ is complete or empty and *heterogeneous* otherwise. An edge $\{X, Y\}$ of the cell graph is called *isotropic* if the bipartite graph $G[X, Y]$ is either complete or empty and *anisotropic* otherwise. By an *anisotropic component* of the cell graph $C(G)$ we mean a maximal connected subgraph of $C(G)$ whose edges are all anisotropic. Note that if a vertex of $C(G)$ has no incident anisotropic edges, it forms a single-vertex anisotropic component.

Theorem 1 (Arvind et al. [2]). *A graph G is amenable if and only if the stable partition $\mathcal{P}_G$ of G fulfils the following three properties:*

(A) *For any cell $X \in \mathcal{P}_G$, $G[X]$ is an empty graph, a complete graph, a matching graph mK_2, the complement of a matching graph, or the 5-cycle;*

(B) *For any two cells $X, Y \in \mathcal{P}_G$, $G[X, Y]$ is an empty graph, a complete bipartite graph, a disjoint union of stars $sK_{1,t}$ where X and Y are the set of s central vertices and the set of st leaves, or the bipartite complement of the last graph.*

(C) *Every anisotropic component A of the cell graph $C(G)$ of G is a tree and contains at most one heterogeneous vertex. If A contains a heterogeneous vertex, it has minimum cardinality among the vertices of A. Let R be any vertex of A having minimum cardinality and let A_R be the rooted directed tree obtained from A by rooting it at R. Then $|X| \leq |Y|$ for any directed edge (X, Y) of A_R.*

3   Amenable Graphs Are Compact

An $n \times n$ real matrix X is *doubly stochastic* if its elements are nonnegative and all its rows and columns sum up to 1. Doubly stochastic matrices are closed under products and convex combinations. The set of all $n \times n$ doubly stochastic matrices forms the *Birkhoff polytope* $B_n \subset \mathbb{R}^{n^2}$. *Permutation matrices* are exactly 0-1 doubly stochastic matrices. By Birkhoff's Theorem (see, e.g. [6]), the $n!$ permutation matrices form precisely the set of all extreme points of B_n. Equivalently, every doubly stochastic matrix is a convex combination of permutation matrices.

Let G and H be graphs with vertex set $\{1, \ldots, n\}$. An isomorphism π from G to H can be represented by the permutation matrix $P_\pi = (p_{ij})$ such that $p_{ij} = 1$ if and only if $\pi(i) = j$. Denote the set of matrices P_π for all isomorphisms π by $\mathrm{Iso}(G, H)$, and let $\mathrm{Aut}(G) = \mathrm{Iso}(G, G)$.

Let A and B be the adjacency matrices of graphs G and H respectively. If the graphs are uncolored, a permutation matrix X is in $\mathrm{Iso}(G, H)$ if and only if $AX = XB$. For vertex-colored graphs, X must additionally satisfy the condition $X[u, v] = 0$ for all pairs of differently colored u and v, i.e., this matrix must be block-diagonal with respect to the color classes. We say that (vertex-colored) graphs G and H are *fractionally isomorphic* if $AX = XB$ for a doubly stochastic matrix X, where $X[u, v] = 0$ if u and v are of different colors. The matrix X is called a *fractional isomorphism*.

Denote the set of all fractional isomorphisms from G to H by $S(G, H)$ and note that it forms a polytope in $\mathbb{R}^{n^2}$. The set of isomorphisms $\mathrm{Iso}(G, H)$ is contained in $\mathrm{Ext}(S(G, H))$, where $\mathrm{Ext}(Z)$ denotes the set of all extreme points of a set Z. Indeed, $\mathrm{Iso}(G, H)$ is the set of integral extreme points of $S(G, H)$.

The set $S(G) = S(G, G)$ is the polytope of *fractional automorphisms* of G. A graph G is called *compact* [17] if $S(G)$ has no other extreme points than $\mathrm{Aut}(G)$, i.e., $\mathrm{Ext}(S(G)) = \mathrm{Aut}(G)$. Compactness of G can equivalently be defined by any of the following two conditions:

- The polytope $S(G)$ is integral;
- Every fractional automorphism of G is a convex combination of automorphisms of G, i.e., $S(G) = \langle \text{Aut}(G) \rangle$, where $\langle Z \rangle$ denotes the convex hull of a set Z.

Example 2. *Complete graphs are compact as a consequence of Birkhoff's theorem. The compactness of trees and cycles is established in [17]. Matching graphs mK_2 are also compact. This is a particular instance of a much more general result by Tinhofer [19]: If G is compact, then mG is compact for any m. Tinhofer [19] also observes that compact graphs are closed under complement.*

For a negative example, note that the graph $C_3 + C_4$ is not compact. This follows from a general result in [19]: All regular compact graphs must be vertex-transitive (and $C_3 + C_4$ is not).

Tinhofer [19] noted that, if G is compact, then for any graph H, either all or none of the extreme points of the polytope $S(G, H)$ are integral. As mentioned in the introduction, this yields a linear-programming based polynomial-time algorithm to test if a compact graph G is isomorphic to any other given graph H. The following result shows that Tinhofer's approach works for all amenable graphs.

Theorem 3. *All amenable graphs are compact.*

Theorem 3 unifies and extends several earlier results providing examples of compact graphs. In particular, it gives another proof of the fact that almost all graphs are compact, which also follows from a result of Godsil [9, Corollary 1.6]. Indeed, while Babai, Erdős, and Selkow [4] proved that almost all graphs are discrete, we already mentioned in Sect. 1 that all discrete graphs are amenable.

Furthermore, Theorem 3 reproves Tinhofer's result that trees are compact.[1] Since also forests are amenable [2], we can extend this result to forests. This extension is not straightforward as compact graphs are not closed under disjoint union; see Example 2. In [18], Tinhofer proves compactness for the class of *strong tree-cographs*, which includes forests only with pairwise non-isomorphic connected components. To the best of our knowledge, compactness of unigraphs, which also follows from Theorem 3, has not been observed earlier. Summarizing, we note the following result.

Corollary 4. *Discrete graphs, forests, and unigraphs are compact.*

In the rest of this section we prove Theorem 3. We will use a known fact on the structure of fractional automorphisms. For a partition $V_1, \ldots, V_m$ of $\{1, \ldots, n\}$ let $X_1, \ldots, X_m$ be matrices, where the rows and columns of X_i are indexed by elements of V_i. Then we denote the block-diagonal matrix with blocks $X_1, \ldots, X_m$ by $X_1 \oplus \cdots \oplus X_m$.

Lemma 5 (Ramana et al. [15]). *Let G be a (vertex-colored) graph on vertex set $\{1, \ldots, n\}$ and assume that the elements $V_1, \ldots, V_m$ of the stable partition $\mathcal{P}_G$ of G are intervals of consecutive integers. Then any fractional automorphism X of G has the form $X = X_1 \oplus \cdots \oplus X_m$.*

---

[1] The proof of Theorem 3 uses only compactness of complete graphs, matching graphs, and the 5-cycle.

Note that the assumption of the lemma can be ensured for any graph by appropriately renaming its vertices. An immediate consequence of Lemma 5 is that a graph G is compact if and only if it is compact with respect to its stable coloring.

Given an amenable graph G and a fractional automorphism X of G, we have to express X as a convex combination of permutation matrices in $\text{Aut}(G)$. Our proof strategy consists in exploiting the structure of amenable graphs as described by Theorem 1. Given an anisotropic component A of the cell graph $C(G)$, we define the *anisotropic component G_A of G* as the subgraph of G induced by the union of all cells belonging to A. Our overall idea is to prove the claim separately for each anisotropic component G_A, applying an inductive argument on the number of cells in A. A key role will be played by the fact that, according to Theorem 1, A is a tree with at most one heterogeneous cell.

We can assume that G is colored by the stable coloring because, by Lemma 5, the colored version has the same polytope of fractional automorphisms. We first consider the case when G consists of a single anisotropic component A. By Theorem 1, the corresponding cell graph $C(G)$ has at most one heterogeneous vertex, and A forms a spanning tree of $C(G)$. Without loss of generality, we can number the cells $V_1, \ldots, V_m$ of G so that V_1 is the unique heterogeneous cell if it exists; otherwise V_1 is chosen among the cells of minimum cardinality. Moreover, we can suppose that, for each $i \leq m$, the cells $V_1, \ldots, V_i$ induce a connected subgraph in the tree A.

We will prove by induction on $i = 1, \ldots, m$ that the graphs $G_i = G[V_1 \cup \cdots \cup V_i]$ are compact. In the base case of $i = 1$, the graph $G_1 = G[V_1]$ is one of the graphs listed in Condition **A** of Theorem 1. All of them are known to be compact; see Example 2. As induction hypothesis, assume that the graph G_{i-1} is compact. For the induction step, we have to show that also G_i is compact.

Denote $D = V_i$. Since G has no more than one heterogeneous cell, $G[D]$ is complete or empty. It will be instructive to think of D as a "leaf" cell having a unique anisotropic link to the remaining part G_{i-1} of G_i. Let $C \in \{V_1, \ldots, V_{i-1}\}$ be the unique cell such that $\{C, D\}$ is an anisotropic edge of $C(G_i)$. To be specific, suppose that $G[C, D] \cong sK_{1,t}$. If $G[C, D]$ is the bipartite complement of $sK_{1,t}$, we can consider the complement of G_i, using the fact that the polytope of fractional automorphisms is the same for a graph and its complement. By the monotonicity property stated in Condition **C** of Theorem 1, $|C| = s$ and $|D| = st$. Let $C = \{c_1, c_2, \ldots, c_s\}$ and, for each j, $N(c_j) \subseteq D$ be the neighborhood of c_j in $G[C, D]$. Thus, $D = \bigcup_{j=1}^{s} N(c_j)$.

Let X be a fractional automorphism of G_i. It is convenient to break it up into three blocks $X = X' \oplus Y \oplus Z$, where Y and Z correspond to C and D respectively, and X' is the rest. By induction hypothesis we have the convex combination

$$X' \oplus Y = \sum_{P' \oplus P \in \text{Aut}(G_{i-1})} \alpha_{P',P}\, P' \oplus P, \tag{3}$$

where $P' \oplus P$ are permutation matrices corresponding to automorphisms π of the graph G_{i-1}, such that the permutation matrix block P denotes the action of π on the color class C and P' the action on the remaining color classes of G_{i-1}.

We need to show that X is a convex combination of automorphisms of G_i. Let A denote the adjacency matrix of G_i, and $A_{S,T}$ denote the submatrix of A row-indexed by $S \subset V(G_i)$ and column-indexed by $T \subset V(G_i)$. Since X is a fractional automorphism of G_i, we have $XA = AX$. Recall that Y and Z are blocks of X corresponding to color classes C and D. Looking at the corner fragments of the matrices XA and AX, we get

$$\begin{pmatrix} Y & 0 \\ 0 & Z \end{pmatrix} \begin{pmatrix} A_{C,C} & A_{C,D} \\ A_{D,C} & A_{D,D} \end{pmatrix} = \begin{pmatrix} A_{C,C} & A_{C,D} \\ A_{D,C} & A_{D,D} \end{pmatrix} \begin{pmatrix} Y & 0 \\ 0 & Z \end{pmatrix},$$

which implies

$$Y A_{C,D} = A_{C,D}\, Z, \tag{4}$$
$$A_{D,C}\, Y = Z\, A_{D,C}. \tag{5}$$

Consider Z as an $st \times st$ matrix whose rows and columns are indexed by the elements of sets $N(c_1), N(c_2), \ldots, N(c_r)$ in that order. We can thus think of Z as an $s \times s$ block matrix of $t \times t$ matrix blocks $Z^{(k,\ell)}, 1 \le k, \ell \le s$. The next claim is a consequence of Eqs. (4) and (5).

Claim 6. *Each block $Z^{(k,\ell)}$ in Z is of the form*

$$Z^{(k,\ell)} = y_{k,\ell} W^{(k,\ell)}, \tag{6}$$

where $y_{k,\ell}$ is the $(k,\ell)^{th}$ entry of Y, and $W^{(k,\ell)}$ is a doubly stochastic matrix.

Proof. We first note from Eq. (4) that the $(k,j)^{th}$ entry of the $s \times st$ matrix $Y A_{C,D} = A_{C,D} Z$ can be computed in two different ways. In the left hand side matrix, it is $y_{k,\ell}$ for each $j \in N(c_\ell)$. On the other hand, the right hand side matrix implies that the same $(k,j)^{th}$ entry is also the sum of the jth column of the $N(c_k) \times N(c_\ell)$ block $Z^{(k,\ell)}$ of the matrix Z.

We conclude, for $1 \le k, \ell \le s$, that each column in $Z^{(k,\ell)}$ adds up to $y_{k,\ell}$. By a similar argument, applied to Eq. (5) this time, it follows, for each $1 \le k, \ell \le s$, that each *row* of any block $Z^{(k,\ell)}$ of Z adds up to $y_{k,\ell}$.

We conclude that, if $y_{k,\ell} \ne 0$, then the matrix $W^{(k,\ell)} = \frac{1}{y_{k,\ell}} Z^{(k,\ell)}$ is doubly stochastic. If $y_{k,\ell} = 0$, then (6) is true for any choice of $W^{(k,\ell)}$. ∎

For every $P = (p_{k\ell})$ appearing in an automorphism $P' \oplus P$ of G_{i-1} (see Eq. (3)), we define the $st \times st$ doubly stochastic matrix W_P by its $t \times t$ blocks indexed by $1 \le k, \ell \le s$ as follows:

$$W_P^{(k,\ell)} = \begin{cases} W^{(k,\ell)} & \text{if } p_{k\ell} = 1, \\ 0 & \text{if } p_{k\ell} = 0. \end{cases} \tag{7}$$

Equations (3) and (6) imply that

$$X = X' \oplus Y \oplus Z = \sum_{P' \oplus P \in \mathrm{Aut}(G_{i-1})} \alpha_{P',P}\, P' \oplus P \oplus W_P. \tag{8}$$

In order to see this, on the left hand side consider the $(k, \ell)^{th}$ block $Z^{(k,\ell)}$ of Z. On the right hand side, note that the corresponding block in each $P' \oplus P \oplus W_P$ is the matrix $W^{(k,\ell)}$. Clearly, the overall coefficient for this block equals the sum of $\alpha_{P',P}$ over all P' and P such that $p_{k,\ell} = 1$, which is precisely $y_{k,\ell}$ by Eq. (3).

Since each $W^{(k,\ell)}$ is a doubly stochastic matrix, by Birkhoff's theorem we can write it as a convex combination of $t \times t$ permutation matrices $Q_{j,k,\ell}$, whose rows are indexed by elements of $N(c_k)$ and columns by elements of $N(c_\ell)$:

$$W^{(k,\ell)} = \sum_{j=1}^{t!} \beta_{j,k,\ell} Q_{j,k,\ell}.$$

Substituting the above expression in Eq. (7), that defines the doubly stochastic matrix W_P, we express W_P as a convex combination of permutation matrices $W_P = \sum_Q \delta_{Q,P} Q$ where Q runs over all $st \times st$ permutation matrices indexed by the vertices in color class D. Notice here that $\delta_{Q,P}$ is nonzero only for those permutation matrices Q that have structure similar to that described in Eq. (7): The block $Q^{(k,\ell)}$ is a null matrix if $p_{k\ell} = 0$ and it is some $t \times t$ permutation matrix if $p_{k\ell} = 1$. For each such Q, the $(s + st) \times (s + st)$ permutation matrix $P \oplus Q$ is an automorphism of the subgraph $G_i[C, D] = sK_{1,t}$ (because Q maps $N(c_i)$ to $N(c_j)$ whenever P maps c_i to c_j). Since $P \in \mathrm{Aut}(G_i[C])$ and D is a homogeneous set in G_i, we conclude that, moreover, $P \oplus Q$ is an automorphism of the subgraph $G_i[C \cup D]$.

Now, if we plug the above expression for each W_P in Eq. (8), we will finally obtain the desired convex combination

$$X = \sum_{P',P,Q} \gamma_{P',P,Q} \, P' \oplus P \oplus Q.$$

It remains to argue that every $P' \oplus P \oplus Q$ occurring in this sum is an automorphism of G_i. Recall that a pair P', P can appear here only if $P' \oplus P \in \mathrm{Aut}(G_{i-1})$. Moreover, if such a pair is extended to a matrix $P' \oplus P \oplus Q$, then $P \oplus Q \in \mathrm{Aut}(G_i[C \cup D])$. Since $G_i[B, D]$ is isotropic for every color class $B \neq D$ of G_i, we conclude that $P' \oplus P \oplus Q \in \mathrm{Aut}(G_i)$. This completes the induction step and finishes the case when G has one anisotropic component.

Next, we consider the case when $C(G)$ has several anisotropic components $T_1, \ldots, T_k$, $k \geq 2$. Let $G_1, \ldots, G_k$, where $G_i = G[\bigcup_{U \in V(T_i)} U]$, be the corresponding anisotropic components of G. By the proof of the previous case we already know that G_i is compact for each i.

Claim 7. *The automorphism group $\mathrm{Aut}(G)$ of G is the product of the automorphism groups $\mathrm{Aut}(G_i)$, $1 \leq i \leq k$.*

Proof. Recall that any automorphism of G must map each color class of G, which is a cell of the underlying amenable graph, onto itself. Thus, any automorphism π of G is of the form $(\pi_1, \ldots, \pi_k)$, where π_i is an automorphism of the subgraph G_i. Now, for any two subgraphs G_i and G_j, we examine the edges between $V(G_i)$ and $V(G_j)$. For any color classes $U \subseteq V(G_i)$ and $U' \subseteq V(G_j)$, the edge

$\{U, U'\}$ is isotropic because it is not contained in any anisotropic component of $C(G)$. Therefore, the bipartite graph $G[U, U']$ is either complete or empty. It follows that for any automorphisms π_i of G_i, $1 \leq i \leq k$, the permutation $\pi = (\pi_1, \ldots, \pi_k)$ is an automorphism of the graph G. ∎

As follows from Lemma 5, any fractional automorphism X of G is of the form $X = X_1 \oplus \cdots \oplus X_k$, where X_i is a fractional automorphism of G_i for each i. As each G_i is compact we can write each X_i as a convex combination $X_i = \sum_{\pi \in \mathrm{Aut}(G_i)} \alpha_{i,\pi} P_\pi$. This implies

$$I \oplus \cdots \oplus I \oplus X_i \oplus I \oplus \cdots \oplus I = \sum_{\pi \in \mathrm{Aut}(G_i)} \alpha_{i,\pi} I \oplus \cdots \oplus I \oplus P_\pi \oplus I \oplus \cdots \oplus I, \quad (9)$$

where block diagonal matrices in the above expression have X_i and P_π respectively in the ith block (indexed by elements of $V(G_i)$) and identity matrices as the remaining blocks.

We now decompose the fractional automorphism X as a matrix product of fractional automorphisms of G

$$X = X_1 \oplus \cdots \oplus X_k = (X_1 \oplus I \oplus \cdots \oplus I) \cdot (I \oplus X_2 \oplus \cdots \oplus I) \cdot \cdots \cdot (I \oplus \cdots \oplus I \oplus X_k).$$

Substituting for $I \oplus \cdots \oplus I \oplus X_i \oplus I \oplus \cdots \oplus I$ from Eq. (9) in the above expression and writing the product of sums as a sum of products, we see that X is a convex combination of permutation matrices of the form $P_{\pi_1} \oplus \cdots \oplus P_{\pi_k}$ where $\pi_i \in \mathrm{Aut}(G_i)$ for each i. By Claim 7, all the terms $P_{\pi_1} \oplus \cdots \oplus P_{\pi_k}$ correspond to automorphisms of G. Hence, G is compact, completing the proof of Theorem 3.

4   A Color-Refinement Based Hierarchy of Graphs

Let $u \in V(G)$ and $v \in V(H)$ be vertices of two graphs G and H. By *individualization* of u and v we mean assigning the same *new color* to u and v, which makes them distinguished from the remaining vertices of G and H. Tinhofer [19] proved that, if G is compact, then the following polynomial-time algorithm correctly decides if G and H are isomorphic.

1. Run CR on G and H until the coloring of $V(G) \cup V(H)$ stabilizes.
2. If the multisets of colors in G and H are different, then output "non-isomorphic" and stop. Otherwise,
   (a) if all color classes are singletons in G and H, then if the map $u \mapsto v$ matching each vertex $u \in V(G)$ to the vertex $v \in V(H)$ of the same color is an isomorphism, output "isomorphic" and stop. Else output "non-isomorphic" and stop.
   (b) pick any color class with at least two vertices in both G and H, select an arbitrary $u \in V(G)$ and $v \in V(H)$ in this color class and individualize them. Goto Step 1.

If G and H are any two non-isomorphic graphs, then Tinhofer's algorithm will always output "non-isomorphic". However, it can fail for isomorphic input graphs, in general. We call G a *Tinhofer graph* if the algorithm works correctly on G and every H for all choices of vertices to be individualized. Thus, the result of [19] can be stated as the inclusion Compact $\subseteq$ Tinhofer. If G is a Tinhofer graph, then the above algorithm can be used to even find a canonical labeling of G (i.e. a relabeling of the vertices such that isomorphic graphs become equal). Using Theorem 3, we obtain the following fact.

Corollary 8. *Amenable and, more generally, compact graphs admit canonical labeling in polynomial time.*

A partition $\mathcal{P}$ of the vertex set of a graph G is *equitable* if, for any two elements X and Y of $\mathcal{P}$, every vertex $x \in X$ has the same number of neighbors in Y. Note that the stable partition $\mathcal{P}_G$ produced by CR on input G is equitable.

Let A be a subgroup of the automorphism group $Aut(G)$ of a graph G. Then the partition of $V(G)$ into the A-orbits is called an *orbit partition* of G. Any orbit partition of G is equitable, but the converse is not true, in general. However, Godsil [9, Corollary 1.3] has shown that the converse holds for compact graphs. We define *Godsil graphs* as the graphs for which the two notions of an equitable and an orbit partition coincide, that is, every equitable partition is the orbit partition of some group of automorphisms A. Thus, the result of [9] can be stated as the inclusion Compact $\subseteq$ Godsil. Now, the inclusion Compact $\subseteq$ Tinhofer can easily be strengthened as follows.

Lemma 9. *Any Godsil graph is a Tinhofer graph.*

The orbit partition of G with respect to $Aut(G)$ is always a refinement of the stable partition $\mathcal{P}_G$ of G. We call G *refinable* if $\mathcal{P}_G$ is the orbit partition of $Aut(G)$. It is easy to show that Tinhofer graphs are refinable.

Lemma 10. *Any Tinhofer graph is refinable.*

Summarizing Theorem 3, Lemmas 9 and 10, and [9, Corollary 1.3], we state the following hierarchy result.

Theorem 11. *The classes of graphs under consideration form the inclusion chain*

$$\text{Discrete} \subset \text{Amenable} \subset \text{Compact} \subset \text{Godsil} \subset \text{Tinhofer} \subset \text{Refinable}. \tag{10}$$

Moreover, all of the inclusions are strict.

It is worth of noting that the hierarchy (10) collapses to Discrete if we restrict ourselves to only rigid graphs, i.e., graphs with trivial automorphism group.

Finally, we show that testing membership in each of the graph classes in the hierarchy (10) is P-hard.

Theorem 12. *The recognition problem of each of the classes in the hierarchy (10) is P-hard under uniform AC^0 many-one reductions.*

References

1. Arvind, V., Köbler, J., Rattan, G., Verbitsky, O.: Graph isomorphism, color refinement, and compactness. ECCC TR15-032 (2015)
2. Arvind, V., Köbler, J., Rattan, G., Verbitsky, O.: On the power of color refinement. In: Proceedings of the 20th International Symposium on Fundamentals of Computation Theory (FCT), Lecture Notes in Computer Science. Springer (2015, to appear)
3. Atserias, A., Maneva, E.N.: Sherali-Adams relaxations and indistinguishability in counting logics. SIAM J. Comput. **42**(1), 112–137 (2013)
4. Babai, L., Erdös, P., Selkow, S.M.: Random graph isomorphism. SIAM J. Comput. **9**(3), 628–635 (1980)
5. Brualdi, R.A.: Some applications of doubly stochastic matrices. Linear Algebra Appl. **107**, 77–100 (1988)
6. Brualdi, R.A.: Combinatorial Matrix Classes. Cambridge University Press, Cambridge (2006)
7. Evdokimov, S., Karpinski, M., Ponomarenko, I.N.: Compact cellular algebras and permutation groups. Discrete Math. **197–198**, 247–267 (1999)
8. Evdokimov, S., Ponomarenko, I.N., Tinhofer, G.: Forestal algebras and algebraic forests (on a new class of weakly compact graphs). Discrete Math. **225**(1–3), 149–172 (2000)
9. Godsil, C.: Compact graphs and equitable partitions. Linear Algebra Appl. **255**(13), 259–266 (1997)
10. Grohe, M.: Equivalence in finite-variable logics is complete for polynomial time. Combinatorica **19**(4), 507–532 (1999)
11. Grohe, M., Otto, M.: Pebble games and linear equations. In: Computer Science Logic (CSL 2012), vol. 16. LIPIcs, pp. 289–304 (2012)
12. Immerman, N., Lander, E.: Describing graphs: a first-order approach to graph canonization. In: Selman, A.L. (ed.) Complexity Theory Retrospective, pp. 59–81. Springer, New York (1990)
13. Kiefer, S., Schweitzer, P., Selman, E.: Graphs identified by logics with counting. In: Italiano, G.F., et al. (eds.) MFCS 2015, Part I. LNCS, vol. 9234, pp. 319–330. Springer, Heidelberg (2015)
14. Malkin, P.N.: Sherali-Adams relaxations of graph isomorphism polytopes. Discrete Optim. **12**, 73–97 (2014)
15. Ramana, M.V., Scheinerman, E.R., Ullman, D.: Fractional isomorphism of graphs. Discrete Math. **132**(1–3), 247–265 (1994)
16. Schreck, H., Tinhofer, G.: A note on certain subpolytopes of the assignment polytope associated with circulant graphs. Linear Algebra Appl. **111**, 125–134 (1988)
17. Tinhofer, G.: Graph isomorphism and theorems of Birkhoff type. Computing **36**, 285–300 (1986)
18. Tinhofer, G.: Strong tree-cographs are Birkhoff graphs. Discrete Appl. Math. **22**(3), 275–288 (1989)
19. Tinhofer, G.: A note on compact graphs. Discrete Appl. Math. **30**(2–3), 253–264 (1991)
20. Tinhofer, G., Klin, M.: Algebraic combinatorics in mathematical chemistry. Methods and algorithms III. Graph invariants and stabilization methods. Technical report TUM-M9902, Technische Universität München (1999)
21. Wang, P., Li, J.S.: On compact graphs. Acta Math. Sinica **21**(5), 1087–1092 (2005)

On the Complexity of Noncommutative Polynomial Factorization

V. Arvind[1], Gaurav Rattan[1](✉), and Pushkar Joglekar[2]

[1] The Institute of Mathematical Sciences, Chennai 600 113, India
{arvind,grattan}@imsc.res.in
[2] Vishwakarma Institute of Technology, Pune, India
joglekar.pushkar@gmail.com

Abstract. In this paper we study the complexity of factorization of polynomials in the free noncommutative ring $\mathbb{F}\langle x_1, x_2, \ldots, x_n \rangle$ of polynomials over the field $\mathbb{F}$ and noncommuting variables $x_1, x_2, \ldots, x_n$. Our main results are the following:

- Although $\mathbb{F}\langle x_1, \ldots, x_n \rangle$ is not a unique factorization ring, we note that *variable-disjoint* factorization in $\mathbb{F}\langle x_1, \ldots, x_n \rangle$ has the uniqueness property. Furthermore, we prove that computing the variable-disjoint factorization is polynomial-time equivalent to Polynomial Identity Testing (both when the input polynomial is given by an arithmetic circuit or an algebraic branching program). We also show that variable-disjoint factorization in the black-box setting can be efficiently computed (where the factors computed will be also given by black-boxes, analogous to the work [12] in the commutative setting).
- As a consequence of the previous result we show that homogeneous noncommutative polynomials and multilinear noncommutative polynomials have unique factorizations in the usual sense, which can be efficiently computed.
- Finally, we discuss a polynomial decomposition problem in $\mathbb{F}\langle x_1, \ldots, x_n \rangle$ which is a natural generalization of homogeneous polynomial factorization and prove some complexity bounds for it.

1 Introduction

Let $\mathbb{F}$ be any field and $X = \{x_1, x_2, \ldots, x_n\}$ be a set of n free noncommuting variables. Let X^* denote the set of all free words (which are monomials) over the alphabet X with concatenation of words as the monoid operation and the empty word ϵ as identity element.

The *free noncommutative ring* $\mathbb{F}\langle X \rangle$ consists of all finite $\mathbb{F}$-linear combinations of monomials in X^*, where the ring addition $+$ is coefficient-wise addition and the ring multiplication $*$ is the usual convolution product. More precisely, let $f, g \in \mathbb{F}\langle X \rangle$ and let $f(m) \in \mathbb{F}$ denote the coefficient of monomial m in polynomial f. Then we can write $f = \sum_m f(m)m$ and $g = \sum_m g(m)m$, and in the product polynomial fg for each monomial m we have

$$fg(m) = \sum_{m_1 m_2 = m} f(m_1)g(m_2).$$

© Springer-Verlag Berlin Heidelberg 2015
G.F. Italiano et al. (Eds.): MFCS 2015, Part II, LNCS 9235, pp. 38–49, 2015.
DOI: 10.1007/978-3-662-48054-0_4

The *degree* of a monomial $m \in X^*$ is the length of the monomial m, and the degree $\deg f$ of a polynomial $f \in \mathbb{F}\langle X \rangle$ is the degree of a largest degree monomial in f with nonzero coefficient. For polynomials $f, g \in \mathbb{F}\langle X \rangle$ we clearly have $\deg(fg) = \deg f + \deg g$.

A *nontrivial factorization* of a polynomial $f \in \mathbb{F}\langle X \rangle$ is an expression of f as a product $f = gh$ of polynomials $g, h \in \mathbb{F}\langle X \rangle$ such that $\deg g > 0$ and $\deg h > 0$. A polynomial $f \in \mathbb{F}\langle X \rangle$ is *irreducible* if it has no nontrivial factorization and is *reducible* otherwise. For instance, all degree 1 polynomials in $\mathbb{F}\langle X \rangle$ are irreducible. Clearly, by repeated factorization every polynomial in $\mathbb{F}\langle X \rangle$ can be expressed as a product of irreducibles.

In this paper we study the algorithmic complexity of polynomial factorization in the free ring $\mathbb{F}\langle X \rangle$.

Polynomial Factorization Problem. The problem of polynomial factorization in the *commutative* polynomial ring $\mathbb{F}[x_1, x_2, \ldots, x_n]$ is a classical well-studied problem in algorithmic complexity culminating in Kaltofen's celebrated efficient factorization algorithm [11]. Kaltofen's algorithm builds on efficient algorithms for univariate polynomial factorization; there are deterministic polynomial-time algorithms over rationals and over fields of unary characteristic and randomized polynomial-time over large characteristic fields (the textbook [9] contains a comprehensive excellent treatment of the subject). The basic idea in Kaltofen's algorithm is essentially a randomized reduction from multivariate factorization to univariate factorization using Hilbert's irreducibility theorem. Thus, we can say that Kaltofen's algorithm uses randomization in two ways: the first is in the application of Hilbert's irreducibility theorem, and the second is in dealing with *univariate* polynomial factorization over fields of large characteristic. In a recent paper Kopparty et al. [13] have shown that the first of these requirements of randomness can be eliminated, assuming an efficient algorithm as subroutine for the problem of *polynomial identity testing* for small degree polynomials given by circuits. More precisely, it is shown in [13] that over finite fields of unary characteristic (or over rationals) polynomial identity testing is deterministic polynomial-time equivalent to multivariate polynomial factorization.

Thus, in the commutative setting it turns out that the complexity of multivariate polynomial factorization is closely related to polynomial identity testing (whose deterministic complexity is known to be related to proving superpolynomial size arithmetic circuit lower bounds).

Noncommutative Polynomial Factorization. The study of noncommutative arithmetic computation was initiated by Nisan [14] in which he showed exponential size lower bounds for algebraic branching programs that compute the noncommutative permanent or the noncommutative determinant. Noncommutative polynomial identity testing was studied in [5,15]. In [5] a randomized polynomial time algorithm is shown for identity testing of polynomial degree noncommutative arithmetic circuits. For algebraic branching programs [15] give a deterministic polynomial-time algorithm. Proving superpolynomial size lower bounds for noncommutative arithmetic circuits computing the noncommutative permanent is open. Likewise, obtaining a deterministic polynomial-time identity test for polynomial degree noncommutative circuits is open.

In this context, it is interesting to ask if we can relate the complexity of noncommutative factorization to noncommutative polynomial identity testing. However, there are various mathematical issues that arise in the study of non-commutative polynomial factorization.

Unlike in the commutative setting, the noncommutative polynomial ring $\mathbb{F}\langle X \rangle$ is *not* a unique factorization ring. A well-known example is the polynomial

$$xyx + x$$

which has two factorizations: $x(yx + 1)$ and $(xy + 1)x$. Both $xy + 1$ and $yx + 1$ are irreducible polynomials in $\mathbb{F}\langle X \rangle$.

There is a detailed theory of factorization in noncommutative rings [7,8]. We will mention an interesting result on the structure of polynomial factorizations in the ring $R = \mathbb{F}\langle X \rangle$. Two elements $a, b \in R$ are *similar* if there are elements $a', b' \in R$ such that $ab' = a'b$, and (i) a and a' do not have common nontrivial left factors, (ii) b and b' do not have common nontrivial right factors. Among other results, Cohn [8] has shown the following interesting theorem about factorizations in the ring $R = \mathbb{F}\langle X \rangle$.

Theorem 1.1 (Cohn's Theorem). *For a polynomial $a \in \mathbb{F}\langle X \rangle$ let*

$$a = a_1 a_2 \ldots a_r \text{ and } a = b_1 b_2 \ldots b_s$$

be any two factorizations of a into irreducible polynomials in $\mathbb{F}\langle X \rangle$. Then $r = s$, and there is a permutation π of the indices $\{1, 2, \ldots, r\}$ such that a_i and $b_{\pi(i)}$ are similar polynomials for $1 \leq i \leq r$.

For instance, consider the two factorizations of $xyx + x$ above. We note that polynomials $xy + 1$ and $yx + 1$ are similar. It is easy to construct examples of degree d polynomials in $\mathbb{F}\langle X \rangle$ that have $2^{\Omega(d)}$ distinct factorizations. Cohn [7] discusses a number of interesting properties of factorizations in $\mathbb{F}\langle X \rangle$. But it is not clear how to algorithmically exploit these to obtain an efficient algorithm in the general case.

Our Results. In this paper, we study some *restricted* cases of polynomial factorization in the ring $\mathbb{F}\langle X \rangle$ and prove the following results. In this extended abstract all missing proofs are given in the appendix. The full version of the paper is available at [3].

- We consider *variable-disjoint* factorization of polynomials in $\mathbb{F}\langle X \rangle$ into *variable-disjoint irreducibles*. It turns out that such factorizations are unique and computing them is polynomial-time equivalent to polynomial identity testing (for both noncommutative arithmetic circuits and algebraic branching programs).
- It turns out that we can apply the algorithm for variable-disjoint factorization to two special cases of factorization in $\mathbb{F}\langle X \rangle$: homogeneous polynomials and multilinear polynomials. These polynomials do have unique factorizations and we obtain efficient algorithms for computing them.

- We also study a natural polynomial decomposition problem for noncommutative polynomials and obtain complexity results.

2 Variable-Disjoint Factorization Problem

In this section we consider the problem of factorizing a noncommutative polynomial $f \in \mathbb{F}\langle X \rangle$ into variable disjoint factors.

For a polynomial $f \in \mathbb{F}\langle X \rangle$ let $Var(f) \subseteq X$ denote the set of all variables occurring in nonzero monomials of f.

Definition 2.1. *A* nontrivial *variable-disjoint factorization of a polynomial $f \in \mathbb{F}\langle X \rangle$ is a factorization*

$$f = gh$$

such that $\deg g > 0$ *and* $\deg h > 0$, *and* $Var(g) \cap Var(h) = \emptyset$.

A polynomial f is variable-disjoint irreducible *if it does not have a nontrivial variable-disjoint factorization.*

Clearly, all irreducible polynomials are also variable-disjoint irreducible. But the converse is not true. For instance, the familiar polynomial $xyx + x$ is variable-disjoint irreducible but not irreducible. Furthermore, all univariate polynomials in $\mathbb{F}\langle X \rangle$ are variable-disjoint irreducible.

We will study the complexity of variable-disjoint factorization for noncommutative polynomials. First of all, it is interesting that although we do not have the usual unique factorization in the ring $\mathbb{F}\langle X \rangle$, we can prove that every polynomial in $\mathbb{F}\langle X \rangle$ has a *unique* variable-disjoint factorization into variable-disjoint irreducible polynomials.[1]

We can exploit the properties we use to show uniqueness of variable-disjoint factorization for computing the variable-disjoint factorization. Given $f \in \mathbb{F}\langle X \rangle$ as input by a noncommutative arithmetic circuit the problem of computing arithmetic circuits for the variable-disjoint irreducible factors of f is polynomial-time reducible to PIT for noncommutative arithmetic circuits. An analogous result holds for f given by an algebraic branching programs, abbreviated as ABPs. (The reader is referred to the excellent survey of [17] for technical definitions). Hence, there is a deterministic polynomial-time algorithm for computing the variable-disjoint factorization of f given by an ABP. Also in the case when the polynomial $f \in \mathbb{F}\langle X \rangle$ is given as input by a black-box (appropriately defined) we give an efficient algorithm that gives black-box access to each variable-disjoint irreducible factor of f.

Remark 2.2. Factorization of *commutative* polynomials into variable-disjoint factors is studied by Shpilka and Volkovich in [16]. They show a deterministic reduction to polynomial identity testing. However, the techniques used in their paper are specific to commutative rings, involving scalar substitutions, and do not appear useful in the noncommutative case. Our techniques for factorization are simple, essentially based on computing left and right partial derivatives of noncommutative polynomials given by circuits or branching programs.

---

[1] Uniqueness of the factors is up to scalar multiplication.

2.1   Uniqueness of Variable-Disjoint Factorization

Although the ring $\mathbb{F}\langle X \rangle$ is not a unique factorization domain we show that factorization into variable-disjoint irreducible factors is unique.

For a polynomial $f \in \mathbb{F}\langle X \rangle$ let $mon(f)$ denote the set of monomials which occur in f with non-zero coefficients.

Lemma 2.3. *Let $f = gh$ such that $Var(g) \cap Var(h) = \emptyset$ and $|Var(g)|, |Var(h)| \geq 1$. Then*

$$mon(f) = \{mw | m \in mon(g), w \in mon(h)\}.$$

Moreover, the coefficient of mw in f is the product of the coefficients of m in g and w in h.

Lemma 2.4. *Let $f = g.h$ and $f = u.v$ be two nontrivial variable-disjoint factorizations of f. That is,*

$$Var(g) \cap Var(h) = \emptyset$$
$$Var(u) \cap Var(v) = \emptyset.$$

Then either $Var(g) \subseteq Var(u)$ and $Var(h) \supseteq Var(v)$ or $Var(u) \subseteq Var(g)$ and $Var(v) \supseteq Var(h)$.

Lemma 2.5. *Let $f \in \mathbb{F}\langle X \rangle$ and suppose $f = gh$ and $f = uv$ are two variable-disjoint factorizations of f such that $Var(g) = Var(u)$. Then $g = \alpha u$ and $h = \beta v$ for scalars $\alpha, \beta \in \mathbb{F}$.*

Using the lemmas above we prove the uniqueness of variable-disjoint factorizations in $\mathbb{F}\langle X \rangle$.

Theorem 2.6. *Every polynomial in $\mathbb{F}\langle X \rangle$ has a unique variable-disjoint factorization as a product of variable-disjoint irreducible factors, where the uniqueness is upto scalar multiples of the irreducible factors.*

2.2   Equivalence with PIT

Theorem 2.7. *Let $f \in \mathbb{F}\langle X \rangle$ be a polynomial as input instance for variable-disjoint factorization. Then*

1. *If f is input by an arithmetic circuit of degree d and size s there is a randomized $\mathrm{poly}(s,d)$ time algorithm that factorizes f into variable-disjoint irreducible factors.*
2. *If f is input by an algebraic branching program there is a deterministic polynomial-time algorithm that factorizes f into its variable-disjoint irreducible factors.*

Next we consider variable-disjoint factorization of polynomials input in sparse representation and show that the problem is solvable in deterministic logspace (even by constant-depth circuits). Recall that AC^0 circuits mean a family of circuits $\{C_n\}$, where C_n for length n inputs, such that C_n has polynomially bounded size and constant-depth and is allowed unbounded fanin AND and OR gates. The class of TC^0 circuits is similarly defined, but is additionally allowed unbounded fanin majority gates. The logspace uniformity condition means that there is a logspace transducer that outputs C_n on input 1^n for each n.

Theorem 2.8. *Let $f \in \mathbb{F}\langle X \rangle$ be a polynomial input instance for variable-disjoint factorization given in sparse representation.*

(a) When $\mathbb{F}$ is a fixed finite field the variable-disjoint factorization is computable in deterministic logspace (more precisely, even by logspace-uniform AC^0 circuits).

(b) When $\mathbb{F}$ is the field of rationals the variable-disjoint factorization is computable in deterministic logspace (even by logspace-uniform TC^0 circuits).

We now observe that PIT for noncommutative arithmetic circuits is also deterministic polynomial-time reducible to variable-disjoint factorization, making the problems polynomial-time equivalent.

Theorem 2.9. *Polynomial identity testing for noncommutative polynomials $f \in \mathbb{F}\langle X \rangle$ given by arithmetic circuits (of polynomial degree) is deterministic polynomial-time equivalent to variable-disjoint factorization of polynomials given by noncommutative arithmetic circuits.*

2.3   Black-Box Variable-Disjoint Factorization Algorithm

In this subsection we give an algorithm for variable-disjoint factorization when the input polynomial $f \in \mathbb{F}\langle X \rangle$ is given by black-box access. We explain the black-box model below:

In this model, the polynomial $f \in \mathbb{F}\langle X \rangle$ can be evaluated on any n-tuple of matrices $(M_1, M_2, \ldots, M_n)$ where each M_i is a $t \times t$ matrix over $\mathbb{F}$ (or a suitable extension field of $\mathbb{F}$) and get the resulting $t \times t$ matrix $f(M_1, M_2, \ldots, M_n)$ as output.

The algorithm for black-box variable-disjoint factorization takes as input such a black-box access for f and outputs black-boxes for each variable-disjoint irreducible factor of f. More precisely, the factorization algorithm, on input i, makes calls to the black-box for f and works as black-box for the ith variable-disjoint irreducible factor of f, for each i.

The efficiency of the algorithm is measured in terms of the number of calls to the black-box and the size t of the matrices. We will design a variable disjoint factorization algorithm that makes polynomially many black-box queries to f on matrices of polynomially bounded dimension. First we state the theorem formally.

Theorem 2.10. *Suppose $f \in \mathbb{F}\langle X \rangle$ is a polynomial of degree bounded by D, given as input via black-box access. Let $f = f_1 f_2 \ldots f_r$ be the variable-disjoint factorization of f. Then there is a polynomial-time algorithm that, on input i, computes black-box access to the ith factor f_i.*

Remark 2.11. The variable-disjoint irreducible factors of f are unique only upto scalar multiples. However, we note that the algorithm in Theorem 2.10 computes as black-box some *fixed* scalar multiple for each variable-disjoint irreducible factor f_i. However, it can be ensured that the product $f_1 f_2 \ldots f_r$ equals f by appropriate scaling in the algorithm.

Proof of the Theorem 2.10 closely follows the white-box algorithm as in Sect. 2.2. The complete proof of Theorem 2.10 is in the Appendix. For the proof, we need to compute a largest degree monomial of f and create efficient black-box access for left and right partial derivatives of f with respect to any given monomial m'. We create such an efficient black-box access in the next two lemmas.

Lemma 2.12. *Given a polynomial f of degree d with black-box access, we can compute a degree-d nonzero monomial of f, if it exists, with at most nd calls to the black-box on $(d+1)2d \times (d+1)2d$ matrices.*

In the next lemma we show that, there is an efficient algorithm to create black box access for left and right partial derivatives of f w.r.t. monomials. Let $m \in X^*$ be a monomial. We recall that the left partial derivative $\frac{\partial^\ell f}{\partial m}$ of f w.r.t. m is the polynomial

$$\frac{\partial^\ell f}{\partial m} = \sum_{f(mm') \neq 0} f(mm')m'.$$

Similarly, the right partial derivative $\frac{\partial^r f}{\partial m}$ of f w.r.t. m is the polynomial

$$\frac{\partial^r f}{\partial m} = \sum_{f(m'm) \neq 0} f(m'm)m'.$$

Lemma 2.13. *Given a polynomial f of degree d with black-box access, there are efficient algorithms that give black-box access to the polynomials $\frac{\partial^\ell f}{\partial m_1}$, $\frac{\partial^r f}{\partial m_2}$ for any monomials $m_1, m_2 \in X^*$. Furthermore, there is also an efficient algorithm giving black-box access to the polynomial $\frac{\partial^\ell}{\partial m_1} \left(\frac{\partial^r f}{\partial m_2} \right)$.*

3   Factorization of Multilinear and Homogeneous Polynomials

In this section we briefly discuss two interesting special cases of the standard factorization problem for polynomials in $\mathbb{F}\langle X \rangle$. Namely, the factorization of multilinear polynomials and the factorization of homogeneous polynomials. It turns out, as we show, that factorization of multilinear polynomials coincides with their variable-disjoint factorization. In the case of homogeneous polynomials, it turns out that by renaming variables we can reduce the problem to variable-disjoint factorization.

A polynomial $f \in \mathbb{F}\langle X \rangle$ is *multilinear* if in every nonzero monomial of f every variable in X occurs at most once. Let $\text{Var}(f)$ denote the set of all indeterminates from X which appear in some nonzero monomial of f.

Lemma 3.1. *Let $f \in \mathbb{F}\langle X \rangle$ be a multilinear polynomial and $f = gh$ be any nontrivial factorization of f. Then, $Var(g) \cap Var(h) = \emptyset$.*

Thus, factorization and variable-disjoint factorization of multilinear polynomials coincide. Hence, by Theorem 2.6, multilinear polynomials in $\mathbb{F}\langle X \rangle$ have unique factorization. Furthermore, the algorithms described in Sect. 2.2 can be applied to efficiently factorize multilinear polynomials.

We now briefly consider factorization of homogeneous polynomials in $\mathbb{F}\langle X \rangle$.

Definition 3.2. *A polynomial $f \in \mathbb{F}\langle X \rangle$ is said to be* homogeneous *of degree d if every nonzero monomial of f is of degree d.*

Homogeneous polynomials do have the unique factorization property. This is attributed to J.H. Davenport in [6]. However, we argue this by reducing the problem to variable-disjoint factorization.

Given a degree-d homogeneous polynomial $f \in \mathbb{F}\langle X \rangle$, we apply the following simple transformation to f: For each variable $x_i \in X$ we introduce d variables $x_{i1}, x_{i2}, \ldots, x_{id}$. For each monomial $m \in mon(f)$, we replace the occurrence of variable x_i in the jth position of m by variable x_{ij}. The new polynomial f' is in $\mathbb{F}\langle\{x_{ij}\}\rangle$. The crucial property of homogeneous polynomials we use is that for any factorization $f = gh$ both g and h must be homogeneous.

Lemma 3.3. *Let $f \in \mathbb{F}\langle X \rangle$ be a homogeneous degree d polynomial and f' be the polynomial in $\mathbb{F}\langle\{x_{ij}\}\rangle$ obtained as above. Then*

- *The polynomial f' is variable-disjoint irreducible iff f is irreducible.*
- *If $f' = g'_1 g'_2 \ldots g'_t$ is the variable-disjoint factorization of f', where each g'_k is variable-disjoint irreducible then, correspondingly $f = g_1 g_2 \ldots g_t$ is a factorization of f into irreducibles g_k, where g_k is obtained from g'_k by replacing each variable x_{ij} in g'_k by x_i.*

It follows that factorization of homogeneous polynomials can be reduced to variable-disjoint factorization. Using Theorems 2.7 and 2.8, we thus obtain

Theorem 3.4. *Homogeneous polynomials $f \in \mathbb{F}\langle X \rangle$ have unique factorizations into irreducible polynomials. Moreover, this factorization can be efficiently computed:*

- *Computing the factorization of a homogeneous polynomial f given by an arithmetic circuit of polynomial degree is polynomial-time reducible to computing the variable-disjoint factorization of a polynomial given by an arithmetic circuit.*

- *Factorization of f given by an ABP is constant-depth reducible to variable-disjoint factorization of polynomials given by ABPs.*
- *Factorization of f given in sparse representation is constant-depth reducible to variable-disjoint factorization of polynomials given by sparse representation.*

Homogeneous polynomials in $\mathbb{F}\langle X \rangle$ given by black-box access can also be efficiently factorized by suitably adapting the algorithm for variable-disjoint factorization in the previous section.

4   A Polynomial Decomposition Problem

Given a degree d homogeneous noncommutative polynomial $f \in \mathbb{F}\langle X \rangle$, a number k in unary as input we consider the following decomposition problem, denoted by SOP (for sum of products decomposition):

Does f admit a decomposition of the form

$$f = g_1 h_1 + \cdots + g_k h_k?$$

where each $g_i \in \mathbb{F}\langle X \rangle$ is a homogeneous polynomial of degree d_1 and each $h_i \in \mathbb{F}\langle X \rangle$ is a homogeneous polynomial of degree d_2. Notice that this problem is a generalization of homogeneous polynomial factorization. Indeed, homogeneous factorization is simply the case when $k = 1$.

Remark 4.1. As mentioned in [2], it is interesting to note that for *commutative polynomials* the complexity of SOP is open even in the case $k = 2$. However, when f is of constant degree then it can be solved efficiently by applying a very general algorithm [2] based on a regularity lemma for polynomials.

When the input polynomial f is given by an arithmetic circuit, we show that SOP is in MA ∩ coNP. On the other hand, when f is given by an algebraic branching program then SOP can be solved in deterministic polynomial time by some well-known techniques [4]. Moreover, we can also compute ABPs for the g_i and h_i for the minimum k.

Theorem 4.2. *Suppose a degree d homogeneous noncommutative polynomial $f \in \mathbb{F}\langle X \rangle$, and positive integer k encoded in unary are the input to SOP:*

(a) If f is given by a polynomial degree arithmetic circuit then SOP is in MA ∩ coNP.

(b) If f is given by an algebraic branching program then SOP is in deterministic polynomial time (even in randomized NC^2).

(c) If f is given in the sparse representation then SOP is equivalent to the problem of checking if the rank of a given matrix is at most k. In particular, if $\mathbb{F}$ is the field of rationals, SOP is complete for the complexity class $C_{=}L$.[2]

---

[2] The logspace counting class $C_{=}L$ captures the complexity of matrix rank over rationals [1].

We first focus on proving part (a) of the theorem. If (f, k) is a "yes" instance to SOP, then we claim that there exist small arithmetic circuits for the polynomials $g_i, h_i, i \in [k]$.

We define the *partial derivative matrix* A_f for the polynomial f as follows. The rows of A_f are indexed by degree d_1 monomials and the columns of A_f by degree d_2 monomials (over variables in X). For the row labeled m and column labeled m', the entry $A_{m,m'}$ is defined as

$$A_{m,m'} = f(mm').$$

The key to analyzing the decomposition of f is the rank of the matrix A_f.

Claim 4.3. *Let $f \in \mathbb{F}\langle X \rangle$ be a homogeneous degree d polynomial.*

(a) *Then f can be decomposed as $f = g_1 h_1 + \cdots + g_k h_k$ for homogeneous degree d_1 polynomials g_i and homogeneous degree d_2 polynomials h_i if and only if the rank of A_f is bounded by k.*

(b) *Furthermore, if f is computed by a noncommutative arithmetic circuit C then if the rank of A_f is bounded by k there exist polynomials $g_i, h_i \in \mathbb{F}\langle X \rangle$, $i \in [k]$, such that $f = g_1 h_1 + \cdots + g_k h_k$, where g_i and h_i have noncommutative arithmetic circuits of size $poly(|C|, n, k)$ satisfying the above conditions.*

An NP-Hard Decomposition Problem. We now briefly discuss a generalization of SOP. Given a polynomial $f \in \mathbb{F}\langle X \rangle$ as input along with k in unary, can we decompose it as a k-sum of products of *three* homogeneous polynomials:

$$f = a_1 b_1 c_1 + a_2 b_2 c_2 + \cdots + a_k b_k c_k,$$

where each a_i is degree d_1, each b_i is degree d_2, and each c_i is degree d_3?

It turns out that even in the simplest case when f is a cubic polynomial and the a_i, b_i, c_i are all homogeneous linear forms, this problem is NP-hard. The tensor rank problem: given a 3-dimensional tensor $A_{ijk}, 1 \leq i, j, k \leq n$ checking if the tensor rank of A is bounded by k, which is known to be NP-hard [10] is easily shown to be polynomial-time reducible to this decomposition problem.

Indeed, we can encode a three-dimensional tensor A_{ijk} as a homogeneous cubic noncommutative polynomial $f = \sum_{i,j,k \in [n]} A_{ijk} x_i y_j z_k$, such that any summand in the decomposition, which is product of three homogeneous linear forms, corresponds to a rank-one tensor. This allows us to test whether a tensor can be decomposed into at most k rank-one tensors, which is equivalent to testing whether the rank of the tensor is at most k.

5   Concluding Remarks and Open Problems

The main open problem is the complexity of noncommutative polynomial factorization in the general case. Even when the input polynomial $f \in \mathbb{F}\langle X \rangle$ is given in sparse representation we do not have an efficient algorithm nor any nontrivial complexity-theoretic upper bound. Although polynomials in $\mathbb{F}\langle X \rangle$ do not have unique factorization, there is interesting structure to the factorizations [7,8] which can perhaps be exploited to obtain efficient algorithms.

In the case of irreducibility testing of polynomials in $\mathbb{F}\langle X \rangle$ we have the following observation that contrasts it with commutative polynomials. Let $\mathbb{F}$ be a fixed finite field. We note that checking if $f \in \mathbb{F}\langle X \rangle$ given in sparse representation is *irreducible* is in coNP. To see this, suppose f is s-sparse of degree D. If f is reducible and $f = gh$ is any factorization then each monomial in g or h is either a prefix or a suffix of some monomial of f. Hence, both g and h are sD-sparse polynomials. An NP machine can guess g and h (since coefficients are constant-sized) and we can verify if $f = gh$ in deterministic polynomial time.

On the other hand, it is an interesting contrast to note that given an s-sparse polynomial f in the *commutative* ring $\mathbb{F}[x_1, x_2, \ldots, x_n]$ we do not know if checking irreducibility is in coNP. However, checking irreducibility is known to be in RP (randomized polynomial time with one sided-error) as a consequence of the Hilbert irreducibility criterion [11]. If the polynomial f is irreducible, then if we assign random values from a suitably large extension field of $\mathbb{F}$ to variables $x_2, \ldots, x_n$ (say, $x_i \leftarrow r_i$) the resulting univariate polynomial $f(x_1, r_2, \ldots, r_n)$ is irreducible with high probability.

Another interesting open problem that seems closely related to noncommutative sparse polynomial factorization is the problem of finite language factorization [18]. Given as input a finite list of words $L = \{w_1, w_2, \ldots, w_s\}$ over the alphabet X the problem is to check if we can factorize L as $L = L_1 L_2$, where L_1 and L_2 are finite sets of words over X and $L_1 L_2$ consists of all strings uv for $u \in L_1$ and $v \in L_2$. This problem can be seen as noncommutative sparse polynomial factorization problem where the coefficients come from the *Boolean ring* $\{0,1\}$. No efficient algorithm is known for this problem in general, neither is any nontrivial complexity bound known for it [18]. On the other hand, analogous to factorization in $\mathbb{F}\langle X \rangle$, we can solve it efficiently when L is *homogeneous* (i.e. all words in L are of the same length). Factorizing L as $L_1 L_2$, where L_1 and L_2 are variable-disjoint can also be efficiently done by adapting our approach from Sect. 2. It would be interesting if we can relate language factorization to sparse polynomial factorization in $\mathbb{F}\langle X \rangle$ for a field $\mathbb{F}$. Is one efficiently reducible to the other?

References

1. Allender, E., Beals, R., Ogihara, M.: The complexity of matrix rank and feasible systems of linear equations. Comput. Complex. **8**(2), 99–126 (1999)
2. Bhattacharyya, A.: Polynomial decompositions in polynomial time. In: Schulz, A.S., Wagner, D. (eds.) ESA 2014. LNCS, vol. 8737, pp. 125–136. Springer, Heidelberg (2014)
3. Arvind, V., Joglekar, P.S., Rattan, G.: On the complexity of noncommutative polynomial factorization. Electronic Colloquium on Computational Complexity, TR15-004 (2015)

4. Beimel, A., Bergadano, F., Bshouty, N.H., Kushilevitz, E., Varricchio, S.: Learning functions represented as multiplicity automata. J. ACM **47**(3), 506–530 (2000)
5. Bogdanov, A., Wee, H.: More on noncommutative polynomial identity testing. In: Proceedings of 20th Annual Conference on Computational Complexity (CCC), pp. 92–99 (2005)
6. Caruso, F.: Factorization of noncommutative polynomials. CoRR abs/1002.3180 (2010)
7. Cohn, P.M.: Free rings and their relations. Academic Press, London Mathematical Society Monograph No. 19 (1985)
8. Cohn, P.M.: Noncommutative unique factorization domains. Trans. Am. Math. Soc. **109**, 313–331 (1963)
9. Gathen, J., Gerhard, J.: Modern Computer Algebra 2nd Edition. Cambridge University Press, Cambridge (2003)
10. Hastad, J.: Tensor rank is NP-complete. J. Algorithms **11**(4), 644–654 (1990)
11. Kaltofen, E.: Factorization of polynomials given by straight-line programs. In: Micali, S. (ed.) Randomness in Computation. Advances in Computing Research, pp. 375–412. JAI Press, Greenwhich (1989)
12. Kaltofen, E., Trager, B.: Computing with polynomials given by black-boxes for their evaluations: greatest common divisors, factorization, separation of numerators and denominators. J. Symbolic Comput. **9**(3), 301–320 (1990)
13. Kopparty, S., Saraf, S., Shpilka, A.: Equivalence of polynomial identity testing and deterministic multivariate polynomial factorization. Electronic Colloquium on Computational Complexity, TR14-001 (2014)
14. Nisan, N.: Lower bounds for noncommutative computation. In: Proceedings of 23rd ACM Symposium on Theory of Computing (STOC), pp. 410–418 (1991)
15. Raz, R., Shpilka, A.: Deterministic polynomial identity testing in non commutative models. Comput. Complex. **14**(1), 1–19 (2005)
16. Shpilka, A., Volkovich, I.: On the relation between polynomial identity testing and finding variable disjoint factors. In: Abramsky, S., Gavoille, C., Kirchner, C., Meyer auf der Heide, F., Spirakis, P.G. (eds.) ICALP 2010. LNCS, vol. 6198, pp. 408–419. Springer, Heidelberg (2010)
17. Shpilka, A., Yehudayoff, A.: Arithmetic circuits: a survey of recent results and open questions. Found. Trends Theor. Comput. Sci. **5**(3), 217–388 (2010)
18. Salomaa, A., Yu, S.: On the decomposition of finite languages. In: Rozenberg, G., Thomas, W. (eds.) Proceedings of Developments in Language Theory, Foundations, Applications, and Perspectives, pp. 22–31. World Scientific, Singapore (2000)

An Algebraic Proof of the Real Number PCP Theorem

Martijn Baartse and Klaus Meer[(✉)]

Computer Science Institute, BTU Cottbus-Senftenberg,
Platz der Deutschen Einheit 1, 03046 Cottbus, Germany
baartse@tu-cottbus.de, meer@b-tu.de

Abstract. The PCP theorem has recently been shown to hold as well in the real number model of Blum, Shub, and Smale [3]. The proof given there structurally closely follows the proof of the original PCP theorem by Dinur [7]. In this paper we show that the theorem also can be derived using algebraic techniques similar to those employed by Arora et al. [1,2] in the first proof of the PCP theorem. This needs considerable additional efforts. Due to severe problems when using low-degree algebraic polynomials over the reals as codewords for one of the verifiers to be constructed, we work with certain trigonometric polynomials. This entails the necessity to design new segmentation procedures in order to obtain well structured real verifiers appropriate for applying the classical technique of verifier composition.

We believe that designing as well an algebraic proof for the real PCP theorem on one side leads to interesting questions in real number complexity theory and on the other sheds light on which ingredients are necessary in order to prove an important result like the PCP theorem in different computational structures.

1 Introduction

The PCP theorem was an important break-through in discrete complexity theory in the early 1990s. The theorem states that all languages in NP have a proof that can be checked by a randomized verifier which queries only a constant number of components in the proof. Around the same time Blum, Shub, and Smale introduced a uniform model of computation for more general structures than finite alphabets with particular focus on computations over $\mathbb{R}$ and $\mathbb{C}$. One line of research in the BSS model since then has been guided by the question whether deep and important statements in the Turing model hold as well in other structures, and if so, by what reason. This ideally might lead to a better understanding of why the theorems hold, what the heart of their proofs is, and how major open questions over different structures are related. The last conveys hope to find new techniques for attacking them—see [4] for a discussion of a few examples.

M. Baartse, K. Meer—Supported under projects ME 1424/7-1 and ME 1424/7-2 by the Deutsche Forschungsgemeinschaft DFG. We gratefully acknowledge the support.

G.F. Italiano et al. (Eds.): MFCS 2015, Part II, LNCS 9235, pp. 50–61, 2015.
DOI: 10.1007/978-3-662-48054-0_5

The classical PCP theorem for the class NP defined in the Turing model so far has two intrinsically different proofs. The first one was worked out in two papers by Arora and co-authors [1,2]. It is based on a lot of deep algebraic techniques dealing with multivariate polynomials over finite fields. Dinur [7] later on gave another proof based on more combinatorial techniques related to graph theory. In view of transfering such an important theorem to the real number model and the algebraic character of the $NP_\mathbb{R}$-complete QPS problem of deciding the solvability of real quadratic polynomial systems, one might expect that the original proof by Arora et al. is a more appropriate starting point for showing a real version. Nevertheless, this turned out not to be the case. In [3] the authors started from Dinur's proof, showing that with a bunch of sometimes more and sometimes less intricate arguments a real version of the theorem can be obtained along this line.

In this paper we come back to the first proof. The motivation for doing so is at least twofold: as mentioned already we believe it to be interesting to fully understand which parts of proofs build the core when studying different computational structures. Second, along the way of studying such a question new interesting problems occur. In the present setting such problems arise when dealing with trigonometric functions as codes, objects that so far have not occured to the best of our knowledge in BSS complexity theory. And finally, it would be disturbing to see that a combinatorial proof would work whereas an algebraic proof does not. Nevertheless, it turns out that the algebraic proof we present here needs considerable additional efforts. This is due to structural obstacles certain fundamental test procedures on algebraic polynomials used in the classical proof evoke when we try to transform them to real domains. This is solved by using instead trigonometric polynomials as codewords and develop a complete test-scenario for them. Already testing such polynomials requires a lot of work; this was done in [5] and we rely as starting point on the main result therein —see the explanation below.

1.1   Problems with the Classical Proof and Outline

Let us first outline the structure of the classical proof by Arora et al. and where it creates severe problems over the reals. At the same time we refer to intermediate results that have already been shown in the BSS framework and indicate how we shall solve the difficulties arising because of the classical proof not being applicable to the reals.

The Arora et al. proof is built up as follows. Using truth tables of corresponding linear functions as codewords for strings of length n a linearity test is designed to construct a $(\mathrm{poly}(n), 1)$-restricted verifier for NP. This so-called long transparent proof verifier in a suitable way has also been constructed in the (real and complex number) BSS framework [4,9] already and we shall use it below.

Next, a $(\log(n), \mathrm{polylog}(n))$-restricted verifier for NP is designed. It uses as codewords low-degree algebraic polynomials over suitable finite fields and designs a low-degree test together with a sum-check procedure to verify whether a codeword represents a satisfying assignment of a 3-SAT formula through a low-degree polynomial. Such verifiers have been shown to exist for $NP_\mathbb{R}$ as well [10], but their structure is not appropriate to be used further on in the proof.

The low-degree test is not only important for itself, but used in [1] to put arbitrary verifiers into a more structured segmented form. Here, segmentation refers to the aspect of how the verifier inspects the proof. It means that instead of arbitrarily exploiting the query resources it has, the verifier asks the information from the proof in a highly structured form by querying a constant number of blocks of length $\mathrm{polylog}(n)$. With this technique the $(\log(n), \mathrm{polylog}(n))$-restricted verifier is turned into an equivalent segmented verifier which makes it suitable for being used in a procedure called verifier composition. During this process, the above verifier first is composed with itself once and then with the $(\mathrm{poly}(n), 1)$-restricted verifier. The outcome of these two compositions is the required $(\log(n), 1)$-restricted verifier for NP.

It is here where further major difficulties arise. First, we need a low-degree test which at the same time respects the $(\log(n), \mathrm{polylog}(n))$ resources and can be used to achieve a kind of segmentation procedure for other verifiers. Such a test has been designed in [5] using trigonometric polynomials; there it is also explained why algebraic polynomials as codewords seem not appropriate. However, because of the use of trigonometric polynomials the test lacks the kind of structure that allows to prove a full segmentation lemma like the one in [1]. The difficulties for turning any real verifier into a segmented one with the low-degree test arise because the composition of a trigonometric polynomial with an algebraic polynomial in general needs not to be a (low-degree) polynomial anymore. We solve this problem by proving a weak segmentation lemma. With this lemma at hand, we develop a variant of a sum-check procedure that can be segmented. So even if we are not able to give a full segmentation for any real verifier, our techniques solve the problem for a special sum-check sufficient in our setting. This finally clears the way to apply a real version of verifier composition and to obtain the $PCP_\mathbb{R}$ theorem.

The paper is structured as follows. After recalling main notions and concepts about real number computations and PCPs we briefly describe in Sect. 2.1 trigonometric polynomials and the low-degree test from [5]. Sections 3 and 4 are the main parts of the paper. The main task to solve is the design of a segmented almost transparent proof for all problems in $NP_\mathbb{R}$. We show how to extend the low-degree test to check correctness of a given function in a fixed point. This is necessary for proving a weak segmentation lemma. The lemma then is used in order to design a sum-check test in segmented form. Both results are combined to obtain a segmented and $(\log(n), \mathrm{polylog}(n))$-restricted verifier for $NP_\mathbb{R}$. These results finally enter a verifier composition procedure. Segmentation makes it work and leads to the full $PCP_\mathbb{R}$ theorem. Due to space limitation all proofs are postponed to the full version.

2 Basic Notions

We assume the reader to be familiar with BSS machines, real number complexity classes $P_\mathbb{R}$ and $NP_\mathbb{R}$ as well as $NP_\mathbb{R}$-completeness, see [6]. Briefly, a BSS machine over $\mathbb{R}$ is a machine that can add, subtract, multiply and divide two real numbers and can perform tests of the form $x \geq 0$ at unit cost. Complexity classes $P_{\mathbb{R}}$ and $NP_{\mathbb{R}}$ are defined analogously to P and NP. In order to define as well real $PCP_{\mathbb{R}}$ classes we need to introduce real probabilistic verifiers.

Definition 1. *(a) Let $r, q : \mathbb{N} \to \mathbb{N}$ be two resource functions. A real probabilistic $(r(n), q(n))$-restricted verifier V is a randomized real BSS machine; on input x of size n it first generates uniformly and independently a string ρ of $O(r(n))$ random bits. Using x and ρ it makes $O(q(n))$ queries into a proof π. A query is made by writing an address on a query tape and then in one step the real number stored at that address in π is returned. The verifier uses x, ρ and the results of the queries and computes in time polynomial in n its decision 'accept' or 'reject'.*

(b) V accepts (x, π) with probability α if for an α-fraction of sequences of random bits ρ (of the length required by V for x) the computation of V results in acceptance.

V accepts a language $L \subseteq \mathbb{R}^{\infty} := \bigsqcup_{i \geq 1} \mathbb{R}^i$ iff

(i) for every $x \in L$ there exists π such that V accepts (x, π) with probability 1 and

(ii) for every $x \notin L$ and for every π, V rejects (x, π) with probability at least $\frac{3}{4}$.

(c) The class $PCP_{\mathbb{R}}(r(n), q(n))$ is defined as the set of languages that are accepted by an $(r(n), q(n))$-restricted verifier.

The goal of this paper is to prove by algebraic means the real $PCP_{\mathbb{R}}$ theorem

Theorem 1. $NP_{\mathbb{R}} = PCP_{\mathbb{R}}(\log(n), 1)$

To prove this theorem it suffices to prove the existence of a corresponding verifier for a $NP_{\mathbb{R}}$-complete problem. The one we use is QPS, the problem of deciding whether a system of real quadratic polynomials has a common real zero. We can assume each polynomial to depend on at most three variables [6].

2.1 Testing Trigonometric Polynomials

The construction of a well structured $(\log(n), \text{polylog}(n))$-restricted verifier for $NP_{\mathbb{R}}$ is based on using trigonometric polynomials as code words for potential zeros of a polynomial system and a test procedure for such polynomials.

Definition 2. *Let $d \in \mathbb{N}, q$ be a prime and $F := \{0, 1, \ldots, q-1\}$ be a finite field.*

(a) A univariate trigonometric polynomial $f : F \mapsto \mathbb{R}$ of degree d is a function of form

$$f(x) = a_0 + \sum_{m=1}^{d} a_m \cdot \cos(2\pi m \frac{x}{q}) + b_m \cdot \sin(2\pi m \frac{x}{q}),$$

where $a_0, \ldots, a_d, b_1, \ldots, b_d \in \mathbb{R}$.

(b) For $k \in \mathbb{N}$ a multivariate trigonometric polynomial $f : F^k \mapsto \mathbb{R}$ of max-degree d is defined recursively via

$$f(x_1, \ldots, x_k) = a_0(x_1, \ldots, x_{k-1}) + \sum_{m=1}^{d} a_m(x_1, \ldots, x_{k-1}) \cdot \cos(2\pi m \frac{x_k}{q}) +$$

$$+ \sum_{m=1}^{d} b_m(x_1, \ldots, x_{k-1}) \cdot \sin(2\pi m \frac{x_k}{q}),$$

where the a_i, b_j are trigonometric polynomials of max-degree d in $k-1$ variables. We will usually drop the term 'trigonometric' if this is clear from the context.

Remark 1. Note the following technical detail: if below a verifier is used as BSS algorithm for inputs of varying size, then for different cardinalities q of the finite field F it needs to work with different constants $\cos\frac{2\pi}{q}, \sin\frac{2\pi}{q}$. It is not hard to see that given q one could add two real numbers to the verification proof which in the ideal case represent the real and imaginary part of a complex primitive q-th root of unity. The verifier in question deterministically checks in polynomial time whether this is the case and then continues to use these constants for evaluating trigonometric polynomials.

As starting point for the construction of a $(\log(n), \text{polylog}(n))$-restricted verifier for $\text{NP}_\mathbb{R}$ suitable to be used in a composition step the main result from [5] is crucial. After citing the statement we illuminate its main features for the further ongoing.

Theorem 2 ([5]). *Let $d \in \mathbb{N}$, $h = 10^{15}$, $k \geq \frac{3}{2}(2h + 1)$ and let F be a finite field with $q:=|F|$ being a prime number larger than $10^4(2hkd + 1)^3$. There exists a probabilistic verification algorithm in the BSS-model of computation over the reals with the following properties:*

(i) *The verifier gets as input a function value table of a multivariate function $f : F^k \to \mathbb{R}$ and a proof string π consisting of at most $|F|^{2k}$ segments (blocks). Each segment consists of $2hkd + k + 1$ real components. The verifier generates $O(k \cdot \log q)$ random bits and inspects $O(1)$ many positions in the table for f as well as $O(1)$ many segments in the proof string π in order to make its decision. The running time of the verifier is polynomially bounded in the quantity kd.*

(ii) *For every function value table representing a trigonometric max-degree d polynomial there exists a proof string such that is accepted with probability 1.*

(iii) *For any $0 < \epsilon < 10^{-19}$ and for every function value table whose distance to a closest max-degree $\tilde{d}:=2hkd$ polynomial is at least 2ϵ the probability that the verifier rejects is at least ϵ, no matter what proof string is given. Here, for two functions $f, g : F^k \mapsto \mathbb{R}$ their distance is defined as $dist(f, g):=\frac{1}{|F^k|} \cdot |\{x \in F^k | f(x) \neq g(x)\}|$.*

Remark 2. Note that in contrast to most other low degree tests, this test is not sharp in the sense that it accepts every max-degree d polynomial but it does not necessarily reject every function which has a large distance to any max-degree d polynomial. Hence, if the test accepts we only know that with high probability the given function is close to a max-degree $2 \cdot 10^{15}kd$ polynomial and not necessarily close to a polynomial of max-degree d. Nevertheless, this will be sufficient for our purposes.

To avoid confusion below we would like to point out already here that in some situations we shall still use the parameter d in subsequent tests instead of $2 \cdot 10^{15}kd$; the assumption that the given function is close to a max-degree $2 \cdot 10^{15}kd$ polynomial will make these tests work. The reason for still using the parameter d is that although we cannot test it sharply with the low degree test, if the verifier expects additional information about f from an ideal prover, the latter still should fit to a max-degree d polynomial. A typical example would be that the verifier expects a univariate restriction of an f that has passed the test. Then for a prover working correctly the degree of such a restriction should result from a max-degree d polynomial and not from a max-degree $\tilde{d}$ polynomial.

The theorem says that there exists a $(k \log |F|, kd)$-restricted verification procedure that receives as input a function value table $f : F^k \to \mathbb{R}$ and a proof π, accepts if f is a max-degree d polynomial and rejects with high probability if it is not close to a max-degree $2 \cdot 10^{15}kd$ polynomial. Note that in the statement above we do not only count the queries into π but also those into the table of f. The reason is that in the later use of the test both f and π are parts of the (expected) proof that the trigonometric polynomial f codes a zero of a given polynomial system. The crucial property of the procedure is that (f, π) are queried in a very structured form, i.e., only constantly many blocks of size at most $O(kd)$) are inspected.[1] Though the test presented in [10] achieves the same total bound on queries into a pair (f, π) verifying that π proves f to be close to an algebraic low-degree polynomial, the latter is not structured because it makes $O(kd)$ questions also into the table for f. The improved structure is essentially necessary for using the verifier in the composition part. Let us already mention here that below a potential zero $x \in \mathbb{R}^n$ of a QPS instance is coded via a trigonometric polynomial with the parameters k, d, q specified as $k := \lceil \frac{\log n}{\log \log n} \rceil, d := O(\log n)$ and $|F| := O(\log^6 n)$, respectively. Then the above verifier is $(\log(n), polylog(n))$-restricted.

3   The Correctness Test

Our goal in this section is to design a $(\log(n), polylog(n))$-restricted segmented verifier for $\mathrm{NP}_{\mathbb{R}}$. It will then be an outer verifier in the verifier composition procedure (to be developed as well) to prove the $\mathrm{PCP}_{\mathbb{R}}$ theorem. So far we have a segmented low-degree test for trigonometric polynomials at hand. In order to use it for the final verifier a sum-check procedure will be designed.

---

[1] In f each such block only has one real component because the test asks a constant number of function values only.

As said already also this procedure must be in segmented form. In order to achieve this task we use the low-degree verifier for segmentation. However, a direct segmentation of arbitrary real verifiers using it fails because composing a low degree trigonometric polynomial with a low degree algebraic polynomial does not necessarily result in a low-degree polynomial whereas the composition of two low degree algebraic polynomials does. We overcome this restriction by first designing a segmented correctness test. The correctness test allows to achieve a weak form of segmentation also for a sum-check procedure. This finally leads to a segmented $(\log(n), \text{polylog}(n))$-restricted verifier.

Suppose throughout the following that f has passed the test of Theorem 2. Thus, f is close to a trigonometric polynomial $\tilde{f}$ of max-degree $\tilde{d} = 2 \cdot 10^{15} kd$; therefore, for a randomly chosen point x with high probability $f(x)$ and $\tilde{f}(x)$ coincide. However, if we fix an x_0 it still might be the case that $f(x_0)$ is false. Error correction now means to be able to correct the value with high probability; for example, the Walsh-Hadamard code used in the design of long transparent proofs allows such a correction.

For our purposes in the low-degree setting it will be sufficient to detect such an error with high probability for any fixed x_0. Thus we want to design a verifier V which works on a given *fixed* x_0, a table for f and an additional proof string and rejects with high probability if $f(x_0) \neq \tilde{f}(x_0)$.

An easy approach for designing such a V is the following. Assume that $f(x_0) \neq \tilde{f}(x_0)$. The verifier chooses a random line ℓ passing through x_0 and expects the proof string to contain a univariate trigonometric polynomial p for ℓ. The verifier checks whether p and f have the same value in x_0. If not V rejects directly. If yes, then it means that p is different from the restriction of $\tilde{f}$ to ℓ and since they are both polynomials of low degree they actually disagree on most points on ℓ. With high probability this will be exposed by the following procedure. The verifier chooses a random point $x_{rand} \in \ell$ different from x_0 and checks whether p and f have the same value in x_{rand}. We already know that with high probability $\tilde{f}$ disagrees with p on x_{rand} and since x_{rand} is uniformly distributed over $F^k \setminus \{x_0\}$ and f is close to $\tilde{f}$ it follows that with high probability $f(x_{rand}) = \tilde{f}(x_{rand})$. Thus with high probability it holds that f and p disagree on x_{rand}, in which case the verifier rejects.

The problem with this approach is the following: Since above a random point was chosen the proof string needs to contain univariate restrictions with respect to all lines through some point in F^k. But opposed to the case of algebraic polynomials in the trigonometric setting the degrees of such restrictions can get much larger than the degree of the corresponding multivariate polynomials because they might get an additional factor of size $\Omega(kq)$. So we need a set of lines through F^k that result in univariate restrictions of lower degree only. In fact, the resulting degrees will be bounded by $O(kd\sqrt{q})$ which turns out to suffice. A new problem occurs that way: How do we get a randomly distributed point if the set of used directions is restricted? This is solved as follows. V sets up a random process by choosing finitely many times lines randomly from this set together with random points x_i on those lines. The above argument concerning x_0 and x_{rand} now is iteratively applied to the univariate polynomials along the lines through (x_i, x_{i+1}). The key point for V to work correctly is the observation that after finitely many steps (actually, 24 steps) the resulting point x_{24} in the process is almost randomly distributed in F^k. Most of the work below is devoted to demonstrate that this holds for the set of lines that we will construct.

Now towards the details. First, the correctness test shall be described. Let parameters $k, d, \tilde{d}, q$ be as in Theorem 2. The following objects are needed in the test procedure.

Definition 3. *(a)   For $F:=\{0, 1, \ldots, q-1\}$ define the set $W \subset F$ as*

$$W:=\{r \in F \mid |r| \le \lfloor \sqrt{q} \rfloor\}.$$

Here, the absolute value of $r \in F$ is defined as $|r|:=\begin{cases} r & \text{if } r < \frac{q}{2} \\ q-r & \text{if } r > \frac{q}{2} \end{cases}$.

(b)   For $x, v \in F^k$ denote by $\ell_{x,v}:=\{x + tv | t \in F\}$ the line in F^k through x with direction v.

The set of lines that we will use is the set of lines $\ell_{x,v}$ with $x \in F^k$ and $v \in W^k$, so that if f is a max-degree d polynomial, then the restriction of f to any such line has degree at most $k\lfloor\sqrt{q}\rfloor d$ which is much smaller than q. Since $|W^k| = (2\lfloor\sqrt{q}\rfloor + 1)^k$ the set contains $q^{k-1} \cdot (2\lfloor\sqrt{q}\rfloor + 1)^k$ lines. Here we consider $\ell_{x,v}$ to be the same as $\ell_{x+v,v}$, but to be different from $\ell x, cv$ for some $c \in F \setminus \{1\}$.

Correctness Test: Let $k, q, d, \tilde{d}, F, W$ be as above.

Input: • A point $x_0 \in F^k$ and a function $f : F^k \to \mathbb{R}$, given by a table of its values. We suppose that f has passed without error the test behind Theorem 2 and denote by $\tilde{f}$ the unique polynomial of max-degree $\tilde{d} = 2 \cdot 10^{15} kd$ that is δ-close to f for small enough $0 < \delta$;

• a proof string containing a list of univariate trigonometric polynomials $p_{x,v}$ of degree $k\lfloor\sqrt{q}\rfloor d$ defined for each line $\ell_{x,v}$ with direction in W^k and specified by its $2k\lfloor\sqrt{q}\rfloor d + 1$ coefficients.[2] Recall from Remark 2 that although we can only assume f to be close to a max-degree $2 \cdot 10^{15}d$ polynomial, we can still use the parameter d because an ideal prover should specify a max-degree d polynomial.

1. For all $0 \le i \le 23$ pick uniformly and independently a direction $v_i \in W^k$ and a random point $t_i \in F \setminus \{0\}$; set $x_{i+1}:=x_i + t_i \cdot v_i$;
2. check whether $f(x_i) = p_{x_i, v_i}(0)$ and $f(x_{i+1}) = p_{x_i, v_i}(t_i)$. If for all i equality holds accept, otherwise reject.

The same line might have several representations using elements from W^k. We are interested in the uniform distribution among all those representations;

---

[2] There is a certain ambiguity in representing such a polynomial $p_{x,v}$ because different points on the line can be used and different vectors from W^k might result in the same line. This is not a problem since one can efficiently switch between those representations. Below, when we evaluate such a polynomial in a point t^* we silently assume the parametrization induced by the x, v mentioned.

thus, the probability of picking one is proportional to the number of representa-tions it has with respect to different elements in W^k.

The verifier performing the correctness test makes a random walk through F^k along lines with direction in W^k. Each x_i is a random point on a random line from W^k. Then it is checked whether the table for f corresponds to the value of the univariate polynomial p_{x_i,v_i} given as part of the input. In an ideal situation $f = \tilde{f}$ is a polynomial of max-degree d and the p_{x_i,v_i} are the univariate restrictions of f to the lines in W^k. Our aim is to generate from an arbitrary x_0 a point that is almost uniformly distributed in F^k, but only using a restricted set of lines during the corresponding random walk. As the calculations then show the point x_{24} has this property. Thus, finally the test nearly does the same as the straightforward approach described above which directly picks a random x_{rand}. However, this time the degrees of the corresponding univariate polynomials remain small.

Theorem 3 (Correctness Test). *With the above notations and assumptions a verifier performing the Correctness Test satisfies the following:*

(i) *if $f \equiv \tilde{f}$ is a max-degree d polynomial and the $p_{x,v}$ represent the correct restrictions of f to line $\ell_{x,v}$, then the verifier accepts with probability 1;*

(ii) *if $f(x_0) \neq \tilde{f}(x_0)$ the verifier accepts with probability at most $\delta + \frac{25}{\sqrt[4]{q}} + \frac{48\sqrt{q}k\tilde{d}}{q-1}$;*

(iii) *the verifier is $(k\log(q), k\lfloor\sqrt{q}\rfloor d)$-restricted; it inspects $O(1)$ many values of f and $O(1)$ many polynomials $p_{x,v}$, i.e., segmented blocks of length $2k\lfloor\sqrt{q}\rfloor d + 1 = O(kd)$. Its running time is polynomially bounded in $k\lfloor\sqrt{q}\rfloor d$.*

4   Segmented Almost Transparent Proofs for NP$_{\mathbb{R}}$

In the Arora et al. proof the next important step in designing a $(\log n, poly \log n)$-restricted verifier for NP by using a low-degree test is a so-called sum-checking procedure [8]. Very roughly, such a procedure probabilistically checks whether an expression of the form $\sum_{z_1,\ldots,z_k=0}^{h} \tilde{g}(z)$ for some specified $h \in F$ vanishes. Below we apply all the above to an instance of size n for an NP$_{\mathbb{R}}$-complete problem and get a similar problem where $\tilde{g}$ is a trigonometric polynomial of (low) degree $O(\log n)$ and $h = \lfloor \log n \rfloor, k = \lceil \frac{\log n}{\log \log n} \rceil$. Then the sum has $h^k = \Omega(n)$ many terms, so computing it exactly would require to query $\tilde{g}$ in too many arguments. Sum-checking does the evaluation randomly in order to reduce the query complexity to $poly \log n$. In addition, this randomized evaluation has to be segmented for using it in the rest of the proof. This creates serious problems.

In order to do a sum-check the idea is as follows. We do not work directly with $\tilde{g}$, but with a polynomial $\tilde{f} : F^{2k} \mapsto \mathbb{R}$ that is related to $\tilde{g}$ in a very special manner through a summation property. This technical property allows to express the sum-checking problem for $\tilde{g}$ as evaluation of $\tilde{f}$ in a single point. The disadvantage is that it must be checked as well whether $\tilde{f}$ has the summation property. Of course, this verification needs to be done in segmented form as well.

We continue as follows: First, the summation property relating two polynomials $\tilde{g}$ and $\tilde{f}$ is defined; if satisfied it directly transfers sum-checking for $\tilde{g}$ to a single-point evaluation of $\tilde{f}$. Next, we prove why for polynomials $\tilde{g}$ such an $\tilde{f}$ always exists. Finally, it is proved that the summation property for $\tilde{f}$ can be tested within the allowed resources.

Definition 4. (a) For a k-tuple $z = (z_1, \ldots, z_k)$, a component $1 \leq j \leq k$ and a $t \in F$ define $P_t^j(z) := (z_1, \ldots, z_{j-1}, t, z_{j+1}, \ldots z_k)$, i.e., the j-th component is assigned the value t.

(b) If $\tilde{g} : F^k \to \mathbb{R}$ is a max-degree d polynomial and $h \in F$. We say that a max-degree d polynomial $\tilde{f} : F^{2k} \to \mathbb{R}$ satisfies the summation property with respect to $\tilde{g}$ and h if the following conditions hold:

(i) for all $x, y \in F^k, j \in \{1, \ldots, k\}$ it holds $\frac{1}{h+1} \sum\limits_{t=0}^{h} \tilde{f}(P_t^j(x), y) = \tilde{f}(x, P_1^j(y))$;

(ii) the restriction of $\tilde{f}$ to $F^k \times \{0\}^k$ equals $\tilde{g}$.

The summation property can be used to replace the sum-check for $\tilde{g}$ by an equality test for a single argument:

Lemma 1. Let $h, \tilde{f}, \tilde{g}$ be as above. If the summation property holds for $\tilde{f}$ with respect to $\tilde{g}$ and h, then

$$\frac{1}{(h+1)^k} \sum_{z_1, \ldots, z_k = 0}^{h} \tilde{g}(z) = \tilde{f}(0^k, 1^k) = \tilde{f}(x, 1^k) \text{ for every } x \in F^k.$$

Thus it is sufficient to develop a test for verifying condition (i) of the summation property and do a Schwartz-Zippel equality test for condition (ii). We shall first show why all this is useful under certain assumptions. In the next subsection we then show why those assumptions can be satisfied.

The scenario needed further on below is as follows: We are given a black-box for computing a function $g : F^k \mapsto \mathbb{R}$ which is close to a max-degree $\tilde{d}$ trigonometric polynomial $\tilde{g} : F^k \mapsto \mathbb{R}$. In the ideal case we have that $g \equiv \tilde{g}$ is a max-degree d polynomial, cf. again Remark 2. Given suitable $h \in F$ an equation $\sum\limits_{z_1=0}^{h} \cdots \sum\limits_{z_k=0}^{h} \tilde{g}(z_1, \ldots, z_k) = 0$ should be verified. We shall specify later how g and $\tilde{g}$ arise in the general framework of this paper.

The following easy theorem shows how finally a sum-check can be done if an $\tilde{f}$ is available that satisfies the summation property with respect to $\tilde{g}$ and h. The difficult part will be to prove that the assumptions of the theorem can be satisfied.

Theorem 4. Let $q, k, h, d, \tilde{d}, g, \tilde{g}$ be as above. Suppose there exists a max-degree d polynomial $\tilde{f} : F^{2k} \mapsto \mathbb{R}$ such that $\tilde{f}$ satisfies the summation property with respect to $\tilde{g}$ and h. Suppose furthermore that there exists a segmented test for verifying condition (i) of the summation property for $\tilde{f}$. Then there is a segmented verification procedure for the sum check $\sum\limits_{z_1=0}^{h} \cdots \sum\limits_{z_k=0}^{h} \tilde{g}(z_1, \ldots, z_k) = 0$.

The resources the latter verifier needs are those for the test verifying condition (i) plus $O(k \log q)$ random bits plus $O(k \lfloor \sqrt{q} \rfloor d)$ proof components.

It remains to be shown that the assumptions of the theorem can be satisfied. The next technical lemma shows that for an arbitrary polynomial $\tilde{g}$ and $h \in F$ such an $\tilde{f}$ exists.

Lemma 2. *Let $\tilde{g} : F^k \to \mathbb{R}$ be a max-degree d polynomial and let $h \in F$. There exists a max-degree d polynomial $\tilde{f} : F^{2k} \to \mathbb{R}$ satisfying the summation property with respect to $\tilde{g}$ and h.*

Our test for checking whether a given function $\tilde{f}$ satisfies condition (i) from Definition 4 is based on the next lemma.

Lemma 3. *Suppose $|F| =: q \in \mathbb{N}, \tilde{d}, h < q$ and let $f : F^{2k} \to \mathbb{R}$ be 0.1-close to a max-degree $\tilde{d}$ polynomial $\tilde{f} : F^{2k} \to \mathbb{R}$. For every $i \in \{1, \dots, 2k\}$ let $\tilde{f}_i : F^{2k} :\to \mathbb{R}$ be a degree d polynomial in its i-th variable; this means that $\tilde{f}_i$ restricted to a line in the i-th paraxial direction e_i is a polynomial of degree d.*
If for every $j \leq k$ the following three conditions hold:

$$\Pr_{x,y \in F^k} \left(\frac{1}{h+1} \sum_{t=0}^{h} \tilde{f}_j(P_t^j x, y) = \tilde{f}_{k+j}(x, P_1^j y) \right) \geq 0.9 \tag{1}$$

$$\Pr_{x,y \in F^k} \left(f(x, y) = \tilde{f}_j(x, y) \right) \geq 0.9 \tag{2}$$

$$\Pr_{x,y \in F^k} \left(f(x, y) = \tilde{f}_{k+j}(x, y) \right) \geq 0.9, \tag{3}$$

then $\tilde{f}$ satisfies condition (i) of the summation property for h.

Remember Remark 2: Below f typically has passed the low-degree test. Though we then cannot assume that it is close to a max-degree d polynomial because the low-degree test is not sharp, it is no problem to just reject if we find out that f is not a max-degree d polynomial. Thus the verifier still presumes $f \equiv \tilde{f}$ to have max degree d and rejects in case that certain univariate restrictions $\tilde{f}_i$ do not have the required degree d. Of course, concerning probability estimation involving $\tilde{f}$ we still can only assume it to be of max degree $\tilde{d}$.

The above lemma implies a test whether a given function $f : F^{2k} \to \mathbb{R}$ is close to a max-degree $\tilde{d}$ polynomial $\tilde{f}$ which satisfies condition (i) of the summation property: use the low-degree test to verify closeness, then expect the proof in addition to contain the univariate polynomials $\tilde{f}_i$ and check for all $1 \leq j \leq k$ whether (1), (2) and (3) hold. The problem is that this is not a segmented procedure because these checks have to be made for every j. Therefore we will now turn this test into a segmented one.

Lemma 4. *There exists a segmented procedure to test whether a given function $f : F^{2k} \to \mathbb{R}$ is close to a low degree polynomial $\tilde{f}$ satisfying the condition (i) of summation property with respect to h.*

Putting all the above together the following theorem can be proved. The setup for the sum-check is the same as for example in [10], so below we describe it briefly only in the proof.

Theorem 5. *For every problem $L \in \mathrm{NP}_\mathbb{R}$ there is a $(\log n, poly \log n)$-restricted real verifier accepting L. The verifier is segmented, i.e., it reads $O(1)$ many blocks of length $poly \log n$ from the proof certificate.*

The final step will combine the segmented verifier from Theorem 5 with the verification procedure in [9] using long transparent proofs in order to obtain the full $\mathrm{PCP}_\mathbb{R}$-theorem. Having segmented verifiers at hand, the way achieving this by verifier composition is very similar to the corresponding final step in the original proof. We finally obtain the real PCP theorem via algebraic proof methods.

Theorem 6. $NP_\mathbb{R} = PCP_\mathbb{R}(\log(n), 1)$.

Acknowledgment. We thank an anonymous referee for his/her very careful reading and comments.

References

1. Arora, S., Lund, C., Motwani, R., Sudan, M., Szegedy, M.: Proof verification and hardness of approximation problems. J. ACM **45**(3), 501–555 (1998)
2. Arora, S., Safra, S.: Probabilistic checking proofs: A new characterization of NP. J. ACM **45**(1), 70–122 (1998)
3. Baartse, M., Meer, K.: The PCP theorem for NP over the reals. Foundations of Computational Mathematics **15**(3), 651–680 (2015). (Springer)
4. Baartse, M., Meer, K.: Topics in real and complex number complexity theory. In: Montana, J.L., Pardo, L.M. (eds.) Recent Advances in Real Complexity and Computation, Contemporary Mathematics, vol. 604, pp. 1–53. American Mathematical Society (2013)
5. Baartse, M., Meer, K.: Testing low degree trigonometric polynomials. In: Hirsch, E.A., Kuznetsov, S.O., Pin, J.É., Vereshchagin, N.K. (eds.) CSR 2014. LNCS, vol. 8476, pp. 77–96. Springer, Heidelberg (2014)
6. Blum, L., Cucker, F., Shub, M., Smale, S.: Complexity and Real Computation. Springer, Heidelberg (1998)
7. Dinur, I.: The PCP theorem by gap amplification. J. ACM **54**(3), 12 (2007)
8. Lund, C., Fortnow, L., Karloff, H., Nisan, N.: Algebraic methods for interactive proof systems. J. ACM **39**(4), 859–868 (1992)
9. Meer, K.: Transparent long proofs: A first PCP theorem for NP$_\mathbb{R}$. Found. Comput. Math. 5(3), 231–255 (2005). (Springer)
10. Meer, K.: Almost transparent short proofs for NP$_\mathbb{R}$. In: Owe, O., Steffen, M., Telle, J.A. (eds.) FCT 2011. LNCS, vol. 6914, pp. 41–52. Springer, Heidelberg (2011)

On the Complexity of Hub Labeling (Extended Abstract)

Maxim Babenko[1], Andrew V. Goldberg[2], Haim Kaplan[3],
Ruslan Savchenko[4(✉)], and Mathias Weller[5]

[1] Yandex and Higher School of Economics, Moscow, Russia
maxim.babenko@gmail.com
[2] Amazon.com, Inc., E. Palo Alto, USA
avg@alum.mit.edu
[3] School of Computer Science, Tel Aviv University, Tel Aviv-Yafo, Israel
haimk@post.tau.ac.il
[4] Yandex, Moscow, Russia
ruslan.savchenko@gmail.com
[5] LIRMM, Université Montpellier II, Montpellier, France
mathias.weller@lirmm.fr

Abstract. Hub Labeling (HL) is a data structure for distance oracles. Hierarchical HL (HHL) is a special type of HL, that received a lot of attention from a practical point of view. However, theoretical questions such as NP-hardness and approximation guarantees for HHL algorithms have been left aside. We study the computational complexity of HL and HHL. We prove that both HL and HHL are NP-hard, and present upper and lower bounds on the approximation ratios of greedy HHL algorithms that are used in practice. We also introduce a new variant of the greedy HHL algorithm that produces small labels for graphs with small highway dimension.

1 Introduction

The point-to-point shortest path problem is a classical optimization problem with many applications. The input to the problem is a graph $G = (V, E)$, a length function $\ell : E \rightarrow \mathbb{R}$, and a pair $s, t \in V$. The goal is to find $\mathrm{dist}(s, t)$, the length of the shortest s–t path in G, where the length of a path is the sum of the lengths of its arcs. We consider the data structure version of the problem where the goal is to construct a compact data structure that can efficiently answer point to point distance queries. We assume that the length function is non-negative and that there are no zero-length cycles. We define $n = |V|$ and $m = |E|$.

The hub labeling algorithm (HL) [9,13] is a shortest path algorithm that computes vertex labels during a preprocessing stage and answers s, t queries using only the labels of s and t; the input graph is not used for queries [16].

A.V. Goldberg—Part of the work done while the author was at Microsoft Research.
R. Savchenko—Part of the work done while the author was at Department of Mech. and Math., Moscow State University.

© Springer-Verlag Berlin Heidelberg 2015
G.F. Italiano et al. (Eds.): MFCS 2015, Part II, LNCS 9235, pp. 62–74, 2015.
DOI: 10.1007/978-3-662-48054-0_6

For a directed graph, a *label* $L(v)$ for a vertex $v \in V$ consists of the *forward label* $L_f(v)$ and the *backward label* $L_b(v)$. The forward label $L_f(v)$ consists of a sequence of pairs $(w, \text{dist}(v, w))$, where $\text{dist}(v, w)$ is the distance (in G) from v to w. The backward label $L_b(v)$ is similar, with pairs $(u, \text{dist}(u, v))$. Vertices $w \in L_f(v)$ and $u \in L_b(v)$ are called the *hubs* of v. In undirected graphs we define $L(v)$ to be a (single) set of pairs $(w, \text{dist}(v, w))$.

The labels must obey the *cover property*: for any two vertices s and t, the set $L_f(s) \cap L_b(t)$ ($L(s) \cap L(t)$ in undirected graphs) must contain at least one hub v that is on a shortest s–t path (we say that v *covers* the pair $[s, t]$). Given the labels, HL queries are straightforward: to find $\text{dist}(s, t)$, simply find the hub $v \in L_f(s) \cap L_b(t)$ that minimizes $\text{dist}(s, v) + \text{dist}(v, t)$.

Query time and space complexity depend on the label size. The size of a label, $|L(v)|$, is the number of hubs it contains. For a directed graph $|L(v)| = |L_f(v)| + |L_b(v)|$: the sum of the sizes of the forward and backward labels. Unless mentioned otherwise, preprocessing algorithms attempt to minimize the *total labeling size* $|L| = \sum_{v \in V} |L(v)|$.

Cohen et al. [9] give an $O(\log n)$ approximation algorithm for constructing HL. This algorithm was generalized in [7] and sped up in [11]. These approximation algorithms compute small labels but, although polynomial, do not scale to large input instances [11].

A special case of HL is *hierarchical hub labeling* (HHL) [4], where only labelings are valid for which the binary relation "is in the label of" is a partial order. That is, valid labelings allow a global ranking by "importance" and the label for a vertex v can only contain vertices u that are at least as important as v (formally, there exists a bijection $\pi : V \to \{1, \ldots, |V|\}$ such that $u \in L_f(v) \cup L_b(v)$ implies $\pi(u) \geq \pi(v)$). HHL implementations are faster in practice than general HL ones. For several important graph classes, such as road and complex networks, HHL implementations find small labeling and scale to large input instances [3,4,6,10]. However, for the algorithms used in practice such as hierarchical greedy (g-HHL) and hierarchical weighted greedy (w-HHL) there was no theoretical guarantee on the approximation ratio.[1]

Previous work on the computational complexity of HHL algorithms is *purely experimental*. The work on the complexity of HL is mostly experimental; the exceptions are approximation algorithms for HL mentioned above, and upper bounds for HL in case of low highway dimension [1,2,5]. However, prior to our work it was not even known if finding an optimal HL is NP-hard. The NP-hardness was implicitly conjectured in [9], motivating their $O(\log n)$-approximation algorithm. In addition, Cohen et al. [9] prove that a more general problem in which the paths to cover are part of the input is NP-hard (which does not imply NP-hardness of the original problem).

We obtain several natural results on the complexity of HL and HHL.

NP-hardness: We show that both finding an optimal HL and finding an optimal HHL is NP-hard. This solves a long-standing open problem, and the proof

---

[1] g-HHL and w-HHL are introduced in [4] under names TDc and TDwc respectively.

is non-trivial. In particular, proofs for the directed and undirected cases use different constructions.

Greedy HHL *Algorithms:* We prove both upper and lower bounds on the approximation ratio of several greedy HHL algorithms. The greedy algorithms include g-HHL and w-HHL, which are used in practice and give the fastest known distance oracle implementations for several classes of problems. We also introduce a new d-HHL algorithm that works well on networks with small highway dimension. We denote the highway dimension of a network by h and the diameter by D.

- For g-HHL, we prove
  - an $O(\sqrt{n}\log n)$-approximation ratio with respect to the optimal HL.
  - an $\Omega(\sqrt{n})$ lower bound on the approximation ratio with respect to the optimal HHL.
- For w-HHL, we prove
  - an $O(\sqrt{n}\log n)$-approximation ratio with respect to the optimal HL.
  - $\Omega(\sqrt[3]{n})$ lower bound on the approximation ratio with respect to the optimal HHL.
- For d-HHL, we prove
  - an $O(h\log n\log D)$ bound for every label size,
  - an $O(\sqrt{n}\log n\log D)$-approximation ratio with respect to the optimal HL (and therefore the optimal HHL),
  - an $\Omega(\sqrt{n})$ lower bound on the approximation ratio with respect to the optimal HHL.

Other Results: In addition, we show that in a network of highway dimension h and diameter D, there is an HHL such that every label is of size $O(h\log D)$, matching the HL bound of [1–3]. We also give an example showing that hierarchical labelings can be $\Omega(\sqrt{n})$ bigger than general labelings, improving and simplifying [14].

Our lower bounds show that the greedy algorithms do not have a polylogarithmic approximation ratio, leaving the question of the existence of a polylogarithmic approximation algorithm open. This is an interesting theoretical problem that may have practical consequences. Our results on d-HHL give a theoretical justification for using greedy HHL preprocessing algorithms on road networks.

This is an extended abstract that omits many proofs and technical details. These can be found in [8].

2   Preliminaries

2.1   HL Approximation Algorithm

Cohen et al. obtain their $O(\log n)$ approximation algorithm for HL by formulating it as a weighted set cover problem and applying the well known greedy approximation algorithm for set cover. In the weighted set cover problem there is a *universe* set U, a family $\mathcal{F}$ of some subsets of U, a cost function $c : \mathcal{F} \to \mathbb{R}_+$, and the goal is to find a collection $\mathcal{C} \subseteq \mathcal{F}$ such that $\cup_{S \in \mathcal{C}} S = U$ and $\sum_{S \in \mathcal{C}} c(S)$ is minimized. The greedy set cover algorithm starts with an empty $\mathcal{C}$, then iteratively picks a set S which maximizes the ratio of the number of newly covered elements in U to the cost of S and adds S to $\mathcal{C}$.

The elements to cover in the equivalent set cover instance are vertex pairs $[u, v]$. For a directed graph pairs in U are ordered and for an undirected graph pairs are unordered. We first discuss directed graphs, then undirected ones. Every possible set P of vertex pairs such that there exists a vertex u which hits a shortest path between every pair in P is an element of $\mathcal{F}$ in the corresponding set cover instance. (There are exponentially many sets P, but they are not used explicitly.) The cost of a set P is the number of vertices that appear in the first component of a pair in P plus the number of vertices that appear in the second component of a pair in P.

The greedy approximation algorithm for set cover as applied to this set cover instance is as follows. The algorithm maintains the set U of *uncovered* vertex pairs: $[u, w] \in U$ if $L_f(u) \cap L_b(w)$ does not contain a vertex on a shortest u–w path. Initially U contains all vertex pairs $[u, w]$ such that w is reachable from u. The algorithm terminates when U becomes empty. Starting with an empty labeling, in each iteration, the algorithm adds a vertex v to forward labels of vertices in a set $S' \subseteq V$ and to backward labels of the vertices in $S'' \subseteq V$ such that the ratio of the number of newly-covered pairs over the total increase in the size of the labeling is (approximately) maximized. Formally, let $U(v, S', S'')$ be the set of pairs in U which are covered if v is added to $L_f(u) : u \in S'$ and $L_b(w) : w \in S''$. The algorithm maximizes $|U(v, S', S'')|/(|S'| + |S''|)$ over all $v \in V$ and $S', S'' \subseteq V$.

To find the triples (v, S', S'') efficiently the algorithm uses center graphs defined as follows. A *center graph* of v, $G_v = (X, Y, A_v)$, is a bipartite graph with $X = V$, $Y = V$, and an arc $(u, w) \in A_v$ if $[u, w] \in U$ and some shortest path from u to w goes through v. The algorithm finds (v, S', S'') that maximizes $|U(v, S', S'')|/(|S'| + |S''|)$ by computing a densest subgraph among all the subgraphs of the center graphs G_v. The *density* of a graph $G = (V, A)$ is $|A|/|V|$. The *maximum density subgraph (MDS)* problem is the problem of finding an (induced) subgraph of a given graph G of maximum density. This problem can be solved in polynomial time using parametric flows (e.g., [12]). For a vertex v, the arcs of a subgraph of G_v induced by $S' \subseteq X$ and $S'' \subseteq Y$ correspond to the pairs of vertices in U that become covered if v is added to $L_f(u) : u \in S'$ and $L_b(w) : w \in S''$. Therefore, the MDS of G_v maximizes $|U(v, S', S'')|/(|S'| + |S''|)$ over all S', S''.

For undirected graphs we have $L_f(v) = L_b(v) = L(v)$ by definition. Pairs $[u, v] \in U$ are unordered and the cost of a set P of unordered vertex pairs is the number of vertices that appear in a pair in P. Let $U(v, S)$ be the set of unordered vertex pairs that become covered if we add v to $L(u) : u \in S$. We want to maximize $U(v, S)/|S|$. To find such a tuple, we use another type of a center graph of v, $G_v = (V, E_v)$. G_v is an undirected graph with vertex set V and with an edge $\{u, w\} \in E_v$ if $[u, w] \in U$ and some shortest path between u and w goes trough v. (For a pair $[v, v]$ there is a self-loop $\{v, v\}$ in E_v.) Note that G_v is not necessarily bipartite. As in the directed case, MDS of G_v maximizes $U(v, S)/|S|$ over all S.

The following is a folklore lemma about the greedy set cover algorithm.

Lemma 1. *If we run the greedy set cover algorithm where in each iteration we pick a set whose coverage to cost ratio is at least $1/f(n)$ fraction of the maximum coverage to cost ratio, then we get a cover of cost within an $O(f(n) \log n)$ factor of optimal.*

Cohen et al. [9] used this lemma and instead of finding the MDS exactly they used a linear-time 2-approximation algorithm for MDS [15]. The result is an $O(\log n)$-approximation algorithm running in $O(n^5)$ time. Delling et al. [11] improve the running time to $O(n^3 \log n)$.

2.2   Greedy HHL Algorithms

In this section we describe greedy HHL algorithms in terms of center graphs. For an alternative description and efficient implementation of these algorithms, see [4,10].

A greedy HHL algorithm maintains the center graphs $G_v = (X, Y, A_v)$ defined in Sect. 2.1. At each iteration, the algorithm selects a center graph of a vertex v and adds v to $L_f(u)$ for all non-isolated vertices $u \in X$ and to $L_b(w)$ for all non-isolated vertices $w \in Y$. Note that after the labels are augmented this way, all vertex pairs $[u, w]$ for which there is a u-w shortest path passing through v are covered. Therefore, the center graph of every vertex is chosen once, and the labeling is hierarchical.

Greedy algorithms differ by the criteria used to select the next center graph to process. The *greedy* HHL *(g-HHL)* algorithm selects the center graph with most edges. The *weighted greedy* HHL *(w-HHL)* algorithm selects a center graph with the highest density (the number of edges divided by the number of non-isolated vertices).

We propose a new *distance greedy* HHL *(d-HHL)* algorithm. To every vertex pair $[u, v]$ we assign a weight

$$W(u, v) = \begin{cases} 0, & \text{if } \mathrm{dist}(u, v) = 0 \\ n^{2\lfloor \log_2(\mathrm{dist}(u,v)) \rfloor}, & \text{otherwise} \end{cases}$$

and use W to weight the corresponding edges in center graphs. At each iteration, d-HHL selects a center graph with the largest sum of edge weights.

We define the *level* of $[u, v]$ as $\lfloor \log_2(\mathrm{dist}(u, v)) \rfloor$ (if $\mathrm{dist}(u, v) = 0$ the level of $[u, v]$ is $-\infty$). The definition of W insures that if $[u, v]$ is the maximum level uncovered vertex pair, $W(u, v)$ is greater than the total weight of all lower-level uncovered pairs. Therefore d-HHL primarily maximizes the number of uncovered maximum level pairs that become covered, and other pairs that become covered are used essentially as tie-breakers.

We say that a vertex w has *level i* if at the iteration when w is selected by d-HHL, the maximum level of an uncovered vertex pair is i. As the algorithm proceeds, the levels of vertices it selects are monotonically decreasing.

3 HHL and Highway Dimension

Motivated by computing driving directions, several distance (and shortest path) oracles have been developed which are particularly efficient for road networks. In order to formalize what properties of road networks make these distance oracles so efficient Abraham et al. [5] (see also [1,2]) defined the notion of highway dimension.

Intuitively, a graph has highway dimension h if for every $r > 0$, there is a sparse set of vertices S_r such that every shortest path of length greater than r includes a vertex from S_r. The set S_r is sparse if $|S_r \cap B_{2r}(v)| \le h$ for all $v \in V$, where $B_{2r}(v)$ is a ball of radius $2r$ around v. We omit the exact technical definition of highway dimension.

Recall that in general, HHL can be significanly bigger than HL. However, this doesn't happen for the special case of networks with small highway dimension, for which both HL and HHL are small: Abraham et al. show that a network with highway dimension h and diameter D has an HL with $|L(v)| = O(h \log D)$, and that, if shortest paths are unique, in polynomial time one can find an HL with $|L(v)| = O(h \log h \log D)$. We show similar results for HHL.

Theorem 1. *A network with highway dimension h and diameter D has an HHL with $|L(v)| = O(h \log D)$ for all $v \in V$. If shortest paths are unique one can find in polynomial time an HHL with $|L(v)| = O(h \log h \log D)$.*

Theorem 2. *In a network with highway dimension h and diameter D d-HHL finds a labeling with $|L(v)| = O(h \log n \log D)$, for all $v \in V$.*

4 Upper Bounds

In this section we assume that isolated vertices are deleted from the center graphs, so their density is the number of edges divided by the number of (non-isolated) vertices.

At first, we show that g-HHL finds an HHL of size that is within an $O(\sqrt{n} \log n)$ factor of the optimal HL size. We prove this by bounding the ratio of the density of the center graph picked by g-HHL and the density of the MDS of a center graph.

Theorem 3. *g-HHL produces an HHL of size no larger than $O(\sqrt{n} \log n)$ times the size of the optimal HL (and therefore also the optimal HHL).*

Proof. Suppose that at some iteration, the algorithm picks a center graph with m' arcs and n' vertices. Then by the definition of g-HHL all center graphs have at most m' arcs, so the density of the maximum density subgraph (over all center graphs) is at most $\sqrt{m'}$. The ratio of the density of the maximum density subgraph to that of the chosen center graph is at most

$$\frac{\sqrt{m'}}{m'/n'} = \frac{n'}{\sqrt{m'}} \leq \frac{n'}{\sqrt{n'/2}} = \sqrt{2n'} \leq \sqrt{2n}.$$

Here we use the fact that the chosen graph has no isolated vertices, so $m' \geq n'/2$. It follows that the density of the chosen center graph is a $\sqrt{2n}$-approximation of the maximum density of any subgraph. By Lemma 1 we have that the labeling size is larger than the size of the optimal HL by a factor of $O(\sqrt{n}\log n)$.

Now we show that d-HHL finds an HHL of size within an $O(\sqrt{n}\log n \log D)$ factor of the optimal HL size. But first we need to extend our concept of hub labels. Cohen et al. [9] defined a more general notion of *hub labels for a given set U of vertex pairs*. Such labels are required to have a vertex $w \in L(u) \cap L(v)$ which is on a shortest path between u and v for each $[u, v] \in U$. The $O(\log n)$ approximation algorithm described in Sect. 2.1 works for this more general notion of HL. Moreover, the above Theorem 3 is naturally extended and still holds.

Theorem 4. *d-HHL produces an HHL of size no larger than $O(\sqrt{n}\log n \log D)$ times the size of the optimal HL (and therefore also the optimal HHL).*

Proof. Let OPT be the size of the optimal HL. Let U_i be a set of vertex pairs at level i which are not covered by vertices at higher levels when we run d-HHL. Let HL_i be the optimal HL to cover vertex pairs from U_i and let OPT_i be size of HL_i. Since U_i is a subset of all vertex pairs, OPT_i doesn't exceed OPT. By Theorem 3 we can use g-HHL to find $O(\sqrt{n}\log n)$ approximation for HL_i.

Now let's return to d-HHL. Since every two pairs at the same level have the same weight and weights of all lower-level vertex pairs are negligible, at the consecutive set of iterations in which d-HHL covers U_i it picks the same vertices as g-HHL when we run it on U_i. So the labels found by d-HHL have size

$$\sum_{i=0}^{\lfloor \log D \rfloor} O(\sqrt{n}\log n)\mathrm{OPT}_i \leq \sum_{i=0}^{\lfloor \log D \rfloor} O(\sqrt{n}\log n)\mathrm{OPT} = O(\sqrt{n}\log n \log D)\mathrm{OPT}.$$

Finally, we study w-HHL. Although w-HHL is motivated by the approximation algorithm of Cohen et al., it does not achieve $O(\log n)$ approximation. We show that w-HHL finds an HHL of size larger than that of the optimal HL by an $O(\sqrt{n}\log n)$ factor. The key to the analysis is the following lemma.

Lemma 2. *If $G(V, E)$ is a graph with no isolated vertices, then G itself is an $O(\sqrt{n})$-approximation of the maximum density subgraph of G.*

Proof. Consider a subgraph (V', E') of G. Let $|V| = n$, $|E| = m$, $|V'| = n'$, $|E'| = m'$. Then

$$m' \leq \min(m, n'^2) = n' \min\left(\frac{m}{n'}, n'\right) \leq n'\sqrt{m},$$

where the last step follows since if $n' \leq \sqrt{m}$, $\min\left(\frac{m}{n'}, n'\right) = n' \leq \sqrt{m}$, and if $n' > \sqrt{m}$, $\min\left(\frac{m}{n'}, n'\right) = \frac{m}{n'} \leq \sqrt{m}$.

Since G goes not have isolated vertices, $m \geq n/2$, so we have

$$\frac{m'}{n'} \leq \sqrt{m} = \frac{m}{n}\frac{n}{\sqrt{m}} \leq \frac{m}{n}\frac{n}{\sqrt{n/2}} \leq \frac{m}{n}\sqrt{2n}.$$

Theorem 5. *w-HHL produces an HHL of size no larger than $O(\sqrt{n}\log n)$ times the size of the optimal HL (and therefore also the optimal HHL).*

Proof. At each iteration, w-HHL picks the center graph with the maximum ratio of the number of edges divided by the number of vertices. By Lemma 2 the density of this graph is smaller than the density of the densest subgraph of a center graph by at most $O(\sqrt{n})$. Therefore by Lemma 1 w-HHL produces an HHL of size within an $O(\sqrt{n}\log n)$ factor of the size of the optimal HL.

5   Lower Bounds

In the full paper [8] we present graphs for which g-HHL, d-HHL and w-HHL find a labeling worse than the optimal HHL by a polynomial factor. Here is the summary of our results. We include the upper bounds from Sect. 4 and Theorem 2 for comparison.

	g-HHL	w-HHL	d-HHL
approximation ratio (upper bound)	$O(\sqrt{n}\log n)$	$O(\sqrt{n}\log n)$	$O(\sqrt{n}\log n\log D)$
approximation ratio (lower bound)	$\Omega(\sqrt{n})$	$\Omega(n^{1/3})$	$\Omega(\sqrt{n})$
maximum label size (upper bound)	?	?	$O(h\log n\log D)$
maximum label size (lower bound)	?	?	$\Omega(h\log D)$

Our lower bounds on the approximation guarantees show that our upper bounds for g-HHL and d-HHL are tight up to a logarithmic factor. For w-HHL there is still an intriguing gap of $O(\sqrt[6]{n}\log n)$ between the lower and the upper bounds.

We also present a graph for which the largest label in the labeling computed by d-HHL is of size $\Omega(h\log D)$. This shows that the upper bound on the maximum label size of Theorem 2 is tight up to a logarithmic factor. Finding non-trivial bounds on the size of the largest label produced by g-HHL and w-HHL is an interesting open question.

6   NP-Hardness Proofs

The question whether computing an optimal HL is NP-hard was open for twelve years [9]. We resolve this question and prove that computing optimal HL or HHL is NP-hard. Note that although a standard reduction maps an undirected shortest path problem with non-negative arc lengths to a symmetric directed problem, this reduction does not preserve labeling optimality (see Sect. 6.2). Therefore NP-hardness of the directed case does not follow easily from the NP-hardness of the undirected case.

In this section we omit some technical details and try to present the key ideas of the proof. For a rigorous description see the full paper [8].

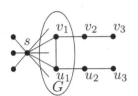

Fig. 1. Graph G'.   **Fig. 2.** G'_v labels.

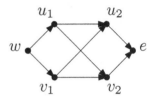

Fig. 3. Directed case.

6.1   Undirected Graphs

We prove that the problems of finding an optimal HL and an optimal HHL are NP-hard by a reduction from Vertex Cover (VC). The reduction takes an instance of VC consisting of a graph G and an integer k and produces an undirected graph G' and an integer k' such that the following conditions are equivalent

1. There is a VC of size k in G.
2. There is an HL of size k' in G'.
3. There is an HHL of size k' in G'.

This clearly implies the NP-hardness of HL and HHL in undirected graphs.
    We construct $G' = (V', E')$ from $G = (V, E)$ as follows.

1. For each vertex $v \in V$ we add three vertices, v_1, v_2, and v_3 to V' and two edges $\{v_1, v_2\}$ and $\{v_2, v_3\}$ to E'.
2. For each edge $\{u, v\} \in E$ we add an edge $\{u_1, v_1\}$ to E'.
3. We add a star S with $3|V|$ leaves and a root s to G' and add $\{s, v_1\}$ to E' for every $v \in V$.

    The graph G' is shown in Fig. 1. All edges have length 1. The purpose of the step 3 is to cover shortest paths between all $[u_i, v_j]$ pairs where $\{u, v\} \notin E$ (hereinafter when we write v_i we mean v_i for all $i \in \{1, 2, 3\}$). Intuitively, the star S is large enough to force every vertex in G' to have s in its label.

For each vertex $v \in V$ we have a subgraph G'_v in G' which is a path (v_1, v_2, v_3). Any labeling must cover all $[v_i, v_j]$ pairs of G'_v. Figure 2 shows two ways to assign labels to G'_v. A dashed arc (x, y) represents that $y \in L(x)$. We also assume $x \in L(x)$ for all vertices $x \in G'_v$ (and do not draw a self-loop at each vertex). We show that w.l.g. we may assume that the labeling covers paths in G'_v either as in Fig. 2 (upper) in which case we say that v is a vertex of *type 1* or as in Fig. 2 (lower) in which case we say that v is a vertex of *type 2*. We prove this by arguing that, if $v_1 \in L(v_2)$ then the labels of v_i $(1 \le i \le 3)$ can be rearranged such that v is of type 2. Similarly, if $v_1 \notin L(v_2)$, then the labels can be rearranged such that v is of type 1. Thus, we assume that all vertices in G are either of type 1 or of type 2.

Note that a vertex of type 2 uses one more hub in the labeling, so to reduce the labeling size we want to minimize the number of vertices of type 2. We will show, however, that the vertices of type 2 must form a vertex cover for the labeling to be valid.

For an edge $\{u, v\} \in E$ let G'_{uv} be the subgraph of G' corresponding to this edge. G'_{uv} contains all shortest paths between u_i and v_j. Note that no vertex of G' other than u_i and v_j hits these paths. We say that a hub $u_i \in L(v_j)$ or $v_i \in L(u_j)$ is a $\{u, v\}$-*crossing* hub. Clearly we must have a $\{u, v\}$-crossing hub in each of $L(u_i) \cup L(v_i)$, $1 \le i \le 3$ so the labels of u_i, v_i, $1 \le i \le 3$ contain at least 3 $\{u, v\}$-crossings.

Finally, we can prove that covering all $[u_i, v_j]$ pairs by 3 $\{u, v\}$-crossings is possible if and only if u or v is of type 2. Thus, computing an optimal HL for G' is equal to selecting type 2 vertices such that all $[u_i, v_j]$ pairs with $\{u, v\} \in E$ can be covered by 3 $\{u, v\}$ crossings, which is equal to selecting type 2 vertices according to an optimal VC of G.

Now consider $\{u, v\} \in E$ and let v be of type 2. By putting v_1 into $L(u_i)$, all $[u_i, v_j]$ pairs can be covered by 3 $\{u, v\}$-crossings. Thus, given a vertex cover X in G, we can construct a valid labeling by making each $v \in X$ a type 2 vertex and then, for each edge $\{u, v\} \in E$, covering the pairs in G'_{uv} using only 3 $\{u, v\}$-crossings. Taking into account that the label of each vertex v contains s and v itself, this gives a labeling of size $k' = 12|V| + 1 + 3|E| + k$. Conversely, an HL of size k' can make at most k vertices type 2 and for each $\{u, v\} \in E$, at least one of u,v must be type 2. Thus, the type 2 vertices form a size-k vertex cover. NP-hardness of HL in undirected graphs follows.

With a slightly more technical effort one can show that this labeling is actually hierarchical. Consider the following order. The most important vertex is s. The next important vertices are v_1 followed by v_2 for all vertices $v \in V$ of type 2. These are followed by v_2 for all vertices $v \in V$ of type 1. All other vertices form the rest of the ordering. One may verify that the resulting HHL is exactly the optimal HL. So NP-hardness of HHL in undirected graphs follows.

Our reduction shows that both HL and HHL are NP-hard in undirected graphs with unit lengths. If we change length of edges $\{s, v_1\}$ for $v \in V$ from 1 to 0.9 our proof is not affected. However, the shortest paths in G' become unique. This shows that HL and HHL are NP-hard in undirected graphs even when shortest paths are unique.

6.2   Directed Graphs

Here we show that both optimal HL and HHL are NP-hard in directed graphs. We begin with HHL, for which there is a simple reduction from the undirected case.

Let G be an undirected graph. We transform G to a directed graph G' by replacing each edge $\{u, v\}$ with two arcs (u, v) and (v, u). It is not hard to show that the graph G has an HHL of size k if and only if G' has an HHL of size $2k$. It follows that finding optimal HHL in directed graphs is NP-hard.

Unfortunately, the above simple transformation does not work for HL. In the full paper [8] we give an example showing a directed graph in which the optimal HL is not symmetric (and of size strictly less than twice the size of optimal HL in the underlying undirected graph).

Next, we present another reduction from VC to HL in a directed graph. For a VC instance $G = (V, E)$ we construct an HL instance $G' = (V', A')$,

$$
\begin{aligned}
V' =& \{w\} \cup \{v_1, v_2 \mid v \in V\} \cup \{e \mid e \in E\}, \\
A' =& \{(w, v_1), (v_1, v_2) \mid v \in V\} \\
& \cup \{(u_1, v_2), (v_1, u_2), (u_2, e), (v_2, e) \mid e = \{u, v\} \in E\}.
\end{aligned}
$$

All arcs have length 1. Figure 3 shows a gadget for an edge $e = \{u, v\}$.

For any labeling, each vertex $x \in V'$ is in both $L_f(x)$ and $L_b(x)$ and, for each arc (x, y), either $x \in L_b(y)$ or $y \in L_f(x)$. Let us call such hubs *mandatory* and all other hubs *non-mandatory*. Mandatory hubs cover all pairs $[x, y]$ such that $\mathrm{dist}(x, y) \leq 1$. Any labeling for G' has at least $M(G') = 2|V'| + |A'|$ mandatory hubs.

We claim that the graph G has a VC of size k if and only if G' has an HL of size $M(G') + k$. The key idea required to prove this claim is to carefully assign mandatory hubs to labels. We show that $M(G')$ mandatory hubs are enough to cover all pairs at distance at most 2. In fact, the following labeling suffices: u_1 is in $L_f(w)$ and $L_b(u_2)$ for all $u \in V$, u_2 is in $L_f(v_1)$ and $L_b(e)$ (also v_2 is in $L_f(u_1)$ and $L_b(e)$) for all $e = \{u, v\} \in E$, finally x is in $L_f(x)$ and $L_b(x)$ for all $x \in V'$.

So with only $M(G')$ labels we cover all pairs except $[w, e]$, $e \in E$. Observe, that to cover $[w, e]$ where $e = \{u, v\}$ it is sufficient to add v_2 or u_2 to $L_f(w)$. By adding v_2 to $L_f(w)$ for every $v \in$ VC we make the labeling complete. On the other hand, an arbitrary labeling can be rearranged such that $u_1, v_1 \in L_f(w)$ and $u_2, v_2 \in L_b(e)$ for each $e = \{u, v\} \in E$. It follows directly that the set $\{v \mid v_2 \in L_f(w)\} \cup \{v \mid v_1 \in L_b(e), e = \{u, v\}\}$ is a VC. We get that HL in directed graphs is NP-hard.

7   HL vs HHL

In [14], it is shown that the gap between the size of the optimal HHL and the size of the optimal HL can be $\Omega(n^{0.26})$. In the full paper [8], we show that the gap can be as large as $\Omega(n^{0.5})$. The results of Sect. 4 imply that this gap is within $O(\log n)$ factor of the best possible.

Theorem 6. *There is a graph family for which the optimal* HHL *size is* $\Omega(\sqrt{n})$ *times larger than the optimal* HL *size.*

8   Concluding Remarks

Our lower bounds for greedy algorithms show that in contrast with HL the greedy algorithm does not give a poly-log approximation for HHL. This motivates the question of whether a poly-log approximation algorithm for HHL exists. Our lower bound for w-HHL is $\Omega(\sqrt[6]{n})$ factor away from the upper bound, we leave open the problem of closing this gap.

On many problem classes g-HHL and w-HHL find labelings of size close to that found by the $O(\log n)$-approximation algorithm for HL [11]. It would be interesting to find a theoretical explanation for this phenomena, for example a better approximation ratio for g-HHL or w-HHL on natural classes of graphs.

References

1. Abraham, I., Delling, D., Fiat, A., Goldberg, A.V., Werneck, R.F.: Highway dimension and provably efficient shortest path algorithms. TR MSR-TR-2013-91, Microsoft Research (2013)
2. Abraham, I., Delling, D., Fiat, A., Goldberg, A.V., Werneck, R.F.: VC-dimension and shortest path algorithms. In: Aceto, L., Henzinger, M., Sgall, J. (eds.) ICALP 2011, Part I. LNCS, vol. 6755, pp. 690–699. Springer, Heidelberg (2011)
3. Abraham, I., Delling, D., Goldberg, A.V., Werneck, R.F.: A hub-based labeling algorithm for shortest paths in road networks. In: Pardalos, P.M., Rebennack, S. (eds.) SEA 2011. LNCS, vol. 6630, pp. 230–241. Springer, Heidelberg (2011)
4. Abraham, I., Delling, D., Goldberg, A.V., Werneck, R.F.: Hierarchical hub labelings for shortest paths. In: Epstein, L., Ferragina, P. (eds.) ESA 2012. LNCS, vol. 7501, pp. 24–35. Springer, Heidelberg (2012)
5. Abraham, I., Fiat, A., Goldberg, A.V., Werneck, R.F.: Highway dimension, shortest paths, and provably efficient algorithms. In: Proceedings of the 21st ACM-SIAM Symposium on Discrete Algorithms, pp. 782–793 (2010)
6. Akiba, T., Iwata, Y., Yoshida, Y.: Fast exact shortest-path distance queries on large networks by pruned landmark labeling. In: SIGMOD 2013, pp. 349–360. ACM, New York (2013)
7. Babenko, M., Goldberg, A.V., Gupta, A., Nagarajan, V.: Algorithms for hub label optimization. In: Fomin, F.V., Freivalds, R., Kwiatkowska, M., Peleg, D. (eds.) ICALP 2013, Part I. LNCS, vol. 7965, pp. 69–80. Springer, Heidelberg (2013)
8. Babenko, M., Goldberg, A.V., Kaplan, H., Savchenko, R., Weller, M.: On the Complexity of Hub Labeling (2015). http://arxiv.org/abs/1501.02492 [cs.DS]
9. Cohen, E., Halperin, E., Kaplan, H., Zwick, U.: Reachability and distance queries via 2-hop labels. SIAM J. Comput. **32**(5), 1338–1355 (2003)
10. Delling, D., Goldberg, A.V., Pajor, T., Werneck, R.F.: Robust exact distance queries on massive networks. TR MSR-TR-2014-12, Microsoft Research (2014)
11. Delling, D., Goldberg, A.V., Savchenko, R., Werneck, R.F.: Hub labels: theory and practice. In: Gudmundsson, J., Katajainen, J. (eds.) SEA 2014. LNCS, vol. 8504, pp. 259–270. Springer, Heidelberg (2014)

12. Gallo, G., Grigoriadis, M.D., Tarjan, R.E.: A fast parametric maximum flow algorithm and applications. SIAM J. Comput. **18**, 30–55 (1989)
13. Gavoille, C., Peleg, D., Pérennes, S., Raz, R.: Distance labeling in graphs. J. Algorithms **53**(1), 85–112 (2004)
14. Goldberg, A.V., Razenshteyn, I., Savchenko, R.: Separating hierarchical and general hub labelings. In: Chatterjee, K., Sgall, J. (eds.) MFCS 2013. LNCS, vol. 8087, pp. 469–479. Springer, Heidelberg (2013)
15. Kortsarz, G., Peleg, D.: Generating sparse 2-spanners. J. Alg. **17**, 222–236 (1994)
16. Peleg, D.: Proximity-preserving labeling schemes. J. Gr. Th. **33**(3), 167–176 (2000)

On the Complexity of Speed Scaling

Neal Barcelo[1], Peter Kling[2], Michael Nugent[1(✉)],
Kirk Pruhs[1], and Michele Scquizzato[3]

[1] Department of Computer Science, University of Pittsburgh,
Pittsburgh, USA
mpn1@pitt.edu
[2] School of Computing Science, Simon Fraser University,
Burnaby, Canada
[3] Department of Computer Science, University of Houston,
Houston, USA

Abstract. The most commonly studied energy management technique
is speed scaling, which involves operating the processor in a slow, energy-
efficient mode at non-critical times, and in a fast, energy-inefficient mode
at critical times. The natural resulting optimization problems involve
scheduling jobs on a speed-scalable processor and have conflicting dual
objectives of minimizing energy usage and minimizing waiting times. One
can formulate many different optimization problems depending on how
one models the processor (e.g., whether allowed speeds are discrete or
continuous, and the nature of relationship between speed and power), the
performance objective (e.g., whether jobs are of equal or unequal importance, and whether one is interested in minimizing waiting times of jobs
or of work), and how one handles the dual objective (e.g., whether they
are combined in a single objective, or whether one objective is transformed into a constraint). There are a handful of papers in the algorithmic literature that each give an efficient algorithm for a particular
formulation. In contrast, the goal of this paper is to look at a reasonably
full landscape of all the possible formulations. We give several general
reductions which, in some sense, reduce the number of problems that
are distinct in a complexity theoretic sense. We show that some of the
problems, for which there are efficient algorithms for a fixed speed processor, turn out to be NP-hard. We give efficient algorithms for some of the
other problems. Finally, we identify those problems that appear to not be
resolvable by standard techniques or by the techniques that we develop
in this paper for the other problems.

N. Barcelo—This material is based upon work supported by the National Science
Foundation Graduate Research Fellowship under Grant No. DGE-1247842.
P. Kling—Supported by a fellowship within the Postdoc-Programme of the
German Academic Exchange Service (DAAD). Work done while at the University
of Pittsburgh.
K. Pruhs—Supported by NSF grants CCF-1115575, CNS-1253218, CCF-1421508,
and an IBM Faculty Award.
M. Scquizzato—Work done while at the University of Pittsburgh.

G.F. Italiano et al. (Eds.): MFCS 2015, Part II, LNCS 9235, pp. 75–89, 2015.
DOI: 10.1007/978-3-662-48054-0_7

1   Introduction

The most commonly studied energy management technique is speed scaling. It involves operating the processor in a slow, energy-efficient mode at non-critical times, and in a fast, energy-inefficient mode at critical times. The natural resulting optimization problems involve scheduling jobs on such a processor and have conflicting dual objectives of minimizing both energy usage and waiting times. This leads to many different optimization problems, depending on how one models the processor, the performance objective, and how one handles the dual objectives. There are several papers in algorithmic literature that give an efficient algorithm for a particular formulation. In contrast, we strive to look at a reasonably full landscape of all the possible formulations. We give several general reductions which reduce the number of problems that are distinct in a complexity theoretic sense. We show that some of the problems, for which there are efficient algorithms for a fixed speed processor, turn out to be NP-hard. We give efficient algorithms for some of the other problems. Finally, we identify those problems that appear to not be resolvable by standard techniques or by the techniques that we develop in this paper for the other problems.

Models. We now describe the different models that have been considered in the literature. Given the multitude of problems, we use a succinct representation, which we introduce in parenthesis.

- *Energy Budget (B) vs. Flow plus Energy (FE):* In an energy budget problem, the energy objective is turned into a constraint that the total energy used is at most some budget B. This setting is most appropriate when the energy is limited to some finite supply, such as the battery of a laptop. In a flow plus energy problem, the performance and energy objectives are linearly combined into a single objective. We use a constant coefficient β for the energy objective that, intuitively, represents the desired trade-off between the value of energy and the value of performance.
- *Integral Flow (I) vs. Fractional Flow (F):* In an integral flow problem, the objective is total (weighted) flow/waiting time of the jobs. In a fractional flow problem, the objective is total (weighted) flow of the work (job parts). If there is no benefit in partially completing a job, then integral flow is the right performance metric. Otherwise, fractional flow may be more appropriate.
- *Continuous Speeds (C) vs. Discrete Speeds (D):* In the discrete speed setting, the processor has a finite collection of allowable speeds and corresponding powers at which it may run. In the continuous speed setting, the allowable speeds are the nonnegative real numbers. While the discrete speed model is more realistic, it is often mathematically convenient to assume continuous speeds.
- *Weighted (W) vs. Unweighted (U):* In the unweighted setting, each job is of equal importance and is weighted equally in the performance objective. However, the raison d'être for power heterogeneous technologies, such as speed-scalable processors, is ubiquity of heterogeneity in the jobs. In the weighted case, the flow of jobs/work is weighted by their importance.

– *Arbitrary Size (A) vs. Unit Size (U):* In the unit size setting, each job has the same amount of work. Similar sized jobs occur in many information technology settings (e.g., for name servers). In the arbitrary size setting, the jobs may have different sizes.
– *Power Function:* In the continuous speed setting, one needs to model how a speed s maps to power. There are two common assumptions: Most commonly one assumes $P(s) = s^\alpha$ for a constant α, slightly generalizing the well-known cube-root rule that speed is approximately the cube-root of dynamic power. The second common assumption is that $P(s)$ is a general "nice" convex function. Intuitively, the complexity of speed scaling should not come from the power function's complexity. Foreshadowing slightly, our results support this intuition.

Previous Results. We now summarize the known complexity theoretic and offline algorithmic results using the succinct notation we just introduced. The format of our description is essentially a 5-tuple of the form *-****. The first entry captures the objective (**B**udget or **F**low plus **E**nergy). The remaining entries are **I**ntegral or **F**ractional flow, **C**ontinuous or **D**iscrete speed, **W**eighted or **U**nweighted, and **A**rbitrary or **U**nit size. A * represents a "don't care" entry. See Table 1 for an overview that puts all these results into the context of the full range of possible problems.

– B-ICUU: [17] gave a polynomial-time homotopic optimization algorithm for the problem of minimizing integral flow (I) with continuous speeds (C) subject to an energy budget (B) for unweighted jobs (U) of unit size (U). They also used the assumption that $P(s) = s^\alpha$. The key insights were that jobs should be scheduled in FIFO order and that the KKT conditions for the natural convex program can be used to guide the homotopic search.
– FE-ICUU: [2] gave a polynomial-time dynamic programming algorithm for the problem of minimizing integral flow (I) with continuous speeds (C) for unweighted jobs (U) of unit size (U) and the objective of flow plus β energy (FE). Again, they were under the assumption that the power function was $P(s) = s^\alpha$, and again the key insight was that jobs should be scheduled in FIFO order.
– FE-FCWA: [10] gave a (not necessarily polynomial-time) homotopic optimization algorithm for the problem of minimizing fractional flow (F) with continuous speeds (C) for weighted jobs (W) of arbitrary size (A) and the objective of flow plus β energy (FE). The algorithm guides its search via the KKT conditions of the natural convex program.
– FE-FDWA: [4] gave a polynomial-time algorithm for the problem of minimizing fractional flow (F) with discrete speeds (D) for weighted jobs (W) of arbitrary size (A) and the objective of flow plus β energy (FE). The algorithm constructed an optimal schedule job by job, using the duality conditions of the natural linear program to find a new optimal schedule when a new job is added.
– *-I*WA: NP-hardness for integral flow (I) and weighted jobs (W) of arbitrary size (A) for the objective of flow plus β energy (FE) follows from NP-hardness

of weighted integral flow for fixed speed processors [14] (via a suitable power function). The same holds for weighted integral flow (I) for weighted jobs (W) of arbitrary size (A) subject to a budget (B).

Our Results. The goal in this paper is to more fully map out the landscape of complexity and algorithmic results for the range of problems reflected in Table 1. In particular, for each setting our aim is to either give a combinatorial algorithm (i.e., without the use of a convex program) or show it is NP-hard. Let us summarize our results:

Hardness Results:

- `B-IDUA` *is NP-hard:* The reduction is from the subset sum problem. The basic idea is to associate several high density and low density jobs with each number in the subset sum instance, and show that for certain parameter settings, there are only two possible choices for this set of jobs, with the difference in energy consumption being this number.
- `B-IDWU` *is NP-hard:* A reduction similar to `B-IDUA`, but more technical.

These results are a bit surprising as, unlike the previous NP-hardness results for speed scaling in Table 1, these problems are either open or can be solved in polynomial time on a fixed speed processor.

Polynomial Time Algorithms:

- `FE-ICUU` *is in P:* We extend [2] to general power functions. This follows by noticing that a certain set of equations can be solved for general "nice" power functions.
- `FE-IDUU` *is in P:* The algorithm utilizes the structure of the unit size unweighted case. Here, discrete speeds allow for a much simpler algorithm than for `FE-ICUU`.
- `FE-FCWA` *is in P:* We generalize [4]'s algorithm to continuous speeds. The main hurdle is a more complicated equation system at certain points in the algorithm.

Equivalence Reductions:

- *Reduction from* `B-FC**` *to* `FE-FC**`: We reduce any energy budget problem with fractional flow and continuous speeds to the corresponding flow plus β energy problem using binary search.
- *Reduction from* `B-ICUU` *to* `FE-ICUU`: The difficulty here stems from the fact that there may be multiple optimal flow plus energy schedules for a β (so binary search over β does not suffice).
- *Reduction from* `*-*D**` *to* `*-*C**`: We give a reduction from any discrete speed problem to the corresponding continuous speed problem.

While not explicitly needed for our main results, we also provide the other directions for the first two reductions, in order to improve the understanding of structural similarities.

Table 1 summarizes our results and sets them into context of previous work. Note that for some of the problems shown to be solvable in polynomial time we give a direct algorithm, while others follow by one of our reductions. For problems that are solvable by reduction to linear/convex programming, our algorithms are faster and simpler than general linear/convex programming algorithms. The key takeaways from this more holistic view of the complexity of speed scaling problems are:

Certain parameters are sufficient for determining complexity:

- *Fractional Flow:* Looking at the first two rows, we see that any problem involving fractional flow can be solved in polynomial time. This generally follows from the ability to write the problem as a convex program, although we can give simpler and more efficient algorithms.
- *Integral Flow:* For integral flow there is a more fine grained distinction:
  - Weighted & Arbitrary Size: Everything with these settings is NP-hard. Given the NP-hardness of weighted flow for a fixed speed processor, this is not surprising.
  - Unweighted & Unit Size: Everything with these settings can be solved in polynomial time largely because FIFO is the optimal ordering of the jobs.
  - Unweighted & Arbitrary or Weighted & Unit size: These seem to be the most interesting settings (w.r.t. complexity). We show their hardness for a budget, but flow plus energy remains open.

Complexity of budget problem vs. flow plus energy: For every setting for which the complexity of each is known, the complexities (in terms of membership in P or NP-hardness) match. This might be seen as circumstantial evidence that the resolution to the remaining open complexity questions is that they are NP-hard. If these open problems do indeed have polynomial algorithms, it will require new insights as there are clear barriers to applying known techniques to these problems.

Other Related Work. There is a fair number of papers that study approximately computing optimal trade-off schedules, both offline and online. [16] also gives PTAS's for minimizing total flow without release times subject to an energy budget in both the continuous and discrete speed settings. [2,3,5-7,11,12,15] consider online algorithms for optimal total flow and energy, [5,7,12] considers online algorithms for fractional flow and energy. In particular, [7] show that there are $O(1)$-competitive algorithms for all of the flow plus β energy problems that we consider (with arbitrary power functions). For a survey on energy-efficient algorithms, see [1]. For a fixed speed processor, all the fractional problems can be solved by running the job with highest density (=weight/size). Turning to integral flow, if all jobs are unit size, then always running the job of highest weight is optimal. The complexity of the problem if all jobs have the same (not unit) size is open [8,9]. The complexity of FE-I*WU seems at least as hard (but perhaps not much harder) than this problem. If all jobs have unit weight, then Shortest Remaining Processing Time is optimal for total flow.

Outline of the Paper. Section 2 provides basic definitions. In Sect. 3 we show that B-IDWU and B-IDUA are NP-hard. In Sect. 4 we give several polynomial time

Table 1. Summary of known and new results. Each cell's upper-half refers to the flow $+ \beta \cdot$ energy objective and the lower-half refers to flow minimization subject to an energy constraint. Results of this paper are indicated by [⋆], and $\equiv$ indicates that two problems are computationally equivalent

		Unweighted Jobs		Weighted Jobs	
		Unit Sizes	Arbitrary Sizes	Unit Sizes	Arbitrary Sizes
Fractional Flow	Discrete Speeds	P [4] / P [⋆]	P [4] / P [⋆]	P [4] / P [⋆]	P [4] / P [⋆]
	Continuous Speeds	P [⋆] / P [⋆]	P [⋆] / P [⋆]	P [⋆] / P [⋆]	P [⋆] / P [⋆]
Integral Flow	Discrete Speeds	P [⋆] / P [⋆]	? / NP-hard [⋆]	? / NP-hard [⋆]	NP-hard [14] / NP-hard [16]
	Continuous Speeds	P [2][⋆] / P [17][⋆]	? / NP-hard [⋆]	? / NP-hard [⋆]	NP-hard [14] / NP-hard [16]

algorithms. Finally, in Sect. 5, we give the reductions between budget and flow plus β energy problems. Due to space constraints, omitted proofs are left to the full version of the paper.

2   Model and Notation

We consider n jobs $J = 1, 2, \ldots, n$ to be processed on a single, speed-scalable processor. In the *continuous* setting, the processor's energy consumption is modeled by a power function $P \colon \mathbb{R}_{\geq 0} \to \mathbb{R}_{\geq 0}$ mapping a speed s to a power $P(s)$. We require P to be continuous, convex, and non-decreasing. Other than that, we merely assume P to be "nice" in the sense that we can solve basic equations involving the power function and, in particular, its derivative and inverse. In the *discrete* setting, the processor features only k distinct speeds $0 < s_1 < s_2 < \cdots < s_k$, where a speed s_i consumes energy at the rate $P_i \geq 0$. Even in the discrete case, we will often use $P(s)$ to refer to the power consumption when "running at a speed $s \in (s_i, s_{i+1})$" in between the discrete speeds. This is to be understood as interpolating the speed $s = s_i + \gamma(s_{i+1} - s_i)$ (running for a γ fraction at speed s_{i+1} and a $1 - \gamma$ fraction at speed s_i), yielding an equivalent discrete schedule. Each job $j \in J$ has a release time r_j, a processing volume p_j, and a weight w_j. The density of j is w_j/p_j. For each time t, a schedule S must decide which job to process at what speed. Preemption is allowed, so that a job may be suspended and resumed later on. We model a schedule S by a speed function $\mathcal{V} \colon \mathbb{R}_{\geq 0} \to \mathbb{R}_{\geq 0}$ and a scheduling policy $\mathcal{J} \colon \mathbb{R}_{\geq 0} \to J$.

Here, $\mathcal{V}(t)$ denotes the speed at time t, and $\mathcal{J}(t)$ the job that is scheduled at time t. Jobs can be processed only after they have been released. For job j let $I_j = \mathcal{J}^{-1}(j) \cap [r_j, \infty)$ be the set of times during which it is processed. A feasible schedule must finish the work of all jobs. That is, the inequality $\int_{I_j} \mathcal{V}(t)dt \geq p_j$ must hold for all jobs j.

We measure the quality of a given schedule S by means of its energy consumption and its fractional or integral flow. The energy consumption of a job j is $E_j = \int_{I_j} P(\mathcal{V}(t))dt$, and the energy consumption of schedule S is $\sum_{j \in J} E_j$. The *integral flow* $F_j = w_j(C_j - r_j)$ of a job j is the weighted difference between its completion time C_j and release time r_j. The integral flow of schedule S is $F(S) = \sum_{j \in J} F_j$. In contrast, the *fractional flow* can be seen as the flow on a per workload basis (instead of per job). More formally, if $p_j(t)$ denotes the work remaining on job j at time t, the fractional flow time of job j is $w_j \int_{r_j}^{\infty} \frac{p_j(t)}{p_j} dt$. Our goal is to find energy-efficient schedules that provide a good (low) flow. We consider two different ways to combine these conflicting goals. In the *budget setting*, we fix an *energy budget* $B \geq 0$ and seek the minimal (fractional or integral) flow achievable with this energy. In the *flow plus energy setting*, we want to minimize a linear combination $F(S) + \beta E(S)$ of energy and (fractional or integral) flow.

3   Hardness Results

This section proves NP-hardness for the problems B-IDUA and B-IDWU. The reductions are from the subset sum problem, where we are given n elements $a_1 \geq a_2 \geq \cdots \geq a_n$ with $a_i \in \mathbb{N}$ as well as a target value $A \in \mathbb{N}$ with $a_1 < A < \sum_{i=1}^{n} a_i$. The goal is to decide whether there is a subset $L \subseteq [n]$ such that $\sum_{i \in L} a_i = A$.

Basic Idea. For both reductions, we define for each element a_i a job set J_i such that jobs of different sets will not influence each other. Each J_i contains one low density job and one/several high density jobs. Starting from a *base schedule*, we choose the parameters such that investing roughly a_i energy into J_i improves its flow by roughly a_i. More precisely, when J_i gets a_i energy, additional energy can be used to decreases the flow at a rate $\gg 1/2$ per energy unit. Given substantially more or less energy, additional energy decreases the flow at a rate of only $1/2$. We achieve this by ensuring that at about a_i energy, the schedule switches from finishing the low density job after the high density jobs to finishing it before them. For an energy budget of A, we can define a target flow that is reached if and only if there is an $L \subseteq [n]$ such that $\sum_{i \in L} a_i = A$ (corresponding to job sets that are given about a_i extra energy).

Remarks. We assume a processor with two speeds $s_1 = 1$ and $s_2 = 2$ and power consumption rates $P_1 = 1$ and $P_2 = 4$. For an isolated job of weight w, this means that increasing a workload of x from speed s_1 to s_2 increases the energy by x and decreases the flow by $w \cdot \frac{x}{2}$. To ensure that jobs of different job sets do not influence each other, one can increase all release times of the job set J_i by the total workload of all previous job sets. For ease of exposition, we consider each job group J_i in isolation, and assume its first job is released at time 0. Due to space constraints, the reduction for B-IDWU is deferred to the full version.

3.1 Hardness of B-IDUA

For $i \in [i]$, we define a job set $J_i = (i,1),(i,2)$ of two unit weight jobs and set $\delta = \frac{1}{a_1 n^2}$. The release time of job $(i,1)$ is $r_{i1} = 0$ and its size is $p_{i1} = a_i$. The release time of job $(i,2)$ is $r_{i2} = a_i/2$ and its size is $p_{i2} = 2\delta a_i$.

Definition 1 (Base Schedule). *The* base schedule BS_i *schedules job* $(i,1)$ *at speed 1 and job* $(i,2)$ *at speed 2. It finishes job* $(i,1)$ *after* $(i,2)$, *has energy consumption* $E(\mathrm{BS}_i) = a_i + 4\delta a_i$, *and flow* $F(\mathrm{BS}_i) = a_i + 2\delta a_i$.

Note that BS_i is optimal for the energy budget $E(\mathrm{BS}_i)$. Consider an optimal schedule S for the jobs $J = \bigcup_{i=1}^n J_i$ (release times shifted such that they do not interfere) for the energy budget $B = \sum_{i=1}^n E(\mathrm{BS}_i) + A$. Let $L \subseteq [n]$ be such that $i \in L$ if and only if J_i gets at least $E(\mathrm{BS}_i) + a_i - 4\delta a_i = 2a_i$ energy in S.

Lemma 1. S *has flow at most* $F = \sum_{i=1}^n F(\mathrm{BS}_i) - (\frac{1}{2} + \delta)A$ *iff* $\sum_{i \in L} a_i = A$.

Proof. For the first direction, given that $\sum_{i \in L} a_i = A$, note that the schedule that gives each job set J_i with $i \in L$ exactly $E(\mathrm{BS}_i) + a_i$ energy and each J_i with $i \notin L$ exactly $E(\mathrm{BS}_i)$ energy adheres to the energy budget and has flow exactly F. For the other direction, consider $i \in [n]$, let E_i be the total energy used to schedule J_i in S, and let $\Delta_i = E_i - E(\mathrm{BS}_i)$ the additional energy used with respect to the base schedule. Then, for $i \notin L$, the flow of J_i is $F(\mathrm{BS}_i) - \frac{1}{2}\Delta_i$, yielding an average flow gain per energy unit of $1/2$. For $i \in L$, the flow gain per energy unit is 1 for the interval $[2a_i, 2a_i + 2\delta a_i)$ and $1/2$ otherwise. Thus, the maximum average flow gain is achieved for $E_i = 2a_i + 2\delta a_i$, where the energy usage is $E(\mathrm{BS}_i) + a_i - 2\delta a_i$ and the flow is $F(\mathrm{BS}_i) - a_i/2$. This yields a maximum average flow gain per energy unit of $\frac{a_i/2}{a_i - 2\delta a_i} = \frac{1}{2 - 4\delta}$. Using these observations, we now show that, if $\sum_{i \in L} a_i \neq A$, the schedule has either too much flow or uses too much energy. Let us distinguish two cases:

Case 1: $\sum_{i \in L} a_i < A$: Using $a_i, A \in \mathbb{N}$ and our observations, the flow decreases by at most (w.r.t. $\sum_{i=1}^n \mathrm{BS}_i$)

$$\frac{1}{2 - 4\delta}\sum_{i \in L} a_i + \frac{1}{2}\left(A - \sum_{i \in L} a_i\right) = \frac{1}{2}A + \frac{\delta}{1 - 2\delta}\sum_{i \in L} a_i \leq \frac{1}{2}A + \frac{\delta}{1 - 2\delta}(A-1) < \left(\frac{1}{2} + \delta\right)A.$$

The last inequality follows from $\delta = \frac{1}{a_1 n^2} < \frac{1}{2A}$.

Case 2: $\sum_{i \in L} a_i > A$: This implies $\sum_{i \in L} a_i \geq A + 1$. Note that even if all jobs $(i,2)$ with $i \in 1,2,\ldots,n$ are run at speed 1 instead of speed 2, the total energy saved with respect to the base schedules is at most $\sum_{i=1}^n 2\delta a_i \leq \frac{2}{n}$. By this and the previous observations, the additional energy used by S with respect to the base schedules is at least $(1 - 4\delta)\sum_{i \in L} a_i - \frac{2}{n} \geq \sum_{i \in L} a_i - \frac{6}{n} \geq A + 1 - \frac{6}{n} > A.$ □

Theorem 1. *B-IDUA is NP-hard.*

4   Polynomial Time Algorithms

In this section we provide polynomial time algorithms for FE-IDUU, FE-ICUU, and FE-FCWA. The algorithm for FE-ICUU generalizes and makes slight modifications to the algorithm in [2] to handle arbitrary power functions. We also provide a new, simple, combinatorial algorithm for FE-IDUU. While by the results of Sect. 5.1 we could use the algorithm for FE-ICUU to solve FE-IDUU, the algorithm we provide has the advantages of not having the numerical qualifications of the algorithm for FE-ICUU, as well as providing some additional insight into the open problem FE-IDUA. The algorithm for FE-FCWA generalizes and makes slight modifications to the algorithm in [4] to handle arbitrary power functions.

4.1   An Algorithm for FE-IDUU

Here we give a polynomial time algorithm for FE-IDUU. We describe the algorithm for two speeds; it is straightforward to generalize it to k speeds. The algorithm relies heavily upon the fact that, when jobs are of unit size, the optimal completion ordering is always FIFO (since any optimal schedule uses the SRPT (shortest remaining processing time) scheduling policy).[1] Before describing the algorithm, we provide the necessary optimality conditions in Lemma 2. They are based on the following definitions, capturing how jobs may affect each other.

Definition 2 (Lower Affection). *For a fixed schedule, a job j_1 lower affects a job j_2 if there is an $\varepsilon > 0$ such that decreasing the speed of j_1 by any value in $(0, \varepsilon]$ increases the flow of j_2.*

Definition 3 (Upper Affection). *For a fixed schedule, a job j_1 upper affects a job j_2 if there is some $\varepsilon > 0$ such that increasing the speed of j_1 by any value in $(0, \varepsilon]$ decreases the flow of j_2.*

Lemma 2. *Be S an optimal schedule and s_1 and s_2 consecutive speeds. Define $\alpha = \frac{P_2 - P_1}{s_2 - s_1}$ and $\kappa = -(P_1 - \alpha s_1) \geq 0$. For job j with (interpolated) speed $s_j \in [s_1, s_2]$: (a) $s_j > s_1 \Rightarrow j$ lower affects at least κ jobs, and (b) $s_j < s_2 \Rightarrow j$ upper affects at most $\kappa - 1$ jobs.*

Proof. We start with (a). To get a contradiction, assume $s_j > s_1$ but j lower affects less than κ jobs. Thus, for any $\varepsilon > 0$, increasing j's completion time by ε increases the flow of at most $\kappa - 1$ jobs by ε. If the resulting schedule is S'. For $t = \frac{1}{s_j}$, the energy from S to S' decreases by

---

[1] In fact, a slightly more general result yields an optimal FE-IDUA schedule given an optimal completion ordering.

$$tP\left(\frac{1}{t}\right) - (t+\varepsilon)P\left(\frac{1}{(t+\varepsilon)}\right)t\left(\frac{\alpha}{t}+P_1-\alpha s_1\right) - (t+\varepsilon)\left(\frac{\alpha}{(t+\varepsilon)}+P_1-\alpha s_1\right)$$
$$= \alpha + tP_1 - t\alpha s_1 - \alpha - (t+\varepsilon)P_1 + (t+\varepsilon)\alpha s_1 = -\varepsilon(P_1-\alpha s_1) = \kappa\varepsilon.$$

So, the total change in the objective function is at most $(\kappa - 1)\varepsilon - \kappa\varepsilon < 0$, contradicting the optimality of S. Statement (b) follows similarly by decreasing the completion time of j by ε.   □

Observation 2. *Consider two arbitrary jobs j and j' in an arbitrary schedule S.*

(a) If j upper affects $j' \neq j$ and j does not run at s_2, j' must run at s_1.
(b) While raising j's speed, the number of its lower and upper affections can only decrease.
(c) If j upper affects j', then changing the speed of j' will not change j's affection on j'.
(d) Assume j runs at speed s_j and upper affects m jobs. Then, in any schedule where j's speed is increased (and all other jobs remain unchanged), j lower affects at most m jobs.

Our algorithm GREEDYAFFECTION initializes each job with speed s_1. Consider jobs in order of release times and let j denote the current job. While j upper affects at least κ jobs and is not running at s_2, increase its speed. Otherwise, update j to the next job (or terminate if $j = n$).

Theorem 3. GREEDYAFFECTION *solves FE-IDUU in polynomial time.*

Proof. Assume A is not optimal and let O be an optimal schedule agreeing with A for the most consecutive job speeds (in order of release times). Let j be the first job that runs at a different speed and let s_A and s_O be the job's speeds in A and O. We consider two cases: If $s_A > s_O$, Observation 2(a) implies that every job that is upper affected by j in O other than j itself is run at s_1. Consider the time of A's execution when the speed of j was at s_O. Since A continued to raise j's speed, j upper affected at least κ jobs. Let J be this set of jobs. By Observation 2(c), j still upper affects all jobs $j' \in J$ in O. This contradicts the optimality of O (Lemma 2). For the second case, assume $s_A < s_O$. By Lemma 2, j upper affects less than κ jobs in A. When A stops raising j's speed, all jobs to the right run at s_1. Observations 2(b) and (d) imply that j lower affects less than κ jobs in O, contradicting O's optimality (Lemma 2).   □

4.2   An Algorithm for FE-ICUU

In this subsection we show that FE-ICUU is in P. Essentially, it is possible to modify the algorithm from [2] to work with arbitrary power functions. The main alteration is that, for certain power functions that would yield differential equations too complicated for the algorithm to solve, we use binary search to find solutions to a these equations rather than solve the equations analytically.

Theorem 4. *There is a polynomial time algorithm for solving FE-ICUU.*

4.3   An Algorithm for FE-FCWA

This subsection shows that FE-FCWA is in P. The basic idea is to modify the algorithm from [4] to work with arbitrary power functions, under some mild assumptions. In order to maintain polynomial running time, the algorithm must efficiently find the next occurrence of certain *events*. In [4], this is done by analytically solving a series of differential equations, which are too complicated to solve for arbitrary power functions. Instead, our algorithm finds the occurrence of the next event by using binary search to "guess" its occurrence, and then (numerically) solve a (simpler) set of equations to determine if an event did, in fact, occur. Our only assumption is that it is possible to numerically find a solution to the involved equations.

Theorem 5. *There is a polynomial time algorithm for solving FE-FCWA.*

5   Equivalence Reductions

Here we provide the reductions to obtain the hardness and algorithmic results that are not proven explicitly. First, we reduce B-ICUU to FE-ICUU. Combined with the algorithm from Sect. 4.2, this shows that B-ICUU is in P. The second reduction is from any problem in the discrete power setting to the corresponding continuous variant. As a result, the hardness proofs from Sect. 3 for B-IDWU and B-IDUA imply that B-ICWU and B-ICUA are NP-hard. Our final reduction is from B-FCWA to FE-FCWA. As a result of the algorithm in Sect. 4.3, this shows that B-FCWA is in P.

5.1   Reducing B-ICUU to FE-ICUU

We show that, given an algorithm for the flow plus energy variant, we can solve the energy budget variant of ICUU. The basic idea is to modify the coefficient β in the flow plus energy objective until we find a schedule that fully utilizes the energy budget B. This schedule gives the minimum flow for B. The major technical hurdles to overcome are that the power function P may be non-differentiable, and may lead to multiple optimal flow plus energy schedules, each using different energies. Thus, we may not find a corresponding schedule for the given budget, even if there is one. To overcome this, we define the *affectance* ν_j of a job j. Intuitively, ν_j represents how many jobs' flows will be affected by a speed change of j. We show that a job's affectance is, in contrast to its energy and speed, unique across optimal schedules and changes continuously in β. This will imply that job speeds change continuously in β (i.e., for small enough changes, there are two optimal schedules with speeds arbitrarily close). We also give a continuous transformation process between any two optimal schedules. This eventually allows us to apply binary search to find the correct β.

Definitions and Notation. We start with some formal definitions for this section and a small overview of what they will be used for in the remainder.

Definition 6 (affectance) will be most central to this section, as it will be shown in Lemma 3 and Corollary 1 to characterize optimal schedules. It uses the *sub-differential*[2] $\partial P(s)$ to handle non-differentiable power functions P.

Definition 4 (Total Weight of Lower/Upper Affection). *In any schedule, l_j and u_j are the total weight of jobs lower and upper affected by j, respectively (see Definitions 2 and 3).*

Definition 5 (Job Group). *A job group is a maximal subset of jobs such that each job in the subset completes after the release of the next job. Let J_i denote the job group with the i-th earliest release time and W_i the total weight of J_i ($J_i = \emptyset$ and $W_i = 0$ if J_i does not exist). Job groups J_i and J_{i+1} are consecutive if the last job in J_i ends at the release time of the first job in J_{i+1}. We set the indicator $\zeta_i = 1$ if and only if J_{i+1} exists and J_i and J_{i+1} are consecutive.*

Definition 6 (Affectance Property). *The ith job group of a schedule satisfies the* affectance property *if either $\zeta_{i+1} = 0$ or the $i + 1$st job group also satisfies the affectance property, and there exists $\mathcal{N}^i$ such that for all $v^i \in \mathcal{N}^i$ and $j \in J_i$*

$$v^i \in [0, \zeta_{i+1}(\nu^{i+1} + W_{i+1})], \tag{1}$$

$$v_j = v^i + u_j, \text{ and} \tag{2}$$

$$v_j = s_j d - P(s_j) \text{ for some } d \in \partial P(s). \tag{3}$$

Here, $\nu^i = \max \mathcal{N}^i$ if job group i exists, and $\nu^i = 0$ otherwise. A schedule satisfies the affectance property if all job groups in the schedule satisfy the affectance property.

Definition 7 (Affectance of a Job). *The set of speeds satisfying Eq. (3) for $v_j = \nu$ is $\mathcal{S}(\nu)$. For each job j in group i with the affectance property, the* affectance *of job j is $\nu_j = \nu^i + u_j$.*

Characterizing Optimal Schedules. We first prove that the affectance property characterizes optimal schedules. Lemma 3 shows that this property is necessary, Lemma 4 shows that affectance is unique across optimal schedules, and Corollary 1 shows that the affectance property is sufficient for optimality.

Lemma 3. *Any optimal schedule for FE-ICUU satisfies the affectance property.*

Lemma 4. *Let S_1 and S_2 be schedules with the affectance property and let ν_j^i denote the affectance of job j in the corresponding schedule. Then $\nu_j^1 = \nu_j^2$ for all j.*

Next, we show how to transform any schedule that has the affectance property into any other such schedule without changing the flow plus energy value.

---

[2] Subdifferentials generalize the derivative of convex functions. $\partial P(s)$ is the set of slopes of lines through $(s, P(s))$ that lower bound P. It is closed, convex on the interior of P's domain, and non-decreasing if P is increasing [13].

Together with Lemma 3, this immediately implies that the affectance property is sufficient for optimality (Corollary 1). Also, Lemma 3 is a nice algorithmic tool, as it allows us to find schedules "in between" any two optimal schedules with arbitrary precision. We will make use of that in the proof of Theorem 6.

Lemma 5. *Let S_1 and S_2 be schedules with the affectance property. We can transform S_1 to S_2 without changing its flow plus energy. All intermediate schedules satisfy the affectance property and we can make the speed changes between intermediate schedules arbitrarily small.*

Corollary 1. *Any schedule satisfying the affectance property is optimal.*

Binary Search Algorithm. We now provide the main technical result of this section, a polynomial time algorithm for B-ICUU based on any such algorithm for FE-ICUU (Theorem 6). In order to state the algorithm and its correctness, we need two more auxiliary lemmas. Lemma 6 proves that the affectance of jobs is continuous in β, while Lemma 7 does the same for job speeds.

Lemma 6. *For $\beta > 0$ and $\varepsilon > 0$, there exists $\delta > 0$ such that for all jobs j and $\beta' \in [\beta - \delta, \beta + \delta]$, any optimal FE-ICCU schedules S for β and S' for β' adhere to $\nu'_j \in [\nu_j - \varepsilon, \nu_j + \varepsilon]$.*

Lemma 7. *For $\beta > 0$ and $\varepsilon > 0$, there exists $\delta > 0$ such that for all jobs j and $\beta' \in [\beta - \delta, \beta + \delta]$, any optimal FE-ICUU schedules S for β and S' for β' adhere to $s'_j \in [s_j - \varepsilon, s_j + \varepsilon]$.*

Theorem 6. *Given a polynomial time algorithm for the continuous flow plus energy problem with unit size unit weight jobs, there is a polynomial time algorithm for the budget variant.*

Proof. Suppose we are given an energy budget B, and an algorithm to solve FE-ICUU. As we formally show in the proof of Theorem 8, the energy of optimal schedules increases as β decreases (even though we are considering here integral flow rather than fractional flow). Thus, the first step of the algorithm is to binary search over β until we find a schedule that fully utilizes B. If we find such a β, we are done (any optimal FE-ICUU schedule must minimize flow for the energy it consumes). Otherwise, we consider three cases:

Case 1: We find a β for which the optimal FE-ICUU schedule runs every job at the lowest speed used by any optimal schedule and uses $> B$ energy. Here, this lowest speed is (if it exists) the largest speed s such that for all $s' < s$ we have $\frac{P(s)}{s} \leq \frac{P(s')}{s'}$. In this case, no solution exists, since running a job at a lower speed increases its flow but does not decrease its energy.

Case 2: We find a β for which the optimal FE-ICUU schedule runs every job at the highest speed used by any optimal schedule and uses $\leq B$ energy. Here, this highest speed is (if it exists) the largest speed s such that for all $s' > s$ we have $P(s') = \infty$. In this case, β yields the optimal budget solution, since running any job at a higher speed uses infinite energy.

Case 3: There is $\varepsilon > 0$ such that for any β, the computed optimal FE-ICUU schedule uses at least $B + \varepsilon$ or at most $B - \varepsilon$ energy. Since job speeds are continuous in β (Lemma 7) and the energy increases as β decreases, we know that there is some β such that the corresponding FE-ICUU solutions contain schedules using both $B + \varepsilon_1$ energy and $B - \varepsilon_2$ energy $(\varepsilon_1, \varepsilon_2 > 0)$. Fix such a β and let S_1 and S_2 be the corresponding schedules using $B - \varepsilon_1$ and $B + \varepsilon_2$, respectively. By Lemma 5, we can continuously change the speeds (and, thus, energy) of S_1 to obtain S_2. During this process, we obtain an intermediate optimal FE-ICUU schedule that uses exactly B energy. As described above, this schedule is also optimal for B-ICUU. □

5.2 Reducing the Discrete to the Continuous Setting

The main result of this subsection is a reduction from the discrete to the continuous setting. Using mild computational power assumptions, Theorem 7 shows how to use an algorithm for the continuous variant of one of our problems (*-*C**) to solve the corresponding discrete variant (*-*D**). It is worth noting that our reduction makes use of arbitrary continuous power functions (especially power functions with a maximum speed).

Theorem 7. *Given a polynomial time algorithm for any budget or flow plus energy variant in the continuous setting, there is a polynomial time algorithm for the discrete variant.*

5.3 Reducing from Budget to Flow Plus Energy for Fractional Flow

This subsection gives a reduction from the budget to the flow plus energy objective. The reduction given in Theorem 8 is for fractional flow, assumes the most general setting (weighted jobs of arbitrary size), and preserves unit size and unit weight jobs, making it applicable to reduce B-FC** to FE-FC**.

Theorem 8. *Given a polynomial time algorithm for the budget variant and fractional flow, there is a polynomial time algorithm for the corresponding flow plus energy variant.*

References

1. Albers, S.: Energy-efficient algorithms. Commun. of the ACM **53**(5), 86–96 (2010)
2. Albers, S., Fujiwara, H.: Energy-efficient algorithms for flow time minimization. ACM Trans. Algorithms 3(4) (2007)
3. Andrew, L.L.H., Wierman, A., Tang, A.: Optimal speed scaling under arbitrary power functions. SIGMETRICS Perform. Eval. Rev. **37**(2), 39–41 (2009)
4. Antoniadis, A., Barcelo, N., Consuegra, M., Kling, P., Nugent, M., Pruhs, K., Scquizzato, M.: Efficient computation of optimal energy and fractional weighted flow trade-off schedules. In: Proceedings of the 31st International Symposium on Theoretical Aspects of Computer Science (STACS), pp. 63–74 (2014)

5. Bansal, N., Chan, H.-L., Lam, T.-W., Lee, L.-K.: Scheduling for speed bounded processors. In: Aceto, L., Damgård, I., Goldberg, L.A., Halldórsson, M.M., Ingólfsdóttir, A., Walukiewicz, I. (eds.) ICALP 2008, Part I. LNCS, vol. 5125, pp. 409–420. Springer, Heidelberg (2008)
6. Bansal, N., Pruhs, K., Stein, C.: Speed scaling for weighted flow time. SIAM J. Comput. 39(4), 1294–1308 (2009)
7. Bansal, N., Chan, H.-L., Pruhs, K.: Speed scaling with an arbitrary power function. ACM Trans. Algorithms 9(2), 18:1–18:14 (2013)
8. Baptiste, P.: Polynomial time algorithms for minimizing the weighted number of late jobs on a single machine with equal processing times. J. Sched. 2(6), 245–252 (1999)
9. Baptiste, P.: Scheduling equal-length jobs on identical parallel machines. Discrete Appl. Math. 103(1–3), 21–32 (2000)
10. Barcelo, N., Cole, D., Letsios, D., Nugent, M., Pruhs, K.: Optimal energy trade-off schedules. Sustainable Comput.: Inform. Syst. 3, 207–217 (2013)
11. Chan, S.-H., Lam, T.-W., Lee, L.-K.: Non-clairvoyant speed scaling for weighted flow time. In: de Berg, M., Meyer, U. (eds.) ESA 2010, Part I. LNCS, vol. 6346, pp. 23–35. Springer, Heidelberg (2010)
12. Devanur, N.R., Huang, Z.: Primal dual gives almost optimal energy efficient online algorithms. In: Proceedings of the Twenty-Fifth Annual ACM-SIAM Symposium on Discrete Algorithms (SODA), pp. 1123–1140 (2014)
13. Hiriart-Urruty, J.-B., Lemaréchal, C.: Fundamentals of Convex Analysis. Springer, Heidelberg (2001)
14. Labetoulle, J., Lawler, E.L., Lenstra, J.K., Kan, A.H.G.R.: Preemptive scheduling of uniform machines subject to release dates. In: P.H.R. (eds.) Progress in Combinatorial Optimization, pp. 245–261. Academic Press (1984)
15. Lam, T.-W., Lee, L.-K., To, I.K.K., Wong, P.W.H.: Speed scaling functions for flow time scheduling based on active job count. In: Halperin, D., Mehlhorn, K. (eds.) ESA 2008. LNCS, vol. 5193, pp. 647–659. Springer, Heidelberg (2008)
16. Megow, N., Verschae, J.: Dual techniques for scheduling on a machine with varying speed. In: Fomin, F.V., Freivalds, R., Kwiatkowska, M., Peleg, D. (eds.) ICALP 2013, Part I. LNCS, vol. 7965, pp. 745–756. Springer, Heidelberg (2013)
17. Pruhs, K., Uthaisombut, P., Woeginger, G.J.: Getting the best response for your erg. ACM Trans. Algorithms 4(3) (2008)

Almost All Functions Require Exponential Energy

Neal Barcelo[1], Michael Nugent[1]([✉]), Kirk Pruhs[1], and Michele Scquizzato[2]

[1] Department of Computer Science, University of Pittsburgh,
Pittsburgh, USA
{ncb30,mpn1,kirk}@pitt.edu
[2] Department of Computer Science, University of Houston,
Houston, USA
michele@cs.uh.edu

Abstract. One potential method to attain more energy-efficient circuits with the current technology is Near-Threshold Computing, which means using less energy per gate by designing the supply voltages to be closer to the threshold voltage of transistors. However, this energy savings comes at a cost of a greater probability of gate failure, which necessitates that the circuits must be more fault-tolerant, and thus contain more gates. Thus achieving energy savings with Near-Threshold Computing involves properly balancing the energy used per gate with the number of gates used. The main result of this paper is that almost all Boolean functions require circuits that use exponential energy, even if allowed circuits using heterogeneous supply voltages. This is not an immediate consequence of Shannon's classic result that almost all functions require exponential sized circuits of faultless gates because, as we show, the same circuit layout can compute many different functions, depending on the value of the supply voltages. The key step in the proof is to upper bound the number of different functions that one circuit layout can compute. We also show that the Boolean functions that require exponential energy are exactly the Boolean functions that require exponentially many faulty gates.

1 Introduction

The threshold voltage of a transistor is the minimum supply voltage at which the transistor starts to conduct current. However, if the designed supply voltage was exactly the ideal threshold voltage, some transistors would likely fail to operate as designed due to manufacturing and environmental variations. In the traditional approach to circuit design the supply voltages for each transistor/gate are set

---

N. Barcelo—This material is based upon work supported by the National Science Foundation Graduate Research Fellowship under Grant No. DGE-1247842.

K. Pruhs—Supported in part by NSF grants CCF-1115575, CNS-1253218, CCF-1421508, and an IBM Faculty Award.

M. Scquizzato—Work done while at the University of Pittsburgh.

G.F. Italiano et al. (Eds.): MFCS 2015, Part II, LNCS 9235, pp. 90–101, 2015.
DOI: 10.1007/978-3-662-48054-0_8

sufficiently high so that with sufficiently high probability no transistor fails, and thus the designed circuits need not be fault-tolerant. One potential method to attain more energy-efficient circuits is *Near-Threshold Computing*, which simply means that the supply voltages are designed to be closer to the threshold voltage. As the power used by a transistor/gate is roughly proportional to the square of the supply voltage [4], Near-Threshold Computing can potentially significantly decrease the energy used per gate. However, this energy savings comes at a cost of a greater probability of functional failure, which necessitates that the circuits must be more fault-tolerant, and thus contain more gates. For an example of this tradeoff in an SRAM cell, see [7].

1.1   Our Contributions

As the total energy used by a circuit is roughly the sum over all gates of the energy used by that gate, achieving energy savings with Near-Threshold Computing involves properly balancing the energy used per gate with the number of gates used. In principle, for every function f there exists a circuit C computing f with probability of error at most δ that uses minimum energy. It is natural to ask questions about the minimum energy required for various functions. Pippenger showed that all Boolean functions with n inputs can be computed by circuit layouts with $O(2^n/n)$ noisy gates (i.e., gates that fail independently with some known, fixed probability) [12]. Using that construction, it immediately follows that all Boolean functions can be computed by some circuit C that uses $O(2^n/n)$ energy when δ is a fixed constant. Our main result, which we state somewhat informally below, is that this result is tight for almost all functions.

Theorem 1. *Almost all Boolean functions on n variables require circuits that use $\Omega(2^n/n)$ energy.*

The main component of the proof is to show that most functions require circuit layouts with exponentially many gates. Note that in this setting, this is not an immediate consequence of Shannon's classic result [15] that most functions require circuit layouts with exponentially many faultless gates. To understand this point better, let us consider the simple counting-based proof of the following somewhat informal statement of Shannon's classic result:

Theorem 2 (Shannon [15]). *Almost all Boolean functions on n inputs require circuits with faultless gates of size $\Omega(2^n)$ bits (and of size $\Omega(2^n/n)$ gates).*

Proof. We will associate circuit layouts with their binary representation in some standard form. Each string of k bits specifies at most one circuit layout. There are 2^k bit strings of length k. Thus using k or less bits, at most $\sum_{i=0}^{k} 2^i \leq 2^{k+1}$ different circuit layouts can be specified. But there are 2^{2^n} Boolean functions with n input bits, hence $k = 2^n - \ell$ bits are only sufficient to specify a $2^{2^n-\ell+1}/2^{2^n} = 1/2^{\ell-1}$ fraction of all the possible Boolean functions. The bound on the number of gates follows by noting that the number of bits per gate is logarithmic in the number of gates. $\qquad\square$

The reason that this proof does not work in a Near-Threshold Computing setting is because a circuit now not only consists of a layout, but also of a set of supply voltages. Thus in principle a circuit may compute different functions for different settings of the supply voltages. We start by showing that, perhaps somewhat surprisingly until one sees the trick, this can in fact actually happen. In Sect. 2 we show that when supply voltages must be homogeneous, that is every gate of the circuit is supplied with the same voltage, there are simple circuits with n inputs and $O(n)$ gates that compute $\Omega\big(\log n/\log(\frac{1}{\delta}\log n)\big)$ different functions with probability of error at most δ, and when heterogeneous supply voltages are allowed, there are circuits with n inputs and $O(n^2)$ gates that compute $\Omega(3^n)$ different functions. Here by heterogeneous voltages we simply mean that gates could be supplied with different voltages.

In contrast, in Sect. 3 we show that, for each $\delta < 1/2$, every homogeneous circuit with n inputs and s faulty gates computes at most $s2^n + 1$ different functions, and every heterogeneous circuit with s faulty gates computes at most $(8e2^n)^s$ different functions. These upper bounds are then sufficient to prove our main result using the same counting-based technique as in the proof of Shannon's classic result. Since a homogeneous voltage setting is also a heterogeneous voltage setting, the result that almost all functions require heterogeneous circuits using exponential energy is strictly stronger than the corresponding result for homogeneous circuits. Nevertheless, we include the latter as it demonstrates how much simpler homogeneous supply voltages are, and as we are able to obtain a slightly stronger bound in terms of the required error probability δ.

These results leave open the possibility that some Boolean functions that do not require circuits with exponentially many gates still require exponential energy. For example, it could be the case that for some function the energy-optimal circuit has sub-exponentially many gates, with many of them requiring exponential energy. We show in Sect. 4 that this is not the case, i.e., the Boolean functions that require exponential energy are exactly the Boolean functions that require exponentially many faulty gates.

1.2   Related Work

The study of fault-tolerant circuits started with the seminal paper by von Neumann [16]. Subsequent work can be found in [5,6,8–10,12–14]. The results of [6,12,16] show that any circuit layout of s faultless gates can be simulated by a circuit with $O(s \log s)$ noisy gates. As already mentioned, Pippenger [12] showed that all Boolean functions can be computed by circuit layouts with $O(2^n/n)$ noisy gates. In fact, he proved this result in a stronger model in which the error probabilities of the gates could be adversarially set in the range $[0, \epsilon]$. In this model, the fact that almost all functions require $\Omega(2^n/n)$ noisy gates immediately follows from the classic result of Shannon that most functions require $\Omega(2^n/n)$ faultless gates and noting that the circuit must compute correctly if there are no gate failures. It is also known that functions with sensitivity m (roughly, the number of bits that affect the output on any input) require $\Omega(m \log m)$ noisy gates [5,9,13]. A more detailed history can be found in [8,9].

The general idea of trading accuracy of a hardware circuit and computing architecture for energy savings dates back to at least [11]. An excellent survey on Near-Threshold Computing can be found in [7]. A theoretical study of Near-Threshold Computing was initiated in [3]. The four main results in [3] are: (1) to compute a function with sensitivity m requires a circuit that uses energy $\Omega(m \log m)$, (2) if a function can be computed by a circuit with s faultless gates, then it can be computed by a circuit with energy $O(s \log s)$ when δ is a fixed constant, (3) there are circuits where there is a feasible heterogeneous setting of the supply voltages which uses much less energy than any feasible homogeneous setting of the supply voltages, and (4) there are functions where there are nearly optimal energy circuits that have a homogeneous setting of the supply voltages when δ is a fixed constant. [2] considered the problem of setting the supply voltage of a given circuit in such a way that the circuit has a specified reliability with the objective of minimizing energy. [2] showed that obtaining a significantly better approximation ratio than the traditional approach, which sets the voltage sufficiently high so that with sufficiently high probability no gate fails, is NP-hard.

1.3   Formal Model

A *Boolean function* f is a function from $\{0,1\}^n$ to $\{0,1\}$. A *gate* is a function $g : \{0,1\}^{n_g} \to \{0,1\}$, where n_g is the number of inputs (i.e., the *fan-in*) of the gate. We assume that the maximum fan-in is at most a constant. A *Boolean circuit* C with n inputs is a directed acyclic graph in which every node is a gate. Among them there are n gates with fan-in zero, each of which outputs one of the n inputs of the circuit. The *size* of a circuit, denoted by s, is the number of gates it contains. For any $I \in \{0,1\}^n$, we denote by $C(I)$ the output of the Boolean function computed by Boolean circuit layout C.

In this paper we consider circuits $(C, \bar{v})$ that consist of both a traditional circuit layout C as well as a vector of supply voltages $\bar{v}$, one for each gate of C. Every gate g is supplied with a voltage v_g. We say that the supply voltages are *homogeneous* when every gate of the circuit is supplied with the same voltage, and *heterogeneous* otherwise. A circuit is said to be homogeneous when its supply voltages are homogeneous, and heterogeneous otherwise. We say that a gate *fails* when it produces an incorrect output, that is, when given an input x it produces an output other than $g(x)$.

Each (*faulty*[1]) non-input gate g fails independently with probability $\epsilon(v_g)$, where $\epsilon : \mathbb{R}^+ \to (0, 1/2)$ is a decreasing function. The voltage supplied to a gate determines both its energy usage and its failure probability, thus we define $\epsilon_g := \epsilon(v_g)$ and drop all future formal reference to supply voltages. Finally we assume there is a failure-to-energy function $E(\epsilon)$ that maps the failure probability ϵ to

---

[1] In previous work *faulty* and *noisy* are often used as synonyms, however, in order to provide additional clarity in regards to which model is currently being referred to, we use noisy when referring to gates in the fault-tolerant model, and faulty when referring to gates in the near-threshold model.

the energy used by a gate. The only constraints we impose on $E(\epsilon)$ are that it is decreasing and $\lim_{x \to 0^+} E(1/2 - x) > 0$. In practice $E(\epsilon)$ is observed to be roughly $\Theta(\log^2(1/\epsilon))$ [3,7]. The energy used by a circuit C is simply the aggregate energy used by the gates, $\sum_{g \in C} E(\epsilon_g)$ in our notation.

A gate that never fails is said to be *faultless*. Given a value $\delta \in (0, 1/2)$ (δ may not be constant), a circuit $(C, \bar{\epsilon})$ that computes a Boolean function f is said to be $(1 - \delta)$-*reliable* if for every input I, $C(I)$ equals $f(I)$ with probability at least $1 - \delta$. We say that C can compute ℓ different Boolean functions $(1 - \delta)$-reliably if there exist $\bar{\epsilon}_1, \bar{\epsilon}_2, \ldots, \bar{\epsilon}_\ell \in (0, 1/2)^{|C|}$ and different Boolean functions $f_1, f_2, \ldots, f_\ell$ such that $(C, \bar{\epsilon}_i)$ computes f_i $(1 - \delta)$-reliably, for each $i \in 1, 2, \ldots, \ell$.

2   A Lower Bound on the Number of Functions Computable by a Circuit

In this section we show that, in both the homogeneous case and the heterogeneous case, a single circuit can $(1 - \delta)$-reliably compute many different functions, by changing the supply voltage(s). Both of these lower bounds demonstrate that Shannon's counting argument will not be sufficient to show that almost all functions require exponential energy.

2.1   Homogeneous Supply Voltages

We start with the homogeneous case, giving an explicit construction of a circuit that computes approximately $\log n$ different functions. The key concept used throughout is that for a large enough perfect binary tree of AND gates (referred to as an AND tree) there is some ϵ such that, regardless of the input, the tree will output 0 with high probability. By combining such trees of different sizes into a single circuit we can essentially ignore different parts of the input depending on ϵ. The statement and proof are formalized below.

Theorem 3. *For any $\delta \in (0, 1/2)$ and $n \in \mathbb{N}$, there exists a homogeneous circuit C with n inputs and size $O(n)$ that computes $\Omega\left(\frac{\log n}{\log(\frac{1}{\delta} \log n)}\right)$ different Boolean functions $(1 - \delta)$-reliably.*

Proof (Sketch). The circuit, which we indicate with C, consists of k perfect binary trees of AND gates, which we refer to as $\mathrm{AND}_1, \ldots, \mathrm{AND}_k$, and of a complete binary tree of OR gates, denoted OR_1. The size of AND_i, which will be determined later but decreases exponentially as i increases, is denoted by s_i, and the size of OR_1 is $k - 1$. Each AND tree receives its own set of input bits. The outputs of these k trees are fed into the tree of OR gates, and the output of the latter tree is the output of the circuit. Thus, when $\epsilon = 0$, the circuit C computes $\mathrm{OR}(\mathrm{AND}_1, \ldots, \mathrm{AND}_k)$ (see Fig. 1).

The high level approach is to show that as ϵ grows larger, the larger AND trees switch from computing the AND function to computing the 0 function. In other words, the result is completely determined by the remaining functional

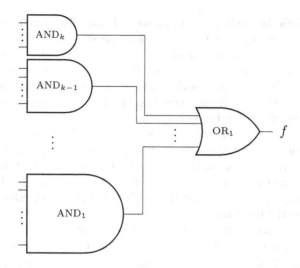

Fig. 1. The circuit used in the proof of Theorem 3.

AND trees. By choosing the sizes s_i to be sufficiently different, we can show that each AND_i will switch to computing the 0 function at a different ϵ, and further, when this switch occurs all of the smaller trees will still be functioning correctly with high probability. The details are left to the full version.   □

2.2 Heterogeneous Supply Voltages

We now show that with heterogeneous voltage settings, we can construct a circuit that computes exponentially many functions $(1 - \delta)$-reliably. We leverage the power of heterogeneity to ensure that certain parts of the circuit compute correctly with high probability, while other parts can fail with high probability. In particular, we build a circuit for a conjunctive normal form (CNF) Boolean formula where the literals of the formula can be determined dynamically by forcing certain gates to fail while preserving the correctness of the CNF calculation. This allows a single circuit to compute all possible functions representable by CNF formulas with n variables and a fixed number of fixed-sized clauses.

Theorem 4. *For any constant $\delta \in (0, 1/2)$ and $n \in \mathbb{N}$, there exists a heterogeneous circuit C of size $O(n^2)$ that computes $\Omega(3^n)$ different Boolean functions $(1 - \delta)$-reliably.*

Proof (Sketch). We give a circuit that computes at least 3^n different functions. We delay the discussion of voltages and correctness until we have completely described the circuit. Consider a 3CNF formula Φ with n variables and m clauses, i.e., $\Phi(x)$ is 1 if x satisfies all the clauses and 0 otherwise. To build a circuit that computes Φ, for each clause $(\ell_1 \vee \ell_2 \vee \ell_3)$ we have a single OR gate the inputs of which are variables ℓ_1, ℓ_2, ℓ_3 (note these need not be different and we are ignoring negations here). The output of each such OR gate is fed into an AND tree which outputs the conjunction of all such clauses. This circuit computes f_Φ, the Boolean function computed by 3CNF formula Φ.

We now give the construction of the circuit C. Consider a generic 3CNF formula $\Phi = (\ell_1 \vee \ell_2 \vee \ell_3) \wedge \cdots \wedge (\ell_{3m-2} \vee \ell_{3m-1} \vee \ell_{3m})$, and the corresponding series of OR and AND gates as described above, however with input wires coming into each ℓ_i removed. We will use a selection circuit to dynamically connect each ℓ_i to some x_j, depending on the supply voltages.

We define the selection circuit for ℓ_i, S_i as follows. This circuit takes as input $\log 2n$ bits as selectors as well as the $2n$ bits $(x_1, \neg x_1, \ldots, x_n, \neg x_n)$. The output of S_i is the bit corresponding to the location determined by the first $\log 2n$ bits. Note that Pippenger provides such a circuit of size $O(n)$ in [12]. Hence for all possible Φ, by appropriately setting the $\log 2n$ bits of each selection circuit, this circuit computes the function f_Φ.

The last piece necessary to define C is describing how the $\log 2n$ input bit b_k of each selection circuit are set. For each such b_k, we have a tree of AND gates with $\Theta\left(\log \frac{m \log n}{\delta}\right)$ inputs, the output of which is fed into b_k. The input to these AND gates are constant 0's that go through a single NOT gate, which have failure probability close to 0 if we want b_k to be 1, and close to $1/2$ if we want b_k to be 0. We leave to the full version the details of showing that for any fixed Φ there are voltage settings such that, for all x, with probability at least $1 - \delta$, $C(x) = f_\Phi(x)$.

Consider the case where $m = n$. We now compute the size of C. The size of the 3CNF circuit is at most $3n$. For each of the $3n$ literals, there is a circuit of size $O(n)$ that uses $\log 2n$ bits to map an input or its negation to that literal. Each of the $O(n \log n)$ bits is created by a tree of size $O(\log(n \log(n)/\delta))$. Thus C has size $O(n^2 + n \log(n) \log(1/\delta))$.

The last step is to show that there are $\Omega(3^n)$ unique functions $f_\Phi(x)$ with m clauses. Consider some subset $S = \{s_1, \ldots, s_{|S|}\} \subseteq [n]$ and some assignment $x = (x_{s_1}, x_{s_2}, \ldots, x_{s_{|S|}})$ for the variables x_i such that $i \in S$. Then, for each such x_i, if $x_i = 1$ create the clause $(x_i \vee x_i \vee x_i)$ and if $x_i = 0$ create the clause $(\neg x_i \vee \neg x_i \vee \neg x_i)$. Create $n - |S|$ additional clauses that are a duplicate of one of these clauses. Note that the resulting formula Φ returns 1 exactly when the input bits S are set to x, regardless of the value of the rest of the input bits, and 0 otherwise. Thus for each unique assignment of x and each unique S we obtain a new function. Since there are $\binom{n}{|S|}$ ways to choose S and $2^{|S|}$ possible assignments for x, by the binomial theorem we have that the sum over $0 \leq |S| \leq n$ of $\binom{n}{|S|} 2^{|S|}$ is 3^n. □

3   Almost All Functions Require Exponential Energy

In this section we show that, despite the ability of a single circuit to compute multiple functions, an upper bound on the number of such functions and an adaptation of Shannon's argument allows us to show that almost all functions require exponential energy, both in the homogeneous and heterogeneous case. In some sense, this is evidence that the advantages heterogeneity provides are somewhat limited, as even though some heterogeneous circuits can compute many more functions than any homogeneous circuit of the same size, this advantage is not sufficient to reduce the minimal circuit size by more than a constant for almost all functions.

3.1   Adaptation of Shannon's Argument

Inspired by Shannon's counting argument that almost all Boolean functions require exponentially-sized circuits, we show first that, in circuit models where circuits can compute multiple functions, as long as the number of functions a single circuit can compute is not too many, almost all functions still require exponentially-sized circuits. We will combine this with upper bounds on the number of functions homogeneous and heterogeneous circuits can compute to obtain our main results. Note that the following lemma assumes gates have fan-in at most two, and thus all of our results assume gate fan-in is at most two; It is straightforward to generalize this lemma and our results to any setting where the fan-in of the gates is a constant.

Lemma 1. *Suppose a circuit of size s can compute at most $f(s)$ Boolean functions in some circuit model where gates have fan-in at most two. If there exists some constant $c > 0$ such that $s^{4s} f(s) = o\big(2^{2^n}\big)$ for $s = 2^n/cn$, then almost all Boolean functions require $\Omega(2^n/n)$ gates in that model.*

Proof. Consider the set of circuits with at most s gates. A standard counting argument shows that any circuit in this set can be represented with $4s \log s$ bits, and therefore there are at most s^{4s} circuits with size at most s. Thus, if for some $c > 0$ and $s = 2^n/cn$ it holds that $s^{4s} f(s) = o\big(2^{2^n}\big)$, then almost all Boolean functions require circuits of size at least $2^n/cn = \Omega(2^n/n)$.   □

3.2   Homogeneous Supply Voltages

In this subsection we show that almost all functions require exponential-energy homogeneous circuits. In some sense, this result is a corollary of the later result that almost all functions require exponential-energy heterogeneous circuits; However, we include this result as it illustrates how homogeneous circuits are simpler than heterogeneous circuits, and we are able to obtain a slightly stronger lower bound on the energy used by almost all functions. Our proof aims to bound the number of functions a circuit of size s can compute, which is necessary, since, as we showed in the previous section, a single circuit can compute many functions.

Lemma 2. *For any circuit C on n inputs with s gates, and any $\delta > 0$, let $\mathcal{F}$ be the set of all Boolean functions f for which there exists some $\epsilon \in (0, 1/2)$ such that (C, ϵ) is $(1 - \delta)$-reliable for f. Then, $|\mathcal{F}| \leq s2^n + 1$.*

Proof. Fix some circuit C and input I, and let $C_I(\epsilon)$ be the probability that (C, ϵ) outputs a 1 on input I. Note that by definition for (C, ϵ) to compute some function f we must have that for all inputs I, either $C_I(\epsilon) \geq 1 - \delta$ or $C_I(\epsilon) \leq \delta$.

Fix some input I and consider how the output of C changes as we vary ϵ. Note that the above observation implies that C will only switch the function it is computing due to input I if $C_I(\epsilon) = 1 - \delta$ and $C_I(\epsilon)$ is decreasing or $C_I(\epsilon) = \delta$ and $C_I(\epsilon)$ is increasing. However note that $C_I(\epsilon)$ is a polynomial in ϵ of degree s,[2] and therefore there are at most s such points since between any two of them the function must change at least once from increasing to decreasing or vice versa. This means that each input I can cause C to switch the function it is computing at most s times. Since there are 2^n distinct inputs, this means that C can switch functions at most $s2^n$ times, and therefore it is able to compute at most $s2^n + 1$ different functions.   □

Since $E(\epsilon) = \Omega(1)$ for $\epsilon > 1/2$, we need only show that almost all functions require exponentially many gates in this model to show that almost all functions require exponential energy. However, the following lemma will allow us to strengthen our theorem statement.

Lemma 3. *Let C be a homogeneous circuit that is $(1-\delta)$-reliable. Then, $\epsilon \le \delta$.*

Proof. Let f be the function C is trying to compute, and fix some input I. It suffices to show that the output gate, g_o, must fail with probability less than δ. Let p be the probability that g_o receives an input I' such that $g_o(I') = f(I)$. Then, note that $\Pr[g_o(I') = f(I)] = p(1-\epsilon)+(1-p)\epsilon \le 1-\epsilon$. Since by hypothesis $\Pr[C(I) = f(I)] \ge 1 - \delta$, it follows that $\epsilon \le \delta$.   □

We can now prove the desired theorem.

Theorem 5. *For any $\delta \in (0, 1/2)$, almost all Boolean functions on n variables require homogeneous circuits using $\Omega(E(\delta)2^n/n)$ energy.*

Proof. From Lemma 2 we know that each circuit of size s computes at most $s2^n + 1$ different functions. We now show that for $s = 2^n/4n$, the quantity $s^{4s}(s2^n + 1)$ is asymptotically smaller than 2^{2^n}, the number of functions on n inputs. Plugging in and simplifying we have

$$\left(\frac{2^n}{4n}\right)^{4\frac{2^n}{4n}} \left(\frac{2^n}{4n}2^n + 1\right) \le \frac{2^{2^n}}{n^{\frac{2^n}{n}}}2^{2n} = o(2^{2^n}).$$

Hence, Lemma 1 implies that almost all homogeneous circuits require $\Omega(2^n/n)$ gates. By Lemma 3, we have $\epsilon \le \delta$, so each gate uses at least $E(\delta)$ energy.   □

3.3   Heterogeneous Supply Voltages

In this section we show that almost all functions require exponential energy, even when allowed circuits with heterogeneous voltages. The approach is similar to the one for the homogeneous case, however the bound on the number of functions a heterogeneous circuit can compute requires a technical result from real algebraic geometry, which was proved by Alon [1].

---

[2] If we fix which gates fail, then the output of C on I is fixed to either 1 or 0. A fixed set of q gates fail with probability $\epsilon^q(1 - \epsilon)^{s-q}$, a polynomial of degree s in ϵ. $C_I(\epsilon)$ can be viewed as the sum over the sets of gates that, when failing, cause C to output 1 on I, of the probability of that set failing.

Lemma 4. *For any circuit C on n inputs with s gates, and any $\delta > 0$, let $\mathcal{F}$ be the set of all Boolean functions f for which there exists some $\bar{\epsilon} \in (0, 1/2)^{|C|}$ such that $(C, \bar{\epsilon})$ is $(1 - \delta)$-reliable for f. Then, $|\mathcal{F}| \leq (8e2^n)^s$.*

Proof. Let $\mathcal{P} \subset \mathbb{R}[X_1, \ldots, X_k]$ be a finite set of p polynomials with degree at most d. A *sign condition* on $\mathcal{P}$ is an element of $\{0, 1, -1\}^p$. The *realization* of the sign condition σ in $\mathbb{R}^k$ is the semi-algebraic set

$$\mathcal{R}(\sigma) = \left\{ x \in \mathbb{R}^k : \bigwedge_{P \in \mathcal{P}} \text{sign}\,(P(x)) = \sigma(P) \right\}.$$

Let $N(p, d, k)$ be the number of realizable sign conditions, i.e., the cardinality of the set $\{\sigma : \mathcal{R}(\sigma) \neq \emptyset\}$. The following theorem is due to Alon.

Theorem 6 (Proposition 5.5 in [1]). *If $2p > k$, then $N(p, d, k) \leq \left(\frac{8edp}{k} \right)^k$.*

Let $I \in \{0, 1\}^n$ be some input to C, and let $P_I(\epsilon_1, \ldots, \epsilon_s)$ be the probability that C outputs 1 on I, when gate i fails with probability ϵ_i. Observe that $P_I \in \mathbb{R}[\epsilon_1, \ldots, \epsilon_s]$ and that P_I has degree at most s, since we can compute P_I by summing over all possible subsets of gates that could fail and cause C to output a 1, of the probability that exactly those gates fail and no others (which is a polynomial is $\epsilon_1, \ldots, \epsilon_s$, where each ϵ_i has exponent 1).

Let $\mathcal{P} = \{P_I - (1 - \delta) | I \in \{0, 1\}^n\}$. Clearly, the cardinality of $\mathcal{P}$ is at most 2^n. Observe that every different function f that C calculates must correspond to a unique realizable sign condition of $\mathcal{P}$, in the sense that there is some setting of $\bar{\epsilon} = (\epsilon_1, \ldots, \epsilon_s)$ such that

1. $P(\bar{\epsilon}) - (1 - \delta) > 0$ on inputs I such that $f(I) = 1$, and
2. $P(\bar{\epsilon}) - (1 - \delta) < 0$ on inputs I such that $f(I) = 0$ (in fact, we need $P(\bar{\epsilon}) - \delta < 0$, an even stronger condition).

By Theorem 6, if the size of $\mathcal{P}$ is at least $n/2$, the number of realizable sign conditions of $\mathcal{P}$ is at most $(8e2^n)^s$. Otherwise, if the size of $\mathcal{P}$ is at most $n/2$, the total number of sign conditions is at most $3^{n/2} = o((8e2^n)^s)$. Thus, we have obtained an upper bound on the number of different functions C can compute. $\square$

We can now prove our main theorem.

Theorem 7. *For any $\delta \in (0, 1/2)$, almost all Boolean functions on n variables require heterogeneous circuits using $\Omega(2^n/n)$ energy.*

Proof. From Lemma 4 we know that each circuit of size s computes at most $(8e2^n)^s$ different functions. We now show that for $s = 2^n/8n$, the quantity $s^{4s}(8e2^n)^s$ is asymptotically smaller than 2^{2^n}, the number of functions on n inputs. Plugging in and simplifying we have

$$\left(\frac{2^n}{8n} \right)^{4 \frac{2^n}{8n}} (8e2^n)^{\frac{2^n}{8n}} \leq 2^{\frac{2^n}{2}} 2^{\frac{2^n(3+2-12-4\log n)}{8n}} 2^{\frac{2^n}{8}} \leq 2^{\frac{5 \cdot 2^n}{8}} = o(2^{2^n}).$$

Hence, Lemma 1 implies that almost all heterogeneous circuits require $\Omega(2^n/n)$ gates. The theorem follows since $E(1/2) = \Omega(1)$ and E is decreasing in the interval $(0, 1/2)$. □

4   Relating Energy and the Number of Faulty Gates

In this section, we show that the Boolean functions that require exponential energy are exactly the Boolean functions that require exponentially many faulty gates. Before formalizing this notion we introduce some additional notation. For any Boolean function f on n variables and any reliability parameter δ, let $NG(f, \delta)$ denote the minimum size of any (heterogeneous) circuit that $(1 - \delta)$-reliably computes f, and $\widetilde{NG}(f, \delta)$ denote the minimum size of any homogeneous circuit that $(1 - \delta)$-reliably computes f. Similarly define $\mathcal{E}(f, \delta)$ to be the minimum energy used by any (heterogeneous) circuit that $(1 - \delta)$-reliably computes f, and $\widetilde{\mathcal{E}}(f, \delta)$ the minimum energy used by any homogeneous circuit that $(1-\delta)$-reliably computes f. We are now ready to state the main result of this section.

Lemma 5. *For all Boolean functions* f, *and for all* $\delta < 1/2$,

$$E(1/2)NG(f,\delta) \le \mathcal{E}(f,\delta) \le \widetilde{\mathcal{E}}(f,\delta) \le E\left(\frac{\delta}{\widetilde{NG}(f,\delta)}\right)\widetilde{NG}(f,\delta).$$

Proof. First observe that $\mathcal{E}(f,\delta) \le \widetilde{\mathcal{E}}(f,\delta)$. We now prove the leftmost inequality. Let $(C, \bar{\epsilon})$ be the circuit achieving $\mathcal{E}(f, \delta)$ and note that by definition $\mathcal{E}(f,\delta) = \sum_{g \in C} E(\epsilon_g)$. Since E is decreasing, it follows that $E(\epsilon_g) \ge E(1/2)$ for all $g \in C$. Additionally, by definition, $|C| \ge NG(f, \delta)$, and the result follows.

To show the rightmost inequality, fix some Boolean function f, and some δ. Let C be a circuit of size $s = \widetilde{NG}(f, \delta)$, and ϵ the failure probability, such that (C, ϵ) is $(1 - \delta)$-reliable on f. If $\epsilon \ge \delta/s$, we are done, since E is decreasing. Note that for a circuit of size s, if gates fail with probability at most δ/s, then by the union bound, the probability that any gate fails is at most δ. Thus, if $\epsilon < \delta/s$, the probability that any gate fails is at most δ. However, this implies that $(C, \delta/s)$ is $(1 - \delta)$-reliable on f as well, and thus can use energy $E\left(\frac{\delta}{\widetilde{NG}(f,\delta)}\right)\widetilde{NG}(f,\delta)$. □

If $E(1/2)$ is $\Omega(1)$, and $E(\delta/\widetilde{NG}(f,\delta))$ is bounded above by a polynomial in $\widetilde{NG}(f,\delta)$ and $1/\delta$ (recall that in current CMOS technologies $E(\epsilon) = \Theta(\log^2(1/\epsilon))$), this implies that any function that requires exponential energy requires exponential circuit size and vice versa.

References

1. Alon, N.: Tools from higher algebra. In: Graham, R.L., Grötschel, M., Lovász, L. (eds.) Handbook of Combinatorics, vol. 2, pp. 1749–1783. MIT Press (1995)

2. Antoniadis, A., Barcelo, N., Nugent, M., Pruhs, K., Scquizzato, M.: Complexity-theoretic obstacles to achieving energy savings with near-threshold computing. In: Proceedings of the 5th International Green Computing Conference (IGCC), pp. 1–8 (2014)
3. Antoniadis, A., Barcelo, N., Nugent, M., Pruhs, K., Scquizzato, M.: Energy-efficient circuit design. In: Proceedings of the 5th conference on Innovations in Theoretical Computer Science (ITCS), pp. 303–312 (2014)
4. Butts, J., Sohim, G.: A static power model for architects. In: Proceedings of the 33rd Annual ACM/IEEE International Symposium on Microarchitecture (MICRO), pp. 191–201 (2000)
5. Dobrushin, R.L., Ortyukov, S.I.: Lower bound for the redundancy of self-correcting arrangements of unreliable functional elements. Probl. Inf. Transm. **13**, 59–65 (1977)
6. Dobrushin, R.L., Ortyukov, S.I.: Upper bound for the redundancy of self-correcting arrangements of unreliable functional elements. Probl. Inf. Transm. **13**, 203–218 (1977)
7. Dreslinski, R.G., Wieckowski, M., Blaauw, D., Sylvester, D., Mudge, T.N.: Near-threshold computing: Reclaiming Moore's law through energy efficient integrated circuits. Proc. IEEE **98**(2), 253–266 (2010)
8. Gács, P.: Reliable Computation In: Algorithms in Informatics, vol. 2, ELTE Eötvös Kiadó, Budapest (2005)
9. Gács, P., Gál, A.: Lower bounds for the complexity of reliable Boolean circuits with noisy gates. IEEE Trans. Inf. Theory **40**(2), 579–583 (1994)
10. Kleitman, D.J., Leighton, F.T., Ma, Y.: On the design of reliable Boolean circuits that contain partially unreliable gates. J. Comput. Syst. Sci. **55**(3), 385–401 (1997)
11. Palem, K.V.: Energy aware computing through probabilistic switching: A study of limits. IEEE Trans. Comput. **54**(9), 1123–1137 (2005)
12. Pippenger, N.: On networks of noisy gates. In: Proceedings of the 26th Symposium on Foundations of Computer Science (FOCS), pp. 30–38 (1985)
13. Pippenger, N., Stamoulis, G.D., Tsitsiklis, J.N.: On a lower bound for the redundancy of reliable networks with noisy gates. IEEE Trans. Inf. Theory **37**(3), 639–643 (1991)
14. Reischuk, R., Schmeltz, B.: Reliable computation with noisy circuits and decision trees-A general $n \log n$ lower bound. In: Proceedings of the 32nd Symposium on Foundations of Computer Science (FOCS), pp. 602–611 (1991)
15. Shannon, C.E.: The synthesis of two-terminal switching circuits. Bell Syst. Tech. J. **28**, 59–98 (1949)
16. von Neumann, J.: Probabilistic logics and the synthesis of reliable organisms from unreliable components. In: Shannon, C.E., McCarthy, J. (eds.) Automata Studies, pp. 329–378. Princeton University Press (1956)

On Dynamic DFS Tree in Directed Graphs

Surender Baswana and Keerti Choudhary[✉]

Department of CSE, IIT Kanpur, Kanpur 208016, India
{sbaswana,keerti}@cse.iitk.ac.in
http://www.cse.iitk.ac.in

Abstract. Let $G = (V, E)$ be a directed graph on n vertices and m edges. We address the problem of maintaining a depth first search (DFS) tree efficiently under insertion/deletion of edges in G.

1. We present an efficient randomized decremental algorithm for maintaining a DFS tree for a directed acyclic graph. For processing any arbitrary online sequence of edge deletions, this algorithm takes expected $O(mn \log n)$ time.
2. We present the following lower bound results.
   (a) Any decremental (or incremental) algorithm for maintaining the ordered DFS tree *explicitly* requires $\Omega(mn)$ total update time in the worst case.
   (b) Any decremental (or incremental) algorithm for maintaining the ordered DFS tree is at least as hard as computing all-pairs reachability in a directed graph.

Keywords: Dynamic · Decremental · Directed · Graph · Depth first search

1 Introduction

Depth First Search (DFS) is a well known graph traversal technique. Tarjan, in his seminal work [14], demonstrated the power of DFS traversal for solving various fundamental graph problems, namely, connected components, topological sorting and strongly connected components.

A DFS traversal is a recursive algorithm to traverse a graph. Let $G = (V, E)$ be a directed graph on $n = |V|$ vertices and $m = |E|$ edges. The DFS traversal carried out in G from a vertex $r \in V$ produces a tree, called DFS tree rooted at r. This tree spans all the vertices reachable from r in G. It takes $O(m + n)$

Full version of this article is available at http://www.cse.iitk.ac.in/users/sbaswana/ Papers-published/Dynamic-DFS-digraph.pdf.

S. Baswana—This research was partially supported by the India-Israel joint research project on dynamic graph algorithms, and the Indo-German Max Planck Center for Computer Science (IMPECS).

K. Choudhary—This research was partially supported by Google India under the Google India PhD Fellowship Award.

© Springer-Verlag Berlin Heidelberg 2015
G.F. Italiano et al. (Eds.): MFCS 2015, Part II, LNCS 9235, pp. 102–114, 2015.
DOI: 10.1007/978-3-662-48054-0_9

time to perform a DFS traversal and generate its DFS tree. A DFS tree leads to a classification of all non-tree edges into three categories, namely, *back* edges, *forward* edges, and *cross* edges. Most of the applications of a DFS tree exploit the relationship among the edges based on this classification. For a given graph, there may exist many DFS trees rooted at a vertex r. However, if the DFS traversal is performed strictly according to the adjacency lists, then there will be a unique DFS tree rooted at r. The ordered DFS problem is to compute the order in which the vertices get visited during this restricted traversal.

Most of the graph applications in real life deal with a graph that is not static. Instead, the graph keeps changing with time - some edges get deleted while some new edges get inserted. The dynamic nature of these graphs has motivated researchers to design efficient algorithms for various graph problems in a dynamic environment. Any algorithmic graph problem can be modeled in the dynamic environment as follows. There is an online sequence of insertion and/or deletion of edges, and the aim is to update the solution of the graph problem efficiently after each edge insertion/deletion. There exist efficient dynamic algorithms for various fundamental problems in graphs [4, 7, 12, 13, 15]. In this article we address the problem of maintaining a DFS tree in a dynamic graph.

Though an efficient algorithm is known for the static version of the DFS tree problem, the same is not true for its dynamic counterpart. Reif [9, 10] and Miltersen et al. [8] addressed the complexity of the ordered DFS problem in a dynamic environment. Miltersen et al. [8] introduced a class of problems called non-redundant polynomial (NRP) complete. They showed that if the solution of any NRP-complete problem is updatable in $O(\text{polylog}(n))$ time, then the solution of every problem in the class P is updatable in $O(\text{polylog}(n))$ time. The ordered DFS tree problem was shown to be NRP-complete. So it is highly unlikely that any $O(\text{polylog}(n))$ update time algorithm would exist for the ordered DFS problem in the dynamic setting.

Apart from showing the hardness of the ordered DFS tree problem, very little work has been done on the design of any non-trivial algorithm for the problem of maintaining any DFS tree in a dynamic environment. Franciosa et al. [6] designed an algorithm for maintaining a DFS tree in a DAG under insertion of edges. For any arbitrary sequence of edge insertions, this algorithm takes $O(mn)$ total time to maintain the DFS tree from a given source. This incremental algorithm is the only non-trivial result known for the dynamic DFS tree problem in directed graphs. For the related problem of maintaining a breadth first search (BFS) tree in directed graphs, Franciosa et al. [5] designed a decremental algorithm that achieves $O(n)$ amortized time per edge deletion.

In this article, we complement the existing upper and lower bound results for the dynamic DFS tree problem in a directed graph. Our main result is the first non-trivial decremental algorithm for a DFS tree in a DAG.

1.1   An Efficient Decremental Algorithm for a DFS Tree in a DAG

We present a decremental algorithm that maintains a DFS tree in a DAG and takes expected $O(mn \log n)$ time to process any arbitrary online sequence of edge deletions. Hence the expected amortized update time per edge deletion is $O(n \log n)$. We now provide an overview of this randomized algorithm.

Consider the deletion of a tree edge, say (a, b), in a DFS tree. It may turn out that each vertex of subtree $T(b)$ may have to be *hung* from a different parent in the new DFS tree (see Fig. 1). This is because the parent of a vertex in a DFS tree depends upon the order in which its in-neighbors get visited during the DFS traversal.

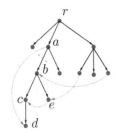

Fig. 1. Deletion of edge (a, b) results in the change of parent of every vertex in $T(b)$.

In order to achieve an efficient update time, we use randomization in the DFS traversal. Once the traversal reaches a vertex v, the next vertex to be traversed is selected randomly uniformly among all unvisited neighbors of v. Our decremental algorithm maintains this *randomized* DFS tree at each stage. This randomization plays a crucial role in fooling the adversary that generates the sequence of edge deletions: For a pair of vertices, say u and x, the deletion of very few (in expectation) outgoing edges of u will disconnect the tree path from root to x. Hence a vertex will have to search for a new parent *fewer* times during any given sequence of edge deletions. We analyze this algorithm using probability tools and some fundamental properties of a DFS tree in a DAG. This leads us to achieve an upper bound of $O(mn \log n)$ on the expected running time of this algorithm for any arbitrary sequence of edge deletions.

The DFS tree maintained (based on the random bits) by the algorithm at any stage is not known to the adversary for it to choose the updates adaptively. This oblivious adversarial model is no different from randomized data structures like universal hashing.

1.2   Lower Bounds on Dynamic DFS Tree

We get the following lower bound results for maintaining an ordered DFS tree.

1. The most common way of storing any rooted tree is by keeping a parent pointer for each vertex in the tree. We call this representation an *explicit* representation of a tree. We establish a worst case lower bound of $\Omega(mn)$ for maintaining the ordered DFS tree explicitly under deletion (or insertion) of edges. This lower bound holds for undirected as well as directed graphs. Recently, an $O(n^2)$ time incremental algorithm [2] has been designed for maintaining a DFS tree explicitly in any undirected graph. Therefore, our lower bound result implies that for dense graphs (when $m = \Theta(n^2)$) maintaining a DFS tree explicitly in the incremental environment is provably faster than maintaining an ordered DFS tree by a factor of $\Omega(n)$.

2. We show that the dynamic DFS tree problem in a directed graph is closely related to the static all-pairs reachability problem. We provide an $O(m + n)$ time reduction from the static all-pairs reachability problem to the decremental (or incremental) maintenance of the ordered DFS tree in a graph G'

with $O(m)$ edges and $O(n)$ vertices. This reduction is similar to the reduction technique used by Roditty and Zwick [11] for the decremental (or incremental) BFS tree problem. This reduction implies conditional lower bounds of $\Omega(\min(mn, n^\omega))$ on the total update time and $\Omega(\min(m, n^{\omega-1}))$ on the worst case update time per edge deletion for any decremental (or incremental) algorithm for the ordered DFS tree problem. Here ω is the exponent of the fastest matrix multiplication algorithm, and currently $\omega < 2.373$ [16].

1.3   Organization of the Paper

We describe notations and terminologies related to a DFS tree in Sect. 2. In Sect. 3, we first describe a deterministic algorithm for maintaining a DFS tree in a DAG under deletion of edges. This algorithm is very simple. However, in the worst case it can take $\Theta(m^2)$ time. In Sect. 4, we show that by adding a small amount of randomization to this algorithm, its expected running time gets reduced to $O(mn \log n)$. In Sect. 5, we provide lower bounds on the dynamic DFS tree problem.

2   Preliminaries

Given a directed acyclic graph $G = (V, E)$ on $n = |V|$ vertices and $m = |E|$ edges, and a given root vertex $r \in V$, the following notations will be used throughout the paper.

- T : A DFS tree of G rooted at r.
- START-TIME(x): The time at which the traversal reaches x for the first time when we carry out the DFS traversal associated with T.
- FINISH-TIME(x): The time at which the DFS traversal finishes for vertex x.
- $\deg(x)$: The number of edges entering into vertex x.
- $IN(x)$: The list of vertices of T having an outgoing edge into x; this list is sorted in the topological ordering.
- $OUT(x)$: The list of vertices of T having an incoming edge from x.
- $par(x)$: The parent of x in T.
- $path(x)$: The path from r to x in T.
- LEVEL(x) : The level of a vertex x in T such that LEVEL$(r) = 0$ and LEVEL$(x) =$ LEVEL$(par(x)) + 1$.
- $T(x)$: The subtree of T rooted at a vertex x.
- $LCA(x, y)$: The Lowest Common Ancestor of x and y in tree T.
- $LA(x, k)$: The ancestor of x at level k in tree T.

Our algorithm uses the following results for the dynamic version of the Lowest Common Ancestor (LCA) and the Level Ancestors (LA) problems.

Theorem 1 (Cole and Hariharan 2005 [3]). *There exists a data structure for maintaining a rooted tree using linear space that can answer any LCA query in $O(1)$ time and can handle insertion or deletion of a leaf node in $O(1)$ time.*

Theorem 2 (Alstrup and Holm 2000 [1]). *There exists a data structure for maintaining a rooted tree using linear space that can answer any level ancestor query in $O(1)$ time and can handle insertion of a leaf node in $O(1)$ time.*

The data structure for the Level Ancestor problem can be easily extended to handle deletion of a leaf node in amortized $O(1)$ time using the standard technique of periodic rebuilding. The following lemma, with a straight forward proof, states two properties of the DFS traversal in a DAG. This lemma will play a crucial role in our randomized decremental algorithm for DFS tree.

Lemma 1. *Let y and z be any two vertices in a given directed acyclic graph G. If there is a path from y to z, then the following two properties hold true for each DFS traversal carried out in G.*

P1 : FINISH-TIME(z) < FINISH-TIME(y).
P2 : *If* START-TIME(y) < START-TIME(z), *then z must be a descendant of y in the DFS tree.*

3   A Simple and Deterministic Decremental Algorithm

We shall now present a deterministic and very simple algorithm for maintaining the ordered DFS tree defined by a fixed permutation of vertices.

Let σ be a permutation of V (a bijective mapping from V to $[1..n]$). We sort the adjacency list (i.e. $OUT(x)$) of each vertex x in the increasing order of σ value. T is initialized as the ordered DFS tree of G with respect to these sorted adjacency lists. Throughout the sequence of edge deletions, our algorithm maintains the following invariant.

$\mathcal{I}$: At any instant of time, T is the ordered DFS tree corresponding to σ.

Consider the deletion of an edge (u, v). We first remove u from $IN(v)$ and v from $OUT(u)$. If (u, v) is not a tree edge, then this is all that is required. Otherwise we scan the vertices of $T(v)$ in the topological ordering, and process each vertex $x \in T(v)$ as follows.

1. If $IN(x)$ is empty, then we set $par(x)$ to null, and for each out-neighbor y of x, delete x from $IN(y)$.
2. If $IN(x)$ is non-empty then we hang x appropriately from its new parent in the DFS tree by executing procedure Hang(x) explained below.

Note: The processing of vertices of $T(v)$ in the topological ordering is very crucial because it ensures that before the processing of any vertex x begins, all its in-neighbors are already hung appropriately in the new DFS tree.

We now explain procedure Hang(x) (see the pseudocode on the next page). Let w be the vertex from $IN(x)$ with minimum finish time. Then the in-neighbors of x visited before w must lie on $path(w)$. Now consider any in-neighbor y of x lying on $path(w)$. Let z be its child on $path(w)$. Then at the time when DFS traversal reaches w, vertex y would have scanned only upto those vertices in its adjacency list whose σ value is less than or equal to $\sigma(z)$. Thus if $\sigma(x) > \sigma(z)$, then x would not have been scanned by y and thus cannot be its child. But if $\sigma(x) < \sigma(z)$, then x would be a child of y in case it is unvisited at the time when DFS traversal reaches y. Based on this observation we can conclude the following. Let b be the first vertex in $IN(x)$ such that b is ancestor of w and the child of b on $path(w)$ has σ value greater than $\sigma(x)$ (see Fig. 2). Then b is assigned as the parent of x. However, if no such vertex exists, then x is hung from w. Hanging x in this manner ensures that the invariant $\mathcal{I}$ is maintained.

Procedure Hang(x): hangs vertex x appropriately in T to preserve the invariant $\mathcal{I}$.	**Procedure** MinFinish(x): computing x's in-neighbor having minimum finish time.
1 $par(x) \leftarrow null$; 2 $w \leftarrow$ MinFinish(x); 3 **for** $y \in IN(x) \backslash \{w\}$ **do** 4 **if** $y = LCA(y, w)$ **then** 5 $z \leftarrow LA(w, 1 + \text{LEVEL}(y))$; 6 **if** $\sigma(x) < \sigma(z)$ **then** 7 $par(x) \leftarrow y$; break; 8 **end** 9 **end** 10 **end** 11 **if** $par(x) = \emptyset$ **then** $par(x) \leftarrow w$; 12 $\text{LEVEL}(x) \leftarrow 1 + \text{LEVEL}(par(x))$; 13 Return;	1 $w \leftarrow$ First vertex in $IN(x)$; 2 **for** $y \in IN(x) \backslash \{w\}$ **do** 3 $z \leftarrow LCA(y, w)$; 4 **if** $w = z$ **then** $w \leftarrow y$; 5 **else if** $y \neq z$ **then** 6 $a \leftarrow LA(y, 1 + \text{LEVEL}(z))$; 7 $b \leftarrow LA(w, 1 + \text{LEVEL}(z))$; 8 **if** $\sigma(a) < \sigma(b)$ **then** 9 $w \leftarrow y$; 10 **end** 11 **end** 12 **end** 13 Return w;

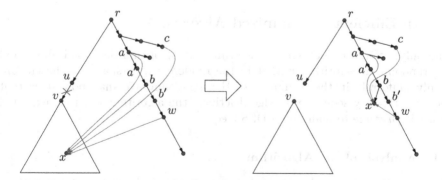

Fig. 2. Illustration of procedure Hang(x) : (i) w is vertex in $IN(x)$ having minimum finish time; (ii) x does not hang from c as it is not on $path(w)$; (iii) x does not hang from a as $\sigma(a') < \sigma(x)$; (iv) x hangs from b as $\sigma(x) < \sigma(b')$.

We now explain procedure MinFinish(x) that computes the in-neighbor w of x with minimum finish time. w is first initialized to be any arbitrary vertex in $IN(x)$. Then we scan $IN(x)$, and for each $y \in IN(x)$ if FINISH-TIME(y) < FINISH-TIME(w) then we update w to y. The FINISH-TIME of vertices y and w can be compared as follows. Let $z = LCA(y, w)$. If w is ancestor of y, i.e. $w = z$, then FINISH-TIME of w would be greater than that y, and vice-versa. Otherwise if there is no ancestor-child relationship between them, then let a, b be respectively the children of z on $path(y)$ and $path(w)$. Now FINISH-TIME of y would be less than that of w if and only if $\sigma(a) < \sigma(b)$. This is because a would be visited before b if $\sigma(a) < \sigma(b)$, and then START-TIME (as well as FINISH-TIME) of all the vertices in $T(a)$ would be less than that of vertices in $T(b)$, and vice versa.

We shall now analyze the time complexity of this decremental algorithm. The procedures Hang(x) and MinFinish(x) scan $IN(x)$ and use only LA/LCA queries that can be answered in $O(1)$ time (see Theorems 1 and 2). So, the time taken by each of them is $O(\deg(x))$. The scanning of the vertices of $T(v)$ in topological ordering can be done in linear time by using any integer sorting algorithm. Thus the time taken to handle deletion of tree edge (u, v) is $\Theta(\sum_{x \in T(v)} \deg(x))$. Now it follows from the discussion given above that a vertex x is processed during an edge deletion if that edge lies on the tree path from root to x. Suppose for a given sequence of edge deletions, this event happens $c(x)$ times for a vertex x. Then we get the following theorem.

Theorem 3. *There exists a decremental algorithm for maintaining a DFS tree T that takes $\Theta(\sum_x \deg(x) \cdot c(x))$ time where $c(x)$ is the number of times vertex x loses an edge on its path from the root in T during a given sequence of edge deletions.*

The algorithm for handling deletion of an edge described above is simple. However, this algorithm can be as inefficient asymptotically as recomputing the DFS tree from scratch after each edge deletion (see Observation 1 in Sect. 5).

4   An Efficient Randomized Algorithm

The only difference between our deterministic algorithm described above and the randomized algorithm is that in the randomized version σ is chosen randomly uniformly in the beginning of the algorithm. We shall now show that for any arbitrary sequence of edge deletions, the algorithm will take expected $O(mn \log n)$ time to maintain a DFS tree.

4.1   Analysis of the Algorithm

Consider vertices u and x such that x is reachable from u in G. x may be reachable from u by a direct edge or through a path from some of its outgoing neighbors. Let S_u be the set consisting of all those vertices $v \in OUT(u)$ such that at the time of deletion of edge (u, v), x was reachable from v. Note that x may also be present in set S_u if $(u, x) \in E$. It can be observed that for vertex $v \in$

$OUT(u)\backslash S_u$, the deletion of (u,v) will have no influence on x, as at that time x was already unreachable from v. For each vertex $v \in S_u$, define DELETE-TIME(v) as the time at which (u,v) is deleted. Let $\langle v_1, v_2, \ldots, v_k \rangle$ be the sequence of vertices from S_u arranged in increasing order of DELETE-TIME. Observe that every edge from $\{u\} \times S_u$, at the time of its deletion, could potentially be present on the path from root to x in T. However, we will now prove that if σ is selected randomly uniformly, then this may happen only $O(\log k)$ times on expectation for any given sequence of edge deletions. For this purpose, we first state a lemma that shows the relationship between any two vertices from set S_u. This lemma crucially exploits the fact the G is acyclic.

Lemma 2. *Suppose (u, v_i) is present in the DFS tree T. If there is any $j > i$ with $\sigma(v_j) < \sigma(v_i)$, then (i) START-TIME$(v_j) <$ START-TIME(v_i), and (ii) there is no path from v_j to v_i in the present graph.*

Proof. (u, v_i) belongs to the DFS tree. So at the moment DFS visited u, v_i must be unvisited. Since $\sigma(v_j) < \sigma(v_i)$ and v_j has an incoming edge from u, so v_j will be visited (if not already visited) before v_i. So START-TIME$(v_j) <$ START-TIME(v_i).

If there is a path from v_j to v_i in the present graph, then it follows from property **P2** of DFS traversal in DAG (stated in Lemma 1) that v_i must be a descendant of v_j in the DFS tree T. But this would imply that u is also descendant of v_j in T since u is parent of v_i in T. This tree path from v_j to u, if exists, and the edge $(u, v_j) \in E$ would form a cycle. This is a contradiction since the graph is acyclic.   □

The following lemma precisely states the conditions under which the deletion of (u, v_i) will have no influence on x.

Lemma 3. *Suppose (u, v_i) is a tree edge at some time. If there is any $j > i$ with $\sigma(v_j) < \sigma(v_i)$, then x does not belong to $T(v_i)$ at that time.*

Proof. As stated above vertex x is reachable from v_j, or x is vertex v_j itself. So from property **P1** stated in Lemma 1 it follows that FINISH-TIME$(x) \leq$ FINISH-TIME(v_j). It also follows from Lemma 2 that START-TIME$(v_j) <$ START-TIME(v_i) and there is no path from v_j to v_i in the graph. Therefore, FINISH-TIME$(v_j) <$ START-TIME(v_i). Hence we get the following relations.

$$\text{FINISH-TIME}(x) \leq \text{FINISH-TIME}(v_j) < \text{START-TIME}(v_i)$$

So v_i can not be ancestor of x in T. Hence $x \notin T(v_i)$.   □

Lemma 3 leads to the following corollary.

Corollary 1. *If there exists j satisfying $i < j$ and $\sigma(v_j) < \sigma(v_i)$, then the deletion of the edge (u, v_i) will have no influence on the path from root to x in the DFS tree.*

Lemma 4. *On expectation, there are $O(\log k)$ edges from $\{(u, v_i) | 1 \leq i \leq k\}$ that may lie on the tree path from u to x at the time of their deletion.*

Proof. For each vertex $v_i, 1 \leq i \leq k$, we define a random variable Y_i as follows. Y_i is 1 if x belongs to $T(v_i)$ at the time of deletion of edge (u, v_i), and 0 otherwise. Let $Y = \sum_{i=1}^{k} Y_i$. It follows from Corollary 1 that if $Y_i = 1$ then for all $j > i$, $\sigma(v_j)$ is greater than $\sigma(v_i)$. As σ is a uniformly random permutation, so

$$P[\sigma(v_j) > \sigma(v_i), \; \forall j, \; i < j \leq k] = \frac{1}{k - i + 1}$$

Thus $E[Y_i] \leq \frac{1}{k-i+1}$, and so using linearity of expectation, $E[Y] \leq \sum_{i=1}^{k} \frac{1}{k-i+1} = O(\log k)$. □

Hence, if x is reachable from u in the initial graph, then for any arbitrary sequence of edge deletions, we can conclude the following. Upon deletion of expected $O(\log n)$ outgoing edges of u, x will have to be re-hung in the DFS tree. Therefore, using Theorem 3, the expected time complexity of the randomized algorithm for processing any arbitrary sequence of edge deletions is $O(mn \log n)$.

Theorem 4. *Let G be a DAG on n vertices and m edges and let $r \in V$. A DFS tree rooted at r can be maintained under deletion of edges using an algorithm that takes expected $O(mn \log n)$ time for any arbitrary sequence of edge deletions.*

5   Lower Bounds for Dynamic DFS Tree Problem

5.1   Maintaining Ordered DFS Tree Explicitly

We show that the problem of maintaining the ordered DFS tree explicitly may require $\Omega(mn)$ time in total for directed as well as undirected graphs. For this we provide construction of graphs with $\Theta(n)$ vertices and then present a sequence of edge deletions and an ordering σ such that each edge deletion results in changing the parent of $\Theta(n)$ vertices in the ordered DFS tree defined by σ.

First we prove the result for the undirected graphs. Let $t = \lfloor m/n \rfloor$. Let H_1 be a graph with $n + t + 1$ vertices - $r, a_1, .., a_n, b_1, .., b_t$. The edges of H_1 are the pairs (r, a_i) and (a_i, b_j) for all i, j such that $1 \leq i \leq n$ and $1 \leq j \leq t$. Let H_2 be another graph with $2n$ vertices - $y_1, .., y_n, z_1, .., z_n$, such that each y_k has edges to z_{k-1}, z_k and z_{k+1} (see Fig. 3). Now graph G is obtained by adding an edge between b_j and y_1 if j is odd, and between b_j and z_1 if j is even. This completes the construction of graph G. Let $\sigma = < r, y_1, .., y_n, z_1, .., z_n, a_1, .., a_n, b_1, .., b_t >$. We delete the edges in n stages. In stage i, we delete the edges incident on vertex a_i and the sequence of deletions is - $(a_i, b_1), .., (a_i, b_t)$.

Lemma 5. *Deletion of each edge (a_i, b_j) leads to the change of parent of all vertices in H_2 in the ordered DFS tree T defined by σ.*

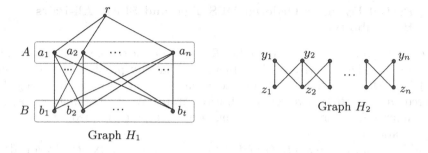

Fig. 3. Subgraphs H_1 and H_2

Proof. Consider the DFS tree just before the deletion of edge (a_i, b_j). The DFS traversal starts with r and visit vertices a_1 to a_i. After a_i it visits b_j, and then visits y_1 or z_1 depending on whether j is odd or even. If y_1 is the first vertex visited in H_2 then the sequence of vertices visited in H_2 would be $y_1, z_1, y_2, z_2, ..., y_n, z_n$. And if z_1 is visited first then the sequence would be $z_1, y_1, z_2, y_2, ..., z_n, y_n$. Thus if j is odd then for each k, y_k is parent of z_k in T, and if j is even then for each k, y_k is child of z_k in T. Hence each edge deletion (a_i, b_j) results in changing of parent of all vertices in H_2.   □

We now prove the result for the directed graphs. Let H_1 be the same graph as described earlier except that (r, a_i) and (a_i, b_j) are now directed edges. Let H_3 be a graph on n vertices - $x_1, .., x_n$ without any edge. The directed graph G' is obtained by adding edges from each b_j to all the vertices in H_3. The ordering is $\sigma = < r, x_1, .., x_n, a_1, .., a_n, b_1, .., b_t >$ and we use the same sequence of edge deletions as for the undirected case.

Lemma 6. *Deletion of each edge (a_i, b_j) leads to the change of parent of all vertices in H_3 in the ordered DFS tree T defined by σ.*

Proof. Note that at any instant of time all the vertices of H_3 would be hanging from the same vertex in H_1. Now after deletion of (a_i, b_j) edge, each vertex of H_3 will hang from vertex b_{j+1} if $j < n$, and b_1 otherwise. Thus on each edge deletion, the parent of each vertex in H_3 changes.

Theorem 5. *For each pair of positive integers n, m $(m > 3n)$, there exists a directed (undirected) graph with $\Theta(n)$ vertices and $\Theta(m)$ edges such that any decremental algorithm for maintaining the ordered DFS tree explicitly in this graph requires $\Omega(mn)$ time.*

Observation 1. *In the directed graph G' described above, each vertex x_k in H_3 changes its parent on deletion of (a_i, b_j) edge. So the total number of times x_k changes its parent is $\Theta(m)$. Notice that $\deg(x_k)$ remains $t = \lfloor m/n \rfloor$ throughout the sequence of edge deletions. Thus Theorem 3 implies that the total time taken by the deterministic algorithm in Sect. 3 for this graph is $\Theta(m^2)$.*

5.2 Partial Dynamic Ordered DFS Tree and Static All-Pairs Reachability

Let $G = (V, E)$ be a directed graph. Let $\mathcal{R}(v)$ denote the set of vertices reachable from $v \in V$. The objective of the all-pairs reachability problem is to compute $\mathcal{R}(v)$ for each $v \in V$. Let $\mathcal{A}$ be any decremental algorithm for maintaining the ordered DFS tree from a vertex in a directed graph. We shall now show that we can compute $\mathcal{R}(v)$ for all v by executing $\mathcal{A}$ on a graph G' formed by a suitable augmentation of G.

Add a dummy vertex s to G and connect it to each vertex of G by an edge. This completes the description of graph G'. Let $v_1, \ldots, v_n$ be any arbitrary sequence of V and let σ be the ordering defined by this sequence. We sort all the edges in the adjacency lists of G' according to σ. This takes $O(m)$ time using any integer sorting algorithm.

We now execute $\mathcal{A}$ on graph G' to maintain the ordered DFS tree from s. The sequence of edge deletions is $(s, v_1), (s, v_2), \ldots, (s, v_{n-1})$. The definition of ordered DFS tree implies the following. Just before the deletion of edge (s, v_i), the set of vertices in the subtree $T(v_i)$ is exactly equal to $R(v_i)$. So we can compute $R(v_i)$ by traversing $T(v_i)$ just before the deletion of edge (s, v_i). In this manner, at the end of deletion of edge (s, v_{n-1}), we have obtained entire reachability information of the graph. Since $\sum_i |R(v_i)| = O(n^2)$, we can state the following theorem.

Theorem 6. *Let $f(m, n)$ be the total time taken by any decremental algorithm for maintaining the ordered DFS tree in a directed graph for any sequence of n edge deletions. If the algorithm can report DFS tree at any stage in $O(n)$ time, we can compute all-pairs reachability of a directed graph on n vertices and m edges in $O(f(m, n) + n^2)$ time.*

Static all-pairs reachability is a classical problem and its time complexity is $O(\min(mn, n^\omega))$ [16]. This bound has remained unbeaten till now. So it is natural to believe that $O(\min(mn, n^\omega))$ is the best possible bound on the time complexity of the static all-pairs reachability. Therefore, Theorem 6 implies the following conditional lower bounds on the dynamic complexity of ordered DFS tree.

Theorem 7. *Let G be a directed graph on n vertices and m edges. Let $\mathcal{A}$ be a decremental algorithm for maintaining the ordered DFS tree in G with $O(n)$ query time. The following lower bounds hold for $\mathcal{A}$ unless we have an $o(\min(mn, n^\omega))$ time algorithm for the static all-pairs reachability problem.*

- *The total update time for any sequence of edge deletions is $\Omega(\min(mn, n^\omega))$.*
- *The worst case time to handle an edge deletion is $\Omega(\min(m, n^{\omega-1}))$.*

For an algorithm that takes more than $O(n)$ query time our reduction implies the following lower bound.

Theorem 8. *Let $q(m, n)$ be the worst case query time to report the DFS tree and $u(m, n)$ be the worst case time to handle an edge deletion by a decremental algorithm for maintaining the ordered DFS tree. Then either $q(m, n)$ or $u(m, n)$ must be $\Omega(\min(m, n^{\omega-1}))$ unless we have an $o(\min(mn, n^{\omega}))$ time algorithm for the static all-pairs reachability problem.*

Remark 1. We can obtain exactly same conditional lower bounds as in Theorems 7 and 8 for any incremental algorithm for the ordered DFS tree in a digraph. The graph G' is the same except that we insert the edges in the order $(s, v_n), \ldots, (s, v_1)$, and traverse the subtree $T(v_i)$ in the DFS tree just after the insertion of (s, v_i).

6   Conclusion

We presented the first decremental algorithm for maintaining a DFS tree in a DAG. This decremental algorithm as well as the incremental algorithm of Franciosa [6] crucially exploit the acyclic condition of a DAG in order to achieve efficient update time. In fact it can be shown that there exists a directed graph on which these algorithms perform no better than rebuilding the DFS tree form scratch after each update (see the full version of this article for the details). So we would like to conclude with the following open question whose answer will surely add significantly to our understanding of the dynamic DFS tree problem.

– Does there exist an incremental/decremental algorithm with $O(mn)$ update time for DFS tree in general directed graphs?

References

1. Alstrup, S., Holm, J.: Improved algorithms for finding level ancestors in dynamic trees. In: Welzl, E., Montanari, U., Rolim, J.D.P. (eds.) ICALP 2000. LNCS, vol. 1853, p. 73. Springer, Heidelberg (2000)
2. Baswana, S., Khan, S.: Incremental algorithm for maintaining DFS tree for undirected graphs. In: Esparza, J., Fraigniaud, P., Husfeldt, T., Koutsoupias, E. (eds.) ICALP 2014. LNCS, vol. 8572, pp. 138–149. Springer, Heidelberg (2014)
3. Cole, R., Hariharan, R.: Dynamic LCA queries on trees. SIAM J. Comput. **34**(4), 894–923 (2005)
4. Demetrescu, C., Italiano, G.F.: A new approach to dynamic all pairs shortest paths. J. ACM **51**(6), 968–992 (2004)
5. Franciosa, P.G., Frigioni, D., Giaccio, R.: Semi-dynamic breadth-first search in digraphs. Theor. Comput. Sci. **250**(1–2), 201–217 (2001)
6. Franciosa, P.G., Gambosi, G., Nanni, U.: The incremental maintenance of a depth-first-search tree in directed acyclic graphs. Inf. Process. Lett. **61**(2), 113–120 (1997)
7. Holm, J., de Lichtenberg, K., Thorup, M.: Poly-logarithmic deterministic fully-dynamic algorithms for connectivity, minimum spanning tree, 2-edge, and biconnectivity. J. ACM **48**(4), 723–760 (2001)
8. Miltersen, P.B., Subramanian, S., Vitter, J.S., Tamassia, R.: Complexity models for incremental computation. Theor. Comput. Sci. **130**(1), 203–236 (1994)

9. Reif, J.H.: Depth-first search is inherently sequential. Inf. Process. Lett. **20**(5), 229–234 (1985)
10. Reif, J.H.: A topological approach to dynamic graph connectivity. Inf. Process. Lett. **25**(1), 65–70 (1987)
11. Roditty, L., Zwick, U.: On Dynamic Shortest Paths Problems. In: Albers, S., Radzik, T. (eds.) ESA 2004. LNCS, vol. 3221, pp. 580–591. Springer, Heidelberg (2004)
12. Roditty, L., Zwick, U.: Improved dynamic reachability algorithms for directed graphs. SIAM J. Comput. **37**(5), 1455–1471 (2008)
13. Roditty, L., Zwick, U.: Dynamic approximate all-pairs shortest paths in undirected graphs. SIAM J. Comput. **41**(3), 670–683 (2012)
14. Robert Endre Tarjan: Depth-first search and linear graph algorithms. SIAM J. Comput. **1**(2), 146–160 (1972)
15. Thorup, M.: Fully-dynamic min-cut. Combinatorica **27**(1), 91–127 (2007)
16. Williams, V.V.: Multiplying matrices faster than Coppersmith-Winograd. In: STOC, pp. 887–898 (2012)

Metric Dimension of Bounded Width Graphs

Rémy Belmonte[1], Fedor V. Fomin[2],
Petr A. Golovach[2], and M.S. Ramanujan[2](✉)

[1] Department of Architecture and Architectural Engineering,
Kyoto University, Kyoto, Japan
remybelmonte@gmail.com
[2] Department of Informatics, University of Bergen, Bergen, Norway
{fedor.fomin,petr.golovach,ramanujan.sridharan}@ii.uib.no

Abstract. The notion of *resolving sets* in a graph was introduced by
Slater (1975) and Harary and Melter (1976) as a way of uniquely identifying every vertex in a graph. A set of vertices in a graph is a resolving
set if for any pair of vertices x and y there is a vertex in the set which
has distinct distances to x and y. A smallest resolving set in a graph is
called a *metric basis* and its size, the *metric dimension* of the graph. The
problem of computing the metric dimension of a graph is a well-known
NP-hard problem and while it was known to be polynomial time solvable
on trees, it is only recently that efforts have been made to understand its
computational complexity on various restricted graph classes. In recent
work, Foucaud et al. (2015) showed that this problem is NP-complete
even on interval graphs. They complemented this result by also showing
that it is *fixed-parameter tractable (FPT)* parameterized by the metric
dimension of the graph. In this work, we show that this FPT result can
in fact be extended to all graphs of bounded tree-length. This includes
well-known classes like chordal graphs, AT-free graphs and permutation
graphs. We also show that this problem is FPT parameterized by the
modular-width of the input graph.

1 Introduction

A vertex v of a connected graph G *resolves* two distinct vertices x and y of
G if $\text{dist}_G(v, x) \neq \text{dist}_G(v, y)$, where $\text{dist}_G(u, v)$ denotes the length of a shortest path between u and v in the graph G. A set of vertices $W \subseteq V(G)$ is a
resolving (or *locating*) set for G if for any two distinct $x, y \in V(G)$, there is
$v \in V(G)$ that resolves x and y. The *metric dimension* $\text{md}(G)$ is the minimum
cardinality of a resolving set for G. This notion was introduced independently
by Slater [22] and Harary and Melter [15]. The task of the MINIMUM METRIC
DIMENSION problem is to find the metric dimension of a graph G. Respectively,

Supported by the European Research Council (ERC) via grant Rigorous Theory of
Preprocessing, reference 267959 and the the ELC project (Grant-in-Aid for Scientific
Research on Innovative Areas, MEXT Japan).

© Springer-Verlag Berlin Heidelberg 2015
G.F. Italiano et al. (Eds.): MFCS 2015, Part II, LNCS 9235, pp. 115–126, 2015.
DOI: 10.1007/978-3-662-48054-0_10

METRIC DIMENSION
Input: A connected graph G and a positive integer k.
Question: Is $\mathrm{md}(G) \leq k$?

is the decision version of the problem.

The problem was first mentioned in the literature by Garey and Johnson [12] and the same authors later proved it to be NP-complete in general. Khuller et al. [18] have also shown that this problem is NP-complete on general graphs while more recently Diaz et al. [4] showed that the problem is NP-complete even when restricted to planar graphs. In this work, Diaz et al. also showed that this problem is solvable in polynomial time on the class of outer-planar graphs. Prior to this, not much was known about the computational complexity of this problem except that it is polynomial-time solvable on trees (see [18,22]), although there are also results proving combinatorial bounds on the metric dimension of various graph classes [3]. Subsequently, Epstein et al. [7] showed that this problem is NP-complete on split graphs, bipartite and co-bipartite graphs. They also showed that the *weighted version* of METRIC DIMENSION can be solved in polynomial time on paths, trees, cycles, co-graphs and trees augmented with k-edges for fixed k. Hoffmann and Wanke [17] extended the tractability results to a subclass of unit disk graphs and most recently, Foucaud et al. [9] showed that this problem is NP-complete on interval graphs.

The NP-hardness of the problem in general as well as on several special graph classes raises the natural question of resolving its parameterized complexity. Parameterized complexity is a two dimensional framework for studying the computational complexity of a problem. One dimension is the input size n and another one is a parameter k. It is said that a problem is *fixed parameter tractable* (or FPT), if it can be solved in time $f(k) \cdot n^{O(1)}$ for some function f. We refer to the books of Downey and Fellows [6], Flum and Grohe [8], and Niedermeier [21] for detailed introductions to parameterized complexity. The parameterized complexity of METRIC DIMENSION under the standard parameterization – the metric dimension of the input graph, on general graphs was open until 2012, when Hartung and Nichterlein [16] proved that it is W[2]-hard. The next natural step in understanding the parameterized complexity of this problem is the identification of special graph classes which permit fixed-parameter tractable algorithms (FPT). Recently, Foucaud et al. [9] showed that when the input is restricted to the class of interval graphs, there is an FPT algorithm for this problem parameterized by the metric dimension of the graph. However, as Foucaud et al. note, it is far from obvious how the crucial lemmas used in their algorithm for interval graphs might extend to natural superclasses like chordal graphs and charting the actual boundaries of tractability of this problem remains an interesting open problem.

In this paper, we identify two width-measures of graphs, namely *tree-length* and *modular-width* as two parameters under which we can obtain FPT algorithms for METRIC DIMENSION. The notion of tree-length was introduced by Dourisboure and Gavoille [5] in order to deal with tree-decompositions whose quality is measured not by the size of the bags but the *diameter* of the bags.

Essentially, the *length* of a tree-decomposition is the maximum diameter of the bags in this tree-decomposition and the tree-length of a graph is the minimum length over all tree-decompositions. The class of bounded tree-length graphs is an extremely rich graph class as it contains several well-studied graph classes like interval graphs, chordal graphs, AT-free graphs, permutation graphs and so on. As mentioned earlier, out of these, only interval graphs were known to permit FPT algorithms for METRIC DIMENSION. This provides a strong motivation for studying the role played by the tree-length of a graph in the computation of its metric dimension. Due to the obvious generality of this class, our results for METRIC DIMENSION on this graph class significantly expand the known boundaries of tractability of this problem (see Fig. 1). Modular-width was introduced

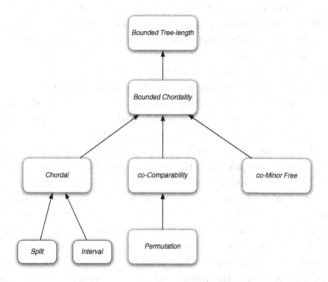

Fig. 1. Some well known graph classes which are subclasses of the class of bounded tree-length graphs. Out of these, METRIC DIMENSION was previously known to be FPT only on Split graphs and Interval graphs. Our results imply FPT algorithms parameterized by metric dimension on all other graph classes in the figure.

by Gallai [11] in the context of comparability graphs and transitive orientations. A module in a graph is a set X of vertices such that each vertex in $V \setminus X$ adjacent all or none of X. A partition of the vertex set into modules defines a *quotient graph* with the set of modules as the vertex set. Roughly speaking, the modular decomposition tree is a rooted tree that represents the graph by recursively combining modules and quotient graphs. The modular-width of the decomposition is the size of the largest *prime* node in this decomposition, that is, a node which cannot be partitioned into a set of non-trivial modules. Modular-width is a larger parameter than the more general clique-width and has been used in the past as a parameterization for problems where choosing clique-width as a parameter leads to W-hardness [10].

Our main result is an FPT algorithm for METRIC DIMENSION parameterized by the *maximum degree* and the tree-length of the input graph.

Theorem 1. METRIC DIMENSION *is* FPT *when parameterized by* $\Delta + \mathbf{tl}$, *where* Δ *is the max-degree and* $\mathbf{tl}$ *is the tree-length of the input graph.*

It follows from (Theorem 3.6, [18]) that for any graph G, $\Delta(G) \le 2^{\mathrm{md}(G)} + \mathrm{md}(G) - 1$. Therefore, one of the main consequences of this theorem is the following.

Corollary 1. METRIC DIMENSION *is* FPT *when parameterized by* $\mathbf{tl} + k$, *where* k *is the metric dimension of the input graph.*

Further, it is known that chordal graphs and permutation graphs have tree-length at most 1 and 2 respectively. This follows from the definition in the case of chordal graphs. In the case of permutation graphs it is known that their chordality is bounded by 4 (see for example [2]) and by using the result of Gavoille et al. [13] for any h-chordal graph G, $\mathbf{tl}(G) \le h/2$ and a tree decomposition of length at most $h/2$ can be constructed in polynomial time. Therefore, we obtain FPT algorithms for METRIC DIMENSION parameterized by the solution size on chordal graphs and permutation graphs. This answers a problem posed by Foucaud et al. [9] who proved a similar result for the case of interval graphs.

The algorithm behind Theorem 1 is a dynamic programming algorithm on a bounded *width* tree-decomposition. However, it is not sufficient to have bounded tree-width (indeed it is open whether METRIC DIMENSION is polynomial time solvable on graphs of treewidth 2). This is mainly due to the fact that pairs of vertices can be resolved by a vertex 'far away' from them hence making the problem extremely non-local. However, we use delicate distance based arguments using the tree-length and degree bound on the graph to show that most pairs are trivially resolved by any vertex that is sufficiently far away from the vertices in the pair and furthermore, the pairs that are not resolved in this way must be resolved 'locally'. We then design a dynamic programming algorithm incorporating these structural lemmas and show that it is in fact an FPT algorithm for METRIC DIMENSION parameterized by max-degree and tree-length.

Our second result is an FPT algorithm for METRIC DIMENSION parameterized by the modular-width of the input graph.

Theorem 2. METRIC DIMENSION *is* FPT *when parameterized by the modular-width of the input graph.*

2   Basic Definitions and Preliminaries

Graphs. We consider finite undirected graphs without loops or multiple edges. The vertex set of a graph G is denoted by $V(G)$, the edge set by $E(G)$. We typically use n and m to denote the number of vertices and edges respectively. For a set of vertices $U \subseteq V(G)$, $G[U]$ denotes the subgraph of G induced by U. and by $G - U$ we denote the graph obtained form G by the removal of all the vertices of U, i.e., the subgraph of G induced by $V(G) \setminus U$. A set of vertices $U \subset V(G)$ is a *separator* of a connected graph G if $G - U$ is disconnected. Let G be a graph. For a vertex v, we denote by $N_G(v)$ its *(open) neighborhood*, that is, the set of vertices which are adjacent to v. The *distance* $\mathrm{dist}_G(u, v)$ between two vertices u and v in a connected graph G is the number of edges in a shortest (u, v)-path. For a positive integer r, $N_G^r[v] = \{u \in V(G) \mid \mathrm{dist}_G(u, v) \leq r\}$. For a vertex $v \in V(G)$ and a set $U \subseteq V(G)$, $\mathrm{dist}_G(v, U) = \min\{\mathrm{dist}_G(v, u) \mid u \in U\}$. For a set of vertices $U \subseteq V(G)$, its *diameter* $\mathrm{diam}_G(U) = \max\{\mathrm{dist}_G(u, v) \mid u, v \in U\}$. The *diameter of a graph* G is $\mathrm{diam}(G) = \mathrm{diam}_G(V(G))$. A vertex $v \in V(G)$ is *universal* if $N_G(v) = V(G) \setminus \{v\}$. For two graphs G_1 and G_2 with $V(G_1) \cap V(G_2)$, the *disjoint union* of G_1 and G_2 is the graph with the vertex set $V(G_1) \cup V(G_2)$ and the edge set $E(G_1) \cup E(G_2)$, and the *join* of G_1 and G_2 is the graph the vertex set $V(G_1) \cup V(G_2)$ and the edge set $E(G_1) \cup E(G_2) \cup \{uv \mid u \in V(G_1), v \in V(G_2)\}$. For a positive integer k, a graph G is *k-chordal* if the length of the longest induced cycle in G is at most k. The *chordality* of G is the smallest integer k such that G is k-chordal. It is usually assumed that forests have chordality 2; *chordal* graphs are 3-chordal graphs. We say that a set of vertices $W \subseteq V(G)$ *resolves* a set of vertices $U \subseteq V(G)$ if for any two distinct vertices $x, y \in U$, there is a vertex $v \in W$ that resolves them. Clearly, W is a resolving set for G if W resolves $V(G)$.

Modular-Width. A set $X \subseteq V(G)$ is a *module* of graph G if for any $v \in V(G) \setminus X$, either $X \subseteq N_G(v)$ or $X \cap N_G(v) = \emptyset$. The *modular-width* of a graph G introduced by Gallai in [11] is the maximum size of a prime node in the modular decomposition tree. For us, it is more convenient to use the following recursive definition. The modular-width of a graph G is at most t if one of the following holds:

(i) G has one vertex,
(ii) G is disjoint union of two graphs of modular-width at most t,
(iii) G is a join of two graphs of modular-width at most t,
(iv) $V(G)$ can be partitioned into $s \leq t$ modules $X_1, \ldots, X_s$ such that $\mathbf{mw}(G[X_i]) \leq t$ for $i \in \{1, \ldots, s\}$.

The modular-width of a graph can be computed in linear time by the algorithm of Tedder et al. [23] (see also [14]). Moreover, this algorithm outputs the algebraic expression of G corresponding to the described procedure of its construction.

Tree Decompositions. A *tree decomposition* of a graph G is a pair $(\mathcal{X}, T)$ where T is a tree and $\mathcal{X} = \{X_i \mid i \in V(T)\}$ is a collection of subsets (called *bags*) of $V(G)$ such that:

1. $\bigcup_{i \in V(T)} X_i = V(G)$,
2. for each edge $xy \in E(G)$, $x, y \in X_i$ for some $i \in V(T)$, and
3. for each $x \in V(G)$ the set $\{i \mid x \in X_i\}$ induces a connected subtree of T.

The *width* of a tree decomposition $(\{X_i \mid i \in V(T)\}, T)$ is $\max_{i \in V(T)} |X_i| - 1$. The *length* of a tree decomposition $(\{X_i \mid i \in V(T)\}, T)$ is $\max_{i \in V(T)} \mathrm{diam}_G(X_i)$.

The *tree-length* if a graph G denoted as $\mathbf{tl}(G)$ is the minimum length over all tree decompositions of G.

The notion of tree-length was introduced by Dourisboure and Gavoille [5]. Lokshtanov proved in [20] that it is NP-complete to decide whether $\mathbf{tl}(G) \leq \ell$ for a given G for any fixed $\ell \geq 2$, but it was shown by Dourisboure and Gavoille in [5] that the tree-length can be approximated in polynomial time within a factor of 3.

We say that a tree decomposition $(\mathcal{X}, T)$ of a graph G with $\mathcal{X} = \{X_i \mid i \in V(T)\}$ is *nice* if T is a rooted binary tree such that the nodes of T are of four types:

(i) a *leaf node* i is a leaf of T and $|X_i| = 1$;
(ii) an *introduce node* i has one child i' with $X_i = X_{i'} \cup \{v\}$ for some vertex $v \in V(G) \setminus X_{i'}$;
(iii) a *forget node* i has one child i' with $X_i = X_{i'} \setminus \{v\}$ for some vertex $v \in X_{i'}$; and
(iv) a *join node* i has two children i' and i'' with $X_i = X_{i'} = X_{i''}$ such that the subtrees of T rooted in i' and i'' have at least one forget vertex each.

By the same arguments as were used by Kloks in [19], it can be proved that every tree decomposition of a graph can be converted in linear time to a nice tree decomposition of the same length and width such that the size of the obtained tree is $O(n)$. Moreover, for an arbitrary vertex $v \in V(G)$, it is possible to obtain such a nice tree decomposition with the property that v is the unique vertex of the root bag. Due to space constraints we have omitted some proofs although we attempt to give proof sketches wherever appropriate. The appended full version contains all the missing proofs.

3   METRIC DIMENSION on Graphs of Bounded Tree-Length and Max-Degree

In this section we prove that METRIC DIMENSION is FPT when parameterized by the max-degree and tree-length of the input graph. Throughout the section we use the following notation. Let $(\mathcal{X}, T)$, where $\mathcal{X} = \{X_i \mid i \in V(T)\}$, be a nice tree decomposition of a graph G. Then for $i \in V(T)$, T_i is the subtree of T rooted in i and G_i is the subgraph of G induced by $\cup_{j \in V(T_i)} X_j$. We first begin with a subsection where we prove the required structural properties of graphs of bounded tree-length and max-degree.

3.1   Properties of Graphs of Bounded Tree-Length and Max-Degree

We need the following lemma from [1], bounding the treewidth of graphs of bounded tree-length and degree.

Lemma 1. [1] *Let G be a connected graph with $\Delta(G) = \Delta$ and let $(\mathcal{X}, T)$ be a tree decomposition of G with the length at most ℓ. Then the width of $(\mathcal{X}, T)$, is at most $w(\Delta, \ell) = \Delta(\Delta - 1)^{(\ell-1)}$.*

We also need the next lemma which essentially bounds the number of bags of $(\mathcal{X}, T)$ a particular vertex of the graph appears in. We then use this lemma to prove Lemma 3, which states that the 'distance between a pair of vertices in the tree-decomposition' in fact approximates the distance between these vertices in the graph by a factor depending only on Δ and ℓ.

Lemma 2. *Let G be a connected graph with $\Delta(G) = \Delta$, and let $(\mathcal{X}, T)$, where $\mathcal{X} = \{X_i \mid i \in V(T)\}$, be a nice tree decomposition of G of length at most ℓ. Furthermore, let P be a path in T such that for some vertex $z \in V(G)$, $z \in X_i$ for every $i \in V(P)$. Then $|V(P)| \leq \alpha(\Delta, \ell) = 2(\Delta^\ell(\Delta + 2) + 4)$.*

Using Lemma 2, we obtain the following.

Lemma 3. *Let G be a connected graph with max-degree $\Delta(G) = \Delta$, and let $(\mathcal{X}, T)$, where $\mathcal{X} = \{X_i \mid i \in V(T)\}$, be a nice tree decomposition of G with the length at most ℓ. Then for $i, j \in V(T)$ and any $x \in X_i$ and $y \in X_j$,*

$$dist_T(i, j) \leq \alpha(\Delta, \ell)(dist_G(x, y) + 1) - 1.$$

The following lemma is the main structural lemma based on which we design our algorithm.

Lemma 4 (Locality Lemma). *Let $(\mathcal{X}, T)$, where $\mathcal{X} = \{X_i \mid i \in V(T)\}$, be a nice tree decomposition of length at most ℓ of a connected graph G such that T is rooted in r, $X_r = \{u\}$. Let $\Delta = \Delta(G)$ be the max-degree of G and let $s = \alpha(\Delta, \ell)(2\ell + 1)$. Then the following holds:*

(i) If $i \in V(G)$ is an introduce node with the child i' and v is the unique vertex of $X_i \setminus X_{i'}$, then for any $x \in V(G_j)$ for a node $j \in V(T_i)$ such that $dist_T(i, j) \geq s$, u resolves v and x.

(ii) If $i \in V(G)$ is a join node with the children i', i'' and $x \in V(G_j) \setminus X_j$ for $j \in T_{i'}$ such that $dist_T(i', j) \geq s - 1$ and $y \in V(G_{i''}) \setminus X_{i''}$, then u or an arbitrary vertex $v \in (V(G_j) \setminus X_j)$ resolves x and y.

Proof. To show (i), consider $x \in V(G_j)$ for some $j \in V(T_{i'})$ such that $dist_T(i', j) \geq s$. As either $u \in X_i$ or X_i separates u and x,

$$dist_G(u, x) = \min\{dist_G(u, y) + dist_G(y, z) + dist_G(z, x) \mid y \in X_i, z \in X_j\}.$$

Let $y \in X_i$ and $z \in X_j$ be vertices such that $dist_G(u, x) = dist_G(u, y) + dist_G(y, z) + dist_G(z, x)$. Then by Lemma 3,

$$dist_G(u, x) \geq dist_G(u, y) + dist_G(y, z) \geq dist_G(u, y) + \frac{s + 1}{\alpha(\Delta, \ell)} - 1.$$

Because $v \in X_i$ and $diam_G(X_i) \leq \ell$,

$$dist_G(u, v) \leq dist_G(u, y) + dist_G(y, v) \leq dist_G(u, y) + \ell.$$

Because $s = \alpha(\Delta, \ell)(2\ell+1)$, we obtain that $\mathrm{dist}_G(u, v) < \mathrm{dist}_G(u, x)$, completing the proof of the first statement.

To prove (ii), let $x \in V(G_j)$ for $j \in T_{i'}$ such that $\mathrm{dist}_T(i', j) \geq s - 1$, and let $y \in V(G_{i''}) \setminus X_{i''}$. Assume also that $v \in V(G_j) \setminus X_j$. Suppose that u does not resolve x and y. It means that $\mathrm{dist}_G(u, x) = \mathrm{dist}_G(u, y)$. Because either $u \in X_i$ or X_i separates u and $\{x, y\}$, there are $x', y' \in X_i$ such that $\mathrm{dist}_G(u, x) = \mathrm{dist}_G(u, x') + \mathrm{dist}_G(x', x)$ and $\mathrm{dist}_G(u, y) = \mathrm{dist}_G(u, y') + \mathrm{dist}_G(y', y)$. As $\mathrm{dist}_G(u, x) = \mathrm{dist}_G(u, y)$ and $\mathrm{diam}_G(X_i) \leq \ell$,

$$\mathrm{dist}_G(x', x) - \mathrm{dist}_G(y', y) = \mathrm{dist}_G(u, y') - \mathrm{dist}_G(u, x') \leq \ell.$$

Notice that $\mathrm{dist}_G(x, X_i) \leq \mathrm{dist}_G(x, x')$ and $\mathrm{dist}_G(y, X_i) \geq \mathrm{dist}_G(y, y') - \ell$, because $\mathrm{diam}_G(X_i) \leq \ell$. Hence, $\mathrm{dist}_G(x, X_i) - \mathrm{dist}_G(y, X_i) \leq 2\ell$. There are $z, z' \in X_j$ such that $\mathrm{dist}_G(x, X_i) = \mathrm{dist}_G(x, z) + \mathrm{dist}_G(z, X_i)$ and $\mathrm{dist}_G(v, X_i) = \mathrm{dist}_G(v, z') + \mathrm{dist}_G(z', X_i)$. Because $\mathrm{diam}_G(X_j) \leq \ell$, $\mathrm{dist}_G(v, z) \leq \mathrm{dist}_G(v, z') + \ell$ and $\mathrm{dist}_G(z, X_i) \leq \mathrm{dist}_G(z', X_i) + \ell$. Hence,

$$\mathrm{dist}_G(v, z) + \mathrm{dist}_G(z, X_i) \leq \mathrm{dist}_G(v, z') + \mathrm{dist}_G(z', X_i) + 2\ell \leq \mathrm{dist}_G(v, X_i) + 2\ell.$$

Since X_i separates v and y,

$$
\begin{aligned}
\mathrm{dist}_G(v, y) &\geq \mathrm{dist}_G(v, X_i) + \mathrm{dist}_G(y, X_i) \\
&\geq \mathrm{dist}_G(v, z) + \mathrm{dist}_G(z, X_i) - 2\ell + \mathrm{dist}_G(y, X_i) \\
&\geq \mathrm{dist}_G(v, z) + \mathrm{dist}_G(z, X_i) - 2\ell + \mathrm{dist}_G(x, X_i) - 2\ell \\
&\geq \mathrm{dist}_G(v, z) + 2\mathrm{dist}_G(z, X_i) + \mathrm{dist}_G(x, z) - 4\ell
\end{aligned}
$$

Clearly, $\mathrm{dist}_G(v, x) \leq \mathrm{dist}_G(x, z) + \mathrm{dist}_G(v, z)$. Hence,

$$
\begin{aligned}
\mathrm{dist}_G(v, y) - \mathrm{dist}_G(v, x) &\geq (\mathrm{dist}_G(v, z) + 2\mathrm{dist}_G(z, X_i) + \mathrm{dist}_G(x, z) - 4\ell) \\
&\quad - (\mathrm{dist}_G(x, z) + \mathrm{dist}_G(v, z)) \\
&\geq 2\mathrm{dist}_G(z, X_i) - 4\ell.
\end{aligned}
$$

It remains to observe that $\mathrm{dist}_G(z, X_i) \geq \frac{s+1}{\alpha(\Delta,\ell)} - 1 > 2\ell$, and we obtain that $\mathrm{dist}_G(v, y) - \mathrm{dist}_G(v, x) > 0$, i.e., v resolve x and y. $\square$

3.2   The Algorithm

Now we are ready to prove the main result of the section.

Theorem 1. METRIC DIMENSION *is* FPT *when parameterized by* $\Delta + \mathbf{tl}$, *where* Δ *is the max-degree and* **tl** *is the tree-length of the input graph.*

Proof (Sketch). Let (G, k) be an instance of METRIC DIMENSION. Recall that the tree-length of G can be approximated in polynomial time within a factor of 3 by the results of Dourisboure and Gavoille [5]. Hence, we assume that a tree-decomposition $(\mathcal{X}, T)$ of length at most $\ell \leq 3\mathbf{tl}(G) + 1$ is given. By Lemma 1, the width of $(\mathcal{X}, T)$ is at most $w(\Delta, \ell)$. We consider at most n choices of a vertex $u \in V(G)$, and for each u, we check the existence of a resolving set W of size at most k that includes u.

From now on, we assume that $u \in V(G)$ is given. We use the techniques of Kloks from [19] and construct from $(\mathcal{X}, T)$ a nice tree decomposition of the same width and the length at most ℓ such that the root bag is $\{u\}$. To simplify notations, we assume that $(\mathcal{X}, T)$ is such a decomposition and T is rooted in r. By Lemma 2, for any path P in T, any $z \in V(G)$ occurs in at most $\alpha(\Delta, \ell)$ bags X_i for $i \in V(P)$.

We now design a dynamic programming algorithm over the tree decomposition that checks the existence of a resolving set of size at most k that includes u. For simplicity, we only solve the decision problem. However, the algorithm can be modified to find such a resolving set (if exists). Due to space constraints we only give a rough intuition behind the design of the algorithm. The formal description of the algorithm is given in the appended full version.

Let i be a node in the tree-decomposition and suppose that it is an introduce node where the vertex v is introduced. The case when i is a join node can be argued analogously by appropriate applications of the statements of Lemma 4. Since any vertex outside G_i has at most $\ell + 1$ possible distances to the vertices of X_i, the resolution of any pair in G_i by a vertex outside can be expressed in a 'bounded' way. The same holds for a vertex in $G_i - X_i$ which resolves a pair in $G - V(G_i)$. The tricky part is when a vertex in G_i resolves a pair with at least one vertex in G_i. Now, consider pairs of vertices in G which are necessarily resolved by a vertex of the solution in G_i. Let a, b be such a pair. Now, for those pairs a, b such that both are contained in $G_{i'}$, either v resolves them or we may inductively assume that these resolutions have been handled during the computation for node i'. We now consider other possible pairs. Now, if a is v, then by Lemma 4, if b is in $V(G_j)$ for any j which is at a distance at least s from i, then this pair is trivially resolved by u. Therefore, any 'interesting pair' containing v is contained within a distance of s from X_i in the tree-decomposition induced on G_i. However, due to Lemma 2 and the fact that G has bounded degree, the number of such vertices which form an interesting pair with v is bounded by a function of Δ and ℓ. Now, suppose that a is in $V(G_i)$ and b is a vertex in $V(G) \setminus V(G_i)$ and there is an introduce node on the path from i to the root which introduces b. Then, if $a \in V(G_j)$ where j is at a distance at least s from i, then this pair is trivially resolved by u. By the same reasoning, if the bag containing a is within a distance of s from i then the node where b is introduced must be within a distance of s from i. Otherwise this pair is again trivially resolved by u. Again, there are only a bounded number of such pairs. Finally, suppose that $a \in V(G_i)$ and b is not introduced on a bag on the path from i to the root. In this case, there is a join node, call it l, on the path from i to the root with children l' and l'' such that l' lies on the path from i to the root and b is contained in $V(G_{l''})$. In this case, we can use statement (ii) of Lemma 4 to argue that if a lies in $V(G_j)$ where j is at a distance at least s from i then it lies at a distance at least s from l and hence either u or a vertex in G_j resolves this and in the latter case, *any* arbitrary vertex achieves this. Therefore, we simply compute solutions corresponding to both cases. Otherwise, the bag containing a lies at a distance at most s from i. In this case, if l is at a distance greater than s from i then the previous argument based on statement (ii) still holds. Therefore, it only remains to consider the case when l is at a distance at most s from i. However, in this case, due to Lemma 3, if u does not resolve this pair, it must be the case that even b lies in a bag which is at a distance at most s from l. Hence, the number of such pairs is also bounded and we conclude that at any node i of the dynamic program, the number of interesting pairs we need to consider is bounded by a function of Δ and ℓ and hence we can perform a bottom up parse of the tree-decomposition and compute the appropriate solution values at each node. $\square$

4   METRIC DIMENSION on Graphs of Bounded Modular-Width

In this section we prove that the metric dimension can be computed in linear time for graphs of bounded modular-width. Let X be a module of a graph G and $v \in V(G) \setminus X$. Then the distances in G between v and the vertices of X are the same. This observation immediately implies the following lemma.

Lemma 5. *Let $X \subset V(G)$ be a module of a connected graph G and $|X| \geq 2$. Let also H be a graph obtained from $G[X]$ by the addition of a universal vertex. Then any $v \in V(G)$ resolving $x, y \in X$ is a vertex of X, and if $W \subseteq V(G)$ is a resolving set of G, then $W \cap X$ resolves X in H.*

Theorem 2. *The metric dimension of a connected graph G of modular-width at most t can be computed in time $O(t^3 4^t \cdot n + m)$.*

Proof (Sketch). To compute $\mathrm{md}(G)$, we consider auxiliary values $w(H, p, q)$ defined for a (not necessarily connected) graph H of modular-width at most t with at least two vertices and boolean variables p and q as follows. Let H' be the graph obtained from H by the addition of a universal vertex u. Notice that $\mathrm{diam}_{H'}(V(H)) \leq 2$. Then $w(H, p, q)$ the minimum size of a set $W \subseteq V(H)$ such that

(i) W resolves $V(H)$ in H',
(ii) H has a vertex x such that $\mathrm{dist}_{H'}(x, v) = 1$ for every $v \in W$ if and only if $p = true$, and
(iii) H has a vertex x such that $\mathrm{dist}_{H'}(x, v) = 2$ for every $v \in W$ if and only if $q = true$.

We assume that $w(H, p, q) = +\infty$ if such a set does not exists. The intuition behind the definition of the function $w(.)$ is as follows. Let X be a module in the graph G, $H = G[X]$ and let $H_1, \ldots, H_s$ be the partition of H into modules, of which $H_1, \ldots, H_t$ are trivial. Let Z be a hypothetical optimal resolving set and let $Z' = Z \cap X$. By Lemma 5, we know that every pair of vertices in H

must be resolved by a vertex in Z'. Therefore, we need to compute a set which, amongst satisfying other properties must be a resolving set for the vertices in X. However, since these vertices are all in the same module and G is connected, any pair of vertices are either adjacent or at a distance exactly 2 in G. Hence, we ask for W (condition (i)) to be a resolving set of $V(H)$ in H', the graph obtained by adding a universal vertex to H. Further, it could be the case that a vertex z in Z' is required to resolve a pair of vertices, one contained in X say x and the other disjoint from X, say y. Now, if x is at a distance 1 in G (and hence H') from every vertex in Z' then for any vertex $x' \in X$ which is also at a distance exactly 1 from every vertex of Z', z is also required to resolve x' and y. The same argument holds for vertices at distance exactly 2 from every vertex of Z'. Therefore, in order to keep track of such resolutions, it suffices to know *whether* exists a vertex in X which is at a distance exactly 1 (respectively 2) from every vertex of Z'. This is precisely what is captured by the boolean variables p and q.

Recall that since H has modular-width at most t, it can be constructed from single vertex graphs by the disjoint union and join operation and decomposing H into at most t modules and H has at least two vertices. In the formal proof, we we formally describe our algorithm to compute $w(H, p, q)$ given the modular decomposition of H and the values computed for the 'child' nodes. As the base case corresponds to graphs of size at most t we may compute the values for the leaf nodes by brute force and execute a bottom up dynamic program.   □

5   Conclusions

We have essentially shown that METRIC DIMENSION can be solved in polynomial time on graphs of constant degree and tree-length. For this, amongst other things, we used the fact that such graphs have constant treewidth. Therefore, the most natural step forward would be to attempt to extend these results to graphs of constant treewidth which do not necessarily have bounded degree or tree-length. In fact, we point out that it is not known whether METRIC DIMENSION is polynomial-time solvable even on graphs of treewidth at most 2.

References

1. Bodlaender, H.L., Thilikos, D.M.: Treewidth for graphs with small chordality. Discrete Appl. Math. **79**(1–3), 45–61 (1997)
2. Chandran, L.S., Lozin, V.V., Subramanian, C.R.: Graphs of low chordality. Discrete Math. & Theor. Comput. Sci. **7**(1), 25–36 (2005). www.dmtcs.org/volumes/abstracts/dm070103.abs.html
3. Chartrand, G., Eroh, L., Johnson, M.A., Oellermann, O.: Resolvability in graphs and the metric dimension of a graph. Discrete Appl. Math. **105**(1–3), 99–113 (2000)
4. Díaz, J., Pottonen, O., Serna, M., van Leeuwen, E.J.: On the complexity of metric dimension. In: Epstein, L., Ferragina, P. (eds.) ESA 2012. LNCS, vol. 7501, pp. 419–430. Springer, Heidelberg (2012)
5. Dourisboure, Y., Gavoille, C.: Tree-decompositions with bags of small diameter. Discrete Math. **307**(16), 2008–2029 (2007)

6. Downey, R.G., Fellows, M.R.: Fundamentals of Parameterized Complexity. Texts in Computer Science. Springer, London (2013)
7. Epstein, L., Levin, A., Woeginger, G.J.: The (weighted) metric dimension of graphs: hard and easy cases. In: Golumbic, M.C., Stern, M., Levy, A., Morgenstern, G. (eds.) WG 2012. LNCS, vol. 7551, pp. 114–125. Springer, Heidelberg (2012)
8. Flum, J., Grohe, M.: Parameterized Complexity Theory. Texts in Theoretical Computer Science. An EATCS Series. Springer-Verlag, Berlin (2006)
9. Foucaud, F., Mertzios, G.B., Naserasr, R., Parreau, A., Valicov, P.: Identification, location-domination and metric dimension on interval and permutation graphs. In: Workshop on Graph-Theoretic Concepts in Computer Science, WG 2015 to appear
10. Gajarský, J., Lampis, M., Ordyniak, S.: Parameterized algorithms for modular-width. In: Gutin, G., Szeider, S. (eds.) IPEC 2013. LNCS, vol. 8246, pp. 163–176. Springer, Heidelberg (2013)
11. Gallai, T.: Transitiv orientierbare Graphen. Acta Math. Acad. Sci. Hungar **18**, 25–66 (1967)
12. Garey, M.R., Johnson, D.S.: Computers and Intractability: A Guide to the Theory of NP-Completeness. W. H. Freeman, San Franciso (1979)
13. Gavoille, C., Katz, M., Katz, N.A., Paul, C., Peleg, D.: Approximate distance labeling schemes. In: Meyer auf der Heide, F. (ed.) ESA 2001. LNCS, vol. 2161, pp. 476–487. Springer, Heidelberg (2001)
14. Habib, M., Paul, C.: A survey of the algorithmic aspects of modular decomposition. Comput. Sci. Rev. **4**(1), 41–59 (2010)
15. Harary, F., Melter, R.A.: On the metric dimension of a graph. Ars Combinatoria **2**, 191–195 (1976)
16. Hartung, S., Nichterlein, A.: On the parameterized and approximation hardness of metric dimension. In: Proceedings of the 28th Conference on Computational Complexity, CCC 2013, pp. 266–276. K.lo Alto, California, USA, 5–7 June 2013
17. Hoffmann, S., Wanke, E.: METRIC DIMENSION for gabriel unit disk graphs is NP-complete. In: Bar-Noy, A., Halldórsson, M.M. (eds.) ALGOSENSORS 2012. LNCS, vol. 7718, pp. 90–92. Springer, Heidelberg (2013)
18. Khuller, S., Raghavachari, B., Rosenfeld, A.: Landmarks in graphs. Discrete Appl. Math. **70**(3), 217–229 (1996)
19. Kloks, T.: Treewidth, Computations and Approximations. LNCS, vol. 842. Springer, Heidelberg (1994)
20. Lokshtanov, D.: On the complexity of computing treelength. Discrete Appl. Math. **158**(7), 820–827 (2010)
21. Niedermeier, R.: Invitation to Fixed-parameter Algorithms. Oxford Lecture Series in Mathematics and its Applications, vol. 31. Oxford University Press, Oxford (2006)
22. Slater, P.J.: Leaves of trees. In: Proceedings of the Sixth Southeastern Conference on Combinatorics, Graph Theory, and Computing (Florida Atlantic Univ., Boca Raton, Fla., 1975). pp. 549–559. Congressus Numerantium, No. XIV. Utilitas Math., Winnipeg, Man (1975)
23. Tedder, M., Corneil, D.G., Habib, M., Paul, C.: Simpler linear-time modular decomposition via recursive factorizing permutations. In: Aceto, L., Damgård, I., Goldberg, L.A., Halldórsson, M.M., Ingólfsdóttir, A., Walukiewicz, I. (eds.) ICALP 2008, Part I. LNCS, vol. 5125, pp. 634–645. Springer, Heidelberg (2008)

Equality, Revisited

Ralph Bottesch[1], Dmitry Gavinsky[2]([⊠]), and Hartmut Klauck[3,4]

[1] Division of Mathematical Sciences,
Nanyang Technological University, Singapore City, Singapore
[2] Institute of Mathematics, Academy of Sciences, Žitna 25,
Prague 1, Czech Republic
gavinsky@math.cas.cz
[3] Division of Mathematical Sciences,
Nanyang Technological University, Singapore City, Singapore
[4] Centre for Quantum Technologies,
National University of Singapore, Singapore City, Singapore

Abstract. We develop a new lower bound method for analysing the complexity of the Equality function (EQ) in the Simultaneous Message Passing (SMP) model of communication complexity. The new technique gives tight lower bounds of $\Omega(\sqrt{n})$ for both EQ and its negation NE in the *non-deterministic* version of quantum-classical SMP, where Merlin is also quantum – this is the strongest known version of SMP where the complexity of both EQ and NE remain high (previously known techniques seem to be insufficient for this).

Besides, our analysis provides to a unified view of the communication complexity of EQ and NE, allowing to obtain tight characterisation in all previously studied and a few newly introduced versions of SMP, including all possible combination of either quantum or randomised Alice, Bob and Merlin in the non-deterministic case.

Some of our results highlight that NE is easier than EQ in the presence of classical proofs, whereas the problems have (roughly) the same complexity when a quantum proof is present.

1 Introduction

The Equality function (EQ) and the Simultaneous Message Passing model (SMP) are among the longest-studied objects in communication complexity: When in 1979 Yao published his seminal paper [18] that introduced communication complexity, EQ was repeatedly used as an example (under the name of "the identification function"), SMP was introduced (referred to as "$1 \rightarrow 3 \leftarrow 2$"), and determining the SMP complexity of EQ was posed as an open problem.

D. Gavinsky—Partially funded by the grant P202/12/G061 of GA ČR and by RVO: 67985840.

H. Klauck—This work is funded by the Singapore Ministry of Education (partly through the Academic Research Fund Tier 3 MOE2012-T3-1-009) and by the Singapore National Research Foundation.

Being one of the weakest models that has been studied in communication complexity, SMP is, probably, the most suitable for studying EQ. While in several natural variants of SMP the complexity of EQ varies from constant to $\Omega(\sqrt{n})$, in the most of more powerful setups EQ can be solved by very efficient protocols of at most logarithmic cost.[1] That happens due to the fact that EQ becomes very easy (even for SMP) in the presence of shared randomness, and virtually all commonly studied stronger models can emulate shared randomness at the cost of at most $O(\log n)$ additional bits of communication.

The standard SMP setting involves three participants: the *players* Alice and Bob, and the *referee*; each of them can use private randomness. The input consists of two parts, one is given to Alice and the other to Bob; upon receiving their parts, the players send one message each to the referee; upon receiving the messages, the referee outputs the answer. Depending on the considered model, each player can be deterministic, randomised or quantum. In the non-deterministic regime, there is a third player called Merlin, who knows both parts of the input and sends his message to the referee, but is a dishonest person whose goal is to convince the referee to accept.[2] All players are assumed to be computationally unlimited, and the cost of a protocol is the (maximum) number of (qu)bits that the players send to the referee. Unless stated otherwise, we consider the *worst-case setting*, where a protocol should give correct answer on every possible input with probability at least $2/3$.

Yao's question about the SMP complexity of EQ was answered seventeen years later, in 1996, by Ambainis [3], who gave a protocol of cost $O(\sqrt{n})$, and by Newman and Szegedy [15], giving the matching lower bound $\Omega(\sqrt{n})$. Babai and Kimmel [4] showed the same lower bound shortly afterwards using a simpler argument. The above results address the "basic" version of SMP, where all the participants are randomised and no shared randomness is allowed (recall that otherwise SMP becomes "too powerful" for EQ).

In 2001, Buhrman, Cleve, Watrous and de Wolf [5] considered the version of SMP with quantum players and gave a very efficient and surprising protocol solving EQ at cost $O(\log n)$, and showed its optimality. In 2008, Gavinsky, Regev and de Wolf [8] studied the "quantum-classical" version of SMP, where only one of the players could send a quantum message, and they showed that the complexity of EQ in that model was $\Omega\left(\sqrt{n/\log n}\right)$ (which was, tight up to the multiplicative factor $\sqrt{\log n}$ by [15], and was improved to $\Omega(\sqrt{n})$ in [11]) .

---

[1] The communication complexity of EQ becomes n in the most of *deterministic* models, but those results are usually trivial and we do not consider the deterministic setup in this work (except for one special case where a "semi-deterministic" protocol has complexity $O(\sqrt{n})$ – that situation will be analysed in one of our lower bound proofs).

[2] Note that Merlin's message is only seen by the referee, and not by Alice and Bob. Letting the players receive messages from Merlin prior to sending their own messages would contradict the "simultaneous flavour" of the SMP model. Practically, that would make *NE* trivial; while the case of EQ is less obvious, we believe that the techniques developed in this work would be useful there as well.

2   Our Results

In this work we revisit the question about the SMP complexity of the Equality function EQ and its negation NE (the two cases are different in the context of non-deterministic models). We give a complete characterisation of the complexity of EQ and NE in a number of new non-deterministic SMP models corresponding to all possible combinations of either classical or quantum Alice, Bob, and Merlin. Moreover, our characterisation also covers the "asymmetric" scenarios when the (qu)bits from (untrusted) Merlin are either cheaper or more expensive than those from the trusted parties.

Let us denote the *type* of a non-deterministic SMP model by three letters, like QRQ, RRQ, RRR etc., where the letter in the corresponding position determines whether, respectively, Alice's, Bob's or Merlin's message is quantum or randomised. We will say that a protocol is "(a, b, m)" if Alice, Bob and Merlin sends at most a, b and m (qu)bits, respectively.

Table 1. Summary of results: complexity of (a, b, m) protocols.

Type of SMP	Task	Lower bound (assuming $ab = o(n)$)	Non-trivial upper bound (skipping O)	Tight bound on $a + b + m$
RRR	EQ	$m = \Omega(n)$	$(\log n, \log n, n)$	$\Theta(\sqrt{n})$
	NE	$m = \Omega(n/a + n/b)$	$(a, a, n/a)$	$\Theta(\sqrt{n})$
QRR	EQ	$m = \Omega(n)$	$(\log n, \log n, n)$	$\Theta(\sqrt{n})$
	NE	$m = \Omega(n/a + n/b)$	$(a, a, n/a)$	$\Theta(\sqrt{n})$
RRQ	EQ	$m = \Omega(n/a + n/b)$	$(a \log a, a \log a, n/a \cdot \log n)$	$\Theta(\sqrt{n})$
	NE	$m = \Omega(n/a + n/b)$	$(a, a, n/a)$	$\Theta(\sqrt{n})$
QRQ	EQ	$m = \Omega(\min\{n/a, n/b\})$	$(\log n, b \log b, n/b \cdot \log n)$	$\Theta(\sqrt{n})$
	NE	$m = \Omega(\min\{n/a, n/b\})$	$(\log n, b \log b, n/b \cdot \log n)$	$\Theta(\sqrt{n})$

Our results for all possible types of protocols are summarised in Table 1. As it was shown in [3] that both EQ and NE can be computed without any message from Merlin if $ab = const \cdot n$ and $a, b \geq \log n$, we present our hardness results via lower bounds on m as a function of a and b, assuming $ab = o(n)$. Note that we are closing the gap left by [8], as our results about non-deterministic SMP complexity of EQ imply that its quantum-classical complexity is also in $\Omega(\sqrt{n})$. This result has also recently been obtained by Klauck and Podder [11].

One of the key ingredients in our lower bound method is a tight analysis of a new communication primitive that we call *"One out of two"*, which might be of independent interest:

- Alice receives X_1 and X_2, Bob receives Y.
- It is promised that either $X_1 = Y$ or $X_2 = Y$.
- The referee has to distinguish the two cases.

The problem closely resembles *EQ*, but can be shown to be easier in some situations: We show that its quantum-classical SMP complexity is $\Omega(\sqrt{n})$, and here the interesting case is when Alice's message is quantum (the case of "quantum Bob" can be handled relatively easily by previously known techniques). A new method developed for its analysis can be viewed as the main technical contribution of this work.

In all cases we are showing tightness of our bounds even for arbitrary trade-off between a, b and m (Table 1). For that we demonstrate two protocols: one for *NE* in RRR, and one for *EQ* in QRQ and RRQ. For the latter, we combine the protocol of [5] for *EQ* with a new primitive in quantum communication, which might be of independent interest:

Assume that both Alice and Merlin know the classical description of a quantum state on $\log n$ qubits. Another player, the referee, wants to obtain one approximate copy of this state. Alice is trusted, but can send only classical information; Merlin can send quantum messages, but not be trusted. For which message lengths a and m from, respectively, Alice and Merlin can this be achieved? We call this problem *"Untrusted quantum state transfer"* and give a tight analysis, up to log-factors. We show that, essentially, $am = \tilde{\Theta}(n)$ is both enough and required.

3   The New Lower Bound Technique

Let us illustrate the idea of our method by applying it in the regular (randomised) SMP regime. Consider any randomised mapping $A(X)$ that Alice uses to create her message, and similarly $B(Y)$. The goal of the referee is to distinguish messages between the cases $X = Y$ and $X \neq Y$. In our method we create (depending on a given protocol) a distribution μ on $\{0,1\}^n$ of high entropy, such that X, X chosen according to μ leads to messages $A(X), B(X)$ with small mutual information $\mathbf{I}[A : B]$. This implies that the message state on inputs X, X is close to the product of its marginal states, which is the message state that would result from input strings X, Y chosen from μ, but independently. Since independent X, Y satisfy $X \neq Y$ with high probability and the referee cannot distinguish these two situations, he cannot decide *EQ*.

The basic argument is quite simple. Suppose Alice sends k bits and Bob l bits. Assume we create $100 \cdot k$ independent samples B_i of Bob's message (for the same input X, but using different random choices each time). Then

$$k \geq \mathbf{I}[A : B_1, \ldots, B_{100k}]$$
$$= \sum_i \mathbf{I}[A : B_i | B_1, \ldots, B_{i-1}].$$

Hence, for a random i we can make the information $\mathbf{I}[A : B_i] \leq 0.01$ by conditioning on a random value of the $i - 1 \leq k$ previous samples of Bob's message and observing that all B_i have the same distribution. Fixing those other $i - 1$ messages leaves us with a distribution μ on X that has entropy at least $n - O(kl)$ (on average) and makes the information between Alice's and Bob's messages small, i.e., their messages are close to the situation where Alice and Bob send messages for independent strings (X, Y). Hence, choosing pairs (X, X) with X from μ is not distinguishable (for the referee) from choosing (X, Y) independently, each from μ.

Let us compare our technique with the one used in [4] (and later in [8]). There the main idea was to amplify the success probability by making one player's message longer, until that message can be made deterministic – thus forcing communication $\Omega(n)$ from that player's side. An advantage of this was that it could be applied to any function; the main disadvantage for the special case of EQ was that this approach was "too demanding", and could not work in the stronger modifications of SMP.

4   Extensions

4.1   QR Protocols

Note that in the proof sketched above we never utilise the classical nature of Alice's message. The reason why we need Bob to be classical is that in order to construct the distribution μ we interpret $\mathbf{I}\big[B_{t_0+1} : A \big| B_1 = B'_1, \ldots, B_{t_0} = B'_{t_0}\big]$ as an expectation over fixing the random variables in the condition, something that is impossible in the quantum case. Alice's message is not required to be classical, and the whole argument goes through unchanged for QR protocols, where we use the quantum version of Pinsker's inequality ([9], see also [10]) in the last step.

Another generalisation is to extend the lower bound to a model, where Alice and the referee share entanglement (no other two players share entanglement). Without loss of generality (at the expense of a factor of two in the message length) in this case Alice can replace her quantum message by a classical message through teleportation. Nevertheless we can argue that the referee's part of the entangled state plus the message from Alice together carry little information about Alice's input, and leave the remaining argument unchanged. So even in this potentially stronger model we obtain the same $\Omega(\sqrt{n})$ lower bound.

4.2   The *"One Out of Two"* Problem

We are now going to extend the lower bound to the "One out of two" problem described in Sect. 1. Recall that now Alice (the quantum player) has two inputs X_1, X_2 and the task is to decide which one is equal to Y. To show that this problem is still hard we want to construct a distribution ρ_1 on triples of strings X_1, X_2, Y, such that X_1, Y always have the same value, X_2 is distributed like the marginal of ρ on Y, and given inputs under ρ the messages of Alice and Bob are almost independent. Then we repeat the argument with the roles of X_1, X_2 exchanged. The distributions seen by the referee are close to each other, yet the first one is almost completely on 1-inputs, the second one on 0-inputs.

We now define a distribution on triples of strings. Start by choosing X_0 uniformly at random from $\{0,1\}^n$. Let B_i be random variables taking values $\beta(X_0, R_i)$ for random strings R_i with i running from 1 to some value t. Choose t_0 uniformly at random between 1 and t. We now fix the values of the random variables B_i: $B_1 = \beta_1, \ldots, B_{t_0} = \beta_{t_0}$. Note that we keep X_0 random conditional on the fixed B_i's.

Now choose a random variable X as follows: There is an induced conditional distribution $\mu = \mu_{\beta_1, \ldots, \beta_{t_0}}$ on $\{0,1\}^n$. Choose X according to this distribution. As the resulting distribution of the input random variables X_1, X_2, Y we set $X_1 = X_0, X_2 = X, Y = X_0$. Call this distribution ρ_1.

Now we define another distribution ρ as follows. Choose X as above, but pick another sample Z distributed like X, but independent of X and of X_0 conditional on $B_1 = \beta_1, \ldots, B_{t_0} = \beta_{t_0}$. Set $X_1 = Z, X_2 = X, Y = X_0$. Now all three random variables will usually have different values. Note that inputs from ρ almost certainly fall outside of the promise of the problem we study.

Finally, ρ_2 is like ρ_1, only with $X_1 = X, X_2 = X_0, Y = X_0$.

Our goal is now to show that the messages produced by Alice and Bob on ρ_1 and on ρ are very similar, and by symmetry the same for the messages on ρ_2 and ρ, which implies that the referee cannot distinguish ρ_1 and ρ_2, which almost certainly have different function values, hence the error is close to $1/2$. The full details are in the full version of the paper.

Theorem 1. *In any QR protocol for the "One out of two" problem, the product of the message lengths must be $\Omega(n)$.*

We note that there is a protocol for the problem (as described in the full version), in which Alice sends a deterministic message and Bob a randomised message, both of length $O(\sqrt{n})$, and indeed other trade-offs of the form $ab = O(n)$ are possible by arranging the inputs as non-square matrices.

5   Lower Bounds

5.1   Bounds for *NE*

Theorem 2. *Any QRQ protocol for NE, in which the message lengths of Alice, Bob, Merlin are a, b, m satisfies $b(a + m) \geq \Omega(n)$.*

Corollary 1. *For QRQ protocols, if $ab = o(n)$, then $bm = \Omega(n)$, and indeed $m = \Omega(n/b) \geq \Omega(\min\{n/a, n/b\})$ as claimed in Table 1.*

Proof of Theorem 2. Take any QRQ (a, b, m)-protocol for *NE*. We show that this implies a QR $(a + m, b)$-protocol for the "One out of two" problem and we are done by Theorem 1. Instead of sending a quantum proof/message of length m to the referee we let Merlin simply provide Alice with a string Z claimed to be equal to Y, Bob's real input. Alice can now provide Referee with her own message (depending on X only) and the proof $\rho_{X,Z}$ that would maximise Merlin's success probability if Bob's input is really Z. Clearly the communication is $a + m$ from Alice, and still b from (unchanged) Bob.

The new protocol is as good as the previous at distinguishing $X \neq Y$ and $X = Y$: in both situations its maximum acceptance is the same as in the QRQ protocol.

We claim that the new protocol also solves the "One out of two" problem: given inputs X, Y, Z and the promise that $X = Y$ or $Z = Y$ (and $X \neq Z$) then in the first case the protocol will reject, in the second case accept (with high probability).

Hence by Theorem 1 , we get $b(a + m) \geq \Omega(n)$.   □

The following corollary is easy by symmetry.

Corollary 2. *Any RRQ or QRR (a, b, m)-protocol for NE satisfies $b(a + m), a(b + m) \geq \Omega(n)$, and hence when $ab = o(n)$ we have $m \geq \Omega(\max\{n/a, n/b\})$.*

Proof. The RRQ case is trivial since, for the same protocol, we can use the above argument with the role of Alice and Bob exchanged, and not exchanged (they both send classical messages), leading to lower bounds $b(a + m) \geq \Omega(n)$ and $a(b + m) \geq \Omega(n)$. For the QRR case this can also be done: now Merlin's message can be sent by either the classical player or the quantum player, since it not quantum itself, and in any case the protocol stays of type QR.   □

5.2   Bounds for *EQ*

Theorem 3. *Any QRQ protocol for EQ, in which the message lengths of Alice, Bob, Merlin are a, b, m satisfies $b(a + m) \geq \Omega(n)$.*

Corollary 3. *For QRQ protocols, if $ab = o(n)$, then $bm = \Omega(n)$, and indeed $m = \Omega(n/b) \geq \Omega(\min\{n/a, n/b\})$ as claimed in Table 1.*

Proof of Theorem 3. In the case of $X = Y$ there is a proof/message from Merlin ρ_X that makes the protocol accept with high probability. Note that Alice knows this proof, since she knows X.

To solve the "One out of two" problem again, Alice will simply choose the proof ρ_X, and send it together with her original message. Bob, again remains unchanged. If the referee accepts in the original protocol, then he will now announce that $X = Y$, otherwise that $Z = Y$.

Note that when $X = Y$ then the proof ρ_X makes the Referee accept with high probability, and if not, then $Z = Y$ via the promise, and the referee will reject, because the proof ρ_Z on inputs $X \neq Y$ must lead to rejection with high probability in the original protocol.   □

Again by symmetry we get the following corollary:

Corollary 4. *Any RRQ a, b, m-protocol for EQ satisfies $b(a + m), a(b + m) \geq \Omega(n)$, and hence when $ab = o(n)$ we have $m \geq \Omega(\max\{n/a, n/b\})$.*

We now turn to the case of QRR protocols, which exhibits a fundamentally different trade-off compared to the same case for *NE*: while for *NE* a proof from Merlin of length $n^{2/3}$ requires messages of length $n^{1/3}$ from Alice and Bob, for *EQ* any sub-linear proof requires $ab = \Omega(n)$, at which point the problem can be solved without Merlin.

Theorem 4. *Any QRR or RRR protocol for EQ, in which the message lengths of Alice, Bob, Merlin are a, b, m satisfies $ab + m \geq \Omega(n)$.*

Proof. Given a QRR protocol for *EQ* first observe that the classical message by Merlin can be assumed to be deterministic. He knows the whole protocol (although not the internal or measurement randomness of Alice, Bob, Referee). For any probability distribution on messages his winning probability (Merlin's goal is to make the referee accept) is a convex combination of winning probabilities over deterministic messages, and he can as well pick the best of those without losing anything.

Assuming that $m \leq n/3$ (otherwise we are done) we can find a message for Merlin such that the message maximises acceptance for at least $2^{2n/3}$ inputs X, X while still rejecting inputs X, Y with $X \neq Y$ with high probability. Hence we can find a QR protocol that solves a sub-problem of *EQ* that by renaming is equivalent to *EQ* on $2n/3$ bit inputs, and we get that $ab = \Omega(n)$. □

In fact we also have a more general statement for any function f.

Theorem 5. *For any Boolean function f: if we have a QRR (a, b, m)-protocol for f, then $ab + m \geq \Omega(N(f))$, for the non-deterministic communication complexity $N(f)$ of f.*

Proof. SKETCH. Again we may assume that Merlin's message is deterministic. The idea is to fix proof messages M of Merlin, and thus obtain partial functions f_M accepting all 1-inputs for which M is a good proof, while rejecting all 0-inputs. Using a result of Klauck and Podder [11] we can find a deterministic one-way protocol for f_M of cost $O(ab)$, and putting Merlin's proofs back in gives us a non-deterministic protocol. □

6 Protocols

In this section we give two protocols that illustrate the tightness of most of our bounds. For details see the full version of the paper.

6.1 Tightness for *NE* and RRR Protocols

We start with the case of *NE*. Here RRR, QRR, RRQ protocols all have the same complexity, as shown by our lower bounds and the following upper bound.

Theorem 6. *There is a RRR $(a + \log n, a + \log n, m)$-protocol for NE for all a, m such that $am \geq c \cdot n$ for some constant c.*

Proof. Consider the following protocol.

- Alice, Bob, Merlin fix an error-correcting code $C : \{0,1\}^n \to \{0,1\}^N$, where $N = O(n)$ and $C(x)$ differs from $C(y)$ on at least $N/3$ positions for every $x \neq y$. Additionally, let $N = am$ for some $a, m \in \mathbb{N}$.
- Viewing the code-words of C as $a \times m$ Boolean matrices, Alice chooses a uniformly random $i \in [m]$, and Bob uniformly random $j \in [m]$. Merlin (if honest) chooses some $k \in [a]$ such that $C(x)$ and $C(y)$ differ on at least $m/3$ positions of the k'th row.
- Alice and Bob send columns i resp. j of the encodings of x, y plus the numbers i, j.
- Merlin sends row k of both of the encodings of x, y plus the number k.
- If the k'th entry of the column sent by Alice is not equal to the i'th entry of the row sent by Merlin, then the referee rejects. Similarly, if the k'th entry of the column sent by Bob is not equal to the j'th entry of the row sent by Merlin, reject.
- Accept, if the rows sent by Merlin for x, y differ in at least $m/3$ positions, otherwise reject.

If $x \neq y$ then Merlin can proceed as indicated, and the protocol will accept with certainty.

It remains to show that Merlin cannot cheat if $x = y$. Denote by r_k and s_k the two rows sent by Merlin, which coincide in at most $2m/3$ positions. a_i and β_j are the columns sent by Alice and Bob.

If Merlin cheats by changing u entries in r_k, then he will be caught with probability u/m by the test against Alice's message, similarly he will be caught with probability v/m if he changes v entries in s_k. But to pass the last test $u + v \geq m/3$. Hence the total probability with which he will be caught is at least $1/3$.

We thus have an (a, a, m)- protocol with $am = O(n)$ that has perfect completeness and soundness error $2/3$. Repetition can improve this to arbitrarily small error. □

6.2   Tightness for QRQ and RRQ Protocols

In the next section we will show that a quantum message of length $\log n$ can be replaced by a classical message of length $O(\sqrt{n} \log n)$ by a trusted player, and a quantum message of length $O(\sqrt{n} \log n)$ by an untrusted player. This is the new primitive Untrusted Quantum State Transfer (UQST).

Furthermore we need the following fact from [5]: a QQ protocol for EQ (or NE) needs only communication $\log n + O(1)$ from either player. In this protocol, the players send superpositions over the indices and entries of an error-correcting code for x, y, and the swap-test is used by the referee to tell whether $x = y$ or not.

Hence we may replace the $\log n + O(1)$ qubit message from Bob by a $O(\sqrt{n} \log n)$ bit randomised message from Bob together with a $O(\sqrt{n} \log n)$

quantum message from Merlin, leading to a QRQ, $(\log n, \sqrt{n} \log n, \sqrt{n} \log n)$-protocol. Note that this works for both EQ and NE.

The same approach works for other values of Bob's message length, leading to a protocol in which Alice sends $O(\log n)$ qubits, Bob $O(b \log b)$ bits, and Merlin $O(n/b \log b)$ qubits.

We would like to stress that such a protocol is impossible in the QRR or the RRQ case: our lower bounds show that for NE Merlin's message needs to be at least $\Omega(n/\log n)$ (qu-)bits long if another player only sends $O(\log n)$ qubits.

We now turn to RRQ protocols. Informally, the idea is to "de-quantise" both messages from Alice and Bob as above. This leads to a protocol in which Alice and Bob both send $O(a \log a)$ bits, whereas Merlin sends $O(n/a \cdot \log n)$ qubits.

Note that the RRQ protocol also works for NE, however, the protocol in Theorem 6 is simpler, slightly more efficient, and does not use quantum messages.

7   Untrusted Quantum State Transfer

For the task of untrusted quantum state transfer (UQST) players Alice and Merlin, holding the classical description of a pure quantum state $|\phi\rangle$ on $\log n$ qubits, have to provide messages to the referee, such that the referee can get a single copy of a state ρ that is ϵ-close to $|\phi\rangle$ in the trace distance. Here Merlin can send quantum messages, but is untrusted, whereas trusted Alice can send only classical (randomised) messages. We are interested in the lengths of the messages they have to send.

More formally, in a protocol for UQST Alice and Merlin send messages to the referee. The referee produces two outputs: a classical bit meaning acceptance or rejection, and a quantum state ρ on $\log n$ qubits. The protocol is (ϵ, δ)-successful, if

1. For every quantum message from Merlin: the probability of the event that ρ satisfies $||\rho - |\phi\rangle\langle\phi| \,||_t > \epsilon$ and the referee accepts (simultaneously) is at most δ. The probability is over randomness in the (possibly mixed) quantum state ρ as well as Alice's random choices and the referee's measurements.
2. there is a message from Merlin such that the referee will accept with probability at least $1 - \delta$.

Note that in the discrete version of this problem $|\phi\rangle$ is given as a vector of n floating point numbers with (say) $100 \log n$ bits precision each.

7.1   A Protocol

Theorem 7. *For any $a \geq 10 \log(n)/(\epsilon^6 \delta^6)$ Untrusted Quantum State Transfer can be implemented with Alice sending $O(a \log(a/(\epsilon\delta))) + O(\log n)$ bits and Merlin $O((n/a) \log n/(\delta^3 \epsilon^2))$ qubits.*

7.2   A Lower Bound

We now observe that our protocol is optimal within lower order terms.

Theorem 8. *Any protocol for UQST in which Alice sends u bits and Merlin v qubits on states on $\log n$ qubits must satisfy $u(v + \log n) \geq \Omega(n)$.*

Proof. Assume there is a protocol for UQST with message lengths u, v. As shown in Theorem 3 any QRQ protocol for EQ must satisfy $b(a + m) \geq \Omega(n)$. There is a QQ (a, b)-protocol for EQ (see [5]) with $a = b = \log n + O(1)$, which can be turned into a QRQ protocol with message lengths $(\log n + O(1), O(u), O(v))$ by applying UQST to the message of Bob, making Bob randomised and introducing Merlin. Hence $u(v \log n) \geq \Omega(n)$.   □

8   Discussion

The main purpose of this work was to address what we viewed as remaining gaps in the understanding of one of the most basic communication primitives – the Equality function. We have developed a new technique for analysing EQ. Compared to previously used methods, it has the following advantages:

- It seems to be more widely applicable, as it allowed us to show $\Omega(\sqrt{n})$ lower bound for both EQ and its negation NE in the *non-deterministic* version of quantum-classical SMP, where Merlin is also quantum. This is the strongest known version of SMP where the complexity of both EQ and NE remains high, and the previously known lower bound techniques seemed to be insufficient for showing that.[3]
- It provides a unified view upon the complexity of EQ in all the versions of SMP mentioned above. Moreover, it simplifies some previously known bounds (even for the best-understood case of EQ in the randomised SMP, the new proof is, arguably, the simplest known).

We mention here that it is also interesting to study other functions in the SMP model with a prover (and private randomness). For instance, we also have the following upper bound for the Disjointness problem: there is an RRR protocol with communication $O(n^{2/3} \log n)$ from Alice, Bob, and Merlin. We conjecture this to be tight up to the log-factor. For details see the full version of the paper.

References

1. Aaronson, S.: Limitations of quantum advice and one-way communication. In: Proceedings of the 19th IEEE Conference on Computational Complexity, pp. 320–332 (2004)

---

[3] The best lower bound that we were able to prove using combinations of known techniques is $\tilde{\Omega}(n^{1/3})$.

2. Aaronson, S.: QMA/qpoly ⊆ pspace/poly: de-merlinizing quantum protocols. In: Proceedings of 21st IEEE Conference on Computational Complexity (2006)
3. Ambainis, A.: Communication complexity in a 3-computer model. Algorithmica **16**(3), 298–301 (1996)
4. Babai, L., Kimmel, P.: Randomized simultaneous messages: solution of a problem of yao in communication complexity. In: Proceedings of the 12th Annual IEEE Conference on Computational Complexity, p. 239 (1997)
5. Buhrman, H., Cleve, R., Watrous, J., de Wolf, R.: Quantum Fingerprinting. Phys. Rev. Lett. **87**(16), 167902 (2001)
6. Dasgupta, S., Gupta, A.: An elementary proof of a theorem of johnson and lindenstrauss. Random Struct. Algorithms **22**(1), 60–65 (2003)
7. Fuchs, C.A., van de Graaf, J.: Cryptographic distinguishability measures for quantum-mechanical states. IEEE Trans. Inf. Theory **45**(4), 1216–1227 (1999)
8. Gavinsky, D., Regev, O., de Wolf, R.: Simultaneous communication protocols with quantum and classical messages. Chicago J. Theor. Comput. Sci. **7** (2008)
9. Hiai, F., Ohya, M., Tsukada, M.: Sufficiency, KMS condition and relative entropy in von neumann algebras. Pacific J. Math. **96**, 99–109 (1981)
10. Klauck, H., Nayak, A., Ta-Shma, A., Zuckerman, D.: Interaction in quantum communication and the complexity of set disjointness. In: Proceedings of 33rd ACM STOC, pp. 124–133 (2001)
11. Klauck, H., Podder, S.: Two results about quantum messages. In: Csuhaj-Varjú, E., Dietzfelbinger, M., Ésik, Z. (eds.) MFCS 2014, Part II. LNCS, vol. 8635, pp. 445–456. Springer, Heidelberg (2014)
12. Kobayashi, H., Matsumoto, K., Yamakami, T.: Quantum merlin-arthur proof systems: are multiple merlins more helpful to arthur? In: Ibaraki, T., Katoh, N., Ono, H. (eds.) ISAAC 2003. LNCS, vol. 2906, pp. 189–198. Springer, Heidelberg (2003)
13. Kushilevitz, E., Nisan, N.: Communication Complexity. Cambridge University Press, New York (1997)
14. Newman, I.: Private vs. common random bits in communication complexity. Inf. Process. Lett. **39**(2), 67–71 (1991)
15. Newman, I., Szegedy,v: Public vs. private coin flips in one round communication games. In: Proceedings of the 28th Symposium on Theory of Computing, pp. 561–570 (1996)
16. Nielsen, M.A., Chuang, I.L.: Quantum Computation and Quantum Information. Cambridge University Press, New York (2000)
17. de Wolf, R.: Quantum communication and complexity. Theor. Comput. Sci. **287**(1), 337–353 (2002)
18. Yao, A.C-C.: Some complexity questions related to distributed computing. In: Proceedings of the 11th Symposium on Theory of Computing, pp. 209–213 (1979)

Bounding the Clique-Width of H-free Chordal Graphs

Andreas Brandstädt[1], Konrad K. Dabrowski[2(✉)], Shenwei Huang[3], and Daniël Paulusma[2]

[1] Institute of Computer Science, Universität Rostock, Albert-Einstein-Straße 22, 18059 Rostock, Germany
ab@informatik.uni-rostock.de
[2] School of Engineering and Computing Sciences, Durham University, Science Laboratories, South Road, Durham DH1 3LE, UK
{konrad.dabrowski,daniel.paulusma}@durham.ac.uk
[3] School of Computing Science, Simon Fraser University, 8888 University Drive, Burnaby, BC V5A 1S6, Canada
shenweih@sfu.ca

Abstract. A graph is H-free if it has no induced subgraph isomorphic to H. Brandstädt, Engelfriet, Le and Lozin proved that the class of chordal graphs with independence number at most 3 has unbounded clique-width. Brandstädt, Le and Mosca erroneously claimed that the gem and the co-gem are the only two 1-vertex P_4-extensions H for which the class of H-free chordal graphs has bounded clique-width. In fact we prove that bull-free chordal and co-chair-free chordal graphs have clique-width at most 3 and 4, respectively. In particular, we prove that the clique-width is:

(i) bounded for four classes of H-free chordal graphs;
(ii) unbounded for three subclasses of split graphs.

Our main result, obtained by combining new and known results, provides a classification of all but two stubborn cases, that is, with two potential exceptions we determine *all* graphs H for which the class of H-free chordal graphs has bounded clique-width. We illustrate the usefulness of this classification for classifying other types of graph classes by proving that the class of $(2P_1 + P_3, K_4)$-free graphs has bounded clique-width via a reduction to K_4-free chordal graphs. Finally, we give a complete classification of the (un)boundedness of clique-width of H-free weakly chordal graphs.

1 Introduction

Clique-width is a well-studied graph parameter; see for example the surveys of Gurski [29] and Kamiński, Lozin and Milanič [30]. In particular, there are

The research in this paper was supported by EPSRC (EP/K025090/1). The third author is grateful for the generous support of the Graduate (International) Research Travel Award from Simon Fraser University and Dr. Pavol Hell's NSERC Discovery Grant.

© Springer-Verlag Berlin Heidelberg 2015
G.F. Italiano et al. (Eds.): MFCS 2015, Part II, LNCS 9235, pp. 139–150, 2015.
DOI: 10.1007/978-3-662-48054-0_12

numerous graph classes, such as those that can be characterized by one or more forbidden induced subgraphs,[1] for which it has been determined whether or not the class is of *bounded clique-width* (i.e. whether there is a constant c such that the clique-width of every graph in the class is at most c). Clique-width is one of the most difficult graph parameters to deal with and our understanding of it is still very limited. We do know that computing clique-width is NP-hard [25] but we do not know if there exist polynomial-time algorithms for computing the clique-width of even very restricted graph classes, such as unit interval graphs. Also the problem of deciding whether a graph has clique-width at most c for some fixed constant c is only known to be polynomial-time solvable if $c \leq 3$ [13] and is a long-standing open problem for $c \geq 4$. Identifying more graph classes of bounded clique-width and determining what kinds of structural properties ensure that a graph class has bounded clique-width increases our understanding of this parameter. Another important reason for studying these types of questions is that certain classes of NP-complete problems become polynomial-time solvable on any graph class $\mathcal{G}$ of bounded clique-width.[2] Examples of such problems are those definable in Monadic Second Order Logic using quantifiers on vertices but not on edges.

Notation. The *disjoint union* $(V(G) \cup V(H), E(G) \cup E(H))$ of two vertex-disjoint graphs G and H is denoted by $G + H$ and the disjoint union of r copies of a graph G is denoted by rG. The *complement* of a graph G, denoted by $\overline{G}$, has vertex set $V(\overline{G}) = V(G)$ and an edge between two distinct vertices if and only if these vertices are not adjacent in G. If G is a graph, for $S \subseteq V(G)$, we let $G[S]$ denote the *induced* subgraph of G, which has vertex set S and edge set $\{uv \mid u, v \in S, uv \in E(G)\}$. For two graphs G and H we write $H \subseteq_i G$ to indicate that H is an induced subgraph of G. The graphs $C_r, K_r, K_{1,r-1}$ and P_r denote the cycle, complete graph, star and path on r vertices, respectively. The graph $S_{h,i,j}$, for $1 \leq h \leq i \leq j$, denotes the *subdivided claw*, that is the tree that has only one vertex x of degree 3 and exactly three leaves, which are of distance h, i and j from x, respectively. For a set of graphs $\{H_1, \ldots, H_p\}$, a graph G is $(H_1, \ldots, H_p)$-*free* if it has no induced subgraph isomorphic to a graph in $\{H_1, \ldots, H_p\}$. A graph G is *chordal* if it is $(C_4, C_5, \ldots)$-free and *weakly chordal* if both G and $\overline{G}$ are $(C_5, C_6, \ldots)$-free. Every chordal graph is weakly chordal.

Research Goal and Motivation. The class of chordal graphs has unbounded clique-width, as it contains the classes of proper interval graphs and split graphs, both of which have unbounded clique-width as shown by Golumbic and Rotics [28] and Makowsky and Rotics [35], respectively. We want to determine all graphs H for which the class of H-free chordal graphs has *bounded* clique-width. Our motivation for this research is threefold.

---

[1] For a record see also the Information System on Graph Classes and their Inclusions [22].

[2] This follows from results [15,24,31,38] that assume the existence of a so-called c-expression of the input graph $G \in \mathcal{G}$ combined with a result [37] that such a c-expression can be obtained in cubic time for some $c \leq 8^{\mathrm{cw}(G)} - 1$, where $\mathrm{cw}(G)$ is the clique-width of the graph G.

Firstly, as discussed, such a classification might generate more graph classes for which a number of NP-complete problems can be solved in polynomial time. Although many of these problems, such as the COLOURING problem [27], are polynomial-time solvable on chordal graphs, many others stay NP-complete for graphs in this class. Of course, in order to find new "islands of tractability", one may want to consider superclasses of H-free chordal graphs instead. However, already when one considers H-free weakly chordal graphs, one does not obtain any new tractable graph classes. Indeed, the clique-width of the class of H-free graphs is bounded if and only if H is an induced subgraph of P_4 [21], and as we prove later, the induced subgraphs of P_4 are also the only graphs H for which the class of H-free weakly chordal graphs has bounded clique-width. The same classification therefore also follows for superclasses, such as $(H, C_5, C_6, \ldots)$-free graphs (or H-free perfect graphs, to give another example). Since forests, or equivalently, $(C_3, C_4, \ldots)$-free graphs have bounded clique-width it follows that the class of $(H, C_3, C_4, \ldots)$-free graphs has bounded clique-width for every graph H. It is therefore a natural question to ask for which graphs H the class of $(H, C_4, C_5, \ldots)$-free (i.e. H-free chordal) graphs has bounded clique-width.

Secondly, we have started to extend known results [2,5–9,11,17,19,35] on the clique-width of classes of (H_1, H_2)-free graphs in order to try to determine the boundedness or unboundedness of the clique-width of every such graph class [18,21]. This led to a classification of all but 13 open cases (under some equivalence relation, see [21]). An important technique that we used for showing the boundedness of the clique-width of three new graph classes of (H_1, H_2)-free graphs [18] was to reduce these classes to some known subclass of perfect graphs of bounded clique-width (recall that perfect graphs form a superclass of chordal graphs). An example of such a subclass, which we used for one of the three cases, is the class of diamond-free chordal graphs (the diamond is the graph $\overline{2P_1 + P_2}$), which has bounded clique-width [28]. We believe that a full classification of the boundedness of clique-width for H-free chordal graphs would be useful to attack some of the remaining open cases, just as the full classification for H-free bipartite graphs [20] has already proven to be [18,21]. Examples of open cases included the class of $(2P_1 + P_3, K_4)$-free graphs and its superclass of $(2P_1 + P_3, \overline{2P_1 + P_3})$-free graphs [21], the first of which turns out to have bounded clique-width, as we shall prove in this paper via a reduction to K_4-free chordal graphs. The second case is still open.

Thirdly, a classification of those graphs H for which the clique-width of H-free chordal graphs is bounded would complete a line of research in the literature, which we feel is an interesting goal on its own. As a start, using a result of Corneil and Rotics [14] on the relationship between treewidth and clique-width it follows that the clique-width of the class of K_r-free chordal graphs is bounded for all $r \geq 1$. Brandstädt, Engelfriet, Le and Lozin [5] proved that the class of $4P_1$-free chordal graphs has unbounded clique-width. Brandstädt, Le and Mosca [9] considered forbidding the graphs $\overline{P_1 + P_4}$ (gem) and $P_1 + P_4$ (co-gem) as induced subgraphs (see also Fig. 1). They showed that $(P_1 + P_4)$-free chordal graphs have clique-width at most 8 and also observed that $\overline{P_1 + P_4}$-free chordal graphs belong to the class of distance-hereditary graphs, which have clique-width at most 3 (as shown by Golumbic and Rotics [28]). Moreover, the same authors [9] erroneously claimed that the gem and co-gem are the only two 1-vertex P_4-extensions H for which the class of H-free chordal graphs has bounded clique-width. We prove that bull-free chordal graphs have clique-width at most 3, improving a known bound of 8, which was shown by Le [33]. We also prove that $\overline{S_{1,1,2}}$-free chordal graphs have clique-width at most 4, which Le posed as an open problem. Results [28,32,35] for split graphs and proper interval graphs lead to other classes of H-free chordal graphs of unbounded clique-width, as we shall discuss in Sect. 2. However, in order to obtain our almost-full dichotomy for H-free chordal graphs new results also need to be proved.

Our Results. In Sect. 2, in addition to some known results for H-free chordal graphs, we give our result that bull-free chordal graphs have clique-width at most 3. In Sect. 3 we present four new classes of H-free chordal graphs of bounded clique-width,[3] namely when $H \in \{\overline{K_{1,3} + 2P_1}, P_1 + \overline{P_1 + P_3}, P_1 + \overline{2P_1 + P_2}, \overline{S_{1,1,2}}\}$ (see also Fig. 1). We include most of the proof for the $\overline{S_{1,1,2}}$ case, but do not include any other proofs due to space restrictions. In the same section we present three new subclasses of split graphs that have unbounded clique-width, namely $\overline{\mathbb{H}}$-free, $(\overline{3P_1 + P_2})$-free and $(K_3 + 2P_1, K_4 + P_1, P_1 + \overline{P_1 + P_4})$-free split graphs. By combining all these results with a number of previously known results [5,9,28,32,33,35], we obtain an almost-complete classification for H-free chordal graphs, leaving only two open cases (see also Figs. 1 and 2). We omit the proof, which is based on case analysis.

Theorem 1. *Let H be a graph with $H \notin \{F_1, F_2\}$. The class of H-free chordal graphs has bounded clique-width if and only if*

- $H = K_r$ *for some $r \geq 1$;*
- $H \subseteq_i$ *bull;*
- $H \subseteq_i P_1 + P_4$;
- $H \subseteq_i \overline{P_1 + P_4}$;
- $H \subseteq_i \overline{K_{1,3} + 2P_1}$;
- $H \subseteq_i P_1 + \overline{P_1 + P_3}$;
- $H \subseteq_i P_1 + \overline{2P_1 + P_2}$ *or*
- $H \subseteq_i \overline{S_{1,1,2}}$.

We also present our full classification for H-free weakly chordal graphs. We omit the proof.

Theorem 2. *Let H be a graph. The class of H-free weakly chordal graphs has bounded clique-width if and only if H is an induced subgraph of P_4.*

---

[3] In Theorems 8, 9 and 10, we do not specify our upper bounds as this would complicate our proofs for negligible gain. In our proofs we repeatedly apply graph operations that exponentially increase the upper bound on the clique-width, which means that the bounds that could be obtained from our proofs would be very large and far from being tight. We use different techniques to prove Lemma 5 and Theorem 11, and these allow us to give good bounds for these cases.

Finally, we illustrate the usefulness of having a classification for H-free chordal graphs by proving that the class of $(2P_1+P_3, K_4)$-free graphs has bounded clique-width via a reduction to K_4-free chordal graphs, and mention future research directions.

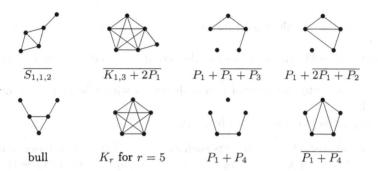

| $S_{1,1,2}$ | $\overline{K_{1,3}+2P_1}$ | $P_1+\overline{P_1+P_3}$ | $P_1+\overline{2P_1+P_2}$ |

| bull | K_r for $r=5$ | P_1+P_4 | $\overline{P_1+P_4}$ |

Fig. 1. The graphs H for which the class of H-free chordal graphs has bounded clique-width; the four graphs at the top are new cases proved in this paper.

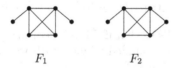

| F_1 | F_2 |

Fig. 2. The graphs H for which boundedness of clique-width of the class of H-free chordal graphs is open.

2   Preliminaries

All graphs considered in this paper are finite, undirected and have neither multiple edges nor self-loops. Let $G = (V,E)$ be a graph. Let $S, T \subseteq V$ with $S \cap T = \emptyset$. We say that S is *complete* to T if every vertex in S is adjacent to every vertex in T, and we say that S is *anti-complete* to T if every vertex in S is non-adjacent to every vertex in T. Similarly, a vertex $v \in V \setminus T$ is *complete* or *anti-complete* to T if it is adjacent or non-adjacent, respectively, to every vertex of T. A set of vertices M is a *module* if every vertex not in M is either complete or anti-complete to M. We say that a vertex v *distinguishes* two vertices x and y if v is adjacent to precisely one of x and y. Note that if a set $M \subseteq V$ is not a module then there must be vertices $x, y \in M$ and a vertex $v \in V \setminus M$ such that v distinguishes x and y.

Let $G = (V,E)$ be a graph. The graph G is a *split graph* if it has a *split partition*, i.e. a partition of V into two (possibly empty) sets K and I, where K is a clique and I is an independent set; if K and I are complete to each other, then G is a *complete* split graph. Every split graph is chordal. It is well known [26] that a graph is split if and only if it is $(C_4, C_5, 2P_2)$-free.

Clique-Width. The *clique-width* of a graph G, denoted by $\mathrm{cw}(G)$, is the minimum number of labels needed to construct G by using the following four operations:

1. creating a new graph consisting of a single vertex v with label i (denoted by $i(v)$);
2. taking the disjoint union of two labelled graphs G_1 and G_2 (denoted by $G_1 \oplus G_2$);
3. joining each vertex with label i to each vertex with label j ($i \neq j$, denoted by $\eta_{i,j}$);
4. renaming label i to j (denoted by $\rho_{i \to j}$).

An algebraic term that represents such a construction of G and uses at most k labels is said to be a *k-expression* of G (i.e. the clique-width of G is the minimum k for which G has a k-expression). For instance, an induced path on four consecutive vertices a, b, c, d has clique-width equal to 3, and the following 3-expression can be used to construct it:

$$\eta_{3,2}(3(d) \oplus \rho_{3 \to 2}(\rho_{2 \to 1}(\eta_{3,2}(3(c) \oplus \eta_{2,1}(2(b) \oplus 1(a)))))).$$

The following lemma tells us that if $\mathcal{G}$ is a hereditary graph class then in order to determine whether $\mathcal{G}$ has bounded clique-width we may restrict ourselves to the graphs in $\mathcal{G}$ that are prime.

Lemma 3 ([16]). *Let G be a graph and let $\mathcal{P}$ be the set of all induced subgraphs of G that are prime. Then $\mathrm{cw}(G) = \max_{H \in \mathcal{P}} \mathrm{cw}(H)$.*

Known Results on H-free Chordal Graphs. To prove our results, we need to use a number of known results. We present these results as lemmas below; a number of relevant graphs are displayed in Figs. 1 and 3. For a graph G, let $\mathrm{tw}(G)$ denote the treewidth of G. Corneil and Rotics [14] showed that $\mathrm{cw}(G) \leq 3 \times 2^{\mathrm{tw}(G)-1}$ for every graph G. Because the treewidth of a chordal graph is equal to the size of a maximum clique minus 1 (see e.g. [1]), this result leads to the following well-known lemma.

Lemma 4. *The class of K_r-free chordal graphs has bounded clique-width for all $r \geq 1$.*

The *bull* is the graph obtained from the cycle $abca$ after adding two new vertices d and e with edges ad, be (see also Fig. 1). In [9], Brandstädt, Le and Mosca erroneously mentioned that the clique-width of $\overline{S_{1,1,2}}$-free chordal graphs and of bull-free chordal graphs is unbounded. Using a general result of De Simone [23], Le [33] proved that every bull-free chordal graph has clique-width at most 8. Using a result of Olariu [36] we can show the following (we omit the proof).

Lemma 5. *Every bull-free chordal graph has clique-width at most 3.*

Lemma 6 ([9]). *Every $P_1 + P_4$-free chordal graph has clique-width at most 8 and every $\overline{P_1 + P_4}$-free chordal graph has clique-width at most 3.*

Lemma 7 ([5, 28, 32, 35]). *The class of H-free chordal graphs has unbounded clique-width if $H \in \{4P_1, K_{1,3}, 2P_2, C_4, C_5, net, \overline{net}\}$.*

3 New Classes of Bounded and Unbounded Clique-Width

We first present four new classes of H-free chordal graphs that have bounded clique-width. We omit the proofs for the first three of these.

Theorem 8. *The class of $\overline{K_{1,3} + 2P_1}$-free chordal graphs has bounded clique-width.*

Theorem 9. *The class of $(P_1 + \overline{P_1 + P_3})$-free chordal graphs has bounded clique-width.*

Theorem 10. *The class of $(P_1 + \overline{2P_1 + P_2})$-free chordal graphs has bounded clique-width.*

To prove Theorem 8, we make use of the celebrated Menger's Theorem and a tool developed by Lozin and Rautenbach, who proved that a graph G has bounded clique-width if and only if every block of G has bounded clique-width [34]. To the best of our knowledge, this technique has not been explored in previous research on clique-width. For Theorem 9, one may get the impression that the class of $(P_1 + \overline{P_1 + P_3})$-free chordal graphs is not much more complicated than the class of $\overline{P_1 + P_3}$-free chordal graphs and therefore expect it to have bounded clique-width (and similarly for the class of $(P_1 + \overline{2P_1 + P_2})$-free chordal graphs). We point out, however, that clique-width has a subtle transition from bounded to unbounded even if the class of graphs under consideration has a "slight" enlargement. For instance, the class of $(2P_1 + \overline{3P_1})$-free chordal (or even split) graphs (see Theorem 17) turns out to have unbounded clique-width. In fact, our proofs for Theorems 9 and 10 are rather involved. We now present a (detailed) proof sketch of our last new result for boundedness.

Theorem 11. *Every $\overline{S_{1,1,2}}$-free chordal graph has clique-width at most 4.*

We first provide a structural description of prime $\overline{S_{1,1,2}}$-free chordal graphs, and then Theorem 11 follows easily from our structural result. To this end, we appeal to the well-developed technique of *prime extension*. Results on prime extension effectively say that a prime graph that contains a particular pattern H as an induced subgraph must contain some extension of H (in the sense of being a supergraph of H) from a prescribed list of graphs (see e.g. [23, 33]). The following two structural lemmas, both of which play fundamental roles in the proof of Theorem 11, are of this flavour. The first is due to Brandstädt, Le and de Ridder and the second is due to Brandstädt.

Lemma 12 ([10]). *If a prime graph G contains an induced subgraph isomorphic to $P_1 + P_4$ then it contains one of the graphs in Fig. 3 as an induced subgraph.*

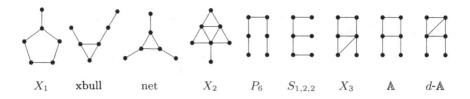

X_1 xbull   net X_2 P_6 $S_{1,2,2}$ X_3 $\mathbb{A}$ $d\text{-}\mathbb{A}$

Fig. 3. The minimal prime extensions of $P_1 + P_4$.

Lemma 13 ([3]). *If a prime graph G contains an induced $\overline{2P_1 + P_2}$ then it contains an induced $\overline{P_1 + P_4}$, $d\text{-}\mathbb{A}$ or d-domino. (See also Figs. 1 and 3. The d-domino is the graph with vertex set $\{x_1, \ldots, x_6\}$ and edge set $\{x_1x_2, x_2x_3, x_3x_4, x_4x_5, x_5x_6, x_6x_1, x_1x_3, x_1x_4\}$.)*

A graph G is a *thin spider* if its vertex set can be partitioned into a clique K, an independent set I and a set R such that $|K| = |I| \geq 2$, the set R is complete to K and anti-complete to I and the edges between K and I form an induced matching (that is, every vertex of K has a unique neighbour in I and vice versa). Note that if a thin spider is prime then $|R| \leq 1$. A *thick spider* is the complement of a thin spider. A graph is a *spider* if it is either a thin or a thick spider. Spiders play an important role in our result for $\overline{S_{1,1,2}}$-free chordal graphs and we will need the following lemma (due to Brandstädt and Mosca).

Lemma 14 ([12]). *If G is a prime $S_{1,1,2}$-free split graph then it is a spider.*

We now show that the clique-width of $\overline{S_{1,1,2}}$-free chordal graphs is bounded. Switching to the complement, we study $S_{1,1,2}$-free co-chordal graphs which are a subclass of $(2P_2, C_5, S_{1,1,2})$-free graphs. The main step consists of the following structural result.

Lemma 15. *If a prime $(2P_2, C_5, S_{1,1,2})$-free graph G contains an induced subgraph isomorphic to the net (see Fig. 3) then G is a thin spider.*

Proof. Suppose that G is a prime $(2P_2, C_5, S_{1,1,2})$-free graph and suppose that G contains a net, say N with vertices $a_1, a_2, a_3, b_1, b_2, b_3$ such that a_1, a_2, a_3 is an independent set (the *end-vertices* of N), b_1, b_2, b_3 is a clique (the *mid-vertices* of N), and the only edges between a_1, a_2, a_3 and b_1, b_2, b_3 are $a_ib_i \in E(G)$ for $i \in \{1, 2, 3\}$.

Let $M = V(G) \setminus V(N)$. We partition M as follows: For $i \in \{1, \ldots, 5\}$, let M_i be the set of vertices in M with exactly i neighbours in $V(N)$. Let U be the set of vertices in M adjacent to every vertex of $V(N)$. Let Z be the set of vertices in M with no neighbours in $V(N)$. Note that Z is an independent set in G, since G is $2P_2$-free. We now analyse the structure of G through a series of claims. The proofs of these claims have been omitted.

Claim 1. $M_1 \cup M_2 \cup M_5 = \emptyset$.

Next, we show that vertices in $M_3 \cup M_4$ have a restricted type of neighbourhood in $V(N)$:

Claim 2. *Every $x \in M_3$ is adjacent to either exactly one end-vertex a_i and its two opposite mid-vertices b_j and b_k ($j \neq i$, $k \neq i$) or to all three mid-vertices of N.*

The situation for M_4 is similar to that of M_3, as shown in the following claim.

Claim 3. *If $x \in M_4$ then it is adjacent to exactly one end-vertex and all mid-vertices.*

Let Mid_3 denote the set of vertices in M_3 that are adjacent to all three mid-vertices of N (and non-adjacent to any end-vertex of N).

Claim 4. *U is complete to $(M_3 \cup M_4)$.*

Let Z_1 denote the set of vertices in Z that have a neighbour in $M_3 \cup M_4$, and let $Z_0 = Z \setminus Z_1$. The next two claims show the adjacency between Z_1 and other subsets of $V(G)$.

Claim 5. *Z_1 is anti-complete to $((M_3 \cup M_4) \setminus Mid_3)$.*

Claim 6. *U is complete to Z_1.*

Let $X = V(N) \cup M_3 \cup M_4 \cup Z_1$. Then X is a module: every vertex in U is complete to X (due to the definition of U, together with Claims 4 and 6) and every vertex in Z_0 is anti-complete to X (due to the definitions of Z, Z_0 and Z_1, together with the fact that Z is an independent set). Since G is prime, X must be a trivial module. Since X contains more than one vertex, it follows that $V(G) = X = V(N) \cup M_3 \cup M_4 \cup Z_1$. Hence $U \cup Z_0 = \emptyset$. It remains to show that $G = G[V(N) \cup M_3 \cup M_4 \cup Z_1]$ is a thin spider. For $i \in \{1,2,3\}$ let $M_i' = (M_3 \cup M_4) \cap N(a_i)$. Note that $M_3 \cup M_4 = Mid_3 \cup M_1' \cup M_2' \cup M_3'$. The next two claims show how each M_i' is connected to other subsets of $V(G)$.

Claim 7. *For $i \neq j$, M_i' is complete to M_j'.*

Claim 8. *For every $i = 1,2,3$, M_i' is complete to Mid_3.*

By Claims 2, 3, 5, 7 and 8 we find that, for every $i \in \{1,2,3\}$, $M_i' \cup \{b_i\}$ is a module, so $M_i' = \emptyset$ (since G is prime). Consequently, $V(G) = V(N) \cup Mid_3 \cup Z_1$. Next, we show the following:

Claim 9. *Mid_3 is a clique.*

By Claim 9 and the definition of Mid_3, we find that $\{b_1, b_2, b_3\} \cup Mid_3$ is a clique. By the definition of Z and the fact that Z is independent, $\{a_1, a_2, a_3\} \cup Z_1$ is an independent set. Therefore G is a split graph. By Lemma 14, since G is prime and $S_{1,1,2}$-free, it must be a spider. Since G contains an induced net, it must be a thin spider. $\qquad\square$

The following is our new structural theorem on prime chordal $\overline{S_{1,1,2}}$-free graphs.

Theorem 16. *If G is a prime chordal $\overline{S_{1,1,2}}$-free graph then it is either a $\overline{2P_1 + P_2}$-free graph or a thick spider.*

Proof. Let G be a prime $\overline{S_{1,1,2}}$-free chordal graph. Note that since G is $\overline{S_{1,1,2}}$-free, it cannot contain d-$\mathbb{A}$ (see also Fig. 3) or d-domino as an induced subgraph. If G is $\overline{P_1 + P_4}$-free then, by Lemma 13, it must therefore be $\overline{2P_1 + P_2}$-free.

Now suppose that G contains an induced copy of $\overline{P_1 + P_4}$. Since G is prime, $\overline{G}$ is also prime. Furthermore, $\overline{G}$ is $(2P_2, C_5, S_{1,1,2})$-free. By Lemma 12, $\overline{G}$ must contain one of the graphs in Fig. 3. The only graph in Fig. 3 which is $(2P_2, C_5, S_{1,1,2})$-free is the net, so $\overline{G}$ must contain a net. By Lemma 15, $\overline{G}$ is a thin spider, so G is a thick spider. □

We are now ready to prove Theorem 11.

Theorem 11 (restated). *Every $\overline{S_{1,1,2}}$-free chordal graph has clique-width at most 4.*

Proof. Let G be an $\overline{S_{1,1,2}}$-free chordal graph. By Lemma 3, we may assume that G is prime. If G is $\overline{2P_1 + P_2}$-free then it has clique-width at most 3 by Lemma 6. By Theorem 16, we may therefore assume that G is a thick spider. Note that since a thick spider is the complement of a thin spider (see also the definition of a thin spider), K is an independent set, I is a clique and R is complete to I and anti-complete to K. Every vertex in K has exactly one non-neighbour in I and vice versa. Since G is prime and R is a module, R contains at most one vertex.

Let $i_1, \ldots, i_p$ be the vertices in I and let $k_1, \ldots, k_p$ be the vertices in K such that for each $j \in \{1, \ldots, p\}$, the vertex i_j is the unique non-neighbour of k_j in I. Let G_j be the labelled copy of $G[\{i_1, \ldots, i_j, k_1, \ldots, k_j\}]$ where every i_h is labelled 1 and every k_h is labelled 2. Now $G_1 = 1(i_1) \oplus 2(k_1)$ and for $j \in \{1, \ldots, p-1\}$ we can construct G_{j+1} from G_j as follows:

$$G_{j+1} = \rho_{3\to1}(\rho_{4\to2}(\eta_{1,3}(\eta_{1,4}(\eta_{2,3}(G_j \oplus 3(i_{j+1}) \oplus 4(k_{j+1})))))).$$

If $R = \emptyset$ then using the above recursively we get a 4-expression for G_p and therefore for G. If $R = \{x\}$ then we obtain a 4-expression for G using $\eta_{1,4}(G_p \oplus 4(x))$. Therefore G indeed has clique-width at most 4. This completes the proof. □

We now present three new subclasses of H-free split graphs that have unbounded clique-width. (The graph $\overline{\mathbb{H}}$ is the graph on six vertices whose complement looks like a capital letter "H".) We omit the proofs.

Theorem 17. *The following classes have unbounded clique-width:*

- $\overline{\mathbb{H}}$-*free split graphs*
- $3P_1 + P_2$-*free split graphs*
- $(K_3 + 2P_1, K_4 + P_1, P_1 + \overline{P_1 + P_4})$-*free split graphs.*

4   Concluding Remarks

Using our new results and a significant amount of non-trivial case analysis, we are able to prove our classification theorems. As an application of this, we can prove the following theorem via a reduction to K_4-free chordal graphs (we omit the proof). This has reduced the number of open problems posed in [21] to 13.

Theorem 18. *The class of $(K_4, 2P_1 + P_3)$-free graphs has bounded clique-width.*

We still need to determine whether or not the classes of F_i-free chordal graphs have bounded clique-width when $i \in \{1, 2\}$. For this purpose, we recently managed to show that for $i \in \{1, 2\}$, the class of F_i-free split graphs has bounded clique-width [4] and we are currently exploring whether it is possible to generalize the proof of this result to the class of F_i-free chordal graphs.

References

1. Bodlaender, H.L.: A tourist guide through treewidth. Acta Cybernetica **11**(1–2), 1–21 (1993)
2. Boliac, R., Lozin, V.V.: On the clique-width of graphs in hereditary classes. In: Bose, P., Morin, P. (eds.) ISAAC 2002. LNCS, vol. 2518, pp. 44–54. Springer, Heidelberg (2002)
3. Brandstädt, A.: (P_5, diamond)-free graphs revisited: structure and linear time optimization. Discrete Appl. Math. **138**(1–2), 13–27 (2004)
4. Brandstädt, A., Dabrowski, K.K., Huang, S., Paulusma, D.: Bounding the clique-width of H-free split graphs. In: Proceedings of EuroComb ENDM (to appear, 2015)
5. Brandstädt, A., Engelfriet, J., Le, H.-O., Lozin, V.V.: Clique-width for 4-vertex forbidden subgraphs. Theor. Comput. Sys. **39**(4), 561–590 (2006)
6. Brandstädt, A., Klembt, T., Mahfud, S.: P_6- and triangle-free graphs revisited: structure and bounded clique-width. Discrete Math. Theor. Comput. Sci. **8**(1), 173–188 (2006)
7. Brandstädt, A., Kratsch, D.: On the structure of (P_5, gem)-free graphs. Discrete Appl. Math. **145**(2), 155–166 (2005)
8. Brandstädt, A., Le, H.-O., Mosca, R.: Gem- and co-gem-free graphs have bounded clique-width. Int. J. Found. Comput. Sci. **15**(1), 163–185 (2004)
9. Brandstädt, A., Le, H.-O., Mosca, R.: Chordal co-gem-free and (P_5, gem)-free graphs have bounded clique-width. Discrete Appl. Math. **145**(2), 232–241 (2005)
10. Brandstädt, A., Le, V.B., de Ridder, H.N.: Efficient robust algorithms for the maximum weight stable set problem in chair-free graph classes. Inf. Process. Lett. **89**(4), 165–173 (2004)
11. Brandstädt, A., Mahfud, S.: Maximum weight stable set on graphs without claw and co-claw (and similar graph classes) can be solved in linear time. Inf. Process. Lett. **84**(5), 251–259 (2002)
12. Brandstädt, A., Mosca, R.: On the structure and stability number of P_5- and co-chair-free graphs. Discrete Appl. Math. **132**(1–3), 47–65 (2003)
13. Corneil, D.G., Habib, M., Lanlignel, J.-M., Reed, B.A., Rotics, U.: Polynomial-time recognition of clique-width ≤ 3 graphs. Discrete Appl. Math. **160**(6), 834–865 (2012)
14. Corneil, D.G., Rotics, U.: On the relationship between clique-width and treewidth. SIAM J. Comput. **34**, 825–847 (2005)
15. Courcelle, B., Makowsky, J.A., Rotics, U.: Linear time solvable optimization problems on graphs of bounded clique-width. Theor. Comput. Sys. **33**(2), 125–150 (2000)
16. Courcelle, B., Olariu, S.: Upper bounds to the clique width of graphs. Discrete Appl. Math. **101**(1–3), 77–114 (2000)

17. Dabrowski, K.K., Golovach, P.A., Paulusma, D.: Colouring of graphs with Ramsey-type forbidden subgraphs. Theoret. Comput. Sci. **522**, 34–43 (2014)
18. Dabrowski, K.K., Huang, S., Paulusma, D.: Bounding clique-width via perfect graphs. In: Dediu, A.-H., Formenti, E., Martín-Vide, C., Truthe, B. (eds.) LATA 2015. LNCS, vol. 8977, pp. 676–688. Springer, Heidelberg (2015)
19. Dabrowski, K.K., Lozin, V.V., Raman, R., Ries, B.: Colouring vertices of triangle-free graphs without forests. Discrete Math. **312**(7), 1372–1385 (2012)
20. Dabrowski, K.K., Paulusma, D.: Classifying the clique-width of H-free bipartite graphs. Discrete Appl. Math. (to appear)
21. Dabrowski, K.K., Paulusma, D.: Clique-width of graph classes defined by two forbidden induced subgraphs. In: Paschos, V.T., Widmayer, P. (eds.) CIAC 2015. LNCS, vol. 9079, pp. 167–181. Springer, Heidelberg (2015)
22. de Ridder, H.N., et al.: Information System on Graph Classes and their Inclusions (2001–2013). www.graphclasses.org
23. De Simone, C.: On the vertex packing problem. Graphs Comb. **9**(1), 19–30 (1993)
24. Espelage, W., Gurski, F., Wanke, E.: How to solve NP-hard graph problems on clique-width bounded graphs in polynomial time. In: Brandstädt, A., Le, V.B. (eds.) WG 2001. LNCS, vol. 2204, pp. 117–128. Springer, Heidelberg (2001)
25. Fellows, M.R., Rosamond, F.A., Rotics, U., Szeider, S.: Clique-width is NP-Complete. SIAM J. Discrete Math. **23**(2), 909–939 (2009)
26. Földes, S., Hammer, P.L.: Split graphs. Congressus Numerantium **XIX**, 311–315 (1977)
27. Golumbic, M.C.: Algorithmic Graph Theory and Perfect Graphs. Annals of Discrete Mathematics, vol. 57. North-Holland Publishing Company, Amsterdam (2014)
28. Golumbic, M.C., Rotics, U.: On the clique-width of some perfect graph classes. Int. J. Found. Comput. Sci. **11**(03), 423–443 (2000)
29. Gurski, F.: Graph operations on clique-width bounded graphs. CoRR, abs/cs/0701185 (2007)
30. Kamiński, M., Lozin, V.V., Milanič, M.: Recent developments on graphs of bounded clique-width. Discrete Appl. Math. **157**(12), 2747–2761 (2009)
31. Kobler, D., Rotics, U.: Edge dominating set and colorings on graphs with fixed clique-width. Discrete Appl. Math. **126**(2–3), 197–221 (2003)
32. Korpelainen, N., Lozin, V.V., Mayhill, C.: Split permutation graphs. Graphs Comb. **30**(3), 633–646 (2014)
33. Le, H.-O.: Contributions to clique-width of graphs. Ph.D. thesis, University of Rostock (2003). Cuvillier Verlag Göttingen (2004)
34. Lozin, V.V., Rautenbach, D.: On the band-, tree-, and clique-width of graphs with bounded vertex degree. SIAM J. Discrete Math. **18**(1), 195–206 (2004)
35. Makowsky, J.A., Rotics, U.: On the clique-width of graphs with few P_4's. Int. J. Found. Comput. Sci. **10**(03), 329–348 (1999)
36. Olariu, S.: On the homogeneous representation of interval graphs. J. Graph Theor. **15**(1), 65–80 (1991)
37. Oum, S.-I.: Approximating rank-width and clique-width quickly. ACM Trans. Algorithms **5**(1), 10 (2008)
38. Rao, M.: MSOL partitioning problems on graphs of bounded treewidth and clique-width. Theoret. Comput. Sci. **377**(1–3), 260–267 (2007)

New Bounds for the CLIQUE-GAP Problem Using Graph Decomposition Theory

Vladimir Braverman[1], Zaoxing Liu[1]([✉]), Tejasvam Singh[1],
N.V. Vinodchandran[2], and Lin F. Yang[1]

[1] Johns Hopkins University, Baltimore, MD 21218, USA
zaoxing@jhu.edu
[2] University of Nebraska-Lincoln, Lincoln, NE 68588, USA

Abstract. Halldórsson, Sun, Szegedy, and Wang (*ICALP 2012*) [16] investigated the space complexity of the following problem CLIQUE-GAP(r, s): given a graph stream G, distinguish whether $\omega(G) \geq r$ or $\omega(G) \leq s$, where $\omega(G)$ is the clique-number of G. In particular, they give matching upper and lower bounds for CLIQUE-GAP(r, s) for any r and $s = c \log(n)$, for some constant c. The space complexity of the CLIQUE-GAP problem for smaller values of s is left as an open question. In this paper, we answer this open question. Specifically, for $s = \tilde{O}(\log(n))$ and for any $r > s$, we prove that the space complexity of CLIQUE-GAP problem is $\tilde{\Theta}(\frac{ms^2}{r^2})$. Our lower bound is based on a new connection between graph decomposition theory (Chung, Erdös, and Spencer [11], and Chung [10]) and the multi-party set disjointness problem in communication complexity.

1 Introduction

Graphs are ubiquitous structures for representing real-world data in several scenarios. In particular, when the data involves relationships between entities it is natural to represent it as a graph $G = (V, E)$ where V represents entities and E represents the relationships between entities. Examples of such entity-relationship pairs include webpages-hyperlinks, papers-citations, IP addresses-network flows, and people-friendships. Such graphs are usually very large in size, e.g. the people-friendships "Facebook graph" [24] has 1 billion nodes. Because of the massive size of such graphs, analyzing them using classical algorithmic approaches is challenging and often infeasible. A natural way to handle such

V. Braverman—This material is based upon work supported in part by the National Science Foundation under Grant No. 1447639, by the Google Faculty Award and by DARPA grant N660001-1-2-4014. Its contents are solely the responsibility of the authors and do not represent the official view of DARPA or the Department of Defense.

Z. Liu—This work is supported in part by DARPA grant N660001-1-2-4014.

N.V. Vinodchandran—Research supported in part by National Science Foundation grant CCF-1422668.

L.F. Yang—This work is supported in part by NSF grant No. 1447639.

© Springer-Verlag Berlin Heidelberg 2015
G.F. Italiano et al. (Eds.): MFCS 2015, Part II, LNCS 9235, pp. 151–162, 2015.
DOI: 10.1007/978-3-662-48054-0_13

massive graphs is to process them under the *data streaming* model. When dealing with graph data, algorithms in this model have to process the input graph as a stream of edges. Such an algorithm is expected to produce an approximation of the required output while using only a limited amount of memory for any ordering of the edges. This streaming model has become one of the most widely accepted models for designing algorithms over large data sets and has found deep connections with a number of areas in theoretical computer science including communication complexity [3,9] and compressed sensing [14].

While most of the work in the data streaming model is for processing numerical data, processing large graphs is emerging as one of the key topics in this area. Graph problems considered so far in this model include counting problems such as triangle counting [4,6,12,17,18,23], MAX-CUT [19] and small graph minors [8], and classical graph problems such as bipartite matching [15], shortest path [13], and graph sparsification [1]. We refer the reader to a recent survey by McGregor for more details on streaming algorithms for graph problems [22]. Recently, Halldórsson, Sun, Szegedy, Wang [16] considered the problem of approximating the size of maximum clique in a graph stream. In particular, they introduced the CLIQUE-GAP(r, s) problem:

Definition 1. CLIQUE-GAP(r, s): *given a graph stream G, integer r and s with $0 \le s \le r$, output "1" if G has a r-clique or "0" if G has no $(s + 1)$-clique. The output can be either 0 or 1 if the size of the max-clique $w(G)$ is in $[s + 1, r]$.*

In this paper we further investigate the space complexity of the CLIQUE-GAP problem and its relation to other well studied topics including multiparty communication, graph decomposition theory, and counting triangles. We establish several new results including a solution to an open question raised in [16].

1.1 Our Results

In this paper, we establish a new connection between graph decomposition theory [10,11] and the multi-party set disjointness problem of the communication complexity theory. Using this connection, we prove new lower bounds for CLIQUE-GAP(r, s) when $s = O(\log n)$ and complement the results of [16]. Our main technical results are Theorems 1, 2, 3, and 4. We summarize our results below.

The Upper Bound : We give a one-pass streaming algorithm that solves CLIQUE-GAP(r, s) using $\tilde{O}(ms^2/r^2)$ space. Note that our results do not contradict the lower bounds in [16], since their results apply for dense graphs with $m = \Theta(n^2)$.

Theorem 1. *For any r and s where $r \ge 100s$, there is a one-pass streaming algorithm (Algorithm 1) that, on any streaming graph G with m edges and n vertices, answers CLIQUE-GAP(r, s) correctly with probability ≥ 0.99, using $\tilde{O}(ms^2/r^2)$ space.*[1]

---

[1] In this and following theorems, the constants we choose are only for demonstrative convenience.

Lower Bounds : We give a matching lower bound of $\tilde{\Omega}(ms^2/r^2)$ on the space complexity of CLIQUE-GAP(r, s) when $s = O(\log n)$.

Theorem 2. *For any $0 < \delta < 1/2$ there exists a global constant $c > 0$ such that for any $0 < s < r$, $M > 0$, there exists graph families $\mathcal{G}_1$ and $\mathcal{G}_2$ that satisfy the following:*

- *for all graph $G_1 \in \mathcal{G}_1$, $|E(G_1)| = m \geq M$, G_1 has a r-clique;*
- *for all graph $G_2 \in \mathcal{G}_2$; $|E(G_2)| = m \geq M$, G_2 has no $(s+1)$-clique;*
- *any randomized one-pass streaming algorithm $\mathcal{A}$ that distinguishes whether $G \in \mathcal{G}_1$ or $G \in \mathcal{G}_2$ with probability at least $1 - \delta$ uses at least $cm/(r^2 \log_s^2 r)$ memory bits.*

For $s = O(\log n)$ our lower bound matches, up to polylogarithmic factors, the upper bound of Theorem 1. Using the terminology from graph decomposition theory [10,11] we extend our results to a lower bound theorem for the general promise problem GAP$(\mathcal{P}, \mathcal{Q})$, which distinguishes between any two graph properties $\mathcal{P}$ and $\mathcal{Q}$ satisfying the following restrictions. Note that $\alpha_*(G_0, \mathcal{Q})$ is a parameter denotes the minimum decomposition of G_0 by graphs in $\mathcal{Q}$, first defined in [10]. Please refer to Eq. 5 for details.

Theorem 3. *Let $\mathcal{P}, \mathcal{Q}$ be two graph properties such that*

- $\mathcal{P} \cap \mathcal{Q} = \emptyset$;
- *If $G'' \in \mathcal{P}$ and G'' is a subgraph of G', then $G' \in \mathcal{P}$;*
- *If $G', G'' \in \mathcal{Q}$ and $V(G') \cap V(G'') = \emptyset$, then $\tilde{G} = (V(G') \cup V(G''), E(G') \cup E(G'')) \in \mathcal{Q}$;*

Let G_0 be an arbitrary graph in $\mathcal{P}$. Given any graph G with m edges and n vertices, if a one-pass streaming algorithm $\mathcal{A}$ solves GAP$(\mathcal{P}, \mathcal{Q})$ correctly with probability at least 3/4, then $\mathcal{A}$ requires $\Omega(\frac{n}{|V(G_0)|} \frac{1}{\alpha_*^2(G_0, \mathcal{Q})})$ space in the worst case.

We use the tools we develop for the CLIQUE-GAP problem to give a new two-pass algorithm to distinguish between graphs with at least T triangles and triangle-free graphs. For $T = n^{2+\beta}$, the space complexity of our algorithm is $o(m/\sqrt{T})$ for $\beta > 2/3$. Cormode and Jowhari [12] give a two-pass algorithm using $O(m/\sqrt{T})$ space. Also, for $T \leq n^2$ they provide a matching lower bound. Our results demonstrate that for some $T > n^2$, it might be possible to refine the lower bound of Cormode and Jowhari. We state our results in Theorem 4.

Theorem 4. *Let $\mathcal{G}_1$ be a class of graphs of n vertices that has at least $T = n^{2+\beta}$ triangles for some $\beta \in [0, 1]$. Let $\mathcal{G}_2$ be a class of graphs of n vertices and triangle-free. Given graph $G = (V, E)$ with n nodes and m edges, there is a two-pass streaming algorithm that distinguishes whether $G \in \mathcal{G}_1$ or $G \in \mathcal{G}_2$ with constant probability using $\tilde{O}(\frac{mn^{2-\beta}}{T})$ space. In particular, for $\beta > 2/3$, the algorithm uses $o(m/\sqrt{T})$ space.*

Incidence Model : We also give a new lower bound for the space complexity of CLIQUE-GAP$(r, 2)$ in the *incidence model* of graph streams (Theorem 5).

Theorem 5. *If one-pass streaming algorithm solves* CLIQUE-GAP$(r, 2)$ *in the incidences model for any G with m edges and n vertices with probability at least* $3/4$*, it requires* $\Omega(m/r^3)$ *space in the worst case.*

We omit the proofs of Theorems 3, 4 and 5 due to space limitation. The readers who are interested can find the proofs in the full version, which is available on arXiv.

1.2   Related Work

Prior work that is closest to our work is the above-mentioned paper of Halldórsson et al. [16]. They show that for any $\epsilon > 0$, any randomized streaming algorithm for approximating the size of the maximum clique with approximation ratio $cn^{1-\epsilon}/\log n$ requires $n^{2\epsilon}$ space (for some constant c). To prove this result they show a lower bound of $\Omega(n^2/r^2)$ for CLIQUE-GAP(r, s) (using the two-party communication complexity of the set disjointness problem) when $r = n^{1-\epsilon}$ and $s = 100 \cdot 2^{1/\epsilon} \log n$.

The problem related to cliques that has received the most attention in the streaming setting is approximately counting the number of triangles in a graph. Counting the number of triangles is usually an essential part of obtaining important statistics such as the clustering coefficient and transitivity coefficient [5,20] of a social network. Starting with the work of Bar-Yossef, Kumar and Sivakumar [4], triangle counting in the streaming model has received sustained attention by researchers [6,12,18,23]. Researchers have also considered counting other substructures such as $K_{3,3}$ subgraphs [7] and cycles [5,21].

The problem of clique identification in a graph has also been considered in other models. For example, Alon, Krivelevich, and Sudakov [2] considered the problem of finding a large *hidden clique* in a random graph.

2   Definitions and Results

2.1   Notations and Definitions

We give notations and definitions that are necessary to explain our results. For a graph $G = (V, E)$ with vertex set V and edge set E, we use m to denote the number of edges, n to denote the number of vertices, T to denote the number of triangles in G, Δ to denote the maximum degree of G, and $\omega(G)$ to denote the size of the maximum clique (also known as the clique number). We use $\tilde{O}$ and $\tilde{\Omega}$ to suppress logarithmic factors in the asymptotics.

We consider *the adjacency streaming model* for processing graphs [4,6]. In this model the graph G is presented as a stream of edges $\langle e_1, e_2, ..., e_m \rangle$. We process edges under the cash register model: edge deletion is not allowed.

A *k-pass* streaming algorithm can access the stream k times and should work correctly irrespective of the order in which the edges arrive (the ordering is fixed for all passes).

2.2 Lower Bound Techniques

To establish our lower bounds on the CLIQUE-GAP(r, s) problem for arbitrarily small s, we use the well known approach of reducing a communication complexity problem to CLIQUE-GAP(r, s). For the reduction, we make use of graph decomposition theory [10,11]. The communication complexity problem we use is the set disjointness problem in the one-way multi-party communication model.

The set disjointness problem in the one-way k-party communication model, denoted by DISJ$_k^n$, is the following promise problem. The input to the problem is a collection of k sets $S_1, \ldots, S_k$ over a universe $[n]$, with the promise that either all the sets are pairwise disjoint or there is a *unique* intersection (that is there is a unique $a \in [n]$ so that $a \in S_i$ for all $1 \le i \le n$). There are k players with unlimited computational power and with access to randomness. Player i has the input S_i. Player i can only send information to Player $(i + 1)$. After all the communication between players, the last player (Player k) outputs "0" if the k sets are pairwise disjoint or outputs "1" if the sets uniquely intersect. For instances that do not meet the promise the last player can output "0" or "1" arbitrarily. The communication complexity of such a protocol is the total number of bits communicated by all players. This problem was first introduced by [3] to prove lower bounds on the space complexity of approximating the frequency moments. In [9], it is shown that the communication complexity of DISJ$_k^n$ is $\Omega(n/k)$.

We review basics of graph decomposition [10,11]. An $\mathcal{H}$-decomposition of graph G is a family of subgraphs $\{G_1, G_2, \ldots, G_t\}$ such that each edge of G is exactly in one of the G_is and each G_i belongs to a specified class of graphs $\mathcal{H}$. Let f be a nonnegative cost function on graphs. The cost of a decomposition with respect to f is defined as $\alpha_f(G, \mathcal{H}) \equiv \min_D \sum_{i=1}^{t} f(G_i)$, where $D = \{G_1, G_2, \ldots, G_t\}$ is an $\mathcal{H}$-decomposition of G. Two functions that have received attention are $f_0(G) \equiv 1$ and $f_1(G) \equiv |V(G)|$. The former one counts the minimum number of subgraphs among all decompositions; and the later one counts the total number of nodes in the minimum decomposition. Many interesting problems in graph theory are related to this framework. For example $\alpha_{f_0}(G, \mathcal{P})$ is the thickness of G, for $\mathcal{P}$ the set of planar graphs; $\alpha_{f_1}(G, \mathcal{B})$, where $\mathcal{B}$ is the set of complete bipartite graphs, arises in the study of network contacts realizing certain symmetric monotone Boolean functions. Refer to [10,11] for more details on graph decomposition.

We are interested in the cost function f_0. $\alpha_{f_0}(G, \mathcal{H})$ is typically denoted as $\alpha_*(G, \mathcal{P})$ which is what we use in this paper. For the class $\mathcal{B}$, the class of complete bipartite graphs, it is known that $\alpha_*(K_n, \mathcal{B}) = \lceil \log_2 n \rceil$ [10].

To illustrate the reduction, consider CLIQUE-GAP$(r, 2)$. Let $k = \lceil \log_2 r \rceil$. Let $\{H_1, H_2, \ldots, H_k\}$ be a decomposition of G so that H_i's are bipartite and $\cup H_i$ is K_r. We will reduce an instance $S_1, \ldots, S_k$ of DISJ$_k^{n/r}$ to a graph G on n vertices as follows. The graph G has n/r groups of r vertices each. The players collectively and independently build the graph G as follows. Consider Player i and her input $S_i \subseteq [n/r]$. For an $a \in S_i$, Player i puts the graph H_i on r vertices of group a into the stream. It is clear that if S_is are disjoint then the graph G

is a collection of disjoint bipartite graphs and if there is a unique intersection a, the group a forms $\cup H_i = K_r$. Using standard arguments, we can show that the space complexity of CLIQUE-GAP$(r, 2)$ is $\Omega(n/r \log_2^2 r)$. Details are given in Sect. 4.

This proof can be generalized. In particular, we prove Theorem 2 by choosing $\mathcal{H}$ as set of s-partite graphs and prove Theorem 5 by choosing $\mathcal{H}$ as set of k-star graphs.

3   An Upper Bound

In this section we give an algorithm for CLIQUE-GAP(r, s) that uses $\tilde{O}(ms^2/r^2)$ space. Note that for $s = \Omega(r)$, the trivial algorithm that stores the entire graph has the required space complexity. Hence we will assume $s = o(r)$.

Algorithm 1. Algorithm for CLIQUE-GAP(r, s)

1: **Input:**
   Graph edge stream $\langle e_1, e_2, \ldots, e_m \rangle$ of graph $G = (V, E)$, positive
   integers r, s.
2: **Output:**
   "1" if a clique of order r is detected in G; "0" if G is
   $(s + 1)$-clique free.
3: **Initialize:**
   Set $p = 40(s + 1)/r$.
   Set memory buffer M empty.
   Compute n pairwise independent bits $\{Q_v | \text{for all } v \in V\}$ using
   $O(\log n)$ space such that for each $v \in V$, $Pr[Q_v = 1] = p$.
4: **while not** the end of the stream **do**
5:    Read an edge $e = (a, b)$.
6:    Insert e into M if $Q_a = 1$ and $Q_b = 1$.
7:    **If** there is an $(s + 1)$-clique in M, **then** output "1".
8: **output** "0".

Theorem 1. *For any r and s where $r \geq 100s$, there is a one-pass streaming algorithm (Algorithm 1) that, on any streaming graph G with m edges and n vertices, answers CLIQUE-GAP(r, s) correctly with probability ≥ 0.99, using $\tilde{O}(ms^2/r^2)$ space.*[2]

Proof. If $s < 2$, it is trivial to detect an edge. So let us assume $s \geq 2$. If the input graph G has no $(s + 1)$-clique, the algorithm always outputs "0" since the algorithm outputs "1" only if there is an $(s + 1)$-clique on a sampled subgraph of G. Consider the case where G has a r-clique. Let $K_r = (V_K, E_K)$ be such a clique.

---

[2] In this and following theorems, the constants we choose are only for demonstrative convenience.

Let the random variable Z denote the number of nodes 'sampled' from V_K. That is, $Z = \sum_{v \in V_K} Q_v$. The probability that $Q_v = 1$ is p and $Var(Q_v) = p(1 - p)$. Hence $E(Z) = rp$ and since each Q_v is pairwise independent, $Var(Z) = rp(1-p)$. Thus for $s \geq 2$, by Chebyshev's bound, we have

$$
\begin{aligned}
Pr(Z \leq s) &= Pr(Z - E(Z) < s + 1 - E(Z)) \\
&\leq Pr(|Z - E(Z)| \geq |s + 1 - E(Z)|) \\
&\leq \frac{Var(Z)}{(s + 1 - E(Z))^2} \\
&= \frac{rp(1 - p)}{(s + 1 - rp)^2} \leq \frac{40(s + 1)}{39^2(s + 1)^2} \leq 1/100.
\end{aligned}
\tag{1}
$$

The probability of sampling an edge (u, v) is p^2, given by the probability of sampling both u and v. Thus the expected memory used by the above algorithm is $\tilde{O}(ms^2/r^2)$.

4 Lower Bounds

In this section we present our lower bounds on the space complexity of the CLIQUE-GAP problem. Our main theorem is the following.

Theorem 2. *For any $0 < \delta < 1/2$ there exists a global constant $c > 0$ such that for any $0 < s < r$, $M > 0$, there exists graph families $\mathcal{G}_1$ and $\mathcal{G}_2$ that satisfy the following:*

- *for all graph $G_1 \in \mathcal{G}_1$, $|E(G_1)| = m \geq M$, G_1 has a r-clique;*
- *for all graph $G_2 \in \mathcal{G}_2$; $|E(G_2)| = m \geq M$, G_2 has no $(s + 1)$-clique;*
- *any randomized one-pass streaming algorithm $\mathcal{A}$ that distinguishes whether $G \in \mathcal{G}_1$ or $G \in \mathcal{G}_2$ with probability at least $1 - \delta$ uses at least $cm/(r^2 \log_s^2 r)$ memory bits.*

For $s = O(\log n)$, this matches our $\tilde{O}(ms^2/r^2)$ upper bound up to polylogarithmic factors and solves the open question of obtaining lower bounds for CLIQUE-GAP(r, s) for small values of s (from [16]). Our main technical contribution is a reduction from the multi-party set disjointness problem (DISJ$_k^n$) in communication complexity to the CLIQUE-GAP problem. The reduction employs efficient graph decompositions.

We use the following optimal bound on the communication complexity of DISJ$_k^n$ proved in [9].

Theorem 6 ([9]). *Any randomized one-way communication protocol that solves DISJ$_k^n$ correctly with probability $> 3/4$ requires $\Omega(n/k)$ bits of communication.*

Before we prove Theorem 2 in detail, we will give the construction for CLIQUE-GAP$(4, 2)$. The reduction is from DISJ$_2^{n/4}$ to CLIQUE-GAP$(4, 2)$ (for the general case it will be from DISJ$_{\lceil \log_s r \rceil}^{n/r}$ to CLIQUE-GAP(r, s)). For any instance of $\mathrm{DISJ}_2^{n/4}$, where Player 1 holds a set $S_1 \subset [n/4]$ and Player 2 holds a set $S_2 \subset [n/4]$, we construct an instance G with n vertices of CLIQUE-GAP(4,2) as follows. The n vertices are denoted by $\{v_{i,j} | i = 1, 2, 3, \ldots, n/4, j = 0, 1, 2, 3\}$. This notation partitions the vertex set to be $n/4$ groups, each of size 4, denoting as $V_i \equiv \{v_{i,0}, v_{i,1}, v_{i,2}, v_{i,3}\}$ for $i = 1, 2, 3, \ldots, n/4$. We partition $V_i = V_{i,0} \cup V_{i,1}$, where $V_{i,0} = \{v_{i,0}, v_{i,1}\}$ and $V_{i,1} = \{v_{i,2}, v_{i,3}\}$. Further partition $V_{i,0} = V_{i,0,0} \cup V_{i,0,1}$ and $V_{i,1} = V_{i,1,0} \cup V_{i,1,1}$, where $V_{i,0,0} = \{v_{i,0}\}$, $V_{i,0,1} = \{v_{i,1}\}$, $V_{i,1,0} = \{v_{i,2}\}$ and $V_{i,1,1} = \{v_{i,3}\}$.

Player 1 places all edges of the complete bipartite graphs between $V_{i,0}$ and $V_{i,1}$ if $i \in S_1$.

Player 2 places all edges between $V_{i,0,0}$ and $V_{i,0,1}$ and edges between $V_{i,1,0}$, $V_{i,1,1}$ if $i \in S_2$.

The edges and partitions are shown in Fig. 1a.

If $S_1 \cap S_2 = \{i\}$, then there is a clique on vertex set V_i (which is of size 4). If $S_1 \cap S_2 = \emptyset$, since both Player 1 and Player 2 have only bipartite graph edges on disjoint vertex sets, the output graph is triangle free.

If there is a one-pass streaming algorithm A for CLIQUE-GAP(4, 2) that distinguishes whether the input graph G has clique of size 4 or triangle-free, the players can use this algorithm to solve $\mathrm{DISJ}_2^{n/4}$ as follows: Player 1 runs A on his edge set and communicates the content of the working memory at the end of his computation to Player 2. Player 2 continues to run the algorithm on his edge set and outputs the result of the algorithm as the answer of the DISJ problem. Hence if A uses space M, then total communication between players $\leq M$ (in general if there are k players we have the inequality: total communication $\leq (k-1)M$). This leads to the required lower bound.

The edge decomposition for the reduction from $\mathrm{DISJ}_3^{n/8}$ to CLIQUE-GAP(8, 2) is shown in Fig. 1b.

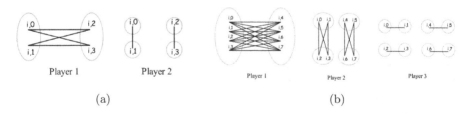

Player 1   Player 2   Player 1   Player 2   Player 3

(a)   (b)

Fig. 1. (a) The decomposition of K_4 to $\log_2 4 = 2$ bipartite graphs. (b) The decomposition of K_8 to $\log_2 8 = 3$ bipartite graphs.

For obtaining a lower bound on the space complexity of CLIQUE-GAP(r, s), we will reduce $\mathrm{DISJ}_{\lceil \log_s r \rceil}^{n/r}$ to CLIQUE-GAP(r, s) and use the lower bound stated in Theorem 6. For the reduction, we give an extension of the bipartite graph decomposition result. In particular, we show (implicitly) that $\alpha_*(K_r, \mathcal{H}) \leq \lceil \log_s r \rceil$ where $\mathcal{H}$ is the class of all s-partite graphs.

*Proof (**of Theorem** 2)*. We will reduce $\text{DISJ}_t^{n/r}$ to CLIQUE-GAP(r,s) where $t = \lceil \log r / \log s \rceil$. Consider an instance of $\text{DISJ}_t^{n/r}$, where Player l holds a set $S_l \subset [n/r]$ for $l = 1, 2, \ldots, t$. To construct an instance G on n vertices of CLIQUE-GAP(r,s), for $l = 1, \ldots, t$, Player l places an edge set E_l as described below.

The Construction of E_l: The construction follows the same pattern as in the figures above. To explain it precisely we need to structure the vertex set of the graph in certain way. W.l.o.g set $r = s^t$ and $n = 0 \mod r$. We will denote an integer in $[r]$ by its s-ary representation using a t-tuple. We denote the n vertices by $V = \{v_{i,[j_1,j_2,\ldots,j_t]} | i = 1, 2, 3, \ldots, n/r, \text{ for all } j_1, j_2, \ldots, j_t \in [s]\}$ ($[j_1, j_2, \ldots, j_t]$ represents an integer in $[r]$ uniquely). This notation partitions the set V into n/r subsets, each of size r. We denote them as $V_1, V_2, \ldots, V_{n/r}$. That is, for each fixed $i = 1, 2, \ldots, n/r$, $V_i = \{v_{i,[j_1,j_2,\ldots,j_t]} | \text{for all } j_1, j_2, \ldots, j_t \in [s]\}$. Next we define a series of s partitions of each V_i where l^{th} partition is a refinement of the $(l-1)^{th}$ partition.

Partition 1: $V_i = V_{i,0} \cup V_{i,1} \ldots \cup V_{i,s-1}$, where for each fixed $j_1 \in [s]$

$$V_{i,j_1} \equiv \{v_{i,[j_1,j_2,j_3,\ldots,j_t]} | \text{ for all } j_2, j_3, \ldots, j_t \in [s]\}. \tag{2}$$

Partition l: For each set $V_{i,j_1,j_2,\ldots,j_{l-1}}$ in Partition $(l-1)$, partition $V_{i,j_1,j_2,\ldots,j_{l-1}} = V_{i,j_1,j_2,\ldots,j_{l-1},0} \cup V_{i,j_1,j_2,\ldots,j_{l-1},1} \ldots \cup V_{i,j_1,j_2,\ldots,j_{l-1},s-1}$ as s subsets, each of which is of size s^{t-l}. Here, for each fixed $i = 1, 2, \ldots, n/r$ and for each fixed $j_1, j_2, \ldots, j_l \in [s]$, we have

$$V_{i,j_1,j_2,\ldots,j_l} \equiv \{v_{i,[j_1,j_2,j_3,\ldots,j_l,j_{l+1},\ldots,j_t]} | \text{ for all } j_{l+1}, j_{l+2}, \ldots, j_t \in [s]\}. \tag{3}$$

With this structuring of vertices, we can now define E_l for each Player l. If an element i is in the set S_l, then for all $j_1, j_2, \ldots, j_{l-1} \in [s]$, Player l has all the s-partite graph edges between the s partitions of the vertex set $V_{i,j_1,j_2,\ldots,j_{l-1}}$, namely, $V_{i,j_1,j_2,\ldots,j_{l-1},0}$, $V_{i,j_1,j_2,\ldots,j_{l-1},1}$, $V_{i,j_1,j_2,\ldots,j_{l-1},3}, \ldots$ and $V_{i,j_1,j_2,\ldots,j_{l-1},s-1}$. Formally, $E_l = \cup_{i \in S_l} \cup_{j_1,j_2,\ldots,j_{l-1} \in [s]} E(i, j_1, j_2, \ldots, j_{l-1})$, where

$$E(i, j_1, j_2, \ldots, j_{l-1})$$
$$\equiv \cup_{j_l,j_l' \in [s], j_l \neq j_l'} \{(a,b)| \text{ for all } a \in V_{i,j_1,j_2,\ldots,j_{l-1},j_l}, b \in V_{i,j_1,j_2,\ldots,j_{l-1},j_l'}\}. \tag{4}$$

Note that each edge appears only in one of the edge set. **End of Construction of E_l.**

Correctness of the Reduction: On a negative instance, players' input sets $S_1, S_2 \ldots S_t$ are pairwise disjoint. The above construction builds all the s-partite graphs on disjoint sets of vertices, hence the output graph is s-partite and hence $(s+1)$-clique free.

On a positive instance, players' input sets have a unique intersection, $S_1 \cap S_2 \ldots \cap S_t = \{i\}$. For each Player l, the edge set E_l includes all the s-partite graph edges on each vertex set $V_{i,j_1,j_2,\ldots,j_{l-1}}$, i.e. $\cup_{j_1,j_2,\ldots,j_{l-1} \in [s]} E(i, j_1, j_2, \ldots, j_{l-1})$. We claim that there is a r-clique on vertex set V_i. Consider any two distinct vertices

$u, v \in V_i$, where $u = v_{i,[j_1,j_2,\ldots,j_t]}$, $v = v_{i,[j'_1,j'_2,\ldots,j'_t]}$. Since $u \neq v$, $(j_1, j_2, \ldots, j_t) \neq (j'_1, j'_2, \ldots, j'_t)$. Let q be first integer such that $j_q \neq j'_q$. By the definition of the partitions, $u \in V_{i,j_1,j_2,\ldots,j_{q-1},j_q}$ and $v \in V_{j,j_1,j_2,\ldots,j_{q-1},j'_q}$. Therefore, there is an edge (u, v) in the edge set output by Player q.

Proof of the Bound: Suppose there is a one-pass streaming algorithm $\mathcal{A}$ that solves CLIQUE-GAP(r, s) in $M(n, r, s)$ space. Then consider the following one-way protocol for DISJ$_t^{n/r}$. For each $1 \leq l < t$, Player l simulates $\mathcal{A}$ on his edge set E_l and communicates the memory content to Player $(l+1)$. Finally Player t simulates $\mathcal{A}$ on E_t and outputs the result of $\mathcal{A}$. The total communication $\leq (t-1)M(n, r, s)$. Hence from the known lower bound on DISJ$_t^{n/r}$, we have that $M(n, r, s) = \Omega(n/rt^2) = \Omega(n/r \log_s^2 r)$. Now consider the hard instance of DISJ$_t^{n/r}$, any player holds a non-empty set (otherwise this is an easy instance). From the construction, for each hard instance we know $m = \Omega(r^2 \times n/r) = \Omega(nr)$. Hence any one-pass streaming algorithm that solves CLIQUE-GAP(r, s) requires $\Omega(m/r^2 \log_s^2 r)$ space. We further justify this argument by the following modification of the reduction.

Constructing Graphs with m Edges: Suppose we are given m, r, s. For an instance of DISJ$_t^{m/2r}$ we can construct a graph on m/r vertices with m edges as follows. Without loss of generality, assume $r = o(\sqrt{m})$, otherwise the bound is trivially $\Omega(1)$. Divide the set of vertices into two groups each with $m/2r$ nodes. For the first $m/2r$ nodes, construct the graph as discribed above with $m' \leq m/2$ edges. For the second group of $m/2r$ nodes, the last player outputs a graph with $(m - m')$ edges incident on this group that does not have an s-clique. This can be done since by Turán's theorem, an $(m/2r)$-vertices graph can have up to $(1 - 1/s)m^2/8r^2 = \omega(m)$ edges without creating an s-clique. The analysis of the lower bound is the same as the previous analysis. By picking up graphs constructed for the hard instances for DISJ problem, we construct the graph class as required by the theorem.

A Lower Bound for The General GAP Problem

Using the terminology from graph decomposition theory we prove a general lower bound theorem for the promise problem GAP$(\mathcal{P}, \mathcal{Q})$ which is defined as follows.

Definition 2. *Let $\mathcal{P}$ and $\mathcal{Q}$ be two graph properties (equivalently, $\mathcal{P}$ and $\mathcal{Q}$ are two sets of graphs) such that $\mathcal{P} \cap \mathcal{Q} = \emptyset$. Given an input graph G, an algorithm for GAP$(\mathcal{P}, \mathcal{Q})$ should output "1" if $G \in \mathcal{P}$ and '0' if $G \in \mathcal{Q}$. For $G \notin \mathcal{P} \cup \mathcal{Q}$, the algorithm can output "1" or "0".*

We first recall the necessary definitions. Let $\mathcal{H}$ be a specified class of graphs. An $\mathcal{H}$-decomposition[3] of a graph G is the decomposition of G into subgraphs $G_1, G_2, \ldots, G_t$ such that any edge in G is an edge of exactly one of the G_i's and

---

[3] Note that some papers define the decomposition on connected graph. We here use a more general statement.

all G_is belong to $\mathcal{H}$. Define $\alpha_*(G, \mathcal{H})$ as:

$$\alpha_*(G, \mathcal{H}) \equiv \min_D |D| \tag{5}$$

where $D = \{G_1, G_2, \ldots, G_t\}$ is an $\mathcal{H}$-decomposition of G. For convenience, we define $\alpha_*(G, \mathcal{H}) = \infty$ if the $\mathcal{H}$-decomposition of G is not defined.

Theorem 3. *Let $\mathcal{P}, \mathcal{Q}$ be two graph properties such that*

- $\mathcal{P} \cap \mathcal{Q} = \emptyset$;
- *If $G'' \in \mathcal{P}$ and G'' is a subgraph of G', then $G' \in \mathcal{P}$;*
- *If $G', G'' \in \mathcal{Q}$ and $V(G') \cap V(G'') = \emptyset$, then $\tilde{G} = (V(G') \cup V(G''), E(G') \cup E(G'')) \in \mathcal{Q}$;*

Let G_0 be an arbitrary graph in $\mathcal{P}$. Given any graph G with m edges and n vertices, if a one-pass streaming algorithm $\mathcal{A}$ solves $\mathrm{GAP}(\mathcal{P}, \mathcal{Q})$ correctly with probability at least $3/4$, then $\mathcal{A}$ requires $\Omega(\frac{n}{|V(G_0)|} \frac{1}{\alpha_*^2(G_0, \mathcal{Q})})$ space in the worst case.

Remark 1. We note that in the above statement G_0 is an arbitrary graph. To get the optimal bound, we can select a G_0 such that the denominator $|V_0|\alpha_*^2(G, \mathcal{Q})$ of the bound is minimized. We also note that this theorem is indeed a generalization of Theorem 2. Let $\mathcal{P} = \{G \mid G$ has a r-clique $\}$ and $\mathcal{Q} = \{G \mid G$ has no $(s + 1)$-clique $\}$. In the proof of Theorem 2 we use $G_0 = K_r$ and shows that $\alpha_*(K_r, \mathcal{Q}) \leq \log_s r$ (in this case $m = O(nr)$).

References

1. Ahn, K.J., Guha, S.: Graph sparsification in the semi-streaming model. In: Albers, S., Marchetti-Spaccamela, A., Matias, Y., Nikoletseas, S., Thomas, W. (eds.) ICALP 2009, Part II. LNCS, vol. 5556, pp. 328–338. Springer, Heidelberg (2009)
2. Alon, N., Krivelevich, M., Sudakov, B.: Finding a large hidden clique in a random graph. In: Proceedings of the Ninth Annual ACM-SIAM Symposium on Discrete Algorithms, pp. 594–598. ACM/SIAM (1998). http://dl.acm.org/citation.cfm?id=314613.315014
3. Alon, N., Matias, Y., Szegedy, M.: The space complexity of approximating the frequency moments. In: Proceedings of the Twenty-eighth Annual ACM Symposium on Theory of Computing. pp. 20–29. ACM (1996)
4. Bar-Yossef, Z., Kumar, R., Sivakumar, D.: Reductions in streaming algorithms, with an application to counting triangles in graphs. In: Proceedings of the Thirteenth Annual ACM-SIAM Symposium on Discrete Algorithms, pp. 623–632. SODA, Society for Industrial and Applied Mathematics (2002)
5. Buriol, L.S., Frahling, G., Leonardi, S., Marchetti-Spaccamela, A., Sohler, C.: Computing clustering coefficients in data streams. In: European Conference on Complex Systems (ECCS) (2006)
6. Buriol, L.S., Frahling, G., Leonardi, S., Marchetti-Spaccamela, A., Sohler, C.: Counting triangles in data streams. In: Proceedings of the Twenty-fifth ACM SIGMOD-SIGACT-SIGART Symposium on Principles of Database Systems (PODS), pp. 253–262. ACM (2006)

7. Buriol, L.S., Frahling, G., Leonardi, S., Sohler, C.: Estimating clustering indexes in data streams. In: Arge, L., Hoffmann, M., Welzl, E. (eds.) ESA 2007. LNCS, vol. 4698, pp. 618–632. Springer, Heidelberg (2007)

8. Buriol, L.S., Frahling, G., Leonardi, S., Spaccamela, A.M., Sohler, C.: Counting graph minors in data streams. Technical report, DELIS - Dynamically Evolving, Large-Scale Information Systems (2005)

9. Chakrabarti, A., Khot, S., Sun, X.: Near-optimal lower bounds on the multi-party communication complexity of set disjointness. In: IEEE Conference on Computational Complexity, pp. 107–117. IEEE Computer Society (2003)

10. Chung, F.: On the decomposition of graphs. SIAM J. Algebraic Discrete Methods **2**(1), 1–12 (1981)

11. Chung, F., Erdős, P., Spencer, J.: On the decomposition of graphs into complete bipartite subgraphs. In: Erdős, P., Alpár, L., Halász, H., SárkÖz, A. (eds.) Studies in Pure Mathematics, pp. 95–101. Birkhäuser, Basel (1983)

12. Cormode, G., Jowhari, H.: A second look at counting triangles in graph streams. Theoret. Comput. Sci. **552**, 44–51 (2014)

13. Demetrescu, C., Finocchi, I., Ribichini, A.: Trading off space for passes in graph streaming problems. In: Proceedings of ACM-SIAM Symposium on Discrete Algorithms. pp. 714–723. ACM (2006)

14. Donoho, D.L.: Compressed sensing. IEEE Trans. Inf. Theory **52**, 1289–1306 (2006)

15. Goel, A., Kapralov, M., Khanna, S.: On the communication and streaming complexity of maximum bipartite matching. In: Proceedings of the Twenty-third Annual ACM-SIAM Symposium on Discrete Algorithms (SODA), pp. 468–485. SIAM (2012)

16. Halldórsson, M.M., Sun, X., Szegedy, M., Wang, C.: Streaming and communication complexity of clique approximation. In: Czumaj, A., Mehlhorn, K., Pitts, A., Wattenhofer, R. (eds.) ICALP 2012, Part I. LNCS, vol. 7391, pp. 449–460. Springer, Heidelberg (2012)

17. Jha, M., Seshadhri, C., Pinar, A.: When a graph is not so simple: Counting triangles in multigraph streams. CoRR abs/1310.7665 (2013). http://arxiv.org/abs/1310.7665

18. Jowhari, H., Ghodsi, M.: New Streaming algorithms for counting triangles in graphs. In: Wang, L. (ed.) COCOON 2005. LNCS, vol. 3595, pp. 710–716. Springer, Heidelberg (2005)

19. Kapralov, M., Khanna, S., Sudan, M.: Streaming lower bounds for approximating MAX-CUT. CoRR abs/1409.2138 (2014). http://arxiv.org/abs/1409.2138

20. Kutzkov, K., Pagh, R.: On the streaming complexity of computing local clustering coefficients. In: Proceedings of the Sixth ACM International Conference on Web Search and Data Mining (WSDM), pp. 677–686. ACM (2013)

21. Manjunath, M., Mehlhorn, K., Panagiotou, K., Sun, H.: Approximate counting of cycles in streams. In: Demetrescu, C., Halldórsson, M.M. (eds.) ESA 2011. LNCS, vol. 6942, pp. 677–688. Springer, Heidelberg (2011)

22. McGregor, A.: Graph stream algorithms: a survey. SIGMOD Rec. **43**(1), 9–20 (2014). http://doi.acm.org/10.1145/2627692.2627694

23. Pavan, A., Tangwongsan, K., Tirthapura, S., Wu, K.L.: Counting and sampling triangles from a graph stream. Proc. VLDB Endowment **6**(14), 1870–1881 (2013)

24. Ugander, J., Karrer, B., Backstrom, L., Marlow, C.: The anatomy of the facebook social graph. CoRR abs/1111.4503 (2011). http://arxiv.org/abs/1111.4503

QMA with Subset State Witnesses

Alex Bredariol Grilo[1]([⊠]), Iordanis Kerenidis[1,2], and Jamie Sikora[2]

[1] LIAFA, CNRS, Université Paris Diderot, Paris, France
abgrilo@gmail.com
[2] Centre for Quantum Technologies, National University of Singapore,
Singapore City, Singapore

Abstract. The class QMA plays a fundamental role in quantum complexity theory and it has found surprising connections to condensed matter physics and in particular in the study of the minimum energy of quantum systems. In this paper, we further investigate the class QMA and its related class QCMA by asking what makes quantum witnesses potentially more powerful than classical ones. We provide a definition of a new class, SQMA, where we restrict the possible quantum witnesses to the "simpler" subset states, i.e. a uniform superposition over the elements of a subset of n-bit strings. Surprisingly, we prove that this class is equal to QMA, hence providing a new characterisation of the class QMA. We also describe a new complete problem for QMA and a stronger lower bound for the class QMA_1.

1 Introduction

One of the notions at the heart of classical complexity theory is the class NP and the fact that deciding whether a boolean formula is satisfiable or not is NP-complete [8,18]. The importance of NP-completeness became evident through the plethora of combinatorial problems that can be cast as constraint satisfaction problems and shown to be NP-complete. Moreover, the famous PCP theorem [4,5] provided a new, surprising description of the class NP: any language in NP can be verified efficiently by accessing probabilistically a constant number of bits of a polynomial-size witness. This opened the way to showing that in many cases, approximating the solution of NP-hard problems remains as hard as solving them exactly. An equivalent definition of the PCP theorem states that it remains NP-hard to decide whether an instance of a constraint satisfaction problem is satisfiable or any assignment violates at least a constant fraction of the constraints.

Not surprisingly, the quantum analog of the class NP, defined by Kitaev [16] and called QMA, has been the subject of extensive study in the last decade. Many important properties of this class are known, including a strong amplification property and an upper bound of PP [19], as well as numerous complete problems related to the ground state energy of different types of Hamiltonians [9,12,15, 16,20]. Nevertheless, there are still many open questions about the class QMA, including whether it admits perfect completeness or not.

© Springer-Verlag Berlin Heidelberg 2015
G.F. Italiano et al. (Eds.): MFCS 2015, Part II, LNCS 9235, pp. 163–174, 2015.
DOI: 10.1007/978-3-662-48054-0_14

Moreover, it is still wide-open if a quantum PCP theorem exists. One way to phrase the quantum PCP theorem is that any problem in QMA can be verified efficiently by a quantum verifier accessing a constant number of qubits of a polynomial-size quantum witness. Another way would be that the problem of approximating the ground state energy of a local Hamiltonian within a constant is still QMA-hard. There have been a series of results, mostly negative, towards the goal of proving or disproving the quantum PCP theorem, but there is still no conclusive evidence [2,3,7].

Another important open question about the class QMA is whether the witness really need be a quantum state or it is enough for the polynomial time quantum Verifier to receive a classical witness. In other words, whether the class QMA is equal to the class QCMA, which is the class of problems that are decidable by a polynomial-time quantum verifier who receives a polynomial-size classical witness. Needless to say, resolving this question can also have implications to the quantum PCP theorem, since in case the two classes are the same, the quantum witness can be replaced by a classical one, which may be more easily checked locally. In addition, we know that perfect completeness is achievable for the class QCMA [14].

In this paper, we investigate QMA by asking the following simple, yet fundamental question: what makes a quantum witness potentially more powerful than a classical one? Is it the fact that to describe a quantum state one needs to specify an exponential number of possibly different amplitudes? Is it the different relative phases in the quantum state? Or is it something else altogether?

QMA with Subset State Witnesses. We provide a definition of a new class, where we restrict the quantum witnesses to be as "classical" as possible, without having by definition an efficient classical description (otherwise our class would be trivially equal to QCMA). All definitions and statements of the results are made formal in their respective sections.

For any subset $S \subseteq [d]$, we define the subset state $|S\rangle \in \mathbb{C}^d$, as the uniform superposition over the elements of S, namely $|S\rangle := \frac{1}{\sqrt{|S|}} \sum_{i \in S} |i\rangle$.

The Class SQMA. A promise problem $A = (A_{\text{yes}}, A_{\text{no}})$ is in SQMA if for every $x \in A_{\text{yes}} \cup A_{\text{no}}$, there exists a polynomial time quantum verifier V_x, such that

- *(completeness)* for all $x \in A_{\text{yes}}$, there exists a subset state witness $|S\rangle$, such that the verifier accepts with probability at least $2/3$.
- *(soundness)* for all $x \in A_{\text{no}}$ and all quantum witnesses $|\psi\rangle$, the verifier accepts with probability at most $1/3$.

The only difference from QMA is that in the yes-instances, we ask that there exists a *subset state witness* that makes the quantum verifier accept with high probability. In other words, an honest prover need only provide such subset states, which in principle are conceptually simpler.

Notice, nevertheless, that the Group Non-Membership Problem is in SQMA, since the witness in the known QMA protocol is a subset state [21]. Moreover, we can define a version of our class with two non-entangled provers, similarly to QMA(2), and we can again see that the protocol of Blier and Tapp [6] which shows that any language in NP has a QMA(2) proof system with logarithmic-size quantum messages uses such subset states. Hence, even though the witnesses we consider are quite restricted, some of the most interesting containments still hold for our class.

More surprisingly, we show that SQMA is, in fact, equal to QMA.

Result 1. SQMA = QMA and SQMA(2) = QMA(2).

Hence, for any problem in QMA, the quantum witness can be a subset state. This provides a new way of looking at QMA and shows that if quantum witnesses are more powerful than classical ones, then this relies solely on the fact that a quantum witness can, in some sense, convey information about an arbitrary subset of classical strings through a uniform superposition of its elements. On the other hand, one way to prove that classical witnesses are as powerful as quantum witnesses, is to find a way to replace such subset states with a classical witness, possibly by enforcing more structure on the accepting subset states.

Our proof relies on a geometric lemma, which shows, for instance, that for any unit vector in $\mathbb{C}^{2^n}$, there exists a subset state, such that their inner product is $\Omega(1/\sqrt{n})$. This lemma, in conjunction with standard amplification techniques for QMA imply our main result.

Complete Problems. The canonical QMA-complete problem is the following: Given a Hamiltonian acting on an n-qubit system, which is a sum of "local" Hamiltonians each acting on a constant number of qubits, decide whether the ground state energy is at most a or all states have energy at least b, where $b - a \geq 1/poly(n)$. The first question is whether we can show that the same problem is complete if we look at the energy of any subset state instead of the ground state. In fact, we do not know how to show that this problem is complete: when we try to follow Kitaev's proof of completeness and approximate his *history state* with a subset state, we cannot retain a sufficient energy gap. Moreover, there exist Hamiltonians with a low energy ground state, but the energy of all subset states is close to 1.

In this work, we provide one new complete problem for QMA related to subset states. This problem is based on the QCMA-complete problem Identity Check on Basis States [23]:

Result 2. The Basis State Check on Subset States problem is QMA-complete:

- Input: Let x be a classical description of a quantum circuit Z_x on m input qubits and a ancilla qubits and y be an m'-bit string, where $n := |x|$ and $m' \leq m + a$. Given the promise that x satisfies one of the following cases for some polynomial[1] q, decide which is true:
- Yes: there is a subset S such that $\|(\langle y| \otimes I)Z_x |S\rangle |0\rangle\|_2^2 \geq 1 - 1/q(n)$,
- No: for all subsets S, we have $\|(\langle y| \otimes I)Z_x |S\rangle |0\rangle\|_2^2 \leq 1/q(n)$.

---

[1] This polynomial needs to have degree at least that of m (see Theorem 3 for a formal statement).

Perfect Completeness. Another important open question about QMA is if it is closed under perfect completeness. Using our characterisation, if SQMA is closed under perfect completeness, then so is QMA. On one hand, the result of [1] can be used to show that there exists a quantum oracle A relative to which these two classes are not equal, i.e., $\text{SQMA}^A \neq \text{SQMA}_1^A$. On the other hand, proving perfect completeness for SQMA may be an easier problem to solve, since unlike general QMA witnesses, the amplitudes involved in subset states may be much easier to handle. Even though we are unable to prove perfect completeness for SQMA, we prove perfect completeness for the following closely related class.

The Class oSQMA. A promise problem $A = (A_{\text{yes}}, A_{\text{no}})$ is in oSQMA if for all $x \in A_{\text{yes}} \cup A_{\text{no}}$, there exists a polynomial time quantum verifier V_x, such that

- *(completeness)* for all $x \in A_{\text{yes}}$, there exists a subset state witness $|S\rangle$ that maximizes the probability the verifier accepts and this probability is at least $2/3$.
- *(soundness)* for all $x \in A_{\text{no}}$ and all quantum witnesses $|\psi\rangle$, the verifier accepts with probability at most $1/3$.

It can be shown that this class still contains the Group Non-Membership problem, while its two-prover version has short proofs for NP. It remains open to understand whether demanding that a subset state is the optimal witness, instead of just an accepting one, reduces the computational power of the class. Moreover, these two classes coincide in the case of perfect completeness, since all accepting witnesses are also optimal. We prove that the class oSQMA admits perfect completeness, which implies a stronger lower bound for the class QMA_1 than the previously known QCMA bound.

Result 3. $\text{SQMA}_1 = \text{oSQMA}_1 = \text{oSQMA}$, hence $\text{QCMA} \subseteq \text{oSQMA} \subseteq \text{QMA}_1$.

The fact that for the class oSQMA there exists a subset state which is an optimal witness implies that the maximum acceptance probability is rational (using a specific gate set) and moreover, it is the maximum eigenvalue of the verifier's operator. These two facts enable us to use the Rewinding Technique used by Kobayashi, Le Gall and Nishimura [17] and prove our result.

2   Preliminaries

2.1   Definitions

For $n \in \mathbb{N}$, we define $[n] := \{1, ..., n\}$. The Hilbert-Schmidt or trace inner product between two operators A and B is defined as $\langle A, B \rangle = \text{Tr}(A^\dagger B)$. For a complex number $x = a + ib$, $a, b \in \mathbb{R}$, we define its norm $|x|$ by $\sqrt{a^2 + b^2}$. For a vector $|v\rangle \in \mathbb{C}^d$, its p-norm is defined as $\| |v\rangle \|_{\text{p}} := \left(\sum_{1 \leq i \leq d} |v_i|^p \right)^{\frac{1}{p}}$. For an operator A, the trace norm is $\|A\|_{\text{tr}} := \text{Tr}\sqrt{A^\dagger A}$, which is the sum of the singular values of A.

We now state two identities which we use in our analysis. For normalized $|v\rangle, |w\rangle \in \mathbb{C}^d$, we have

$$\max_{0 \preceq C \preceq I} |\langle C, |v\rangle \langle v| - |w\rangle \langle w| \rangle| = \frac{1}{2} \||v\rangle \langle v| - |w\rangle \langle w|\|_{\mathrm{tr}}, \qquad (1)$$

since $|v\rangle \langle v| - |w\rangle \langle w|$ has largest eigenvalue $\lambda \geq 0$, smallest eigenvalue $-\lambda$, and the rest are 0, and the trace norm of a Hermitian matrix is the sum of the absolute values of its eigenvalues. We also have that for $|v\rangle, |w\rangle \in \mathbb{C}^d$,

$$\||v\rangle \langle v| - |w\rangle \langle w|\|_{\mathrm{tr}} = 2\sqrt{1 - |\langle v|w\rangle|^2}. \qquad (2)$$

2.2 Complexity Classes and Complete Problems

We start by defining the known quantum complexity classes we will study and a complete problem.

Definition 1 (QMA). *A promise problem $A = (A_{\mathrm{yes}}, A_{\mathrm{no}})$ is in QMA if and only if there exist polynomials p, q and a polynomial-time uniform family of quantum circuits $\{Q_n\}$, where Q_n takes as input a string $x \in \Sigma^*$ with $|x| = n$, a $p(n)$-qubit quantum state, and $q(n)$ ancilla qubits in state $|0\rangle^{\otimes q(n)}$, such that:*

- *(completeness) If $x \in A_{\mathrm{yes}}$, then there exists a $p(n)$-qubit quantum state $|\psi\rangle$ such that Q_n accepts $(x, |\psi\rangle)$ with probability at least $2/3$.*
- *(soundness) If $x \in A_{\mathrm{no}}$, then for any $p(n)$-qubit quantum state $|\psi\rangle$, Q_n accepts $(x, |\psi\rangle)$ with probability at most $1/3$.*

We can restrict QMA in order to always accept yes-instances, a property called *perfect completeness*.

Definition 2 (QMA$_1$). *A promise problem $A = (A_{\mathrm{yes}}, A_{\mathrm{no}})$ is in QMA$_1$ if and only if there exist polynomials p, q and a polynomial-time uniform family of quantum circuits $\{Q_n\}$, where Q_n takes as input a string $x \in \Sigma^*$ with $|x| = n$, a $p(n)$-qubit quantum state, and $q(n)$ ancilla qubits in state $|0\rangle^{\otimes q(n)}$, such that:*

- *(completeness) If $x \in A_{\mathrm{yes}}$, then there exists a $p(n)$-qubit quantum state $|\psi\rangle$ such that Q_n accepts $(x, |\psi\rangle)$ with probability exactly 1.*
- *(soundness) If $x \in A_{\mathrm{no}}$, then for any $p(n)$-qubit quantum state $|\psi\rangle$, Q_n accepts $(x, |\psi\rangle)$ with probability at most $1/3$.*

Another way we can restrict QMA is only allowing classical witnesses, resulting in the definition of the class QCMA (sometimes it is also referred to as MQA [10,22]).

Definition 3 (QCMA). *A promise problem $A = (A_{\mathrm{yes}}, A_{\mathrm{no}})$ is in QCMA if and only if there exist polynomials p, q and a polynomial-time uniform family of quantum circuits $\{Q_n\}$, where Q_n takes as input a string $x \in \Sigma^*$ with $|x| = n$, a $p(n)$-bit string, and $q(n)$ ancilla qubits in state $|0\rangle^{\otimes q(n)}$, such that:*

- *(completeness)* If $x \in A_{\text{yes}}$, then there exists a $p(n)$-bit string y such that Q_n accepts (x, y) with probability at least $2/3$.
- *(soundness)* If $x \in A_{\text{no}}$, then for any $p(n)$-bit string y, Q_n accepts (x, y) with probability at most $1/3$.

We state here one QCMA-complete problem, the *Identity Check on Basis States* problem [23].

Definition 4 (Identity Check on Basis States [23]). *Let x be a classical description of a quantum circuit Z_x on m qubits. Given the promise that Z_x satisfies one of the following cases for $\mu - \delta \geq 1/poly(|x|)$, decide which one is true:*

- *there is a binary string z such that $|\langle z| Z_x |z\rangle|^2 \leq 1 - \mu$, i.e., Z_x does not act as the identity on the basis states,*
- *for all binary strings z, $|\langle z| Z_x |z\rangle|^2 \geq 1 - \delta$, i.e., Z_x acts "almost" as the identity on the basis states.*

3   Subset State Approximations

In this section, we prove the Subset State Approximation Lemma which intuitively says that any quantum state can be well-approximated by a subset state, defined below.

Definition 5. *For a subset $S \subseteq [d]$, a subset state, denoted here as $|S\rangle \in \mathbb{C}^d$, is a uniform superposition over the elements of S. More specifically, it has the form*

$$|S\rangle := \frac{1}{\sqrt{|S|}} \sum_{i \in S} |i\rangle .$$

We now state and prove a useful technical lemma.

Lemma 1 (Geometric Lemma). *For a vector $v \in \mathbb{C}^d$, there exists a subset $S \subseteq [d]$ such that*

$$\frac{1}{\sqrt{|S|}} \left| \sum_{j \in S} v_j \right| \geq \frac{\|v\|_2}{8\sqrt{\log_2(d) + 3}}.$$

Proof. If $v = 0$, the lemma statement is trivially true. Suppose $v \in \mathbb{C}^d$ is nonzero and decompose v into real and imaginary components as $v = u + iw$, where $u, w \in \mathbb{R}^d$. Note that $\|v\|_2 \leq \|u\|_2 + \|w\|_2$, by the triangle inequality, implying at least one has norm at least $\|v\|_2 / 2$. Let us say it is u (the argument for w proceeds analogously). We now partition u into positive and negative entries such that $u = x - y$ where $x, y \geq 0$ and are orthogonal. By the same argument as above, we know at least one has norm at least $\|v\|_2 / 4$. Without loss of generality, suppose it is x.

Let T denote the support of x, i.e., $j \in T$ if and only if $x_j > 0$. The idea is to partition T into a small number of sets, where the entries x_j that belong to each set are roughly the same size, and the sum of the entries corresponding to one set is a large enough fraction of the norm of the entire vector.

More precisely, for $\gamma := \left\lceil \frac{\log_2(d)+1}{2} \right\rceil$ and $k \in [\gamma]$, let us partition T into the following sets

$$T_k := \left\{ j \in T : \frac{\|x\|_2}{2^k} < x_j \leq \frac{\|x\|_2}{2^{k-1}} \right\} \text{ and } T_{\gamma+1} := \left\{ j \in T : 0 < x_j \leq \frac{\|x\|_2}{2^\gamma} \right\}.$$

It follows that

$$\sum_{j \in \cup_{k \in [\gamma]} T_k} (x_j)^2 = \|x\|_2^2 - \sum_{j \in T_{\gamma+1}} (x_j)^2 \geq \|x\|_2^2 - d\frac{\|x\|_2^2}{2^{2\gamma}} = \|x\|_2^2 \left(1 - \frac{d}{2^{2\gamma}}\right).$$

This implies that for some $k' \in [\gamma]$,

$$\sum_{j \in T_{k'}} (x_j)^2 \geq \frac{\|x\|_2^2}{\gamma} \left(1 - \frac{d}{2^{2\gamma}}\right) \text{ and } |T_{k'}| \frac{\|x\|_2^2}{2^{2(k'-1)}} \geq \sum_{j \in T_{k'}} (x_j)^2 \geq \frac{\|x\|_2^2}{\gamma} \left(1 - \frac{d}{2^{2\gamma}}\right).$$

This implies the following lower bound for the size of $T_{k'}$

$$|T_{k'}| \geq \frac{2^{2(k'-1)}}{\gamma} \left(1 - \frac{d}{2^{2\gamma}}\right). \tag{3}$$

Using the definition of $T_{k'}$ and Eq. (3), we have

$$\frac{1}{\sqrt{|T_{k'}|}} \sum_{j \in T_{k'}} x_j \geq \frac{\sqrt{|T_{k'}|}\|x\|_2}{2^{k'}} \geq \frac{\|x\|_2 2^{(k'-1)}}{2^{k'}\sqrt{\gamma}} \sqrt{1 - \frac{d}{2^{2\gamma}}} \geq \frac{\|v\|_2}{8\sqrt{\log_2(d)+3}}.$$

Let $S := T_{k'}$ and s be the vector where $s_j = \frac{1}{\sqrt{|S|}}$ if $j \in S$ and 0 otherwise. We have

$$\frac{1}{\sqrt{|S|}} \left| \sum_{j \in S} v_j \right| = |\langle s, v \rangle| = |\langle s, u \rangle + i\langle s, w \rangle| \geq |\langle s, u \rangle| = \left| \frac{1}{\sqrt{|S|}} \sum_{j \in S} x_j \right|,$$

which gives us the bound $\frac{1}{\sqrt{|S|}} \left| \sum_{j \in S} v_j \right| \geq \frac{\|v\|_2}{8\sqrt{\log_2(d)+3}}$ as desired.

The technique used above of splitting the amplitudes into sets has also been used in some form in a proof in [13] which showed a result for approximating bipartite states by a uniform superposition of their Schmidt basis vectors. Note that our result holds for any state and, since we are concerned with a particular fixed basis, we need to deal with arbitrary complex amplitudes.

Lemma 2 (Subset State Approximation Lemma). *For any n-qubit state* $|\psi\rangle$, *there is a subset* $S \subseteq [2^n]$ *such that* $|\langle S|\psi\rangle| \geq \dfrac{1}{8\sqrt{n+3}}$.

Remark 1. We can further restrict the size of the subsets to always be a power of 2 and lose at most a constant factor in the approximation (equal to $\frac{1}{2}$).

We state now a lemma showing that this approximation factor is optimal up to multiplicative constants. We defer the proof of this lemma to the full version of the paper [11].

Lemma 3. *For any n, define the n-qubit state* $|\psi_n\rangle := \sum_{1 \leq i \leq 2^n - 1} \dfrac{1}{\sqrt{n}\sqrt{2^{\lfloor \log i \rfloor}}} |i\rangle$.

Then we have that $\langle \psi_n | S \rangle \leq \frac{2+\sqrt{2}}{\sqrt{n}}$, *for all* $S \subseteq [2^n]$.

4   Alternative Characterisations of QMA

In this section, we prove that QMA can be characterized as accepting subset states. We start by defining formally the new complexity class

Definition 6 (SQMA). *A promise problem $A = (A_{\text{yes}}, A_{\text{no}})$ is in SQMA if and only if there exist polynomials p, q and a polynomial-time uniform family of quantum circuits $\{Q_n\}$, where Q_n takes as input a string $x \in \Sigma^*$ with $|x| = n$, a $p(n)$-qubit quantum state, and $q(n)$ ancilla qubits in state $|0\rangle^{\otimes q(n)}$, such that:*

- *(completeness) If $x \in A_{\text{yes}}$, then there exists a subset $S \subseteq [2^{q(n)}]$ such that Q_n accepts $(x, |S\rangle)$ with probability at least $2/3$.*
- *(soundness) If $x \in A_{\text{no}}$, then for any $p(n)$-qubit quantum state $|\psi\rangle$, Q_n accepts $(x, |\psi\rangle)$ with probability at most $1/3$.*

Remark. Note that we restricted the witness only in the completeness criterion. In fact, it is straightforward to adapt any QMA protocol to have a subset state being an optimal witness in the soundness criterion. For example, the prover can send an extra qubit with the original witness and the verifier can measure it in the computational basis. If the outcome is 0, he continues verifying the proof. If it is 1, he flips a coin and accepts with probability, say, $1/3$. It is easy to see that an optimal witness for the soundness probability is the string of all 1's, which is classical, hence a subset state! Therefore, restricting the proofs in the completeness criterion is the more natural and interesting case.

We prove now that this restriction does not change the power of QMA.

Theorem 1. QMA = SQMA.

Proof. By definition, SQMA $\subseteq$ QMA, thus we need to show QMA $\subseteq$ SQMA.

Suppose we have a QMA protocol which verifies a $p(n)$-qubit proof $|\psi\rangle$ with the two-outcome POVM measurement $\{C, I - C\}$. More precisely, without loss of generality, we assume that there exists a polynomial r such that if $x \in A_{\text{yes}}$, there exists a state $|\psi\rangle$ such that $\langle \psi | C | \psi \rangle \geq 1 - 2^{-r(n)}$ and if $x \in A_{\text{no}}$, we have for every $|\psi\rangle$, that $\langle \psi | C | \psi \rangle \leq 2^{-r(n)}$. We show that the same verification above accepts a subset state with probability at least $\Omega(1/p(n))$, if $x \in A_{\text{yes}}$, from which we conclude that the same instance can be decided with a SQMA protocol using standard error reduction techniques.

If $x \in A_{no}$ there is nothing to show (since the soundness condition for QMA and SQMA coincide). Suppose $x \in A_{yes}$ and let $|\psi\rangle$ be a proof which maximizes the acceptance probability. We then use the Subset State Approximation Lemma (Lemma 2) to approximate $|\psi\rangle$ with $|S\rangle$, where $S \subseteq [2^{p(n)}]$, satisfies:

$$|\langle\psi|S\rangle| \geq \frac{1}{8\sqrt{p(n)+3}}. \tag{4}$$

We now show that the acceptance probability of $|S\rangle$ is not too small. Note that

$$\langle S|\,C\,|S\rangle = \langle C, |S\rangle\langle S|\rangle = \langle C, |\psi\rangle\langle\psi|\rangle - \langle C, |\psi\rangle\langle\psi| - |S\rangle\langle S|\rangle \tag{5}$$

and since $\langle C, |\psi\rangle\langle\psi|\rangle \geq 1 - 2^{-r(n)}$, we can concentrate now on bounding the value of $\langle C, |\psi\rangle\langle\psi| - |S\rangle\langle S|\rangle$.

Since $\langle C, |\psi\rangle\langle\psi| - |S\rangle\langle S|\rangle \leq \max_{0 \preceq C \preceq I} |\langle C, |\psi\rangle\langle\psi| - |S\rangle\langle S|\rangle|$, it follows from Eq. (1) that

$$\langle C, |\psi\rangle\langle\psi| - |S\rangle\langle S|\rangle \leq \frac{1}{2} \||\psi\rangle\langle\psi| - |S\rangle\langle S|\|_{tr}. \tag{6}$$

We now have

$$\||\psi\rangle\langle\psi| - |S\rangle\langle S|\|_{tr} = 2\sqrt{1 - |\langle\psi|S\rangle|^2} \leq 2 - |\langle\psi|S\rangle|^2, \tag{7}$$

where the equality follows from Eq. (2) and the inequality from the fact that, for $x \geq 0$, we have $\sqrt{1 - x^2} \leq 1 - x^2/2$. Combining Eqs. (4), (5), (6), and (7), we have

$$\langle S|\,C\,|S\rangle \geq 1 - 2^{-r(n)} - \frac{1}{2}\left(2 - |\langle\psi|S\rangle|^2\right)$$

$$= \frac{1}{2}|\langle\psi|S\rangle|^2 - 2^{-r(n)}$$

$$\geq \frac{1}{128(p(n)+3)} - 2^{-r(n)}.$$

Thus, $|S\rangle$ is accepted with probability $\Omega\left(\frac{1}{p(n)}\right)$, as required. We end the proof by noting that the standard error reduction techniques involve sending multiple copies of a proof which preserves the subset state property.

We also define the class SQMA(2), an analog of QMA(2) where the verifier receives two non-entangled subset state proofs.

Theorem 2. QMA(2) = SQMA(2).

The proof of Theorem 2 can be found in the full version of the paper [11].

5   A QMA-Complete Problem Based on Subset States

In this section, we give a complete problem for QMA based on circuits mapping subset states to a basis state. This is similar to the QCMA-complete problem Identity Check on Basis States (see Definition 4).

Definition 7 (Basis State Check on Subset States (BSCSS(α))). *Let x be a classical description of a quantum circuit Z_x on $m(n)$ input qubits and $a(n)$ ancilla qubits, and y be an $m'(n)$-bit string, such that $n := |x|$, m, a, and m' are bounded by polynomials and $m' \leq m + a$. Given the promise that x satisfies one of the following cases, decide which is true:*

- *there exists a subset $S \subseteq [2^{m(n)}]$ such that $\left\| (\langle y| \otimes I) Z_x |S\rangle |0\rangle^{\otimes a(n)} \right\|_2^2 \geq 1 - \alpha;$*
- *for all subsets $S \subseteq [2^{m(n)}]$, we have $\left\| (\langle y| \otimes I) Z_x |S\rangle |0\rangle^{\otimes a(n)} \right\|_2^2 \leq \alpha.$*

Theorem 3. *For any polynomial r, the problem BSCSS is QMA-complete for*

$$2^{-r(n)} \leq \alpha \leq \frac{1}{257(m(n) + 3)}.$$

At first glance, this problem looks very similar to a trivial complete problem for QMA. However, BSCSS only considers subset states in both the yes and no-instances, as opposed to arbitrary states for the version for QMA. Moreover, one may ask what happens to the computational power of SQMA if one were to restrict to only rejecting subset states in the soundness condition in the definition. The bounds on α in the theorem above give bounds on the completeness-soundness gap required for this modified definition of SQMA to still be equivalent to QMA. We defer the proof of Theorem 3 to the full version of the paper [11].

6   On the Perfectly Complete Version of SQMA

In this section, we study the perfectly complete version of SQMA, which is called SQMA_1. Even though we do not prove here that SQMA admits perfect completeness (i.e., $\text{SQMA} = \text{SQMA}_1$), we characterise SQMA_1 showing that it is equal to a variant of SQMA where there is an optimal subset state witness.

Definition 8 (oSQMA). *A promise problem $A = (A_{\text{yes}}, A_{\text{no}})$ is in oSQMA if and only if there exist polynomials p, q and a polynomial-time uniform family of quantum circuits $\{Q_n\}$, where Q_n takes as input a string $x \in \Sigma^*$ with $|x| = n$, a $p(n)$-qubit quantum state, and $q(n)$ ancilla qubits in state $|0\rangle^{\otimes q(n)}$, such that:*

- *(completeness) If $x \in A_{\text{yes}}$, then there exists a subset $S \subseteq [2^{p(n)}]$ such that Q_n accepts $(x, |S\rangle)$ with probability at least 2/3 and this subset state maximizes the acceptance probability over all states.*
- *(soundness) If $x \in A_{\text{no}}$, then for any $p(n)$-qubit quantum state $|\psi\rangle$, Q_n accepts $(x, |\psi\rangle)$ with probability at most 1/3.*

We remark that the perfectly complete versions $SQMA_1$ and $oSQMA_1$ coincide, since in both cases there is an optimal subset state witness for yes-instances which leads to acceptance probability 1. They are also contained in QMA_1. We prove perfect completeness for the class oSQMA.

Theorem 4. $SQMA_1 = OSQMA_1 = OSQMA$.

As a corollary, we get a new lower bound for QMA_1 stronger than the previously known lower bound of QCMA. Further details can be found in the full version of the paper [11].

Corollary 1. $QCMA \subseteq oSQMA \subseteq QMA_1$.

7 Conclusions

Our results provide a new way of looking at the class QMA and provide some insight on the power of quantum witnesses. It shows that all quantum witnesses can be replaced by the "simpler" subset states, a fact that may prove helpful both in the case of a quantum PCP and for proving that QMA admits perfect completeness, towards which we have provided some more partial results. Of course, the main question remains open: Are quantum witnesses more powerful than classical ones and if so, why? What we know now, are some things that do not make the quantum witnesses more powerful, for example arbitrary amplitudes or relative phases.

We conclude by stating some open problems. First, can we restrict the quantum witnesses even further? Can we use our characterization of QMA(2) as SQMA(2) in order to show a better upper bound than NEXP for QMA(2)? Also, can we prove perfect completeness for QMA through our new characterisation? Last, can we obtain other complete problems for QMA, possibly related to finding the energy of subset states of local Hamiltonians?

Acknowledgements. We acknowledge support from a Government of Canada NSERC Postdoctoral Fellowship, ANR RDAM (ANR-12-BS02-005), ERC QCC and FP7 QAlgo. Research at CQT at NUS is partially funded by the Singapore Ministry of Education and the National Research Foundation, also through the Tier 3 Grant "Random numbers from quantum processes," (MOE2012-T3-1-009).

References

1. Aaronson, S.: On perfect completeness for QMA. Quantum Inf. Comput. **9**, 81–89 (2009)
2. Aharonov, D., Arad, I., Landau, Z., Vazirani, U.: The detectability lemma and quantum gap amplification. In: Proceedings of the Forty-First Annual ACM Symposium on Theory of Computing, STOC 2009, pp. 417–426. ACM, New York (2009)
3. Aharonov, D., Arad, I., Vidick, T.: Guest column: the quantum PCP conjecture. SIGACT News **44**(2), 47–79 (2013)

4. Arora, S., Lund, C., Motwani, R., Sudan, M., Szegedy, M.: Proof verification and the hardness of approximation problems. J. ACM **45**(3), 501–555 (1998)
5. Arora, S., Safra, S.: Probabilistic checking of proofs: a new characterization of NP. J. ACM **45**(1), 70–122 (1998)
6. Blier, H., Tapp, A.: All languages in NP have very short Quantum proofs. In: Proceedings of the 2009 Third International Conference on Quantum, Nano and Micro Technologies, ICQNM 2009, pp. 34–37. IEEE Computer Society, Washington, DC (2009)
7. Brandão, F.G.S.L., Harrow, A.W.: Product-state approximations to quantum ground states. In: Proceedings of the Forty-Fifth Annual ACM Symposium on Theory of Computing, STOC 2013, pp. 871–880. ACM, New York (2013)
8. Cook, S.A.: The complexity of theorem proving procedures. In: Proceedings of the Third Annual ACM Symposium, pp. 151–158. ACM, New York (1971)
9. Cubitt, T., Montanaro, A.: Complexity classification of local hamiltonian problems. arXiv.org e-Print quant-ph/1311.3161 (2013)
10. Gharibian, S., Sikora, J., Upadhyay, S.: QMA variants with polynomially many provers. Quantum Inf. Comput. **13**(1–2), 135–157 (2013)
11. Grilo, A.B., Kerenidis, I., Sikora, J.: QMA with subset state witnesses. CoRR, abs/1410.2882 (2014)
12. Hallgren, S., Nagaj, D., Narayanaswami, S.: The local hamiltonian problem on a line with eight states is QMA-complete. Quantum Info. Comput. **13**(9–10), 721–750 (2013)
13. Jain, R., Upadhyay, S., Watrous, J.: Two-message quantum interactive proofs are in PSPACE. In: 50th Annual IEEE Symposium on Foundations of Computer Science, FOCS 2009, October 25–27, 2009, Atlanta, Georgia, USA, pp. 534–543. IEEE Computer Society (2009)
14. Jordan, S.P., Kobayashi, H., Nagaj, D., Nishimura, H.: Achieving perfect completeness in classical-witness quantum Merlin-Arthur proof systems. Quantum Info. Comput. **12**(5–6), 461–471 (2012)
15. Kempe, J., Regev, O.: 3-local Hamiltonian is QMA-complete. Quantum Inf. Comput. **3**(3), 258–264 (2003)
16. Kitaev, A., Shen, A., Vyalyi, M.N.: Classical and quantum computation. Graduate studies in mathematics. American mathematical society, Providence (R.I.) (2002)
17. Kobayashi, H., Le Gall, F., Nishimura, H.: Stronger methods of making quantum interactive proofs perfectly complete. In: Kleinberg, R.D. (ed.) ITCS, pp. 329–352. ACM, New York (2013)
18. Levin, L.A.: Universal sequential search problems. Probl. Inf. Transm. **9**(3), 265–266 (1973)
19. Marriott, C., Watrous, J.: Quantum arthur-merlin games. Comput. Complex. **14**, 122–152 (2005)
20. Oliveira, R., Terhal, B.M.: The complexity of quantum spin systems on a two-dimensional square lattice. Quantum Inf. Comput. **8**(10), 0900–0924 (2008)
21. Watrous, J.: Succinct quantum proofs for properties of finite groups. In: FOCS, pp. 537–546. IEEE Computer Society (2000)
22. Watrous, J.: Quantum computational complexity. In: Meyers, R.A. (ed.) Encyclopedia of Complexity and Systems Science, pp. 7174–7201. Springer, New York (2009)
23. Wocjan, P., Janzing, D., Beth, T.: Two qcma-complete problems. Quantum Inf. Comput. **3**(6), 635–643 (2003)

Phase Transition for Local Search
on Planted SAT

Andrei A. Bulatov[1]([⊠]) and Evgeny S. Skvortsov[2]

[1] Simon Fraser University, Burnaby, Canada
abulatov@cs.sfu.ca
[2] Google, Toronto, Canada
evgenys@google.com

Abstract. The Local Search algorithm (or Hill Climbing, or Iterative Improvement) is one of the simplest heuristics to solve the Satisfiability and Max-Satisfiability problems. Although it is not the best known Satisfiability algorithm even for the class of problems we study, the Local Search is a part of many satisfiability and max-satisfiability solvers, where it is used to find a good starting point for a more sophisticated heuristics, and to improve a candidate solution. In this paper we give an analysis of Local Search on random planted 3-CNF formulas. We show that a sharp transition of efficiency of Local Search occurs at density $\varrho = \frac{7}{6} \ln n$. Specifically we show that if there is $\kappa < \frac{7}{6}$ such that the clause-to-variable ratio is less than $\kappa \ln n$ (n is the number of variables in a CNF) then Local Search whp does not find a satisfying assignment, and if there is $\kappa > \frac{7}{6}$ such that the clause-to-variable ratio is greater than $\kappa \ln n$ then the local search whp finds a satisfying assignment. As a byproduct we also show that for any constant ϱ there is γ such that Local Search applied to a random (not necessarily planted) 3-CNF with clause-to-variable ratio ϱ produces an assignment that satisfies at least γn clauses less than the maximal number of satisfiable clauses.

The k-SAT and MAX-k-SAT problems are known to be hard both to solve exactly and to approximate. In particular, Håstad [17] proved that the MAX-k-SAT problem is NP-hard to approximate within ratio better than $1 - 2^{-k}$. These worst case hardness results motivate the study of the typical case complexity of those problems, and a quest for probabilistic or heuristic algorithms with satisfactory performance, in the typical case. In this paper we analyze the performance of one of the simplest algorithms for (MAX-)k-SAT, the Local Search algorithm, on random planted instances.

The Distribution. One of the most natural and well studied probability distributions of 3-CNFs is the uniform distribution $\Phi(n, \varrho n)$ on the set of 3-CNFs with a given clauses-to-variables ratio. For a given number of variables n and *density* $\varrho = \varrho(n)$ a formula is constructed and sampled as follows. The formula contains $m(n) = \varrho n$ 3-clauses over variables $x_1, \ldots, x_n$. These clauses are chosen uniformly at random from the set of all possible clauses. So the probability of every 3-CNF from $\Phi(n, \varrho n)$ is the same.

© Springer-Verlag Berlin Heidelberg 2015
G.F. Italiano et al. (Eds.): MFCS 2015, Part II, LNCS 9235, pp. 175–186, 2015.
DOI: 10.1007/978-3-662-48054-0_15

The typical case complexity for this distribution is not very complex except for a limited range of densities. The reason is that the random 3-SAT under this distribution demonstrates a sharp satisfiability threshold in the density [2,12]. A random 3-CNF with density below the threshold (estimated to be around 4.2) is satisfiable whp (with high probability, meaning that the probability tends to 1 as n goes to infinity), and a 3-CNF with density above the threshold is unsatisfiable whp. Therefore the trivial algorithm outputting yes or no by just counting the density of a 3-CNF gives a right answer to 3-SAT whp. For more results on the threshold see [1,10,11]. It is also known that, as the density grows, the number of clauses satisfied by a random assignment differs less and less from the maximal number of satisfiable clauses. If the density is *infinite* (i.e. it is an unbounded function of n), then whp this difference becomes negligible, i.e. $o(n)$.

An interesting and useful distribution $\Phi^{\mathsf{sat}}(n, \varrho n)$ is obtained from $\Phi(n, \varrho n)$ by conditioning on satisfiability: such distribution is uniform and its elements are the satisfiable 3-CNFs. Then the problem is to find or approximate a satisfying assignment knowing it exists. If a satisfiable formula has sufficiently high density, the set of satisfying assignments of such formula whp satisfies strong restrictions on its structure, which makes it possible to find a solution in polynomial time [9], see also [24]. In some aspects, however, the satisfiable distribution is not very convenient, for instance, it is not clear how to sample the satisfiable distribution (see, e.g., [6]). An approximation for such a distribution is the planted distribution $\Phi^{\mathsf{plant}}(n, \varrho n)$, which is obtained from $\Phi(n, \varrho n)$ by conditioning on satisfiability by a specific "planted" assignment. To construct an element of a planted distribution we select an assignment of a set of n variables and then uniformly at random include ϱn clauses satisfied by the assignment selected. Coja-Oghlan et al. [9] argued that when density is sufficiently high satisfiable formulas whp have satisfying assignments tightly clustered, and consequently the planted distribution is statistically close to the satisfiable one. They also managed to transfer certain results for $\Phi^{\mathsf{plant}}(n, \varrho n)$ to $\Phi^{\mathsf{sat}}(n, \varrho n)$.

Another interesting feature of the planted distribution (and the satisfiable one, see [9]) is that there is a hope to design an algorithm that solves all planted instances whp [6,14,19]. Algorithms from [14] and [19] use different approaches to solve planted 3-SAT of high but constant density. Note that if the density of formulas is low, properties of the planted and satisfiable distributions may be significantly different. Due to the phase transition phenomenon, almost all 3-CNFs of density below the threshold are satisfiable, while finding a solution of such a formula seems to be hard. On the other hand, experiments show that the algorithm from [5] achieves the goal, but a rigorous analysis of this algorithm is not yet made. For a survey on SAT algorithms the reader is referred to [7].

The Algorithm. The Local Search algorithm (LS) is one of the oldest heuristics for SAT that has been around since the eighties. Numerous variations of LS have been proposed since then, see, e.g., [22]. We study one of the most basic versions of LS, which, given a CNF, starts with a random assignment to its variables, and then on each step chooses at random a variable such that flipping this variable increases the number of satisfied clauses, or stops if no such variable exists. Thus LS finds a random local optimum accessible from the initial assignment.

LS has been studied before. The worst-case performance of pure LS appears to be not very good: the only lower bound for local optima of a k-CNF is $\frac{k}{k+1}m$ of clauses satisfied, where m is the number of all clauses [16]. Local Search was proven to solve 3-SAT in the typical case by Koutsoupias and Papadimitriou [18] for formulas of linear density, that is, $m = \Omega(n^2)$. Finally, in [8], we gave an estimation of the dependence of the number of clauses LS typically satisfies and the density of the formula.

Often visualization of the number of clauses satisfied by an assignment is useful: Assignments can be thought of as points of a landscape, and the elevation of a point corresponds to the number of clauses unsatisfied, the higher the point is, the less clauses it satisfies. It is suspected that 'topographic' properties of such a landscape are responsible for many complexity properties of satisfiability instances [7,20]. As we shall see the performance of LS is closely related to geometric properties of the assignments.

The behavior of other SAT/MAXSAT algorithms has been studied as well. In some cases the algorithms in question are basic heuristics used as a part or prototypes of practical Satisfiability solvers. For example, the random walk has been analyzed in [21] and then in [3], and the performance of GSAT, an improved version of LS, has been studied in [23]. A number of algorithms have been carefully examined to show that they solve whp a certain class of SAT problems. These include, e.g., [5,6,9,14,19,24] for the planted and satisfiable distributions. Message passing type algorithm such as Survey Propagation are known be very efficient for solving 3-SAT of density close to the threshold [7]. Although this algorithm still eludes a rigorous analysis, a similar message passing algorithm, Warning Propagation, is studied in [13].

Our Contribution. We classify the performance of LS for all densities higher than an arbitrary constant. In particular, we demonstrate that LS has a sharp threshold (see [15]) in its performance. The main result is the following theorem.

Theorem 1

(1) Let $\varrho \geq \kappa \cdot \ln n$, and $\kappa > \frac{7}{6}$. Then LS whp finds a solution of an instance from $\Phi^{\text{plant}}(n, \varrho n)$.

(2) Let $\sigma \leq \varrho \leq \kappa \cdot \ln n$, where σ is an arbitrary positive constant, and $0 < \kappa < \frac{7}{6}$. Then LS whp does not find a solution of an instance from $\Phi^{\text{plant}}(n, \varrho n)$.

To prove part (1) of Theorem 1 we show that under those conditions all the local optima of a 3-CNF whp are either satisfying assignments, that is, global optima, or are far from the planted solution, and so are located on the opposite side of the set of assignments. In the former case LS finds a satisfying assignment, while whp it does not reach the local optima of the second type.

For part (2) we prove that if the density is insufficient, the formula whp contains sufficiently many local minima so that LS inevitably falls into one of them, rather than proceeding into the planted global minimum. We also show that for any constant density ϱ there is γ such that the assignment produced by

LS on an instance from $\Phi^{\mathsf{plant}}(n, \varrho n)$ or $\Phi(n, \varrho n)$ satisfies at least γn clauses less than the maximal number of satisfiable clauses.

Another region where LS can find a solution of a random planted 3-CNF is the case of very low density. Methods similar to Lemma 3 and Theorem 2 show that this low density transition happens around $\varrho \approx n^{-1/4}$. However, we do not go into details here.

1   Preliminaries

SAT. A 3-CNF is a conjunction of *3-clauses*. As we consider only 3-CNFs, we will always call them just clauses. Depending on the number of negated literals, we distinguish 4 types of clauses: $(-, -, -), (+, -, -), (+, +, -)$, and $(+, +, +)$. If φ is a 3-CNF over variables $x_1, \ldots, x_n$, an *assignment* of these variables is a Boolean n-tuple $\boldsymbol{u} = (u_1, \ldots, u_n)$, so the value of x_i is u_i. The *density* of a 3-CNF φ is the number $\frac{m}{n}$ where m is the number of clauses, and n is the number of variables in φ.

The *uniform* distribution of 3-CNFs of density ϱ (density may be a function of n), $\Phi(n, \varrho n)$, is the set of all 3-CNFs containing n variables and ϱn clauses equipped with the uniform probability distribution on this set. To sample a 3-CNF accordingly to $\Phi(n, \varrho n)$ one chooses uniformly and independently ϱn clauses out of the $2^3 \binom{n}{3}$ possible clauses. Thus, we allow repetitions of clauses, but not repetitions of variables within a clause. *Random 3-SAT* is the problem of deciding the satisfiability of a 3-CNF randomly sampled accordingly to $\Phi(n, \varrho n)$. For short, we will call such a random formula a 3-CNF from $\Phi(n, \varrho n)$.

The *uniform planted* distribution of 3-CNFs of density ϱ is constructed as follows. First, choose at random a Boolean n-tuple $\boldsymbol{u}$, a *planted* satisfying assignment. Then let $\Phi^{\mathsf{plant}}(n, \varrho n, \boldsymbol{u})$ be the uniform probability distribution over the set of all 3-CNFs over variables $x_1, \ldots, x_n$ with density ϱ and such that $\boldsymbol{u}$ is a satisfying assignment. For our goals we can always assume that $\boldsymbol{u}$ is the all-ones tuple, that is a 3-CNF belongs to $\Phi^{\mathsf{plant}}(n, \varrho n, \boldsymbol{u})$ if and only if it contains no clauses of the type $(-, -, -)$. We also simplify the notation $\Phi^{\mathsf{plant}}(n, \varrho n, \boldsymbol{u})$ by $\Phi^{\mathsf{plant}}(n, \varrho n)$. To sample a 3-CNF accordingly to $\Phi^{\mathsf{plant}}(n, \varrho n)$ one chooses uniformly and independently ϱn clauses out of $7 \binom{n}{3}$ possible clauses of types $(+, -, -), (+, +, -)$, and $(+, +, +)$. *Random Planted 3-SAT* is the problem of deciding the satisfiability of a 3-CNF from $\Phi^{\mathsf{plant}}(n, \varrho n)$.

The problems *Random MAX-3-SAT* and *Random Planted MAX-3-SAT* are the optimization versions of Random 3-SAT and Random Planted 3-SAT. The goal in these problems is to find an assignment that satisfies as many clauses as possible. Although the two problems usually are treated as maximization problems, it will be convenient for us to consider them as problems of minimizing the number of unsatisfied clauses. Since we always evaluate the absolute error of our algorithms, not the relative one, such transformation does not affect the results.

Local Search. A description of the Local Search algorithm (LS) is given below.

Algorithm 1. Local Search

Require: 3-SAT formula φ over variables $x_1, \ldots, x_n$.
Ensure: Boolean n-tuple v, which is a local minimum of φ.
 1: **choose** uniformly at random a Boolean n-tuple u
 2: **let** U be the set of all variables x_i such that the number of clauses that can be
    made satisfied by flipping the value of x_i is strictly greater than the number of
    those made unsatisfied
 3: **while** U is not empty **do**
 4:     pick uniformly at random a variable x_j from U
 5:     change the value of x_j
 6:     recompute U
 7: **end while**

Observe that LS stops when reaches a local minimum of the number of unsatisfied clauses.

Given an assignment u and a clause c it will be convenient to say that c *votes* for a variable x_i to have value 1 if c contains literal x_i and its other two literals are unsatisfied. Similarly, we say that c votes for x_i to have value 0 if c contains the negation of x_i and its other two literals are not satisfied. Using this terminology we can define set U (see Algorithm 1) as the set of all variables such that the number of votes received to change the current value is greater than the number of those to keep it.

Random Graphs. Probabilistic tools we use are fairly standard and can be found in the book [4].

Let φ be a 3-CNF with variables $x_1, \ldots, x_n$. The *primal graph* $G(\varphi)$ of φ is the graph with vertex set $\{x_1, \ldots, x_n\}$ and edge set $\{x_i x_j \mid$ literals containing x_i, x_j appear in the same clause$\}$. Note that $G(\varphi)$ is not a random graph.

We will need the following properties that a graph $G(\varphi)$ has, provided its density is sufficiently low.

Lemma 1. *Let $\varrho < \kappa \ln n$ for a certain constant κ, and let $\varphi \in \Phi^{\text{plant}}(n, \varrho n)$.*

*(1) For any $\alpha < 1$, whp all the subgraphs of $G(\varphi)$ induced by at most $O(n^\alpha)$
     vertices have the average degree less than 4.*
(2) The probability that $G(\varphi)$ has a vertex of degree greater than $\ln^2 n$ is $o(n^{-3})$.

2   Success of Local Search

In this section we prove Theorem 1(1). This will be done as follows. First, we show that if a 3-CNF has high density, that is, greater than $\kappa \log n$ for some $\kappa > \frac{7}{6}$ then whp all the local minima that do not satisfy the CNF — we call such minima *proper* — concentrate very far from the planted assignment. This

Fig. 1. Caps and crowns

is the statement of Proposition 1 below. Then we use Lemma 2 to prove that starting from a random assignment LS whp does not go to that remote region. Therefore the algorithm does not get stuck to a local minimum that is not a solution. Recall that the planted solution is the all-ones one.

Lemma 2. *Let* $\varrho \geq \kappa \ln n$ *for some constant* $\kappa > 0$, *and let constants* q_0, q_1 *be such that* $0 \leq q_0 < q_1 \leq 1$. *Whp any assignment with less than* $q_0 n$ *zeros satisfies more clauses than any assignment with more than* $q_1 n$ *zeros.*

Proposition 1. *Let* $\varrho \geq \kappa \cdot \ln n$, *and* $\kappa > \frac{7}{6}$. *Then whp proper local minima of a 3-CNF from* $\Phi^{\text{plant}}(n, \varrho n)$ *have at most* $\frac{n}{10}$ *ones.*

Proof (of Theorem 1(1)). By Lemma 2 for a $\varphi \in \Phi^{\text{plant}}(n, \varrho n)$ whp any assignment with dn variables equal to 1, where $\frac{1}{3} \leq d \leq \frac{2}{3}$, satisfies more clauses than any assignment with less than $\frac{n}{10}$ variables equal to 1. Then, whp a random initial assignment for LS assigns between $\frac{1}{3}$ and $\frac{2}{3}$ of all variables to 1. Therefore, whp LS never arrives to a proper local minimum with less than $\frac{n}{10}$ variables equal to 1, and, by Proposition 1, to any proper local minimum.

3   Failure of Local Search

We now prove statement (2) of Theorem 1. The overall strategy is the following. First, we show, Proposition 2, that in contrast to the previous case there are many proper local minima in the close proximity of the planted assignment. Then we show, Proposition 3, that those local minima are located so that they intercept almost every run of LS, and thus almost every run is unsuccessful.

A variable of a CNF is called *k-isolated* if it appears positively in at most k clauses of the type $(+, -, -)$. The *distance* between variables of a CNF φ is the length of the shortest path in $G(\varphi)$ connecting them. A pair of clauses $c_1 = (x_1, \overline{x}_2, \overline{x}_3)$, $c_2 = (\overline{x}_1, \overline{x}_4, x_5)$ is called a *cap* over variables $x_1, \dots, x_5$ if x_1, x_5 are 1-isolated, that is, they do not appear positively in any clause of the type $(+, -, -)$ except for c_1 and c_2, respectively, and x_2, x_3 are not 0-isolated (see Fig. 1(a)). If pair c_1, c_2 is a cap, then clause c_2 is called *cap support*.

The proof of the following lemma is fairly standard, see, e.g. the proof of Theorem 4.4.4 in [4].

Lemma 3. *Let* $n^{-\frac{1}{4}} < \varrho \leq \kappa \cdot \ln n$, *and* $\kappa < \frac{7}{6}$. *There is* α, $0 < \alpha < 1$, *such that whp a random planted CNF* $\varphi \in \Phi^{\text{plant}}(n, \varrho n)$ *contains at least* n^{α} *caps.*

Proposition 2. *Let $\sigma > 0, 0 < \kappa < \frac{7}{6}$, and density ϱ be such that $\sigma < \varrho \leq \kappa \cdot \ln n$. Then there is α, $0 < \alpha \leq 1$, such that a 3-CNF from $\Phi^{\text{plant}}(n, \varrho n)$ whp has at least n^{α} proper local minima.*

Indeed, let $c_1 = (x_1, \overline{x}_2, \overline{x}_3)$, $c_2 = (\overline{x}_1, \overline{x}_4, x_5)$ be a cap and $\boldsymbol{u}$ an assignment such that $u_1 = u_5 = 0$, and $u_i = 1$ otherwise. It is straightforward that $\boldsymbol{u}$ is a proper local minimum because c_1 is the only unsatisfied clause but Local Search does not flip any variables. By Lemma 3, there is α such that whp the number of such minima is at least n^{α}.

Before proving Theorem 1(2), we note that a construction similar to caps helps evaluate the approximation rate of Local Search applied to random and random planted CNFs. A subformula $c = (x_1, x_2, x_3), c_1 = (\overline{x}_1, x_4, x_5), c_2 = (\overline{x}_2, x_6, x_7), c_3 = (\overline{x}_3, x_8, x_9)$ is called a *crown* if the variables $x_1, \ldots, x_9$ do not appear in any clauses other than c, c_1, c_2, c_3 (see Fig. 1(b)). The crown is satisfiable, while the all-zero assignment is a proper local minimum. For a CNF φ and an assignment $\boldsymbol{u}$ to its variables, by $\mathsf{OPT}(\varphi)$ and $\mathsf{sat}(\boldsymbol{u})$ we denote the maximal number of simultaneously satisfiable clauses and the number of clauses satisfied by $\boldsymbol{u}$, respectively.

Theorem 2. *If the density ϱ is such that $\sigma \leq \varrho \leq \kappa \ln n$ for some $\sigma > 0, 0 < \kappa < 1/27$, then there is $\gamma_{\varrho}(n) > 0$ such that whp Local Search on a 3-CNF $\varphi \in \Phi(n, \varrho n)$ ($\varphi \in \Phi^{\text{plant}}(n, \varrho n)$) returns an assignment $\boldsymbol{u}$ such that $\mathsf{OPT}(\varphi) - \mathsf{sat}(\boldsymbol{u}) \geq \gamma_{\varrho} \cdot n$. If ϱ is constant then γ_{ϱ} is also constant.*

Proof of Theorem 2 is similar to that of Lemma 3. It can be shown that for ϱ that satisfies conditions of this theorem there is $\gamma'(n) > 0$ such that whp a random [random planted] formula has at least $\gamma' n$ crowns. If ϱ is a constant, γ' is also a constant. For a random assignment $\boldsymbol{u}$, whp the variables of at least $\frac{\gamma'}{1024} n$ crowns are assigned zeroes. Such an all-zero assignment of a crown cannot be changed by the local search.

Proposition 3. *Let $\sigma > 0, 0 < \kappa < \frac{7}{6}$, and density ϱ be such that $\sigma < \varrho \leq \kappa \cdot \ln n$. Then Local Search on a 3-CNF from $\Phi^{\text{plant}}(n, \varrho n)$ whp ends up in a proper local minimum.*

In what follows we prove Proposition 3.

If $\varrho = o(\ln n)$ then Proposition 3 follows from Theorem 2. So we assume that $\varrho > \kappa' \cdot \ln n$. The main tool of proving Proposition 3 is coupling of local search (LS) with the algorithm STRAIGHT DESCENT (SD) that on each step chooses at random a variable assigned to 0 and changes its value to 1. Obviously SD is not a practical algorithm, since to apply it we need to know the solution. For the purposes of our analysis we modify SD as follows. At each step SD chooses a variable at random, and if it is assigned 0 changes its value (see Algorithm 2). The algorithm LS is modified in a similar way. It chooses a variable at random and flips it if it increases the number of satisfied clauses. Modified Local Search is identical to Local Search in the sense that it flips the same sequence of variables.

Since actions of SD do not depend on the formula, its position at each step is random, in a sense that under the condition that SD at step t arrived to a vector with m ones, each vector with m ones has the same probability of being the location of SD.

Algorithm 2. Straight Descent

Require: $\varphi \in \Phi^{\mathsf{plant}}(n, \varrho n)$ with the all-ones solution, Boolean tuple $\boldsymbol{u}$,
Ensure: The all-ones Boolean tuple.
1: **while** there is a variable assigned 0 **do**
2:     pick uniformly at random variable x_j from the set of all variables
3:     **if** $u_j = 0$ **then**
4:         $u_j = 1$
5:     **end if**
6: **end while**

Given 3-CNF φ and an assignment $\boldsymbol{u}$ we say that a variable x_i is k-*righteous* if the number of clauses voting for it to be one is greater by at least k than the number of clauses voting for it to be zero. Let $\varphi \in \Phi^{\mathsf{plant}}(n, \varrho n)$ and $\boldsymbol{u}$ be a Boolean tuple. The *ball* of radius m with the center at $\boldsymbol{u}$ is the set of all tuples of the same length as $\boldsymbol{u}$ at Hamming distance at most m from $\boldsymbol{u}$. Let $f(n)$ and $g(n)$ be arbitrary functions and d be an integer constant. We say that a set S of n-tuples is $(g(n), d)$-*safe*, if for any $\boldsymbol{u} \in S$ the number of variables that are not d-righteous does not exceed $g(n)$. A run of SD is said to be $(f(n), g(n), d)$-*safe* if at each step of this run the ball of radius $f(n)$ with the center at the current assignment is $(g(n), d)$-safe.

Lemma 4. *Let $\varrho > \kappa \cdot \ln n$ for some $\kappa, \kappa > 0$. For any positive constants γ and d there is a constant $\alpha_1 < 1$ such that, for any α with $\alpha_1 < \alpha < 1$, whp a run of SD on $\varphi \in \Phi^{\mathsf{plant}}(n, \varrho n)$ and a random initial assignment $\boldsymbol{u}$ is $(\gamma n^{\alpha}, n^{\alpha}, d)$-safe.*

For CNFs ψ_1, ψ_2 we denote by $\psi_1 \wedge \psi_2$ their conjunction.

We will need formulas that are obtained from a random formula by adding some clauses in an 'adversarial' manner. Following [19] we call distributions for such formulas *semi-random*. However, the type of semi-random distributions we need is different from that in [19]. Let $\eta < 1$ be some constant. A formula φ is sampled according to semi-random distribution $\Phi_{\eta}^{\mathsf{plant}}(n, \varrho n)$ if $\varphi = \varphi' \wedge \psi$, where φ' is sampled according to $\Phi^{\mathsf{plant}}(n, \varrho n)$ and ψ contains at most n^{η} clauses and is given by an adversary. That is, by definition an event E whp takes place for $\varphi = \varphi' \wedge \psi \in \Phi_{\eta}^{\mathsf{plant}}(n, \varrho n)$ if whp φ' is such that $E(\varphi)$ is true for any ψ. It is not difficult to see that if $\eta' \leq \eta$ then any event that occurs whp for $\varphi \in \Phi_{\eta}^{\mathsf{plant}}(n, \varrho n)$, also occurs whp for $\varphi \in \Phi_{\eta'}^{\mathsf{plant}}(n, \varrho n)$.

We say that a variable *plays d-righteously in a run of LS* if every time it is considered for flipping it is d-righteous. For semi-random distributions we then have the following

Lemma 5. *Let $\varrho > \kappa \cdot \ln n$ for some $\kappa, \kappa > 0$. For any d there is $\alpha_2 < 1$ such that, for a run of LS on $\varphi \in \Phi_{\eta}^{\mathsf{plant}}(n, \varrho n)$ whp the number of variables that do not play d-righteously is bounded above by n^{α_2}.*

For a formula ψ we denote the set of variables that occur in it by $\mathrm{var}(\psi)$. For a set of clauses K we denote by $\bigwedge K$ a CNF formula constructed by conjunction of

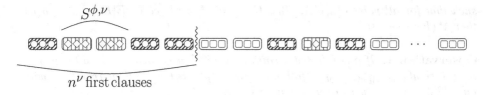

$S^{\phi,\nu}$ ... n^{ν} first clauses

Fig. 2. A scheme of a 3-CNF. Every clause is shown as a rectangle with its literals represented by squares, circles, or diamonds inside the rectangle. Literals corresponding to variables from $L^{\phi\nu}$ and from var$([n^{\nu}] \setminus S^{\varphi,\nu})$ are shown as diamonds and circles, respectively. Shaded rectangles with vertical and diagonal lines represent clauses from $T^{\varphi\nu}$ and $U^{\varphi\nu}$, respectively.

the clauses. For the sake of simplicity we will write var(K) instead of var$(\bigwedge K)$. In what follows it will be convenient to view a CNF as a sequence of clauses. Note that such representation of a CNF is quite natural when we sample a random CNF by generating random clauses. This way every clause occupies certain position in the formula. For a set of positions P we denote the formula obtained from φ by removing all clauses except for occupying positions P by $\varphi \downarrow_P$. The set of variables occurring in the clauses in positions from P will be denoted by var(P). For a set of variables V the set of positions of clauses which contain a variable from V is denoted by clp(V). By $[k]$ we denote the set of the first k positions of clauses in φ. Recall that a clause $(\overline{x}, \overline{y}, z)$ is called a *cap support* if there are w_1, w_2 such that $(x, \overline{w}_1, \overline{w}_2), (\overline{x}, \overline{y}, z)$ is a cap in φ.

For a real constant $\nu, 0 < \nu < 1$ we will recognize the following subsets of clauses and variables of φ. Let $S^{\varphi,\nu}$ be the set of positions from $[n^{\nu}]$ occupied by clauses that are cap supports in φ; let $L^{\varphi,\nu} = \text{var}(S^{\varphi,\nu})$; let $T^{\varphi,\nu} = \text{clp}(L^{\varphi,\nu})$, $U^{\varphi,\nu} = \text{clp}(\text{var}([n^{\nu}] \setminus S^{\varphi,\nu}))$; finally, let $M^{\varphi,\nu} = \text{var}(T^{\varphi,\nu})$ and $N^{\varphi,\nu} = \text{var}(U^{\varphi,\nu})$. Figure 2 pictures the notation just introduced.

We denote equality $f(n) = g(n)(1 + o(1))$ by $f(n) \sim g(n)$.

Lemma 6. *Let $\sigma > 0, 0 < \kappa < \frac{7}{6}$, and density ϱ be such that $\sigma < \varrho \le \kappa \cdot \ln n$. Then there is $\mu_0 > 0$ such that for any $\mu < \mu_0$ there is $\nu < 1$ such that whp:*

(1) $|S^{\varphi,\nu}| \sim n^{\mu}$;
(2) $M^{\varphi,\nu} \cap N^{\varphi,\nu} = \varnothing$, i.e. variables from clauses in $U^{\varphi,\nu}$ do not appear in the same clauses with variables from $S^{\varphi,\nu}$;
(3) $|M^{\varphi,\nu}| = 3|T^{\varphi,\nu}|$, that is no variable occurs twice in the clauses from $T^{\varphi,\nu}$.

Let n be the number of variables, let ϱ be the density, let ν be a real constant such that $0 < \nu < 1$, let T_0 and U_0 be subsets of $[\varrho n]$ such that $T_0 \cap U_0 = \varnothing$, $[n^{\nu}] \subseteq T_0 \cup U_0$, and let $S_0 = T_0 \cap [n^{\nu}]$. We denote by $H_{T_0 U_0 \nu}$ a hypothesis stating that φ is such that $S^{\varphi,\nu} = S_0$, $T^{\varphi,\nu} = T_0$, $U^{\varphi,\nu} = U_0$ and also $M^{\varphi,\nu} \cap N^{\varphi,\nu} = \varnothing$, $|M^{\varphi,\nu}| = 3|T^{\varphi,\nu}|$. The following lemma allows us to focus on small sets U_0, T_0.

Lemma 7. *Let $\phi \in \Phi^{\text{plant}}(n, \varrho n)$, and $\mu_0, \mu < \mu_0$, and ν be as in Lemma 6; also let $\varepsilon > 0$ be such that $\nu + \varepsilon < 1$. If for an event E there is a sequence $\delta(n) \xrightarrow[n \to \infty]{} 0$*

such that for all pairs (T_0, U_0), $|T_0 \cup U_0| < n^{\nu + \varepsilon}$ we have $\mathbf{P}\left(E|H_{T_0 U_0 \nu}\right) \leq \delta(n)$ then $\mathbf{P}\left(E\right) \underset{n \to \infty}{\longrightarrow} 0$.

Observation 1. *If φ is selected according to $\Phi^{\mathtt{plant}}(n, \varrho n)$ conditioned to $H_{T_0 U_0 \nu}$ then formula $\varphi \downarrow_{[\varrho n] \setminus (T_0 \cup U_0)}$ follows random planted 3-CNF distribution with $\varrho n - |T_0| - |U_0|$ clauses over variables $V \setminus \text{var}([n^\nu])$.*

Proof (of Proposition 3). For this proof we shall need a small enough μ. Here is how to pick it. We start by applying Lemma 6 to $\varrho \in \Phi^{\mathtt{plant}}(n, n\varrho)$ and get $\tilde{\mu}$ and $\tilde{\nu}$ that satisfy conditions (1)–(3) of Lemma 6. Let η be such that $\tilde{\nu} < \eta < 1$ and α_2 be picked from Lemma 5 for $\Phi^{\mathtt{plant}}_\eta(n, n\varrho)$, so the number of variables that are not playing d-righteously is bounded by n^{α_2}. Now we set $\mu = \min(\tilde{\mu}, \frac{1 - \alpha_2}{4})$ and by Lemma 6 there is $\nu \leq \tilde{\nu}$ which makes conditions (1)–(3) true for μ.

We fix an arbitrary pair (T_0, U_0) of subsets of $[\varrho n]$ with $T_0 \cap U_0 = \varnothing$, $[n^\nu] \subseteq T_0 \cup U_0$, $|T_0 \cup U_0| < n^\eta$. We will bound the probability of success of LS under a hypothesis of the form $H_{T_0 U_0 \nu}$ and apply Lemma 7 to get the result.

Let $M = M^{\varphi, \nu}$ and $L = L^{\varphi, \nu}$. We split formula φ into $\varphi_1 = \varphi \downarrow_{T_0}$ and $\varphi_2 = \varphi \downarrow_{[\varrho n] \setminus T_0}$ and first consider a run of LS applied to φ_2 only. Formula φ_2 can in turn be considered as the conjunction of $\varphi_{21} = \varphi \downarrow_{U_0}$ and $\varphi_{22} = \varphi \downarrow_{[\varrho n] \setminus (T_0 \cup U_0)}$. In Fig. 2 formula φ_1 consists of clauses shaded with vertical lines, formula φ_{21} of clauses shaded with diagonal lines and formula φ_{22} of clauses that are not shaded. By Observation 1 formula φ_{22} is sampled according to $\Phi^{\mathtt{plant}}(n - \delta_1(n), n\varrho - \delta_2(n))$ modulo the names of variables where $\delta_1(n)$ and $\delta_2(n)$ are $o(n)$. So formula φ_2 is sampled according to $\Phi^{\mathtt{plant}}_\eta(n - \delta_1(n), n\varrho - \delta_2(n))$ and by the choice of α_2 the number of variables that do not play 2-righteously during a run of LS on φ_2 is bounded from above by n^{α_2}.

We consider coupling $(LS_\varphi, LS_{\varphi_2})$ of runs of LS on φ and φ_2. The assignments obtained by the runs of the algorithm at step t we denote by $\boldsymbol{u}_\varphi(t)$ and $\boldsymbol{u}_{\varphi_2}(t)$ respectively. Let K be the set of those variables which do not belong to L (squares and circles in Fig. 2). Formula φ_2 is a 3-CNF containing only variables from K. If $\boldsymbol{u}$ is an assignment of values to all variables, then $\boldsymbol{u}|_K$ will denote the restriction of $\boldsymbol{u}$ onto variables from K. We make process LS_φ start with a random assignment $\boldsymbol{u}_\varphi(0) = \boldsymbol{u}^0_\varphi$ to all variables, and LS_{φ_2} with a random assignment $\boldsymbol{u}_{\varphi_2}(0) = \boldsymbol{u}^0_{\varphi_2}$ to variables in K, such that $\boldsymbol{u}^0_\varphi|_K = \boldsymbol{u}^0_{\varphi_2}$. Now the algorithms work as follows. At every step a random variable x_i is chosen. Process LS_φ makes its step, and process LS_{φ_2} makes its step if $x_i \in K$.

Let W denote the set of variables from $\text{var}(\varphi_2)$ that do not play 2-righteously during the run of LS_{φ_2}. By choice of α_2 whp we have $|W| \leq n^{\alpha_2}$. Variables in formula φ_1 are selected uniformly at random and $\alpha_2 + 2\mu < 1$ so whp set M does not intersect with W. Hence, every time LS_φ considers some variable from M it is 2-righteous in φ_2 and belongs to at most one clause of φ_1. Therefore such a variable is at least 1-righteous in φ and is flipped to 1, or stays 1, whichever is to happen for LS_{φ_2}. Thus whp at every step of $(LS_\varphi, LS_{\varphi_2})$ we have $\boldsymbol{u}_\varphi(t)|_K = \boldsymbol{u}_{\varphi_2}(t)$. In the rest of the proof we consider only this highly probable case.

Consider some cap support $c_i = (\overline{x}_1, \overline{x}_4, x_5)$ occupying a position $i \in [n^\nu]$ and such that $x_1 = 0, x_4 = 1, x_5 = 0$ at time 0, and a set P_{c_i} of variables occurring in clauses that contain variables $var(c_i)$ (obviously $var(c_i) \subseteq P_{c_i}$). Let c_j be the clause that forms a cap with c_i. We say that a variable is *discovered* at step t if it is considered for the first time at step t. Let $p_1, \ldots, p_k$ be an ordering of elements of P_{c_i} according to the step of their discovery. In other words if variable p_1 is the first variable from P_{c_i} that is discovered, p_k was the last. If some variables are not considered at all, we place them in the end of the list in a random order. Observe that all variables that play at least 1-righteously are discovered at some step. All orderings of variables are equiprobable, hence, the probability of variables $var(c_i)$ to occupy places p_{k-2}, p_{k-1} and p_k equals $3!/k(k-1)(k-2)$. We will call this ordering *unlucky*.

Let us consider what happens if the order of discovery of P_{c_i} is unlucky. All variables in $P_{c_i} \setminus var(c_i)$ play 1-righteously, therefore once they are discovered by LS_φ they are equal to 1. Thus when x_1, x_4, x_5 are finally considered all clauses they occur in are satisfied, except for c_j. So variables x_1, x_4, x_5 do not change their values and the clause c_j remains unsatisfied by the end of the work of LS_φ.

By Lemma 1(2) whp no vertex has degree greater than $\ln^2 n$, so the size of the set P_{c_i} is bounded above by $3 \ln^2 n$. Thus the probability of event $Unluck(i) =$ "order of discovery of P_{c_i} is unlucky" is greater than $\frac{1}{\ln^6 n}$. Thus, the expectation of $|\{i | Unluck(i)\}|$ equals $\frac{|S_0|}{\ln^6 n} = \frac{n^\mu}{\ln^6 n}$. By definition of $H_{T_0 U_0 \nu}$ any variable whp occurs in clauses from $T^{\varphi, \nu}$ at most once, hence there is no variable that occurs in the same clause with a variable from c_{i_1} and a variable from c_{i_2} for $i_1, i_2 \in S_0$, $i_1 \neq i_2$. This implies that events of the form $Unluck(i)$ are independent. Therefore random variable $|\{i | Unluck(i)\}|$ is Bernoulli and, as its expectation tends to infinity, the probability that it equals to 0 goes to 0. As unlucky ordering of at least one cap support leads to failure of the LS, this proves the result.

Acknowledgment. The fist author was supported by an NSERC Discovery grant.

References

1. Achlioptas, D.: Lower bounds for random 3-SAT via differential equations. Theor. Comput. Sci. **265**(1–2), 159–185 (2001)
2. Achlioptas, D., Friedgut, E.: A sharp threshold for k-colorability. Random Struct. Algorithms **14**(1), 63–70 (1999)
3. Alekhnovich, M., Ben-Sasson, E.: Linear upper bounds for random walk on small density random 3-CNFs. SIAM J. Comput. **36**(5), 1248–1263 (2007)
4. Alon, N., Spencer, J.: The Probabilistic Method. Wiley, New York (2000)
5. Amiri, E., Skvortsov, E.S.: Pushing random walk beyond golden ratio. In: Diekert, V., Volkov, M.V., Voronkov, A. (eds.) CSR 2007. LNCS, vol. 4649, pp. 44–55. Springer, Heidelberg (2007)
6. Ben-Sasson, E., Bilu, Y., Gutfreund, D.: Finding a randomly planted assignment in a random 3-CNF (2002). (manuscript)
7. Braunstein, A., Mézard, M., Zecchina, R.: Survey propagation: an algorithm for satisfiability. Random Struct. Algorithms **27**(2), 201–226 (2005)
8. Bulatov, A.A., Skvortsov, E.S.: Efficiency of local search. In: Biere, A., Gomes, C.P. (eds.) SAT 2006. LNCS, vol. 4121, pp. 297–310. Springer, Heidelberg (2006)

9. Coja-Oghlan, A., Krivelevich, M., Vilenchik, D.: Why almost all k-colorable graphs are easy. In: Thomas, W., Weil, P. (eds.) STACS 2007. LNCS, vol. 4393, pp. 121–132. Springer, Heidelberg (2007)

10. Coja-Oghlan, A., Panagiotou, K.: Going after the k-SAT threshold. In: Boneh, D., Roughgarden, T., Feigenbaum, J. (eds.) STOC, pp. 705–714. ACM (2013)

11. Crawford, J.M., Auton, L.D.: Experimental results on the crossover point in random 3-SAT. Art. Int. **81**(1–2), 31–57 (1996)

12. Ding, J., Sly, A., Sun, N.: Proof of the satisfiability conjecture for large k. In: Servedio, R., Rubinfeld, R. (eds.) STOC, pp. 59–68. ACM (2015)

13. Feige, U., Mossel, E., Vilenchik, D.: Complete convergence of message passing algorithms for some satisfiability problems. In: Díaz, J., Jansen, K., Rolim, J.D.P., Zwick, U. (eds.) APPROX 2006 and RANDOM 2006. LNCS, vol. 4110, pp. 339–350. Springer, Heidelberg (2006)

14. Flaxman, A.: A spectral technique for random satisfiable 3CNF formulas. In: SODA, pp. 357–363. ACM/SIAM (2003)

15. Friedgut, E.: Sharp thresholds of graph properties, and the k-SAT problem. J. Amer. Math. Soc. **12**, 1017–1054 (1999)

16. Hansen, P., Jaumard, B.: Algorithms for the maximum satisfiability problem. Computing **44**, 279–303 (1990)

17. Håstad, J.: Some optimal inapproximability results. J. ACM **48**(4), 798–859 (2001)

18. Koutsoupias, E., Papadimitriou, C.: On the greedy algorithm for satisfiability. Inf. Process. Lett. **43**(1), 53–55 (1992)

19. Krivelevich, M., Vilenchik, D.: Solving random satisfiable 3CNF formulas in expected polynomial time. In: SODA, pp. 454–463. ACM Press (2006)

20. Mézard, M., Mora, T., Zecchina, R.: Clustering of solutions in the random satisfiability problem. CoRR abs/cond-mat/0504070 (2005)

21. Papadimitriou, C.: On selecting a satisfying truth assignment (extended abstract). In: FOCS, pp. 163–169. IEEE Computer Society (1991)

22. Selman, B., Levesque, H., Mitchell, D.: A new method for solving hard satisfiability problems. In: Swartout, W. (ed.) AAAI, pp. 440–446. AAAI Press/The MIT Press (1992)

23. Skvortsov, E.S.: A theoretical analysis of search in GSAT. In: Kullmann, O. (ed.) SAT 2009. LNCS, vol. 5584, pp. 265–275. Springer, Heidelberg (2009)

24. Vilenchik, D.: It's all about the support: a new perspective on the satisfiability problem. JSAT **3**(3–4), 125–139 (2007)

Optimal Bounds for Estimating Entropy with PMF Queries

Cafer Caferov[1]([⊠]), Barış Kaya[1], Ryan O'Donnell[2], and A.C. Cem Say[1]

[1] Computer Engineering Department, Boğaziçi University, Istanbul, Turkey
{cafer.caferov,baris.kaya,say}@boun.edu.tr
[2] Department of Computer Science, Carnegie Mellon University, Pittsburgh, USA
odonnell@cs.cmu.edu

Abstract. Let p be an unknown probability distribution on $[n] := \{1, 2, \ldots n\}$ that we can access via two kinds of queries: A SAMP query takes no input and returns $x \in [n]$ with probability $p[x]$; a PMF query takes as input $x \in [n]$ and returns the value $p[x]$. We consider the task of estimating the entropy of p to within $\pm \Delta$ (with high probability). For the usual Shannon entropy $H(p)$, we show that $\Omega(\log^2 n/\Delta^2)$ queries are necessary, matching a recent upper bound of Canonne and Rubinfeld. For the Rényi entropy $H_\alpha(p)$, where $\alpha > 1$, we show that $\Theta(n^{1-1/\alpha})$ queries are necessary and sufficient. This complements recent work of Acharya et al. in the SAMP-only model that showed $O(n^{1-1/\alpha})$ queries suffice when α is an integer, but $\widetilde{\Omega}(n)$ queries are necessary when α is a non-integer. All of our lower bounds also easily extend to the model where CDF queries (given x, return $\sum_{y \leq x} p[y]$) are allowed.

1 Introduction

The field of statistics is to a large extent concerned with questions of the following sort: How many samples from an unknown probability distribution p are needed in order to accurately estimate various properties of the distribution? These sorts of questions have also been studied more recently within the theoretical computer science framework of *property testing*. In this framework, one typically makes no assumptions about p other than that it is a discrete distribution supported on, say, $[n] := \{1, 2, \ldots, n\}$. There is a vast literature on testing and estimating properties of unknown distributions; for surveys with pointers to the literature, see Rubinfeld [15] and Canonne [5].

One of the most important properties of a probability distribution p is its *Shannon entropy*, $H(p) = \sum_x p[x] \log \frac{1}{p[x]}$.[1] Shannon entropy is a measure of the "amount of randomness" in p. In this work we will be concerned with the associated task of estimating the entropy of an unknown p within a confidence

---

R. O'Donnell—Work performed while the author was at the Boğaziçi University Computer Engineering Department, supported by Marie Curie International Incoming Fellowship project number 626373.
[1] In this paper, log denotes $\log_2$.

© Springer-Verlag Berlin Heidelberg 2015
G.F. Italiano et al. (Eds.): MFCS 2015, Part II, LNCS 9235, pp. 187–198, 2015.
DOI: 10.1007/978-3-662-48054-0_16

interval of $\pm\Delta$ with probability at least $1 - \delta$. (Typically $\Delta = 1$ and $\delta = 1/3$.) We also remark that if $\boldsymbol{p}$ is a distribution on $[n] \times [n]$ representing the joint pmf of random variables $\boldsymbol{X}$ and $\boldsymbol{Y}$, then $H(\boldsymbol{p})$ is related to the *mutual information* $I(\boldsymbol{X};\boldsymbol{Y})$ of $\boldsymbol{X}$ and $\boldsymbol{Y}$ via $H(\boldsymbol{X}) + H(\boldsymbol{Y}) - H(\boldsymbol{X},\boldsymbol{Y})$. Thus additively estimating mutual information easily reduces to additively estimating entropy. For an extended survey and results on the fundamental task of estimating entropy, see Paninski [13]; this survey includes justification of discretization, as well as discussion of applications to neuroscience (e.g., estimating the information capacity of a synapse).

It is known that in the basic "samples-only model" — in which the only access to $\boldsymbol{p}$ is via independent samples — estimation of entropy is a very expensive task. From the works [13, 14, 17–19] we know that estimating $H(\boldsymbol{p})$ to within ± 1 with confidence $2/3$ requires roughly a linear number of samples; more precisely, $\Theta(n/\log n)$ samples are necessary (and sufficient). Unfortunately, for many applications this quantity is too large. E.g., for practical biological experiments it may be infeasible to obtain that many samples; or, for the enormous data sets now available in computer science applications, it may simply take too long to process $\widetilde{\Theta}(n)$ samples.

To combat this difficulty, researchers have considered an extension to the samples-only model, called the "Generation+Evaluation" model in [11] and the "combined model" in [3]. We will refer to it as the SAMP+PMF model[2] because it allows the estimation algorithm two kinds of "queries" to the unknown distribution $\boldsymbol{p}$: a SAMP query, which takes no input and returns $x \in [n]$ with probability $\boldsymbol{p}[x]$; and a PMF query, which takes as input $x \in [n]$ and returns the value $\boldsymbol{p}[x]$. As we will see, in this model entropy can be accurately estimated with just polylog(n) queries, dramatically smaller than the $\Omega(n/\log n)$ queries needed in the samples-only model.

Regarding the relevance of the SAMP+PMF model, an example scenario in which it might occur is the Google n-gram database; the frequency of each n-gram is published, and it is easy to obtain a random n-gram from the underlying text corpus. Another motivation for SAMP+PMF access comes from the *streaming* model of computation [2], where entropy estimation has been well studied [4, 7–10, 12]. The SAMP+PMF testing model and the streaming model are closely related. Roughly speaking, any q-query estimation algorithm in the SAMP+PMF model can be converted to a $q \cdot$ polylog(n)-space streaming algorithm with one or two passes (with precise details depending on the model for how the items in the stream are ordered). More motivation and results for the SAMP+PMF model can be found in Canonne and Rubinfeld [6].

1.1   Our Results, and Comparison with Prior Work

The first works [3,9] on entropy estimation in the SAMP+PMF model were concerned with *multiplicative* estimates of $H(\boldsymbol{p})$. Together they show relatively

---

[2] In this paper, PMF, CDF and SAMP are abbreviations for probability mass function, cumulative distribution function and sampling, respectively.

tight bounds for this problem: estimating (with high probability) $H(\boldsymbol{p})$ to within a multiplicative factor of $1 + \gamma$ requires

$$\text{between} \quad \Omega\left(\frac{\log n}{\max(\gamma, \gamma^2)}\right) \cdot \frac{1}{H(\boldsymbol{p})} \quad \text{and} \quad O\left(\frac{\log n}{\gamma^2}\right) \cdot \frac{1}{H(\boldsymbol{p})} \quad (1)$$

queries. Unfortunately, these bounds depend quantitatively on the entropy $H(\boldsymbol{p})$ itself; the number of queries necessarily scales as $1/H(\boldsymbol{p})$. Further, whereas additive estimation of entropy can be used to obtain additive estimates of mutual information, multiplicative estimates are insufficient for this purpose. Thus in this paper we consider only the problem of additive approximation.

For this problem, Canonne and Rubinfeld [6] recently observed that $O(\log^2 n)$ SAMP+PMF queries are sufficient to estimate $H(\boldsymbol{p})$ to ± 1 with high probability, and $\Omega(\log n)$ queries are necessary. The first main result in this work is an improved, optimal lower bound:

First Main Theorem. *In the* SAMP+PMF *model,* $\Omega(\log^2 n)$ *queries are necessary to estimate (with high probability) the Shannon entropy* $H(\boldsymbol{p})$ *of an unknown distribution* $\boldsymbol{p}$ *on* $[n]$ *to within* ± 1.

Remark 1. Our lower bound and the lower bound from (1) hold even under the promise that $H(\boldsymbol{p}) = \Theta(\log n)$. The lower bound in (1) yields a lower bound for our additive approximation problem by taking $\gamma = \frac{1}{\log n}$, but only a nonoptimal one: $\Omega(\log n)$.

More generally, Canonne and Rubinfeld showed that $O(\frac{\log^2 n}{\Delta^2})$ queries suffice for estimating Shannon entropy to within $\pm \Delta$.[3] Note that this result is trivial once $\Delta \leq \frac{\log n}{\sqrt{n}}$ because of course the entire distribution $\boldsymbol{p}$ can be determined precisely with n PMF queries. In fact, our first main theorem is stated to give a matching lower bound for essentially the full range of Δ: for any $\frac{1}{n^{4999}} \leq \Delta \leq \log n$ we show that $\Omega(\frac{\log^2 n}{\Delta^2})$ queries are necessary in the SAMP+PMF model.

Our second main theorem is concerned with estimation of the *Rényi entropy* $H_\alpha(\boldsymbol{p})$ for various parameters $0 \leq \alpha \leq \infty$. Here

$$H_\alpha(\boldsymbol{p}) = \frac{1}{1-\alpha} \log\left(\sum_x \boldsymbol{p}[x]^\alpha\right),$$

interpreted in the limit when $\alpha = 0, 1, \infty$. The meaning for $\boldsymbol{p}$ is as follows: when $\alpha = 0$ it is the (log of the) support size; when $\alpha = 1$ it is the usual Shannon entropy; when $\alpha = \infty$ it is the min-entropy; when $\alpha = 2$ it is the (negative-log of the) collision probability; and for general integer $\alpha \geq 2$ it is related to higher-order collision probabilities.

A recent work of Acharya, Orlitsky, Suresh, and Tyagi [1] showed that for estimating $H_\alpha(\boldsymbol{p})$ in the samples-only model to within ± 1, $\Theta(n^{1-1/\alpha})$ samples

---

[3] They actually state $O(\frac{\log^2(n/\Delta)}{\Delta^2})$, but this is the same as $O(\frac{\log^2 n}{\Delta^2})$ because the range of interest is $\frac{1}{\sqrt{n}} \leq \Delta \leq \log n$.

are necessary and sufficient when α is an integer greater than 1, and $\widetilde{\Omega}(n)$ queries are necessary when α is a noninteger greater than 1. Our second main result is a tight characterization of the number of SAMP+PMF queries necessary and sufficient for estimating $H_\alpha(\boldsymbol{p})$ for all $\alpha > 1$. It turns out that PMF queries do not help in estimating Rényi entropies for integer α, whereas they *are* helpful for noninteger α.

Second Main Theorem. *Let $\alpha > 1$ be a real number. In the SAMP + PMF model, $\Omega(n^{1-1/\alpha}/2^\Delta)$ queries are necessary and $O\left(\dfrac{n^{1-1/\alpha}}{\left(1-2^{(1-\alpha)\Delta}\right)^2}\right)$ queries are sufficient to estimate (with high probability) the Rényi entropy $H_\alpha(\boldsymbol{p})$ of an unknown distribution $\boldsymbol{p}$ on $[n]$ to within $\pm\Delta$.*

Finally, we mention that our two lower bounds easily go through even when the more liberal "CDF" queries introduced in [6] are allowed. These queries take as input $x \in [n]$ and return the value $\sum_{y \le x} \boldsymbol{p}[y]$.[4] We will also show that the Canonne–Rubinfeld SAMP+PMF lower bound of $\Omega(1/\epsilon^2)$ for estimating support size to within $\pm\epsilon n$ can be extended to the more general SAMP+CDF model.

2   Lower Bound for Estimating Shannon Entropy

Theorem 2. *In the SAMP+PMF model, $\Omega\left(\dfrac{\log^2 n}{\Delta^2}\right)$ queries are necessary to estimate (with high probability) the Shannon entropy $H(\boldsymbol{p})$ of an unknown distribution $\boldsymbol{p}$ on $[n]$ to within $\pm\Delta$, where $\frac{1}{n^{.4999}} \le \Delta \le \log n$.*

Proof. We will show that a hypothetical SAMP+PMF algorithm $\mathcal{E}$ that can estimate the entropy of an unknown distribution on $[n]$ to within $\pm\Delta$ using $o\left(\dfrac{\log^2 n}{\Delta^2}\right)$ queries would contradict the well-known fact that $\Omega(1/\lambda^2)$ coin tosses are necessary to determine whether a given coin is fair, or comes up heads with probability $1/2 + \lambda$.

The idea is to use the given coin to realize the probability distribution that $\mathcal{E}$ will work on. Let n be the smallest one millionth power of a natural number that satisfies $\frac{3 \cdot 10^6 \Delta}{\log n} \le \lambda$. Partition the domain $[n]$ into $M = n^{.999999}$ consecutive blocks $I_1, \ldots, I_M$, each containing $K = \frac{n}{M} = n^{.000001}$ elements. Each block will be labeled either as a tails or a heads block. The internal distribution of each heads block is uniform, i.e. each element has probability mass $\frac{1}{MK} = \frac{1}{n}$. In each tails block, the first element has probability mass $\frac{1}{n^{.999999}}$, while the rest of the elements have probability mass 0. Note that the total probability mass of each block is $K \cdot \frac{1}{MK} = \frac{1}{M} = \frac{1}{n^{.999999}}$, regardless of its label.

We will now describe a costly method of constructing a probability distribution $\boldsymbol{p}$ of this kind, using a coin that comes up heads with probability d:

- Throw the coin M times to obtain the outcomes $\boldsymbol{X_1}, \ldots, \boldsymbol{X_M}$,
- Set the label of block I_m to $\boldsymbol{X_m}$, for all $m \in [M]$.

---

[4] Note that a PMF query can be simulated by two CDF queries.

Let $\boldsymbol{X}$ be the number of heads blocks in $\boldsymbol{p}$. Then $\mu = \mathbf{E}[\boldsymbol{X}] = Md$. Let $\overline{\boldsymbol{X}} = \frac{X}{M}$ denote the proportion of heads blocks in $\boldsymbol{p}$. Then we can calculate the entropy $H[\boldsymbol{p}]$ by calculating the individual entropies of the blocks. For a heads block, the entropy is $K \cdot \frac{1}{MK} \cdot \log(MK) = \frac{1}{M}\log n$. The entropy of a tails block is $\frac{1}{n^{.999999}}\log(n^{.999999}) = \frac{.999999}{M}\log n$. Since there are $M\overline{\boldsymbol{X}}$ heads blocks and $M(1-\overline{\boldsymbol{X}})$ tails blocks, the total entropy becomes $H[\boldsymbol{p}] = M\overline{\boldsymbol{X}} \cdot \frac{1}{M}\log n + M(1-\overline{\boldsymbol{X}}) \cdot \frac{.999999}{M}\log n = \overline{\boldsymbol{X}}\log n + .999999(1-\overline{\boldsymbol{X}})\log n = (.999999 + .000001\overline{\boldsymbol{X}})\log n$. Note that this function is monotone with respect to $\overline{\boldsymbol{X}}$.

Define two families of distributions $\mathcal{P}_1$ and $\mathcal{P}_2$ constructed by the above process, taking d to be $p_1 = \frac{1}{2}$ and $p_2 = \frac{1}{2} + \lambda$, respectively. Let $\boldsymbol{p_1}$ (respectively $\boldsymbol{p_2}$) be a probability distribution randomly chosen from $\mathcal{P}_1$ (respectively $\mathcal{P}_2$).

Proposition 3. $\boldsymbol{p_1}$ has entropy at most $.9999995\log n + \Delta$ with high probability.

Proof. We prove this by using a Chernoff bound on the number of heads blocks in the distribution.

$$\mathbf{Pr}\left[X \geq \left(p_1 + \frac{10^6\Delta}{\log n}\right)M\right] \leq \exp\left(-\frac{\frac{4\cdot10^{12}\Delta^2}{\log^2 n}}{2 + \frac{2\cdot10^6\Delta}{\log n}}\frac{M}{2}\right)$$

$$\leq \exp\left(-\frac{10^{12}\cdot n^{.999999}/n^{.999998}}{\log^2 n(1+10^6)}\right) = o(1).$$

The last term indicates that the proportion of the heads blocks $\overline{\boldsymbol{X}} < \left(p_1 + \frac{10^6\Delta}{\log n}\right)$ with high probability. Thus with high probability $H[\boldsymbol{p_1}] = (.999999 + .000001\overline{\boldsymbol{X}})\log n < .9999995\log n + \Delta$.   □

Proposition 4. $\boldsymbol{p_2}$ has entropy at least $.9999995\log n + 2\Delta$ with high probability.

Proof. We find a similar bound by;

$$\mathbf{Pr}\left[X \leq \left(p_2 - \frac{10^6\Delta}{\log n}\right)M\right] \leq \exp\left(-\frac{\frac{10^{12}\Delta^2}{p_2^2\log^2 n}}{2}p_2 M\right) \leq \exp\left(-\frac{n^{.000001}}{\log^2 n}\right) = o(1).$$

The last term indicates that the proportion of the heads blocks $\overline{\boldsymbol{X}} > \left(p_2 - \frac{10^6\Delta}{\log n}\right)$ with high probability. Thus with high probability $H[\boldsymbol{p_2}] = (.999999 + .000001\overline{\boldsymbol{X}})\log n > .9999995\log n + .000001(\lambda - \frac{10^6\Delta}{\log n})\log n \geq .9999995\log n + .000001(\frac{2\cdot10^6\Delta}{\log n})\log n = .9999995\log n + 2\Delta$.   □

Since the entropies of $\boldsymbol{p_1}$ and $\boldsymbol{p_2}$ are sufficiently far apart from each other, our hypothetical estimator $\mathcal{E}$ can be used to determine whether the underlying coin has probability p_1 or p_2 associated with it. To arrive at the contradiction we want, we must ensure that the coin is not thrown too many times during this process. This is achieved by constructing the distribution "on-the-fly"[6] during the execution of $\mathcal{E}$, throwing the coin only when it is required to determine the label of a previously undefined block:

When $\mathcal{E}$ makes a SAMP query, we choose a block I_m uniformly at random (since each block has probability mass $\frac{1}{M}$), and then flip the coin for I_m to decide its label if it is yet undetermined. We then draw a sample $i \sim \mathbf{d_m}$ from I_m, where $\mathbf{d_m}$ is the normalized distribution of the mth block.

When $\mathcal{E}$ makes a PMF query on $i \in [n]$, we flip the coin to determine the label of the associated block I_m if it is yet undetermined. We then return the probability mass of i.

By this procedure, the queries of $\mathcal{E}$ about the probability distribution $\mathbf{p}$ (known to be either $\mathbf{p_1}$ or $\mathbf{p_2}$) can be answered by using at most one coin flip per query, i.e. $o\left(\frac{\log^2 n}{\Delta^2}\right)$ times in total.

Since we selected n so that $1/\lambda^2 = \Theta(\frac{\log^2 n}{\Delta^2})$, this would mean that it is possible to distinguish between the two coins using only $o(1/\lambda^2)$ throws, which is a contradiction, letting us conclude that no algorithm can estimate the Shannon entropy $H(\mathbf{p})$ of an unknown distribution $\mathbf{p}$ on $[n]$ to within $\pm\Delta$ with high probability making $o\left(\frac{\log^2 n}{\Delta^2}\right)$ queries. $\qquad\square$

We now give a similar lower bound for the SAMP+CDF model.

Corollary 5. *In the SAMP+CDF model, any algorithm estimating (with high probability) the Shannon entropy $H(\mathbf{p})$ of an unknown distribution $\mathbf{p}$ on $[n]$ to within $\pm\Delta$ must make $\Omega\left(\frac{\log^2 n}{\Delta^2}\right)$ queries.*

Proof. The construction is identical to the one in the proof of Theorem 2, except that we now have to describe how the CDF queries of the estimation algorithm must be answered using the coin:

When $\mathcal{E}$ makes a CDF query on $i \in [n]$, we flip the coin to determine the label of the associated block I_m if this is necessary. We then return the sum of the total probability mass of the blocks preceding I_m (which is $\frac{m-1}{M}$, since each block has a total probability mass of $\frac{1}{M}$ regardless of its label) and the probability masses of the elements from the beginning of I_m up to and including i itself. At most one coin flip per CDF query is therefore sufficient. $\qquad\square$

3   Estimating Rényi Entropy

We start by demonstrating a lower bound.

Theorem 6. *For any $\alpha > 1$, $\Omega\left(\dfrac{n^{1-1/\alpha}}{2^\Delta}\right)$ SAMP+PMF queries are necessary to estimate (with high probability) the Rényi entropy $H_\alpha(\mathbf{p})$ of an unknown distribution $\mathbf{p}$ on $[n]$ to within $\pm\Delta$.*

Proof. We will first prove the theorem for rational α, and show that it remains valid for irrationals at the end.

The proof has the same structure as that of Theorem 2. One difference is that we reduce from the problem of distinguishing a maximally biased coin that never comes up tails from a less biased one (instead of the problem of distinguishing a fair coin from a biased one).

Suppose that we are given a coin whose probability of coming up heads is promised to be either $p_1 = 1$ or $p_2 = 1 - \lambda$ for a specified number λ, and we must determine which is the case. It is easy to show that this task requires at least $\Omega(1/\lambda)$ coin throws. We will show that this fact is contradicted if one assumes that there exist natural numbers s and t, where $\alpha = \frac{s}{t} > 1$, such that it is possible to estimate (with high probability) the Rényi entropy $H_\alpha(\boldsymbol{p})$ of an unknown distribution $\boldsymbol{p}$ on $[n]$ to within $\pm\Delta$ using an algorithm, say $\mathcal{R}$, that makes only $o(\frac{n^{1-1/\alpha}}{2^\Delta})$ SAMP+PMF queries.

Let n be the smallest number of the form $\left(\lceil 2^\Delta \rceil j\right)^s$ that satisfies $\frac{5 \cdot \lceil 2^\Delta \rceil}{n^{1-1/\alpha}} \leq \lambda$, where j is some natural number. Partition $[n]$ into $M = \frac{n^{1-1/\alpha}}{\lceil 2^\Delta \rceil}$ consecutive blocks $I_1, I_2, \ldots I_M$, each of size $K = \lceil 2^\Delta \rceil \cdot n^{1/\alpha}$. As in the proof of Theorem 2, a probability distribution $\boldsymbol{p}$ can be realized by throwing a given coin M times to obtain the outcomes $\boldsymbol{X_1}, \ldots, \boldsymbol{X_M}$, and setting the label of block I_m to $\boldsymbol{X_m}$, for all $m \in [M]$, where each member of each heads block again has probability mass $1/n$. The first member of each tails block has probability mass $\frac{\lceil 2^\Delta \rceil}{n^{1-1/\alpha}}$, and the remaining members have probability mass 0. We again have that each block has total probability mass $\frac{K}{n} = \frac{\lceil 2^\Delta \rceil n^{1/\alpha}}{n} = \frac{1}{M}$ regardless of its label, so this process always results in a legal probability distribution.

If the coin is maximally biased, then $\boldsymbol{p}$ becomes the uniform distribution, and $H_\alpha(\boldsymbol{p}) = \log n$. We will examine the probability of the same distribution being obtained using the less biased coin. Let $\mathcal{P}_2$ be the family of distributions constructed by the process described above, using a coin with probability p_2 of coming up heads. Let $\boldsymbol{p_2}$ be a probability distribution randomly chosen from $\mathcal{P}_2$.

The probability of the undesired case where $\boldsymbol{p_2}$ is the uniform distribution is

$$\Pr\left[\boldsymbol{p_2} = \mathcal{U}([n])\right] = p_2^M \leq \left(1 - \frac{5 \cdot \lceil 2^\Delta \rceil}{n^{1-1/\alpha}}\right)^M \leq e^{-\frac{5 \cdot \lceil 2^\Delta \rceil}{n^{1-1/\alpha}} M} = e^{-5} \leq \frac{1}{1000}.$$

That is, with probability $\geq .999$, $\boldsymbol{p_2}$ has at least one element with probability mass $\frac{\lceil 2^\Delta \rceil n^{\frac{1}{\alpha}}}{n}$. Let $\boldsymbol{X}$ be the number of heads outcomes and $\boldsymbol{B}$ and $\boldsymbol{W}$ denote the number of elements with probability mass $\frac{1}{n}$ and $\frac{\lceil 2^\Delta \rceil n^{\frac{1}{\alpha}}}{n}$, respectively. It is not difficult to see that $\boldsymbol{B} = K \cdot \boldsymbol{X}$ and $\boldsymbol{W} = M - \boldsymbol{X}$. We just showed that $\boldsymbol{X} < M$ with high probability.

Then the Rényi entropy of the constructed distribution $\boldsymbol{p_2} \in \mathcal{P}_2$ is, with high probability:

$$H_\alpha(\boldsymbol{p_2}) = \frac{1}{1 - \alpha} \log\left(\frac{K \cdot \boldsymbol{X}/n + (M - \boldsymbol{X})\lceil 2^\Delta \rceil^\alpha}{n^{\alpha-1}}\right)$$

$$\leq \log n - \frac{1}{\alpha - 1} \log\left(\lceil 2^\Delta \rceil^\alpha\right) \leq \log n - \Delta.$$

Because $H_\alpha(\mathcal{U}([n])) - H_\alpha(\boldsymbol{p_2}) \geq \Delta$, $\mathcal{R}$ has to be able to distinguish $\mathcal{U}([n])$ and $\boldsymbol{p_2}$ with high probability.

We can then perform a simulation of $\mathcal{R}$ involving an "on-the-fly" construction of distribution $\boldsymbol{p}$ exactly as described in the proof of Theorem 2. As discussed in Sect. 2, this process requires no more coin throws than the number of SAMP+PMF queries made by $\mathcal{R}$, allowing us to determine the type of the coin using only $o(\dfrac{n^{1-1/\alpha}}{2^\Delta})$, that is, $o(1/\lambda)$ tosses with high probability, a contradiction.

Having thus proven the statement for rational α, it is straightforward to cover the case of irrational α: Note that $H_\alpha(\boldsymbol{p})$ is a continuous function of α for fixed $\boldsymbol{p}$. Given any $\boldsymbol{p}$ and ε, for any irrational number α_i greater than 1, there exists a rational α_r which is so close to α_i such that $H_{\alpha_i}(\boldsymbol{p}) - H_{\alpha_r}(\boldsymbol{p}) < \varepsilon$. An efficient entropy estimation method for some irrational value of α would therefore imply the existence of an equally efficient method for some rational value, contradicting the result obtained above.   □

These results are generalized to the SAMP+CDF model in the same way as in Sect. 2:

Corollary 7. *For any* $\alpha > 1$, $\Omega\left(\dfrac{n^{1-1/\alpha}}{2^\Delta}\right)$ SAMP+PMF *or* SAMP+CDF *queries are necessary to estimate (with high probability) the Rényi entropy* $H_\alpha(\boldsymbol{p})$ *of an unknown distribution* $\boldsymbol{p}$ *on* $[n]$ *to within* $\pm\Delta$.

We now show that PMF queries are useful for the estimation of H_α for non-integer α.

Lemma 8. *For any number* $\alpha > 1$, *there exists an algorithm estimating (with high probability) the Rényi entropy* $H_\alpha(\boldsymbol{p})$ *of an unknown distribution* $\boldsymbol{p}$ *on* $[n]$ *to within* $\pm\Delta$ *with* $O\left(\dfrac{n^{1-1/\alpha}}{\left(1 - 2^{(1-\alpha)\Delta}\right)^2}\right)$ SAMP+PMF *queries.*

Proof. We will prove this statement for rational α. The generalization to irrational α discussed in the proof of Theorem 6 can be applied.

Defining the αth moment of $\boldsymbol{p}$ as

$$\mathcal{M}_\alpha(\boldsymbol{p}) = \sum_{i=1}^{n} (\boldsymbol{p}\,[i])^\alpha,$$

the Rényi entropy can be written as

$$H_\alpha(\boldsymbol{p}) = \frac{1}{1-\alpha} \log \mathcal{M}_\alpha(\boldsymbol{p}).$$

Observe that estimating $H_\alpha(\boldsymbol{p})$ to an additive accuracy of $\pm\Delta$ is equivalent to estimating $\mathcal{M}_\alpha(\boldsymbol{p})$ to a multiplicative accuracy of $2^{\pm\Delta(1-\alpha)}$. Therefore we construct an estimator for $\mathcal{M}_\alpha(\boldsymbol{p})$.

Let $M = \left\lceil \frac{100n^{1-1/\alpha}}{\left(1-2^{(1-\alpha)\Delta}\right)^2} \right\rceil$, and let $\boldsymbol{X_1}, \ldots, \boldsymbol{X_M}$ be i.i.d. random variables drawn from $\boldsymbol{p}$. Define $\boldsymbol{Y_i} = (\boldsymbol{p}\,[\boldsymbol{X_i}])^{\alpha-1}$, where $\boldsymbol{p}\,[\boldsymbol{X_i}]$ can be calculated using a PMF query on $\boldsymbol{X_i}$ for $1 \leq i \leq M$. Note that $\mathbf{E}\,[\boldsymbol{Y_i}] = \sum_{j=1}^{n} \boldsymbol{p}\,[j]\,(\boldsymbol{p}\,[j])^{\alpha-1} = \sum_{j=1}^{n} (\boldsymbol{p}\,[j])^{\alpha} = \mathcal{M}_\alpha(\boldsymbol{p})$. Then $\frac{1}{M}\sum_{i=1}^{M} \boldsymbol{Y_i}$ is an unbiased estimator of $\mathcal{M}_\alpha(\boldsymbol{p})$, because $\mathbf{E}\left[\frac{1}{M}\sum_{i=1}^{M} \boldsymbol{Y_i}\right] = \frac{1}{M}\sum_{i=1}^{M} \mathbf{E}\,[\boldsymbol{Y_i}] = \mathcal{M}_\alpha(\boldsymbol{p})$. Moreover, $\mathbf{Var}\,(\boldsymbol{Y_i}) = \mathbf{E}\,[\boldsymbol{Y_i}^2] - \mathbf{E}^2\,[\boldsymbol{Y_i}] = \sum_{j=1}^{n} \boldsymbol{p}\,[j]\,(\boldsymbol{p}\,[j])^{2\alpha-2} - \mathbf{E}^2[\boldsymbol{Y_i}] = \mathcal{M}_{2\alpha-1}(\boldsymbol{p}) - \mathcal{M}_\alpha^2(\boldsymbol{p})$. Since the $\boldsymbol{Y_i}$'s are also i.i.d. random variables, $\mathbf{Var}\left(\frac{1}{M}\sum_{i=1}^{M} \boldsymbol{Y_i}\right) = \frac{1}{M^2}\sum_{i=1}^{M} \mathbf{Var}\,(\boldsymbol{Y_i}) = \frac{M}{M^2}\mathbf{Var}\,(\boldsymbol{Y}) = \frac{1}{M}\left(\mathcal{M}_{2\alpha-1}(\boldsymbol{p}) - \mathcal{M}_\alpha^2(\boldsymbol{p})\right)$.

We use the following fact from [1] to find an upper bound for the variance of our empirical estimator.

Fact 9 ([1], Lemma 1). *For $\alpha > 1$ and $0 \leq \beta \leq \alpha$*

$$\mathcal{M}_{\alpha+\beta}(\boldsymbol{p}) \leq n^{(\alpha-1)(\alpha-\beta)/\alpha}\mathcal{M}_\alpha^2(\boldsymbol{p}) \ .$$

By taking $\beta = \alpha - 1$, we get

$$\sigma^2 = \mathbf{Var}\left(\frac{1}{M}\sum_{i=1}^{M} \boldsymbol{Y_i}\right) \leq \frac{1}{M}\mathcal{M}_\alpha^2(\boldsymbol{p})\left(n^{1-1/\alpha} - 1\right) \leq \mathcal{M}_\alpha^2(\boldsymbol{p})\frac{\left(1-2^{(1-\alpha)\Delta}\right)^2}{100}.$$

By Chebyshev's inequality we have

$$\mathbf{Pr}\left[\left|\frac{1}{M}\sum_{i=1}^{M} \boldsymbol{Y_i} - \mathcal{M}_\alpha(\boldsymbol{p})\right| \geq 10\sigma\right] \leq \frac{1}{100} \Rightarrow$$

$$\mathbf{Pr}\left[\left|\frac{1}{M}\sum_{i=1}^{M} \boldsymbol{Y_i} - \mathcal{M}_\alpha(\boldsymbol{p})\right| \leq \mathcal{M}_\alpha(\boldsymbol{p})\left(1-2^{(1-\alpha)\Delta}\right)\right] \geq .99$$

Thus we can estimate $\mathcal{M}_\alpha(\boldsymbol{p})$ to a desired multiplicative accuracy with $O\left(\frac{n^{1-1/\alpha}}{\left(1-2^{(1-\alpha)\Delta}\right)^2}\right)$ queries, which ends the proof. $\qquad\square$

Finally, we show a similar upper bound for $\alpha < 1$.

Lemma 10. *Let $\alpha < 1$ and $1 > \epsilon > 0$ be rational numbers. There exists an algorithm estimating (with high probability) the Rényi entropy $H_\alpha(\boldsymbol{p})$ of an unknown distribution $\boldsymbol{p}$ on $[n]$ to within $\pm\Delta$ with $O\left(\frac{n^\epsilon}{\left(1-2^{(1-\alpha)\Delta}\right)^2}\right)$ SAMP+PMF queries.*

The proof is omitted due to space constraints.

4   Lower Bound for Estimating Support Size

For any probability distribution p, $H_0(p) = \log(\mathrm{supp}(p))$, where $\mathrm{supp}(p)$ denotes the support size of p. Canonne and Rubinfeld [6] have shown that $\Omega(1/\epsilon^2)$ SAMP+PMF queries are necessary for estimating $\mathrm{supp}(p)$ to within $\pm \epsilon n$ where $[n]$ is the domain of p. We modify their proof to establish the same lower bound for this task in the SAMP+CDF model.

Theorem 11. $\Omega\left(\dfrac{1}{\epsilon^2}\right)$ SAMP+CDF queries are necessary to estimate (with high probability) the support size of an unknown distribution p on domain $[n]$ to within $\pm \epsilon n$.

Proof. Assume that there exists a program $\mathcal{S}$ which can accomplish the task specified in the theorem statement with only $o\left(\frac{1}{\epsilon^2}\right)$ queries. Let us show how $\mathcal{S}$ can be used to determine whether a given a coin is fair, or comes up heads is with probability $p_2 = \frac{1}{2} + \lambda$.

Set $\epsilon = \frac{\lambda}{6}$, and let n be the smallest even number satisfying $n \geq 10/\epsilon^2$. Partition the domain $[n]$ into $M = \frac{n}{2}$ blocks $I_1, \ldots, I_M$ where $I_m = \{2m-1, 2m\}$ for all $m \in [M]$. The construction of a probability distribution p based on coin flips is as follows:

- Throw the coin M times, with outcomes $X_1, \ldots, X_M$,
- for $m \in [M]$, set $p[2m-1] = \frac{2}{n}$ and $p[2m] = 0$ if X_m is heads, and set $p[2m-1] = p[2m] = \frac{1}{n}$ if X_m is tails.

Note that by construction $p[2m-1] + p[2m] = \frac{2}{n}$ for all $m \in [M]$.

Let $\mathcal{P}_1$ and $\mathcal{P}_2$ be the families of distributions constructed by the above process, using the fair and biased coin, respectively. Let p_1 (respectively p_2) be a probability distribution randomly chosen from $\mathcal{P}_1$ (respectively $\mathcal{P}_2$). Then

$$\mathbf{E}\left[\mathrm{supp}(p_1)\right] = n - M\frac{1}{2} = \frac{3}{4}n, \quad \mathbf{E}\left[\mathrm{supp}(p_2)\right] = n - Mp_2 = \left(\frac{3}{4} - 3\epsilon\right)n.$$

and via additive Chernoff bound,

$$\mathbf{Pr}\left[\mathrm{supp}(p_1) \leq \frac{3}{4}n - \frac{\epsilon}{2}n\right] \leq e^{-\frac{\epsilon^2 n}{2}} \leq e^{-5} < \frac{1}{1000}$$

$$\mathbf{Pr}\left[\mathrm{supp}(p_2) \geq \frac{3}{4}n - \frac{5\epsilon}{2}n\right] \leq e^{-\frac{\epsilon^2 n}{2}} \leq e^{-5} < \frac{1}{1000}$$

In other words, with high probability the resulting distributions will satisfy $\mathrm{supp}(p_1) - \mathrm{supp}(p_2) > 2\epsilon n$, distant enough for $\mathcal{S}$ to distinguish between two families.

As in our previous proofs, we could use $\mathcal{S}$ (if only it existed) to distinguish between the two possible coin types by using the coin for an on-the-fly construction of p. As before, SAMP and CDF queries are answered by picking a block randomly, throwing the coin if the type of this block has not been fixed before, and returning the answer depending on the type of the block. Since $o\left(\frac{1}{\epsilon^2}\right) = o\left(\frac{1}{\lambda^2}\right)$ coin tosses would suffice for this task, we have reached a contradiction.   □

5 Concluding Remarks

Tsallis entropy, defined as [16]

$$S_\alpha(\boldsymbol{p}) = \frac{\mathbf{k}}{\alpha - 1}\left(1 - \sum_{i=1}^{n}(\boldsymbol{p}\,[i])^\alpha\right),$$

where $\alpha \in \mathbb{R}$, and $\mathbf{k}$ is the Boltzmann constant, is a generalization of Boltzmann-Gibbs entropy. Harvey *et al.* [10] gave an algorithm to estimate Tsallis entropy, and used it to estimate Shannon entropy in the most general streaming model. Recalling the link shown in Lemma 8 between the tasks of estimating Rényi entropy $H_\alpha(\boldsymbol{p})$ and the αth moment $\mathcal{M}_\alpha(\boldsymbol{p})$, the results we obtained for Rényi entropy can be extended easily to Tsallis entropy:

Remark 12. Let $\alpha > 1$ be a real number. In both the SAMP+PMF and the SAMP+CDF models, $\Omega(n^{1-1/\alpha}/2^\Delta)$ queries are necessary and $O\left(\frac{n^{1-1/\alpha}}{\left(1-2^{(1-\alpha)\Delta}\right)^2}\right)$ queries are sufficient to estimate (with high probability) the Tsallis entropy $S_\alpha(\boldsymbol{p})$ of an unknown distribution $\boldsymbol{p}$ on $[n]$ to within $\pm\Delta$.

One problem left open by our work is that of optimal lower bounds for estimating the Rényi entropy $H_\alpha(\boldsymbol{p})$ in the SAMP + PMF model for $\alpha < 1$. The work [1] showed that in the model where only SAMP are allowed, $\widetilde{\Omega}(n^{1/\alpha})$ queries are necessary when $0 < \alpha < 1$. It is interesting to ask whether the bound in Lemma 10 is optimal.

Acknowledgments. We thank Clément Canonne for his assistance with our questions about the literature, and an anonymous reviewer for helpful remarks on a previous version of this manuscript.

References

1. Acharya, J., Orlitsky, A., Suresh, A.T., Tyagi, H.: The complexity of estimating Rényi entropy. In: Proceedings of the 26th Annual ACM-SIAM Symposium on Discrete Algorithms (2015)
2. Alon, N., Matias, Y., Szegedy, M.: The space complexity of approximating the frequency moments. J. Comput. Syst. Sci. **58**(1), 137–147 (1999)
3. Batu, T., Dasgupta, S., Kumar, R., Rubinfeld, R.: The complexity of approximating the entropy. SIAM J. Comput. **35**(1), 132–150 (2005)
4. Bhuvanagiri, L., Ganguly, S.: Estimating entropy over data streams. In: Azar, Y., Erlebach, T. (eds.) ESA 2006. LNCS, vol. 4168, pp. 148–159. Springer, Heidelberg (2006)
5. Canonne, C.: A survey on distribution testing: Your data is big. But is it blue? Technical Report TR15-063, ECCC (2015)
6. Canonne, C., Rubinfeld, R.: Testing probability distributions underlying aggregated data. Technical Report 1402.3835, arXiv (2014)

7. Chakrabarti, A., Cormode, G., McGregor, A.: A near-optimal algorithm for computing the entropy of a stream, pp. 328–335 (2007)
8. Chakrabarti, A., Ba, K.D., Muthukrishnan, S.: Estimating entropy and entropy norm on data streams. Internet Math. **3**(1), 63–78 (2006)
9. Guha, S., McGregor, A., Venkatasubramanian, S.: Streaming and sublinear approximation of entropy and information distances. In: Proceedings of the 17th Annual ACM-SIAM Symposium on Discrete Algorithms, pp. 733–742. ACM (2006)
10. Harvey, N., Nelson, J., Onak, K.: Sketching and streaming entropy via approximation theory. In: Proceedings of the 49th Annual IEEE Symposium on Foundations of Computer Science, pp. 489–498 (2008)
11. Kearns, M., Mansour, Y., Ron, D., Rubinfeld, R., Schapire, R., Sellie, L.: On the learnability of discrete distributions. In: Proceedings of the 26th Annual ACM Symposium on Theory of Computing, pp. 273–282 (1994)
12. Lall, A., Sekar, V., Ogihara, M., Xu, J., Zhang, H.: Data streaming algorithms for estimating entropy of network traffic. In: Proceedings of ACM SIGMETRICS, pp. 145–156 (2006)
13. Paninski, L.: Estimation of entropy and mutual information. Neural Comput. **15**(6), 1191–1253 (2003)
14. Paninski, L.: Estimating entropy on m bins given fewer than m samples. IEEE Trans. Inf. Theory **50**(9), 2200–2203 (2004)
15. Rubinfeld, R.: Taming big probability distributions. XRDS: Crossroads ACM Mag. Stud. **19**(1), 24–28 (2012)
16. Tsallis, C.: Possible generalization of Boltzmann-Gibbs statistics. Technical Report CBPF-NF-062/87, CBPF (1987)
17. Valiant, G., Valiant, P.: A CLT and tight lower bounds for estimating entropy. Technical Report TR10-179, Electronic Colloquium on Computational Complexity (2011)
18. Valiant, G., Valiant, P.: Estimating the unseen: an $n/\log(n)$-sample estimator for entropy and support size, shown optimal via new CLTsse. In: Proceedings of the 43rd Annual ACM Symposium on Theory of Computing, pp. 685–694 (2011)
19. Valiant, P.: Testing symmetric properties of distributions. SIAM J. Comput. **40**(6), 1927–1968 (2011)

Mutual Dimension and Random Sequences

Adam Case[✉] and Jack H. Lutz

Department of Computer Science, Iowa State University, Ames, IA 50011, USA
{adamcase,lutz}@iastate.edu

Abstract. If S and T are infinite sequences over a finite alphabet, then the *lower* and *upper mutual dimensions* $mdim(S : T)$ and $Mdim(S : T)$ are the upper and lower densities of the algorithmic information that is shared by S and T. In this paper we investigate the relationships between mutual dimension and *coupled randomness*, which is the algorithmic randomness of two sequences R_1 and R_2 with respect to probability measures that may be dependent on one another. For a restricted but interesting class of coupled probability measures we prove an explicit formula for the mutual dimensions $mdim(R_1 : R_2)$ and $Mdim(R_1 : R_2)$, and we show that the condition $Mdim(R_1 : R_2) = 0$ is necessary but not sufficient for R_1 and R_2 to be independently random.

We also identify conditions under which Billingsley generalizations of the mutual dimensions $mdim(S : T)$ and $Mdim(S : T)$ can be meaningfully defined; we show that under these conditions these generalized mutual dimensions have the "correct" relationships with the Billingsley generalizations of $dim(S)$, $Dim(S)$, $dim(T)$, and $Dim(T)$ that were developed and applied by Lutz and Mayordomo; and we prove a divergence formula for the values of these generalized mutual dimensions.

1   Introduction

Algorithmic information theory combines tools from the theory of computing and classical Shannon information theory to create new methods for quantifying information in an expanding variety of contexts. Two notable and related strengths of this approach that were evident from the beginning [11] are its abilities to quantify the information in and to access the randomness of *individual* data objects.

Some useful mathematical objects, such as real numbers and execution traces of nonterminating processes, are intrinsically infinitary. The randomness of such objects was successfully defined very early [18] but it was only at the turn of the present century [14,15] that ideas of Hausdorff were reshaped in order to define *effective fractal dimensions*, which quantify the densities of algorithmic information in such infinitary objects. Effective fractal dimensions, of which there

---

This research was supported in part by National Science Foundation Grants 0652519, 1143830, and 124705. Part of the second author's work was done during a sabbatical at Caltech and the Isaac Newton Institute for Mathematical Sciences at the University of Cambridge.

G.F. Italiano et al. (Eds.): MFCS 2015, Part II, LNCS 9235, pp. 199–210, 2015.
DOI: 10.1007/978-3-662-48054-0_17

are now many, and their relations with randomness are now a significant part of algorithmic information theory [6].

Many scientific challenges require us to quantify not only the information in an individual object, but also the information *shared* by two objects. The *mutual information* $I(X : Y)$ of classical Shannon information theory does something along these lines, but for two probability spaces of objects rather than for two individual objects [5]. The *algorithmic mutual information* $I(x : y)$, defined in terms of Kolmogorov complexity [13], quantifies the information shared by two individual finite objects x and y.

The present authors recently developed the *mutual dimensions* $mdim(x : y)$ and $Mdim(x : y)$ in order to quantify the density of algorithmic information shared by two infinitary objects x and y [4]. The objects x and y of interest in [4] are points in Euclidean spaces $\mathbb{R}^n$ and their images under computable functions, so the fine-scale geometry of $\mathbb{R}^n$ plays a major role there.

In this paper we investigate mutual dimensions further, with objectives that are more conventional in algorithmic information theory. Specifically, we focus on the lower and upper mutual dimensions $mdim(S : T)$ and $Mdim(S : T)$ between two sequences $S, T \in \Sigma^\infty$, where Σ is a finite alphabet. (If $\Sigma = \{0, 1\}$, then we write **C** for the *Cantor space* Σ^∞.) The definitions of these mutual dimensions, which are somewhat simpler in Σ^∞ than in $\mathbb{R}^n$, are implicit in [4] and explicit in Sect. 2 below.

Our main objective here is to investigate the relationships between mutual dimension and *coupled randomness*, which is the algorithmic randomness of two sequences R_1 and R_2 with respect to probability measures that may be dependent on one another. In Sect. 3 below we formulate coupled randomness precisely, and we prove our main theorem, Theorem 3.4, which gives an explicit formula for $mdim(R_1 : R_2)$ and $Mdim(R_1 : R_2)$ in a restricted but interesting class of coupled probability measures. This theorem can be regarded as a "mutual version" of Theorem 7.7 of [14], which in turn is an algorithmic extension of a classical theorem of Eggleston [2,7]. We also show in Sect. 3 that $Mdim(R_1 : R_2) = 0$ is a necessary, but not sufficient condition for two random sequences R_1 and R_2 to be independently random.

In 1960 Billingsley investigated generalizations of Hausdorff dimension in which the dimension itself is defined "through the lens of" a given probability measure [1,3]. Lutz and Mayordomo developed the effective Billingsley dimensions $dim^\nu(S)$ and $Dim^\nu(S)$, where ν is a probability measure on Σ^∞, and these have been useful in the algorithmic information theory of self-similar fractals [8,17].

In Sect. 4 we investigate "Billingsley generalizations" $mdim^\nu(S : T)$ and $Mdim^\nu(S : T)$ of $mdim(S : T)$ and $Mdim(S : T)$, where ν is a probability measure on $\Sigma^\infty \times \Sigma^\infty$. These turn out to make sense only when S and T are *mutually normalizable*, which means that the normalizations implicit in the fact that these dimensions are *densities* of shared information are the same for S as for T. We prove that, when mutual normalizability is satisfied, the Billingsley mutual dimensions $mdim^\nu(S : T)$ and $Mdim^\nu(S : T)$ are well behaved. We also identify a sufficient condition for mutual normalizability, make some preliminary observations on when it holds, and we prove a divergence formula, analogous to a theorem of [16], for computing the values of the Billingsley mutual dimensions in many cases.

2   Mutual Dimension in Cantor Spaces

In [4] we defined and investigated the mutual dimension between points in Euclidean space. The purpose of this section is to develop a similar framework for the mutual dimension between sequences in Σ^∞. We begin by reviewing the definitions of Kolmogorov complexity and dimension.

Definition. The *conditional Kolmogorov complexity* of $u \in \Sigma^*$ given $w \in \Sigma^*$ is

$$K(u \,|\, w) = \min\{|\pi| \,\big|\, \pi \in \{0,1\}^* \text{ and } U(\pi, w) = u\}.$$

In the above definition, U is a *universal* Turing machine. The *Kolmogorov complexity* of a string $u \in \Sigma^*$ is $K(u) = K(u \,|\, \lambda)$, where λ is the empty string. For a detailed overview of Kolmogorov complexity and its properties, see [13].

Definition. The *lower* and *upper dimensions* of $S \in \Sigma^\infty$ are

$$dim(S) = \liminf_{u \to S} \frac{K(u)}{|u| \log |\Sigma|}$$

and

$$Dim(S) = \limsup_{u \to S} \frac{K(u)}{|u| \log |\Sigma|},$$

respectively.

The rest of this section will be about mutual information and mutual dimension. We now give the definition of the mutual information between strings as defined in [13].

Definition. The *mutual information* between $u \in \Sigma^*$ and $w \in \Sigma^*$ is

$$I(u : w) = K(w) - K(w \,|\, u).$$

Using the above definition, we introduce the upper and lower mutual dimensions between sequences. Note that, for all $S, T \in \Sigma^\infty$, the notation (S, T) represents the sequence in $(\Sigma \times \Sigma)^\infty$ obtained after pairing each symbol in S with the symbol in T located at the same position.

Definition. The *lower* and *upper mutual dimensions* between $S \in \Sigma^\infty$ and $T \in \Sigma^\infty$ are

$$mdim(S : T) = \liminf_{(u,w) \to (S,T)} \frac{I(u : w)}{|u| \log |\Sigma|}$$

and

$$Mdim(S:T) = \limsup_{(u,w)\to(S,T)} \frac{I(u:w)}{|u|\log|\Sigma|},$$

respectively.

(We insist that $|u| = |w|$ in the above limits.) The mutual dimension between two sequences is regarded as the density of algorithmic mutual information between them.

Our main result for this section shows that mutual dimension between sequences is well behaved.

Theorem 2.1. *For all $S, T \in \Sigma^{\infty}$,*

1. $mdim(S:T) \leq Dim(S) + Dim(T) - Dim(S,T)$.
2. $mdim(S:T) \geq dim(S) + dim(T) - Dim(S,T)$.
3. $Mdim(S:T) \leq Dim(S) + Dim(T) - dim(S,T)$.
4. $Mdim(S:T) \geq dim(S) + dim(T) - dim(S,T)$.
5. $mdim(S:T) \leq \min\{dim(S), dim(T)\}$.
6. $Mdim(S:T) \leq \min\{Dim(S), Dim(T)\}$.
7. $0 \leq mdim(S:T) \leq Mdim(S:T) \leq 1$.
8. *If S and T are independently random, then $Mdim(S:T) = 0$.*
9. $mdim(S:T) = mdim(T:S)$, $Mdim(S:T) = Mdim(T:S)$.

3   Mutual Dimension and Coupled Randomness

In this section we investigate the mutual dimensions between coupled random sequences. Because coupled randomness is new to algorithmic information theory, we first review the technical framework for it. Let Σ be a finite alphabet. A *(Borel) probability measure* on the Cantor space Σ^{∞} of all infinite sequences over Σ is (conveniently represented by) a function $\nu : \Sigma^* \to [0,1]$ with the following two properties.

1. $\nu(\lambda) = 1$, where λ is the empty string.
2. For every $w \in \Sigma^*$, $\nu(w) = \displaystyle\sum_{a \in \Sigma} \nu(wa)$.

Intuitively, here, $\nu(w)$ is the probability that $w \sqsubseteq S$ (w is a *prefix* of S) when $S \in \Sigma^{\infty}$ is "chosen according to" the probability measure ν.

Most of this paper concerns a very special class of probability measures on Σ^{∞}. For each $n \in \mathbb{N}$, let $\alpha^{(n)}$ be a probability measure on Σ, i.e., $\alpha^{(n)} : \Sigma \to [0,1]$, with

$$\sum_{a \in \Sigma} \alpha^{(n)}(a) = 1,$$

and let $\vec{\alpha} = (\alpha^{(0)}, \alpha^{(1)}, \ldots)$ be the sequence of these probability measures on Σ. Then the *product* of $\vec{\alpha}$ (or, emphatically distinguishing it from the products

$\nu_1 \times \nu_2$ below, the *longitudinal product* of $\vec{\alpha}$) is the probability measure $\mu[\vec{\alpha}]$ on Σ^* defined by

$$\mu[\vec{\alpha}](w) = \prod_{n=0}^{|w|-1} \alpha^{(n)}(w[n])$$

for all $w \in \Sigma^*$, where $w[n]$ is the nth symbol in w. Intuitively, a sequence $S \in \Sigma^\infty$ is "chosen according to" $\mu[\vec{\alpha}]$ by performing the successive experiments $\alpha^{(0)}, \alpha^{(1)}, \dots$ *independently.*

To extend probability to pairs of sequences, we regard $\Sigma \times \Sigma$ as an alphabet and rely on the natural identification between $\Sigma^\infty \times \Sigma^\infty$ and $(\Sigma \times \Sigma)^\infty$. A probability measure on $\Sigma^\infty \times \Sigma^\infty$ is thus a function $\nu : (\Sigma \times \Sigma)^* \to [0, 1]$. It is convenient to write elements of $(\Sigma \times \Sigma)^*$ as ordered pairs (u, v), where $u, v \in \Sigma^*$ *have the same length.* With this notation, condition 2 above says that, for every $(u, v) \in (\Sigma \times \Sigma)^*$,

$$\nu(u, v) = \sum_{a,b \in \Sigma} \nu(ua, vb).$$

If ν is a probability measure on $\Sigma^\infty \times \Sigma^\infty$, then the first and second *marginal probability measures* of ν (briefly, the first and second *marginals* of ν) are the functions $\nu_1, \nu_2 : \Sigma^* \to [0, 1]$ defined by

$$\nu_1(u) = \sum_{v \in \Sigma^{|u|}} \nu(u, v), \quad \nu_2(v) = \sum_{u \in \Sigma^{|v|}} \nu(u, v).$$

It is easy to verify that ν_1 and ν_2 are probability measures on Σ^*. The probability measure ν here is often called a *joint probability measure* on $\Sigma^\infty \times \Sigma^\infty$, or a *coupling* of the probability measures ν_1 and ν_2.

If ν_1 and ν_2 are probability measures on Σ^∞, then the *product probability measure* $\nu_1 \times \nu_2$ on $\Sigma^\infty \times \Sigma^\infty$ is defined by

$$(\nu_1 \times \nu_2)(u, v) = \nu_1(u)\nu_2(v)$$

for all $u, v \in \Sigma^*$ with $|u| = |v|$. It is well known and easy to see that $\nu_1 \times \nu_2$ is, indeed, a probability measure on $\Sigma^\infty \times \Sigma^\infty$ and that the marginals of $\nu_1 \times \nu_2$ are ν_1 and ν_2. Intuitively, $\nu_1 \times \nu_2$ is the coupling of ν_1 and ν_2 in which ν_1 and ν_2 are *independent*, or *uncoupled.*

We are most concerned here with coupled longitudinal product probability measures on $\Sigma^\infty \times \Sigma^\infty$. For each $n \in \mathbb{N}$, let $\alpha^{(n)}$ be a probability measure on $\Sigma \times \Sigma$, i.e., $\alpha^{(n)} : \Sigma \times \Sigma \to [0, 1]$, with

$$\sum_{a,b \in \Sigma} \alpha^{(n)}(a, b) = 1,$$

and let $\vec{\alpha} = (\alpha^{(0)}, \alpha^{(1)}, \dots)$ be the sequence of these probability measures. Then the longitudinal product $\mu[\vec{\alpha}]$ is defined as above, but now treating $\Sigma \times \Sigma$ as the alphabet. It is easy to see that the marginals of $\mu[\vec{\alpha}]$ are $\mu[\vec{\alpha}]_1 = \mu[\vec{\alpha_1}]$ and $\mu[\vec{\alpha}]_2 = \mu[\vec{\alpha_2}]$, where each $\alpha_i^{(n)}$ is the marginal on Σ given by

$$\alpha_1^{(n)}(a) = \sum_{b \in \Sigma} \alpha^{(n)}(a, b), \quad \alpha_2^{(n)}(b) = \sum_{a \in \Sigma} \alpha^{(n)}(a, b).$$

The following class of examples is useful [20] and instructive.

Example 3.1. Let $\Sigma = \{0, 1\}$. For each $n \in \mathbb{N}$, fix a real number $\rho_n \in [-1, 1]$, and define the probability measure $\alpha^{(n)}$ on $\Sigma \times \Sigma$ by $\alpha^{(n)}(0, 0) = \alpha^{(n)}(1, 1) = \frac{1+\rho_n}{4}$ and $\alpha^{(n)}(0, 1) = \alpha^{(n)}(1, 0) = \frac{1-\rho_n}{4}$. Then, writing $\alpha^{\vec{\rho}}$ for $\vec{\alpha}$, the longitudinal product $\mu[\alpha^{\vec{\rho}}]$ is a probability measure on $\mathbf{C} \times \mathbf{C}$. It is routine to check that the marginals of $\mu[\alpha^{\vec{\rho}}]$ are

$$\mu[\alpha^{\vec{\rho}}]_1 = \mu[\alpha^{\vec{\rho}}]_2 = \mu,$$

where $\mu(w) = 2^{-|w|}$ is the uniform probability measure on $\mathbf{C}$.

It is convenient here to use Schnorr's martingale characterization [6, 13, 19, 21–23] of the algorithmic randomness notion introduced by Martin-Löf [18]. If ν is a probability measure on Σ^{∞}, then a ν–*martingale* is a function $d : \Sigma^* \to [0, \infty)$ satisfying $d(w)\nu(w) = \sum_{a \in \Sigma} d(wa)\nu(wa)$ for all $w \in \Sigma^*$. A ν–martingale d *succeeds* on a sequence $S \in \Sigma^{\infty}$ if $\limsup_{w \to S} d(w) = \infty$. A ν–martingale d is *constructive*, or *lower semicomputable*, if there is a computable function $\hat{d} : \Sigma^* \times \mathbb{N} \to \mathbb{Q} \cap [0, \infty]$ such that $\hat{d}(w, t) \leq (\hat{d})(w, t + 1)$ holds for all $w \in \Sigma^*$ and $t \in \mathbb{N}$, and $\lim_{t \to \infty} \hat{d}(w, t) = d(w)$ holds for all $w \in \Sigma^*$. A sequence $R \in \Sigma^{\infty}$ is *random* with respect to a probability measure ν on Σ^* if no lower semicomputable ν–martingale succeeds on R.

If we once again treat $\Sigma \times \Sigma$ as an alphabet, then the above notions all extend naturally to $\Sigma^{\infty} \times \Sigma^{\infty}$. Hence, when we speak of a *coupled pair* (R_1, R_2) *of random sequences*, we are referring to a pair $(R_1, R_2) \in \Sigma^{\infty} \times \Sigma^{\infty}$ that is random with respect to some probability measure ν on $\Sigma^{\infty} \times \Sigma^{\infty}$ that is explicit or implicit in the discussion. An extensively studied special case here is that $R_1, R_2 \in \Sigma^{\infty}$ are defined to be *independently random* with respect to probability measures ν_1, ν_2, respectively, on Σ^{∞} if (R_1, R_2) is random with respect to the product probability measure $\nu_1 \times \nu_2$ on $\Sigma^{\infty} \times \Sigma^{\infty}$.

When there is no possibility of confusion, we use such convenient abbreviations as "random with respect to $\vec{\alpha}$" for "random with respect to $\mu[\vec{\alpha}]$".

A trivial transformation of Martin-Löf tests establishes the following well known fact.

Observation 3.2. *If ν is a computable probability measure on $\Sigma^{\infty} \times \Sigma^{\infty}$ and $(R_1, R_2) \in \Sigma^{\infty} \times \Sigma^{\infty}$ is random with respect to ν, then R_1 and R_2 are random with respsect to the marginals ν_1 and ν_2.*

Example 3.3. If $\vec{\rho}$ is a computable sequence of reals $\rho_n \in [-1, 1]$, $\alpha^{\vec{\rho}}$ is as in Example 3.1, and $(R_1, R_2) \in \mathbf{C} \times \mathbf{C}$ is random with respect to $\alpha^{\vec{\rho}}$, then Observation 3.2 tells us that R_1 and R_2 are random with respect to the uniform probability measure on $\mathbf{C}$.

In applications one often encounters longitudinal product measures $\mu[\vec{\alpha}]$ in which the probability measures $\alpha^{(n)}$ are all the same (the i.i.d. case) or else converge to some limiting probability measure. The following theorem says that, in such cases, the mutual dimensions of coupled pairs of random sequences are easy to compute.

Theorem 3.4. *If $\vec{\alpha}$ is a computable sequence of probability measures $\alpha^{(n)}$ on $\Sigma \times \Sigma$ that converge to a probability measure α on $\Sigma \times \Sigma$, then for every coupled pair $(R_1, R_2) \in \Sigma^\infty \times \Sigma^\infty$ that is random with respect to $\vec{\alpha}$,*

$$mdim(R_1 : R_2) = Mdim(R_1 : R_2) = \frac{I(\alpha_1 : \alpha_2)}{\log |\Sigma|},$$

where

$$I(\alpha_1 : \alpha_2) = \sum_{a,b \in \Sigma} \alpha(a,b) \log \frac{\alpha(a,b)}{\alpha_1(a)\alpha_2(b)}$$

is the Shannon mutual information between the marginals α_1 and α_2 of α, and the logarithms are base-2.

Example 3.5. Let $\Sigma = \{0,1\}$, and let $\vec{\rho}$ be a computable sequence of reals $\rho_n \in [-1,1]$ that converge to a limit ρ. Define the probability measure α on $\Sigma \times \Sigma$ by $\alpha(0,0) = \alpha(1,1) = \frac{1+\rho}{4}$ and $\alpha(0,1) = \alpha(1,0) = \frac{1-\rho}{4}$, and let α_1 and α_2 be the marginals of α. If $\alpha^{\vec{\rho}}$ is as in Example 3.1, then for every pair $(R_1, R_2) \in \Sigma^\infty \times \Sigma^\infty$ that is random with respect to $\alpha^{\vec{\rho}}$, Theorem 3.4 tells us that

$$mdim(R_1 : R_2) = Mdim(R_1 : R_2)$$
$$= I(\alpha_1 : \alpha_2)$$
$$= 1 - \mathcal{H}(\frac{1+\rho}{2}),$$

where

$$\mathcal{H}(\theta) = \theta \log \frac{1}{\theta} + (1-\theta) \log \frac{1}{1-\theta}$$

is the *Shannon entropy* of $\theta \in [0,1]$. In particular, if the limit ρ is 0, then

$$mdim(R_1 : R_2) = Mdim(R_1 : R_2) = 0.$$

Theorem 3.4 has the following easy consequence, which generalizes the last sentence of Example 3.5.

Corollary 3.6. *If $\vec{\alpha}$ is a computable sequence of probability measures $\alpha^{(n)}$ on $\Sigma \times \Sigma$ that converge to a product probability measure $\alpha_1 \times \alpha_2$ on $\Sigma \times \Sigma$, then for every coupled pair $(R_1, R_2) \in \Sigma^\infty \times \Sigma^\infty$ that is random with respect to $\vec{\alpha}$,*

$$mdim(R_1 : R_2) = Mdim(R_1 : R_2) = 0.$$

Applying Corollary 3.6 to a constant sequence $\vec{\alpha}$ in which each $\alpha^{(n)}$ is a product probability measure $\alpha_1 \times \alpha_2$ on $\Sigma \times \Sigma$ gives the following.

Corollary 3.7. *If α_1 and α_2 are computable probability measures on Σ, and if $R_1, R_2 \in \Sigma^\infty$ are independently random with respect to α_1, α_2, respectively, then*

$$mdim(R_1 : R_2) = Mdim(R_1 : R_2) = 0.$$

We conclude this section by showing that the converse of Corollary 3.7 does *not* hold. This can be done via a direct construction, but it is more instructive to use a beautiful theorem of Kakutani, van Lambalgen, and Vovk. The *Hellinger distance* between two probability measures α_1 and α_2 on Σ is

$$H(\alpha_1, \alpha_2) = \sqrt{\sum_{a \in \Sigma} (\sqrt{\alpha_1(a)} - \sqrt{\alpha_2(a)})^2}.$$

(See [12], for example.) A sequence $\alpha = (\alpha^{(0)}, \alpha^{(1)}, \ldots)$ of probability measures on Σ is *strongly positive* if there is a real number $\delta > 0$ such that, for all $n \in \mathbb{N}$ and $a \in \Sigma$, $\alpha^{(n)}(a) \geq \delta$. Kakutani [10] proved the classical, measure-theoretic version of the following theorem, and van Lambalgen [24,25] and Vovk [26] extended it to algorithmic randomness.

Theorem 3.8. *Let $\vec{\alpha}$ and $\vec{\beta}$ be computable, strongly positive sequences of probability measures on Σ.*

1. If

$$\sum_{n=0}^{\infty} H(\alpha^{(n)}, \beta^{(n)})^2 < \infty,$$

then a sequence $R \in \Sigma^\infty$ is random with respect to $\vec{\alpha}$ if and only if it is random with respect to $\vec{\beta}$.

2. If

$$\sum_{n=0}^{\infty} H(\alpha^{(n)}, \beta^{(n)})^2 = \infty,$$

then no sequence is random with respect to both $\vec{\alpha}$ and $\vec{\beta}$.

Observation 3.9. *Let $\Sigma = \{0, 1\}$. If $\rho = [-1, 1]$ and probability measure α on $\Sigma \times \Sigma$ is defined from ρ as in Example 3.5, then*

$$H(\alpha_1 \times \alpha_2, \alpha)^2 = 2 - \sqrt{1 + \rho} - \sqrt{1 - \rho}.$$

Corollary 3.10. *Let $\Sigma = \{0, 1\}$ and $\delta \in (0, 1)$. Let $\vec{\rho}$ be a computable sequence of real numbers $\rho_n \in [\delta - 1, 1 - \delta]$, and let $\alpha^{\vec{\rho}}$ be as in Example 3.1. If*

$$\sum_{n=0}^{\infty} \rho_n^2 = \infty,$$

and if $(R_1, R_2) \in \Sigma^\infty \times \Sigma^\infty$ is random with respect to $\alpha^{\vec{\rho}}$, then R_1 and R_2 are not independently random with respect to the uniform probability measure on $\mathbf{C}$.

Corollary 3.11. *There exist sequences $R_1, R_2 \in \mathbf{C}$ that are random with respect to the uniform probability measure on $\mathbf{C}$ and satisfy $Mdim(R_1 : R_2) = 0$, but are not independently random.*

4   Billingsley Mutual Dimensions

We begin this section by reviewing the Billingsley generalization of constructive dimension, i.e., dimension with respect to probability measures. A probability measure β on Σ^∞ is *strongly positive* if there exists $\delta > 0$ such that, for all $w \in \Sigma^*$ and $a \in \Sigma$, $\beta(wa) > \delta\beta(w)$. The *Shannon self-information* of $w \in \Sigma^*$ with respect to β is $\ell_\beta(w) = \log \frac{1}{\beta(w)}$. The *Kullback-Leibler divergence* between probability measures α and β on Σ is $\mathcal{D}(\alpha||\beta) = \sum_{a \in \Sigma} \alpha(a) \log \frac{\alpha(a)}{\beta(a)}$.

In [17], Lutz and Mayordomo defined (and usefully applied) constructive Billingsley dimension in terms of gales and proved that it can be characterized using Kolmogorov complexity. Since Kolmogorov complexity is more relevent in this discussion, we treat the following theorem as a definition.

Definition (Lutz and Mayordomo [17]). The *dimension of $S \in \Sigma^\infty$ with respect to* a strongly positive probability measure β on Σ^∞ is

$$dim^\beta(S) = \liminf_{w \to S} \frac{K(w)}{\ell_\beta(w)}.$$

In the above definition the denominator $\ell_\beta(w)$ normalizes the dimension to be a real number in [0,1]. It seems natural to define the Billingsley generalization of mutual dimension in a similar way by normalizing the algorithmic mutual information between u and w by $\log \frac{\beta(u,w)}{\beta_1(u)\beta_2(w)}$ (i.e., the *self-mutual information* or *pointwise mutual information* between u and w [9]) as $(u, w) \to (S, T)$. However, this results in bad behavior. For example, the mutual dimension between any two sequences with respect to the uniform probability measure on $\Sigma \times \Sigma$ is *always* undefined. Other thoughtful modifications to this natural definition resulted in sequences having negative or infinitely large mutual dimension. The main problem here is that, given a particular probability measure, one can construct certain sequences whose prefixes have extremely large positive or negative self-mutual information. In order to avoid undesirable behavior, we restrict the definition of Billingsley mutual dimension to sequences that are mutually normalizable.

Definition. Let β be a probability measure on $\Sigma^\infty \times \Sigma^\infty$. Two sequences $S, T \in \Sigma^\infty$ are *mutually β–normalizable* (in this order) if

$$\lim_{(u,w) \to (S,T)} \frac{\ell_{\beta_1}(u)}{\ell_{\beta_2}(w)} = 1.$$

Definition. Let $S, T \in \Sigma^\infty$ be mutually β–normalizable. The *upper* and *lower* *mutual dimensions* between S and T with respect to β are

$$mdim^\beta(S : T) = \liminf_{(u,w) \to (S,T)} \frac{I(u : w)}{\ell_{\beta_1}(u)} = \liminf_{(u,w) \to (S,T)} \frac{I(u : w)}{\ell_{\beta_2}(w)}$$

and

$$Mdim^\beta(x : y) = \limsup_{(u,w) \to (S,T)} \frac{I(u : w)}{\ell_{\beta_1}(u)} = \limsup_{(u,w) \to (S,T)} \frac{I(u : w)}{\ell_{\beta_2}(w)},$$

respectively.

The above definition has nice properties because β–normalizable sequences have prefixes with asymptotically equivalent self-information. Given the basic properties of mutual information and Shannon self-information, we can see that

$$0 \le mdim^\beta(S : T) \le \min\{dim^{\beta_1}(S), dim^{\beta_2}(T)\} \le 1.$$

Clearly, $Mdim^\beta$ also has a similar property.

The rest of this paper will be primarily concerned with probability measures on alphabets. Our first result of this section is a mutual divergence formula for random, mutually β–normalizable sequences. This can be thought of as a "mutual" version of a divergence formula in [16].

Theorem 4.1 (Mutual Divergence Formula). *If α and β are computable, positive probability measures on $\Sigma \times \Sigma$, then, for every $(R_1, R_2) \in \Sigma^\infty \times \Sigma^\infty$ that is random with respect to α such that R_1 and R_2 are mutually β–normalizable,*

$$mdim^\beta(R_1 : R_2) = Mdim^\beta(R_1 : R_2) = \frac{I(\alpha_1 : \alpha_2)}{\mathcal{H}(\alpha_1) + \mathcal{D}(\alpha_1 || \beta_1)} = \frac{I(\alpha_1 : \alpha_2)}{\mathcal{H}(\alpha_2) + \mathcal{D}(\alpha_2 || \beta_2)}.$$

We conclude this section by making some initial observations regarding when mutual normalizability can be achieved.

Definition. Let α_1, α_2, β_1, β_2 be probability measures over Σ. We say that α_1 is (β_1, β_2)–*equivalent* to α_2 if

$$\sum_{a \in \Sigma} \alpha_1(a) \log \frac{1}{\beta_1(a)} = \sum_{a \in \Sigma} \alpha_2(a) \log \frac{1}{\beta_2(a)}.$$

For a probability measure α on Σ, let $FREQ_\alpha$ be the set of sequences $S \in \Sigma^\infty$ satisfying $\lim_{n \to \infty} n^{-1}|\{i < n \mid S[i] = a\}| = \alpha(a)$ for all $a \in \Sigma$.

Lemma 4.2. *Let α_1, α_2, β_1, β_2 be probability measures on Σ. If α_1 is (β_1, β_2)–equivalent to α_2, then, for all pairs $(S, T) \in FREQ_{\alpha_1} \times FREQ_{\alpha_2}$, S and T are mutually β–normalizable.*

Given probability measures β_1 and β_2 on Σ, we would like to know which sequences are mutually β–normalizable. The following results help to answer this question for probability measures on and sequences over $\{0, 1\}$.

Lemma 4.3. *Let β_1 and β_2 be probability measures on $\{0,1\}$ such that exactly one of the following conditions hold.*

1. $0 < \beta_2(0) < \beta_1(1) < \beta_1(0) < \beta_2(1) < 1$
2. $0 < \beta_2(1) < \beta_1(0) < \beta_1(1) < \beta_2(0) < 1$
3. $0 < \beta_2(0) < \beta_1(0) < \beta_1(1) < \beta_2(1) < 1$
4. $0 < \beta_2(1) < \beta_1(1) < \beta_1(0) < \beta_2(0) < 1$
5. $\beta_1 = \mu$ and $\beta_2 \neq \mu$.

For every $x \in [0,1]$,

$$0 < \frac{x \cdot \log \frac{\beta_1(1)}{\beta_1(0)} + \log \frac{\beta_2(1)}{\beta_1(1)}}{\log \frac{\beta_2(1)}{\beta_2(0)}} < 1.$$

Theorem 4.4. *Let β_1 and β_2 be probability measures on $\{0,1\}$ that satisfy exactly one of the conditions from Lemma 4.3, and let α_1 be an arbitrary probability measure on $\{0,1\}$. Then α_1 is (β_1, β_2)-equivalent to only one probability measure.*

The following corollary follows from Theorem 4.4 and Lemma 4.2.

Corollary 4.5. *Let β_1, β_2, and α_1 be as defined in Theorem 4.4. There exists only one probability measure α_2 such that, for all $(S,T) \in FREQ_{\alpha_1} \times FREQ_{\alpha_2}$, S and T are mutually β-normalizable.*

Acknowledgment. We thank an anonymous reviewer of [4] for posing the question answered by Corollary 3.11. We also thank anonymous reviewers of this paper for useful comments, especially including Observation 3.2.

References

1. Billingsley, P.: Hausdorff dimension in probability theory. Ill. J. Math. **4**, 187–209 (1960)
2. Billingsley, P.: Ergodic Theory and Information. R. E. Krieger Pub., Co., Huntington (1978)
3. Cajar, H.: Billingsley Dimension in Probability Spaces. Lecture Notes in Mathematics, vol. 892. Springer, Heidelberg (1981)
4. Case, A., Lutz, J.H.: Mutual dimension. ACM Trans. Comput. Theory **7**(12) July 2015
5. Cover, T.R., Thomas, J.A.: Elements of Information Theory, 2nd edn. John Wiley & Sons Inc., New York (2006)
6. Downey, R.G., Hirschfeldt, D.R.: Algorithmic Randomness and Complexity. 2010 edn. Springer, Heidelberg (2010)
7. Eggleston, H.G.: The fractional dimension of a set defined by decimal properties. Q. J. Math. **20**, 31–36 (1965)
8. Xiaoyang, G., Lutz, J.H., Elvira Mayordomo, R., Moser, P.: Dimension spectra of random subfractals of self-similar fractals. Ann. Pure Appl. Logic **165**, 1707–1726 (2014)

9. Han, T.S., Kobayashi, K.: Mathematics of Information and Coding. Translations of Mathematical Monographs (Book 203). American Mathematical Society (2007)
10. Kakutani, S.: On equivalence of infinite product measures. Ann. Math. **49**(1), 214–224 (1948)
11. Kolmogorov, A.N.: Three approaches to the quantitative definition of information. Probl. Inf. Transm. **1**(1), 1–7 (1965)
12. Levin, D.A., Peres, Y., Wilmer, E.L.: Markov Chains and Mixing Times. American Mathematical Society (2009)
13. Li, M., Vitányi, P.: An Introduction to Kolmogorov Complexity and Its Applications. 3rd edn. Springer, Heidelberg (2008)
14. Lutz, J.H.: Dimension in complexity classes. SIAM J. Comput. **32**(5), 1235–1259 (2003)
15. Lutz, J.H.: The dimensions of individual strings and sequences. Inf. Comput. **187**(1), 49–79 (2003)
16. Lutz, J.H.: A divergence formula for randomness and dimension. Theor. Comput. Sci. **412**, 166–177 (2011)
17. Lutz, J.H., Mayordomo, E.: Dimensions of points in self-similar fractals. SIAM J. Comput. **38**(3), 1080–1112 (2008)
18. Martin-Löf, P.: The definition of random sequences. Inf. Control **9**, 602–619 (1966)
19. Nies, A.: Computability and Randomness. Oxford University Press, New York (2009)
20. O'Donnell, R.: Analysis of Boolean Functions. Cambridge University Press, New York (2014)
21. Schnorr, C.-P.: A unified approach to the definition of random sequences. Math. Syst. Theory **5**(3), 246–258 (1971)
22. Schnorr, C.-P.: Zuflligkeit und Wahrscheinlichkeit: Eine algorithmische Begrndung der Wahrscheinlichkeitstheorie. 1971 edn. Springer (1971)
23. Schnorr, C.-P.: A survey of the theory of random sequences. In: Butts, R.E., Hintikka, J. (eds.) Basic Problems in Methodology and Linguistics, pp. 193–211. Springer, Netherlands (1977)
24. van Lambalgen, M.: Random Sequences. Ph.D. thesis, University of Amsterdam (1987)
25. van Lambalgen, M.: Von mises' definition of random sequences reconsidered. J. Symbolic Logic **52**(3), 725–755 (1987)
26. Vovk, V.G.: On a criterion for randomness. Dokl. Akad. Nauk SSSR **294**(6), 1298–1302 (1987)

Optimal Algorithms and a PTAS for Cost-Aware Scheduling

Lin Chen[1], Nicole Megow[2], Roman Rischke[1(✉)], Leen Stougie[3],
and José Verschae[4]

[1] Department of Mathematics, Technische Universität Berlin, Berlin, Germany
{lchen,rischke}@math.tu-berlin.de
[2] Center for Mathematics, Technische Universität München, Munich, Germany
nicole.megow@tum.de
[3] Department of Econometrics and Operations Research, Vrije Universiteit
Amsterdam and CWI, Amsterdam, The Netherlands
stougie@cwi.nl
[4] Departamento de Matemáticas and Departamento de Ingeniería Industrial
y de Sistemas, Pontificia Universidad Católica de Chile, Santiago, Chile
jverschae@uc.cl

Abstract. We consider a natural generalization of classical scheduling problems in which using a time unit for processing a job causes some time-dependent cost which must be paid in addition to the standard scheduling cost. We study the scheduling objectives of minimizing the makespan and the sum of (weighted) completion times. It is not difficult to derive a polynomial-time algorithm for preemptive scheduling to minimize the makespan on unrelated machines. The problem of minimizing the total (weighted) completion time is considerably harder, even on a single machine. We present a polynomial-time algorithm that computes for any given sequence of jobs an optimal schedule, i.e., the optimal set of time-slots to be used for scheduling jobs according to the given sequence. This result is based on dynamic programming using a subtle analysis of the structure of optimal solutions and a potential function argument. With this algorithm, we solve the unweighted problem optimally in polynomial time. Furthermore, we argue that there is a $(4+\varepsilon)$-approximation algorithm for the strongly NP-hard problem with individual job weights. For this weighted version, we also give a PTAS based on a dual scheduling approach introduced for scheduling on a machine of varying speed.

1 Introduction

We consider a natural generalization of classical scheduling problems in which occupying a time slot incurs certain cost that may vary over time and which must be paid in addition to the actual scheduling cost. Such a framework has been proposed recently in [11] and [7]. It models additional cost for operating servers or machines that vary over time such as, e.g., labor cost that may vary by the day of the week, or the hour of the day [11], or electricity cost fluctuating over day time [7]. On the one hand, the latter have economically a huge impact on

© Springer-Verlag Berlin Heidelberg 2015
G.F. Italiano et al. (Eds.): MFCS 2015, Part II, LNCS 9235, pp. 211–222, 2015.
DOI: 10.1007/978-3-662-48054-0_18

facilities with enormous power consumption such as large data centers, and on the other hand, these fluctuations reflect the imbalance between generation and consumption of electricity on a daily or weekly basis. Hence, cost aware scheduling is economically profitable and supports an eco-aware usage and generation of energy. Another motivation stems from cloud computing. Users of such services, e.g., Amazon EC2, are offered time-varying pricing schemes for processing jobs on a remote cloud server [1]. It is a challenging task for them to optimize the tradeoff between resource provisioning cost and the scheduling quality.

Problem Definition. We first describe the underlying classical scheduling problems. We are given a set of jobs $J := \{1, \ldots, n\}$ where every job $j \in J$ has given a processing time $p_j \in \mathbb{N}$ and possibly a weight $w_j \in \mathbb{Q}_{\geq 0}$. The task is to find a preemptive schedule on a single machine such that the total (weighted) completion time, $\sum_{j \in J} w_j C_j$, is minimized. Here C_j denotes the completion time of job j. In the standard scheduling notation, this problem is denoted as $1 \mid pmtn \mid \sum (w_j) C_j$. We also consider makespan minimization on unrelated machines, typically denoted as $R \mid pmtn \mid C_{\max}$. Here we are given a set of machines M, and each job $j \in J$ has an individual processing time $p_{ij} \in \mathbb{N}$ for running on machine $i \in M$. The task is to find a preemptive schedule that minimizes the makespan, that is, the completion time of the latest job.

In this paper, we consider a generalization of these scheduling problems within a time-varying reservation cost model. We are given a cost function $e : \mathbb{N} \to \mathbb{R}$, where $e(t)$ denotes the reservation cost for processing job(s) at time t. We assume that e is piecewise constant with given breakpoints at integral time points. We assume, more formally, that time is discretized into unit-size time slots, and the time horizon is partitioned into given intervals $I_k = [s_k, d_k)$ with $s_k, d_k \in \mathbb{N}$, $k = 1, \ldots, K$, within which unit-size time slots have the same *unit reservation cost* e_k. To ensure feasibility, let $d_K \geq \sum_{j \in J} \min_{i \in M} p_{ij}$.

Given a schedule $\mathcal{S}$, let $y(t)$ be a binary variable indicating if any processing is assigned to time slot $[t, t+1)$. The *reservation cost* in $\mathcal{S}$ is $E(\mathcal{S}) = \sum_t e(t) y(t)$. That means, for any time unit that is used in $\mathcal{S}$ we pay the full unit reservation cost, even if the unit is only partially used. We also emphasize that in case of multiple machines, a reserved time slot can be used by all machines. This models applications in which reserving a time unit on a server gives access to all processors on this server.

The overall objective now is to find a schedule that minimizes the scheduling objective, $C_{\max}$ resp. $\sum_{j \in J} w_j C_j$, *plus* the reservation cost E. We refer to the resulting problems as $R \mid pmtn \mid C_{\max} + E$ and $1 \mid pmtn \mid \sum w_j C_j + E$.

Related Work. Scheduling with time-varying reservation cost (aka variable time slot cost) has been studied explicitly in [11] and [7]. The seminal paper [11] is concerned with several *non-preemptive* single machine problems, which are polynomial-time solvable in the classical setting, such as minimizing the total completion time, lateness, and total tardiness, or maximizing the weighted number of on-time jobs. These problems are shown to be strongly NP-hard when taking reservation cost into account, while efficient algorithms exist for restricted reservation cost functions. In particular, the problem $1 \mid \mid \sum C_j + E$ is strongly

NP-hard, and it is efficiently solvable when the reservation cost is increasing or convex non-increasing [11]. The research focus in [7] is on *online* flow-time minimization using resource augmentation. Their main result is a scalable algorithm that obtains a constant performance guarantee when the machine speed is increased by a constant factor and there are only $K = 2$ distinct unit reservation cost. They also show that, in this online setting, for arbitrary K there is no constant speedup-factor that allows for a constant approximate solution.

In this paper, we study the simpler—but not yet well understood—offline problem without release dates. For this problem, the authors in [7] announce the following results: a pseudo-polynomial $(4 + \varepsilon)$-approximation for $1 \mid pmtn \mid \sum w_j C_j + E$, which gives an optimal solution in case that all weights are equal, and a constant approximation in quasi-polynomial time for a constant number of distinct reservation costs or when using a machine that is processing jobs faster by a constant factor.

The general concept of taking into consideration additional (time-dependent) cost for resource utilization when scheduling has been implemented differently in other models. We mention the area of energy-aware scheduling, where the energy consumption is taken into account (see [2] for an overview), or scheduling with generalized non-decreasing (completion-) time dependent cost functions, such as minimizing $\sum_j w_j f(C_j)$, e.g. [5,6,10], or even more general job-individual cost functions $\sum_j f_j(C_j)$, e.g. [3]. Our model differs fundamentally since our cost function may decrease with time, because delaying the processing in favor of cheaper time slots may decrease the overall cost. This is not the case in the above-mentioned models. Thus, in our framework we have the additional dimension in decision-making of choosing the time slots to be reserved.

There is also some similarity between our model and scheduling on a machine of varying speed. Notice that the latter problem (with $\sum_j w_j C_j$ as objective function) can be reformulated as minimizing $\sum_j w_j f(C_j)$ on a single machine with constant speed. Interestingly, the independently studied problem of scheduling with non-availability periods, see e.g. the survey [9], is a special case of both, the time-varying speed and the time-varying reservation cost model. Indeed, machine non/availability can be expressed either by 0/1-speed or equivalently by $\infty/0$ unit reservation cost. Results shown in this context imply that our problem $1 \mid pmtn \mid \sum_j w_j C_j + E$ is strongly NP-hard, even if there are only two distinct unit reservation costs [12].

Our Contribution. We present new optimal algorithms and best-possible approximation results for a generalization of standard scheduling problems to a framework with time-varying reservation cost.

Firstly, we give an optimal polynomial-time algorithm for the problem $R \mid pmtn \mid C_{\max} + E$ (Sect. 2). It relies on a known algorithm for the problem without reservation cost [8] to determine the optimal *number* of time slots to be reserved, together with a procedure for choosing the time slots to be reserved.

Our main results concern single-machine scheduling to minimize the total (weighted) completion time (Sect. 3). We present an algorithm that computes for a given ordered set of jobs an optimal choice of time slots to be used for scheduling. We derive this by first showing structural properties of an optimal schedule, which we then exploit together with a properly chosen potential function in a dynamic program yielding polynomial running time. Based on this algorithm, we show that the unweighted problem $1 \mid pmtn \mid \sum C_j + E$ can be solved in polynomial time and that there is a $(4 + \varepsilon)$-approximation algorithm for the weighted version $1 \mid pmtn \mid \sum w_j C_j + E$. A pseudo-polynomial $(4 + \varepsilon)$-approximation has been known before [7]. While pseudo-polynomial time algorithms are rather easy to derive, it is quite remarkable that our DP's running time is polynomial in the input, in particular, independent of d_K.

Finally, we design for the strongly NP-hard weighted problem variant (Sect. 4) a pseudo-polynomial algorithm that computes for any fixed ε a $(1+\varepsilon)$-approximate schedule for $1 \mid pmtn \mid \sum w_j C_j + E$. If d_K is polynomially bounded, then the algorithm runs in polynomial time for any ε, i.e., it is a polynomial-time approximation scheme (PTAS). In terms of approximation, our algorithm is best possible since the problem is strongly NP-hard even if there are only two different reservation costs [12].

Our approach is inspired by a recent PTAS for scheduling on a machine of varying speed [10] and it uses some of its properties. As discussed above, there is no formal mathematical relation known between these two seemingly related problems which allows to directly apply the result from [10]. The key is a dual view on scheduling: instead of directly constructing a schedule in the time-dimension, we first construct a dual scheduling solution in the weight-dimension which has a one-to-one correspondence to a true schedule. We design an exponential-time dynamic programming algorithm which can be trimmed to polynomial time using techniques known for scheduling with varying speed [10].

For both the makespan and the min-sum problem, job preemption is crucial for obtaining worst-case bounds. For non-preemptive scheduling, a reduction from 2-PARTITION shows that no approximation within a polynomial ratio is possible, unless P=NP, even if there are only two different reservation costs, 0 and ∞.

Finally, we remark that in general it is not clear that a schedule can be encoded polynomially in the input. However, for our completion-time based minimization objective, it is easy to observe that if an algorithm reserves p unit-size time slots in an interval of equal cost, then it reserves the first p slots within this interval, which simplifies the structure and output of an optimal solution.

2   Minimizing the Makespan on Unrelated Machines

The standard scheduling problem without reservation $R \mid pmtn \mid C_{\max}$ can be solved optimally in polynomial time [8]. We show that taking into account time-varying reservation cost does not significantly increase the problem complexity.

Consider the generalized preemptive makespan minimization problem with reservation cost. Recall that we can use every machine in a reserved time slot and pay only once. By [8] it is sufficient to find an optimal reservation decision.

Observation 1. *Given the set of time slots reserved in an optimal solution, we can compute an optimal schedule in polynomial time.*

Given an instance of our problem, let Z be the optimal makespan of the relaxed problem without reservation cost. Notice that Z is not necessarily integral. To determine an optimal reservation decision, we use the following observation.

Observation 2. *Given an optimal makespan $C^*_{\max}$ for $R \mid pmtn \mid C_{\max} + E$, an optimal schedule reserves the $\lceil Z \rceil$ cheapest slots before $\lceil C^*_{\max} \rceil$.*

Note that we must pay full reservation cost for a used time slot, no matter how much it is utilized, and so does an optimal solution. In particular, this holds for the last reserved slot. Hence, it remains to design a procedure for computing an optimal value $C^* := \lceil C^*_{\max} \rceil$.

We compute for every interval $I_k = [s_k, d_k)$, $k = 1, \ldots, K$, an optimal point in time for C^* assuming that $C^* \in I_k$. Start from the value $C^* = s_k$, if feasible, with the corresponding cheapest $\lceil Z \rceil$ reserved time slots. Notice that any of these reserved time slots that has cost e such that $e > e_k + 1$ can be replaced by a time slot from I_k leading to a solution with less total cost. Thus, if such a time slot does not exist, then s_k is the best choice for C^* in I_k. Otherwise, let $R \subseteq \{1, \ldots, k - 1\}$ be the index set of intervals that contain at least one reserved slot. We define I_ℓ to be the interval with $e_\ell = \max_{h \in R} e_h$ and denote by r_h the number of reserved time slots in I_h. Replace $\min\{r_\ell, d_k - s_k - r_k\}$ reserved slots from I_ℓ by slots from I_k and update R, I_ℓ and r_k. This continues until $e_\ell \leq e_k + 1$ or the interval I_k is completely reserved, i.e., $r_k = d_k - s_k$. This operation takes at most $O(K)$ computer operations per interval to compute the best C^*-value in that interval. It yields the following theorem.

Theorem 1. *The scheduling problem $R \mid pmtn \mid C_{\max} + E$ can be solved in polynomial time equal to the running time required to solve an LP for $R \mid pmtn \mid C_{\max}$ without reservation cost plus $O(K^2)$.*

3   Minimizing $\sum_j (w_j) C_j$ on a Single Machine

In this section, we consider the problem $1 \mid pmtn \mid \sum (w_j) C_j + E$. We design an algorithm that computes, for a given (not necessarily optimal) scheduling sequence σ, an optimal reservation decision for σ. We firstly identify structural properties of an optimal schedule, which we then exploit in a dynamic program. Based on this algorithm, we show that the unweighted problem $1 \mid pmtn \mid \sum C_j + E$ can be solved optimally in polynomial time and that there is a $(4 + \varepsilon)$-approximation algorithm for the weighted problem $1 \mid pmtn \mid \sum w_j C_j + E$.

In principle, an optimal schedule may preempt jobs at fractional time points. However, since time slots can only be reserved entirely, any reasonable schedule uses the reserved slots entirely as long as there are unprocessed jobs. The following lemma shows that this is also true if we omit the requirement that time slots must be reserved entirely. (For the makespan problem considered in Sect. 2 this is not true.)

Lemma 1. *There is an optimal schedule S^* in which all reserved time slots are entirely reserved and jobs are preempted only at integral points in time.*

In the following, we assume that we are given a (not necessarily optimal) sequence of jobs, $\sigma = (1, \ldots, n)$, in which the jobs must be processed. We want to characterize an optimal schedule S^* for σ, that is, in particular the optimal choice of time slots for scheduling σ. We first split S^* into smaller sub-schedules, for which we introduce the concept of a *split point*.

Definition 1 (Split Point). *Consider an optimal schedule S^* and the set of potential split points $\mathcal{P} := \bigcup_{k=1}^{K} \{s_k, s_k + 1\} \cup \{d_K\}$. Let S_j and C_j denote the start time and completion time of job j, respectively. We call a time point $t \in \mathcal{P}$ a split point for S^* if all jobs that start before t also finish their processing not later than t, i.e., if $\{j \in J : S_j < t\} = \{j \in J : C_j \le t\}$.*

Given an optimal schedule S^*, let $0 = \tau_1 < \tau_2 < \cdots < \tau_\ell = d_K$ be the *maximal* sequence of split points of S^*, i.e. the sequence containing all split points of S^*. We denote the interval between two consecutive split points τ_x and τ_{x+1} as region $R_x^{S^*} := [\tau_x, \tau_{x+1})$, for $x = 1, \ldots, \ell - 1$.

Consider now any region $R_x^{S^*}$ for an optimal schedule S^* with $x \in \{1, \ldots, \ell - 1\}$ and let $J_x^{S^*} := \{j \in J : S_j \in R_x^{S^*}\}$. Note that $J_x^{S^*}$ might be empty. Among all optimal schedules we shall consider an optimal solution S^* that minimizes the value $\sum_{t=0}^{d_K - 1} t \cdot y(t)$, where $y(t)$ is a binary variable that indicates if time slot $[t, t + 1)$ is reserved or not.

Observation 3. *There is no job $j \in J_x^{S^*}$ with $C_j \in R_x^{S^*} \cap \mathcal{P}$.*

Namely, every C_j with $C_j = s_k \in R_x^{S^*}$ or $C_j = s_k + 1 \in R_x^{S^*}$ would make s_k or $s_k + 1$ a split point, whereas $R_x^{S^*}$ is defined as the interval between two consecutive split points.

We say that an interval I_k is *partially reserved* if at least one slot in I_k is reserved, but not all.

Lemma 2. *There exists an optimal schedule S^* in which at most one interval is partially reserved in $R_x^{S^*}$.*

We are now ready to bound the unit reservation cost spent for jobs in $J_x^{S^*}$. Let $e_{\max}^j$ be the maximum unit reservation cost spent for job j in S^*. Furthermore, let $\Delta_x := \max_{j \in J_x^{S^*}} (e_{\max}^j + \sum_{j' < j} w_{j'})$ and let j_x be the last job (according to sequence σ) that achieves Δ_x. Suppose, there are $b \ge 0$ jobs before and $a \ge 0$ jobs after job j_x in $J_x^{S^*}$. The following lemma gives for every job $j \in J_x^{S^*} \setminus \{j_x\}$ an upper bound on the unit reservation cost spent in the interval $[S_j, C_j)$.

Lemma 3. *Consider an optimal schedule S^*. For any job $j \in J_x^{S^*} \setminus \{j_x\}$ a slot $[t, t + 1) \in [S_j, C_j)$ is reserved if and only if the cost of $[t, t + 1)$ satisfies the upper bound given in the table below.*

$j_x - b$	$\cdots$	$j_x - 1$	$j_x + 1$	$\cdots$	$j_x + a$
$\le e_{\max}^{j_x} + \sum_{j'=j_x-b}^{j_x-1} w_{j'}$	$\cdots$	$\le e_{\max}^{j_x} + w_{j_x-1}$	$< e_{\max}^{j_x} - w_{j_x}$	$\cdots$	$< e_{\max}^{j_x} - \sum_{j'=j_x}^{j_x+a-1} w_{j'}$

Proof. Consider any job $j := j_x - \ell$ with $0 < \ell \leq b$. Suppose there is a job j for which a slot is reserved with cost $e_{\max}^j > e_{\max}^{j_x} + \sum_{j'=j}^{j_x-1} w_{j'}$. Then $e_{\max}^j + \sum_{j'<j} w_{j'} > e_{\max}^{j_x} + \sum_{j'<j_x} w_{j'}$, which is a contradiction to the definition of job j_x. Thus, $e_{\max}^j \leq e_{\max}^{j_x} + \sum_{j'=j}^{j_x-1} w_{j'}$.

Now suppose, there is a slot $[t, t+1) \in [S_j, C_j)$ with cost $e(t) \leq e_{\max}^{j_x} + \sum_{j'=j}^{j_x-1} w_{j'}$ that is not reserved. There must be a slot $[t', t'+1) \in [S_{j_x}, C_{j_x})$ with cost exactly $e_{\max}^{j_x}$. If we reserve slot $[t, t+1)$ instead of $[t', t'+1)$, then the difference in cost is non-positive, because the completion times of at least ℓ jobs ($j = j_x - \ell, \ldots, j_x - 1$ and maybe also j_x) decrease by one. This contradicts either the optimality of $\mathcal{S}^*$ or our assumption that $\mathcal{S}^*$ minimizes $\sum_{t=0}^{d_K-1} t \cdot y(t)$.

The proof of the statement for any job $j_x + \ell$ with $0 < \ell \leq a$ follows a similar argument, but now using the fact that for every job $j := j_x + \ell$ we have $e_{\max}^j < e_{\max}^{j_x} - \sum_{j'=j_x}^{j-1} w_{j'}$, because j_x was the last job with $e_{\max}^j + \sum_{j'<j} w_{j'} = \Delta_x$. □

To construct an optimal sub-schedule, we need the following two lemmas.

Lemma 4. *Let $[t', t'+1) \in [S_{j_x}, C_{j_x})$ be the last time slot with cost $e_{\max}^{j_x}$ that is used by job j_x. If there is a partially reserved interval I_k in $R_x^{\mathcal{S}^*}$, then either (i) I_k is not the last interval of $R_x^{\mathcal{S}^*}$ and I_k contains $[t', t'+1)$ as its last reserved time slot or (ii) I_k is the last interval of $R_x^{\mathcal{S}^*}$.*

Lemma 5. *Let $[t', t'+1) \in [S_{j_x}, C_{j_x})$ be the last time slot with cost $e_{\max}^{j_x}$ that is used by job j_x. There exists an optimal solution $\mathcal{S}^*$ such that if there is a partially reserved interval I_k in $R_x^{\mathcal{S}^*}$ and it is the last one in $R_x^{\mathcal{S}^*}$, then there is no slot $[t, t+1) \in [S_{j_x}, C_{j_x})$ with cost at most $e_{\max}^{j_x}$ that is not reserved.*

We now show how to construct an optimal partial schedule for a given ordered job set in a given region in polynomial time.

Lemma 6. *Given a region R_x and an ordered job set J_x, we can construct in polynomial time an optimal schedule for J_x within the region R_x, which does not contain any other split point than τ_x and τ_{x+1}, the boundaries of R_x.*

Proof. Given R_x and J_x, we guess the optimal combination $(j_x, e_{\max}^{j_x})$, i.e., we enumerate over all nK combinations and choose eventually the best solution.

We firstly assume that a partially reserved interval exists and it is the last one in R_x (case (ii) in Lemma 4). Based on the characterization in Lemma 3 we find in polynomial time the slots to be reserved for the jobs $j_x - b, \ldots, j_x - 1$. This defines $C_{j_x-b}, \ldots, C_{j_x-1}$. Then starting job j_x at time C_{j_x-1}, we check intervals in the order given and reserve as much as needed of each next interval I_h if and only if $e_h \leq e_{\max}^{j_x}$, until a total of p_{j_x} time slots have been reserved for processing j_x. Lemma 5 justifies to do that. This yields a completion time C_{j_x}. Starting at C_{j_x}, we use again Lemma 3 to find in polynomial time the slots to be reserved for processing the jobs $j_x + 1, \ldots, j_x + a$. This gives $C_{j_x+1}, \ldots, C_{j_x+a}$.

Now we assume that there is no partially reserved interval or we are in case (i) of Lemma 4. Similar to the case above, we find in polynomial time the slots that $\mathcal{S}^*$ reserves for the jobs $j_x - b, \ldots, j_x - 1$ based on Lemma 3. This defines

$C_{j_x-b}, \ldots, C_{j_x-1}$. To find the slots to be reserved for the jobs $j_x + 1, \ldots, j_x + a$, in this case, we start at the end of R_x and go backwards in time. We can start at the end of R_x because in this case the last interval of R_x is fully reserved. This gives $C_{j_x+1}, \ldots, C_{j_x+a}$. Job j_x is thus to be scheduled in $[C_{j_x-1}, S_{j_x+1})$. In order to find the right slots for j_x we solve a makespan problem in the interval $[C_{j_x-1}, S_{j_x+1})$, which can be done in polynomial time (Theorem 1) and gives a solution that cannot be worse than what an optimal schedule $\mathcal{S}^*$ does.

If anywhere in both cases the reserved intervals can not be made sufficient for processing the job(s) for which they are intended, or if scheduling the jobs in the reserved intervals creates any intermediate split point, then this $(j_x, e_{\max}^{j_x})$-combination is rejected. Hence, we have computed the optimal schedules over all nK combinations of $(j_x, e_{\max}^{j_x})$ and over both cases of Lemma 4 concerning the position of the partially reserved interval. We choose the schedule with minimum total cost and return it with its value. This completes the proof.   □

Now we are ready to prove our main theorem.

Theorem 2. *Given an instance of* $1 \,|\, pmtn \,|\, \sum w_j C_j + E$ *and an arbitrary processing sequence of jobs* σ, *we can compute an optimal reservation decision for* σ *in polynomial time.*

Proof. We give a dynamic program. We define a state for every possible potential split point $t \in \mathcal{P}$. By definition, there are $2K + 1$ of them. A state also includes the set of jobs processed until time t. Given the sequence σ, this job set can be uniquely identified by the index of the last job, say j, that finished by time t. By relabeling the job set J, we can assume w.l.o.g. that $\sigma = (1, \ldots, n)$.

For each state (j, t) we compute and store recursively the optimal scheduling cost plus reservation cost $Z(j, t)$ by

$$Z(j, t) = \min\left\{Z(j', t') + z(\{j'+1, \ldots, j\}, [t', t)) : t', t \in \mathcal{P}, t' < t, j', j \in J, j' < j\right\},$$

where $z(\{j'+1, \ldots, j\}, [t', t))$ denotes the value of an optimal partial schedule for job set $\{j'+1, j'+2, \ldots, j\}$ in the region $[t', t)$, or ∞ if no such schedule exists. This value can be computed in polynomial time, by Lemma 6. Hence, we compute $Z(j, t)$ for all $O(nK)$ states in polynomial time, which concludes the proof.   □

The following observation follows from a standard interchange argument.

Observation 4. *In an optimal schedule* $\mathcal{S}^*$ *for the problem* $1 \,|\, pmtn \,|\, \sum C_j + E$, *jobs are processed according to the Shortest Processing Time First (SPT) policy.*

Combining this observation with Theorem 2 gives the following corollary.

Corollary 1. *There is a polynomial-time algorithm for* $1 \,|\, pmtn \,|\, \sum C_j + E$.

For the weighted problem $1 \,|\, pmtn \,|\, \sum w_j C_j + E$, there is no sequence that is universally optimal for all reservation decisions [5]. However, in the context of scheduling on an unreliable machine there has been shown a polynomial-time algorithm that computes a universal $(4 + \varepsilon)$-approximation [5]. More precisely, the algorithm constructs a sequence of jobs which approximates the scheduling cost for any reservation decision with a factor at most $4 + \varepsilon$.

Consider an instance of problem $1 \,|\, pmtn \,|\, \sum w_j C_j + E$ and compute such a universally $(4+\varepsilon)$-approximate sequence. Applying Theorem 2 to σ, we obtain a schedule $\mathcal{S}$ with an optimal reservation decision for σ. Let $\mathcal{S}'$ denote the schedule which we obtain by changing the reservation decision of $\mathcal{S}$ to the reservation in an optimal schedule $\mathcal{S}^*$ (but keeping the scheduling sequence σ). The schedule $\mathcal{S}'$ has cost no less than the original cost of $\mathcal{S}$. Furthermore, given the reservation decision in the optimal solution $\mathcal{S}^*$, the sequence σ approximates the scheduling cost of $\mathcal{S}^*$ within a factor of $4 + \varepsilon$. This gives the following result.

Corollary 2. *There is a $(4 + \varepsilon)$-approx. algorithm for $1 \,|\, pmtn \,|\, \sum w_j C_j + E$.*

4   A PTAS for Minimizing Total Weighted Completion Time

The main result of this section is an approximation scheme for minimizing the total weighted completion time with time-varying reservation cost.

Theorem 3. *For any fixed $\varepsilon > 0$, there is a pseudo-polynomial time algorithm that computes a $(1 + \varepsilon)$-approximation for the problem $1 \,|\, pmtn \,|\, \sum_j w_j C_j + E$. This algorithm runs in polynomial time if d_K is polynomially bounded.*

In the remainder of this section we describe some preliminaries, present a dynamic programming (DP) algorithm with exponential running time, and then we argue that it can be trimmed down to (pseudo-)polynomial size. As noted in the introduction, our approach is inspired by a PTAS for scheduling on a machine of varying speed [10], but a direct application does not seem possible.

4.1   Preliminaries and Scheduling in the Weight-Dimension

We describe a schedule $\mathcal{S}$ not in terms of completion times $C_j(\mathcal{S})$, but in terms of the remaining weight function $W^{\mathcal{S}}(t)$ which, for a given schedule $\mathcal{S}$, is defined as the total weight of all jobs not completed by time t. Based on the remaining weight function we can express the cost for any schedule $\mathcal{S}$ as

$$\int_0^\infty W^{\mathcal{S}}(t) = \sum_{j \in J} w_j C_j(\mathcal{S}).$$

This has a natural interpretation in the standard 2D-Gantt chart, which was originally introduced in [4].

For a given reservation decision, we follow the idea of [10] and implicitly describe the completion time of a job j by the value of the function W at the time that j completes. This value is referred to as the *starting weight* S_j^w of job j. In analogy to the time-dimension, the value $C_j^w := S_j^w + w_j$ is called *completion weight* of job j. When we specify a schedule in terms of the remaining weight function, then we call it a *weight-schedule*, otherwise a *time-schedule*. Other terminologies, such as feasibility and idle time, also translate from the time-dimension to the weight-dimension. A weight-schedule is called *feasible* if no two jobs overlap and the machine is called *idle in weight-dimension* if there exists a point w in the weight-dimension with $w \notin \left[S_j^w, C_j^w \right]$ for all jobs $j \in J$.

A weight-schedule together with a reservation decision can be translated into a time-schedule by ordering the job in decreasing order of completion weights and scheduling them in this order in the time-dimension in the reserved time slots. For a given reservation decision, consider a weight-schedule $\mathcal{S}$ with completion weights $C_1^w > \cdots > C_n^w > C_{n+1}^w := 0$ and the corresponding completion times $0 =: C_0 < C_1 < \cdots < C_n$ for the jobs $j = 1, \ldots, n$. We define the *(scheduling) cost of a weight schedule* $\mathcal{S}$ as $\sum_{j=1}^{n} \left(C_{j+1}^w - C_j^w \right) C_j$. This value equals $\sum_{j=1}^{n} \pi_j^{\mathcal{S}} C_j^w$, where $\pi_j^{\mathcal{S}} := C_j - S_j$, if and only if there is no idle weight. If there is idle weight, then the cost of a weight-schedule can only be greater, and we can safely remove idle weight without increasing the scheduling cost [10].

4.2 Dynamic Programming Algorithm

Let $\varepsilon > 0$. Firstly, we apply standard geometric rounding to the weights to gain more structure on the input, i.e., we round the weights of all jobs up to the next integer power of $(1 + \varepsilon)$, by losing at most a factor $(1 + \varepsilon)$ in the objective value. Furthermore, we discretize the weight-space into intervals of exponentially increasing size: we define intervals $WI_u := \left[(1 + \varepsilon)^{u-1}, (1 + \varepsilon)^u \right)$ for $u = 1, \ldots, \nu$ with $\nu := \lceil \log_{1+\varepsilon} \sum_{j \in J} w_j \rceil$.

Consider a subset of jobs $J' \subseteq J$ and a partial weight-schedule of J'. In the dynamic program, the set J' represents the set of jobs at the beginning of a corresponding weight-schedule, i.e., if $j \in J'$ and $k \in J \setminus J'$, then $C_k^w < C_j^w$. As discussed in Sect. 4.1, a partial weight-schedule $\mathcal{S}$ for the jobs in J' together with a reservation decision for all jobs in J can be translated into a time-schedule. Note that the makespan of this time-schedule is completely defined by the reservation decision and the total processing volume $\sum_{j \in J} p_j$. Moreover, knowing the last $p(J') := \sum_{j \in J'} p_j$ reserved slots is sufficient for scheduling the jobs in J' in the time-dimension, since we know that the first job in the weight-schedule finishes last in the time-schedule, i.e., at the makespan. This gives a unique completion time C_j and a unique execution time $\pi_j^{\mathcal{S}} := C_j - S_j$ for each job $j \in J'$. The total scheduling and reservation cost of this partial schedule is $\sum_{j \in J'} \pi_j^{\mathcal{S}} C_j^w + E$.

Let $\mathcal{F}_u := \{ J_u \subseteq J : \sum_{j \in J_u} w_j \leq (1 + \varepsilon)^u \}$. The set $\mathcal{F}_u$ contains all the possible job sets J_u that can be scheduled in WI_u or before. With every possible pair (J_u, t), $J_u \in \mathcal{F}_u$ and $t \in [0, d_K)$, we associate a recursively constructed weight-schedule together with a reservation decision starting at time t so that the current scheduling and reservation cost is a good approximation of the optimal total cost for processing the set J_u starting at t. More precisely, given a $u \in \{1, \ldots, \nu\}$, a set $J_u \in \mathcal{F}_u$, and a time point $t \in [0, d_K)$, we create a table entry $Z(u, J_u, t)$

that represents a $(1 + \mathcal{O}(\varepsilon))$-approximation of the scheduling and reservation cost of an optimal weight-schedule of J_u subject to $C_j^w \leq (1+\varepsilon)^u$ for all $j \in J_u$ and $S_j \geq t$ for all $j \in J_u$. Initially, we create table entries $Z(0, \emptyset, t) := 0$ for all $t = 0, \ldots, d_K$ and we define $\mathcal{F}_0 = \{\emptyset\}$. With this, basically, we control the makespan of our time-schedule.

The table entries in iteration u are created based on the table entries from iteration $u-1$ in the following way. Consider candidate sets $J_u \in \mathcal{F}_u$ and $J_{u-1} \in \mathcal{F}_{u-1}$, a partial weight-schedule S of J_u, in which the set of jobs with completion weight in WI_u is exactly $J_u \setminus J_{u-1}$, and two integer time points t, t' with $t < t'$.

We let $APX_u(J_u \setminus J_{u-1}, [t, t'))$ denote a $(1 + \varepsilon)$-approximation of the minimum total cost (for reservation and scheduling) when scheduling job set $J_u \setminus J_{u-1}$ in the time interval $[t, t')$, i.e.,

$$APX_u(J_u \setminus J_{u-1}, [t, t')) := (1 + \varepsilon)^u \cdot (t' - t) + RES(J_u \setminus J_{u-1}, [t, t')),$$

where $RES(J_u \setminus J_{u-1}, [t, t'))$ denotes the cost of the $p(J_u \setminus J_{u-1})$ cheapest slots in the interval $[t, t')$. If $p(J_u \setminus J_{u-1}) > t' - t$, then we set $RES(J_u \setminus J_{u-1}, [t, t'))$ to infinity to express that we cannot schedule all jobs in $J_u \setminus J_{u-1}$ within $[t, t')$.

Based on this, we compute the table entry $Z(u, J_u, t)$ with $J_u \in \mathcal{F}_u$ according to the following recursive formula

$$Z(u, J_u, t) := \min \{ Z(u - 1, J_{u-1}, t') + APX_u(J_u \setminus J_{u-1}, [t, t')) :$$
$$J_{u-1} \in \mathcal{F}_{u-1}, J_{u-1} \subseteq J_u, t \leq t' \}.$$

We return $Z(\nu, J, 0)$ after iteration ν. From $Z(\nu, J, 0)$ we can construct a schedule and its reservation decision by backtracking.

Notice that the values $APX_u(J_u \setminus J_{u-1}, [t, t'))$ do not depend on the entire schedule, but only on the time interval $[t, t')$ and the total processing volume of jobs in $J_u \setminus J_{u-1}$. Since we approximate the scheduling cost by a factor $1+\varepsilon$ and determine the minimum reservation cost, the dynamic programming algorithm obtains the following result.

Lemma 7. *The DP computes a $(1 + \mathcal{O}(\varepsilon))$-approximate solution.*

4.3   Trimming the State Space

The set $\mathcal{F}_u$, containing all possible job sets J_u, is of exponential size, and so is the DP state space. In the context of scheduling with variable machine speed, it has been shown in [10] how to reduce the set $\mathcal{F}_u$ for a similar DP (without reservation decision, though) to one of polynomial size at only a small loss in the objective value. In general, such a procedure is not necessarily applicable to our setting because of the different objective involving additional reservation cost and the different decision space. However, the compactification in [10] holds *independently of the speed of the machine* and, thus, independently of the reservation decision of the DP (interpret non/reservation as speed 0/1). Hence, we can apply it to our cost aware scheduling framework and obtain a PTAS. For more details on the procedure we refer to the full version.

5 Open Problems

Our PTAS for the weighted problem runs in polynomial time when d_K is polynomially bounded. Otherwise, the scheme is a "PseuPTAS". The question is if, by a careful analysis of the structure of optimal solutions, we can avoid the dependence of the running time on d_K, as in our dynamic program from Sect. 3.

Furthermore, it would be interesting to algorithmically understand other scheduling problems in the model of time-varying reservation cost. An immediate open question concerns the problems considered in this paper when there are release dates present. In the full version we show that the makespan problem $1 \mid r_j, pmtn \mid C_{\max} + E$ can be solved in polynomial time. We can also solve $R \mid pmtn, r_j \mid C_{\max}$ optimally if we allow *fractional* reservation. The seemingly most simple open problem in our (integral) model is $1 \mid r_j, pmtn \mid \sum_j C_j + E$. While the problem without reservation cost can be solved optimally in polynomial time, the complexity status in the time-varying cost model is unclear, even with only two different unit reservation costs.

Machine-individual time-slot reservation opens a different stream of research. While a standard LP can be adapted for optimally solving $R \mid pmtn, r_j \mid C_{\max}$ with fractional reservation cost, the integrality gap is unbounded for our model.

References

1. Amazon EC2 Pricing Options. https://aws.amazon.com/ec2/pricing/
2. Albers, S.: Energy-efficient algorithms. Commun. ACM **53**(5), 86–96 (2010)
3. Bansal, N., Pruhs, K.: The geometry of scheduling. In: Proceedings of the FOCS 2010, pp. 407–414 (2010)
4. Eastman, W.L., Even, S., Isaac, M.: Bounds for the optimal scheduling of n jobs on m processors. Manage. Sci. **11**(2), 268–279 (1964)
5. Epstein, L., Levin, A., Marchetti-Spaccamela, A., Megow, N., Mestre, J., Skutella, M., Stougie, L.: Universal sequencing on an unreliable machine. SIAM J. Comput. **41**(3), 565–586 (2012)
6. Höhn, W., Jacobs, T.: On the performance of smith's rule in single-machine scheduling with nonlinear cost. In: Fernández-Baca, D. (ed.) LATIN 2012. LNCS, vol. 7256, pp. 482–493. Springer, Heidelberg (2012)
7. Kulkarni, J., Munagala, K.: Algorithms for cost-aware scheduling. In: Erlebach, T., Persiano, G. (eds.) WAOA 2012. LNCS, vol. 7846, pp. 201–214. Springer, Heidelberg (2013)
8. Lawler, E.L., Labetoulle, J.: On preemptive scheduling of unrelated parallel processors by linear programming. J. ACM **25**(4), 612–619 (1978)
9. Lee, C.-Y.: Machine scheduling with availability constraints. In: Leung, J.Y.-T. (eds.) Handbook of Scheduling: Algorithms, Models, and Performance Analysis. Chapter 22. CRC Press (2004)
10. Megow, N., Verschae, J.: Dual techniques for scheduling on a machine with varying speed. In: Fomin, F.V., Freivalds, R., Kwiatkowska, M., Peleg, D. (eds.) ICALP 2013, Part I. LNCS, vol. 7965, pp. 745–756. Springer, Heidelberg (2013)
11. Wan, G., Qi, X.: Scheduling with variable time slot costs. Nav. Res. Logistics **57**, 159–171 (2010)
12. Wang, G., Sun, H., Chu, C.: Preemptive scheduling with availability constraints to minimize total weighted completion times. Ann. Oper. Res. **133**, 183–192 (2005)

Satisfiability Algorithms and Lower Bounds for Boolean Formulas over Finite Bases

Ruiwen Chen$^{(\boxtimes)}$

School of Informatics, University of Edinburgh, Edinburgh, UK
rchen2@inf.ed.ac.uk

Abstract. We give a #SAT algorithm for boolean formulas over arbitrary finite bases. Let B_k be the basis composed of all boolean functions on at most k inputs. For B_k-formulas on n inputs of size cn, our algorithm runs in time $2^{n(1-\delta_{c,k})}$ for $\delta_{c,k} = c^{-O(c^2 k 2^k)}$. We also show the average-case hardness of computing affine extractors using linear-size B_k-formulas.

We also give improved algorithms and lower bounds for formulas over finite unate bases, i.e., bases of functions which are monotone increasing or decreasing in each of the input variables.

Keywords: Boolean formula · Satisfiability algorithm · Average-case lower bound · Random restriction

1 Introduction

The random restriction approach was introduced by Subbotovskaya [15] to prove lower bounds for boolean formulas. For formulas over a basis B, we define the *shrinkage exponent* Γ_B to be the least upper bound on γ such that the formula size shrinks (in expection) by a factor of p^γ under random assignments leaving p fraction of the inputs unfixed. Subbotovskaya [15] showed the shrinkage exponent of de Morgan formulas (formulas over the binary basis $\{\neg, \wedge, \vee\}$) is at least 1.5, and this implies an $\Omega(n^{1.5})$ lower bound on the formula size for computing the parity of n variables. Andreev [1] improved this lower bound by constructing an explicit function which requires size $\Omega(n^{\Gamma+1-o(1)})$ for any formulas with shrinkage exponent Γ. The shrinkage exponent of de Morgan formulas was improved in [8,12], and finally, Hastad [6] showed the tight bound $\Gamma = 2 - o(1)$, which gives an $\Omega(n^{3-o(1)})$ lower bound for computing Andreev's function.

Recently, the shrinkage property was strengthened to get both satisfiability algorithms [13] and average-case lower bounds [9,10]. Santhanam [13] gave a simple deterministic #SAT algorithm for de Morgan formulas. The algorithm recursively restricts the most frequent variables, and the number of branches in the recursion tree is bounded via the *concentrated shrinkage* property; that is, along a random branch of the recursion tree, the formula size shrinks by a factor of $p^{1.5}$ with high probability. For cn-size formulas, Santhanam's algorithm runs

© Springer-Verlag Berlin Heidelberg 2015
G.F. Italiano et al. (Eds.): MFCS 2015, Part II, LNCS 9235, pp. 223–234, 2015.
DOI: 10.1007/978-3-662-48054-0_19

in time $2^{n(1-\Omega(1/c^2))}$. Combining with memoization, similar algorithms running in time $2^{n-n^{\Omega(1)}}$ were given for formula size $n^{2.49}$ [2] and $n^{2.63}$ [3].

Seto and Tamaki [14] extended Santhanam's algorithm to formulas over the full binary basis $\{\neg, \wedge, \vee, \oplus\}$. Note that, the shrinkage exponent here is trivially 1 since $\oplus$ is in the basis, and thus Santhanam's algorithm does not apply directly. Instead, Seto and Tamaki [14] showed that, for a small-size formula over the full binary basis, either satisfiability checking is easy by solving systems of linear equations, or a greedy restriction will shrink the formula non-trivially, which is as required by Santhanam's algorithm. In this work, we will further extend Seto and Tamaki's approach to formulas over arbitrary finite bases.

On the other hand, Komargodski, Raz, and Tal [9,10] applied concentrated shrinkage to prove average-case lower bounds for de Morgan formulas. In particular, they constructed a generalized Andreev's function which is computable in polynomial time, but de Morgan formulas of size $n^{2.99}$ can compute correctly on at most $1/2 + 2^{-n^{\Omega(1)}}$ fraction of all inputs. The result in [10] also implies a randomized #SAT algorithm in time $2^{n-n^{\Omega(1)}}$ for de Morgan formulas of size $n^{2.99}$.

1.1   Our Results and Proof Techniques

In this work, we focus on formulas over arbitrary finite bases, and generalize previous results on formulas over binary bases [2,13,14]. For $k \geqslant 2$, let B_k be the basis consisting of all boolean functions on at most k variables. We consider B_k-formulas, i.e., formulas over the basis B_k.

We first give a satisfiability algorithm for B_k-formulas which is significantly better than brute-force search.

Theorem 1. *For n-input B_k-formulas of size cn, there is a deterministic algorithm counting the number of satisfying assignments in time $2^{n(1-\delta_{c,k})}$, where $\delta_{c,k} = c^{-O(c^2 k 2^k)}$.*

The algorithm is based on a structural property of small formulas, similar to Seto and Tamaki's approach [14] for B_2-formulas. That is, for a small B_k-formula, either satisfiability checking can be done by solving systems of linear equations, or a process of greedy restrictions gives non-trivial shrinkage. The technical difficulty is that, since we have both $\oplus$ and functions of arity larger than 2 in B_k, the formula size shrinks trivially even by restricting the most frequent variables. To get nontrivial shrinkage, we need a formula *weight* function which accounts for not only the formula size but also the basis functions used in the formula, and argue that the weight shrinks nontrivially under certain greedy restrictions (which aims at optimally reducing the weight rather than the size). We also require that the formula weight is proportional to the formula size, and thus, in the end of greedy restrictions, the size also shrinks nontrivially.

The weighting technique was used previously for the shrinkage (in expectation) of de Morgan formulas [8,12] and formulas over finite unate bases [4]. We use weight functions similar as in [4] but with dedicated parameterizations since the basis contains non-unate functions.

The algorithm implicitly constructs a *parity decision tree* for the given formula; this implies that any B_k-formula of linear size has a parity decision tree of size $2^{n-\Omega(n)}$. By the fact that affine extractors are hard to approximate by parity decision trees, we immediately get the following average-case lower bound.

Theorem 2. *There is a polynomial-time computable function f_n such that, for any family of B_k-formulas F_n of size cn for a constant c, on random inputs $x \in \{0,1\}^n$,*

$$\mathbf{Pr}[F_n(x) = f_n(x)] \leqslant 1/2 + 2^{-\Omega(n)}.$$

Note that, an average-case lower bound for larger B_k-formulas also follow from [9,10]. That is, B_k-formulas of size $n^{1.99}$ can compute the generalized Andreev's function of [10] correctly on at most $1/2 + 2^{-n^{\Omega(1)}}$ fraction of inputs.

For the more restrictive unate bases, we get nontrivial #SAT algorithms and average-case lower bounds for formulas of super-quadratic size. The results follow easily by extending the expected shrinkage of Chockler and Zwick [4] to concentrated shrinkage, and generalizing the previous algorithms and lower bounds for de Morgan formulas [2,9,10,13].

1.2   Related Work

This work, or the general task of finding better-than exhaustive search satisfiability algorithms, is largely motivated by the connection between satisfiability algorithms and circuit lower bounds [18]. Williams [16,17] showed that nontrivial satisfiability algorithms (running in time $2^n/n^{\omega(1)}$) for various circuit classes would imply circuit lower bounds. This raises the question of which circuit classes have nontrivial satisfiability algorithms. In this work, we show such an algorithm for linear-size formulas over arbitrary finite bases. Our algorithm is also an example of the other direction in the connection (following the works [2,3,13,14]); that is, proof techniques for circuit lower bounds (shrinkage under random restrictions) can be used to design nontrivial satisfiability algorithms.

Shrinkage (in expection) under restrictions was a successful technique for proving formula size lower bounds [4,6,8,12,15]. Recently, this property was generalized to concentrated shrinkage in several works in order to get nontrivial satisfiability algorithms [3,13,14], average-case lower bounds [9,10], and pseudo-random generators [7].

The weighting and shrinkage technique we use, following from [4,8,12,13], is also related to the measure-and-conquer approach [5], which gives improved exact algorithms for graph problems such as maximum independent set. Both approaches wish to bound the recursion branches via a better measure (weight) on problem instances. A major difference might be that, the usual measure-and-conquer approach reduces the measure additively in each recursion (linear recurrences), whereas the shrinkage approach reduces the measure by a multiplicative factor (non-linear recurrences).

2   Preliminaries

A *basis* is a collection of boolean functions. A formula over a basis B, which we call a *B-formula*, is a tree where each internal node is labeled by a function in B, and each leaf is labeled by a literal (a variable x or its negation $\bar{x}$) or a constant (0 or 1). We call each internal node a *gate*, and require the fan-in of a gate matches with the inputs of the labeling function.

For $k \geqslant 2$, let B_k be the basis composed of all boolean functions on at most k variables.

We say a function $f(x_1, \ldots, x_k)$ is *positive (negative) unate* in x_i if, for all $a_j \in \{0, 1\}$ where $j \neq i$,

$$f(a_1, \ldots, a_{i-1}, 0, a_{i+1}, \ldots, a_k) \leqslant (\geqslant) f(a_1, \ldots, a_{i-1}, 1, a_{i+1}, \ldots, a_k).$$

We say f is *unate* if it is either positive unate or negative unate in each of its input variables. For $k \geqslant 2$, let U_k be the basis consisting of all unate functions on at most k variables.

For convenience, we assume all gates have fan-in at least two; that is, all negations are eliminated by merging to their inputs. We also assume all binary gates are labeled by either $\vee, \wedge$ or $\oplus$; the other binary gates can be represented by adding necessary negations, e.g., replacing $x \equiv y$ by $x \oplus \bar{y}$. When we write $\vee, \wedge$ or $\oplus$, we assume they are binary.

We define the *size* $L(F)$ of a formula F to be the number of non-constant leaves in F. Let $G_\oplus(F)$ be the number of $\oplus$ gates in F, let $G_2(F)$ be the number of $\vee$ and $\wedge$ gates, and let $G_i(F)$ be the number of i-ary gates, for $3 \leqslant i \leqslant k$. Then obviously, for non-constant F, we have $L(F) = 1 + G_\oplus(F) + \sum_{i=2}^{k}(i-1)G_i(F)$. We define the *weight* of F to be

$$W(F) = L(F) + \alpha_\oplus G_\oplus(F) + \sum_{i=2}^{k} \alpha_i G_i(F),$$

where $\alpha_\oplus, \alpha_2, \ldots, \alpha_k$ is a sequence of positive weights associated with gates of the corresponding types. As we will explain later, we require that $\alpha_{i-1} + \alpha_\oplus \leqslant \alpha_i \leqslant (i-1)\alpha_\oplus$ for $i \geqslant 3$.

3   Shrinkage Under Restrictions

In this section, we will characterize the shrinkage of formula weights under random restrictions. We first present formula simplification rules which will help removing redundancy and transforming the formula into a normalized representation.

3.1   Formula Simplification

For $i \geqslant 3$, we say an i-ary gate $g(x_1, \ldots, x_i)$ has a *linear representation* in one of its input variables, say x_1, if $g(x_1, \ldots, x_i) = x_1 \oplus h(x_2, \ldots, x_i)$ for some $(i-1)$-ary gate h. We will replace a gate by its linear representation whenever possible.

We recall some definitions from [14]. A node in the formula tree is called *linear* if (1) it is a leaf, or (2) it is labeled by $\oplus$ and both of its children are linear. We say a linear node is *maximal* if its parent is not linear. Let F_v denote the subformula rooted at a linear node v. Denote by $\text{var}(F_v)$ the set of variables appearing in F_v.

Two maximal linear nodes u and v are *mergeable* if they are connected by a path where every node in between is labeled by $\oplus$. They can be merged into one maximal linear node in the following way. Let s be the parent of u, and u' be the sibling of u; let t be the parent of v, and v' be the sibling of v; that is, $F_s = F_u \oplus F_{u'}$ and $F_t = F_v \oplus F_{v'}$. (Note that the parents of both u and v must be $\oplus$ since otherwise they would not be mergeable.) To merge u and v, we replace F_u by $F_u \oplus F_v$, and replace F_t by $F_{v'}$.

The following are the formula simplification rules, which include the rules in $[6, 13, 14]$ as special cases.

Formula Simplification Rules:

1. Constant elimination:
   (a) Eliminate constants feeding into binary gates by the following: $1 \vee x = 1, 1 \wedge x = x, 0 \wedge x = 0, 0 \vee x = x, 1 \oplus x = \overline{x}, 0 \oplus x = x$.
   (b) For $3 \leqslant i \leqslant k$, if an i-ary gate is fed by a constant, replace it by an equivalent $(i-1)$-ary gate.
2. Redundant sibling elimination:
   (a) If an $\wedge$ or $\vee$ gate is fed by a literal x and a subformula G, eliminate x and $\overline{x}$ from G (if possible) by the following: $x \vee G = x \vee G|_{x=0}$, $x \wedge G = x \wedge G|_{x=1}$.
   (b) For $3 \leqslant i \leqslant k$, if an i-ary gate does not depend on one of its input, replace it an equivalent $(i-1)$-ary gate.
   (c) If an i-ary gate is fed by two literals over the same variable, replace it by an equivalent $(i-1)$-ary gate.
3. Linear node transformation:
   (a) If an i-ary gate has a linear representation in one of its inputs, replace it by $\oplus$ over that input and a new $(i-1)$-ary gate.
   (b) If a variable appears more than once under a linear node, eliminate unnecessary leaves by the commutativity of $\oplus$ and the following: $x \oplus x = 0$, $x \oplus \overline{x} = 1$.
   (c) Merge any mergeable pairs of maximal linear nodes.

We call a formula *simplified* if none of the rules above is applicable. It is easy to check that, given a formula F, one can compute an equivalent simplified formula F' in polynomial time, and it holds that $L(F') \leqslant L(F)$.

3.2 Weight Reduction Under Restrictions

In the following, we analyze weight reductions when a single leaf is randomly fixed. Given a simplified formula F, a variable x, and $b \in \{0, 1\}$, we define the weight reduction from $x = b$ as $\sigma_{x=b}(F) = W(F) - W(F|_{x=b})$, where $F|_{x=b}$

is the simplified formula obtained from F under the restriction $x = b$. We also define $\sigma_x(F) = (\sigma_{x=0}(F) + \sigma_{x=1}(F))/2$.

Lemma 1. *Suppose that* $\alpha_{i-1} + \alpha_\oplus \leqslant \alpha_i \leqslant (i-1)\alpha_\oplus$ *for* $i \geqslant 3$. *Let* $F = g(x, G_1, \ldots, G_{i-1})$ *be a simplified formula where the root gate* g *has arity* $i \geqslant 2$ *and its first input is a literal* x. *Let* $\beta^b = \sigma_{x=b}(F) - \sum_{j=1}^{i-1} \sigma_{x=b}(G_j)$.

- *If* g *is* $\vee$ *or* $\wedge$, *then* $\min\{\beta^0, \beta^1\} \geqslant 1 + \alpha_2$, *and* $\max\{\beta^0, \beta^1\} \geqslant 2 + \alpha_2$.
- *If* g *is* $\oplus$, *then* $\beta^b \geqslant 1 + \alpha_\oplus$ *for* $b \in \{0, 1\}$.
- *If* g *is* i-*ary for* $i \geqslant 3$, *then* $\min\{\beta^0, \beta^1\} \geqslant 1 + \alpha_i - (i-2)\alpha_\oplus$, *and* $\max\{\beta^0, \beta^1\} \geqslant 1 + \alpha_i - \alpha_{i-1}$.

Note that, β^b measures the weight reduction from restricting a single leaf.

Proof. If g is $\vee$ or $\wedge$, then x or $\overline{x}$ does not appear in G_1 by the simplification rule 2(a), which means $\sigma_{x=b}(G_1) = 0$. For both assignments of x, we can eliminate x and g, reducing the weight by $1 + \alpha_2$; in one of the assignments, we can also eliminate G_1, which has size at least 1.

If g is $\oplus$, then, for both assignments of x, we can eliminate x and g, reducing the weight by $1 + \alpha_\oplus$. This will not affect the weight reduction in G_1, since G_1 cannot be a single literal x or $\overline{x}$ by the simplification rule 3(b).

If g is an i-ary gate, for $i \geqslant 3$, then none of G_i's is a literal x or $\overline{x}$ by the simplification rule 2(c), and g does not have any linear representation by the simplification rule 3(a). We restrict $x = b$ in two steps. First restrict $x = b$ on all G_i's, and let $F' = g(x, G_1|_{x=b}, \ldots, G_{i-1}|_{x=b})$, which has weight $W(F') \leqslant W(F) - \sum_{j=1}^{i-1} \sigma_{x=b}(G_j)$; note that F' may not be simplified on the top gate g (however, each $G_i|_{x=b}$ is simplified). Then restrict $x = b$ on the first input of g, and simplify the formula. Next we compute the weight reduction $W(F') - W(F'|_{x=b})$.

This weight reduction is upper bounded by the weight reduction from restricting $x = b$ on $g(x, y_1, \ldots, y_{i-1})$ where $x, y_1, \ldots, y_{i-1}$ are distinct variables. The rest of the proof follows from the next Claim.

Claim. Let $G = g(x, y_1, \ldots, y_{i-1})$ where $x, y_1, \ldots, y_{i-1}$ are distinct variables, and g does not have any linear representation. Then $\min_{b \in \{0,1\}} \sigma_{x=b}(G) \geqslant 1 + \alpha_i - (i-2)\alpha_\oplus$, and $\max_{b \in \{0,1\}} \sigma_{x=b}(G) \geqslant 1 + \alpha_i - \alpha_{i-1}$.

Proof. We consider how g simplifies when x is fixed. Let $g_b := g(b, y_1, \ldots, y_{i-1})$ be an $(i-1)$-ary gate on inputs $y_1, \ldots, y_{i-1}$. Note that g_b could have linear representations. In general, each of g_0 and g_1 could be an $(i-1)$-ary gate, or eventually replaced by an $(i-1-j)$-ary gate together with j of $\oplus$ gates, for $1 \leqslant j \leqslant i-2$; when $j = i-2$, this is just $i-2$ of $\oplus$ gates.

However, we argue that, g_0 and g_1 cannot have linear representations in the same input. For the sake of contradiction, suppose both g_0 and g_1 have linear representations in the same input y, then we get $g_0 = y \oplus h_0$ and $g_1 = y \oplus h_1$, for some $(i-2)$-ary gates h_0 and h_1. But this implies $g = y \oplus ((\overline{x} \wedge h_0) \vee (x \wedge h_1))$, which contradicts the assumption that g does not have any linear representation.

Suppose g_0 does not have any linear representation, then the smallest weight reduction for $x = 1$ is obtained when g_1 is replaced by $i - 2$ of $\oplus$ gates. The weight reduction is $1 + \alpha_i - \alpha_{i-1}$ for $x = 0$, and $1 + \alpha_i - (i - 2)\alpha_\oplus$ for $x = 1$.

If g_0 is replaced by $(i - 2)$ of $\oplus$ gates, then g_1 cannot have any linear representation; this is the same as the previous case.

If g_0 is replaced by an $(i - 1 - j)$-ary gate together with j of $\oplus$ gates, for $1 \leqslant j < i - 2$, then g_1 cannot be completely replaced by $i - 2$ of $\oplus$ gates. For both $x = 0$ and $x = 1$, the weight reduction is at least $1 + \alpha_i - \alpha_{i-1}$, since $\alpha_{i-1-j} + j\alpha_\oplus \leqslant \alpha_{i-1}$. $\qquad\square$

3.3 Upper Bounds on Formula Weights

Random restrictions will only affect gates fed by leaves, but the weight function is defined by attaching weights to all gates of the formula. In order to characterize the weight reduction in terms of the total weight, we give an upper bound on the total weight expressed by the numbers of leaves feeding into different types of gates.

Let F be a simplified formula of size L and weight W. Let L_2 be the number of leaves feeding into $\vee$ or $\wedge$ gates, let $L_\oplus$ be the number of leaves feeding into $\oplus$ gates, and let L_i be the number of leaves feeding into i-ary gates, for $3 \leqslant i \leqslant k$. Then we have $L = L_\oplus + \sum_{i=2}^{k} L_i = G_\oplus(F) + \sum_{i=2}^{k}(i - 1)G_i(F) + 1$.

The next lemma gives an upper bound of the formula weight. It is similar to the bound by Chockler and Zwick [4] for formulas over unate bases. We need the following parameters:

$$\gamma_\oplus = 1 + \frac{\alpha_\oplus}{2} + \frac{\alpha_k}{2(k - 1)},$$

$$\gamma_i = 1 + \frac{\alpha_i}{i} + \frac{\alpha_k}{i(k - 1)}, \quad 2 \leqslant i \leqslant k.$$

Lemma 2 ([4]). *Suppose that $\alpha_k/(k - 1) \geqslant \alpha_\oplus$ and $\alpha_k/(k - 1) \geqslant \alpha_i/(i - 1)$ for $2 \leqslant i \leqslant k$. Then the formula weight $W \leqslant \gamma_\oplus L_\oplus + \sum_{i=2}^{k} \gamma_i L_i$.*

3.4 Choice of Weight Parameters

We next choose suitable weight parameters. To get nontrivial shrinkage, we require that if a leaf feeding into $\oplus$ is restricted, then the weight reduces proportionally; for a leaf feeding into other types of gates, the weight should reduce by a factor strictly larger than 1. Also, since we will use greedy restrictions in our algorithm instead of completely random restrictions as in [4], we require an upper bound γ for both $\gamma_\oplus$ and γ_i's; that is γ does not depend on the gate type. This will give an upper bound of the weight $W \leqslant \gamma L$.

We choose the following:

$$\alpha_i = i - 1 - \frac{1}{2^{i-1}}, \quad i = 2, \dots, k,$$

$$\alpha_\oplus = \frac{\alpha_k}{k-1} = 1 - \frac{1}{(k-1)\cdot 2^{k-1}},$$

$$\gamma = 2 - \frac{1}{(k-1)\cdot 2^{k-1}}.$$

It is easy to check that $\alpha_{i-1} + \alpha_\oplus \leqslant \alpha_i \leqslant (i-1)\alpha_\oplus$ for $i \geqslant 3$. We also have $\gamma_i < \gamma_k = \gamma_\oplus = \gamma$ for $2 \leqslant i < k$, and, by Lemma 2, $W \leqslant \gamma L$.

Let $\beta_\oplus$ (β_2, β_i for $i \geqslant 3$) be the average weight reduction from restricting a leaf which feeds into an $\oplus$ gate ($\vee$ or $\wedge$ gate, i-ary gate). Then, by Lemma 1,

$$\beta_\oplus = 1 + \alpha_\oplus = \gamma,$$
$$\beta_2 = 1.5 + \alpha_2 = 2,$$
$$\beta_i = 1 + \alpha_i - \frac{1}{2}\left[\alpha_{i-1} + (i-2)\alpha_\oplus\right]$$
$$= 1 + \left(i - 1 - \frac{1}{2^{i-1}}\right) - \frac{1}{2}\left[\left(i - 2 - \frac{1}{2^{i-2}}\right) + (i-2)\cdot\left(1 - \frac{1}{(k-1)\cdot 2^{k-1}}\right)\right]$$
$$= 2 + \frac{i-2}{(k-1)\cdot 2^k} > 2, \quad \text{for } 3 \leqslant i \leqslant k,$$
$$\beta_{\min} = \min\{1 + \alpha_2, \ 1 + \alpha_i - (i-2)\alpha_\oplus\} = 1.5,$$
$$\beta_{\max} = \max\{2 + \alpha_2, \ 1 + \alpha_i - \alpha_{i-1}\} = 2.5.$$

Note that, we have $\beta_\oplus/\gamma = 1$, and $\beta_i/\gamma \geqslant 2/\gamma = 1 + \frac{1}{(k-1)\cdot 2^k - 1} > 1$, for $2 \leqslant i \leqslant k$. This means when a leaf feeding into $\oplus$ is restricted, the weight reduces by γ, and when a leaf feeding into other types of gates is restricted, the weight reduces nontrivially (by $\gamma(1 + \frac{1}{(k-1)\cdot 2^k - 1})$ on average).

4   A Satisfiability Algorithm

Seto and Tamaki [14] observed that, for a linear-size formula over B_2, either satisfiability checking is easy by solving systems of linear equations, or one can restrict a constant number of variables such that the formula shrinks non-trivially. This non-trivial shrinkage property leads to a satisfiability algorithm similar to Santhanam's algorithm [13]. The next lemma generalizes this property to B_k-formulas. The proof is essentially the same as the proof for B_2-formulas in [14].

Lemma 3 (Seto and Tamaki [14]). *Let F be a simplified n-input B_k-formula of size cn for $c > 0$. Then one of the following cases must be true:*

1. *The total number of maximal linear nodes is at most $3n/4$.*
2. *There exists a variable appearing at least $c + \frac{1}{8c}$ times.*

3. *There exists a maximal linear node v such that (1) the parent of v is not $\oplus$, (2) there are at most $8c$ variables under v, and (3) each variable under v appears at least c times in F.*

Theorem 3 (Theorem 1 Restated). *For n-input B_k-formulas of size cn, there is a deterministic algorithm counting the number of satisfying assignments in time $2^{n(1-\delta_{c,k})}$, where $\delta_{c,k} = c^{-O(c^2 k 2^k)}$.*

Proof. The algorithm first simplifies the given formula, and then runs recursively. At each recursive step, suppose we have a simplified formula F with n free variables and size cn. Consider the following cases as in Lemma 3:

1. If there are at most $3n/4$ maximal linear nodes, we enumerate all possible assignments to the maximal linear nodes, which will generate at most $2^{3n/4}$ branches; for each branch, solve a system of linear equations using Gaussian elimination.
2. If a variable x appears at least $c + 1/8c$ times, build two branches for $x = 0$ and $x = 1$; for each branch, simplify the restricted formula and recurse.
3. Otherwise, find the *smallest* maximal linear node, say v, such that it has a non-$\oplus$ parent and at most $8c$ variables in $\mathrm{var}(F_v)$, and each variable in $\mathrm{var}(F_v)$ appears at least c times in F. We also find all maximal linear nodes $\{v_j\}$ which are over exactly the same set of variables as v, that is, $\mathrm{var}(F_{v_j}) = \mathrm{var}(F_v)$.
   (a) If all v_j's are feeding into different gates, we choose all variables in $\mathrm{var}(F_v)$.
   (b) If there are two v_j's feeding into the same gate, we choose all but one (arbitrary) variable in $\mathrm{var}(F_v)$.
   Enumerate all assignments to the chosen variables; for each assignment, restrict the formula, and recurse.

Each branch ends when the restricted formula becomes a constant. For each branch, we count the number of assignments consisting with the restrictions along the branch. The final answer is the summation of the counts for all branches where the restricted formula becomes 1. In the rest of the proof, we bound the number of branches, and thus the running time of the algorithm.

In cases 2, 3(a) and 3(b), we wish to restrict a constant number of variables such that the formula weight/size reduces non-trivially. (A trivial weight reduction from restricting one variable is $c\gamma$, where c is the average number of appearances and γ is the smallest average weight reduction for each leaf.)

In case 2, we get non-trivial reduction since the selected variable appears more than the average.

In case 3, suppose there are d variables in $\mathrm{var}(F_v)$, and we restrict these variables one by one. For case 3(a), when we restrict each of the first $d - 1$ variables, we always eliminate leaves feeding to $\oplus$ gates; this is guaranteed by the minimality of v. For the last variable, we eliminate at least one leaf feeding into a non-$\oplus$ gate. The average weight reduction together is at least $(d-1)c\gamma + (c-1)\gamma + 2 = dc\gamma + 2 - \gamma$.

For case 3(b), whenever two v_j's feed into the same gate, the gate cannot be $\oplus$ since v_j's are maximal linear nodes. Similar to case 3(a), for each variable restricted, we always eliminate leaves feeding to $\oplus$ gates. At the end, the unrestricted variable in $\text{var}(F_v)$ will appear twice as literals feeding into a non-$\oplus$ gate; then at least one leaf can be eliminated. The average weight reduction will be at least $(d-1)c\gamma + 1$.

Consider a partial branch in the recursion tree up to depth $n - l$, for l to be specified later. If case 1 occurs along the branch, then all extensions of this branch will have depth at most $n - l/4$. We next assume case 1 does not occur. We will argue that most branches at depth $n-l$ have formulas of size at most $l/2$, although there are still l variables unrestricted; then extensions of such branches will have depth at most $n - l/2$.

In the following, we claim that the formula weight shrinks with high probability. Intuitively this is because we restrict a small number of variables in each of the cases 2, 3(a) and 3(b) and this gives non-trivial weight reductions. We omit the proof of this claim; a similar result for B_2-formulas was shown in [2,14].

Claim. Let F be a simplified U_k-formula of size cn and weight W. Let W' be the weight of the formula obtained after restricting $n - l$ variables (according to cases 2, 3(a) and 3(b)). Then for large enough l,

$$\mathbf{Pr}\left[W' \geq 2W\left(\frac{l}{n}\right)^{\Gamma}\right] < 2^{-l/bc^2},$$

for $\Gamma = 1 + \Omega(1/c^2 k 2^k)$ and a constant $b > 0$.

We choose $l = pn$ for $p = (8c)^{-1/(\Gamma-1)}$. Then after restricting $n - l$ variables, there are 2^{n-l} branches, but $1 - 2^{-l/bc^2}$ fraction of the branches end with formulas of size

$$L' \leq W' \leq 2W\left(\frac{l}{n}\right)^{\Gamma} \leq 2 \cdot \gamma cn \cdot p^{\Gamma} = 2\gamma cp^{\Gamma-1} \cdot l < \frac{l}{2}.$$

These "small" formulas depend on at most $l/2$ variables, although there are still l variables unrestricted. Completing such branches will get to depth at most $n - l/2$. The total number of complete branches will be at most $2^{n-l} \cdot 2^{l/2} = 2^{n-l/2}$.

For $2^{-l/bc^2}$ fraction of branches having "large" formulas, although they may extend to depth n, the total number of such complete branches will be at most $2^{n-l/bc^2}$.

Therefore, the algorithm generates at most $2^{n-\Omega(l/c^2)}$ branches; the running time is bounded by $2^{n(1-\delta)}$, for $\delta = c^{-O(c^2 k 2^k)}$. $\qquad\qquad\qquad\square$

5   Average-Case Lower Bounds

The algorithm in Theorem 1 essentially constructs a *parity decision tree*, where at each node of the tree, we can either restrict a variable or a parity function (maximal linear node). The next corollary follows directly.

Corollary 1. *An n-input B_k-formula of size cn has a parity decision tree of size $2^{(1-\delta_{c,k})n}$, where $\delta_{c,k} = c^{-O(c^2 k 2^k)}$.*

An average-case lower bound (also called correlation bound) is a lower bound on the minimal formula size for approximating an explicit function. Santhanam [13] showed that linear-size de Morgan formulas compute the parity function correctly on at most $1/2 + 2^{-\Omega(n)}$ fraction of the inputs. Seto and Tamaki [14] extended this to B_2-formulas by showing that linear-size B_2-formulas can compute affine extractors correctly on at most $1/2 + 2^{-\Omega(n)}$ fraction of the inputs. We next generalize this result for B_k-formulas.

Let F_2 be the finite field with elements $\{0,1\}$. A function $\mathsf{E} \colon F_2^n \to F_2$ is a (k, ϵ)-*affine extractor* if for any uniform distribution X over some k-dimensional affine subspace of F_2^n, it holds that $|\mathbf{Pr}[\mathsf{E}(X) = 1] - 1/2| \leqslant \epsilon$. We will need the following known construction of affine extractors.

Theorem 4 ([11,19]). *For any $\delta > 0$ there exists a polynomial-time computable (k, ϵ)-affine extractor $\mathsf{E}_\delta \colon \{0,1\}^n \to \{0,1\}$ with $k = \delta n$ and $\epsilon = 2^{-\Omega(n)}$.*

By Corollary 1, any B_k-formulas size cn has a parity decision tree of size $2^{n-\Omega(n)}$. Since most branches of the tree have depth $n - \Omega(n)$ (defining affine subspaces of dimension $\Omega(n)$), by the definition of E_δ, the tree has exponentially small correlation with E_δ. Thus Theorem 2 follows immediately. We omit the proof since it is similar to the proof for B_2-formulas in [14].

6   Formulas over Unate Bases

Chockler and Zwick [4] showed that U_k-formulas have shrinkage exponent $\Gamma_k = 1 + \frac{1}{3k-4} > 1$ under random restrictions. This implies that Parity requires U_k-formula size n^{Γ_k} and Andreev's function [1] requires size $n^{1+\Gamma_k - o(1)}$. This result can be easily strengthen to concentrated shrinkage via a properly defined weight function. Then we can give #SAT algorithms and average-case lower bounds for U_k-formulas of super-quadratic size, similar to the results for de Morgan formulas [2,9,10,13]. We state the results below, and leave the proofs in the full version of this paper.

Theorem 5. *There are deterministic algorithms counting satisfying assignments for n-input U_k-formulas of*

- *size cn in time $2^{n(1-\Omega(1/c^{3k-4}))}$;*
- *size $n^{2+\frac{1}{3k-4}-\epsilon}$ in time $2^{n-n^{\Omega(1)}}$, for any constant $\epsilon > 0$.*

Corollary 2. *A U_k-formula of size cn has a decision tree of size $2^{n(1-\Omega(1/c^{3k-4}))}$.*

Corollary 3. *Any family of U_k-formulas of size cn can compute Parity correctly on at most $1/2 + 2^{-\Omega(n/c^{3k-4})}$ fraction of inputs.*

Theorem 6. *There is a polynomial-time computable function H (constructed in [9,10]) such that, any family of U_k-formulas F_n of size $n^{2+\frac{1}{3k-4}-\epsilon}$ for a constant $\epsilon > 0$ can compute H correctly on at most $1/2 + 2^{-n^{\Omega(1)}}$ fraction of inputs.*

7 Open Questions

For B_k-formulas of size cn, we give a #SAT algorithm improving over exhaustive search by a factor of $2^{\delta n}$, for $\delta = c^{-O(c^2 k 2^k)}$. An open question is whether we can improve δ to be polynomially small in c, as for U_k-formulas. Another question is whether we have nontrivial satisfiability algorithms for U_k-formulas of size larger than n^{Γ_k+1}, where Γ_k is the shrinkage exponent. The satisfiability algorithms we have for U_k-formulas of super-linear size use exponential space; it would be interesting to have polynomial-space algorithms.

References

1. Andreev, A.E.: On a method of obtaining more than quadratic effective lower bounds for the complexity of π-schemes. Vestnik Moskovskogo Universiteta. Matematika **42**(1), 70–73 (1987)
2. Chen, R., Kabanets, V., Kolokolova, A., Shaltiel, R., Zuckerman, D.: Mining circuit lower bound proofs for meta-algorithms. In: CCC (2014)
3. Chen, R., Kabanets, V., Saurabh, N.: An improved deterministic #SAT algorithm for small De Morgan formulas. In: Csuhaj-Varjú, E., Dietzfelbinger, M., Ésik, Z. (eds.) MFCS 2014, Part II. LNCS, vol. 8635, pp. 165–176. Springer, Heidelberg (2014)
4. Chockler, H., Zwick, U.: Which bases admit non-trivial shrinkage of formulae? Comput. Complex. **10**(1), 28–40 (2001)
5. Fomin, F., Kratsch, D.: A measure & conquer approach for the analysis of exact algorithms. J. ACM **56**(5), 25:1–25:32 (2009)
6. Håstad, J.: The shrinkage exponent of de Morgan formulae is 2. SIAM J. Comput. **27**, 48–64 (1998)
7. Impagliazzo, R., Meka, R., Zuckerman, D.: Pseudorandomness from shrinkage. In: FOCS (2012)
8. Impagliazzo, R., Nisan, N.: The effect of random restrictions on formula size. Random Struct. Algorithms **4**(2), 121–134 (1993)
9. Komargodski, I., Raz, R.: Average-case lower bounds for formula size. In: STOC (2013)
10. Komargodski, I., Raz, R., Tal, A.: Improved average-case lower bounds for demorgan formula size. In: FOCS (2013)
11. Li, X.: A new approach to affine extractors and dispersers. In: CCC (2011)
12. Paterson, M., Zwick, U.: Shrinkage of de Morgan formulae under restriction. Random Struct. Algorithms **4**(2), 135–150 (1993)
13. Santhanam, R.: Fighting perebor: New and improved algorithms for formula and qbf satisfiability. In: FOCS (2010)
14. Seto, K., Tamaki, S.: A satisfiability algorithm and average-case hardness for formulas over the full binary basis. In: CCC (2012)
15. Subbotovskaya, B.A.: Realizations of linear functions by formulas using and or, not. Soviet Math. Doklady **2**, 110–112 (1961)
16. Williams, R.: Improving exhaustive search implies superpolynomial lower bounds. In: STOC (2010)
17. Williams, R.: Non-uniform ACC circuit lower bounds. In: CCC (2011)
18. Williams, R.: Algorithms for circuits and circuits for algorithms. In: CCC (2014)
19. Yehudayoff, A.: Affine extractors over prime fields. Combinatorica **31**(2), 245–256 (2011)

Randomized Polynomial Time Protocol
for Combinatorial Slepian-Wolf Problem

Daniyar Chumbalov[1] and Andrei Romashchenko[2]([⊠])

[1] Ecole Polytechnique Federale de Lausanne (EPFL), Lausanne, Switzerland
daniyar.chumbalov@epfl.ch
[2] Le Laboratoire d'Informatique, de Robotique et de Microelectronique
de Montpellier (LIRMM), Montpellier, France
andrei.romashchenko@lirmm.fr

Abstract. We consider the following combinatorial version of the Slepian–Wolf coding scheme. Two isolated Senders are given binary strings X and Y respectively; the length of each string is equal to n, and the Hamming distance between the strings is at most αn. The Senders compress their strings and communicate the results to the Receiver. Then the Receiver must reconstruct both strings X and Y. The aim is to minimise the lengths of the transmitted messages.

The theoretical optimum of communication complexity for this scheme (with randomised parties) was found in [6], though effective protocols with optimal lengths of messages remained unknown. We close this gap and present for this communication problem a polynomial time randomised protocol that achieves the optimal communication complexity.

Keywords: Slepian-Wolf coding · Communication complexity · Coding theory · Randomized encoding · Pseudo-random permutations

1 Introduction

The classic Slepian–Wolf coding theorem characterises the optimal rates for the lossless compression of two correlated data sources. In this theorem the correlated data sources (two sequences of correlated random variables) are encoded separately; then the compressed data are delivered to the receiver where all the data are jointly decoded, see the scheme in Fig. 1. The seminal paper [1] gives a very precise characterisation of the profile of accessible compression rates — Slepian and Wolf found a natural and intuitive characterisation in terms of Shannon's entropies of the sources.

It seems instructive to view the Slepian–Wolf coding problem in the general context of information theory. In Kolmogorov's paper *"Three approaches to the quantitative definition of information"*, [8], he considers a *combinatorial* (*cf.* Hartley's combinatorial definition of information, [9]), *probabilistic*

A. Romashchenko—The second author is on leave from the IITP RAS. This work is supported in part by the RFBR with grant 13-01-12458 ofi-m2.

G.F. Italiano et al. (Eds.): MFCS 2015, Part II, LNCS 9235, pp. 235–247, 2015.
DOI: 10.1007/978-3-662-48054-0_20

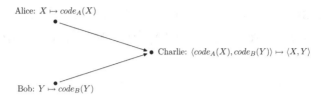

Alice: $X \mapsto code_A(X)$

Charlie: $\langle code_A(X), code_B(Y)\rangle \mapsto \langle X, Y\rangle$

Bob: $Y \mapsto code_B(Y)$

Fig. 1. Distributed sources coding.

(*cf.* Shannon's entropy), and *algorithmic* approach (*cf.* algorithmic complexity a.k.a. Kolmogorov complexity). Many fundamental concepts and constructions in information theory have parallel implementations in all three approaches. An evident example of this parallelism is provided by the formal information inequalities: these can be equivalently represented as linear inequalities for Shannon's entropy, for Kolmogorov complexity, [10], or for (logs of) cardinalities of finite sets, [11,12]. It is remarkable that some results known in one of these approaches look very similar to its homologues from the two other versions of information theory, while mathematical techniques and formal proofs behind them are fairly different.

As for the multi-source coding theory, two parallel versions are known: the probabilistic/Shannon's framework (the Slepian–Wolf coding theory and its generalisations) and the algorithmic/Kolmogorov's one (Muchnik's theorem on conditional coding [13] and its generalisations, respectively). What is missing in this picture is a satisfactory "combinatorial" version of the Slepian–Wolf theorem (though several partial results are known, see blow). We will try to fill this gap and start with formal definitions and some bounds for the combinatorial Slepian–Wolf coding scheme[1].

Thus, we investigate the combinatorial version of the Slepian–Wolf coding problem. To simplify the notation, we focus on the symmetric binary case of this problem. Formally, we consider a communication scheme with two senders (let us call them Alice and Bob) and one receiver (we call him Charlie). We assume Alice is given a string X and Bob is given a string Y. Both strings are of length n, and the Hamming distance between X and Y is not greater than αn. The senders prepare some messages for the receiver (Alice computes her message given X and Bob computes his message given Y). When both messages are delivered to Charlie, he should decode them and reconstruct both strings

---

[1] I. Csiszar and J. Körner described the Slepian–Wolf theorem as "*the visible part of the iceberg*" of the multi-source coding theory; since the seminal paper by Slepian and Wolf, many parts of this "iceberg" were revealed and investigated, see a survey in [14]. Similarly, Muchnik's theorem has motivated numerous researches in the theory of Kolmogorov complexity. Apparently, a similar (probably even bigger) "iceberg" should also exist in the combinatorial version of information theory. However, before we explore this iceberg, we should understand first the very basic multi-source coding models, and the most natural starting point is the combinatorial version of the Slepian–Wolf coding scheme.

X and Y. Our aim is to characterise the optimal lengths of Alice's and Bob's messages.

This is the general scheme of the combinatorial version of the Slepian–Wolf coding problem. Let us place emphasis on the most important points of our setting: (a) the input data are distributed between two senders: Alice knows X but not Y and Bob knows Y but not X; (b) one way comminiction: Alice and Bob send some messages to Charlie without feedback; (c) no communications between Alice and Bob; (d) parameters n and α are known to all parties. It is usual for the theory of communication complexity to consider two types of protocols: deterministic communication protocols (Alice's and Bob's messages are deterministic functions of X and Y respectively) and randomised communication protocol (encoding and decoding procedures are randomised, and for each pair (X, Y) Charlie must get the right answer with only a small probability of error ε). We use the following standard *communication model with private sources of randomness*:

- each party (Alice, Bob, and Charlie) has her/his own "random coin" — a source of random bits,
- the coins are fair, i.e., produce independent and uniformly distributed random bits,
- the sources of randomness are private: each party can access only its own coins.

There is an important difference between the classic probabilistic setting of the Slepian–Wolf coding and randomised protocols for combinatorial version of this problem. In other words, in the probabilistic setting requires to minimise the *average* communication complexity (for *typical* pairs (X, Y)); and in the combinatorial version of the problem we deal with the *worst* case communication complexity (the protocol must succeed with high probability for *each* pair (X, Y) with the bounded Hamming distance).

In terms of the theory of communication complexity, we are looking for an optimal one-round communication protocol. We are interested not only in the total communication complexity (the sum of the lengths of Alice's and Bob's messages) but also in the trade-off between the two sent messages. More formally, we want to characterise the set of *achievable pairs of rates*:

Definition 1. *We say that a pair of integers (k_a, k_b) is* an achievable pair of rates *for the combinatorial Slepian–Wolf problem (in the communication model with private sources of randomness) with parameters (n, α, ε) if there exists a randomised communication protocol such that*

- *the length of Alice's message is equal to k_a,*
- *the length of Bob's message is equal to k_b,*
- *for each pair of inputs $x, y \in \{0, 1\}^n$ such that $\mathrm{dist}(x, y) \leq \alpha n$, the probability of the error is less than ε.*

A simple counting arguments gives very natural lower bounds for lengths of messages in this problem:

Theorem 1 ([6]). *For all $\varepsilon \geq 0$ and $0 < \alpha < 1/2$, a pair (k_a, k_b) can be an achievable pair of rates for the combinatorial Slepian–Wolf problem with parameters (n, α, ε) only if the following three inequalities are satisfied*

- $k_a + k_b \geq (1 + h(\alpha))n - o(n),$
- $k_a \geq h(\alpha)n - o(n),$
- $k_b \geq h(\alpha)n - o(n),$

where $h(\alpha) := -\alpha \log \alpha - (1 - \alpha) \log(1 - \alpha)$ is Shannon's entropy function.

Let us note that Theorem 1 holds also for the model with public randomness.

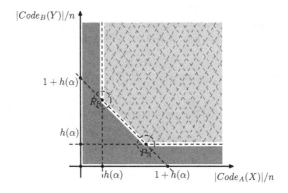

Fig. 2. Achievable rates region.

The asymptotic version of these conditions is shown in Fig. 2: the points in the non-hatched area below the dashed lines are *not achievable*. Notice that these bounds are very similar to the Slepian–Wolf bound known for the classic probabilistic setting, [1]. Indeed, the correspondence is quite straightforward: in Theorem 1 the sum of lengths of two messages is lower-bounded by the "combinatorial entropy of the pair" $(1 + h(\alpha))n$, which is basically the logarithm of the number of possible pairs (X, Y) with the given Hamming distance; in the classic Slepian–Wolf theorem the sum of two channel capacities is bounded by the Shannon entropy of the pair. Similarly, in Theorem 1 the lengths of both messages are bounded by $h(\alpha)n$, which is the "combinatorial conditional entropy" of X conditional on Y or Y conditional on X, i.e., the logarithm of the maximal number of X's compatible with a fixed Y and vice-versa; in the classic Slepian–Wolf theorem these quantities are bounded by the two conditional Shannon entropies.

For the deterministic version of the combinatorial Slepian–Wolf encoding problem, the complete characterisation of the set of achievable pairs remains unknown. Only some partial (negative) results are proven in [7], see also a discussion in [6]. Namely, it is proven that in some $\Theta(n)$-neighborhood of points $(n, h(\alpha)n)$ and $(h(\alpha)n, n)$ (i.e., in the dashed circles around points P_A and P_B in Fig. 2) there are no achievable pairs. Hence, for the case $\varepsilon = 0$ the bound from

Theorem 1 does not provide the exact characterisation of the set of achievable pairs. This result is in sharp contrast with the classic Slepian–Wolf coding.

The case of randomised protocols for this communication problem is somewhat simpler. It is known that the sufficient conditions for achievable pairs are very close to the bound from Theorem 1. More precisely, for every $\varepsilon > 0$, all pairs in the hatched area in Fig. 2 are achievable for the combinatorial Slepian–Wolf problem with parameters (n, α, ε), see [6]. The gap between known necessary and sufficient conditions (the hatched and non-hatched areas in the figure) is negligibly small. This result is similar to the classic Slepian–Wolf theorem.

An annoying shortcoming of the result in [6] was computational complexity. The protocols in [6] require exponential computations on the senders and the receiver sides. In this paper we improve computational complexity without degrading communication complexity. We propose a communication protocol with (i) optimal trade-off between the lengths of senders messages and (ii) polynomial time algorithms for all parties. Technically, we prove the following theorem[2]:

Theorem 2 (main result). *There exists a real $d > 0$ and a function $\delta(n) = o(n)$ such that for all $0 < \alpha < 1/2$ and all integers n, every pair (k_a, k_b) that satisfies three inequalities*

- $k_a + k_b \geq (1 + h(\alpha))n + \delta(n)$,
- $k_a \geq h(\alpha)n + \delta(n)$,
- $k_b \geq h(\alpha)n + \delta(n)$,

is achievable for the combinatorial Slepian–Wolf coding problem with parameters $(n, \alpha, \varepsilon(n) = 2^{-\Omega(n^d)})$ (in the standard communication model with private sources of randomness). Moreover, all the computations in the communication protocol are done in polynomial time.

Protocols achieving the marginal pairs $(n, h(\alpha)n + o(n))$ and $(h(\alpha)n + o(n), n)$ (even for poly-time protocols) were originally proposed in [3] and later in [5]. In our paper we generalise these results: we construct effective protocols for all points in hatched hatched area in Fig. 2. In fact, our construction uses the techniques proposed in [3] and [4].

Our argument employs the following technical tools: reduction of one global coding problem with strings of length n to many local problems with strings of length $\log n$ (similar to the classic technique of concatenated codes); Reed–Solomon checksums; pseudo-random permutations; universal hashing. Due to space limitation of the conference proceedings, the complete proof is deferred to the full version of the paper. In what follows we present the complete description of the communication protocol (encoding and decoding algorithms) and explain the intuition behind the proof.

---

[2] Not surprisingly, the statements of Theorems 1 and 2 are very similar. The gap between necessary and sufficient conditions for achievable pairs is only $o(n)$.

2   Preliminaries

Notation:

- We denote by dist(v, w) the Hamming distance between bit strings v and w.
- For an n-bits string $X = x_1 \ldots x_n$ and a tuple of indices $I = \langle i_1, \ldots, i_s \rangle$ we denote $X_I := x_{i_1} \ldots x_{i_s}$.

Pseudo-Random Permutations: A distribution on the set S_n of permutations of $\{1, \ldots, n\}$ is called *almost t-wise independent* if for every tuple of indices $1 \le i_1 < i_2 < \ldots < i_t \le n$, the distribution of $(\pi(i_1), \pi(i_2), \ldots, \pi(i_t))$ for π chosen according to this distribution has distance at most 2^{-t} from the uniform distribution on t-tuples of t distinct elements from $\{1, \ldots, n\}$.

Proposition 1 ([2]). *For all $1 \le t \le n$, there exists $T = O(t \log n)$ and an explicit map $\Pi : \{0, 1\}^T \to S_n$, computable in time $\mathrm{poly}(n)$, such that the distribution $\Pi(s)$ for random $s \in \{0, 1\}^T$ is almost t-wise independent.*

3   Auxiliarely Communication Models: Shared and Imperfect Randomness

The complete proof of Theorem 2 involves a few different technical tricks. To make the construction more modular and intuitive, we split it in several possibly independent parts. To this end, we introduce several auxiliary communication models. The first two models are somewhat artificial; they are of no independent interest make sense only as intermediate steps of the proof of the main theorem. Here is the list of our communication model:

Model 1. The Model with Partially Shared Sources of Perfect Randomness: Alice and Bob have their own sources of independent uniformly distributed random bits. Charlie has a free access to Alice's and Bob's sources of randomness (these random bits are not included in the communication complexity); but Alice and Bob cannot access the random bits of each other.

Model 2. The Model with Partially Shared Sources of T-Non-perfect Randomness: Alice and Bob have their own (independent of each other) sources of randomness. However these sources are not perfect: they can produce T-independent sequences of bits and T-*wise almost independent* permutations on $\{1, \ldots, n\}$. Charlie has a free access to Alice's and Bob's sources of randomness, while Alice and Bob cannot access the random bits of each other.

Model 3. The Standard Model with Private Sources of Perfect Randomness (Our Principal Model). In this model Alice and Bob have their own sources of independent uniformly distributed random bits. Charlie cannot access random bits of Alice and Bob unless they include these bits in their messages.

In all these models the profile of achievable pairs of rates is the same as in Theorem 1 (the hatched area in Fig. 2). We start with an effective protocol for Model 1, and then extend it to Model 2, and at last to Model 3.

4   An Effective Protocol for Model 1 (Partially Shared Sources of Perfect Randomness)

In this section we show that all pairs of rates from the hatched area in Fig. 2 are achievable for Model 1. Technically, we prove the following statement.

Proposition 2. *The version of Theorem 2 holds for the Communication Model 1.*

Remark 1. Our protocol involves random objects of different kinds: randomly chosen permutations and random hash functions from a universal family. In this section we assume that the used randomness is perfect. This means that all permutations are chosen with the uniform distribution, and all hash functions are chosen independently.

4.1   Parameters of the Construction

Our construction has some "degrees of freedom"; it involves several parameters, and values of these parameters can be chosen in rather broad intervals. In what follows we list these parameters, with some short comments.

– λ is any fixed number between 0 and 1 (this parameter controls the ratio between the lengths of messages sent by Alice and Bob);
– κ_1, κ_2 (some absolute constants that control the asymptotic of communication complexity hidden in the $o(\cdot)$-terms in the statements of Theorem 2 and Proposition 3);
– $k(n) = \log n$ (we will cut strings of Alice and Bob in "blocks" of length k; we can afford the brute force search over all binary strings of length k, since 2^k is polynomial in n);
– $m(n) = n/k(n)$ (when we split n-bits strings into blocks of length k, we get m blocks);
– $r(n) = O(\log k) = O(\log \log n)$ (this parameter controls the chances to get a collision in hashing; we choose $r(n)$ so that $1 \ll r(n) \ll k$);
– $\delta(n) = k^{-0.49} = (\log n)^{-0.49}$ (the threshold for deviation of the relative frequency from the probability involved in the law of large numbers; notice that we choose $\delta(n)$ s.t. $\frac{1}{\sqrt{k}} \ll \delta(n) \ll k$);
– $\sigma = \Theta(\frac{1}{(\log n)^c})$ for some constant $c > 0$ (σn is the length of the Reed-Solomon checksum; we chose σ such that $\sigma \to 0$);
– t (this parameter characterise the quality of the random bits used by Alice and Bob; accordingly, this parameter is involved in the law(s) of large numbers used to bound the probability of the error; we let $t(n) = m^c$ for some $c > 0$).

4.2   The Scheme of the Protocol

Alice's Parts of the Protocol

(1_A) Select at random a tuple of λn indices $I = \{i_1, i_2, \ldots, i_{\lambda n}\} \subset \{1, \ldots, n\}$. Technically, we may assume that Alice chooses at random a permutation π_I on the set $\{1, 2, \ldots, n\}$ and lets $I := \pi_I(\{1, 2, \ldots, \lambda n\})$.

(2_A) Send to the receiver the bits $X_I = x_{i_1} \ldots x_{i_{\lambda n}}$.

(3_A) Choose another random permutation $\pi_A : \{1, \ldots, n\} \to \{1, \ldots, n\}$ and permute the bits of X, i.e., let[3] $X' = x'_1 \ldots x'_n := x_{\pi_A(1)} \ldots x_{\pi_A(n)}$. Further, divide X' into blocks of length $k(n)$, i.e., represent X' as a concatenation $X' = X'_1 \ldots X'_m$, where $X'_j := x'_{(j-1)k+1} x'_{(j-1)k+2} \cdots x'_{jk}$ for each j.

(4_A) Then Alice computes hash values of these blocks. More technically, we consider a universal family of hash functions

$$\mathrm{hash}_l^A \ : \ \{0,1\}^k \to \{0,1\}^{h(\alpha)(1-\lambda)k+\kappa_1\delta k+\kappa_2 \log k+r}.$$

With some standard universal hash family, we may assume that these hash functions are indexed by bit strings l of length $O(k)$. Alice choses at random m indices $l_1, \ldots, l_m$ of hash functions. (We may assume that the sequence of l_i is (T/k)-independent). Then Alice applies each hash_{l_j} to the corresponding block X'_j and sends to Charlie the resulting hash values

$$\mathrm{hash}_{l_1}^A(X'_1), \ldots, \mathrm{hash}_{l_m}^A(X'_m).$$

(5_A) Compute the Reed-Solomon checksums of the sequence $X'_1, \ldots, X'_m$ that are enough to reconstruct all blocks X'_j if most σm of them are corrupted, and send them to Charlie These checksums make a string of $O(\sigma m k)$ bits.

Bob's Parts of the Protocol

(1_B) Choose at random permutation $\pi_B : \{1, \ldots, n\} \to \{1, \ldots, n\}$ and use it to permute the bits of Y, i.e., let[4] $Y'' = y''_1 \ldots y''_n := y_{\pi_B(1)} \ldots y_{\pi_B(n)}$. Further, divide Y'' into blocks of length k, and represent Y'' as a concatenation $Y'' = Y''_1 \ldots Y''_m$, where $Y''_j := y''_{(j-1)k+1} y''_{(j-1)k+2} \cdots y''_{jk}$ for each j.

(2_B) Then choose at random m hash functions $\mathrm{hash}_{l_j}^B$ from a universal family of hash functions

$$\mathrm{hash}_l^B \ : \ \{0,1\}^k \to \{0,1\}^{(1-\lambda)k+h(\alpha)\lambda k+\kappa_1\delta\cdot k+\kappa_2 \log k+r}.$$

(we assume that l_j are (T/k)-independent) and send to Charlie random hash values

$$\mathrm{hash}_{l_1}^B(Y''_1), \ldots, \mathrm{hash}_{l_m}^B(Y''_m).$$

---

[3] In what follows we consider also the π_A-permutation of bits in Y and denote it $Y' = y'_1 \ldots y'_n := y_{\pi_A(1)} \ldots y_{\pi_A(n)}$. Thus, the prime in notation (e.g., X' and Y') implies that we permuted the bits of the original strings by π_A.

[4] Similarly, in what follows we apply this permutation to the bits of X and denote $X'' = x''_1 \ldots x''_n := x_{\pi_B(1)} \ldots x_{\pi_B(n)}$. Thus, the double prime in notation (e.g., X'' and Y'') implies that we permuted the bits of the original strings by π_B.

Similarly to (4_A), we may assume that these hash functions are indexed by bit strings l of length $O(k)$.

(3_B) Compute the Reed-Solomon checksums of the sequence $Y_1'', \ldots, Y_m''$, that are enough to reconstruct all blocks Y_j'', if at most σm of them are corrupted, and send them to Charlie. These checksums should be a string of length $O(\sigma m k)$ bits.

Charlie's Parts of the Protocol

(1_C) Apply Bob's permutation π_B to the positions of bits selected by Alice, and denote the result by I'', i.e., $I'' = \{\pi_B(i_1), \ldots, \pi_B(i_{\lambda n})\}$. Then split indices of I'' into m disjoint parts corresponding to the different intervals $Int_j = \{(j-1)k+1, (j-1)k+2, \ldots, jk\}$, and $I_j'' := I'' \cap Int_j$. Further, for each $j = 1, \ldots, m$ denote by $X_{I_j''}$ the bits sent by Alice, that appear in the interval Int_j after permutation π_B.

(2_C) For each $j = 1, \ldots, m$ try to reconstruct Y_j''. To this end, find all bit strings $Z = z_1 \ldots z_k$ that satisfy a pair of conditions (Cond$_1$) and (Cond$_2$) that we formulate below.

We abuse notation and denote by $Z_{I_j''}$ the subsequence of bits from Z that appear at the positions determined by I_j''. That is, if $I_j'' = \{(j-1)k + s_1, \cdots, (j-1)k + s_l\}$, where

$$(j-1)k + s_1 < (j-1)k + s_2 < \cdots < (j-1)k + s_l,$$

then $Z_{I_j''} = z_{s_1} z_{s_2} \ldots z_{s_l}$. With this notation we can specify the required property of Z:

(Cond$_1$) $\mathrm{dist}(X_{I_j''}, Z_{I_j''}) \leq (\alpha + \delta)|I_j''|$,

(Cond$_2$) $\mathrm{hash}_{l_j}^B(Z)$ must coincide with the hash value $\mathrm{hash}_{l_j}^B(Y_j'')$ received from Bob.

If there is a unique Z that satisfies these two conditions, then take it as a candidate for Y_j''; otherwise (if there is no such Z or if there exist more than one Z that satisfy these conditions) we say that reconstruction of Y_j'' fails.

(3_C) Use Reed-Solomon checksums received from Bob to correct the blocks Y_j'' that we failed to reconstruct or reconstructed incorrectly at step (2_C).

(4_C) Apply permutation π_B^{-1} to the bits of Y'' and obtain Y.

(5_C) Permute bits of Y and X_I using permutation π_A.

(6_C) For each $j = 1, \ldots, m$ try to reconstruct X_j'. To this end, find all bit strings $W = w_1 \ldots w_k$ such that

(Cond$_3$) at each position from $I' \cap Int_j$ the bit from X' (in the j-th block) sent by Alice coincides with the corresponding bit in W,

(Cond$_4$) $\mathrm{dist}(Y_{Int_j \setminus I_j'}', W_{Int_j \setminus I_j'}) \leq (\alpha + \delta)|Int_j \setminus I_j'|$

(Cond$_5$) $\mathrm{hash}_{l_j}^A(W)$ coincides with the hash value $\mathrm{hash}_{l_j}^A(X_j')$ received from Alice. If there is a unique W that satisfies these conditions, then take this string as a candidate for X_j'; otherwise (if there is no such W or if there exist more than one W satisfying these conditions) we say that reconstruction of X_j' fails.

(7_C) Use Reed-Solomon checksums received from Alice to correct the blocks X'_j that were incorrectly decoded at step (6_C).

(8_C) Apply permutation π_A^{-1} to the positions of bits of X' and obtain X.

Lemma 1. *In Communication Model 1, the protocol described above fails with probability at most $O(2^{-m^d})$ for some $d > 0$.*

(The proof is deferred to the full version of the paper.)

4.3   Communication Complexity of the Protocol

Alice sends λn bits at step (2_A), $h(\alpha)(1-\lambda)k + O(\delta)k + O(\log k) + r$ for each block $j = 1, \ldots, m$ at step (3_A), and $\sigma m k$ bits of the Reed-Solomon checksums at step (4_A). So the total length of Alice's message is

$$\lambda n + (h(\alpha)(1-\lambda)k + O(\delta)k + O(\log k) + r) \cdot m + \sigma n.$$

For the values of parameters that we have chosen above, this sum can be estimated as $\lambda n + h(\alpha)(1-\lambda)n + o(n)$. Bob sends $(1-\lambda)k + h(\alpha)\lambda k + O(\delta)k + O(\log k) + r$ bits for each block $j = 1, \ldots, m$ at step (1_B) and $\sigma m k$ bits of the Reed-Solomon checksums at step (2_B). This sums up to

$$((1-\lambda)k + h(\alpha)\lambda k + O(\delta)k + O(\log k) + r) \cdot m + \sigma n$$

bits. For the chosen values of parameters this sum is equal to $(1-\lambda)n + h(\alpha)\lambda n + o(n)$. When we vary parameter λ between 0 and 1, we variate accordingly the lengths of both messages from $h(\alpha)n + o(n)$ to $(1+h(\alpha))n + o(n)$, while the sum of Alice's and Bob's messages always remains equal to $(1+h(\alpha))n + o(n)$. Thus, varying λ from 0 to 1, we move in Fig. 2 from P_B to P_A.

It remains to notice that algorithms of all participants require only poly(n)-time computations. Indeed, all manipulations with Reed-Solomon checksums (encoding and error-correction) can be done in time poly(n), with standard encoding and decoding algorithms. The brute force search used in the decoding procedure requires only the search over sets of size $2^k = \text{poly}(n)$). Thus, Proposition 2 is proven.

5   An Effective Protocol for Model 2

In this section we prove that the pairs of rates from Fig. 2 are achievable for the Communication Model 2. Now the random sources of Alice and Bob are not perfect: the random permutations are t-wise almost independent and the chosen hash functions are t-independent (for a suitable t).

Proposition 3. *The version of Theorem 2 holds for Communication Model 2 (with parameter $T = \Theta(n^c \log n)$).*

To prove Proposition 3 we do not need a new communication protocol — in fact, the protocol for the Model 1 works for the Model 2 as well. The only difference between Proposition 2 and Proposition 3 is a more general statement about the estimation of the error probability:

Lemma 2. *For the Communication Model 2 with parameter $T = \Theta(n^c \log n)$ the communication protocol described in Sect. 4 fails with probability at most $O(2^{-m^d})$ for some $d > 0$.*

(The proof is deferred to the full version of the paper. The argument employs a version of the law of large numbers suitable for t-wise almost independent sequences of random variables.) Since the protocol remains the same, the bounds for the communication and computational complexity, proven in Proposition 2, remain valid in the new setting. With Lemma 2 we get the proof of Proposition 3.

6   The Model with Private Sources of Perfect Randomness

Proposition 3 claims that the protocol from Sect. 4 works well for the artificial Communication Model 2 (with non-perfect and partially private randomness). Now we want to modify this protocol and adapt it to the Communication Model 3. Technically, we have to get rid of (partially) shared randomness. That is, in Model 3 we cannot assume that Charlie gets Alice's and Bob's random bits for free. Moreover, Alice and Bob cannot just send their random bits to Charlie (this would dramatically increase the communication complexity). However, we can use a standard idea: we require that Alice and Bob use pseudo-random bits instead of truly uniformly random bits. Alice and Bob take at random (with the truly uniform distribution) short seeds for pseudo-random generators and expand them to longer sequences of pseudo-random bits, and feed these *pseudo-random* bits in the protocol described in the previous sections. Then Alice and Bob send the random seeds of generators to Charlie (the seeds are rather short, so they do not increase communication complexity substantially); and Charlie (using the same generators) expands the seeds to the same long pseudo-random sequences and plug them in into his side of the protocol.

It remains to choose some specific pseudo-random generators that suits our plan: we need two different pseudo-random generators — one to generate indices of hash functions and another to generate permutations.

Constructing a suitable sequence of pseudo-random hash-functions is simple. Both Alice and Bob needs m random indices l_i of hash functions, and the size of each family of hash functions is $2^{O(k)} = 2^{O(\log n)}$. We need the property of t-independency of l_i for $t = m^c$ (for a small enough c). So we can choose a random polynomial of degree $t - 1$ over $\mathbb{F}_{2^{O(\log n)}}$ and take the values of this polynomial at m different points of the field. Then the seed of this generator is just the tuple of all its coefficients (it consists of $O(t \log n) = o(n)$ bits).

A construction of a pseudo-random permutation is more involved. We need t-wise almost independent pseudo-random permutations. By Proposition 1 such a permutation can be effectively produced by a pseudo-random generator with a seed of length $O(t \log n)$ bits. Again, Alice and Bob chose seeds for random permutation at random, with the uniform distribution. The seeds of the generators involved in our protocol are much shorter than n, so Alice and Bob can send these seeds to Charlie without essentially increasing communication complexity.

The failure probability in Proposition 3 is bounded by $2^{-\Omega(n/\log n)^c}$, which is less than $2^{-\Omega(n^d)}$ for $d < c$. This concludes the proof of Theorem 2.

7 Conclusion

Practical Implementation. The coding and decoding procedures in our protocol run in polynomial time. However, the protocol does not seem very practical (mostly due to the use of the KNR generator from Proposition 1, which requires quite sophisticated computations). A simpler and more practical protocol can be implemented if we substitute t-wise almost independent permutations (KNR generator) by 2-independent permutation (e.g., a random affine mapping). The price for this simplification is a weaker bound for the probability of error, since with 2-independent permutations we have to employ only Chebyshev's inequality instead of stronger versions of the law of large numbers (applicable to n^c-wise almost independent series of random variables). (A similar technique was used in [5] to simplify the protocol from [3].) In this version of the protocol we can conclude that the probability of error $\varepsilon(n)$ tends to 0, but the convergence is rather slow.

Open Question: To characterize the set of all achievable pairs of rates for deterministic communication protocols.

References

1. Slepian, D., Wolf, J.K.: Noiseless coding of correlated information sources. IEEE Trans. Inf. Theory **19**, 471–480 (1973)
2. Kaplan, E., Naor, M., Reingold, O.: Derandomized constructions of k-wise (almost) independent permutations. In: Chekuri, C., Jansen, K., Rolim, J.D.P., Trevisan, L. (eds.) APPROX 2005 and RANDOM 2005. LNCS, vol. 3624, pp. 354–365. Springer, Heidelberg (2005)
3. Smith, A.: Scrambling adversarial errors using few random bits, optimal information reconciliation, and better private codes. In: Proceedings of the 18th ACM-SIAM Symposium on Discrete Algorithms (SODA), pp. 395–404 (2007)
4. Guruswami, V., Smith, A.: Codes for computationally simple channels: explicit constructions with optimal rate. In: Proceedings of the 51st IEEE Symposium on Foundations of Computer Science (FOCS), pp. 723–732 (2010)
5. Chuklin, A.: Effective protocols for low-distance file synchronization (2011). arXiv:1102.4712
6. Chumbalov, D.: Combinatorial version of the Slepian-Wolf coding theorem for binary strings. Siberian Electron. Math. Rep. **10**, 656–665 (2013)
7. Orlitsky, A.: Interactive communication of balanced distributions and of correlated files. SIAM J. Discrete Math. **6**, 548–564 (1993)

8. Kolmogorov, A.N.: Three approaches to the quantitative definition of information. Prob. Inf. Transm. **1**(1), 1–7 (1965)
9. Hartley, R.V.L.: Transmission of information. Bell Syst. Tech. J. **7**(3), 535–563 (1928)
10. Hammer, D., Romashchenko, A., Shen, A., Vereshchagin, N.: Inequalities for Shannon entropy and Kolmogorov complexity. J. Comput. Syst. Sci. **60**(2), 442–464 (2000)
11. Romanshchenko, A., Shen, A., Vereshchagin, N.: Combinatorial interpretation of Kolmogorov complexity. Theoret. Comput. Sci. **271**(1–2), 111–123 (2002)
12. Chan, T.H.: A combinatorial approach to information inequalities. Commun. Inf. Syst. **1**(3), 1–14 (2001)
13. Muchnik, A.: Conditional complexity and codes. Theoret. Comput. Sci. **271**(1–2), 97–109 (2002)
14. Csiszar, I., Körner, J.: Information Theory: Coding Theorems for Discrete Memoryless Systems, 2nd edn. Cambridge University Press, Cambridge (2011)

Network Creation Games: Think Global – Act Local

Andreas Cord-Landwehr[1]([⊠]) and Pascal Lenzner[2]([⊠])

[1] Heinz Nixdorf Institute and Department of Computer Science,
University of Paderborn, Paderborn, Germany
`andreas.cord-landwehr@uni-paderborn.de`
[2] Department of Computer Science, Friedrich-Schiller-University Jena,
Jena, Germany
`pascal.lenzner@uni-jena.de`

Abstract. We investigate a non-cooperative game-theoretic model for the formation of communication networks by selfish agents. Each agent aims for a central position at minimum cost for creating edges. In particular, the general model (Fabrikant et al., PODC'03) became popular for studying the structure of the Internet or social networks. Despite its significance, locality in this game was first studied only recently (Bilò et al., SPAA'14), where a worst case locality model was presented, which came with a high efficiency loss in terms of quality of equilibria. Our main contribution is a new and more optimistic view on locality: agents are limited in their knowledge and actions to their local view ranges, but can probe different strategies and finally choose the best. We study the influence of our locality notion on the hardness of computing best responses, convergence to equilibria, and quality of equilibria. Moreover, we compare the strength of local versus non-local strategy changes. Our results address the gap between the original model and the worst case locality variant. On the bright side, our efficiency results are in line with observations from the original model, yet we have a non-constant lower bound on the Price of Anarchy.

1 Introduction

Many of today's networks are formed by selfish and local decisions of their participants. Most prominently, this is true for the Internet, which emerged from the uncoordinated peering decisions of thousands of autonomous subnetworks. Yet, this process can also be observed in social networks, where participants selfishly decide with whom they want to interact and exchange information. *Network Creation Games* (NCGs) are known as a widely adopted model to study the evolution and outcome of such networks. In the last two decades, several such game variants were introduced and analyzed in the fields of economics, e.g. Jackson

---

A. Cord-Landwehr—This work was partially supported by the German Research Foundation (DFG) within the Collaborative Research Centre "On-The-Fly Computing" (SFB 901).

© Springer-Verlag Berlin Heidelberg 2015
G.F. Italiano et al. (Eds.): MFCS 2015, Part II, LNCS 9235, pp. 248–260, 2015.
DOI: 10.1007/978-3-662-48054-0_21

and Wolinsky [12], Bala and Goyal [3], and theoretical computer science, e.g. Fabrikant et al. [10], Corbo and Parkes [6].

In all of these models, the acting agents are assumed to have a global knowledge about the network structure on which their decisions are based. Yet, due to the size and dynamics of those networks, this assumption is hard to justify. Only recently, Bilò et al. [5] introduced the first variant of the popular (and mathematically beautiful) model by Fabrikant et al. [10] that explicitly incorporates a locality constraint. In this model, the selfish agents are nodes in a network, which can buy arbitrary incident edges. Every agent strives to maximize her service quality in the resulting network at low personal cost for creating edges. The locality notion by Bilò et al. [5] incorporates a worst case view on the network, which limits agents to know only their neighborhood within a bounded distance. The network structure outside of this view range is assumed to be worst possible. In particular, the assumed resulting cost for any agent's strategy-change is estimated as the worst case over all possible network structures outside of this view. In a follow-up work [4], this locality notion was extended by enabling agents to obtain information about the network structure by traceroute based strategies. Interestingly, in both versions the agents' service quality still depends on the whole network, which is actually a realistic assumption. As their main result, Bilò et al. show a major gap in terms of efficiency loss caused by the selfish behavior when compared to the original non-local model.

In this paper, we extend the investigation of the influence of locality in NCGs by studying a more optimistic but still very natural model of locality. Our model allows us to map the boundary of what locally constrained agents can actually hope for. Thereby, we close the gap between the non-local original version and the worst case locality models by Bilò et al. Besides studying the impact of locality on the outcomes' efficiency, we also analyze the impact on the computation of best response strategies and on the dynamic properties of the induced network creation process and we compare the strength of local versus non-local strategy-changes from an agent's perspective.

Due to space constraints we refer to [7] for all omitted proofs and details.

Our Locality Approach. We assume that an agent u in a network only has complete knowledge of her k-neighborhood. Yet, in contrast to Bilò et al. [4,5], besides knowing the induced subnetwork of all the agents that have distance of at most k to u, the agent can, e.g. by sending messages, judge her actual service quality that would result from a strategy-change. That is, we assume rational agents that "probe" different strategies, get direct feedback on their cost and finally select the best strategy. It is easy to see that allowing the agents to probe all available strategy-changes within their respective k-neighborhood and then selecting the best of them is equivalent to providing the agents with a global view, but restricting their actions to k-local moves. Here, a k-*local move* is any combination of (1) removing an own edge, (2) swapping an own edge towards an agent in the k-neighborhood, and (3) buying an edge towards an agent in the k-neighborhood.

Depending on the size of the neighborhood, probing all k-local moves may be unrealistic since there can be exponentially many such strategy-changes. To address this issue, we will also consider k-*local greedy moves*, which are k-local moves consisting only of exactly one of the options (1)–(3). It is easy to see that the number of such moves is quadratic in the number of vertices in the k-neighborhood and especially for small k and sparse networks this results in a small number of probes.

We essentially investigate the trade-off between the cost for eliciting a good local strategy by repeated probing and the obtained network quality for the agents. For this, we consider the extreme cases where agents either probe all k-local moves or only a polynomial fraction of them. Note that the former sheds light on the networks created by the strongest possible locally constrained agents.

Model and Notation. We consider the NCGs as introduced by Fabrikant et al. [10], where n agents V want to create a connected network among themselves. Each agent selfishly strives for minimizing her cost for creating network links, while maximizing her own service quality in the network. All edges in the network are undirected, have unit length and agents can create any incident edge for the price of $\alpha > 0$, where α is a fixed parameter of the game. (Note that in our illustrations, we depict the edge-ownerships by directing the edges away from their owners, yet still understand them as undirected.) The strategy $S_u \subseteq V \setminus \{u\}$ of an agent u determines which edges are bought by this agent, that is, agent u is willing to create (and pay for) all the edges ux, for all $x \in S_u$. Let $\mathcal{S}$ be the n-dimensional vector of the strategies of all agents, then $\mathcal{S}$ determines an undirected network $G_{\mathcal{S}} = (V, E_{\mathcal{S}})$, where for each edge $uv \in E_{\mathcal{S}}$ we have $v \in S_u$ or $u \in S_v$. If $v \in S_u$, then we say that agent u is the *owner* of edge uv, otherwise, if $u \in S_v$, then agent v owns the edge uv. We assume throughout the paper that each edge in $E_{\mathcal{S}}$ has a unique owner, which is no restriction, since no edge can have two owners in any equilibrium network. In particular, we assume that the cost of an edge cannot be shared and every edge is fully paid by its owner. With this, it follows that there is a bijection between strategy-vectors $\mathcal{S}$ and networks G with edge ownership information. Thus, we will use networks and strategy-vectors interchangeably, i.e., we will say that a network (G, α) is in equilibrium meaning that the corresponding strategy-vector $\mathcal{S}$ with $G = G_{\mathcal{S}}$ is in equilibrium for edge-price α. The edge-price α will heavily influence the equilibria of the game, which is why we emphasize this by using (G, α) to denote a network G with edge ownership information and edge-price α. Let $N_k(u)$ in a network G denote the set of all nodes in G with distance of at most k to u. The subgraph of G that is induced by $N_k(u)$ is called the k-*neighborhood* of u.

There are two versions for the cost function of an agent, which will yield two different games called the SUM-NCG [10] and the MAX-NCG [8]. The cost of an agent u in the network $G_{\mathcal{S}}$ with edge-price α is $\text{cost}_u(G_{\mathcal{S}}, \alpha) = \text{edge}_u(G_{\mathcal{S}}, \alpha) + \text{dist}_u(G_{\mathcal{S}}) = \alpha|S_u| + \text{dist}_u(G_{\mathcal{S}})$, where in the SUM-NCG we have that $\text{dist}_u(G_{\mathcal{S}}) = \sum_{w \in V} d_{G_{\mathcal{S}}}(u, w)$, if $G_{\mathcal{S}}$ is connected and $\text{dist}_u(G_{\mathcal{S}}) = \infty$, otherwise. Here, $d_{G_{\mathcal{S}}}(u, w)$ denotes the length of the shortest path between u and w in $G_{\mathcal{S}}$. Since all edges have unit length, $d_{G_{\mathcal{S}}}(u, w)$ is the hop-distance between u and w. In the MAX-NCG, the sum-operator in $\text{dist}_u(G_S)$ is replaced by a max-operator.[1]

A network (G_S, α) is in *Pure Nash Equilibrium* (NE), if no agent can unilaterally change her strategy to strictly decrease her cost. Since the NE has undesirable computational properties, researchers have considered weaker solution concepts for NCGs. G_S is in *Asymmetric Swap Equilibrium* (ASE) [19], if no agent can strictly decrease her cost by swapping one own edge. Here, a swap of agent u is the replacement of one incident edge uv by any other new incident edge uw, abbreviated by $uv \to uw$. Note that this solution concept is independent of the parameter α since the number of edges per agent cannot change. A network (G_S, α) is in *Greedy Equilibrium* (GE) [16], if no agent can buy, swap or delete exactly one own edge to strictly decrease her cost. These solution concepts induce the following k-local solution concepts: (G, α) is in *k-local Nash Equilibrium* (k-NE) if no agent can improve by a k-local move, (G, α) is in *k-local Greedy Equilibrium* (k-GE) if no agent can improve by a k-local greedy move. By slightly abusing notation, we will use the names of the above solution concepts to also denote the sets of instances that satisfy the respective solution concept, i.e., NE denotes the set of all networks (G, α) that are in Pure Nash Equilibrium. The notions GE, k-NE, and k-GE are used respectively. With this, we have the following:

Observation 1. *For $k \geq 1$: $NE \subseteq k\text{-}NE \subseteq k\text{-}GE$ and $NE \subseteq GE \subseteq k\text{-}GE$.*

We will also consider approximate equilibria and say that a network (G, α) is in *β-approximate Nash Equilibrium*, if no strategy-change of an agent can decrease her cost to less than a β-fraction of her current cost in (G, α). Similarly, we say (G, α) is in *β-approximate Greedy Equilibrium*, if no agent can decrease her cost to less than a β-fraction of her current cost by buying, deleting or swapping exactly one own edge.

The *social cost* of a network (G_S, α) is $\text{cost}(G_S, \alpha) = \sum_{u \in V} \text{cost}_u(G_S)$. Let OPT_n be the minimum social cost of an n agent network. Let maxNE_n be the maximum social cost of any NE network on n agents and let minNE_n be the minimum social cost of any NE network on n agents. Then, the *Price of Anarchy* (PoA) [14] is the maximum over all n of the ratio $\frac{\text{maxNE}_n}{\text{OPT}_n}$, whereas the *Price of Stability* (PoS) [2] is the maximum over all n of the ratio $\frac{\text{minNE}_n}{\text{OPT}_n}$.

Known Results. SUM-NCGs were introduced in [10], where the authors proved the first PoA upper bounds, among them a constant bound for $\alpha \geq n^2$ and for trees. Later, by different authors and papers [1,8,17,18], the initial PoA bounds were improved for several ranges of α, resulting in the currently best known bounds of PoA $= \mathcal{O}(1)$, in the range of $\alpha = \mathcal{O}(n^{1-\varepsilon})$, for any fixed $\varepsilon \geq \frac{1}{\log n}$, and the range of $\alpha \geq 65n$, as well as the bound $o(n^\varepsilon)$, for α between $\Omega(n)$ and $\mathcal{O}(n \log n)$. Fabrikant et al. [10] also showed that for $\alpha < 2$ the network having minimum social cost is the complete network and for $\alpha \geq 2$ it is the

---

[1] Throughout this paper, we will only consider connected networks as they are the only ones which induce finite costs.

spanning star which yields a constant PoS for all α. Since it is known that computing the optimal strategy change is NP-hard [10], Lenzner [16] studied the effect of allowing only single buy/delete/swap operations, leading to efficiently computable best responses and 3-approximate NEs. NCG versions where the cost of edges can be shared have been studied by Corbo and Parkes [6] and Albers et al. [1].

For the MAX-NCG, Demaine et al. [8] showed that the PoA is at most 2 for $\alpha \geq n$, for α in range $2\sqrt{\log n} \leq \alpha \leq n$ it is $\mathcal{O}(\min\{4^{\sqrt{\log n}}, (n/\alpha)^{1/3}\})$, and $\mathcal{O}(n^{2/\alpha})$ for $\alpha < 2\sqrt{\log n}$. For $\alpha > 129$, Mihalák and Schlegel [18] showed, similarly to the SUM-NCG, that all equilibria are trees and the PoA is constant.

Kawald and Lenzner [13,15] studied convergence properties of the sequential versions of many NCG-variants and provided mostly negative results. The agents in these variants are myopic in the sense that they only optimize their next step which is orthogonal to our locality constraint.

To the best of our knowledge, the only models in the realm of NCGs that consider locality constraints are [4,5]. As discussed above, both model the local knowledge in a very pessimistic way. Hence, it is not surprising that in [5] the authors lower bound the PoA by $\Omega(n/(1+\alpha))$ for MAX-NCG and by $\Omega(n/k)$ for SUM-NCG, when $k = o(\sqrt[3]{\alpha})$. In particular, for MAX-NCG they show that their lower bound is still $\Omega(n^{1-\varepsilon})$ for every $\varepsilon > 0$, even if k is poly-logarithmic and $\alpha = \mathcal{O}(\log n)$. On the bright side, they provide several PoA upper bounds that match with their lower bounds for different parameter combinations. In their follow-up paper [4], they equip agents with knowledge about either all distances to other nodes, a shortest path tree, or a set of all shortest path trees. However, the MAX-NCG PoA bounds are still $\Theta(n)$ for $\alpha > 1$, while the bounds for SUM-NCG improve to $\Theta(\min\{1 + \alpha, n\})$. Note, that this is in stark contrast to the known upper bounds for the non-local version.

Apart from NCGs the influence of locality has already been studied for different game-theoretic settings, e.g. in the local matching model by Hoefer [11], where agents only know their 2-neighborhood and choose their matching partners from this set.

Our Contribution. We introduce a new locality model for Network Creation Games which allows to explore the intrinsic limits induced by a locality constraint and apply it to one of the mostly studied NCG versions, namely the SUM-NCG introduced by Fabrikant et al. [10]. In Sect. 2, we prove that constraining the agents' actions to their k-neighborhood has no influence on the hardness of computing good strategies and on the game dynamics, even for very small k. In Sect. 3, we explore the impact of locality from the agents' perspective by studying the strength of local versus global strategy-changes obtaining an almost tight general approximation gap of $\Omega(\frac{\log n}{k})$, which is tight for trees. Finally, in Sect. 4 we provide drastically improved PoA upper bounds compared to [4], which are in line with the known results about the non-local SUM-NCG. In contrast to this, we also prove a non-constant lower bound on the PoA, which proves that even in the most optimistic locality model the social efficiency deteriorates significantly.

2   Computational Hardness and Game Dynamics

In this section, we study the effect of restricting agents to k-local (greedy) moves on the hardness of computing a best response and the convergence to equilibria. We start with the observation that for any $k \geq 1$ computing a best possible k-local move is not easier than in the general setting. See [7] for omitted proofs.

Theorem 2. *Computing a best possible k-local move is NP-hard for all $k \geq 1$.*

Proof (Sketch). We reduce from DOMINATING SET [22] and we focus on an agent u who owns edges to any other agent, thus who can only delete edges to improve. For $1 < \alpha < 2$ this yields that u's best possible k-local move only keeps edges to nodes belonging to a minimum dominating set in the original instance.   □

Clearly, the best possible k-local greedy move of an agent can be computed in polynomial time by simply trying all possibilities. Similarly to [16], it is true that the best k-local greedy move is a 3-approximation of the best k-local move. This yields:

Theorem 3. *For any $k \geq 1$, every network in k-GE is in 3-approximate k-NE.*

The same construction as in [16] yields:

Corollary 1. *For $k \geq 2$ there exist k-GE networks that are in $\frac{3}{2}$-approximate k-NE.*

In the following, we analyze the influence of k-locality on the dynamics of the network creation process. For this, we consider several sequential versions of the SUM-NCG, which were introduced in [13,15], and refer to the corresponding papers for further details. Our results only cover the SUM-NCG but we suspect similar results for corresponding versions of MAX-NCG, which would be in line with the non-local version [13].

  In short, we consider the following network creation process: Starting with any connected network having n agents, edge price α, and arbitrary edge ownerships. Now agents move sequentially, that is, at any time exactly one agent is active and checks if an improvement of her current cost is possible. If this is the case, then this agent will perform the strategy-change towards her best possible new strategy. After every such move this process is iterated until there is no agent who can improve by changing her current strategy, i.e., the network is in equilibrium. If there is a cyclic sequence of such moves, then clearly this process is not guaranteed to stop and we call such a situation a *best response cycle* (BR-cycle). Note that the existence of a BR-cycle, even if the participating active agents are chosen by an adversary, implies the strong negative result that no ordinal potential function [20] can exist.

  In Table 1, we summarize our results for four types of possible strategy-changes: If agents can only swap own edges in their k-neighborhood, then this is called the *k-local Asymmetric Swap Game* (k-ASG). If agents are allowed to swap any incident edge within their k-neighborhood, then we have the *k-local Swap Game* (k-SG). If agents are allowed to perform k-local greedy moves only, then we have the *k-local Greedy Buy Game* (k-GBG), and if any k-local move is allowed, then we have the *k-local Buy Game* (k-BG).

Table 1. Overview of convergence speeds in the sequential versions. Note that the existence of a BR-cycle implies that there is no convergence guarantee. Here a move is a strategy-change by one agent.

k	Sum k-SG	Sum k-ASG	Sum k-GBG	Sum k-BG
$k = 1$	no moves	no moves	$\Theta(n^2)$ moves	$\Theta(n)$ moves
$k = 2$	BR-cycle	OPEN	BR-cycle	BR-cycle
$k \geq 3$	BR-cycle	BR-cycle	BR-cycle	BR-cycle
on trees	$\mathcal{O}(n^3)$ moves	$\mathcal{O}(n^3)$ moves	BR-cycle for $k \geq 2$	BR-cycle for $k \geq 2$

3   Local versus Global Moves

In this section, we investigate the agents' perspective. We ask how much agents lose by being forced to act locally. That is, we compare k-local moves to arbitrary non-local strategy-changes. We have already shown that the best possible k-local greedy move is a 3-approximation of the best possible k-local move. The same was shown to be true for greedy moves versus arbitrary strategy-changes [16]. Thus, if we ignore small constant factors, it suffices to compare k-local greedy moves with arbitrary greedy moves. All omitted details of this section can be found in [7].

We start by providing a high lower bound on the approximation ratio for k-local greedy moves versus arbitrary greedy moves. Corresponding to this lower bound, we provide two different upper bounds. The first one is a tight upper bound for tree networks. The second upper bound holds for arbitrary graphs, but is only tight for constant k. Here, the structural difference between trees and arbitrary graphs is captured in the difference of Lemmas 1 and 2. Whereas for tree networks only edge-purchases have to be considered, for arbitrary networks edge-swaps are critical as well.

Theorem 4 (Locality Lower Bound). *For any n' there exist tree networks on $n \geq n'$ vertices having diameter $\Theta(\log n)$ which are in k-GE but only in $\Omega\left(\frac{\log n}{k}\right)$-approximate GE.*

For the first upper bound, let (T, α) be a tree network in k-GE and assume that it is not in GE, that is, we assume there is an agent u, who can decrease her cost by buying or swapping an edge. Note that we can rule out single edge-deletions, since these are 1-local greedy moves and assume that no improving k-local greedy move is possible for any agent. First, we will show that we only have to analyze the cost decrease achieved by single edge-purchases.

Lemma 1. *Let T be any tree network. If an agent can decrease her cost by performing any single edge-swap in T, then this agent also can decrease her cost by performing a k-local single edge-swap in T, for any $k \geq 2$.*

Now we analyze the maximum cost decrease achievable by buying one edge in T. We show that our obtained lower bound from Theorem 4 is actually tight.

Theorem 5 (Tree Network Locality Upper Bound). *Any tree network on n vertices in k-GE is in $\mathcal{O}\left(\frac{\log n}{k}\right)$-approximate GE.*

We now prove a slightly inferior approximation upper bound for arbitrary networks. For this, we show that for any k the diameter of any network is an upper bound on the approximation ratio. For constant k this matches the $\Omega(\mathrm{diam}(G)/k)$ lower bound from Theorem 4 and is almost tight otherwise.

Lemma 2. *For any $k \geq 2$ there is a non-tree network in which some agent cannot improve by performing a k-local swap but by a $k + 1$-local swap.*

It turns out that both buying and swapping operations yield the same upper bound on the approximation ratio, which is even independent of k.

Theorem 6 (General Locality Upper Bound). *Let G' be the network after some agent u has performed a single improving edge-swap or -purchase in a network G, then $\frac{\mathrm{cost}_u(G)}{\mathrm{cost}_u(G')} \leq \mathrm{diam}(G)$. Thus, for any k any network G in k-GE is in $\mathcal{O}(\mathrm{diam}(G))$-approximate GE.*

4 The Quality of k-Local Equilibria

In this section we prove bounds on the Price of Anarchy and on the diameter of equilibrium networks. All omitted proofs can be found in [7].

We start with a theorem about the relationship of β-approximate Nash equilibria and the social cost. This may be of independent interest, since it yields that the PoA can be bounded by using upper bounds on approximate best responses. In fact, we generalize an argument by Albers et al. ([1], Proof of Theorem 3.2).

Theorem 7. *If for $\alpha \geq 2$, (G, α) is in β-approximate NE, then the ratio of social cost and social optimum is at most $\beta(3 + \mathrm{diam}(G))$.*

However, for our bounds we will use another theorem, which directly carries over from the original non-local model.

Theorem 8 ([10, 21] Lemma 19.4). *For any $k \geq 1$ and $\alpha \geq 2$, if a k-NE network (G, α) has diameter D, then its social cost is $\mathcal{O}(D)$ times the optimum social cost.*

We obtain several PoA bounds by applying Theorem 8 and the diameter bounds shown later in this section. Summarizing them, the next theorem provides our main insight into the PoA: The larger k grows, the better the bounds get, providing a remarkable gap between $k = 1$ and $k > 1$. In other words, the theorem tells the "price" in terms of minimal required k-locality to obtain a specific PoA upper bound.

Theorem 9. *For k-NE with $k \geq 1$ the following PoA upper bounds hold (cf. Fig. 1):*

$k = 1$: PoA $= \Theta(n)$, *which can be seen by considering a line graph.*

$k \geq 2$: *Theorem 12 gives* PoA $= \begin{cases} \mathcal{O}((\alpha/k) + k) & k < 2\sqrt{\alpha}, \\ \mathcal{O}(\sqrt{\alpha}) & k \geq 2\sqrt{\alpha}. \end{cases}$

$k \geq 6$: *Theorem 13 gives* PoA $= \mathcal{O}\left(n^{1-\varepsilon(\log(k-3)-1)}\right)$, *provided* $1 \leq \alpha \leq n^{1-\varepsilon}$ *with $\varepsilon \geq 1/\log n$. Note, this yields a constant PoA bound for $\varepsilon \geq 1/(\log(k-3)-1)$.*

For tree networks and $k \geq 2$, Corollary 2 gives PoA $= \mathcal{O}(\log n)$.

4.1 Conformity of k-Local and Nash Equilibria

We stated in Observation 1 that $k\text{-}NE \subseteq NE$ holds. In the following, we discuss the limits of known proof techniques to identify the parameters for which both equilibria concepts coincide, i.e., $k\text{-}NE = NE$.

Theorem 10. *The equilibrium concepts k-NE and NE coincide for the following parameter combinations and yield the corresponding PoA results (cf. Fig. 1):*

1. *For $0 < \alpha < 1$ and $2 \leq k$, $k\text{-}NE = NE$ and* PoA $= \mathcal{O}(1)$.
2. *For $1 \leq \alpha \leq \sqrt{n/2}$ and $6 \leq k$, $k\text{-}NE = NE$ and* PoA $= \mathcal{O}(1)$.
3. *For $1 \leq \alpha \leq n^{1-\varepsilon}$, $\varepsilon \geq 1/\log(n)$ and $4.667 \cdot 3^{\lceil 1/\varepsilon \rceil} + 8 \leq k$, $k\text{-}NE = NE$ and* PoA $= \mathcal{O}(3^{\lceil 1/\varepsilon \rceil})$.
4. *For $1 \leq \alpha \leq 12n \log n$ and $2 \cdot 5^{1+\sqrt{\log n}} + 24 \log(n) + 3 \leq k$, $k\text{-}NE = NE$ and* PoA $= \mathcal{O}(5^{\sqrt{\log n}} \log n)$.
5. *For $12n \log n \leq \alpha$ and $2 \leq k$, $k\text{-}NE = NE$ and* PoA $= \mathcal{O}(1)$.

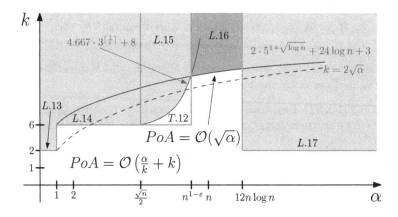

Fig. 1. Overview of our results for both parameters k and α. The shaded areas show where NE and k-NE coincide. In the light gray areas we have a constant PoA, whereas in the dark gray area the PoA is bounded by $\mathcal{O}(5^{\sqrt{\log n}} \log n)$. (L=Lemma, T=Theorem) (Color figure online)

4.2   The Price of Anarchy for Tree Networks

We consider the quality of tree networks that are in k-NE. First of all, by Lemma 1, we have that any tree network in k-GE must be in ASE and therefore, as shown in [9,19], has diameter $\mathcal{O}(\log n)$. Together with Theorem 8 this yields the following upper bound.

Corollary 2. *For k-NE tree networks with $2 \leq k \leq \log n$: PoA $= \mathcal{O}(\log n)$.*

Next, we show a non-constant lower bound, which is tight if $k \in \mathcal{O}(1)$ or if $k \in \Omega(\log n)$. The latter is true, since trees in k-NE have diameter $\mathcal{O}(\log n)$ and thus the technique in [10] yields constant PoA. In particular, this bound states that even with the most optimistic model of locality, there still exists a fundamental difference in the efficiency of the local versus the non-local NCG.

Theorem 11. *For k-NE tree networks with $2 \leq k \leq \log n$: PoA $= \Omega\left(\frac{\log n}{k}\right)$.*

Proof. For any $2 \leq k \leq d$ consider a complete binary tree T_d of depth d with root r, where every edge is owned by the incident agent that is closer to the root r of T. Clearly, independent of α no agent in T_d can delete or swap edges to improve. Thus, towards proving that T_d is in k-NE, we only have to choose α high enough so that no agent can improve by buying any number of edges in her k-neighborhood. Any edge-purchase within the k-neighborhood of some agent u can decrease u's distances to other agents by at most $k-1$. Thus, if $\alpha = (k-1)n$, then no agent can buy one single edge to decrease her cost. Moreover, by buying more than one edge, agent u cannot decrease her distance cost by more than $(k-1)n$, which implies that (T_d, α) is in k-NE for $\alpha = (k-1)n$.

Now we consider the social cost ratio of T_d and the spanning star OPT on $n = 2^{d+1} - 1$ agents, which is the social cost minimizing network, since $\alpha \geq 2$. We will use $\mathrm{cost}(T_d) \geq \alpha(n-1) + n\,\mathrm{dist}_r(T_d)$, which is true, since the root r has minimum distance cost among all agents in T_d. Note that $\mathrm{dist}_r(T_d) > \frac{n}{4}\log n$, since r has at least $\frac{n}{2}$ many agents in distance $\frac{\log n}{2}$. Moreover, the spanning star on n vertices has a total distance cost of less than $2n^2$, since it has diameter 2. With $\alpha = (k-1)n$ this yields:

$$\frac{\mathrm{cost}(T_d)}{\mathrm{cost}(\mathrm{OPT})} > \frac{(k-1)\frac{n^2}{4} + \frac{n^2}{4}\log n}{(k-1)n^2 + 2n^2} > \frac{\frac{n^2}{4}((k-1) + \log n)}{n^2(k+1)} \in \Omega\left(\frac{\log n}{k}\right). \quad \square$$

4.3   The Price of Anarchy for Non-tree Networks

We first provide a simple diameter upper bound that holds for any combination of $k \geq 2$ and $\alpha > 0$.

Theorem 12. *Given a k-NE network (G, α) for $k \geq 2$, then it holds:*

$$\mathrm{diam}(G) \leq \begin{cases} \alpha/(k-1) + k^{\frac{3}{2}} + 1 & k < 2\sqrt{\alpha}, \\ 2\sqrt{\alpha} & k \geq 2\sqrt{\alpha}. \end{cases}$$

In the following, we provide a more involved upper bound for $1 \leq \alpha < n^{1-\varepsilon}$ by modifying an approach by [8]. First, we give three lemmas that lower bound the number of agents in specific sized neighborhoods for k-NE networks. Then, considering the locality parameter k, we look at the maximal neighborhoods for which these lower bounds apply and present an estimation on the network's diameter.

First, we restate Lemma 3 from [8], which holds for any $k \geq 2$ since only 2-local operations are considered.

Lemma 3 (Lemma 3 from [8]). *For any $k \geq 2$ and any k-NE network G with $\alpha \geq 0$ it holds $|N_2(v)| > n/(2\alpha)$ for every agent $v \in V$.*

Lemma 4. *For $k \geq 6$, let G be a k-NE network and $d \leq k/3 - 1$ an integer. If there is a constant $\lambda > 0$ such that $|N_d(u)| > \lambda$ holds for every $u \in V$, then*

(1) either $|N_{2d+3}(u)| > n/2$ for some agent $u \in V$
(2) or $|N_{3d+3}(v)| > \lambda \frac{n}{\alpha}$ for every agent $v \in V$.

Lemma 5. *For $k \geq 4$, let G be a k-NE network with $\alpha < n/2$ and $d \leq k/2 - 1$ an integer. If there is an agent $u \in V$ with $|N_d(u)| \geq n/2$, then $|N_{2d+1}(u)| \geq n$.*

Theorem 13. *For $k \geq 6$, $n \geq 4$, and $1 \leq \alpha \leq n^{1-\varepsilon}$ with $\varepsilon \geq 1/\log n$, the maximal diameter of any k-NE network is $\mathcal{O}\left(n^{1-\varepsilon(\log(k-3)-1)}\right)$.*

Proof. Let G be a k-NE network. We define a sequence $(a_i)_{i \in \mathbb{N}}$ by $a_1 := 2$ and for any $i \geq 2$ with $a_i := 3a_{i-1} + 3$. We want to apply Lemma 4 iteratively with $\lambda_i := (n/\alpha)^i/2$. Lemma 3 ensures that with $|N_2(v)| > n/(2\alpha) = \lambda_1$ for all $v \in V$, we have a start for this.

Let m be the highest sequence index with $a_m \leq k/4 - 3$. If there is a $j \leq m$ such that case (1) of Lemma 4 applies, then there is an agent $u \in V$ with $|N_{2a_m+3}(u)| > n/2$. With $\alpha \leq n^{1-\varepsilon} < n/2$ we get with Lemma 5 that $|N_{4a_m+7}| \geq n$ holds and hence the diameter is at most $a_m < k$. Else, case (2) applies for all $i \leq m$ and we know that for every $v \in V$ it holds $|N_{3a_m+3}(v)| > (n/\alpha)^{m-1}/2$. Using $a_i = \frac{7}{6}3^i - 3/2$ and $a_m \leq k/4 - 3$, we get $m \geq \log(k-3) - 1$.

Let D be the diameter of G and p a longest shortest path. We define a set C by selecting the first agent of p as c_1 and than along the path selecting every further agent with distance of $2k$ to the last selected agent. Now consider the operation of c_1 buying an edge to a agent at distance k in the direction of c_2. Using $k \geq 3a_m + 3$, $|N_{3a_m+3}(c)| > (n/\alpha)^{m-1}/2$ for all $c \in C$ and that G forms an equilibrium:

$$\alpha \geq (k-1)(|C|-1)(n/\alpha)^{\log(k-3)-2}/2 \geq \frac{k-1}{2}\left(\frac{D}{2k}-1\right)(n/\alpha)^{\log(k-3)-2}$$

This gives: $D \leq \frac{4k}{k-1}\alpha\left(\frac{\alpha}{n}\right)^{\log(k-3)-2} + 2k \leq 5n^{1-\varepsilon(\log(k-3)-1)} + 2k$. By using Theorem 10 from [8], we get the claimed diameter upper bound for any $k \geq 6$. $\square$

5   Conclusion and Open Problems

Our results show a major gap in terms of social efficiency between the worst case locality model by Bilò et al. [5] and our more optimistic locality assumption. This gap is to be expected since agents in our model can base their decisions on more information. Interestingly, since most of our upper bounds on the PoA are close to the non-local model, this shows that the natural approach of probing different local strategies is quite convenient for creating socially efficient networks. On the other hand, the non-constant lower bound on the PoA and our negative results concerning the approximation of non-local strategies by local strategies show that the locality constraint does have a significant impact on the game, even in the most optimistic setting. Moreover, our negative results on the hardness of computing best responses and on the dynamic properties show that these problems seem to be intrinsically hard and mostly independent of locality assumptions.

The exact choice of the distance cost function seems to have a strong impact in our model. For the MAX-NCG, which was extensively studied by Bilò et al. [5], it is easy to see that a cycle of odd length, where every agent owns exactly one edge, yields even in our model a lower bound of $\Omega(n)$ on the PoA for $k = 2$ and $\alpha > 1$. This seems to be an interesting contrast between the SUM-NCG and the MAX-NCG, which should be further explored. It might also be interesting to further study what happens if agents are limited to probe only a certain number of strategies. So far, we only considered the cases of probing all strategies and of probing the quadratic number of greedy strategies. Also, applying our locality approach to other models, in particular those motivated from the economics perspective, like [3,12], seems a natural next step of inquiry.

References

1. Albers, S., Eilts, S., Even-Dar, E., Mansour, Y., Roditty, L.: On nash equilibria for a network creation game. ACM Trans. Econ. Comput. **2**(1), 2 (2014)
2. Anshelevich, E., Dasgupta, A., Kleinberg, J., Tardos, E., Wexler, T., Roughgarden, T.: The price of stability for network design with fair cost allocation. SIAM J. Comput. **38**(4), 1602–1623 (2008)
3. Bala, V., Goyal, S.: A noncooperative model of network formation. Econometrica **68**(5), 1181–1229 (2000)
4. Bilò, D., Gualà, L., Leucci, S., Proietti, G.: Network creation games with traceroute-based strategies. In: Halldórsson, M.M. (ed.) SIROCCO 2014. LNCS, vol. 8576, pp. 210–223. Springer, Heidelberg (2014)
5. Bilò, D., Gualà, L., Leucci, S., Proietti, G.: Locality-based network creation games. In: SPAA 2014, pp. 277–286. ACM, New York (2014)
6. Corbo, J., Parkes, D.: The price of selfish behavior in bilateral network formation. In: PODC 2005 Proceedings, pp. 99–107. ACM, New York (2005)
7. Cord-Landwehr, A., Lenzner, P.: Network creation games: Think global - act local. CoRR, abs/1506.02616 (2015)
8. Demaine, E.D., Hajiaghayi, M.T., Mahini, H., Zadimoghaddam, M.: The price of anarchy in network creation games. ACM Trans. Algorithms **8**(2), 13 (2012)

9. Ehsani, S., Fazli, M., Mehrabian, A., Sadeghabad, S.S., Safari, M., Saghafian, M., ShokatFadaee, S.: On a bounded budget network creation game. In: SPAA 2011, pp. 207–214. ACM (2011)

10. Fabrikant, A., Luthra, A., Maneva, E., Papadimitriou, C.H., Shenker, S.: On a network creation game. In: PODC 2003 Proceedings, pp. 347–351. ACM (2003)

11. Hoefer, M.: Local matching dynamics in social networks. In: Aceto, L., Henzinger, M., Sgall, J. (eds.) ICALP 2011, Part II. LNCS, vol. 6756, pp. 113–124. Springer, Heidelberg (2011)

12. Jackson, M.O., Wolinsky, A.: A strategic model of social and economic networks. J. Econ. Theory **71**(1), 44–74 (1996)

13. Kawald, B., Lenzner, P.: On dynamics in selfish network creation. In: Proceedings of the 25th ACM Symposium on Parallelism in Algorithms and Architectures, pp. 83–92. ACM (2013)

14. Koutsoupias, E., Papadimitriou, C.: Worst-case equilibria. In: Meinel, C., Tison, S. (eds.) STACS 1999. LNCS, vol. 1563, p. 404. Springer, Heidelberg (1999)

15. Lenzner, P.: On dynamics in basic network creation games. In: Persiano, G. (ed.) SAGT 2011. LNCS, vol. 6982, pp. 254–265. Springer, Heidelberg (2011)

16. Lenzner, P.: Greedy selfish network creation. In: Goldberg, P.W. (ed.) WINE 2012. LNCS, vol. 7695, pp. 142–155. Springer, Heidelberg (2012)

17. Mamageishvili, A., Mihalák, M., Müller, D.: Tree nash equilibria in the network creation game. In: Bonato, A., Mitzenmacher, M., Prałat, P. (eds.) WAW 2013. LNCS, vol. 8305, pp. 118–129. Springer, Heidelberg (2013)

18. Mihalák, M., Schlegel, J.C.: The price of anarchy in network creation games is (Mostly) constant. In: Kontogiannis, S., Koutsoupias, E., Spirakis, P.G. (eds.) SAGT 2010. LNCS, vol. 6386, pp. 276–287. Springer, Heidelberg (2010)

19. Mihalák, M., Schlegel, J.C.: Asymmetric swap-equilibrium: a unifying equilibrium concept for network creation games. In: Rovan, B., Sassone, V., Widmayer, P. (eds.) MFCS 2012. LNCS, vol. 7464, pp. 693–704. Springer, Heidelberg (2012)

20. Monderer, D., Shapley, L.S.: Potential games. Games Econ. Behav. **14**(1), 124–143 (1996)

21. Nisan, N., Roughgarden, T., Tardos, E., Vazirani, V.V.: Algorithmic Game Theory. Cambridge University Press, New York (2007)

22. Williamson, D.P., Shmoys, D.B.: The Design of Approximation Algorithms. Cambridge University Press, New York (2011)

Oblivious Transfer from Weakly Random Self-Reducible Public-Key Cryptosystem

Claude Crépeau and Raza Ali Kazmi[✉]

School of Computer Science, McGill University, Montreal, Canada
crepeau@cs.mcgill.ca, raza-ali.kazmi@mail.mcgill.ca

Abstract. We define a new notion of *Weakly Random-Self-Reducibile* cryptosystems and show how it can be used to implement secure Oblivious Transfer. We also show that two recent Post-quantum cryptosystems (based on Learning With Errors and Approximate Integer GCD) can be viewed as Weakly Random-Self-Reducible.

1 Introduction

An oblivious transfer is a protocol by which a sender transfers one of many pieces of information to a receiver, but remains oblivious as to which piece has been transferred. The early implementations of Oblivious Transfer of Rabin [1], and Even, Goldreich and Lempel [2] were very innovative but did not offer very strong security. The first OT protocols that may be considered secure were introduced by Ficher, Micali and Rackoff [3], and Berger, Peralta and Tedrick [4].

Later, two methodologies where introduced. A first set of results by Brassard, Crépeau and Robert [18] relied on *Random-Self-Reducibility* (**RSR** for short) of certain number theoretic assumptions such as the Quadratic Residuosity assumption, the RSA assumption or the Discrete log assumption. These results were not extended to very general computational assumptions because the **RSR** property, which was at the heart of the construction, is not very common. In a second set of results by Goldreich, Micali and Wigderson [19], secure Oblivious Transfer protocols were constructed from the generic assumption that (enhanced)[1] Trap-door One-Way permutations exist.

Unfortunately, all constructions that are used to implement secure OT under either of these methodologies fall apart when faced with a quantum computer [5]: none of the so-called Post-Quantum Cryptosystems can directly implement secure OT under these methodologies. Nevertheless, some small modifications to the GMW methodology have led to proposals for OT under the Learning with error LWE assumption [6]. Similarily, Dowsley, van de Graaf, Müller-Quade and Nascimento [7] as well as Kobara, Morozov and Overbeck [8] have proposed Oblivious Transfer protocols based on assumptions related to the McEliece

---

C. Crépeau and R.A. Kazmi—Supported in part by Québec's FRQNT, Canada's NSERC, CIFAR, and QuantumWorks.
[1] The enhanced property is not very restrictive, but some examples of candidates Trap-door One-Way permutations seem to escape it [20].

G.F. Italiano et al. (Eds.): MFCS 2015, Part II, LNCS 9235, pp. 261–273, 2015.
DOI: 10.1007/978-3-662-48054-0_22

public-key cryptosystem [9]. Both of these papers use generalization of the GMW methodology. However both of them also require an extra computational assumption on top of McEliece's to conclude security. Those were the first proposals for OT protocols believed secure against a quantum computer[2].

More recently, a new methodology has been proposed by Peikert, Vaikuntanathan and Waters [6] using the notion of dual-mode cryptosystems. Their approach can be instantiated using a couple number theoretic assumptions, and LWE in the Post-Quantum case. However, this methodology does not seems to extend to any other post-quantum assumptions such as Approximate Integer GCD, for instance.

In this work, we first formalize the results by Brassard, Crépeau and Robert [18] which relied on the **RSR** property of certain number theoretic assumptions (we now have five candidates) in order to introduce a new notion of weakly random-self-reducible encryption scheme **wRSR**. We then show how it is possible to construct a secure Oblivious Transfer under the sole assumption that a secure **wRSR** encryption scheme exists. We show that encryption schemes from two Post-Quantum computational assumptions, LWE [17] and AIGCD [13], have this weaker property. We hope that in the future, our methodology may be used for various new computational assumptions as well.

2   Previous Work

Random Self-Reducible Encryption Scheme. Informally speaking an encryption scheme is Random-Self-Reducible (**RSR**) if an arbitrary ciphertext c may be efficiently transformed to a uniformly distributed ciphertext c' by a user who only knows the public-key from that system. Moreover, upon learning the decryption m' of c', the user is able to efficiently compute m, the decryption of c, from knowledge of the relation between c and c'.

RSR encryption schemes are generally implemented from a homomorphic encryption scheme. Notice however that the **RSR** property is very strong in its uniformity requirement. Homomorphic encryption schemes may fail to satisfy that extra constraint completely. It is worth mentioning that a fully homomorphic circuit private encryption scheme (see [12–16]) is inherently a **RSR** encryption scheme. However, to guarantee that the homomorphic property is used properly extra work would be required, using Zero-knowledge proofs for instance. In the end, this approach also works but the overhead is similar to ours, while the computational assumptions to obtain fully homomorphic encryption schemes are significantly stronger than ours (see Sect. 6.2).

OT from Random-Self-Reducible Encryption Schemes. Brassard, Crépeau and Robert [18] showed how zero-knowledge protocols may be combined with Random-Self-Reducible assumptions such as Quadratic Residuosity

---

[2] Earlier results accomplished a similar security level using only a One-Way function and Quantum Communication. The motivation of the papers cited above and of the current work is to avoid quantum communication altogether [10].

or RSA to obtain a secure 1-out-of-n OT. The latter has the following structure: a sender encrypts n secret messages using its own public-key cryptosystem; the receiver, upon reception of these n encryptions, picks one of them at its choosing and randomizes it using the **RSR** property of the cryptosystem; the sender receives a ciphertext that could equally come from any of the original encryptions; its decryption is obtained from the sender and returned to the receiver; the receiver obtains its chosen message from the decrypted message and the randomness involved in its self-reduction. Zero-knowledge proofs are used to make sure the receiver constructed the random ciphertext properly.

OT from Trap-Door One-Way Permutations. In an effort to obtain OT from a more general assumption, an alternate approach was introduced by Goldreich, Micali and Wigderson [19]. Suppose the sender knows the trap-door to a one-way permutation. The receiver constructs two messages from the domain of the permutation, one using the one-way algorithm, the other sampling directly from the image. Using the trap-door, the sender will find the pre-image of both of these elements and use them to transmit two messages. The receiver who knows only one pre-image, is able to recover only one of the two messages. Zero-knowledge proofs are used to make sure the receiver does not know the pre-image of both.

It was realized many years after the fact that not all trap-door one-way permutations could be used in the above protocol, because it is not always easy to sample from the image without knowing a pre-image [20]. Thus, the notion of *enhanced* trap-door one-way permutation was introduced to remedy to this difficulty. Worth noticing is the work of Haitner [21] who weakened the requirements to implement OT using a collection of dense trapdoor permutations.

Bit Commitments and Zero-Knowledge Protocols. Under the assumption that we have a One-Way Function, it is possible to get a statistically binding, computationally concealing Bit Commitment scheme by the general constructions of Håstad, Impegliazzo, Levin and Luby's [22] and Naor [23]. These constructions and their proofs of security are also valid when the adversary is a quantum computer.[3] In this paper, ZK protocols will be used to prove that certain values are constructed properly from a public-key. Invoking general techniques developed in [24] and [25], we can prove any relation among committed bits in a computational zero-knowledge fashion.

Both parties may publish a public-key using a similar computational assumption, use its own to construct a Bit Commitment Scheme and prove arbitrary polynomially verifiable statements about its commitments.

---

[3] Under the same assumption, it is also possible to get a computationally binding, statistically concealing, Bit Commitment scheme by the general construction of Haitner, Nguyen, Ong, Reingold, and Vadhan [26]. However their proof technique does not appear to extend to quantum adversaries.

Compairasion with Homomorphic Cryptosystems. The notions of **RSR** and **wRSR** cryptosystems are in some sense similar to that of Homomorphic Cryptosystems. Part of the **wRSR** properties are usually implemented using the Homomorphic property of their related cryptosystem. However, contrary to general constructions such as Fully Homomorphic Cryptosystems where *two* homomorphic operations are necessary, **wRSR** relies on a single operation. Moreover, **wRSR** does not need the cryptosystem to tolerate application of several successive homomorphic operations in a row, which is typical of constructions for Fully Homomorphic Encryptions. As a result, and although we use similar assumptions (LWE [17] and AIGCD [13]) as constructions for Fully Homomorphic Encryptions, the versions of these assumptions we need are actually *weaker* than theirs. In the end, it is more likely that our assumptions are hard.

3   Background Material

Notations

We denote a vector v by lower-case bold letters $\mathbf{v}$ and matrices by upper-case bold letters $\mathbf{V}$. We denote the Euclidean norm of a vector $\mathbf{v}$ by $\|\mathbf{v}\|_2$, the largest entry (in absolute value) of a vector or a matrix is denoted by $\|\mathbf{v}\|_\infty$ or $\|\mathbf{V}\|_\infty$ and $\lfloor x \rceil$ denote the nearest integer to x. We denote ϵ the empty string.

Gaussian Distribution and Standard Tail Inequality. For any $\beta \geq 0$ the *Gaussian distribution* with mean 0 is the distribution on $\mathbb{R}$ having density function $D_\beta(x) = \frac{1}{\beta}\exp(-\pi(x/\beta)^2)$, for all $x \in \mathbb{R}$.

A random variable with normal distribution, lies within $\pm\frac{t\cdot\beta}{\sqrt{2\pi}}$ of its mean, except with probability at most $\frac{1}{t}\cdot\exp(-t^2/2)$.

For any integer $q \geq 2$, the *discrete Gaussian distribution* $\overline{\psi}_\beta(q)$ over $\mathbb{Z}_q$ with mean 0 and standard deviation $\pm\frac{q\cdot\beta}{\sqrt{2\pi}}$ is obtained by drawing $y \leftarrow D_\beta$ and outputting $\lfloor q \cdot y \rceil \pmod{q}$.

$\binom{2}{1}$**-Oblivious Transfer.** A One-out-of-Two Oblivious Transfer denoted as $\binom{2}{1}$-OT, is a protocol in which a sender inputs two ordered bits b_0, b_1 and a receiver inputs a choice bit c. The protocol sends b_c to the receiver, without the sender learning c, while the receiver learns nothing other than b_c. A full definition of the security of oblivious transfer can be obtained from the work of [20].

4   Random Self-Reducible Encryption Scheme

In this section we formalize the definition of **RSR** encryption schemes. We further show that two number theoretic schemes satisfy this property.

Definition 1. *Let* $\xi = (KeyGen, Enc, Dec, \mathcal{M}, \mathcal{C})$ *be a public-key cryptosystem and* λ *be the security parameter. The cryptosystem* ξ *is random-self-reducible if there exists a set* $\widehat{\mathcal{M}}$, *a pair of probabilistic polynomial-time algorithms* $(\mathcal{S}, \mathcal{S}')$, *together with a polynomial-time algorithm* $\widehat{Dec}$, *such that for a key pair* $(sk, pk) \leftarrow KeyGen(1^\lambda)$ *and uniformly picked string* $\mathbf{R}$ *from* $\widehat{\mathcal{M}}$,

1. $\mathcal{S}_{pk} : \widehat{\mathcal{M}} \times \mathcal{C} \to \mathcal{C}$, $\mathcal{S}'_{pk} : \widehat{\mathcal{M}} \times \widehat{\mathcal{M}} \to \mathcal{M}$, *and* $\widehat{Dec}_{sk} : \mathcal{C} \to \widehat{\mathcal{M}}$,
2. $\mathcal{S}_{pk}(\mathbf{R}, c)$ *is uniformly distributed over* $\mathcal{C}$, *for all* $c \in \mathcal{C}$,
3. $\mathcal{S}'_{pk}(\mathbf{R}, \widehat{Dec}_{sk}(\mathcal{S}_{pk}(\mathbf{R}, Enc_{pk}(m)))) = m$, *for all messages* $m \in \mathcal{M}$.

Examples of RSRCryptosystem. There are several cryptosystems that satisfy **RSR** property for example the RSA [27], Micali-Goldwasser [28], Paillier cryptosystems [31], Elgamal [29], Elliptic curve [30] are all random self-reducible. We give details of the first two:

Goldwasser-Micali Cryptosystem. Let (x, n) and (p, q) denote the public/private keys and (Enc, Dec) denote the encryption/decryption algorithms.

1. $\widehat{\mathcal{M}} = \{0, 1\}$, $\widehat{Dec} = Dec$
2. $\mathcal{S}_{pk}(\mathbf{R}, Enc(b)) = Enc(\mathbf{R}) \cdot Enc(b) \mod n = Enc(\mathbf{R} \oplus b)$, where bit $\mathbf{R}$ is uniformly chosen.
3. $\mathcal{S}'_{pk}(\mathbf{R}, b) = \mathbf{R} \oplus b$.

Semantically Secure RSA Cryptosystem. Let (e, n) and d denote the public/private keys and $\left(Enc := \left(lsb^{-1}(b)\right)^e \mod n, Dec := lsb(m^d \mod n)\right)$ denote the encryption/decryption algorithms, where $lsb^{-1}(b)$ is a random element r in $\mathbb{Z}_n^*$ such that $lsb(r) = b$.

1. $\widehat{\mathcal{M}} = \mathbb{Z}_n^*$, $\widehat{Dec}(m) = m^d \mod n$
2. $\mathcal{S}_{pk}(\mathbf{R}, Enc(b)) = \mathbf{R}^e \cdot Enc(b) \mod n = \left(\mathbf{R} \cdot lsb^{-1}(b)\right)^e \mod n$ where $\mathbf{R}$ is uniformly chosen from the message space $\widehat{M}$.
3. $\mathcal{S}'_{pk}(\mathbf{R}, m) = lsb\left(\mathbf{R}^{-1} \cdot m \mod n\right)$.

5 $\binom{2}{1}$-OT from a RSR Public-Key Cryptosystem

Let $\xi = (KeyGen, Enc, Dec, \mathcal{M}, \mathcal{C})$ be a **RSR** public-key cryptosystem and λ be the security parameter. Let $(sk, pk) \leftarrow KeyGen(1^\lambda)$ be sender's private and public-keys. The sender encodes his bits so that $Enc_{pk}(b_0)$ and $Enc_{pk}(b_1)$ are semantically secure encryptions of b_0, b_1.

Protocol 1. $\binom{2}{1}$-OT from **RSR** Cryptosystem.

1: The sender computes $c_0 \leftarrow Enc_{pk}(b_0)$ and $c_1 \leftarrow Enc_{pk}(b_1)$.
2: The sender sends the ordered pair (c_0, c_1) to the receiver.
3: The receiver picks a string $\mathbf{R}$ uniformly from $\mathcal{C}$ and computes $c \leftarrow \mathcal{S}_{pk}(\mathbf{R}, c_i)$ for its choice bit i and sends c to the sender.
4: The sender computes $\widehat{m} \leftarrow \widehat{Dec}_{sk}(c)$ and sends $\widehat{m}$ to the receiver.
5: The receiver obtains the bit $b_i \leftarrow \mathcal{S}'_{pk}(\mathbf{R}, \widehat{m})$.

Correctness

We first observe that this protocol correctly computes $\binom{2}{1}$-OT.

$$\mathcal{S}'_{pk}(\mathbf{R}, \widehat{m}) = \mathcal{S}'_{pk}(\mathbf{R}, \widehat{Dec}_{sk}(c)) = \mathcal{S}'_{pk}(\mathbf{R}, \widehat{Dec}_{sk}(\mathcal{S}_{pk}(\mathbf{R}, c_i)))$$
$$= \mathcal{S}'_{pk}(\mathbf{R}, \widehat{Dec}_{sk}(\mathcal{S}_{pk}(\mathbf{R}, Enc_{pk}(b_i)))) = b_i \text{ by definition 1.}$$

Theorem 1. *Protocol* 1 *is a secure oblivious transfer in the semi-honest model.*

Proof. We will present a simulator for each party. These simulators are given the local input (which also includes the security parameter λ) and the local output of the corresponding party.

Simulator for the Sender's View: We will first present a simulator for the sender's view. On input $((b_0, b_1), 1^\lambda, \epsilon)$, this simulator uniformly picks c' from $\mathcal{C}$ and outputs $((b_0, b_1), 1^\lambda, c')$. Clearly this output distribution is identical to the view of the sender in the real execution. This hold because c' is uniformly distributed over the ciphertext space $\mathcal{C}$. Therefore, the receiver's security is perfect.

Simulator for the Receiver's View: On input $(i, b_i, 1^\lambda)$, this simulator generates $(sk', pk') \leftarrow KeyGen(1^\lambda)$ as in protocol 1. It computes $c'_i \leftarrow Enc_{pk'}(b_i)$ and $c'_{1-i} \leftarrow Enc_{pk'}(b)$ (for some $b \in \mathcal{M}$) The simulator then picks a string $\mathbf{R}'$ uniformly. It then computes $c' \leftarrow \mathcal{S}_{pk'}(\mathbf{R}', c'_i)$ and $\widehat{m}' \leftarrow \widehat{Dec}_{sk'}(c')$. The simulator outputs $(i, 1^\lambda, pk', c'_0, c'_1, \widehat{m}')$. Note that except for c'_{1-i}, this output distribution is identical to the view of the receiver in the real execution. Moreover, since ξ is a semantically secure encryption scheme, it is impossible to distinguish between the encryption of b_{1-i} and b for any probabilistic polynomial time adversary except with negligible probability. Therefore, the sender's security is computational.

Malicious Adversaries: Of course we are not only interested in the semi-honest case but also to the situation with malicious adversaries. To handle these cases, zero-knowledge proofs are used by the sender to demonstrate that c_0, c_1 are well formed encryptions and by the receiver to demonstrate that c is indeed constructed from a single c_i and not a combination of both. We leave it as an exercise to demonstrate the full result including zero-knowledge proofs [20]:

Theorem 2. *Protocol* 1 *may be compiled to a secure oblivious transfer in the malicious model.*

Proof. (see [20,24]).

6   Weakly Random Self-Reducible Encryption

The current state of affairs is that we don't know of any **RSR** cryptosystem believed to be resistant to quantum attacks. The **RSR** property may be considered too strong in its uniformity requirement of the output of $\mathcal{S}$. One can weaken this property to statistical indistinguishability for some pair of probabilistic polynomial distributions and can still obtain a secure OT protocol provided we have cryptosystems satisfying this weaker property.

In this section we define the notion of *weakly Random-Self-Reducibile* public-key cryptosystem. Informally speaking a *public-key cryptosystem* is weakly Random-Self-Reducible if it is possible efficiently (using the public key) to re-encrypt a ciphertext c_i in a way to make it unrecognizable, regardless of the plaintext it carries. After obtaining decryption of the re-encrypted ciphertext $\widehat{c}$, it is possible to recover the plaintext hidden by the original encryption c_i. We accept that the unrecognizability property be statistical indistinguishability instead of perfect indistinguishability as in **RSR** .

Our definition is motivated by the fact that many post-quantum encryption schemes use random errors in the process of encrypting the plaintext. Many of these schemes provide a fair amount of flexibility in choosing the size of the error for a fixed pair of public and private keys. Due to this flexibility, one can easily convert these cryptosystem into a **wRSR** scheme. The encryption algorithm *Enc* involves relatively small errors, while the re-encryption process uses relatively large errors that will hide the original error. The definition is formally stated below.

Definition 2. *A public-key cryptosystem* $\xi = (KeyGen, Enc, Dec, \mathcal{M}, \mathcal{C})$ *is weakly random-self-reducible if there exist sets* $\widehat{\mathcal{M}}, \widehat{\mathcal{C}}$, *a pair of probabilistic polynomial-time algorithms* $(\mathcal{S}, \mathcal{S}')$, *together with a probabilistic polynomial-time algorithm* $\widehat{Dec}$, *and a probabilistic polynomial-time distribution* χ *on* $\widehat{\mathcal{C}}$ *such that for all* $c_1, c_2 \in \mathcal{C}$, *key pair* $(sk, pk) \leftarrow KeyGen\left(1^\lambda\right)$ *and* $\mathbf{R} \xleftarrow{\chi} \widehat{\mathcal{C}}$:

1. $\mathcal{S}_{pk} : \widehat{\mathcal{C}} \times \mathcal{C} \to \widehat{\mathcal{C}}$, $\mathcal{S}'_{pk} : \widehat{\mathcal{C}} \times \widehat{\mathcal{M}} \to \mathcal{M}$ *and* $\widehat{Dec}_{sk} : \widehat{\mathcal{C}} \to \widehat{\mathcal{M}}$,
2. $\mathcal{S}_{pk}(\mathbf{R}, c_1)$ *and* $\mathcal{S}_{pk}(\mathbf{R}, c_2)$ *are statistically indistinguishable,*
3. $\mathcal{S}'_{pk}\left(\mathbf{R}, \widehat{Dec}_{sk}\left(\mathcal{S}_{pk}(\mathbf{R}, Enc_{pk}\left(m\right))\right)\right) = m$, *for all messages* $m \in \mathcal{M}$.

Note that **RSR** is the sub-case of **wRSR** where $\widehat{\mathcal{M}} = \mathcal{C}$, χ is the uniform distribution over $\widehat{\mathcal{C}}$ and $\mathcal{S}_{pk}(\mathbf{R}, c)$ is uniformly distributed over $\mathcal{C}$. In Sect. 6.1

we show that one can construct a weakly Random-Self-Reducible encryption schemes based on the Approximate Integer GCD assumption [17] or the Learning with Errors assumption [13].

6.1   Instantiation of wRSR Public-Key Cryptosystems

We provide concrete instantiations of **wRSR** schemes from two different post-quantum assumptions.

1. Approximate Integer GCD problem (**AIGCD**)[13].
2. Learning with Errors (**LWE**)[33].

More precisely we show that one can easily construct a **wRSR** from the cryptosystems presented in [13,17]. Please note that for these encryption schemes, operation $(a \mod n)$ means mapping integer a into the interval $[-\lfloor n/2 \rfloor, \lfloor n/2 \rfloor]$, (where n is an odd positive integer).

6.2   Approximate Integer GCD

Let p be a large η-bit odd integer and x_i's are defined as follows

$$x_i = q_i p + r_i, \ \ 0 \leq i \leq \tau$$

where x_i is a γ-bit number which is much larger than p and r_i is a ρ-bit error-term which is much smaller than p in absolute value. W.l.o.g. assume that x_0 is the largest of them, and that x_0 is odd. Under the Approximate Integer GCD assumption the function

$$f_x(s, z, b) = \left(2\overline{z} + b + 2 \sum_{i=1}^{\tau} x_i s_i \right) \mod x_0$$

is one-way for anyone who does not know p, where $b \in \{0, 1\}$, $\mathbf{s} \in \{0, 1\}^\tau$ is a random binary vector and $\overline{z}$ is a random error term of appropriate size (see below).

Public-Key Cryptosystem from AIGCD Problem. Van Dijk, Gentry, Halevi and Vaikuntanathan constructed a fully homomorphic encryption scheme based on the problem of finding an approximate integer gcd [13]. The construction below has many parameters, controlling things like the number of integers in the public-key and the bit-length of the various components. Specifically, we use the following five parameters (all polynomial in the security parameter λ):

η is the bit-length of the secret key p. γ is the bit-length of the integers x_i in the public-key. ρ is the bit-length of the noise r_i. ρ' is the bit-length of the random error z. τ is the number of integers in the public-key, (contrary to the other parameters, this is *not* a bit-size.)

These parameters must be set under the following constraints:

- $\rho \in \omega(\log \lambda)$, to protect against brute-force attacks on the noise.
- $\rho' = \Omega(\rho + \log \tau)$ (τ is a polynomial is λ, e.g. $\tau = \lambda$).
- $\eta \geq \rho \cdot \Theta(\lambda \log^2 \lambda)$ and should satisfy $2^{\eta-2} > 2^{\rho'} + \tau \cdot 2^{\rho}$, to avoid sums of errors passing $p/2$.
- $\gamma \in \omega(\eta^2 \log \lambda)$, to thwart various lattice-based attacks on the underlying approximate-gcd problem.
- $\tau \geq \gamma + \omega(\log \lambda)$, in order to use the leftover hash lemma in the reduction to approximate gcd.

The public-key is the vector $\mathbf{x} = (x_0, x_1, \ldots, x_\tau)$ and the private key is the η bit integer p. To encrypt a bit $b \in \{0, 1\}$ under the public-key $\mathbf{x}$.

- $Enc_{\mathbf{x}}(b)$
  1. Pick uniformly a random bit string $s_1, \cdots, s_\tau$ and pick uniformly a $\bar{p}$-bit error-term $\bar{z}$.

  2. Output the ciphertext $c \leftarrow \left(2\bar{z} + b + 2 \sum_{i=1}^{\tau} x_i s_i \right) \bmod x_0$.

- $Dec_p(c)$
  1. Output bit $b \leftarrow (c \bmod p) \bmod 2$.

The decryption works, provided the overall distance to the nearest multiple of p does not exceed $p/2$, that is $2(\bar{z} + \sum_{i=1}^{\tau} r_i s_i)$ is less than $p/2$ in absolute value. For the above choice of parameters this will always be the case. We rely on the work of [13] to assess that the resulting cryptosystem is a semantically secure encryption scheme.

Weakly RSR Based on Approximate Integer GCD. The cryptosystem based on **AIGCD** can easily be converted to a **wRSR** encryption scheme. Keeping the same notations as above we set

- $\rho = 2\sqrt{\lambda}$ (is the size of $r_i's$ in the public key), $\bar{p} = \rho/2$ (size of the error term $\bar{z}$ in Enc).
- $\rho' = 2\rho$ (size of the error term $\mathcal{Z}$ in $\mathcal{S}_{pk}$)
- $\eta = \Theta(\lambda)$ (size of the private key). This parameter is smaller than suggested in [13].
- $\gamma \in \omega(\eta^2 \log \lambda)$. This parameter is smaller than suggested in [13].
- $\tau = \gamma + \rho$ (number of $x_i's$ in public-key $\mathbf{x}$).
- $\mathcal{I} = \left\{ [-2^{\rho'}, -2^{\rho'-1}] \cup [2^{\rho'-1}, 2^{\rho'}] \right\} \cap \mathbb{Z}$.

- $\mathcal{R}' = \left\{ 2\mathcal{Z} + 2 \sum_{i=1}^{\tau} x_i w_i \bmod x_0 : \mathcal{Z} \in \mathcal{I}, w_i \in \{0, 1\} \right\}$.

- $\widehat{\mathcal{M}} = \{0,1\}$ and $\widehat{\mathcal{C}} = \mathcal{C} \times \widehat{\mathcal{M}}$.
- The distribution χ is induced by picking $\mathbf{r} \xleftarrow{uniform} \mathcal{R}'$, $e \xleftarrow{uniform} \widehat{\mathcal{M}}$ and outputting $\mathbf{R} = (\mathbf{r}, e)$.
- $\mathcal{S}_{pk}(\mathbf{R}, c) := (\mathbf{r} + e + c) \bmod x_0$.
- $\widehat{Dec}_{sk} := Dec_{sk}$.
- $\mathcal{S}'_{pk}(\mathbf{R}, \hat{b}) := (e + \hat{b}) \bmod 2$.

wRSR Encryption Scheme from AIGCD

wRSR Properties(Semi-Honest Case). The scheme clearly satisfies the first and the third properties for the above choice of parameters. For the second property let

$$\mathcal{S}_{pk}(\mathbf{R}, c) = \left(2\mathcal{Z} + e + 2\sum_{i=1}^{\tau} x_i w_i\right) + c \ \bmod \ x_0$$

$$\mathcal{S}_{pk}(\mathbf{R}, c') = \left(2\mathcal{Z} + e' + 2\sum_{i=1}^{\tau} x_i w_i\right) + c' \ \bmod \ x_0.$$

Since $c, c' \in \mathcal{C}$, there exist $\overline{\rho}$ bit integers $\overline{z}, \overline{z}'$, vectors $\mathbf{s}, \mathbf{s}' \in \{0,1\}^{\tau}$ and bits $b, b' \in \{0,1\}$ such that

$$c = \left(2\overline{z} + b + 2\sum_{i=1}^{\tau} x_i s_i\right) \ \bmod \ x_0 \ \& \ c' = \left(2\overline{z}' + b' + 2\sum_{i=1}^{\tau} x_i s_i'\right) \ \bmod \ x_0$$

Note that $\mathcal{S}_{pk}(\mathbf{R}, c)$ and $\mathcal{S}_{pk}(\mathbf{R}, c')$ are perfectly indistinguishable if $\mathbf{r} + b + e$ and $\mathbf{r} + b' + e$ lie in the interval $\mathcal{I}$. Also note that both $\mathbf{r} + b + e$ and $\mathbf{r} + b' + e$ can at most be $2^{\rho'+1} + 2^{\overline{\rho}+1} + \tau \cdot 2^{\rho+2} + 2$ in the absolute value and are guaranteed to lie in $\mathcal{I}$ as far as $\mathcal{Z}$ or $\mathcal{Z}'$ do not lie in

$$\mathbb{Z} \cap \left\{ [-2^{\rho'}, (2^{\overline{\rho}+1} + \tau 2^{\rho+2} + 2) - 2^{\rho'}] \cup [2^{\rho'} - (2^{\overline{\rho}+1} + \tau 2^{\rho+2} + 2), 2^{\rho'}] \right\}.$$

Note that $\overline{\rho}$ is $\rho/2$ bits, $\rho = 2\sqrt{\lambda}$ and $\tau = \tilde{O}(\lambda^2)$. The probability of $\mathcal{Z}$ or $\mathcal{Z}'$ lie in this interval is

$$2 \times \left(\frac{2^{\overline{\rho}+1} + \tau \cdot 2^{\rho+1} + 2}{2^{2\rho-1}}\right) = \left(\frac{2^{\sqrt{\lambda}+1} + \tau \cdot 2^{2\sqrt{\lambda}+1} + 2}{2^{4\sqrt{\lambda}-3}}\right) < 2^{-\sqrt{\lambda}} \cdot \tau$$

which is negligible in the security parameter λ. Hence, $\mathcal{S}_{pk}(\mathbf{R}, c)$ and $\mathcal{S}_{pk}(\mathbf{R}, c')$ are statistically indistinguishable.

6.3   Learning with Errors (LWE)

Due to space limitation, the second instantiation of **wRSR** is only described in the full version of this paper.

7   Conclusion and Open Problem

In this work we introduced a new notion of a **Weakly Random Self-Reducible** public key cryptosystem and a general methodology to obtain secure oblivious transfer under this assumption. We also show that **wRSR** schemes can be constructed from post-quantum assumptions presented in [13,17]. We conclude with two open problems related to our work.

McEliece Assumption: Construct a Weakly Random-Self-Reducible encryption from the McEliece assumption [9].

NTRU Assumption: Construct a Weakly Random-Self-Reducible encryption from NTRU [35].

References

1. Rabin, M.O.: How to exchange secrets by oblivious transfer. Technical Memo TR81, Aiken Computation Laboratory, Harvard University (1981)
2. Even, S., Goldreich, O., Lempel, A.: A randomized protocol for signing contracts. Commun. ACM **28**(6), 637–647 (1985)
3. Fischer, M.J., Micali, S., Rackoff, C.: A secure protocol for the oblivious transfer. J. Cryptol. **9**(3), 191–195 (1996)
4. Berger, R., Peralta, R., Tedrick, T.: A provably secure oblivious transfer protocol. In: Beth, T., Cot, N., Ingemarsson, I. (eds.) EUROCRYPT 1984. LNCS, vol. 209, pp. 379–386. Springer, Heidelberg (1985)
5. Shor, P.W.: Polynomial-time algorithms for prime factorization and discrete logarithms on a quantum computer. SIAM J. Comput. **26**(5), 1484–1509 (1997)
6. Peikert, C., Vaikuntanathan, V., Waters, B.: A framework for efficient and composable oblivious transfer. In: Wagner, D. (ed.) CRYPTO 2008. LNCS, vol. 5157, pp. 554–571. Springer, Heidelberg (2008)
7. Dowsley, R., van de Graaf, J., Müller-Quade, J., Nascimento, A.C.A.: Oblivious transfer based on the McEliece assumptions. In: Safavi-Naini, R. (ed.) ICITS 2008. LNCS, vol. 5155, pp. 107–117. Springer, Heidelberg (2008)
8. Kobara, K., Morozov, K., Overbeck, R.: Coding-Based Oblivious Transfer. In: Calmet, J., Geiselmann, W., Müller-Quade, J. (eds.) Essays in Memory of Thomas Beth. LNCS, pp. 142–156. springer, Heidelberg (2008)
9. Robert, J.: McEliece. A public-key cryptosystem based on algebraic coding theory. Technical memo, California Institute of Technology (1978)
10. Crépeau, C.: Quantum oblivious transfer. J. Modern Opt. Spec. Issue Quantum Commun. Crypt. **41**(12), 2445–2454 (1994)
11. Howgrave-Graham, N.: Approximate integer common divisors. In: Silverman, J.H. (ed.) CaLC 2001. LNCS, vol. 2146, p. 51. Springer, Heidelberg (2001)
12. Gentry, C.: Fully homomorphic encryption using ideal lattices. In: STOC, pp. 169–178 (2009)
13. van Dijk, M., Gentry, C., Halevi, S., Vaikuntanathan, V.: Fully homomorphic encryption over the integers. In: Gilbert, H. (ed.) EUROCRYPT 2010. LNCS, vol. 6110, pp. 24–43. Springer, Heidelberg (2010)
14. Brakerski, Z., Vaikuntanathan, V.: Efficient fully homomorphic encryption from (Standard) LWE. In: FOCS, pp. 97–106 (2011)

15. Brakerski, Z., Vaikuntanathan, V.: Fully homomorphic encryption from ring-lwe and security for key dependent messages. In: Rogaway, P. (ed.) CRYPTO 2011. LNCS, vol. 6841, pp. 505–524. Springer, Heidelberg (2011)
16. Gentry, C., Halevi, S.: Fully homomorphic encryption without squashing using depth-3 arithmetic circuits. In: FOCS, pp. 107–109 (2011)
17. Gentry, C., Halevi, S., Vaikuntanathan, V.: A simple BGN-type cryptosystem from LWE. In: Gilbert, H. (ed.) EUROCRYPT 2010. LNCS, vol. 6110, pp. 506–522. Springer, Heidelberg (2010)
18. Brassard, G., Crépeau, C., Robert, J.M.: All-or-nothing disclosure of secrets. In: Odlyzko, A.M. (ed.) CRYPTO 1986. LNCS, vol. 263, pp. 234–238. Springer, Heidelberg (1987)
19. Goldreich, O., Micali, S., Wigderson, A.: How to play any mental game or a completeness theorem for protocols with honest majority. In: STOC, pp. 218–229 (1987)
20. Goldreich, O.: Foundations of Cryptography, vol. I and II. Cambridge University-Press, Cambridge (2004)
21. Haitner, I.: Semi-honest to malicious oblivious transfer—the black-box way. In: Canetti, R. (ed.) TCC 2008. LNCS, vol. 4948, pp. 412–426. Springer, Heidelberg (2008)
22. Håstad, J., Impagliazzo, R., Levin, L.A., Luby, M.: A Pseudo-random generator from any one-way function. SIAM J. Comput. 28(4), 12–24 (1993)
23. Naor, M.: Bit commitment using pseudorandomness. J. Cryptol. 4(2), 151–158 (1991)
24. Brassard, G., Crépeau, C.: Zero-knowledge simulation of boolean circuits. In: Odlyzko, A.M. (ed.) CRYPTO 1986. LNCS, vol. 263, pp. 223–233. Springer, Heidelberg (1987)
25. Crépeau, C., van de Graaf, J., Tapp, A.: Committed oblivious transfer and private multi-party computation. In: Coppersmith, D. (ed.) CRYPTO 1995. LNCS, vol. 963, pp. 110–123. Springer, Heidelberg (1995)
26. Haitner, I., Nguyen, M.-H., Ong, S.J., Reingold, O., Vadhan, S.: Statistically hiding commitments and statistical zero-knowledge arguments from any one-way function. SIAM J. Comput. 39(3), 1153–1218 (2009)
27. Rivest, R., Shamir, A., Adleman, L.: A method for obtaining digital signatures and public-key cryptosystems. Commun. ACM 21(2), 120–126 (1978)
28. Goldwasser, S., Micali, S.: Probabilistic encryption. J. Comput. Syst. Sci. 28(2), 270–299 (1984)
29. ElGamal, T.: A Public-key cryptosystem and a signature scheme based on discrete logarithms. IEEE Trans. Inf. Theory 31(4), 469–472 (1985)
30. Lawrence, C.: Washington. Number Theory and Cryptography. Discrete Mathematics and Its Applications, Elliptic Curves (2003)
31. Paillier, P.: Public-key cryptosystems based on composite degree residuosity classes. In: Stern, J. (ed.) EUROCRYPT 1999. LNCS, vol. 1592, p. 223. Springer, Heidelberg (1999)
32. Peikert, C., Vaikuntanathan, V.: Noninteractive statistical zero-knowledge proofs for lattice problems. In: Wagner, D. (ed.) CRYPTO 2008. LNCS, vol. 5157, pp. 536–553. Springer, Heidelberg (2008)
33. Regev, O.: On lattices, learning with errors, random linear codes, and cryptography. In: STOC, pp. 84–93 (2005)

34. Crépeau, C.: Equivalence between two flavours of oblivious transfer. In: CRYPTO, pp. 350–354 (1987)
35. Hoffstein, J., Pipher, J., Silverman, J.H.: NTRU: a ring-based public key cryptosystem. In: Buhler, J.P. (ed.) ANTS 1998. LNCS, vol. 1423, pp. 267–288. Springer, Heidelberg (1998)

Efficient Computations over Encrypted Data Blocks

Giovanni Di Crescenzo$^{(\boxtimes)}$, Brian Coan, and Jonathan Kirsch

Applied Communication Sciences, Basking Ridge, NJ, USA
{gdicrescenzo,bcoan,jkirsch}@appcomsci.com

Abstract. Secure computation (i.e., performing computation while keeping inputs private) is a fundamental problem in cryptography. In this paper, we present an efficient and secure 2-party computation protocol for any function computable via a monotone formula over equality statements between data blocks, under standard cryptographic assumptions. Our result bypasses roadblocks in previous general solutions, like Yao's garbled circuits and Gentry's lattice-based fully homomorphic encryption, by performing secure computations over data blocks (instead of bits) and using typical-size (instead of impractically large) cryptographic keys. An important efficiency property achieved is that the number of cryptographic operations in the protocol is *sublinear* in the size of the circuit representing the computed function. Even though not as general as in the two mentioned techniques, the class of formulae in our result contains a large number of well-known computational problems (while previously, only single specific problems were known to satisfy the mentioned sublinear efficiency property). Our main underlying technique is a new cryptographic primitive, perhaps of independent interest, that we call real-or-random conditional transfer, built as a variant of the well-known Rabin's oblivious transfer primitive.

1 Introduction

Secure two-party computation is a fundamental cryptographic primitive with significant application potential. In the formulation of interest for this paper, there are two parties, Alice and Bob, who would like to interactively compute a function f on their inputs x and y, respectively, such that at the end of the protocol: Bob obtains $f(x, y)$; an efficient adversary corrupting Alice learns nothing new about Bob's input y; and an efficient adversary corrupting Bob learns nothing new about Alice's input x, in addition to what is efficiently computable from $f(x, y)$. The first general solution to this problem for any arbitrary function f was presented in [16], assuming that the adversary is semi-honest (i.e., he follows the protocol as the corrupted party but may at the end try any polynomial-time algorithm to learn about the other party's input). Another important general solution for any arbitrary f was given in [10] in the scenario of secure computation among more than 2 parties. The generality of such solutions is so attractive that, even if decades after their introduction, researchers are considering

© Springer-Verlag Berlin Heidelberg 2015
G.F. Italiano et al. (Eds.): MFCS 2015, Part II, LNCS 9235, pp. 274–286, 2015.
DOI: 10.1007/978-3-662-48054-0_23

improvements and optimizations (see, e.g., [11,13]), thus bringing them closer to being usable in practice, at least in some specific scenarios (i.e., with the help of additional servers [1]). An important roadblock in this process is represented by the fact that the 'garbled circuit' technique from [16], using a boolean circuit representation of the function f, requires cryptographic operations for all input bits and gates in the circuit.

Recently, another general and powerful cryptographic primitive, fully homomorphic encryption, has been realized [5]. This primitive allows arbitrary polynomial-time computations over encrypted data and thus can be applied to construct secure 2-party computation protocols for any arbitrary polynomial-size circuit. Even in this case, researchers are recently considering improvements and optimizations, trying to bring it closer to being usable in practice (see, e.g., [6]). The roadblock for garbled circuits does not apply here, since fully homomorphic encryption solutions typically do operate over data blocks (instead of bits). However, another roadblock on the way to efficiency appears here: the security of all known constructions of fully homomorphic encryption is based on problems whose required key lengths are significantly high and the overall scheme is only theoretically efficient, but not in practice.

In this paper, we look for alternative approaches that attempt to combine the best features from both approaches: computing over encrypted data blocks (as in fully homomorphic encryption), limited requirements on key lengths (as in garbled circuits), and achieving solutions for a large class of problems (as in both).

Our Contribution. Our main result is an efficient and secure 2-party protocol for any polynomial-size and equality-based monotone formula, under standard cryptographic assumptions. Here, equality-based monotone formulae can be represented as polynomial-size monotone circuits on top of equality gates whose inputs are data blocks owned by either one of the two parties. The security of our protocol is proved based on the existence of secure 2-party protocols for simpler tasks: pseudo-random function evaluation and scalar product computation, which, in turn, were previously proved secure based on standard number-theoretic assumptions with conventional key lengths (see, e.g., [4,7,12,15]). The main efficiency property is the protocol's time complexity, as it only requires a number of cryptographic operations *sub-linear* in the size of the circuit computing the function. Specifically, it improves over the garbled circuit technique from [16] by a factor equal to the size of a data block. In practice, depending on the block size required by the specific application, this can be anywhere between a small and a very large improvement. Although not as general as the set of all polynomial-size boolean circuits, equality-based polynomial-size monotone formulae are a large class of circuits including important classes of problems, like text/pattern matching, publish/subscribe predicates, set disjointness and operations, search problems, etc.

To establish this result, we introduce a notion of real-or-random conditional transfer evaluation protocols (a variant of Rabin's oblivious transfer protocols [14]), and present one such efficient protocol for equality, and-of-equality, and or-of-equality conditions, which might be results of independent interest. These protocols compose to any monotone formula over equalities. Moreover, they perform computations over data blocks (instead of bits), and thus improve over the solution based on the garbled circuit technique from [16] by a factor equal to the size of a data block.

2   Definitions and Background

In this section we give definitions and background useful in the rest of the document. Definitions in Sect. 2.1 are specific to the main problem of interest in the paper, and include equality-based formulae, secure 2-party function evaluation protocols, and efficiency requirements. Definitions in Sect. 2.2 are specific to our solutions to the main problem considered, and include pseudo-random functions, and secure 2-party protocols for pseudo-random function evaluation, scalar product evaluation, and real-or-random conditional transfer.

2.1   Efficient and Secure 2-Party Evaluation of Equality-Based Formulae

Algebraic Structure. We consider a ring $(R, +, \cdot)$ as the underlying algebraic structure for data blocks input to te two parties Alice and Bob. We assume values in R can be represented as ℓ-bit strings using conventional, one-to-one, and polynomial-time computable and invertible, encoding from R to the set of ℓ-bit binary strings.

Equality-Based Formulae. We study 2-party protocols for the computation of a subclass of boolean circuits, called equality-based formulae, and defined as follows.

Let $x = (x(1), \ldots, x(a))$ and $y = (y(1), \ldots, y(b))$ denote two sequences of values from ring R. An (x, y)-equality gate is a function that takes as input two values $x(i), y(j) \in R$, for some known $i \in \{1, \ldots, a\}$ and $j \in \{1, \ldots, b\}$, and returns 1 if $x(i) = y(j)$ and 0 otherwise. Let $f : \{0, 1\}^n \to \{0, 1\}$ be a boolean formula. We say that f is an *equality-based formula* if there exists an integer m and a boolean formula $g : \{0, 1\}^m \to \{0, 1\}$ such that f can be written as $g(z_1, \ldots, z_m)$, where, for $h = 1, \ldots, m$, each variable z_h is the output of an $(x(i), y(j))$-equality gate, for some $i \in \{1, \ldots, a\}$ and $j \in \{1, \ldots, b\}$. Thus, the length of the input to formula f is $n = O(m\ell)$. In our 2-party formulation of the formula evaluation problem, Alice is given x as input, Bob is given y as input and is returned $f(x, y)$ as output. We say that f is *monotone* if the associated formula g only contains AND and OR gates.

Examples of equality-based monotone formulae include several variations of formulae related to well-known computational problems, such as string or pattern matching, set disjointness, publish/subscribe predicates, etc.

Secure 2-Party Function Evaluation Protocols. Let σ denote a security parameter. A function over the set of natural numbers is *negligible* if for all sufficiently large natural numbers $\sigma \in \mathcal{N}$, it is smaller than $1/p(\sigma)$, for all polynomials p. Two distribution ensembles $\{D_\sigma^0 : \sigma \in \mathcal{N}\}$ and $\{D_\sigma^1 : \sigma \in \mathcal{N}\}$ are *computationally indistinguishable* if for any efficient algorithm A, the quantity $|\text{Prob}[x \leftarrow D_\sigma^0 : A(x) = 1] - \text{Prob}[x \leftarrow D_\sigma^1 : A(x) = 1]|$ is negligible in σ (i.e., no efficient algorithm can distinguish if a random sample came from one distribution or the other). In a 2-party protocol execution, a party's *view* is the sequence containing the party's input, the party's random string, and all messages received during the execution.

We use the simulation-based definition from [8] for security of 2-party function evaluation protocols in the presence of semi-honest adversaries (i.e., adversaries that corrupt one party, follow the protocol as that party and then attempt to obtain some information about the other party's input). According to this definition, a protocol π to evaluate a (possibly probabilistic) function f satisfies *simulation-based security* in the presence of a semi-honest adversary, if there exists two efficient algorithms Sim_A, Sim_B (called the *simulators*), such that:

1. let $out_{S,B}$ be Sim_B's output on input Bob's input and Bob's output (if any); then, it holds that the pair $(out_{S,A}, \text{Bob's output})$ is computationally indistinguishable from the pair (Alice's view, Bob's output); and
2. let $out_{S,B}$ be Sim_A's output on input Alice's input and Alice's output (if any); then, it holds that the pair (Alice's output, $out_{S,B}$) is computationally indistinguishable from the pair (Alice's output, Bob's view).

In the above, the first (resp., second) condition says that a semi-honest adversary's view when corrupting Alice (resp., Bob), can be generated by an efficient algorithm not knowing Bob's (resp., Alice's) input, and thus the adversary does not learn anything about the uncorrupted party's input.

Efficiency Requirements. We will consider the following efficiency metrics, relative to a single execution of a given secure 2-party protocol:

1. *time complexity*: time between the protocol execution's beginning and end;
2. *communication complexity*: length of all messages exchanged; and
3. *round complexity*: number of messages exchanged.

All efficiency metrics are expressed as a function of the security parameter σ, and parameters a, b, ℓ associated with the equality-based function that is input to the protocol. In evaluating protocol latency, we will pay special attention to the number of asymmetric cryptography operations (e.g., modular exponentiations in a large group) and of symmetric cryptographic operations (e.g., block cipher executions), since the former are typically orders of magnitude more expensive than the latter (although the latter might applied a larger number of times). As a comparison result, we will target the general solution from [16] for the 2-party secure evaluation of function $f(x, y)$, where x is Alice's input and y is Bob's input, which requires $O(|y|)$ asymmetric cryptography operations and $O(|C_f|)$ symmetric cryptography operations, if C_f denotes the size of the boolean circuit computing f. Even if we will mainly focus our efficiency analysis on time complexity, our design targets minimization of all the mentioned efficiency metrics.

2.2   Protocols and Cryptographic Primitives Used in Our Solutions

Secure Evaluation Protocols for Specific Functions. In our solutions, we use or build constructions of 2-party secure evaluation protocols for the following functionalities: pseudo-random function, scalar product, and real-or-random conditional transfer.

A *secure pseudo-random function evaluation protocol* (briefly, sPRFeval protocol) is a protocol between two parties: Alice, having as input a key k for a PRF F, and Bob, having as input a string x, where the description of F is known to both parties. The protocol is defined as a secure function evaluation of the value $F(k, x)$, returned to Bob (thus, without revealing any information about x to Alice, or any information about k to Bob in addition to $F(k, x)$). Efficient constructions of sPRFeval protocols, based on the hardness of number-theoretic problems, were given in [4,12].

A *secure scalar product evaluation protocol* (briefly, sSPeval protocol) is a protocol between two parties: Alice, having as input a t-component vector $x = (x_1, \ldots, x_t)$ of values in R, and Bob, having as input a t-component vector $y = (y_1, \ldots, y_t)$ of values in R. The protocol is defined as a secure function evaluation of the value $\sum_{i=1}^{t} x_i y_i$, computed over R and returned to Bob (thus, without revealing any information about the y values to Alice, or any information about vector x to Bob in addition to t and $\sum_{i=1}^{t} x_i y_i$). Efficient constructions of sSPeval protocols, based on the hardness of number-theoretic problems, were given in [7,15].

A *secure real-or-random conditional transfer protocol* for the condition predicate p (briefly, p-srCTeval protocol, or srCTeval protocol when p is clear from the context) is a protocol between two parties: Alice, having as input a message m and a string x, and Bob, having as input a string y. The protocol is defined as a secure function evaluation of the value m', returned to Bob, where $m' = m$ if $p(x, y) = 1$ or m' is computationally indistinguishable from a string random and independent from m, and of the same length as m, if $p(x, y) = 0$. Thus, an execution of the protocol does not reveal any information about y to Alice, or any information about x to Bob in addition to m', and m' only reveals m when $p(x, y) = 1$ or the (possibly padded) length of m when $p(x, y) = 0$. Also, note that if m is a pseudo-random string, then at the end of a p-srCTeval protocol, Bob does not obtain any information about the value of predicate p. The notion of a p-srCTeval protocol is new but a close variant of the conditional oblivious transfer notion in [2,3]. Specifically, it differs in additionally requiring the pseudo-randomness of m' when $p(x, y) = 0$. Thus, an srCTeval protocol is a conditional oblivious transfer protocol in the sense of [2,3], but the converse may not be true. In particular, in a conditional oblivious transfer protocol, Bob may obtain information about the value of predicate p. The notion from [2,3] is, in turn, a variant of the much studied oblivious transfer protocol notion from [14].

Pseudo-Random Function Families. A family of functions $\{r_n : n \in \mathcal{N}\}$ is a *random function family* if, for each value of the security parameter n, the function r_n associated with that value is chosen with distribution uniform across all possible functions of the pre-defined input and output domains. A family of keyed functions $\{F_n(k, \cdot) : n \in \mathcal{N}\}$ is a *pseudo-random function family* (briefly, a PRF family, first defined in [9]) if, after key k is randomly chosen, no efficient algorithm allowed to query an oracle function O_n can distinguish whether O_n is $F_n(k, \cdot)$ or O_n is a random function $R_n(\cdot)$ over the same input and output domain, with probability greater than $1/2$ plus a negligible (in n) quantity. In practice, finite versions of PRF families (for a fixed value of the security parameter n) are implemented using very efficient and standard cryptographic primitives like block ciphers (e.g., AES), which are conjectured to behave like a pseud-random function.

3   Real-or-Random Conditional Transfer Protocols

In this section we present our 2-party srCTeval protocols for secure evaluation of the following 3 predicates: (1) *equality predicate* $p_=$, on input x, y, returns 1 if $x = y$ and 0 otherwise; (2) *OR-of-equalities predicate* $p_{or,=}$, on input x_0, x_1, y_0, y_1, returns 1 if $(x_0 = y_0) \vee (x_1 = y_1)$ and 0 otherwise; (3) *AND-of-equalities predicate* $p_{and,=}$, on input x_0, x_1, y_0, y_1, returns 1 if $(x_0 = y_0) \wedge (x_1 = y_1)$ and 0 otherwise.

3.1   An srCTeval Protocol for the Equality Predicate

In this subsection we show the following result.

Theorem 1. Assume the existence of pseudo-random permutation families F and prF, and of an sPRFeval protocol for prF. There exists a (black-box) construction of a 2-party srCTeval protocol $\pi_=$ for the secure evaluation of predicate $p_=$, only requiring 1 execution of the sPRFeval protocol for prF and $O(1)$ executions of F.

When the sPRFeval protocol is instantiated using the protocol in [12], one execution of prF and of the sPRFeval protocol only require $O(1)$ asymmetric cryptography operations. Moreover, in practice, F can be instantiated using a symmetric cryptography primitive like a block cipher. Thus, $\pi_=$ only requires a total of $O(1)$ asymmetric and $O(1)$ symmetric cryptography operations, while using the general solution from [16] for the same predicate would require $O(\ell)$ asymmetric and $O(\ell)$ symmetric cryptography operations, where ℓ is the length of the input strings x, y. We prove Theorem 1 by describing protocol $\pi_=$, and showing its efficiency and security properties.

Informal and Formal Description. Alice non-interactively computes a 'generated value' v_A for her input string x, as the output of pseudo-random function prF on input x. Bob computes, interactively with Alice, a 'received value' v_B for his input string y, as the output of an sPRFeval protocol for pseudo-random function prF on input y. Finally, Alice sends to Bob an encryption of m using pseudo-random function F and v_A as a key. Bob will return the value m' obtained by decrypting Alice's encryption using pseudo-random function F and v_b as a key. Note that if $x = y$, then $v_A = v_B$ and Bob will be able to invert u into

$m' = m$, while on the other hand if $x \neq y$, then m' is computationally indistinguishable from a random string independent from m, by the properties of the pseudo-random function prF. We now proceed more formally.

Our equality-OT protocol is a conditional OT protocols where the condition is an equality between a string x input to Alice and a string y input to Bob. It uses an arbitrary pseudo-random function family $\{F\}$, (which can be implemented using a block cipher like AES), and the pseudo-random function family $\{prF\}$, along with its associated sPRFeval protocol (which can be implemented using the function family and protocol in [12]). Assuming Alice wants to transfer some ℓ-bit value m to Bob under the condition that Alice's ℓ-bit string x is equal to Bob's ℓ-bit string y, the 2-party sCTeval protocol $\pi_=$ goes as follows:

1. Alice randomly chooses key k and computes $v_A = prF(k, x)$;
2. Alice and Bob run the sPRFeval protocol to return $v_B = prF(k, y)$ to Bob;
3. Alice sets key $k_A = v_A$, randomly chooses string r, computes $u = F(k_A, r) \oplus m$ and sends r, u to Bob;
4. Bob sets key $k_B = v_B$, computes $w' = F(k_B, r)$, $m' = u \oplus w'$ and returns: m'.

Remark. We note that if input values x, y are known to be drawn from a uniform or pseudo-random distribution, then steps 1 and 2 in $\pi_=$ can be replaced by setting $v_A = x$ and $v_B = y$. We denote such simplified protocol by $\pi_=^s$.

Properties of $\pi_=$. We now show the efficiency and security properties of $\pi_=$.

The most interesting efficiency property of $\pi_=$ is its time complexity, in that it only requires 1 application of an sPRFeval protocol and $O(1)$ applications of pseudo-random function F, thus resulting in $O(1)$ asymmetric and $O(1)$ symmetric cryptography operations, instead of $O(|y|)$ modular exponentiations and $O(|x| + |y|)$ block cipher applications, as required by the general solution from [16]. Round and communication complexity of $\pi_=$ are also very efficient; specifically, as for the round (resp., communication) complexity, $\pi_=$ only requires at most one message (resp., $O(\ell)$ communication) more than those required by the sPRFeval protocol.

The security property is obtained by showing two efficient simulator algorithms Sim_A, Sim_B that satisfy the simulation-based security definition in Sect. 2 for protocol $\pi_=$.

Algorithm Sim_A takes as input m, x. First of all, it computes k and v_A as in step 1 of $\pi_=$. Then, it simulates Alice's view in step 2 by running, on input k, x, the analogue simulator associated with the sPRFeval subprotocol. Finally, it computes r, u as in step 3 of $\pi_=$, and returns: $(view_{A,sPRFeval}, r, u)$, where $view_{A,sPRFeval}$ denotes Alice's view during the simulation of the sPRFeval protocol. The correctness of the simulation directly follows from the analogue correctness property of the simulator for the sPRFeval subprotocol.

Algorithm Sim_B takes as input y, m'. First of all, it randomly chooses v_B' and simulates Bob's view in step 2 of $\pi_=$ by running, on input y, v_B', the analogue simulator associated with the sPRFeval subprotocol. Then it simulates Bob's view in step 3 and 4 of $\pi_=$ by randomly choosing r and setting $u' = F(v'_B, r) \oplus m'$. Finally, it returns: $(view_{B,sPRFeval}, r, u')$, where $view_{B,sPRFeval}$ denotes Bob's view during the simulation of the sPRFeval protocol. The correctness of the simulation when $p(x, y) = 0$ follows by using the pseudo-randomness of function prF and the analogue correctness property of the simulator for the sPRFeval subprotocol. To observe that the simulation is also correct when $p(x, y) = 1$, we also note that in this case $x = y$ and therefore v_A, even though computed from a value x unknown to Sim_B, is actually constrained to be equal to v_B. This, together with the pseudo-randomness of function prF, suffices to show that the computation of u' by Sim_B is a good simulation of the value u in $\pi_=$.

3.2 An srCTeval Protocol for the AND-of-Equality Predicate

In this subsection we show the following result.

Theorem 2. Assume the existence of a 2-party srCTeval protocol $\pi_=$ for the secure evaluation of predicate $p_=$. There exists a (black-box) construction of a 2-party srCTeval protocol $\pi_{and,=}$ for the secure evaluation of predicate $p_{and,=}$, which only requires 2 executions of $\pi_=$.

We note that when we use the srCTeval protocol $\pi_=$ from Theorem 1, one execution of this protocol only requires $O(1)$ asymmetric cryptography operations, and thus so does one execution of $\pi_{and,=}$. Another interesting property of this protocol is that it composes $\pi_=$ without any additional cryptographic assumption. (This should be contrasted with the protocol in [16], where each boolean gate requires symmetric cryptography operations.)

Informal and Formal Description of $\pi_{and,=}$. Alice has 'generated values' $m, x_0, x_1 \in R$ as inputs and Bob has 'received values' $y_0, y_1 \in R$ as inputs. Alice splits her input m into random $m_0, m_1 \in R$ such that $m_0 \cdot m_1 = m$, and runs protocol $\pi_=$ twice, first on input m_0 and then on input m_1. Bob computes his output as the product, in R, of the outputs from the two subprotocol executions. Note that if both equalities $x_0 = y_0$ and $x_1 = y_1$ hold, then the values m'_0, m'_1 received by Bob satisfy $m'_0 = m_0$ and $m'_1 = m_1$ and bob can thus compute m from the received values. Instead, if at least one of the two equalities does not hold, Bob receives at least one pseudo-random value m'_i, for some $i \in \{0, 1\}$, and therefore the final value computed by Bob will also be pseudo-random. Formally, the 2-party sCTeval protocol $\pi_{and,=}$ goes as follows:

1. Alice randomly chooses $m_0, m_1 \in R$ such that $m = m_0 \cdot m_1$
2. For $i = 0, 1$, Alice and Bob run srCTeval protocol $\pi_=$ to transfer m_i to Bob, with equality '$x_i = y_i$' as the condition, thus resulting in Bob receiving m'_i;
3. Bob computes $m' = m'_0 \cdot m'_1$ and returns: m'.

Remark. We note that if input values x_0, y_0, x_1, y_1 are known to be drawn from a uniform or pseudo-random distribution, then Alice and Bob can run $\pi_=^s$ instead of $\pi_=$ in the above protocol. We denote such simplified protocol by $\pi_{and,=}^s$.

Properties of $\pi_{and,=}$. Protocol $\pi_{and,=}$ inherits the efficiency and security properties of $\pi_=$. In particular, with respect to time complexity, we note that if $\pi_=$ only requires $O(1)$ asymmetric and symmetric cryptography operations, so does $\pi_{and,=}$.

3.3   An srCTeval Protocol for the OR-of-Equality Predicate

In this subsection we show the following result.

Theorem 3. Assume the existence of an sSPeval protocol. There exists a (black-box) construction of a 2-party srCTeval protocol $\pi_{or,=}$ for the secure evaluation of predicate $p_{or,=}$, only requiring 1 execution of the sSPeval protocol on input 4-component vectors.

We note that when we use the sSPeval protocol from [7,15], one execution of this protocol only requires $O(1)$ asymmetric cryptography operations, and thus so does one execution of $\pi_{or,=}$.

Informal and Formal Description of $\pi_{or,=}$. Alice has 'generated values' $m, x_0, x_1 \in R$ as inputs and Bob has 'received values' $y_0, y_1 \in R$ as inputs. Alice defines expression $f(m, x_0, x_1, y_0, y_1) = m + r \cdot (x_0 - y_0) \cdot (x_1 - y_1)$, where r denotes a random value chosen by Alice from the same set. After some algebraic steps, we see that $f(m, x_0, x_1, y_0, y_1) = m + r \cdot x_0 \cdot x_1 - r \cdot y_0 \cdot x_1 - r \cdot x_0 \cdot y_1 + r \cdot y_0 \cdot y_1$, which can be rewritten as the scalar product of Alice's vector $(m + r \cdot x_0 \cdot x_1, -r \cdot x_1, -r \cdot x_0, r)$ and Bob's vector $(1, y_0, y_1, y_0 \cdot y_1)$. Thus, a single execution of the sSPeval protocol with vector length $t = 4$ allows Bob to securely compute $m' = f(m, x_0, x_1, y_0, y_1)$. Note that if at least one of the equalities $x_0 = y_0$ and $x_1 = y_1$ hold, then $m' = m$. Instead, if both equalities do not hold, m' is a random value, as so is r. Formally, the 2-party sCTeval protocol $\pi_{or,=}$ goes as follows:

1. Alice randomly chooses r and sets $x = (m + r \cdot x_0 \cdot x_1, -r \cdot x_1, -r \cdot x_0, r)$
2. Bob sets $y = (1, y_0, y_1, y_0 \cdot y_1)$
3. Alice and Bob run the sSPeval protocol where Alice uses x as input and Bob uses y as input, thus resulting in Bob receiving $m' = f(m, x_0, x_1, y_0, y_1)$.
4. Bob returns: m'.

Properties of $\pi_{or,=}$. Protocol $\pi_{or,=}$ inherits the efficiency and security properties of the sSPeval protocol used. With respect to time complexity, we note that if the sSPeval protocol only requires $O(1)$ asymmetric computations (as in [15]), so does $\pi_{or,=}$. This still compares favorably with $O(\ell)$ asymmetric computations which would be required by the general solution in [16].

4   Secure Evaluation of Equality-Based Monotone Formulae

In this section we present our 2-party protocol for secure evaluation of any arbitrary equality-based monotone formula over Alice and Bob's input strings. Formally, our protocol satisfies the following result.

Theorem 4. Let f be an equality-based monotone formula with m equality gates over ℓ-bit data blocks. Assume the existence of:

1. pseudo-random function families F and prF,
2. an sPRFeval protocol for the evaluation of prF, and
3. an sSPeval protocol.

It is possible to provide a (black-box) construction of a 2-party protocol π_f for the secure evaluation of f over Alice and Bob's input strings, which only requires $O(m)$ executions of the sPRFeval and sSPEval protocols.

We note that the sPRFeval protocol from [4] and the sSPeval protocol from [15] only require $O(1)$ asymmetric cryptography operations, and thus an execution of π_f based on them only requires $O(m)$ asymmetric cryptography operations, which is linear in the number of equality statements in f, and is thus sublinear in the length $n = O(m\ell)$ of the input to f. We start the proof of Theorem 4 with a description of protocol π_f, then continue by showing its efficiency and security properties.

Informal and Formal Descriptions of π_f. The description of protocol π can be divided into 4 phases: Alice's input processing, Bob's input processing, formula processing and output computation, which we now informally describe. The first 3 phases can be seen as an srCTeval protocol for the transfer of a random string gv with the predicate being equal to the equality-based monotone formula f. This is obtained by composing the 3 srCTeval protocols in Sect. 3. The fourth phase consists of Alice letting Bob check the result of the srCTeval protocol. We now proceed more formally.

Alice's Input Processing. This phase can be seen as a suitable generalization of step 1 in protocol $\pi_=$. Then, Alice randomly chooses a key k and non-interactively generates a set of pseudo-random values in correspondence of all his input values $x(1), \dots, x(a)$, as $gv(i) = prF(k, x(i))$, for $i = 1, \dots, a$.

Bob's Input Processing. This phase can be seen as a suitable generalization of step 2 in protocol $\pi_=$. Specifically, Bob obtains a set of pseudo-random values in correspondence of all her input values $y(1), \dots, y(b)$, by running with Alice one execution of the sPRFeval protocol for each $j = 1, \dots, b$, where Alice uses key k as input, and Bob uses $y(j)$ as input and receives $rv(j) = prF(k_a, y(j))$ as output.

Formula Processing. In this phase, each wire of the monotone circuit representing the equality-based monotone formula will be associated with one 'generated value', computed by Alice, and one 'received value', computed by Bob with Alice's help. First of all, Alice associates one random generated value gv with the output wire and one random generated value with each internal wire. Then, for any gate $g \in \{=, or, and\}$ in the circuit, Alice and Bob run an srCTeval protocol for the transfer of the random generated value associated with g's output wire, where the predicate is determined by g and the generated and received values associated with g's input wires, as follows:

1. if g is the equality gate with generated value x_i and received value y_j associated with its input wires, the predicate is '$x_i = y_j$' and the protocol run by Alice and Bob is $\pi^s_=$, where Alice's input is generated value $gv(i)$ and Bob's input is received value $rv(j)$, and $gv(i), rv(j)$ are as computed in the previous phases;

2. if g is an OR gate with generated values x_{i_0}, x_{i_1} and received values y_{i_0}, y_{i_1} associated with its input wires, the predicate is '$(x_{i_0} = y_{i_0}) \vee (x_{i_1} = y_{i_1})$' and the protocol run by Alice and Bob is $\pi^s_{or,=}$;

3. if g is an AND gate with generated values x_{i_0}, x_{i_1} and received values y_{i_0}, y_{i_1} associated with its input wires, the predicate is '$(x_{i_0} = y_{i_0}) \wedge (x_{i_1} = y_{i_1})$' and the protocol run by Alice and Bob is $\pi^s_{and,=}$.

The value rv returned by Bob at the end of this srCTeval protocol's execution is defined as the received value associated with the output wire of gate g. The execution of the srCTeval protocol for all gates can be performed sequentially or even in parallel, since in each of protocols $\pi^s_=, \pi^s_{or,=}, \pi^s_{and,=}$, the received values are only needed by Bob at the end of the protocol.

Output Computation. In the output computation phase, Alice simply sends gv to Bob, and Bob returns 1 if $gv = rv$ and 0 otherwise.

Properties of π_f. We now show the efficiency and security properties of π_f.

As for the time complexity, π_f only requires $O(m)$ applications of an sPRFeval protocol and $O(m)$ applications of an sSPeval protocol. When these protocols are implemented as in [4,15], this results in $O(m)$ asymmetric and symmetric cryptography operations, instead of $O(m\ell)$, as required in a direct application of the general solution from [16]. The communication complexity of π_f is $O(m\ell)$, also improving over the communication complexity $O(m\ell^2)$ that would be obtained using [16].

The security property is obtained by showing two efficient simulator algorithms Sim_A, Sim_B that satisfy the simulation-based security definition in Sect. 2 for protocol π_f. Towards that goal, we first consider the invariant property maintained by protocol π_f across all gates in the circuit computing the equality-based monotone formula f. Specifically, for each gate g, including the circuit output gate, the generated value gv_g and the received value rv_g associated with gate g, satisfy the following properties:

1. if the subformula with g as root gate is true then $rv_g = gv_g$;
2. if the subformula with g as root gate is false, then rv_g and gv_g are pseudo-independent (i.e., computationally indistinguishable from two random and independent strings).

An alternative way to express this invariant property is that, when restricted to its first three phases only, protocol π_f is a srCTeval protocol for the transfer of random value rv, the predicate being the circuit computing the equality-based monotone formula f on input Alice and Bob's input strings.

This invariant is proved by induction over f, as follows. For the base case, when f is a single equality statement, observe that the first 3 phases of protocol

π_f are identical to protocol $\pi_=$, which is an srCTeval protocol for the equality predicate, and thus by definition satisfies the invariant. Now consider the inductive case, when the root gate of the circuit computing f is an OR gate and let gv_0, rv_0 (resp., gv_1, rv_1) be the generated and received values associated with the left (resp., right) input wire to the OR gate. By the inductive hypothesis, at least one of the equalities '$gv_0 = rv_0$', '$gv_1 = rv_1$' holds if and only if formula f is satisfied. Then the invariant for the entire circuit computing f follows by the fact that in correspondence of this OR gate π_f runs $\pi_{or,=}$, which is an srCTeval protocol for the or-of-equalities predicate. The proof is analogue in the case where the root gate of the circuit computing f is an AND gate.

Given this invariant property, we observe that π_f can be seen as an srCTeval protocol for the predicate determined by f (in its first 3 phases), followed by a single message, containing value gv, from Alice to Bob (in its 4th phase). Then, we construct the two simulator algorithms for π_f, as follows. Simulator Sim_A is exactly the same simulator of Alice's view associated with the srCTeval protocol, since messages received by Alice in π_f are precisely those received in the srCTeval protocol. The construction of simulator Sim_B follows the same approach, but it also needs to simulate the last message from Alice in π_f. This message is simulated by Sim_B as a random string if f is false, or as equal to the received value at the end of the srCTeval protocol if f is true.

5   Conclusions

We have built a framework for the design of more efficient secure function evaluation protocols. In a nutshell, this framework is built as follows: first, we consider atomic circuit operations (i.e., equality) that can be realized through secure protocols that are more efficient than known general solutions; then, we define formulae over such atomic circuits. Our main result is a technique to compose the efficient and secure protocols for atomic circuits into efficient and secure protocols for the entire formula, across a large class of formula compositions (i.e., monotone formulae). This adds a significant amount of generality to the otherwise specialized solutions for atomic circuits. Moreover, this opens the possibility of several directions for future research, including: (a) finding other atomic circuit operations which admit secure protocols that are more efficient than general solutions and well compose with the techniques in this paper; (b) generalizing the composition techniques in this paper to larger circuit classes.

Acknowledgements. This work was supported by the Defense Advanced Research Projects Agency (DARPA) via Air Force Research Laboratory (AFRL), contract number FA8750-14-C-0057. The U.S. Government is authorized to reproduce and distribute reprints for Governmental purposes notwithstanding any copyright annotation hereon. Disclaimer: The views and conclusions contained herein are those of the authors and should not be interpreted as necessarily representing the official policies or endorsements, either expressed or implied, of DARPA, AFRL or the U.S. Government.

References

1. Bogetoft, P., et al.: Secure multiparty computation goes live. In: Dingledine, R., Golle, P. (eds.) FC 2009. LNCS, vol. 5628, pp. 325–343. Springer, Heidelberg (2009)
2. Di Crescenzo, G.: Private selective payment protocols. In: Frankel, Y. (ed.) FC 2000. LNCS, vol. 1962, pp. 72–89. Springer, Heidelberg (2001)
3. Di Crescenzo, G., Ostrovsky, R., Rajagopalan, S.: Conditional oblivious transfer and timed-release encryption. In: Stern, J. (ed.) EUROCRYPT 1999. LNCS, vol. 1592, pp. 74–89. Springer, Heidelberg (1999)
4. Freedman, M.J., Ishai, Y., Pinkas, B., Reingold, O.: Keyword search and oblivious pseudorandom functions. In: Kilian, J. (ed.) TCC 2005. LNCS, vol. 3378, pp. 303–324. Springer, Heidelberg (2005)
5. Gentry, C.: Fully homomorphic encryption using ideal lattices. In: Proceedings of 41st ACM STOC, pp. 169–178 (2009)
6. Gentry, C., Halevi, S., Smart, N.P.: Fully homomorphic encryption with polylog overhead. In: Pointcheval, D., Johansson, T. (eds.) EUROCRYPT 2012. LNCS, vol. 7237, pp. 465–482. Springer, Heidelberg (2012)
7. Goethals, B., Laur, S., Lipmaa, H., Mielikäinen, T.: On private scalar product computation for privacy-preserving data mining. In: Park, C., Chee, S. (eds.) ICISC 2004. LNCS, vol. 3506, pp. 104–120. Springer, Heidelberg (2005)
8. Goldreich, O.: The Foundations of Cryptography. Basic Applications, vol. 2. Cambridge University Press, New York (2004)
9. Goldreich, O., Goldwasser, S., Micali, S.: How to construct random functions. J. ACM 33(4), 792–807 (1986)
10. Goldreich, O., Micali, S., Wigderson, A.: How to play any mental game or a completeness theorem for protocols with honest majority. In: Proceedings of 19th ACM STOC, pp. 218–229 (1987)
11. Huang, Y., Evans, D., Katz, J., Malka, L.: Faster secure two-party computation using garbled circuits. In: Proceedings of 20th USENIX Security Symposium (2011)
12. Jarecki, S., Liu, X.: Efficient oblivious pseudorandom function with applications to adaptive OT and secure computation of set intersection. In: Reingold, O. (ed.) TCC 2009. LNCS, vol. 5444, pp. 577–594. Springer, Heidelberg (2009)
13. Malkhi, D., Nisan, N., Pinkas, B., Sella, Y.: Fairplay - secure two-party computation system. In: Proceedings of 13th USENIX Security Symposium, pp. 287–302 (2004)
14. Rabin, M.O.: How to exchange secrets with oblivious transfer. IACR Cryptology ePrint Archive 2005:187 (2005)
15. Wright, R.N., Yang, Z.: Privacy-preserving bayesian network structure computation on distributed heterogeneous data. In: Proceedings of 10th ACM SIGKDD, pp. 713–718 (2004)
16. Yao, A.C.-C.: How to generate and exchange secrets (extended abstract). In: Proceedings of 27th IEEE FOCS, pp. 162–167 (1986)

Polynomial Kernels for Weighted Problems

Michael Etscheid[✉], Stefan Kratsch, Matthias Mnich,
and Heiko Röglin

Institut für Informatik, Universität Bonn, Bonn, Germany
{etscheid,kratsch,roeglin}@cs.uni-bonn.de
mmnich@uni-bonn.de

Abstract. Kernelization is a formalization of efficient preprocessing for NP-hard problems using the framework of parameterized complexity. Among open problems in kernelization it has been asked many times whether there are deterministic polynomial kernelizations for SUBSET SUM and KNAPSACK when parameterized by the number n of items.

We answer both questions affirmatively by using an algorithm for compressing numbers due to Frank and Tardos (Combinatorica 1987). This result had been first used by Marx and Végh (ICALP 2013) in the context of kernelization. We further illustrate its applicability by giving polynomial kernels also for weighted versions of several well-studied parameterized problems. Furthermore, when parameterized by the different item sizes we obtain a polynomial kernelization for SUBSET SUM and an exponential kernelization for KNAPSACK. Finally, we also obtain kernelization results for polynomial integer programs.

1 Introduction

The question of handling numerical values is of fundamental importance in computer science. Typical issues are precision, numerical stability, and representation of numbers. In the present work we study the effect that the presence of (possibly large) numbers has on weighted versions of well-studied NP-hard problems. In other words, we are interested in the effect of large numbers on the computational complexity of solving hard combinatorial problems. Concretely, we focus on the effect that weights have on the preprocessing properties of the problems, and study this question using the notion of *kernelization* from parameterized complexity. Very roughly, kernelization studies whether there are problem *parameters* such that any instance of a given NP-hard problem can be efficiently reduced to an equivalent instance of small size in terms of the parameter. Intuitively, one may think of applying a set of correct simplification rules, but additionally one has a proven size bound for instances to which no rule applies.

The issue of handling large weights in kernelization has been brought up again and again as an important open problem in kernelization [2,6,7,11]. For example, it is well-known that for the task of finding a vertex cover of at most k

---

Supported by the Emmy Noether-program of the German Research Foundation (DFG), KR 4286/1, and ERC Starting Grant 306465 (BeyondWorstCase).

© Springer-Verlag Berlin Heidelberg 2015
G.F. Italiano et al. (Eds.): MFCS 2015, Part II, LNCS 9235, pp. 287–298, 2015.
DOI: 10.1007/978-3-662-48054-0_24

vertices for a given unweighted graph G one can efficiently compute an equivalent instance (G', k') such that G' has at most $2k$ vertices. Unfortunately, when the vertices of G are additionally equipped with positive rational weights and the chosen vertex cover needs to obey some specified maximum weight $W \in \mathbb{Q}$ then it was long unknown how to encode (and shrink) the vertex weights to bitsize polynomial in k. In this direction, Cheblík and Cheblíková [5] showed that an equivalent graph G' with total vertex weight at most $2w^*$ can be obtained in polynomial time, whereby w^* denotes the minimum weight of a vertex cover of G. This, however, does not mean that the size of G' is bounded, unless one makes the additional assumption that the vertex weights are bounded from below; consequently, their method only yields a kernel with that extra requirement of vertex weights being bounded away from zero. In contrast, we do not make such an assumption.

Let us attempt to clarify the issue some more. The task of finding a polynomial kernelization for a weighted problem usually comes down to two parts: (1) Deriving reduction rules that work correctly in the presence of weights. The goal, as for unweighted problems, is to reduce the number of relevant objects, e.g., vertices, edges, sets, etc., to polynomial in the parameter. (2) Shrinking or replacing the weights of remaining objects such that their encoding size becomes (at worst) polynomial in the parameter. The former part usually benefits from existing literature on kernels of unweighted problems, but regarding the latter only little progress was made.

For a pure weight reduction question let us consider the SUBSET SUM problem. Therein we are given n numbers $a_1, \ldots, a_n \in \mathbb{N}$ and a target value $b \in \mathbb{N}$ and we have to determine whether some subset of the n numbers has sum exactly b. Clearly, reducing such an instance to size polynomial in n hinges on the ability of handling large numbers a_i and b. Let us recall that a straightforward dynamic program solves SUBSET SUM in time $\mathcal{O}(nb)$, implying that large weights are to be expected in hard instances. Harnik and Naor [14] showed that taking all numbers modulo a sufficiently large random prime p of magnitude about 2^{2n} produces an equivalent instance with error probability exponentially small in n. (Note that the obtained instance is with respect to arithmetic modulo p.) The total bitsize then becomes $\mathcal{O}(n^2)$. Unfortunately, this elegant approach fails for more complicated problems than SUBSET SUM.

Consider the SUBSET RANGE SUM variant of SUBSET SUM where we are given not a single target value b but instead a lower bound L and an upper bound U with the task of finding a subset with sum in the interval $\{L, \ldots, U\}$. Observe that taking the values a_i modulo a large random prime faces the problem of specifying the new target value(s), in particular if $U - L > p$ because then every remainder modulo p is possible for the solution. Nederlof et al. [21] circumvented this issue by creating not one but in fact a polynomial number of small instances. Intuitively, if a solution has value close to either L or U then the randomized approach will work well (possibly making a separate instance for target values close to L or U). For solutions sufficiently far from L or U there is no harm in losing a little precision and dividing all numbers by 2; then the argument iterates. Overall, because the number of iterations is bounded by the logarithm of the numbers (i.e., their encoding size), this creates a number of instances that is polynomial in the input size, with each instance having size $\mathcal{O}(n^2)$; if the initial input is "yes" then at least one of the created instances is "yes".[1]

To our knowledge, the mentioned results are the only positive results that are aimed directly at the issue of handling large numbers in the context of kernelization. Apart from these, there are of course results where the chosen parameter bounds the variety of feasible weights and values, but this only applies to integer domains; e.g., it is easy to find a kernel for WEIGHTED VERTEX COVER when all weights are positive integers and the parameter is the maximum total weight k. On the negative side, there are a couple of lower bounds that rule out polynomial kernelizations for various weighted and ILP problems, see, e.g., [3,18]. Note, however, that the lower bounds appear to "abuse" large weights in order to build gadgets for lower bound proofs that also include a super-polynomial number of objects as opposed to having just few objects with weights of super-polynomial encoding size. In other words, the known lower bounds pertain rather to the first step, i.e. finding reduction rules that work correctly in the presence of weights, than to the inherent complexity of the numbers themselves. Accordingly, since 2010 the question for a deterministic polynomial kernelization for SUBSET SUM or KNAPSACK with respect to the number of items can be found among open problems in kernelization [2,6,7,11].

Recently, Marx and Végh [20] gave a polynomial kernelization for a weighted connectivity augmentation problem. As a crucial step, they use a technique of Frank and Tardos [12], originally aimed at obtaining strongly polynomial-time algorithms, to replace rational weights by sufficiently small and equivalent integer weights. They observe and point out that this might be a useful tool to handle in general the question of getting kernelizations for weighted versions of parameterized problems. It turns out that, more strongly, Frank and Tardos' result can also be used to settle the mentioned open problems regarding KNAPSACK and SUBSET SUM. We point out that this is a somewhat circular statement since Frank and Tardos had set out to, amongst others, improve existing algorithms for ILPs, which could be seen as *very general* weighted problems.

Our Work. We use the theorem of Frank and Tardos [12] to formally settle the open problems, i.e., we obtain deterministic kernelizations for SUBSET SUM(n) and KNAPSACK(n), in Sect. 3. Generally, in the spirit of Marx and Vegh's observation, this allows to get polynomial kernelizations whenever one is able to first reduce the *number of objects*, e.g., vertices or edges, to polynomial in the parameter. The theorem can then be used to sufficiently shrink the weights of all objects such that the *total size* becomes polynomial in the parameter.

Motivated by this, we consider weighted versions of several well-studied parameterized problems, e.g., d-HITTING SET, d-SET PACKING, and MAX CUT, and show how to reduce the number of relevant structures to polynomial in the parameter. An application of Frank and Tardos' result then implies polynomial kernelizations. We present our small kernels for weighted problems in Sect. 4.

---

[1] This is usually called a (disjunctive) Turing kernelization.

Next, we consider the KNAPSACK problem and its special case SUBSET SUM, in Sect. 5. For SUBSET SUM instances with only k item sizes, we derive a kernel of size polynomial in k. This way, we are improving the exponential-size kernel for this problem due to Fellows et al. [10]. We also extend the work of Fellows et al. in another direction by showing that the more general KNAPSACK problem is fixed-parameter tractable (i.e. has an exponential kernel) when parameterized by the number k of item sizes, even for unbounded number of item values. On the other hand, we provide quadratic kernel size lower bounds for general SUBSET SUM instances assuming the Exponential Time Hypothesis [15].

Finally, as a possible tool for future kernelization results we show that the weight reduction approach also carries over to polynomial ILPs so long as the maximum degree and the domains of variables are sufficiently small, in Sect. 6. Due to space constraints, proofs of statements marked by $\star$ are deferred to the full version.

2   Preliminaries

A *parameterized problem* is a language $\Pi \subseteq \Sigma^* \times \mathbb{N}$, where Σ is a finite alphabet; the second component k of instances $(I, k) \in \Sigma^* \times \mathbb{N}$ is called the *parameter*. A problem $\Pi \subseteq \Sigma^* \times \mathbb{N}$ is *fixed-parameter tractable* if it admits a *fixed-parameter algorithm*, which decides instances (I, k) of Π in time $f(k) \cdot |I|^{\mathcal{O}(1)}$ for some computable function f. The class of fixed-parameter tractable problems is denoted by FPT. Evidence that a problem Π is unlikely to be fixed-parameter tractable is that Π is W[t]-hard for some $t \in \mathbb{N}$ or W[P]-hard, where FPT $\subseteq$ W[1] $\subseteq$ W[2] $\subseteq \ldots \subseteq$ W[P]. To prove hardness of Π, one can give a *parameterized reduction* from a W[$\cdot$]-hard problem Π' to Π that maps every instance I' of Π' with parameter k' to an instance I of Π with parameter $k \leq g(k')$ for some computable function g such that I can be computed in time $f(k') \cdot |I'|^{\mathcal{O}(1)}$ for some computable function f, and I is a "yes"-instance if and only if I' is. If f and g are polynomials, such a reduction is called a *polynomial parameter transformation*. A problem Π that is NP-complete even if the parameter k is constant is said to be *para-NP-complete*.

A *kernelization* for a parameterized problem Π is an efficient algorithm that given any instance (I, k) returns an instance (I', k') such that $(I, k) \in \Pi$ if and only if $(I', k') \in \Pi$ and such that $|I'| + k' \leq f(k)$ for some computable function f. The function f is called the *size* of the kernelization, and we have a polynomial kernelization if $f(k)$ is polynomially bounded in k. It is known that a parameterized problem is fixed-parameter tractable if and only if it is decidable and has a kernelization. Nevertheless, the kernels implied by this fact are usually of superpolynomial size. (The size matches the $f(k)$ from the run time, which for NP-hard problems is usually exponential as typical parameters are upper bounded by the instance size.) On the other hand, assuming FPT $\neq$ W[1] no W[1]-hard problem has a kernelization. Further, there are tools for ruling out polynomial kernels for some parameterized problems [1,8] under an appropriate complexity assumption (namely that NP $\not\subseteq$ coNP/poly). Such lower bounds can be transferred by the mentioned polynomial parameter transformations [4].

3   Settling Open Problems via the Frank-Tardos Theorem

3.1   Frank and Tardos' Theorem

Frank and Tardos [12] describe an algorithm which proves the following theorem.

Theorem 1 ([12]). *There is an algorithm that, given a vector $w \in \mathbb{Q}^r$ and an integer N, in polynomial time finds a vector $\overline{w} \in \mathbb{Z}^r$ with $\|\overline{w}\|_\infty \leq 2^{4r^3} N^{r(r+2)}$ such that $\mathrm{sign}(w \cdot b) = \mathrm{sign}(\overline{w} \cdot b)$ for all vectors $b \in \mathbb{Z}^r$ with $\|b\|_1 \leq N - 1$.*

This theorem allows us to compress linear inequalities to an encoding length which is polynomial in the number of variables. Frank and Tardos' algorithm runs even in strongly polynomial time. As a consequence, all kernelizations presented in this work also have a strongly polynomial running time.

Example 1. There is an algorithm that, given a vector $w \in \mathbb{Q}^r$ and a rational $W \in \mathbb{Q}$, in polynomial time finds a vector $\overline{w} \in \mathbb{Z}^r$ with $\|\overline{w}\|_\infty = 2^{\mathcal{O}(r^3)}$ and an integer $\overline{W} \in \mathbb{Z}$ with total encoding length $\mathcal{O}(r^4)$, such that $w \cdot x \leq W$ if and only if $\overline{w} \cdot x \leq \overline{W}$ for every vector $x \in \{0,1\}^r$.

Proof. Use Theorem 1 on the vector $(w, W) \in \mathbb{Q}^{r+1}$ with $N = r+2$ to obtain the resulting vector $(\overline{w}, \overline{W})$. Now let $b = (x, -1) \in \mathbb{Z}^{r+1}$ and note that $\|b\|_1 \leq N-1$. The inequality $w \cdot x \leq W$ is false if and only if $\mathrm{sign}(w \cdot x - W) = \mathrm{sign}((w, W) \cdot (x, -1)) = \mathrm{sign}((w, W) \cdot b)$ is equal to $+1$. The same holds for $\overline{w} \cdot x \leq \overline{W}$.

As each $|\overline{w}_i|$ can be encoded with $\mathcal{O}(r^3 + r^2 \log N) = \mathcal{O}(r^3)$ bits, the whole vector $\overline{w}$ has encoding length $\mathcal{O}(r^4)$.   □

3.2   Polynomial Kernelization for Knapsack

A first easy application of Theorem 1 is the kernelization of KNAPSACK with the number n of different items as parameter.

KNAPSACK(n)

    Input: An integer $n \in \mathbb{N}$, rationals $W, P \in \mathbb{Q}$,
          a weight vector $w \in \mathbb{Q}^n$, and a profit vector $p \in \mathbb{Q}^n$.
    Parameter: n.
    Question: Is there a vector $x \in \{0,1\}^n$ with $w \cdot x \leq W$ and $p \cdot x \geq P$?

Theorem 2. KNAPSACK*(n) admits a kernel of size $\mathcal{O}(n^4)$.*   □

As a consequence, also SUBSET SUM(n) admits a kernel of size $\mathcal{O}(n^4)$.

4   Small Kernels for Weighted Parameterized Problems

The result of Frank and Tardos implies that we can easily handle large weights or numbers in kernelization provided that the number of different objects is already sufficiently small (e.g., polynomial in the parameter). In the present section we show how to handle the first step, i.e., the reduction of the number of objects, in the presence of weights for a couple of standard problems. Presumably the reduction in size of numbers is not useful for this first part since the number of different values is still exponential.

4.1 Hitting Set and Set Packing

In this section we outline how to obtain polynomial kernelizations for WEIGHTED d-HITTING SET and WEIGHTED d-SET PACKING. Since these problems generalize quite a few interesting hitting/covering and packing problems, this extends the list of problems whose weighted versions directly benefit from our results. The problems are formally defined as follows.

WEIGHTED d-HITTING SET(k)
  Input: A set family $\mathcal{F} \subseteq \binom{U}{d}$, a function $w \colon U \to \mathbb{N}$, and $k, W \in \mathbb{N}$.
Parameter: k.
  Question: Is there a set $S \subseteq U$ of cardinality at most k and weight $\sum_{u \in S} w(u) \leq W$ such that S intersects every set in $\mathcal{F}$?

WEIGHTED d-SET PACKING(k)
  Input: A set family $\mathcal{F} \subseteq \binom{U}{d}$, a function $w \colon \mathcal{F} \to \mathbb{N}$, and $k, W \in \mathbb{N}$.
Parameter: k.
  Question: Is there a family $\mathcal{F}^* \subseteq \mathcal{F}$ of exactly k disjoint sets of weight $\sum_{F \in \mathcal{F}^*} w(F) \geq W$?

Note that we treat d as a constant. We point out that the definition of WEIGHTED SET PACKING(k) restricts attention to exactly k disjoint sets of weight at least W. If we were to relax to at least k sets then the problem would be NP-hard already for $k = 0$. On the other hand, the kernelization that we present for WEIGHTED SET PACKING(k) holds also if we require $\mathcal{F}^*$ to be of cardinality at most k (and total weight at least W, as before).

Both kernelizations rely on the Sunflower Lemma of Erdős and Rado [9], same as their unweighted counterparts. We recall the lemma.

Lemma 1 (Erdős and Rado [9]). *Let $\mathcal{F}$ be a family of sets, each of size d, and let $k \in \mathbb{N}$. If $|\mathcal{F}| > d! k^d$ then we can find in time $\mathcal{O}(|\mathcal{F}|)$ a so-called $k+1$-sunflower, consisting of $k + 1$ sets $F_1, \dots, F_{k+1} \in \mathcal{F}$ such that the pairwise intersection of any two F_i, F_j with $i \neq j$ is the same set C, called the core.*

For WEIGHTED d-HITTING SET(k) we can apply the Sunflower Lemma directly, same as for the unweighted case: Say we are given $(U, \mathcal{F}, w, k, W)$. If the size of $\mathcal{F}$ exceeds $d!(k + 1)^d$ then we find a $k + 2$-sunflower $\mathcal{F}_s$ in $\mathcal{F}$ with core C. Any hitting set of cardinality at most k must contain an element of C. The same is true for $k + 1$-sunflowers so we may safely delete any set $F \in \mathcal{F}_s$ since hitting the set $C \subseteq F$ is enforced by the remaining $k + 1$-sunflower. Iterating this reduction rule yields $\mathcal{F}' \subseteq \mathcal{F}$ with $|\mathcal{F}'| = \mathcal{O}(k^d)$ and such that $(U, \mathcal{F}, w, k, W)$ and $(U, \mathcal{F}', w, k, W)$ are equivalent.

Now, we can apply Theorem 1. We can safely restrict U to the elements U' present in sets of the obtained set family $\mathcal{F}'$, and let $w' = w|_{U'}$. By Theorem 1 applied to weights w' and target weight W with $N = k + 2$ and $r = \mathcal{O}(k^d)$ we get replacement weights of magnitude bounded by $2^{\mathcal{O}(k^{3d})} N^{\mathcal{O}(k^{2d})}$ and bit size $\mathcal{O}(k^{3d})$. Note that this preserves, in particular, whether the sum of any k weights is at most the target weight W, by preserving the sign of $w_{i_1} + \ldots + w_{i_k} - W$. The total bitsize is dominated by the space for encoding the weight of all elements of the set U'.

Theorem 3. WEIGHTED d-HITTING SET *(k) admits a kernelization to* $\mathcal{O}(k^d)$ *sets and total size bounded by* $\mathcal{O}(k^{4d})$.

For WEIGHTED d-SET PACKING(k) a similar argument works.

Theorem 4 ($\star$). WEIGHTED d-SET PACKING *(k) admits a kernelization to* $\mathcal{O}(k^d)$ *sets and total size bounded by* $\mathcal{O}(k^{4d})$.

4.2 Max Cut

Let us derive a polynomial kernel for WEIGHTED MAX CUT(W), which is defined as follows.

WEIGHTED MAX CUT(W)
  Input: A graph G, a function $w\colon E \to \mathbb{Q}_{\geq 1}$, and $W \in \mathbb{Q}_{\geq 1}$.
Parameter: $\lceil W \rceil$.
  Question: Is there a set $C \subseteq V(G)$ such that $\sum_{e \in \delta(C)} w(e) \geq W$?

Note that we chose the weight of the resulting cut as parameter, which is most natural for this problem. The number k of edges in a solution is not a meaningful parameter: If we restricted the cut to have at least k edges, the problem would again be already NP-hard for $k = 0$. If we required at most k edges, we could, in this example for integral weights, multiply all edge weights by n^2 and add arbitrary edges with weight 1 to our input graph. When setting the new weight bound to $n^2 \cdot W + \binom{n}{2}$, we would not change the instance semantically but there may be no feasible solution left with at most k edges.

The restriction to edge weights at least 1 is necessary as otherwise the problem becomes intractable. This is because when allowing arbitrary positive rational weights, we can transform instances of the NP-complete UNWEIGHTED MAX CUT problem (with all weights equal to 1 and parameter k, which is the number of edges in the cut) to instances of the WEIGHTED MAX CUT problem on the same graph with edge weights all equal to $1/k$ and parameter $W = 1$.

Theorem 5. WEIGHTED MAX CUT *(W) admits a kernel of size* $\mathcal{O}(W^4)$.

Proof. Let T be the total weight of all edges. If $T \geq 2W$, then the greedy algorithm yields a cut of weight at least $T/2 \geq W$. Therefore, all instances with $T \geq 2W$ can be reduced to a constant-size positive instance. Otherwise, there are at most $2W$ edges in the input graph as every edge has weight at least 1. Thus, we can use Theorem 1 to obtain an equivalent (integral) instance of encoding length $\mathcal{O}(W^4)$. $\qquad\square$

4.3   Polynomial Kernelization for Bin Packing with Additive Error

BIN PACKING is another classical NP-hard problem involving numbers. Therein we are given n positive integer numbers $a_1, \ldots, a_n$ (the items), a bin size $b \in \mathbb{N}$, and an integer k; the question is whether the integer numbers can be partitioned into at most k sets, the bins, each of sum at most b. From a parameterized perspective the problem is highly intractable for its natural parameter k, because for $k = 2$ it generalizes the (weakly) NP-hard PARTITION problem.

Jansen et al. [16] proved that the parameterized complexity improves drastically if instead of insisting on exact solutions the algorithm only has to provide a packing into $k+1$ bins or correctly state that k bins do not suffice. Concretely, it is shown that this problem variant is fixed-parameter tractable with respect to k. The crucial effect of the relaxation is that small items are of almost no importance: If they cannot be added greedily "on top" of a feasible packing of big items into $k + 1$ bins, then the instance trivially has no packing into k bins due to exceeding total weight kb. Revisiting this idea, with a slightly different threshold for being a small item, we note that after checking for total weight being at most kb (else reporting that there is no k-packing) we can safely discard all small items before proceeding. Crucially, this cannot turn a no- into a yes-instance because the created $k + 1$-packing could then also be lifted to one for all items (contradicting the assumed no-instance). An application of Theorem 1 then yields a polynomial kernelization because we can have only few large items.

ADDITIVE ONE BIN PACKING(k)
    Input: Item sizes $a_1, \ldots, a_n \in \mathbb{N}$, a bin size $b \in \mathbb{N}$, and $k \in \mathbb{N}$.
    Parameter: k.
    Task: Give a packing into at most $k + 1$ bins of size b,
         or correctly state that k bins do not suffice.

Theorem 6 ($\star$). ADDITIVE ONE BIN PACKING *(k) admits a polynomial kernelization to $\mathcal{O}(k^2)$ items and bit size $\mathcal{O}(k^3)$.*

5   Kernel Bounds for Knapsack Problems

In this section we provide lower and upper bounds for kernel sizes for variants of the KNAPSACK problem.

5.1   Exponential Kernel for Knapsack with Few Item Sizes

First, consider the SUBSET SUM problem restricted to instances with only k distinct item weights, which are not restricted in any other way (except for being non-negative integers). Then the problem can be solved by a fixed-parameter algorithm for parameter k by a reduction to integer linear programming in fixed dimension, and applying Lenstra's algorithm [19] or one of its improvements [12,17]. This was first observed by Fellows et al. [10].

We now generalize the results by Fellows et al. [10] to KNAPSACK with few item weights. More precisely, we are given an instance I of the KNAPSACK problem consisting of n items that have only k distinct item weights; however, the number of item values is unbounded. This means in particular, that the "number of numbers" is not bounded as a function of the parameter, making the results by Fellows et al. [10] inapplicable.

Theorem 7 ($\star$). *The* KNAPSACK *problem with k distinct weights can be solved in time $k^{2.5k+o(k)} \cdot \mathrm{poly}(|I|)$, where $|I|$ denotes the encoding length of the instance.*

5.2 Polynomial Kernel for Subset Sum with Few Item Sizes

We now improve the work of Fellows et al. [10] in another direction. Namely, we show that the SUBSET SUM problem admits a polynomial kernel for parameter the number k of item sizes; this improves upon the exponential-size kernel due to Fellows et al. [10]. To show the kernel bound of $k^{\mathcal{O}(1)}$, consider an instance I of SUBSET SUM with n items that have only k distinct item sizes. For each item size s_i, let μ_i be its multiplicity, that is, the number of items in I of size s_i. Given I, we formulate an ILP for the task of deciding whether some subset S of items has weight exactly t. The ILP simply models for each item size s_i the number of items $x_i \leq \mu_i$ selected from it as to satisfy the subset sum constraint:

$$\left. \begin{array}{r} s_1 x_1 + \ldots + s_k x_k = t, \\ 0 \leq x_i \leq \mu_i, \quad i = 1, \ldots, k, \\ x_i \in \mathbb{N}_0, \quad i = 1, \ldots, k \ . \end{array} \right\} \tag{1}$$

Then (1) is an INTEGER LINEAR PROGRAMMING instance on $m = 1$ relevant constraint and each variable x_i has maximum range bound $u = \max_i \mu_i \leq n$.

Now consider two cases:

- If $\log n \leq k \cdot \log k$, then we apply Theorem 1 to (1) to reduce the instance to an equivalent instance I' of size $\mathcal{O}(k^4 + k^3 \log n) = \mathcal{O}(k^4 + k^3 \cdot (k \log k)) = \mathcal{O}(k^4 \log k)$. We can reformulate I' as an equivalent SUBSET SUM instance by replacing each size s_i by $\mathcal{O}(\log \mu_i)$ new weights $2^j \cdot s_i$ for $0 \leq j \leq \ell_i$ and $\left(\mu_i - \sum_{j=0}^{\ell_i} 2^j \right) \cdot s_i$, where ℓ_i is the largest integer such that $\sum_{j=0}^{\ell_i} 2^j < \mu_i$. Then we have $\mathcal{O}(k \log n) = \mathcal{O}(k^2 \log k)$ items each with a weight which can be encoded in length $\mathcal{O}(k^3 + k^2 \log n + \log n) = \mathcal{O}(k^3 \log k)$, resulting in an encoding length of $\mathcal{O}(k^5 \log^2 k)$.
- If $k \log k \leq \log n$, then we solve the integer linear program (1) by the improved version of Kannan's algorithm [17] due to Frank and Tardos [12] that runs in time $d^{2.5d+o(d)} \cdot s$ for integer linear programs of dimension d and encoding size s. As (1) has dimension $d = k$ and encoding size $s = |I|$, the condition $k^k \leq n$ means that we can solve the ILP (and hence decide the instance I) in time $k^{2.5k+o(k)} \cdot s = n^{\mathcal{O}(1)}$.

In summary, we have shown the following:

Theorem 8. SUBSET SUM *with k item sizes admits a kernel of size $\mathcal{O}(k^5 \log^2 k)$. Moreover, it admits a kernel of size $\mathcal{O}(k^4 \log k)$ if the multiplicities of the item weights can be encoded in binary.*

We remark that this method does not work if the instance I is succinctly encoded by specifying the k distinct item weights w_i in binary and for each item size s_i its multiplicity μ_i in binary: then the running time of Frank and Tardos' algorithm can be exponential in k and the input length of the subset sum instance, which is $\mathcal{O}(k \cdot \log n)$.

5.3   A Kernelization Lower Bound for Subset Sum

In the following we show a kernelization lower bound for SUBSET SUM assuming the *Exponential Time Hypothesis*. The Exponential Time Hypothesis [15] states that there does not exist a $2^{o(n)}$-time algorithm for 3-SAT, where n denotes the number of variables.

Lemma 2 ($\star$). SUBSET SUM *does not admit a $2^{o(n)}$-time algorithm assuming the Exponential Time Hypothesis, where n denotes the number of numbers.*

Theorem 9 ($\star$). SUBSET SUM *does not admit kernels of size $\mathcal{O}(n^{2-\varepsilon})$ for any $\varepsilon > 0$ assuming the Exponential Time Hypothesis, where n denotes the number of numbers.*

6   Integer Polynomial Programming with Bounded Range

Up to now, we used Frank and Tardos' result only for linear inequalities with mostly binary variables. But it also turns out to be useful for more general cases, namely for polynomial inequalities with integral bounded variables. We use this to show that INTEGER POLYNOMIAL PROGRAMMING instances can be compressed if the variables are bounded. As a special case, INTEGER LINEAR PROGRAMMING admits a polynomial kernel in the number of variables if the variables are bounded.

Let us first transfer the language of Theorem 1 to arbitrary polynomials.

Lemma 3 ($\star$). *Let $f \in \mathbb{Q}[X_1, \ldots, X_n]$ be a polynomial of degree at most d with r non-zero coefficients, and let $u \in \mathbb{N}$. Then one can efficiently compute a polynomial $\tilde{f} \in \mathbb{Z}[X_1, \ldots, X_n]$ of encoding length $\mathcal{O}(r^4 + r^3 d \log(ru) + rd \log(nd))$ such that $\mathrm{sign}(f(x) - f(y)) = \mathrm{sign}(\tilde{f}(x) - \tilde{f}(y))$ for all $x, y \in \{-u, \ldots, u\}^n$.*

We use this lemma to compress INTEGER POLYNOMIAL PROGRAMMING instances.

INTEGER POLYNOMIAL PROGRAMMING

   Input: Polynomials $c, g_1, \ldots, g_m \in \mathbb{Q}[X_1, \ldots, X_n]$ of degree at most d
      encoded by the coefficients of the $\mathcal{O}(n^d)$ monomials,
      rationals $b_1, \ldots, b_m, z \in \mathbb{Q}$, and $u \in \mathbb{N}$.
   Question: Is there a vector $x \in \{-u, \ldots, u\}^n$ with $c(x) \leq z$ and
      $g_i(x) \leq b_i$ for $i = 1, \ldots, m$?

Theorem 10 (⋆). *Every* INTEGER POLYNOMIAL PROGRAMMING *instance in which c and each g_i consist of at most r monomials can be efficiently compressed to an equivalent instance with an encoding length that is bounded by* $\mathcal{O}\big(m(r^4 + r^3 d \log(ru) + rd \log(nd))\big)$.

This way, Theorem 10 extends an earlier result by Granot and Skorin-Karpov [13] who considered the restricted variant of $d = 2$.

As r is bounded from above by $\mathcal{O}((n + d)^d)$, Theorem 10 yields a polynomial kernel for the combined parameter (n, m, u) for constant dimensions d. In particular, Theorem 10 provides a polynomial kernel for INTEGER LINEAR PROGRAMMING for combined parameter (n, m, u). This provides a sharp contrast to the result by Kratsch [18] that INTEGER LINEAR PROGRAMMING does not admit a polynomial kernel for combined parameter (n, m) unless the polynomial hierarchy collapses to the third level.

7   Conclusion

In this paper we obtained polynomial kernels for the KNAPSACK problem parameterized by the number of items. We further provide polynomial kernels for weighted versions of a number of fundamental combinatorial optimization problems, as well as integer polynomial programs with bounded range. Our small kernels are built on a seminal result by Frank and Tardos about compressing large integer weights to smaller ones. Therefore, a natural research direction to pursue is to improve the compression quality provided by the Frank-Tardos algorithm.

For the weighted problems we considered here, we obtained polynomial kernels whose sizes are generally larger by some degrees than the best known kernel sizes for the unweighted counterparts of these problems. It would be interesting to know whether this increase in kernel size as compared to unweighted problems is actually necessary (say it could be that we need more space for objects but also due to space for encoding the weights), or whether the kernel sizes of the unweighted problems can be matched.

References

1. Bodlaender, H.L., Downey, R.G., Fellows, M.R., Hermelin, D.: On problems without polynomial kernels. J. Comput. Syst. Sci. **75**(8), 423–434 (2009)
2. Bodlaender, H.L., Fomin, F.V., Saurabh, S.: Open problems posed at WORKER 2010 (2010). http://fpt.wdfiles.com/local-files/open-problems/open-problems.pdf
3. Bodlaender, H.L., Jansen, B.M.P., Kratsch, S.: Kernelization lower bounds by cross-composition. SIAM J. Discrete Math. **28**(1), 277–305 (2014)
4. Bodlaender, H.L., Thomassé, S., Yeo, A.: Kernel bounds for disjoint cycles and disjoint paths. Theoret. Comput. Sci. **412**(35), 4570–4578 (2011)
5. Chlebík, M., Chlebíková, J.: Crown reductions for the minimum weighted vertex cover problem. Discrete Appl. Math. **156**(3), 292–312 (2008)

6. Cygan, M., Fomin, F.V., Jansen, B.M.P., Kowalik, L., Lokshtanov, D., Marx, D., Pilipczuk, M., Pilipczuk, M., Saurabh, S.: Open problems for FPT school 2014 (2014). http://fptschool.mimuw.edu.pl/opl.pdf

7. Cygan, M., Kowalik, L., Pilipczuk, M.: Open problems from workshop on kernels (2013). http://worker2013.mimuw.edu.pl/slides/worker-opl.pdf

8. Drucker, A.: New limits to classical and quantum instance compression. In: Proceedings of the FOCS 2012, pp. 609–618 (2012)

9. Erdős, P., Rado, R.: Intersection theorems for systems of sets. J. London Math. Soc. **35**, 85–90 (1960)

10. Fellows, M.R., Gaspers, S., Rosamond, F.A.: Parameterizing by the number of numbers. Theory Comput. Syst. **50**(4), 675–693 (2012)

11. Fellows, M.R., Guo, J., Marx, D., Saurabh, S.: Data reduction and problem kernels (dagstuhl seminar 12241). Dagstuhl Rep. **2**(6), 26–50 (2012)

12. Frank, A., Tardos, É.: An application of simultaneous Diophantine approximation in combinatorial optimization. Combinatorica **7**(1), 49–65 (1987)

13. Granot, F., Skorin-Kapov, J.: On simultaneous approximation in quadratic integer programming. Oper. Res. Lett. **8**(5), 251–255 (1989)

14. Harnik, D., Naor, M.: On the compressibility of NP instances and cryptographic applications. SIAM J. Comput. **39**(5), 1667–1713 (2010)

15. Impagliazzo, R., Paturi, R., Zane, F.: Which problems have strongly exponential complexity. J. Comput. Syst. Sci. **63**(4), 512–530 (2001)

16. Jansen, K., Kratsch, S., Marx, D., Schlotter, I.: Bin packing with fixed number of bins revisited. J. Comput. Syst. Sci. **79**(1), 39–49 (2013)

17. Kannan, R.: Minkowski's convex body theorem and integer programming. Math. Oper. Res. **12**(3), 415–440 (1987)

18. Kratsch, S.: On polynomial kernels for integer linear programs: covering, packing and feasibility. In: Bodlaender, H.L., Italiano, G.F. (eds.) ESA 2013. LNCS, vol. 8125, pp. 647–658. Springer, Heidelberg (2013)

19. Lenstra Jr., H.W.: Integer programming with a fixed number of variables. Math. Oper. Res. **8**(4), 538–548 (1983)

20. Marx, D., Végh, L.A.: Fixed-parameter algorithms for minimum cost edge-connectivity augmentation. In: Fomin, F.V., Freivalds, R., Kwiatkowska, M., Peleg, D. (eds.) ICALP 2013, Part I. LNCS, vol. 7965, pp. 721–732. Springer, Heidelberg (2013)

21. Nederlof, J., van Leeuwen, E.J., van der Zwaan, R.: Reducing a target interval to a few exact queries. In: Rovan, B., Sassone, V., Widmayer, P. (eds.) MFCS 2012. LNCS, vol. 7464, pp. 718–727. Springer, Heidelberg (2012)

A Shortcut to (Sun)Flowers: Kernels in Logarithmic Space or Linear Time

Stefan Fafianie$^{(\boxtimes)}$ and Stefan Kratsch

University of Bonn, Bonn, Germany
{fafianie,kratsch}@cs.uni-bonn.de

Abstract. We investigate whether kernelization results can be obtained if we restrict kernelization algorithms to run in logarithmic space. This restriction for kernelization is motivated by the question of what results are attainable for preprocessing via simple and/or local reduction rules. We find kernelizations for d-HITTING SET(k), d-SET PACKING(k), EDGE DOMINATING SET(k), and a number of hitting and packing problems in graphs, each running in logspace. Additionally, we return to the question of linear-time kernelization. For d-HITTING SET(k) a linear-time kernel was given by van Bevern [Algorithmica (2014)]. We give a simpler procedure and save a large constant factor in the size bound. Furthermore, we show that we can obtain a linear-time kernel for d-SET PACKING(k).

1 Introduction

The notion of *kernelization* from parameterized complexity offers a framework in which it is possible to establish rigorous upper and lower bounds on the performance of polynomial-time preprocessing for NP-hard problems. Efficient preprocessing is appealing because one hopes to simplify and shrink input instances before running an exact exponential-time algorithm, approximation algorithm, or heuristic. Many intricate techniques have been developed for the field of kernelization and other variants have been considered. For example, the more relaxed notion of Turing kernelization asks whether a problem can be solved by a polynomial-time algorithm that is allowed to query an oracle for answers to instances of small size [10]. In this work we take a more restrictive view. When considering reduction rules for NP-hard problems that a human would come up with quickly, these would often be very simple and probably aimed at local structures in the input. Thus, the matching theoretical question would be whether we can also achieve nice kernelization results when restricted to "simple reduction rules." This is of course a very vague statement and largely a matter of opinion. For local reduction rules this seems much easier: If we restrict a kernelization to running in logarithmic space, then we can no longer perform "complicated" computations like, for example, running a linear program or even just finding a maximal matching in a graph. Indeed, for an instance x, the typical use of $\log |x|$

Supported by the Emmy Noether-program of the German Research Foundation (DFG), research project PREMOD (KR 4286/1).

G.F. Italiano et al. (Eds.): MFCS 2015, Part II, LNCS 9235, pp. 299–310, 2015.
DOI: 10.1007/978-3-662-48054-0_25

bits would rather be to store a counter with values up to $|x|^{\mathcal{O}(1)}$ or to remember a pointer to some position in x.

The main focus of our work is to show that a bunch of classic kernelization results can also be made to work in logarithmic space. To the best of our knowledge such a kernelization was previously only known for VERTEX COVER(k) [1]. Concretely, we show that d-HITTING SET(k) (Sect. 4), d-SET PACKING(k) (Sect. 5), and EDGE DOMINATING SET(k) as well as a couple of implicit hitting set and set packing type problems on graphs admit polynomial kernels that can be computed in logarithmic space. The astute reader will suspect that the well-known sunflower lemma will be behind this, but this is only partially true.

It is well-known that so-called *sunflowers* are very useful for kernelization (they can be used to obtain polynomial kernels for, e.g., d-HITTING SET(k) [5] and d-SET PACKING(k) [2]). A k-sunflower is a collection of k sets $F_1, \ldots, F_k$ such that the pairwise intersection of any two sets is the same set C; called the core. The sets $F_1 \setminus C, \ldots, F_k \setminus C$ are therefore pairwise disjoint. When seeking a k-hitting set S, the presence of a $(k+1)$-sunflower implies that S must intersect the core, or else fail to hit at least one set F_i. The Sunflower Lemma of Erdős and Rado [3] implies that any family with more than $d!k^d$ sets, each of size d, must contain a $(k+1)$-sunflower which can be efficiently found. Thus, so long as the instance is large enough, we will find a core C that can be safely added as a new constraint, and the sets F_i containing C may be discarded.

Crucially, the only point of the disjoint sets $F_1 \setminus C, \ldots, F_k \setminus C$ is to certify that we need at least k elements to hit all sets $F_1, \ldots, F_k$, assuming we refuse to pick an element of C. What if we forgo the disjointness requirement and only request that not picking an element of C incurs a hitting cost of at least k (or at least $k+1$ for the above illustration)? It turns out that the corresponding structure is well-known under the name of a *flower*: A set $\mathcal{F}$ is a k-flower with core C if the collection $\{F \setminus C : F \in \mathcal{F}, F \supseteq C\}$ has minimum hitting set size at least k. Despite the seemingly complicated requirement, Håstad et al. [7] showed that any family with more than k^d sets must contain a $(k+1)$-flower. Thus, by replacing sunflowers with flowers, we can save the extra $d!$ factor in quite a few kernelizations with likely no increase to the running time. We give a formal introduction on (sun)flowers in Sect. 3. In order to meet the space requirements for our logspace kernelizations, we avoid explicitly finding flowers and instead use careful counting arguments to ensure that a $(k+1)$-flower with core $C \subseteq F$ exists when we discard a set F.

Finally, we also return to the question of linear-time kernelization that was previously studied in, e.g., [11,12]. Using flowers instead of sunflowers we can improve a linear-time kernelization for d-HITTING SET(k) by van Bevern [12] from $d! \cdot d^{d+1} \cdot (k+1)^d$ to just $(k+1)^d$ sets (we also save the d^{d+1} factor because of the indirect way in which we use flowers). Similarly, we have a linear-time kernelization for d-SET PACKING(k) to $(d(k-1)+1)^d$ sets. We give these kernels in Sect. 6. We note that for linear-time kernelization the extra applications for hitting set and set packing type problems do not necessarily follow: In logarithmic space we can, for example, find all triangles in a graph and thus kernelize TRIANGLE-FREE VERTEX DELETION(k) and TRIANGLE PACKING(k). In linear time we will typically have no algorithm available that can extract the constraints respectively the feasible sets for the packing that are needed to apply a d-HITTING SET(k) or d-SET PACKING(k) kernelization.

We remark that the kernels for d-HITTING SET(k) and d-SET PACKING(k) via representative sets (cf. [9]) give more savings in the size. For d-HITTING SET(k) this approach yields a kernel with at most $\binom{k+d}{d} = \frac{(k+d)!}{d!+k!} = \frac{1}{d!}(k+1)\cdot\ldots\cdot(k+d) > \frac{k^d}{d!}$ sets, thus saving at most another $d!$ factor. It is however unclear if this approach can be made to work in logarithmic space or linear time. Applying the current fastest algorithm for computing a representative set due to Fomin et al. [6] gives us a running time of $\mathcal{O}(\binom{k+d}{d}|\mathcal{F}|d^\omega + |\mathcal{F}|\binom{k+d}{k}^{\omega-1})$ where ω is the matrix multiplication exponent.

2   Preliminaries

Set Families and Graphs. We use standard notation from graph theory and set theory. Let U be a finite set, let $\mathcal{F}$ be a family of subsets of U, and let $S \subseteq U$. We say that S *hits* a set $F \in \mathcal{F}$ if $S \cap F \neq \emptyset$. In slight abuse of notation we also say that S *hits* $\mathcal{F}$ if for every $F \in \mathcal{F}$ it holds that S hits F. More formally, S is a *hitting set* (or *blocking set*) for $\mathcal{F}$ if for every $F \in \mathcal{F}$ it holds that $S \cap F \neq \emptyset$. If $|S| \leq k$, then S is a k-hitting set. A family $\mathcal{P} \subseteq \mathcal{F}$ is a *packing* if the sets in $\mathcal{P}$ are pairwise disjoint; if $|\mathcal{P}| = k$, then $\mathcal{P}$ is called a k-packing. In the context of instances $(U, \mathcal{F}, k)$ for d-HITTING SET(k) or d-SET PACKING(k) we let $n = |U|$ and $m = |\mathcal{F}|$. Similarly, for problems on graphs $G = (V, E)$ we let $n = |V|$ and $m = |E|$. A *restriction* $\mathcal{F}_C$ of a family $\mathcal{F}$ onto a set C is the family $\{F \setminus C : F \in \mathcal{F}, F \supseteq C\}$, i.e., it is obtained by only taking sets in $\mathcal{F}$ that are a superset of C and removing C from these sets.

Parameterized Complexity. A *parameterized problem* is a language $Q \subseteq \Sigma^* \times \mathbb{N}$; the second component of instances $(x, k) \in \Sigma^* \times \mathbb{N}$ is called the *parameter*. A parameterized problem $Q \subseteq \Sigma^* \times \mathbb{N}$ is *fixed-parameter tractable* if there is an algorithm that, on input $(x, k) \in \Sigma^* \times \mathbb{N}$, correctly decides if $(x, k) \in Q$ and runs in time $\mathcal{O}(f(k)|x|^c)$ for some constant c and any computable function f. A *kernelization algorithm* (or *kernel*) for a parameterized problem $Q \subseteq \Sigma^* \times \mathbb{N}$ is an algorithm that, on input $(x, k) \in \Sigma^* \times \mathbb{N}$, outputs in time $(|x| + k)^{\mathcal{O}(1)}$ an equivalent instance (x', k') with $|x'| + k' \leq g(k)$ for some computable function $g \colon \mathbb{N} \to \mathbb{N}$ such that $(x, k) \in Q \Leftrightarrow (x', k') \in Q$. Here g is called the *size* of the kernel; a *polynomial kernel* is a kernel with polynomial size.

3   Sunflowers and Flowers

The notion of a *sunflower* has played a significant role in obtaining polynomial kernels for the d-HITTING SET(k) and d-SET PACKING(k) problems. These kernelization algorithms return instances of size $\mathcal{O}(k^d)$. However, as a consequence of using the sunflower lemma, there is a hidden $d!$ multiplicative factor in these size bounds (see the full paper [4] for a detailed discussion). We avoid this by considering a relaxed form of sunflower, known as *flower* (cf. Jukna [8]) instead.

Definition 1. *An l-flower with core C is a family $\mathcal{F}$ such that any blocking set for the restriction $\mathcal{F}_C$ of sets in $\mathcal{F}$ onto C contains at least l elements.*

Note that every sunflower with l petals is also an l-flower but not vice-versa. From Definition 1 it follows that the relaxed condition for set disjointness outside of C is still enough to force a k-hitting set S to contain an element of C if there is a $(k+1)$-flower with core C. Similar to the sunflower lemma, Håstad et al. [7] give an upper bound on the size of a family that must contain an l-flower. The next lemma, restates this result. We give a self-contained proof following Jukna's book.

Lemma 1 (cf. Jukna[8, Lemma 7.3]).(★[1]) *Let $\mathcal{F}$ be a family of sets each of cardinality d. If $|\mathcal{F}| > (l-1)^d$, then $\mathcal{F}$ contains an l-flower.*

The proof for Lemma 1 implies that we can find a flower in $\mathcal{O}(|\mathcal{F}|)$ time if $|\mathcal{F}| > (l-1)^d$. However, in order to obtain our logspace and linear-time kernels we avoid explicitly finding a flower. To this end we use Lemma 2. Note that we no longer assume $\mathcal{F}$ to be d-uniform but instead only require that any set in $\mathcal{F}$ is of size at most d, similar to the families that we consider in instances of d-HITTING SET(k) and d-SET PACKING(k). If the required conditions hold, we find that a family $\mathcal{F}$ either contains an l-flower with core C or the set C itself. Thus, any hitting set of size at most $l-1$ for $\mathcal{F}$ must contain an element of C. For our d-HITTING SET(k) kernels we use the lemma with $l = k+1$.

Lemma 2. *For a finite set U, constant d, and a set $C \in \binom{U}{<d}$, let $\mathcal{F} \subseteq \binom{U}{\leq d}$ be a family such that*

(1) there are at least $l^{d-|C|}$ supersets $F \supseteq C$ in $\mathcal{F}$ and
(2) there are at most $l^{d-|C'|}$ supersets $F' \supseteq C'$ in $\mathcal{F}$ for any other $C' \supsetneq C$, $C' \in \binom{U}{\leq d}$.

Then $\mathcal{F}$ contains an l-flower with core C or $C \in \mathcal{F}$.

Proof. Let us consider the restriction $\mathcal{F}_C = \{S \setminus C : S \in \mathcal{F}, S \supseteq C\}$ of sets in $\mathcal{F}$ onto C. If $C \in \mathcal{F}$, then we are done. In the other case, let X be a blocking set for $\mathcal{F}_C$, i.e., $X \cap F \neq \emptyset$ for all $F \in \mathcal{F}_C$ (by assumption $C \notin \mathcal{F}$, thus $\emptyset \notin \mathcal{F}_C$ and a blocking set exists). For every element $x \in X$ consider the number of sets in $\mathcal{F}_C$ that contain x; these correspond to the supersets of $C' = C \cup \{x\}$ in $\mathcal{F}$. We obtain from property (2) that there are at most $l^{d-|C'|} = l^{d-|C|-1}$ such sets. Thus, $|\mathcal{F}_C| \leq |X| \cdot l^{d-|C|-1}$ since every set in $\mathcal{F}_C$ has a non-empty intersection with X while we have previously bounded the number of sets in $\mathcal{F}_C$ that contain at least one element of X. By property (1) of $\mathcal{F}$ we have that $|\mathcal{F}_C| \geq l^{d-|C|}$. Therefore $|X| \geq l$ and consequently $\mathcal{F}' = \{S : S \in \mathcal{F}, S \supseteq C\}$ is an l-flower.   □

---

[1] Proofs of statements marked with ★ are in the full paper [4].

The astute reader may find that Lemma 2 is in a sense not completely tight. Indeed, if we require that there are more than $(l-1)^{d-|C|}$ supersets of C (instead of at least that many) and at most $(l-1)^{d-|C'|}$ supersets for bigger cores C', then an l-flower with core C must exist (if X hits $\mathcal{F}_C$, then $(l-1)^{d-|C|} < |\mathcal{F}_C| \leq |X| \cdot (l-1)^{d-|C|-1}$, therefore $|X| > l-1$). For technical convenience, the present formulation is more suitable for our algorithms.

4 Logspace Kernel for Hitting Set

---

d-HITTING SET(k) **Parameter:** k.
Input: A set U and a family $\mathcal{F}$ of subsets of U each of size at most d, i.e., $\mathcal{F} \subseteq \binom{U}{\leq d}$, and $k \in \mathbb{N}$.
Question: Is there a k-hitting set S for $\mathcal{F}$?

We now present a logspace kernelization algorithm for d-HITTING SET(k). The space requirement prevents the normal approach of finding sunflowers and modifying the family $\mathcal{F}$ in memory (we are basically left with the ability to have a constant amount of pointers and counters in memory). We start with an intuitive attempt for getting around the space restriction and show how it would fail.

The intuitive (but wrong) approach at a logspace kernel works as follows. Process the sets $F \in \mathcal{F}$ one at a time and output F unless we find that the subfamily of sets that where processed before F contains a $(k + 1)$-flower that enforces some core $C \subseteq F$ to be hit. For a single step t, let $\mathcal{F}_t$ be the sets that we have processed so far and let $\mathcal{F}'_t \subseteq \mathcal{F}_t$ be the family of sets in the output. We would like to maintain that a set S is a k-hitting set for $\mathcal{F}_t$ if and only if it a k-hitting set for $\mathcal{F}'_t$. Now suppose that this holds and we want to show that our procedure preserves this property in step $t + 1$ when some set F is processed. This can only fail if we discard F, and only in the sense that some S is a k-hitting set for $\mathcal{F}'_{t+1}$ but not for $\mathcal{F}_{t+1}$ because $\mathcal{F}'_{t+1} \subseteq \mathcal{F}_{t+1}$. However, S is also a k-hitting set for $\mathcal{F}'_t \subseteq \mathcal{F}'_{t+1}$ and, by assumption, also for $\mathcal{F}_t$. Recall that we have discarded F because of a $(k + 1)$-flower in $\mathcal{F}_t$ with core C, so S must intersect C (or fail to be a k-hitting set). Thus, S intersects also $F \supseteq C$, making it a k-hitting set for $\mathcal{F}_t \cup \{F\} = \mathcal{F}_{t+1}$. Unfortunately, while the correctness proof would be this easy, such a straightforward approach fails as a consequence of the following lemma.[2]

Lemma 3. (★) *Given a family of sets $\mathcal{F}$ of size d and $F \in \mathcal{F}$. The problem of finding an l-flower in $\mathcal{F} \setminus F$ with core $C \subseteq F$ is* coNP-*hard.*

---

[2] It is well known that finding a k-sunflower is NP-hard in general. Similarly, finding a k-flower is coNP-hard. For self-contained proofs see the full paper [4]. Both proofs do not apply when the size of the set family exceeds the bounds in the (sun)flower lemma.

Even if we know that the number of sets that where processed exceed the k^d bound of Lemma 1, finding out whether there is a flower with core $C \subseteq F$ is hard.[3] Instead we use an application of Lemma 2 that only ensures that there is some flower with $C \subseteq F$ if two stronger counting conditions are met. Whether condition (1) holds in $\mathcal{F}_t$ can be easily checked, but there is no guarantee that condition (2) holds if $\mathcal{F}_t$ exceeds a certain size bound, i.e., this does not give any guarantee on the size of the output if we process the input once. We fix this by taking a layered approach in which we filter out redundant sets for which there exist flowers that ensure that these sets must be hit, such that in each subsequent layer the size of the cores of these flowers decreases.

We consider a collection of logspace algorithms $A_0, \ldots, A_d$ and families of sets $\mathcal{F}(0), \ldots, \mathcal{F}(d)$ where $\mathcal{F}(l)$ is the output of algorithm A_l. Each of these algorithms *simulates* the next algorithm for decision-making, i.e., if we directly *run* A_l, then it simulates A_{l+1} which in turn simulates A_{l+2}, etc. If we run A_l however, then it is the only algorithm that outputs sets; each of the algorithms that are being simulated as a result of running A_l does not produce output.

We maintain the invariant that for all $C \subseteq U$ such that $l \leq |C| \leq d$, the family $\mathcal{F}(l)$ contains at most $(k+1)^{d-|C|}$ supersets of C. Each algorithm A_l processes sets in $\mathcal{F}$ one at a time. For a single step t, let $\mathcal{F}_t$ be the sets that have been processed so far and let $\mathcal{F}_t(l)$ denote the sets in the output. Note that A_l will not have access to its own output $\mathcal{F}_t(l)$ but we use it for analysis.

Let us first describe how algorithm A_d processes set F in step $t+1$. If $F \notin \mathcal{F}_t$, then A_d decides to output F; in the other case it proceeds with the next step. In other words A_d is a simple algorithm that outputs a single copy of every set in $\mathcal{F}$. This ensures that the kernelization is robust for hitting set instances where multiple copies of a single set appear. If we are guaranteed that this is not the case, then simply outputting each set F suffices. Clearly the invariant holds for $\mathcal{F}(d)$ since any $C \in \binom{U}{\geq d}$ only has $F = C$ as a superset and there is at most $(k+1)^0 = 1$ copy of each set in $\mathcal{F}(d)$.

For $0 \leq l < d$ the procedure in Algorithm 1 describes how A_l processes F in step $t+1$. First observe that lines 2 and 3 ensure that $\mathcal{F}(l) \subseteq \mathcal{F}(l+1)$. Assuming that the invariant holds for $\mathcal{F}(l+1), \ldots, \mathcal{F}(d)$, lines 8 and 9 ensure that the invariant is maintained for $\mathcal{F}(l)$. Crucially, we only need make sure that it additionally holds for $C \in \binom{U}{l}$ since larger cores are covered by the invariant for $\mathcal{F}(l+1)$.

Observation 1. A_0 *commits at most* $(k+1)^d$ *sets to the output during the computation. This follows from the invariant for* $\mathcal{F}(0)$ *when considering* $C = \emptyset$ *which is in* $\binom{U}{\geq l}$ *for* $l = 0$.

Since the algorithms just verify containment of sets of size at most d and only require some pointers and counters we have the following lemma.

Lemma 4. ($\bigstar$) *For* $0 \leq l \leq d$, A_l *can be implemented such that it uses logarithmic space and runs in* $\mathcal{O}(|\mathcal{F}|^{d-l+2})$ *time.*

---

[3] Note that we would run into the same obstacle if we use sunflowers instead of flowers; finding out whether there exists a $(k+1)$-sunflower with core $C \subseteq F$ for a specific set F is NP-hard as we show in the full paper [4].

Algorithm 1: Step $t + 1$ of A_l, $0 \leq l < d$.

1 simulate A_{l+1} up to step $t + 1$;
2 **if** A_{l+1} *decides not to output* F **then**
3   do not output F and end the computation for step $t + 1$;
4 **else**
5   **foreach** $C \subseteq F$ *with* $|C| = l$ **do**
6   simulate A_{l+1} up to step t;
7   count the number of supersets of C that A_{l+1} would output;
8   **if** *the result is at least* $(k+1)^{d-|C|}$ **then**
9   do not output F and end the computation for step $t + 1$;
10   output F;

Let us remark that we could also store each $C \subseteq F$ in line 5, allocate a counter for each of these sets, and simulate A_{l+1} only once instead of starting a new simulation for each subset. This gives us a constant factor trade-off in running time versus space complexity. One might also consider a hybrid approach, e.g., by checking x subsets of F at a time.

For correctness we give the following lemma which shows that the answer to d-HITTING SET(k) is preserved in each layer $\mathcal{F}(0) \subseteq \mathcal{F}(1) \subseteq \ldots \subseteq \mathcal{F}(d)$. We give a brief proof sketch outlining the most interesting case and give a complete proof in the full paper [4]. The more involved proof for our logspace kernelization for d-SET PACKING(k) will be presented in full detail in the next section.

Lemma 5. *Let $0 \leq l < d$ and let S be a set of size at most k. It holds that S is a hitting set for $\mathcal{F}(l)$ if and only if it is a hitting set for $\mathcal{F}(l + 1)$.*

Proof. (sketch). For each $0 \leq l < d$, we prove by induction over $0 \leq t \leq m$ that a set S is a k-hitting set for $\mathcal{F}_t(l)$ if and only if it is a k-hitting set for $\mathcal{F}_t(l + 1)$ (step $t = m$ proves the lemma). For $t = 0$ the statement is trivial. Let us assume that it holds for steps $t \leq i$ and prove that it also holds for step $i + 1$.

Let us consider the direction where a set S is a k-hitting set for $\mathcal{F}_{i+1}(l)$ and we have to show that S is also a hitting set for $\mathcal{F}_{i+1}(l + 1)$ (the other direction is straightforward). Let F be the set that is processed in step $i + 1$. The interesting case is when A_{l+1} decided to output F but A_l decides not to output F.

This implies that it established that there are at least $(k + 1)^{d-|C|}$ supersets of some $C \subseteq F$ with $|C| = l$ in $\mathcal{F}_i(l + 1)$. By the invariant for $\mathcal{F}(l + 1)$, we have that for all sets C' that are larger than C there are at most $(k + 1)^{d-|C|'}$ supersets $F' \subseteq C'$ in $\mathcal{F}_i(l + 1) \subseteq \mathcal{F}(l + 1)$. Consequently, by Lemma 2 we have that $\mathcal{F}_i(l + 1)$ contains C or a $(k + 1)$-flower with core C. Thus, a hitting set for $\mathcal{F}_i(l + 1)$ of size at most k must hit C. Since S is a k-hitting set for $\mathcal{F}_{i+1}(l)$ it must also hit $\mathcal{F}_i(l) \subseteq \mathcal{F}_{i+1}(l)$ and by the induction hypothesis we have that S is also a k-hitting set for $\mathcal{F}_i(l + 1)$. We have just established that S must hit C in order to hit $\mathcal{F}_i(l + 1)$ since it has cardinality at most k. Thus, S also hits $F \supseteq C$ and therefore S is a hitting set for $\mathcal{F}_{i+1}(l + 1) = \mathcal{F}_i(l + 1) \cup \{F\}$. $\qquad\square$

It is easy to see that a set S is a k-hitting set for $\mathcal{F}(d)$ if and only if it is a hitting set for $\mathcal{F}$, because A_d only discards duplicate sets. As a consequence of Lemma 5, a set S is a k-hitting set for $\mathcal{F}(0)$ if and only if it is a k-hitting set for $\mathcal{F}(d)$. Therefore, it follows from Observation 1 and Lemma 4 that A_0 is a logspace kernelization algorithm for d-HITTING SET(k).

Theorem 1. *d-HITTING SET(k) admits a logspace kernelization that runs in time $\mathcal{O}(|\mathcal{F}|^{d+2})$ and returns an equivalent instance with at most $(k+1)^d$ sets.*

Let us remark that technically we still have to reduce the ground set to size polynomial in k. We can reduce the ground set of the output instance to at most $d(k+1)^d$ elements by including one more layer. Let A_{d+1} be an algorithm that simulates A_d. Each time that A_d decides to output a set F, algorithm A_{d+1} determines the new identifier of each element e in F by counting the number of distinct elements that have been output by A_d before the first occurrence of e. This can be done by simulating A_d up to the first step in which A_d outputs e by incrementing a counter each time an element is output for the first time (whether an element occurs for the first time can again be verified via simulation of A_d). We take the same approach for other logspace kernelizations in this paper.

5   Logspace Kernel for Set Packing

---

d-SET PACKING(k) **Parameter:** k.

Input: A set U and a family $\mathcal{F}$ of subsets of U each of size at most d, i.e., $\mathcal{F} \subseteq \binom{U}{\leq d}$, and $k \in \mathbb{N}$.

Question: Is there a k-packing $\mathcal{P} \subseteq \mathcal{F}$?

In this section we present a logspace kernelization for d-SET PACKING(k). The strategy for obtaining such a kernel is similar to that in Sect. 4. However, the correctness proof gets more complicated. We point out the main differences.

We consider a collection of logspace algorithms $B_0, \ldots, B_d$ that perform almost the same steps as the collection of algorithms described in the logspace kernelization for d-HITTING SET(k) such that only the invariant differs. For each $0 \leq l \leq d$ we maintain that for all $C \subseteq U$ such that $l \leq |C| \leq d$, the family $\mathcal{F}(l)$ that is produced by B_l contains at most $(d(k-1)+1)^{d-|C|}$ supersets of C.

The idea for the proof of correctness is similar to that of Lemma 5. However, we need a slightly stronger induction hypothesis to account for the behavior of a solution for d-SET PACKING(k) since it is a subset of the considered family.

Lemma 6. *For $0 \leq l < d$, it holds that $\mathcal{F}(l)$ contains a packing $\mathcal{P}$ of size k if and only if $\mathcal{F}(l+1)$ contains a packing $\mathcal{P}'$ of size at most k.*

Proof. For each $0 \leq l < d$, we prove by induction over $0 \leq t \leq m$ that for any $0 \leq j \leq k$ and any set $S \in \binom{U}{\leq d(k-j)}$, $\mathcal{F}_t(l)$ contains a packing $\mathcal{P}$ of size j such that S does not intersect with any set in $\mathcal{P}$ if and only if $\mathcal{F}_t(l+1)$ contains a packing $\mathcal{P}'$ of size j such that S does not intersect with any set in $\mathcal{P}'$.

This proves the lemma since $\mathcal{F}_m(l) = \mathcal{F}(l)$, $\mathcal{F}_m(l+1) = \mathcal{F}(l+1)$, and for packings of size $j = k$ the set S is empty. It trivially holds for any t if $j = 0$; hence we assume $0 < j \leq k$. For $t = 0$ we have $\mathcal{F}_0(l) = \mathcal{F}_0(l+1) = \emptyset$ and the statement obviously holds. Let us assume that it holds for steps $t \leq i$ and consider step $i + 1$ in which F is processed. If $\mathcal{F}_{i+1}(l)$ contains a packing $\mathcal{P}$ of size j, then $F_{i+1}(l+1)$ also contains $\mathcal{P}$ since $\mathcal{F}_{i+1}(l+1) \supseteq \mathcal{F}_{i+1}(l)$. Thus, the status for avoiding intersection with any set S remains the same.

For the converse direction, let us assume that $\mathcal{F}_{i+1}(l+1)$ contains a packing $\mathcal{P}$ of size j that avoids intersection with a set S of size $d(k-j)$. We must show that $\mathcal{F}_{i+1}(l)$ contains a j-packing that avoids S. Let F be the set that is processed in step $i+1$. Suppose that $F \notin \mathcal{P}$. Then $\mathcal{F}_i(l+1)$ contains $\mathcal{P}$ and by the induction hypothesis we have that $\mathcal{F}_i(l) \subseteq \mathcal{F}_{i+1}(l)$ contains a j-packing $\mathcal{P}'$ that avoids S.

In the other case $F \in \mathcal{P}$. Suppose that $F \in \mathcal{F}_{i+1}(l)$, i.e., B_l decided to output F. We know that $\mathcal{F}_i(l+1)$ contains the $(j-1)$-packing $\mathcal{P} \setminus \{F\}$ which avoids $S \cup F$. By the induction hypothesis we have that $\mathcal{F}_i(l)$ contains a packing $\mathcal{P}''$ of size $j-1$ that avoids $S \cup F$. Thus, $\mathcal{F}_{i+1}(l)$ contains the j-packing $\mathcal{P}' = \mathcal{P}'' \cup \{F\}$ which avoids S.

Now suppose that $F \notin \mathcal{F}_{i+1}(l)$. By assumption, $F \in \mathcal{P} \subseteq \mathcal{F}_{i+1}(l+1)$. This implies that B_l decided not to output F because it has established that there are at least $d(k-1)+1)^{d-|C|}$ supersets of some $C \subseteq F$ with $|C| = l$ in $\mathcal{F}_i(l+1)$. Furthermore, by the invariant for $\mathcal{F}(l+1)$ we have that for all sets C' that are larger than C there are at most $(d(k-1)+1)^{d-|C'|}$ supersets $F' \supseteq C'$ in $F_i(l+1) \subseteq \mathcal{F}(l+1)$. Consequently, by Lemma 2 we have that $\mathcal{F}_i(l+1)$ contains C or a $(d(k-1)+1)$-flower with core C. In the first case, any hitting set for $\mathcal{F}_i(l+1)$ must hit C, and in the second case any hitting set for $\mathcal{F}_i(l+1)$ requires at least $d(k-1)+1$ elements if it avoids hitting C; thus any hitting set of size at most $d(k-1)$ must hit C. By assumption, $\mathcal{P} \setminus \{F\}$ and S both avoid $C \subseteq F$ and therefore they both avoid at least one set F' in $\mathcal{F}_i(l+1)$ since together they contain at most $d(k-1)$ elements. Thus we can obtain a j-packing $\mathcal{P}''$ in $\mathcal{F}_{i+1}(l+1)$ that also avoids S by replacing F with F'. Since $\mathcal{P}''$ no longer contains F we find that $\mathcal{F}_i(l+1)$ contains $\mathcal{P}''$ and by the induction hypothesis there is some packing $\mathcal{P}'$ of size j in $\mathcal{F}_i(l) \subseteq \mathcal{F}_{i+1}(l)$ that avoids S. $\square$

Since B_d only discards duplicate sets it holds that $\mathcal{F}(d)$ has a k-packing if and only if $\mathcal{F}$ has a k-packing. By Lemma 6, we have that $\mathcal{F}(0)$ has a k-packing if and only if $\mathcal{F}(d)$ has a k-packing. The kernel size, and running time and space complexity follow analogous to Observation 1 and Lemma 4 respectively. Therefore, we have that B_0 is a logspace kernelization algorithm for d-SET PACKING(k).

Theorem 2. d-SET PACKING(k) *admits a logspace kernelization that runs in time* $\mathcal{O}(|\mathcal{F}|^{d+2})$ *and returns an equivalent instance with at most* $(d(k-1)+1)^d$ *sets.*

6   Linear-Time Kernels for Hitting Set and Set Packing

We show how our techniques can be used in conjunction with a data-structure and subroutine by van Bevern [12] in order to obtain a smaller linear-time kernel for d-HITTING SET(k). The algorithm processes sets in $\mathcal{F}$ one by one and decides whether they should appear in the final output. For a single step t, let $\mathcal{F}_t$ denote the sets that have been processed by the algorithm so far and let $\mathcal{F}'_t \subseteq \mathcal{F}_t$ be the sets stored in memory for which it has decided positively. The algorithm uses a data structure in which it can store the number of supersets of a set $C \in \binom{U}{\leq d}$ that it has stored in $\mathcal{F}'_t$; let supersets$[C]$ denote the entry for set C. We begin by initializing supersets$[C] \leftarrow 0$ for each $C \subseteq F$ where $F \in \mathcal{F}$.

We maintain the invariant that the number of sets $F \in \mathcal{F}'_t$ that contain $C \in \binom{U}{<d}$ is at most $(k+1)^{d-|C|}$. Now assume that the invariant holds after processing sets $\mathcal{F}_t \subseteq F$ and let F be the next set. Algorithm 2 describes how the algorithm processes F in step $t+1$.

Algorithm 2: Step $t+1$ of the linear-time kernel for d-HITTING SET(k).

---

1 **foreach** $C \subseteq F$ **do**
2   query supersets$[C]$ in order to determine the number of supersets of C in $\mathcal{F}'_t$;
3   **if** *the result is at least $(k+1)^{d-|C|}$* **then**
4   do not store F and end the computation for step $t+1$;

5 store F;
6 **foreach** $C \subseteq F$ **do**
7   supersets$[C] \leftarrow$ supersets$[C]+1$;

Note that Line 3 and 4 ensure that the invariant is maintained. After all sets in $\mathcal{F}$ have been processed, the algorithm returns the family $\mathcal{F}' \subseteq \mathcal{F}$ of sets that it has stored in memory.

Observation 2. *After any step t, the algorithm has stored at most $(k+1)^d$ sets in $\mathcal{F}'_t$. This follows from the invariant when considering $C = \emptyset$.*

Lemma 7. *The algorithm can be implemented such that it runs in linear time.*

Proof. Before we run the algorithm, we sort the sets in $\mathcal{F}$ using the linear-time procedure described in [12]. Implementing the data structure as a trie [12] allows us to query and update supersets$[C]$ in time $\mathcal{O}(|C|)$ for a set C where $|C| \leq d$. These operations occur at most 2^d times in each step (this follows from the number of sets $C \subseteq F$ that we consider). Therefore we spend only constant time for each set that is processed and hence the algorithm runs in time $\mathcal{O}(|\mathcal{F}|)$.   □

We obtain the following lemma in a similar way to Lemma 5. The key difference is that we can now check if there are at least $(k+1)^{d-|C|}$ supersets of a set C in $\mathcal{F}_t$ by using the data structure. Crucially, the algorithm has direct access to the previously processed sets that it plans to output; this simplifies the proof.

Lemma 8. *When a family $\mathcal{F}_t \subseteq \mathcal{F}$ has been processed, the algorithm has a family $\mathcal{F}'_t \subseteq \mathcal{F}_t$ such that for any set S of size at most k, S is a hitting set for $\mathcal{F}_t$ if and only if S is a hitting set for $\mathcal{F}'_t$.*

Proof. We prove the lemma by induction and show for each step t that a set S of size k is a hitting set for $\mathcal{F}'_t$ if and only if S is a hitting set for $\mathcal{F}_t$. This obviously holds for $t = 0$ since we have $\mathcal{F}'_0 = \mathcal{F}_0 = \emptyset$. Now let us assume that it holds for steps $t \leq i$ and prove that it also holds for step $i + 1$. One direction is trivial: If a set S is a k-hitting set for $\mathcal{F}_{i+1}$, then it is a hitting set for $\mathcal{F}'_{i+1} \subseteq \mathcal{F}_{i+1}$.

For the converse direction let us suppose that a set S is a k-hitting set for $\mathcal{F}'_{i+1}$. We must show that S is also a hitting set for $\mathcal{F}_{i+1}$. Let F be the set that is processed in step $i+1$. We must consider the case where the algorithm decides not to output F. Otherwise a hitting set for $\mathcal{F}'_{i+1}$ must hit $\mathcal{F}'_i \cup \{F\}$ and thus by the induction hypothesis it must also hit $\mathcal{F}_i \cup \{F\} = \mathcal{F}_{i+1}$.

Now let us assume $F \notin \mathcal{F}'_{i+1}$. This implies that the algorithm decided not to output F because for some $C \subseteq F$ it has determined that there are at least $(k + 1)^{d-|C|}$ supersets of F in $\mathcal{F}'_i$. Furthermore, by the invariant we have that there are at most $(k+1)^{d-|C'|}$ supersets of C' in $\mathcal{F}'_i$ for any C' that is larger than C. Consequently, by Lemma 2 we have that $\mathcal{F}'_i$ contains C or a $(k + 1)$-flower with core C, i.e., any hitting set of size k for $\mathcal{F}'_i$ must hit C and therefore also hits $F \supseteq C$. Let S be a k-hitting set for $\mathcal{F}'_{i+1}$ and note that S is also a k-hitting set for $\mathcal{F}'_i \subseteq \mathcal{F}'_{i+1}$. By the induction hypothesis, S is also a hitting set for $\mathcal{F}_i$. We have just established that S must hit C in order to hit $\mathcal{F}'_i$. Thus, S hits F and is a hitting set for $\mathcal{F}_{i+1} = \mathcal{F}_i \cup \{F\}$. □

Reducing the size of the ground set is simple since we have the equivalent instance in memory. In one pass we compute a mapping of elements occurring in the equivalent instance to $d(k + 1)^d$ distinct identifiers. The following theorem is a consequence of Observation 2, Lemmas 7 and 8.

Theorem 3. d-HITTING SET(k) *admits a linear-time kernelization which returns an equivalent instance with at most $(k + 1)^d$ sets.*

Running an algorithm that follows the same steps as Algorithm 2 such that only the invariant differs gives us a linear-time kernel for d-SET PACKING(k). In this case we maintain the invariant that the number of sets $F \in \mathcal{F}'_t$ that contain $C \in \binom{U}{<d}$ is at most $(d(k - 1) + 1)^{d-|C|}$. For more details and a proof of correctness see the full paper [4].

Theorem 4. d-SET PACKING(k) *admits a linear time kernelization which returns an equivalent instance with at most $(d(k - 1) + 1)^d$ sets.*

7   Concluding Remarks

In this paper we have presented logspace kernelization algorithms for d-HITTING SET(k) and d-SET PACKING(k). By using flowers instead of sunflowers we save a large hidden constant in the size of these kernels. We can show how these kernels can be used to obtain logspace kernels for hitting and packing problems on graphs, and for EDGE DOMINATING SET(k). We obtain the following theorems, the proofs of which can be found in the full paper [4].

Theorem 5. (★) EDGE DOMINATING SET(k) *admits a logspace kernelization that runs in time* $\mathcal{O}(|E|^4)$ *and returns an equivalent instance with* $\mathcal{O}(k^3)$ *edges.*

Theorem 6. (★) $\mathcal{H}$-FREE VERTEX DELETION(k) *admits a logspace kernelization that runs in time* $\mathcal{O}(|E| \cdot \binom{|V|}{d}^{d+1})$ *and outputs an equivalent instance with at most* $\frac{d(d-1)}{2} \cdot (k+1)^d$ *edges, where d is the maximum number of vertices of any $H \in \mathcal{H}$.*

Theorem 7. (★) H-PACKING(k) *admits a logspace kernelization that runs in time* $\mathcal{O}(|E| \cdot \binom{|V|}{d}^{d+1})$ *and outputs an equivalent instance with at most* $\frac{d(d-1)}{2} \cdot (d(k-1)+1)^d$ *edges, where $d = |V(H)|$.*

Furthermore, we have improved upon a linear-time kernel for d-HITTING SET(k) and have given a linear-time kernel for d-SET PACKING(k). One question to settle is whether a vertex-linear kernel for VERTEX COVER(k) can be found in logspace. While known procedures for obtaining a vertex-linear kernel seem unsuitable for adaptation to logspace, we currently do not have tools for ruling out such a kernel. The problem of finding a vertex-linear kernel for VERTEX COVER(k) in linear time also remains open.

References

1. Cai, L., Chen, J., Downey, R.G., Fellows, M.R.: Advice classes of parameterized tractability. Ann. Pure Appl. Logic **84**(1), 119–138 (1997)
2. Dell, H., Marx, D.: Kernelization of packing problems. In: Proceedings of the Twenty-third Annual ACM-SIAM Symposium on Discrete Algorithms, pp. 68–81. SIAM (2012)
3. Erdős, P., Rado, R.: Intersection theorems for systems of sets. J. Lond. Math. Soc. **1**(1), 85–90 (1960)
4. Fafianie, S., Kratsch, S.: A shortcut to (sun) flowers: Kernels in logarithmic space or linear time. arXiv report 1504.08235 (2015)
5. Flum, J., Grohe, M.: Parameterized complexity theory, volume XIV of texts in theoretical computer science. In: An EATCS Series (2006)
6. Fomin, F.V., Lokshtanov, D., Saurabh, S.: Efficient computation of representative sets with applications in parameterized and exact algorithms. In: Proceedings of the Twenty-Fifth Annual ACM-SIAM Symposium on Discrete Algorithms, pp. 142–151. SIAM (2014)
7. Håstad, J., Jukna, S., Pudlák, P.: Top-down lower bounds for depth-three circuits. Comput. Complex. **5**(2), 99–112 (1995)
8. Jukna, S.: Extremal combinatorics: with applications in computer science. Springer Science & Business Media, Heidelberg (2011)
9. Kratsch, S., et al.: Recent developments in kernelization: a survey. Bull. EATCS **2**(113), 58–97 (2014)
10. Lokshtanov, D.: New methods in parameterized algorithms and complexity. Ph.D thesis, Citeseer (2009)
11. Niedermeier, R., Rossmanith, P.: An efficient fixed-parameter algorithm for 3-hitting set. J. Discrete Algorithms **1**(1), 89–102 (2003)
12. van Bevern, R.: Towards optimal and expressive kernelization for d-hitting set. Algorithmica **70**(1), 129–147 (2014)

Metastability of Asymptotically Well-Behaved Potential Games
(Extended Abstract)

Diodato Ferraioli[1]([✉]) and Carmine Ventre[2]

[1] Dipartimento di Informatica, Università di Salerno, Salerno, Italy
dferraioli@unisa.it
[2] School of Computing, Teesside University, Middlesbrough, UK
C.Ventre@tees.ac.uk

Abstract. One of the main criticisms to game theory concerns the assumption of full rationality. Logit dynamics is a decentralized algorithm in which a level of irrationality (a.k.a. "noise") is introduced in players' behavior. In this context, the solution concept of interest becomes the logit equilibrium, as opposed to Nash equilibria. Logit equilibria are distributions over strategy profiles that possess several nice properties, including existence and uniqueness. However, there are games in which their computation may take exponential time. We therefore look at an approximate version of logit equilibria, called *metastable distributions*, introduced by Auletta et al. [4]. These are distributions which remain stable (i.e., players do not go too far from it) for a large number of steps (rather than forever, as for logit equilibria). The hope is that these distributions exist and can be reached quickly by logit dynamics.

We identify a class of potential games, that we name *asymptotically well-behaved*, for which the behavior of the logit dynamics is not chaotic as the number of players increases, so to guarantee meaningful asymptotic results. We prove that any such game admits distributions which are metastable no matter the level of noise present in the system, and the starting profile of the dynamics. These distributions can be quickly reached if the rationality level is not too big when compared to the inverse of the maximum difference in potential. Our proofs build on results which may be of independent interest, including some spectral characterizations of the transition matrix defined by logit dynamics for generic games and the relationship among convergence measures for Markov chains.

1 Introduction

One of the most prominent assumptions in game theory dictates that people are rational. This is contrasted by many concrete instances of people making irrational choices in certain strategic situations, such as stock markets [20]. This might be due to the incapacity of exactly determining one's own utilities: the strategic game is played with utilities perturbed by some noise. Logit dynamics [5] incorporates this noise in players' actions and then is advocated to be a good

© Springer-Verlag Berlin Heidelberg 2015
G.F. Italiano et al. (Eds.): MFCS 2015, Part II, LNCS 9235, pp. 311–323, 2015.
DOI: 10.1007/978-3-662-48054-0_26

model for people behavior. In more detail, logit dynamics features a rationality level $\beta \geq 0$ (equivalently, a noise level $1/\beta$) and each player is assumed to play a strategy with a probability which is proportional to her corresponding utility and β. So the higher β is, the less noise there is and the more rational players are. Logit dynamics can then be seen as a noisy best-response dynamics.

The natural equilibrium concept for logit dynamics is defined by a probability distribution over the pure strategy profiles of the game. Whilst for best-response dynamics Pure Nash equilibria (PNE) are stable states, in logit dynamics there is a chance, which is inversely proportional to β, that players deviate from such strategy profiles. PNE are then not an adequate solution concept for this dynamics. However, the random process defined by the logit dynamics can be modeled via an ergodic Markov chain. Stability in Markov chains is represented by the concept of stationary distributions. These distributions, dubbed logit equilibria, are suggested as a suitable solution concept in this context due to their properties [3]. For example, from the results known in Markov chain literature, we know that any game possesses a logit equilibrium and that this equilibrium is unique. The absence of either of these guarantees is often considered a weakness of PNE. Nevertheless, as for (P)NE, the computation of logit equilibria may be computationally hard depending on whether the chain mixes rapidly or not [2].

As the hardness of computing (P)NE justifies approximate notions of the concept [15], so Auletta et al. [4] look at an approximation of logit equilibria that they call *metastable distributions*. These distributions aim to *describe* regularities arising during the transient phase of the dynamics before stationarity has been reached. Indeed, they are distributions that remain stable for a time which is long enough for the observer (in computer science terms, this time is assumed to be super-polynomial) rather than forever. Roughly speaking, the stability of the distributions in this concept is measured in terms of the generations living some historical era, while stationary distributions remain stable throughout all the generations. When the convergence to logit equilibria is too slow, then there are generations which are outlived by the computation of the stationary distribution. For these generations, metastable distributions grant an otherwise impossible descriptive power. (We refer the interested reader to [4] for a complete overview of the rationale of metastability.) It is unclear whether and which strategic games possess these distributions and if logit dynamics quickly reaches them.

The focus of this paper is the study of metastable distributions for the class of potential games [17]. Potential games are an important and widely studied class of games modeling many strategic settings. These games satisfy several appealing properties, the existence of PNE being one of them. A general study of metastability of potential games was left open by [4] and assumes particular interest due to the known hardness results (e.g., [9]) which suggest that computing a PNE for them is an intractable problem even for centralized algorithms.

Our Contribution. We aim to prove asymptotic results, in terms of the number of players n of potential games, concerning the super-polynomially long stability of metastable distributions, and the polynomial convergence time to them. This desiderata imposes some requirement on the potential games of interest, for otherwise a chaotic behavior (w.r.t. n) of the logit dynamics run on a game would not allow any meaningful asymptotic guarantee for it. We therefore identify a simple-to-describe class of potential games, termed *asymptotically well-behaved*, for which the behavior of the dynamics is "almost" the same for any number of players. Intuitively, the potential function of a game in this class has a shape which is, in a sense, immaterial from the actual value of n. As an example, consider the potential minimized when all the players agree on a strategy x, maximized when no player plays x, and that increases as the number of players playing x decreases. The technical definition of this notion is given in Sect. 4. We stress that similar assumptions are made in related literature on logit dynamics either implicitly (as in [2,18], where it is assumed that certain properties of the potential function do not change as n changes), or explicitly, by considering specific games that clearly enjoy this property [4]. Moreover, asymptotic results on the mixing time of Markov chains require an assumption on the behavior of the chain (i.e., the minimum bottleneck ratio must either be a polynomial or a super-polynomial) usually implicitly guaranteed. Given that our objective is much more complex than bounding the mixing time (i.e., measuring asymptotically the transient phase of the chain and ascertain stability of *and* convergence time to metastable distributions) we need a similar, yet stronger, requirement.

Together with the formalization of the class of games of interest, we formalize, building upon [4], the concept of asymptotic convergence/closeness to a metastable distribution, as a function of the number of players of a game. We then note, via the construction of an ad-hoc n-player potential game that not all potential games admit metastable distributions (cf. Sect. 2), thus showing formally that some restriction on games under consideration is necessary.

Our main result proves that any asymptotically well-behaved n-player potential game has a metastable distribution for each starting profile of the logit dynamics. These distributions remain stable for a time which is super-polynomial in n, if one is content with being within distance $\varepsilon > 0$ from the distributions. (The distance is defined in this context as the total variation distance, see below.) We also prove that the convergence rate to these distributions, called *pseudo-mixing time*, is polynomial in n for values of β not too big when compared to the (inverse of the) maximum difference in potential of neighboring profiles. Note that when β is very high then logit dynamics is "close" to the best-response dynamics and therefore it is impossible to prove quick convergence results due to the aforementioned hardness results. We then give a picture which is as complete as possible relatively to the class of well-behaved potential games.

The proof of the above results consists of two main steps. We first devise a sufficient property for any n-player (not necessarily potential) game to have, for any starting profile, a distribution that is metastable for a super-polynomial number of steps and reached in polynomial time. The main idea behind this sufficient condition is that when the dynamics starts from a subset from which it is "hard to leave" and in which it is "easy to mix", then the dynamics will stay for a long time close to the stationary distribution restricted to that subset. Moreover, if a subset is "easy-to-leave," then the dynamics will quickly reach a "hard-to-leave" subset. The sufficient property consists of a rather technical definition that is intuitively a partition of the profiles into subsets that are asymptotically "hard-to-leave & easy-to-mix" or "easy-to-leave".

The second step amounts to showing that any asymptotically well-behaved potential game admits such a partition. The proof of this result builds on a number of involved technical contributions, some of which might be of independent interest. They mainly concern Markov chains. The concepts of interest are mixing time (how long the chain takes to mix), bottleneck ratio (intuitively, how hard it is for the stationary distribution to leave a subset of states), hitting time (how long the chain takes to hit a certain subset of states) and spectral properties of the transition matrix of Markov chains. We define a procedure which computes the required partition for these games. It iteratively identifies in the set of pure strategy profiles the "hard-to-leave" subsets. To prove that these subsets are "easy-to-mix", we firstly relate the pseudo-mixing time to the mixing time of a certain family of restricted Markov chains. We then prove that the mixing time of these chains is polynomial by using a spectral characterization of these restricted chains. The proof that the remaining profiles are "easy-to-leave" mainly relies on a connection between bottleneck ratio and hitting time.

We remark that, as a byproduct of our result, we essentially close an open problem of [4] about metastability of the Curie-Weiss game.

Related Works. Logit dynamics is defined in [5]. Early works about this dynamics have focused on its long-term behavior: [5] considered 2×2 coordination games and potential games, whereas a general characterization for wider classes of games is given in [1]. Several works gave bounds on the time the dynamics takes to reach specific Nash equilibria of a game for graphical coordination games on cliques and rings [8] and more general families of graphs [18,19]. The stationary distribution of the logit dynamics Markov chain is proposed as a new equilibrium concept in game theory in [3] and the convergence time to it studied in [2]. The logit response function has also been used for defining another equilibrium concept, known as *quantal response equilibrium* [16]. This differs from the logit equilibrium since it is a product distribution (like Nash equilibrium).

In physics, chemistry, and biology, metastability is a phenomenon related to the evolution of systems under noisy dynamics. In particular, metastability concerns moves between regions of the state spaces and the existence of multiple, well separated time scales: at short time scales, the system appears to be in a quasi-equilibrium, but really explores only a confined region of the available space state, while, at larger time scales, it undergoes transitions between such different regions. Research in physics about metastability aims at expressing typical features of a metastable state and to evaluate the transition time between metastable states. Several monographs on the subject are available in physics literature (see, e.g., [12]). The idea of metastability of probability distributions is introduced in [4] together with the concepts of metastable distribution and pseudo-mixing time for some specific potential games.

2   Preliminary Definitions

A *strategic game* $\mathcal{G}$ is a triple $([n], S_1, \ldots, S_n, \mathcal{U})$, where $[n] = \{1, \ldots, n\}$ is a finite set of players, $(S_1, \ldots, S_n)$ is a family of non-empty finite sets (S_i is the set of strategies available to player i), and $\mathcal{U} = (u_1, \ldots, u_n)$ is a family of utility functions (or payoffs), where $u_i \colon S \to \mathbb{R}$, $S = S_1 \times \ldots \times S_n$ being the set of all strategy profiles, is the utility function of player i. We focus on (exact) *potential games*, i.e., games for which there exists a function $\Phi \colon S \to \mathbb{R}$ such that for any pair of $\mathbf{x}, \mathbf{y} \in S$, $\mathbf{y} = (\mathbf{x}_{-i}, y_i)$, we have $\Phi(\mathbf{x}) - \Phi(\mathbf{y}) = u_i(\mathbf{y}) - u_i(\mathbf{x})$. Note that we use the standard game theoretic notation $(\mathbf{x}_{-i}, s)$ to mean the vector obtained from $\mathbf{x}$ by replacing the i-th entry with s; i.e. $(\mathbf{x}_{-i}, s) = (x_1, \ldots, x_{i-1}, s, x_{i+1}, \ldots, x_n)$. For two vectors $\mathbf{x}, \mathbf{y}$, we denote with $H(\mathbf{x}, \mathbf{y}) = |\{i \colon x_i \neq y_i\}|$ their Hamming distance. For two probability distributions μ and ν on the same state space S, the *total variation distance* $\|\mu - \nu\|_{\mathrm{TV}}$ is defined as $\|\mu - \nu\|_{\mathrm{TV}} = \max_{A \subseteq S} |\mu(A) - \nu(A)| = \frac{1}{2} \sum_{\mathbf{x} \in S} |\mu(\mathbf{x}) - \nu(\mathbf{x})|$, where $\mu(A) = \sum_{\mathbf{x} \in A} \mu(\mathbf{x})$ and $\nu(A) = \sum_{\mathbf{x} \in A} \nu(\mathbf{x})$.

Logit Dynamics. The logit dynamics has been introduced in [5] and runs as follows: at every time step (i) Select one player $i \in [n]$ uniformly at random; (ii) Update the strategy of player i according to the *Boltzmann distribution* with parameter β over the set S_i of her strategies. That is, a strategy $s_i \in S_i$ will be selected with probability $\sigma_i(s_i \mid \mathbf{x}_{-i}) = \frac{1}{Z_i(\mathbf{x}_{-i})} e^{\beta u_i(\mathbf{x}_{-i}, s_i)}$, where $\mathbf{x}_{-i}$ is the profile of strategies played at the current time step by players different from i, $Z_i(\mathbf{x}_{-i}) = \sum_{z_i \in S_i} e^{\beta u_i(\mathbf{x}_{-i}, z_i)}$ is the normalizing factor, and $\beta \geq 0$. One can see parameter β as the inverse of the noise or, equivalently, the *rationality level* of the system: indeed, it is easy to see that for $\beta = 0$ player i selects her strategy uniformly at random, for $\beta > 0$ the probability is biased toward strategies promising higher payoffs, and for β that goes to infinity player i chooses her best response strategy (if more than one best response is available, she chooses one of them uniformly at random). The above dynamics defines a *Markov chain* $\{X_t\}_{t \in \mathbb{N}}$ with the set of strategy profiles as state space, and where the transition probability from profile $\mathbf{x} = (x_1, \ldots, x_n)$ to profile $\mathbf{y} = (y_1, \ldots, y_n)$, denoted $P(\mathbf{x}, \mathbf{y})$, is zero if $H(\mathbf{x}, \mathbf{y}) \geq 2$ and it is $\frac{1}{n} \sigma_i(y_i \mid \mathbf{x}_{-i})$ if the two profiles differ exactly at player i. The Markov chain defined by the logit dynamics is ergodic [5]. Hence, from every initial profile $\mathbf{x}$ the distribution $P^t(\mathbf{x}, \cdot)$ over states of S of the chain X_t starting at $\mathbf{x}$ will eventually converge to a *stationary distribution* π as t tends to infinity. We denote with $\mathbf{P}_{\mathbf{x}}(\cdot)$ the probability on an event given that the logit dynamics starts from profile $\mathbf{x}$. The principal notion to measure the rate of convergence of a Markov chain to its stationary distribution is the *mixing time*. For $0 < \varepsilon < 1/2$, the mixing time of the logit dynamics is defined as $t_{\mathrm{mix}}(\varepsilon) = \min\{t \in \mathbb{N} \colon d(t) \leq \varepsilon\}$, where usually $\varepsilon = 1/4$ or $\varepsilon = 1/(2e)$ and $d(t) = \max_{\mathbf{x} \in S} \|P^t(\mathbf{x}, \cdot) - \pi\|_{\mathrm{TV}}$. Other important concepts related to Markov chains important for our study are the *relaxation time*, which is related to the spectra of the transition matrix P, *hitting time* and *bottleneck ratio*. The *hitting time* τ_L of $L \subseteq S$ is the first time a Markov chain is in a profile of L. For an ergodic Markov chain with finite state space S, transition matrix P, and stationary distribution π, we define the *bottleneck ratio* of a non-empty $L \subseteq S$ as $B(L) = \sum_{\mathbf{x} \in L, \mathbf{y} \in S \setminus L} \pi(\mathbf{x}) P(\mathbf{x}, \mathbf{y}) \cdot (\pi(L))^{-1}$. As in [3], we call the stationary distribution π of the Markov chain defined by the logit dynamics on a game $\mathcal{G}$, the *logit equilibrium* of $\mathcal{G}$. In general, a Markov chain with transition matrix P

and state space S is said to be *reversible* with respect to a distribution π if, for all $\mathbf{x}, \mathbf{y} \in S$, it holds that $\pi(\mathbf{x})P(\mathbf{x}, \mathbf{y}) = \pi(\mathbf{y})P(\mathbf{y}, \mathbf{x})$. If an ergodic chain is reversible with respect to π, then π is its stationary distribution. Therefore when this happens, to simplify our exposition we simply say that the matrix P is reversible. It is known [5] that the logit dynamics for the class of potential games is reversible and the stationary distribution is the *Gibbs measure* $\pi(\mathbf{x}) = \frac{1}{Z}e^{-\beta\Phi(\mathbf{x})}$, where $Z = \sum_{\mathbf{y} \in S} e^{-\beta\Phi(\mathbf{y})}$.

Metastability. We now give formal definitions of *metastable distributions* and *pseudo-mixing time*. For a more detailed description we refer the reader to [4].

Definition 1. *Let P be the transition matrix of a Markov chain with state space S. A probability distribution μ over S is $(\varepsilon, \mathcal{T})$-metastable for P, with $\varepsilon > 0$ and $\mathcal{T} \in \mathbb{N}$, if for every $0 \leq t \leq \mathcal{T}$ it holds that $\|\mu P^t - \mu\|_{\mathrm{TV}} \leq \varepsilon$. Moreover, let $L \subseteq S$ be a non-empty set of states. We define the* pseudo-mixing time $t_\mu^L(\varepsilon)$ *as* $t_\mu^L(\varepsilon) = \inf\{t \in \mathbb{N}\colon \|P^t(\mathbf{x}, \cdot) - \mu\|_{\mathrm{TV}} \leq \varepsilon \text{ for all } \mathbf{x} \in L\}$.

The definition of metastable distribution captures the idea of a distribution that behaves approximately like the stationary distribution: if we start from such a distribution and run the chain we stay close to it for a "long" time.

These notions apply to a single Markov chain. Anyway, [4] adopted them to evaluate the asymptotic behavior of the logit dynamics for parametrized classes of potential game, where the parameter is the number n of players. They consider a *sequence of n-player games* $\mathbf{G}$, one for each number n of players, and analyze the asymptotic properties that the logit dynamics enjoys when run on each $\mathcal{G}_k$, where $\mathcal{G}_k$ is the game in the sequence $\mathbf{G}$ with exactly k players. Thus, we do not have a single Markov chain but a sequence of them, one for each number n of players, and need to consider an asymptotic counterpart of the notions above. [4], in fact, showed that the logit dynamics for specific classes of n-player potential games enjoys the following property, that we name *asymptotic metastability*.

Definition 2. *Let $\mathbf{G}$ be a sequence of n-player strategic games. We say that the logit dynamics for $\mathbf{G}$ is asymptotically metastable for the rationality level β if for any $\varepsilon > 0$ there is a polynomial $p = p_\varepsilon$ and a super-polynomial $q = q_\varepsilon$ such that for each n, the logit dynamics with rationality level β for the game $\mathcal{G}_n$ converges in time at most $p(n)$ from each profile of $\mathcal{G}_n$ to a $(\varepsilon, q(n))$-metastable distribution.*

When the logit dynamics for a game is (not) asymptotically metastable, we say for brevity that the game itself is (not) asymptotically metastable. Unfortunately, asymptotic metastability cannot be proved for every (potential) game.

Lemma 1. *There is a sequence of n-player (potential) games $\mathbf{G}$ which is not asymptotically metastable for any β sufficiently high and any $\varepsilon < \frac{1}{4}$.*

Proof (Idea). Consider pairs (p_j, q_j), where p_j is a polynomial asymptotically greater than p_{j-1} and q_j is a super-polynomial asymptotically smaller than q_{j-1}. Let $\mathcal{T}$ be a function sandwiched for any j between p_j and q_j. Let $\mathbf{G}$ be a sequence of n-player games such that each strategic game $\mathcal{G}_n$ in the sequence assigns to each player exactly two strategies, 0 and 1, and has potential function Φ defined as follows: For $t \leq n-1$ and profile $\mathbf{x}$ wherein exactly t players play strategy 1 we have $\Phi(\mathbf{x}) = n - t$, whereas $\Phi(\mathbf{1}) = 1 + \beta^{-1} \log(\mathcal{T}(n)/\varepsilon - 1)$, with $\mathbf{1} = (1, \ldots, 1)$. Intuitively, (p_j, q_j) will describe the behavior of the game with n_j players and will not be precise enough for the game with more than n_j players. Indeed, we can show that for any $\varepsilon < 1/4$, infinitely many values of n, each polynomial p and each super-polynomial q, the logit dynamics for $\mathcal{G}_n$ does not converge in time $O(p(n))$ from $\mathbf{1}$ to any $(\varepsilon, q(n))$-metastable distribution. $\qquad\square$

3   Asymptotic Metastability and Partitioned Games

Motivated by the result above, we next give a sufficient property for *any* (not necessarily potential) game to be asymptotically metastable. We will introduce the concept of game *partitioned* by the logit dynamics and give examples of games satisfying this notion. Then, we prove that games partitioned by the logit dynamics are asymptotically metastable. Note that we focus only on games and values of β such that the mixing time of the logit dynamics is at least super-polynomial in n, otherwise the stationary distribution enjoys the desired properties of stability and convergence. Throughout the rest of the paper we denote with β_0 the smallest value of β such that the mixing time is not polynomial.

Let $\mathbf{G}$ be a sequence of n-player games. Let P be the transition matrix of the logit dynamics on $\mathcal{G}_n$, for some $n > 0$, and let π be the corresponding stationary distribution. For $L \subseteq S$ non-empty, we define a Markov chain with state space L and transition matrix $\mathring{P}_L$ defined as follows: $\mathring{P}_L(\mathbf{x}, \mathbf{y}) = P(\mathbf{x}, \mathbf{y})$ if $\mathbf{x} \neq \mathbf{y}$, and $\mathring{P}_L(\mathbf{x}, \mathbf{y}) = 1 - \sum_{\substack{\mathbf{z} \in L \\ \mathbf{z} \neq \mathbf{x}}} P(\mathbf{x}, \mathbf{z}) = P(\mathbf{x}, \mathbf{x}) + \sum_{\mathbf{z} \in S \setminus L} P(\mathbf{x}, \mathbf{z})$, otherwise. It easy to check that the stationary distribution of this Markov chain is given by the distribution $\pi_L(\mathbf{x}) = \frac{\pi(\mathbf{x})}{\pi(L)}$, for every $\mathbf{x} \in L$. Note also that the Markov chain defined upon $\mathring{P}_L$ is aperiodic, since the Markov chain defined upon P is, and it will be irreducible if L is a connected set. For a fixed $\varepsilon > 0$, we denote with $t_{\mathrm{mix}}^L(\varepsilon)$ the mixing time of the chain described above. We also denote with $B_L(A)$ the bottleneck ratio of $A \subset L$ in the Markov chain with state space L and transition matrix $\mathring{P}_L$. We now introduce the definition of partitioned games.

Definition 3. *Let $\mathbf{G}$ be a sequence of n-player strategic games. $\mathbf{G}$ is* partitioned *by the logit dynamics for the rationality level β if for any $\varepsilon > 0$ there is a polynomial $p = p_\varepsilon$ and a super-polynomial $q = q_\varepsilon$ s. t. for any n there is a family of connected subsets $R_1, \ldots, R_k$ of the set S of profiles of $\mathcal{G}_n$, with $k \geq 1$, and a partition $T_1, \ldots, T_k, N$ of S, with $T_i \subseteq R_i$ for any $i = 1, \ldots, k$, such that*

1. *the bottleneck ratio of R_i is at most $1/q(n)$, for any $i = 1, \ldots, k$;*
2. *the mixing time $t_{\mathrm{mix}}^{R_i}(\varepsilon)$ is at most $p(n)$, for any $i = 1, \ldots, k$;*
3. *for any $i = 1, \ldots, k$ and for any $\mathbf{x} \in T_i$, it holds that $\mathbf{P}_\mathbf{x}\left(\tau_{S \setminus R_i} \leq t_{\mathrm{mix}}^{R_i}(\varepsilon)\right) \leq \varepsilon$;*
4. *for any $\mathbf{x} \in N$, it holds that $\mathbf{P}_\mathbf{x}\left(\tau_{\bigcup_i T_i} \leq p(n)\right) \geq 1 - \varepsilon$.*

Note that we allow in the above definition that T_i, for some $i = 1, \ldots, k$, or N are empty. Linking back to the intuition discussed in the introduction, $R_1, \ldots, R_k$ represent the "easy-to-mix" subsets of states (condition (2)); these sets play a crucial role in defining distributions that are metastable for very long time (condition (1)). However, when the logit dynamics starts close to the boundary of some R_i, it is likely to leave R_i quickly. Since we are interested in "easy-to-mix & hard-to-leave" subsets of profiles, for each R_i we identify its *core* T_i as the set of profiles from which the logit dynamics takes long time to leave R_i (condition (3)). The distinction between core and non-core profiles will help in proving that metastable distributions are quickly reached from any starting profile.

The main result of this section proves that a game partitioned by the logit dynamics is asymptotically metastable.

Theorem 1. *If a sequence of games* **G** *is partitioned by the logit dynamics for* β, *then the logit dynamics for* **G** *is asymptotically metastable for* β.

We remark that Definition 3 and Theorem 1 do not use any specific property of the game, and then can be extended to consider Markov chains in general.

The details of the proof are deferred to the full version [11]. Here we give examples of an actual game satisfying it (more examples can be found in [11]) as well as comments on a game not enjoying it. Consider, indeed, the following game-theoretic formulation of the *Curie-Weiss model* (the *Ising model* on the complete graph), that we call CW-*game*: each one of n players has two strategies, -1 and $+1$, and the utility of player i for profile $\mathbf{x} = (x_1, \ldots, x_n) \in \{-1, +1\}^n$ is $u_i(\mathbf{x}) = x_i \sum_{j \neq i} x_j$. Observe that for every player i it holds that $u_i(\mathbf{x}_{-i}, +1) - u_i(\mathbf{x}_{-i}, -1) = \mathcal{H}(\mathbf{x}_{-i}, -1) - \mathcal{H}(\mathbf{x}_{-i}, +1)$, where $\mathcal{H}(\mathbf{x}) = -\sum_{j \neq k} x_j x_k$, hence the CW-game is a potential game with potential function $\mathcal{H}$. It is known (see Chap. 15 in [14]) that the logit dynamics for this game (or equivalently the *Glauber dynamics* for the Curie-Weiss model) has mixing time polynomial in n for $\beta < 1/n$ and super-polynomial as long as $\beta > 1/n$. Moreover, [4] describes metastable distributions for $\beta > c \log n/n$ and shows that such distributions are quickly reached from profiles where the number of $+1$ (respectively -1) is a sufficiently large majority, namely if the magnetization k is such that $k^2 > c \log n/\beta$, where the *magnetization* of a profile $\mathbf{x}$ is defined as $M(\mathbf{x}) = \sum_i x_i$.

It has been left open what happens when β lies in the interval $(1/n, c \log n/n)$ and if a metastable distribution is quickly reached when in the starting point the number of $+1$ is close to the number of -1. We observe that next lemma, along with Theorem 1 essentially closes this problem by showing that CW-games are asymptotic metastable for $\beta \geq c/n$ for some constant $c > 1$.

Lemma 2. *Let* **G** *be a sequence of* n-player CW-*games. Then* **G** *is partitioned by the logit dynamics for any* $\beta > c/n$, *for constant* $c > 1$.

Proof (Idea). Fix n and let S_+ (resp., S_-) be the set of profiles with positive (resp., negative) magnetization in $\mathcal{G}_n$. Let us set $R_1 = S_+$ and $R_2 = S_-$. It is known that the bottleneck ratio of these subset is super-polynomial for any $\beta > c/n$, for constant $c > 1$, (see, e.g., Chap. 15 in [14]). Moreover, in [13] it has been proved that the mixing time of the chain restricted to S_+ (resp. S_-) is actually $c_1 n \log n$ for some constant $c_1 > 0$. Moreover, from [6,7] it follows that there are subsets of R_1 (resp., R_2) from which the dynamics hits a profile $\mathbf{y} \in S_-$ (resp., S_+) in a time equivalent to the mixing time of the chain restricted to S_+ with probability at most ε. Thus, these subsets correspond to T_1 and T_2. Finally, observe that from [6], we have that for each profile $\mathbf{x} \in N$, the hitting time of $T_1 \cup T_2$ is polynomial with high probability.   □

Consider now the game of Lemma 1. We can then prove the following lemma.

Lemma 3. *Let* **G** *be the sequence of games defined in Lemma 1. Then, for any β sufficiently large,* **G** *is not partitioned by the logit dynamics.*

Proof. Fix n. The bottleneck of $\mathbf{1} = (1, \ldots, 1)$ is asymptotically larger than any polynomial and smaller than any super-polynomial. Hence, this profile cannot be contained in N and, for all i, it must be that $R_i = L$ for some $L \subseteq S$, with $1 \in L$, and $L \neq \{1\}$. Then, for β sufficiently large $\pi_L(\mathbf{1}) \leq 1/2$. But, since the bottleneck of $\mathbf{1}$ is asymptotically larger than any polynomial, the mixing time of the chain restricted to L is not polynomial.   □

4   Asymptotically Well-Behaved Potential Games

We now ask what class of potential games are partitioned by the logit dynamics. We know already that the answer must differ from the whole class of potential games, due to Lemma 1. However, it is important to understand to what extent it is possible to prove asymptotic metastability for potential games.

Our main aim is to give results, asymptotic in the number n of players, about the behavior of logit dynamics run on potential games. Clearly, it makes sense to give asymptotic results about the property of an object, only if this property is asymptotically well-defined, that is, the object is uniquely defined for infinitely many values of the parameter according to which we compute the asymptotic and the property of this object does not depend "chaotically" on this parameter. For example when we say that a graph has large expansion, we actually mean that there is a sequence of graphs indexed by the number of vertices, such that the expansion of each graph can be bounded by a single function of this number. Similarly, when we say that a Markov Chain has large mixing time, we actually mean that there is a sequence of Markov chains indexed by the number of states, such that the mixing time of each Markov chain can be bounded by a single function of this number. Yet another example arises in algorithm game theory: when we say that the Price of Anarchy of a game is large, we actually mean that there is a sequence of games indexed, for example, by the number of players such that the Price of Anarchy of each game can be bounded by a single function of this number. In this work the object of interest is a potential game and the property of interest is the behavior of the logit dynamics for this game. And thus, in our setting, it makes sense to give asymptotic results only when a potential game is uniquely defined for infinitely many n and the behavior of the logit dynamics for the potential game is not chaotic as n increases. However, giving a formal definition of what this means is not as immediate as in the case of the expansion of a graph or of the Price of Anarchy of a game. Thus, in order to gain insight on how to formalize this concept, let us look at an example of a game for which the behavior of the logit dynamics is evidently "almost the same" as n increases (these include the examples of partitioned games analyzed above) and examples in which this behavior instead changes infinitely often.

The behavior of logit dynamics for the CW-game can be described in a way that is immaterial from the actual value of n. Indeed, the potential function has two equivalent minima when either all players adopt strategy -1 or all players adopt strategy $+1$ and it increases as the number of players adopting a strategy different from the one played by the majority of agents increases. The potential reaches its maximum when each strategy is adopted by the same number of players. Moreover, regardless of the actual value of n it is easy to see that if the number of $+1$ strategies is sufficiently larger than the number of -1 strategies[1], then it must be hard for the logit dynamics to reach a profile with more -1's than $+1$'s, whereas it must be easy for the dynamics to converge to the potential minimizer in which all players are playing $+1$. Thus the logit dynamics for the CW-game is asymptotically well-behaved for our purposes. Indeed, the evolution of the dynamics when the number of players is n can be mapped into the evolution of the dynamics with more or less players, so that the time necessary to some events to happen (e.g., for reaching or leaving certain sets of profiles) can be bounded by the same function of the number of players.

A similar argument holds even for a number of other games considered in literature (i.e., pure coordination games, graphical coordination game on the ring, Pigou's congestion game) [11]. Moreover, it is not too hard to see that the finite opinion games on well-defined classes of graphs (cliques, rings, complete bipartite graphs, etc.) studied in [10] are also asymptotically well behaved. Thus it can be seen how our set of examples covers much of the spectrum of games considered in the logit dynamics literature.

A Game for Which the Logit Dynamics Chaotically Depends on n. Observe that, even though the potential function of the game in Lemma 1 can be easily described solely as a function of n, just as done above, we cannot describe how the logit dynamics for this game and a given rationality parameter β behaves as n changes. In particular, we are forced to describe the time that is necessary for the logit dynamics to leave the profile in which all players are adopting strategy 1 by enumerating infinitely many cases and not through a single function of n (thus preventing us to give asymptotic results in n). Indeed, this time, by construction, changes infinitely often and, for any tentative bound, there will always be a value of n from which that bound will turn out to be incorrect.

Asymptotically Well-Behaved Games: the Definition. From the analysis of these games, it is evident that the behavior of the logit dynamics for potential games is

---

[1] The extent to which the number of $+1$ must be larger than the number of -1 can depend on n, but it can be bounded by a single function F on the number of players.

asymptotically well-defined when profiles of the n-player game can be associated to profiles of the n'-player game such that the probability of leaving associated profiles can be always bounded by the same function of the number of players. Formally, we have the following definition.

Definition 4. *The logit dynamics for a sequence of n-player (potential) games is asymptotically well-behaved if there is n_0 and a small constant $0 < \lambda < 1$ such that for every $n \geq n_0$ and for every $L' \subseteq S_n$, where S_n is the set of profiles in G_n, there is a subset $L \subseteq S_{n_0}$ and a function F_L such that $B(L) = F_L(n_0)$ and $B(L') \in [F_L(n)(1 - \lambda), F_L(n)(1 + \lambda)]$.*

For sake of compactness, we will simply say that the potential game is asymptotically well-behaved whenever the logit dynamics (run on it) is.

The main result of our paper follows.

Theorem 2. *Let $\mathbf{G}$ be an asymptotically well-behaved sequence of n-player potential games. Fix $\Delta(n) := \max\{\Phi_n(\mathbf{x}) - \Phi_n(\mathbf{y}) \colon H(\mathbf{x}, \mathbf{y}) = 1\}$, where Φ_n is the potential function of G_n. Then, for any function ρ at most polynomial, $\mathbf{G}$ is asymptotically metastable for $\beta_0 \leq \beta \leq \frac{\rho(n)}{\Delta(n)}$.*

The dependence on $\Delta(n)$ is a by-product of the fact that the logit dynamics is not invariant to scaling of the utility function, i.e., scaling the utility function of a certain factor requires to inversely scale β to get the same logit dynamics (see [3] for a discussion). In a sense, $\beta\Delta(n)$ is the natural parameter that describes the logit dynamics. Then, according to this point of view, the requirement on β in the above theorem becomes almost natural: we, indeed, require that $\beta\Delta(n)$ is sufficiently large in order for the mixing time being not polynomial, but we also require that $\beta\Delta(n)$ is a polynomial. This assumption on $\beta\Delta(n)$ is in general necessary because when it is high enough logit dynamics roughly behaves as best-response dynamics. Moreover, in this case, the only metastable distributions have to be concentrated around the set of Nash equilibria. This is because for $\beta\Delta(n)$ very high, it is extremely unlikely that a player leaves a Nash equilibrium. Then, the hardness results about the convergence of best-response dynamics for potential games, cf. e.g. [9], imply that the convergence to metastable distributions for high $\beta\Delta(n)$ is similarly computationally hard.

The proof builds upon Theorem 1 and proves that any asymptotic well-behaved potential game is partitioned by the logit dynamics for β not too large.

5   Conclusions and Open Problems

In this work we prove that for any asymptotically well-behaved potential game and any starting point of this game there is a distribution that is metastable for super-polynomial time and it is quickly reached. In the proof we also give a sufficient condition for a game to enjoy this metastable behavior. It is a very interesting open problem to prove that this property is also *necessary*. The main obstacle for this direction consists of the fact that we do not know any tool for proving or disproving metastability of distributions that are largely different from the ones considered in this work. Also, given that our arguments are game-independent, it would be interesting to see whether sufficient and necessary conditions can be refined for specific subclass of games.

Acknowledgments. This work is supported by PRIN 2010-2011 project ARS TechnoMedia and EPSRC grant EP/M018113/1. Authors wish to thank Paul W. Goldberg for many invaluable discussions related to a number of results discussed in this paper, and an anonymous reviewer for the enlightening comments on an earlier version of this work.

References

1. Alós-Ferrer, C., Netzer, N.: The logit-response dynamics. Games Econ. Behav. **68**(2), 413–427 (2010)
2. Auletta, V., Ferraioli, D., Pasquale, F., Penna, P., Persiano, G.: Convergence to equilibrium of logit dynamics for strategic games. In: SPAA, pp. 197–206 (2011)
3. Auletta, V., Ferraioli, D., Pasquale, F., Persiano, G.: Mixing time and stationary expected social welfare of logit dynamics. In: Kontogiannis, S., Koutsoupias, E., Spirakis, P.G. (eds.) SAGT 2010. LNCS, vol. 6386, pp. 54–65. Springer, Heidelberg (2010)
4. Auletta, V., Ferraioli, D., Pasquale, F., Persiano, G.: Metastability of logit dynamics for coordination games. In: SODA, pp. 1006–1024 (2012)
5. Blume, L.E.: The statistical mechanics of strategic interaction. Games Econ. Behav. **5**, 387–424 (1993)
6. Ding, J., Lubetzky, E., Peres, Y.: Censored glauber dynamics for the mean field ising model. Stat. Phys. **137**(3), 407–458 (2009)
7. Ding, J., Lubetzky, E., Peres, Y.: The mixing time evolution of glauber dynamics for the mean-field ising model. Commun. Math. Phys. **289**(2), 725–764 (2009)
8. Ellison, G.: Learning, local interaction, and coordination. Econometrica **61**(5), 1047–1071 (1993)
9. Fabrikant, A., Papadimitriou, C.H., Talwar, K.: The complexity of pure nash equilibria. In: STOC, pp. 604–612 (2004)
10. Ferraioli, D., Goldberg, P.W., Ventre, C.: Decentralized dynamics for finite opinion games. In: Serna, M. (ed.) SAGT 2012. LNCS, vol. 7615, pp. 144–155. Springer, Heidelberg (2012)
11. Ferraioli, D., Ventre, C.: Metastability of potential games. CoRR, abs/1211.2696 (2012)
12. Hollander, F.: Metastability under stochastic dynamics. Stoch. Process. Appl. **114**(1), 1–26 (2004)
13. Levin, D., Luczak, M., Peres, Y.: Glauber dynamics for the mean-field ising model: cut-off, critical power law, and metastability. Probab. Theor. Relat. Fields **146**, 223–265 (2010)
14. Levin, D., Yuval, P., Wilmer, E.L.: Markov Chains and Mixing Times. American Mathematical Society, Providence (2008)
15. Lipton, R.J., Markakis, E., Mehta, A.: Playing large games using simple strategies. In: ACM EC, pp. 36–41 (2003)
16. McKelvey, R.D., Palfrey, T.R.: Quantal response equilibria for normal form games. Games Econ. Behav. **10**(1), 6–38 (1995)

17. Monderer, D., Shapley, L.S.: Potential games. Games Econ. Behav. **14**(1), 124–143 (1996)
18. Montanari, A., Saberi, A.: Convergence to equilibrium in local interaction games. In: FOCS (2009)
19. Young, H.P.: The economy as a complex evolving system. In: The Diffusion of Innovations in Social Networks, vol. III (2003)
20. Schiller, R.J.: Irrational Exuberance. Wiley, New York (2000)

The Shifted Partial Derivative Complexity of Elementary Symmetric Polynomials

Hervé Fournier[1]([✉]), Nutan Limaye[2], Meena Mahajan[3],
and Srikanth Srinivasan[4]

[1] IMJ-PRG, Univ Paris Diderot, Paris, France
`fournier@math.univ-paris-diderot.fr`
[2] Department of Computer Science and Engineering, IIT Bombay, Mumbai, India
`nutan@cse.iitb.ac.in`
[3] The Institute of Mathematical Sciences, Chennai, India
`meena@imsc.res.in`
[4] Department of Mathematics, IIT Bombay, Mumbai, India
`srikanth@math.iitb.ac.in`

Abstract. We continue the study of the *shifted partial derivative measure*, introduced by Kayal (ECCC 2012), which has been used to prove many strong depth-4 circuit lower bounds starting from the work of Kayal, and that of Gupta et al. (CCC 2013).

We show a strong lower bound on the dimension of the shifted partial derivative space of the Elementary Symmetric Polynomials of degree d in N variables for $d < \log N/\log \log N$. This extends the work of Nisan and Wigderson (Computational Complexity 1997), who studied the *partial derivative space* of these polynomials. Prior to our work, there have been no results on the shifted partial derivative measure of these polynomials. Our result implies a strong lower bound for Elementary Symmetric Polynomials in the homogeneous $\Sigma\Pi\Sigma\Pi$ model with bounded bottom fan-in. This strengthens (under our degree assumptions) a lower bound of Nisan and Wigderson who proved the analogous result for homogeneous $\Sigma\Pi\Sigma$ model (i.e. $\Sigma\Pi\Sigma\Pi$ formulas with bottom fan-in 1).

Our main technical lemma gives a lower bound for the ranks of certain inclusion-like matrices, and may be of independent interest.

1 Introduction

Motivation. In an influential paper of Valiant [23] the two complexity classes VP and VNP were defined, which can be thought of as algebraic analogues of Boolean complexity classes P and NP, respectively. Whether VP equals VNP or not is one of the most fundamental problems in the study of algebraic computation. It follows from the work of Valiant [23] that a super-polynomial lower bound for arithmetic circuits computing the Permanent implies VP $\neq$ VNP.

This research was supported by IFCPAR/CEFIPRA Project No 4702-1(A) and research grant compA ANR-13-BS02-0001-01.

G.F. Italiano et al. (Eds.): MFCS 2015, Part II, LNCS 9235, pp. 324–335, 2015.
DOI: 10.1007/978-3-662-48054-0_27

The best known lower bound on uniform polynomials for general arithmetic circuits is $\Omega(N \lg N)$ [3] which is unfortunately quite far from the desired super-polynomial lower bound. Over the years, though there has been no stronger lower bound for general arithmetic circuits, many super-polynomial lower bounds have been obtained for special classes for arithmetic circuits [15–17].

A very interesting such subclass of arithmetic circuits is the class of *bounded-depth* arithmetic formulas[1]. The question of proving lower bounds for bounded-depth formulas and in particular depth 3 and 4 formulas has received a lot of attention subsequent to the recent progress in efficient depth reduction of arithmetic circuits [1,11,22,24]. This sequence of results essentially implies that "strong enough" lower bounds for depth-4 homogeneous formulas suffice to separate VP from VNP. More formally, it proves that any sequence $\{f_N\}_N$ of homogeneous N-variate degree $d = N^{O(1)}$ polynomials in VP has depth-4 homogeneous formulas of size $N^{O(\sqrt{d})}$. Hence, proving an $N^{\omega(\sqrt{d})}$ lower bound for depth-4 homogeneous formulas suffices to separate VP from VNP.

Even more can be said about the depth-4 formulas obtained in the above results. For any integer parameter $t \leq d$, they give a $\Sigma\Pi\Sigma\Pi$ formula for f_N where the layer 1 product gates (just above the inputs) have fan-in at most t and the layer 3 gates are again Π gates with fan-in $O(d/t)$. We will refer to such formulas as $\Sigma\Pi^{[O(d/t)]}\Sigma\Pi^{[t]}$ formulas. The depth-reduction results mentioned above produce a depth-4 homogeneous $\Sigma\Pi^{[O(d/t)]}\Sigma\Pi^{[t]}$ formula of size $N^{O((d/t)+t)}$ and top fan-in $N^{O(d/t)}$; at $t = \lceil\sqrt{d}\rceil$, we get the above depth-reduction result.

The tightness of these results follows from recent progress on lower bounds for the model of $\Sigma\Pi^{[O(d/t)]}\Sigma\Pi^{[t]}$ circuits. A flurry of results followed the groundbreaking work of Kayal [7], who augmented the *partial derivative method* of Nisan and Wigderson [15] to devise a new complexity measure called *the shifted partial derivative measure*, using which he proved an exponential lower bound for a special class of depth-4 circuits. Building on this, the first non-trivial lower bound for $\Sigma\Pi^{[O(d/t)]}\Sigma\Pi^{[t]}$ formulas was proved by Gupta, Kamath, Kayal, and Saptharishi [5] for the determinant and permanent polynomials. This was further improved by Kayal, Saha, and Saptharishi [9] who gave a family of explicit polynomials in VNP the shifted partial derivative complexity of which was (nearly) *as large as possible*[2] and hence showed a lower bound of $N^{\Omega(d/t)}$ for the top fan-in of $\Sigma\Pi^{[O(d/t)]}\Sigma\Pi^{[t]}$ formulas computing these polynomials. Later, a similar result for a polynomial in VP was proved in [4] and this was subsequently strengthened by Kumar and Saraf [12], who gave a polynomial computable by homogeneous $\Pi\Sigma\Pi$ formulas which have no $\Sigma\Pi^{[O(d/t)]}\Sigma\Pi^{[t]}$ formulas of top fan-in smaller than $N^{\Omega(d/t)}$. Finally, using a variant of the shifted partial derivative measure, Kayal et al. [8] and Kumar and Saraf [13] were able to prove similar lower bounds for general depth-4 homogeneous formulas as well.

In this work, we investigate the shifted partial derivative measure of the *Elementary Symmetric Polynomials*, which is a very natural family of polynomials whose complexity has been the focus of many previous works [6,15,20,21].

---

[1] For $O(1)$ depth, the sizes of formulas and circuits are polynomially related.

[2] i.e., as large as it can be for a "generic" or "random" polynomial. (See Remark 3).

Nisan and Wigderson [15] proved tight lower bounds on the depth-3 homogeneous formula complexity of these polynomials. Shpilka and Wigderson [21] and Shpilka [20] studied the general (i.e. possibly inhomogeneous) depth-3 circuit complexity of these polynomials, and showed that for certain degrees, the $O(N^2)$ upper bound due to Ben-Or (see [21]) is tight.

Under some degree constraints, we show strong lower bounds on the dimension of the shifted partial derivative space of these polynomials, which implies that the Elementary symmetric polynomial on N variables of degree d cannot be computed by a $\Sigma\Pi^{[O(d/t)]}\Sigma\Pi^{[t]}$ circuit of top fan-in less than $N^{\Omega(d/t)}$. This strengthens the result of Nisan and Wigderson [15] for these degree parameters.

By the upper bound of Ben-Or mentioned above, this also gives the *first* example of an explicit polynomial with small $\Sigma\Pi\Sigma$ circuits for which such a strong lower bound is known.

Results. We show that, for a suitable range of parameters, the shifted partial derivative measure of the N-variate elementary symmetric polynomial of degree d —denoted S_N^d — is large.

Theorem 1. *Let $N, d, k \in \mathbb{N}$ be such that $d \leq (1/10)\lg N / \lg\lg N$ and $k = \lfloor d/(\tau+1) \rfloor$ for some $\tau \in \mathbb{N}$ satisfying $\tau \equiv 1 \pmod 4$. For $\ell = \lfloor N^{1-1/2\tau} \rfloor$,*
$$\dim\langle \partial_k S_N^d \rangle_{\leq \ell} \geq \frac{(1-o(1))\cdot\binom{N+\ell}{\ell}\cdot\binom{N-\ell}{k}}{(3\sqrt{N}/2)^k \cdot (2d)^\tau}.$$

It was observed by [5] that for *any* homogeneous multilinear polynomial f on N variables of degree d, we have $\dim\langle \partial_k S_N^d \rangle_{\leq \ell} \leq \binom{N+\ell}{\ell} \cdot \binom{N}{k}$, which is close to the numerator in the above expression. The theorem above should be interpreted as saying that the dimension is not too far from this upper bound.

A corollary of our main result is an $N^{\Omega(d/t)}$ lower bound on the top fan-in of any $\Sigma\Pi^{[O(d/t)]}\Sigma\Pi^{[t]}$ formula computing S_N^d.

Theorem 2. *Fix $N, d, D, t \in \mathbb{N}$ and constant $\varepsilon > 0$ s.t. $d \leq \frac{\lg N}{10\lg\lg N}$, $D \leq N^{1-\varepsilon}$. Any $\Sigma\Pi^{[D]}\Sigma\Pi^{[t]}$ circuit of top fan-in s computing S_N^d satisfies $s = N^{\Omega(d/t)}$.*

It is worth noting that in most lower bounds of this flavour, the upper product gates have fanin D bounded by $O(d/t)$. Our lower bound works for potentially much larger values of D.

Corollary 1. *Let parameters N, d, t be as in Theorem 2. Any $\Sigma\Pi^{[O(d/t)]}\Sigma\Pi^{[t]}$ computing S_N^d must have top fan-in at least $N^{\Omega(d/t)}$. In particular[3], any homogeneous $\Sigma\Pi\Sigma\Pi$ circuit C with bottom fan-in bounded by t computing S_N^d must have top fan-in at least $N^{\Omega(d/t)}$.*

By the above depth reduction results, this lower bound is tight up to the constant factor in the exponent. Before our work, [15] proved a lower bound for S_N^d of $N^{\Omega(d)}$ for all d, however with respect to $\Sigma\Pi\Sigma$ circuits (i.e. the case $t = 1$).

---

[3] It is known (see, e.g., [19, Proof of Corollary 5.8]) that any homogeneous $\Sigma\Pi\Sigma\Pi^{[t]}$ circuit C can be converted to a $\Sigma\Pi^{[O(d/t)]}\Sigma\Pi^{[t]}$ circuit with the same top fan-in.

Techniques. The analysis of the shifted partial derivative measure for any polynomial essentially requires the analysis of the rank of a matrix arising from the shifted partial derivative space. In this work, we analyse the matrix arising from the shifted partial derivative space of the symmetric polynomials. Our analysis is quite different from previous works (such as [4,8,13]), which are based on either monomial counting (meaning that we find a large identity or upper triangular submatrix inside our matrix) or an analytic inequality of Alon [2].

In our analysis of the shifted partial derivative space, we define a more complicated version of the *Inclusion matrix* (known to be full rank) and lower bound its rank by using a novel technique, which we describe in the next section.

Disjointness and inclusion matrices arises naturally in other branches of theoretical computer science such as Boolean circuit complexity [18], communication complexity [14, Chapter 2] and also in combinatorics [10,25]. Therefore, we believe that our analysis of the Inclusion-like matrix arising from the symmetric polynomial may find other applications.

Organisation of the Paper. In Sect. 2, we set up basic notation, fix the main parameters, and give a high-level outline of our proof of Theorem 1. In Sect. 3 we give the actual proof. The formula size lower bounds from Theorem 2 is established in Sect. 4. Due to space constraints, several proofs are omitted.

2   Proving Theorem 1: High-Level Outline

Notation: For a positive integer n, we let $[n] = \{1, \ldots, n\}$. Let $X = \{x_1, \ldots, x_N\}$. For $A \subseteq [N]$ we define $X_A = \prod_{i \in A} x_i$. The elementary symmetric polynomial of degree d over the set of variables X is defined as $S_N^d(X) = \sum_{A \subseteq [N], |A| = d} X_A$. In the following $S_N^d(X)$ is abbreviated with S_N^d.

For $k, \ell \in \mathbb{N}$ and a multivariate polynomial $f \in \mathbb{F}[x_1, \ldots, x_n]$, we define

$$\langle \partial_k f \rangle_{\leq \ell} = \text{span} \left\{ x_1^{j_1} \ldots x_n^{j_n} \cdot \frac{\partial^k f}{\partial x_1^{i_1} \ldots \partial x_n^{i_n}} \;\middle|\; i_1 + \ldots + i_n = k, \; j_1 + \ldots + j_n \leq \ell \right\}.$$

Our complexity measure is the dimension of this space, i.e., $\dim(\langle \partial_k f \rangle_{\leq \ell})$ [5,7].

For a monomial $m = \prod_{i=1}^N x_i^{n_i}$, $\deg(m) = n_1 + n_2 + \ldots + n_N$ is the total degree of m. We denote by $\deg_{x_i}(m)$ the degree of the variable x_i in m (here $\deg_{x_i}(m) = n_i$). We define the support of m as $\text{supp}(m) = \{i \in [N] \mid n_i > 0\}$. For a monomial m and $p > 0$, let $\text{supp}_p(m) = \{i \in [N], \deg_{x_i}(m) = p\}$.

Let $\mathcal{M}_N^\ell$ the set of monomials of degree at most ℓ over the variables X. For integers $n_1, \ldots, n_p$, let

$$\mathcal{M}_N^\ell(n_1, \ldots, n_p) = \{m \in \mathcal{M}_N^\ell, \; |\text{supp}_i(m)| = n_i \text{ for } i \in [p]\}.$$

Given $p > 0$, a monomial $m \in \mathcal{M}_N^\ell$ can be uniquely written as $m = \tilde{m} \cdot \prod_{i=1}^p (X_{\text{supp}_i(m)})^i$. We write $m \equiv [\tilde{m}, S_1, \ldots, S_p]$ if $S_i = \text{supp}_i(m)$ for all $i \in [p]$ and $m = \tilde{m} \cdot \prod_{i=1}^p (X_{S_i})^i$.

For a finite set S, let $\mathcal{U}(S)$ denote the uniform distribution over the set S.

We assume that we are working over a field $\mathbb{F}$ of characteristic zero. Our results also hold in non-zero characteristic, but the first step of our proof (Lemma 1) becomes a little more cumbersome (this part is omitted in this version).

Proof Outline. Our lower bound on $\dim(\langle \partial_k S_N^d \rangle_{\leq \ell})$ proceeds in 3 steps.

Step 1: We choose a suitable subset $\mathcal{S}$ of the partial derivative space. It is convenient to work with a set that is slightly different from the set of partial derivatives themselves. To understand the advantage of this, consider the simple setting where we are looking at the partial derivatives of the degree-2 polynomial S_N^2 of order 1. It is not difficult to show that the partial derivative with respect to variable x_i is $r_i := \sum_{j \neq i} x_j$. Over characteristic zero, this set of polynomials is known to be linearly independent. One way to show this is by showing that each polynomial x_i can be written as a linear combination of the r_js; explicitly, one can write $x_i = \frac{1}{n-1} \left(\sum_{j \in [n]} r_j \right) - r_i$. Since the x_is are distinct monomials, they are clearly linearly independent and we are done. This illustrates the advantage in moving to a "sparser" basis for the partial derivative space. We do something like this for larger d and k (Lemma 1).

Step 2: After choosing the set $\mathcal{S}$, we construct the set $\mathcal{P}$ of shifts of $\mathcal{S}$ (actually, we will only consider a subset of $\mathcal{P}$) and lower bound the rank of the corresponding matrix M. To do this, we also prune the set of rows of the matrix M. In other words, we consider a carefully chosen set of monomials $\mathcal{M}$ and project each polynomial in $\mathcal{P}$ down to these monomials. The objective in doing this is to infuse some structure into the matrix while at the same time preserving its rank (up to small losses). Having chosen $\mathcal{M}$, we show that the corresponding submatrix can be block-diagonalized into matrices each of which is described by a simple inclusion pattern between the (tuples of) sets labelling its rows and columns. This is done in Lemmas 4, 5 and 6.

Step 3: The main technical step in the proof is to lower bound the rank of the inclusion pattern matrix mentioned above with an algebraic trick. We first find a full-rank matrix that is closely related to our matrix and then show that the columns of our matrix can (with the aid of just a few other columns) generate the columns of the full-rank matrix.

The Main Parameters. Let N, d and t be fixed. Throughout, we assume that $d \leq (1/10) \lg N / \lg \lg N$. Let $\tau = 4t + 1$, $\delta = 1/(2\tau)$ and $\ell = \lfloor N^{1-\delta} \rfloor$. Finally, let k be such that $d - k = \tau k$. (In particular, we assume that $4t + 2 \leq d$.)

The following are easy to verify for our choice of parameters:

Fact 1. $\tau^2 = o(\ell)$ and $\tau = o(N^\delta)$.

Remark 1. In the above setting of parameters, d has to be divisible by $\tau + 1 = 4t + 2$. For ease of exposition, we present the proof for these parameters. Our proof can be modified so that it works for any d large enough as compared to τ (this part is omitted in this short version).

3   Proving Theorem 1: Details

3.1   Choice of Basis: Step 1 of the Proof

Lemma 1. *Let $k \leq d$. The vector space spanned by the set of k-partial derivatives of S_N^d, that is $\langle \partial_k S_N^d \rangle_{\leq 0}$, contains $\{p_T \mid T \subseteq [N], |T| = k\}$ where $p_T = \sum_{T \subseteq A \subseteq [N], |A| = d-k} X_A$ (that is, $p_T = X_T \cdot S_{N-k}^{d-2k}(X \setminus T))$.*

Let $\mathcal{P} = \{m \cdot p_T \mid T \subseteq [N], |T| = k, m \in \mathcal{M}_N^\ell, \text{supp}(m) \cap T = \emptyset\}$. From Lemma 1, $\mathcal{P} \subseteq \langle \partial_k S_N^d \rangle_{\leq \ell}$. Hence, a lower bound on the dimension of span $\mathcal{P}$ is also a lower bound on $\dim(\langle \partial_k S_N^d \rangle_{\leq \ell})$.

3.2   Choice of Shifts: Step 2 of the Proof

Instead of considering arbitrary shifts m as in the definition of $\mathcal{P}$, we will consider shifts by monomials m with various values of $|\text{supp}_i(m)|$ for $i \in [\tau]$. We first present a technical lemma that is needed to establish the lower bound. It is a concentration bound for support sizes in random monomials.

Definition 1. *For $i \in [\tau]$, $\hat{s}_i$ denotes the average number of variables with degree exactly i; $\hat{s}_i = \mathbb{E}_{m \sim \mathcal{U}(\mathcal{M}_N^\ell)}[|\text{supp}_i(m)|]$.*

Definition 2 (Good signature). *Given $m \in \mathcal{M}_N^\ell$, the signature of m, $s(m)$, is the tuple $(s_1, \ldots, s_\tau)$ such that $m \in \mathcal{M}_N^\ell(s_1, \ldots, s_\tau)$. We call the signature $(s_1, \ldots, s_\tau)$ a good signature if for each $i \in [\tau]$, we have $\hat{s}_i/2 \leq s_i \leq 3\hat{s}_i/2$.*

Lemma 2. *For our choice of the main parameters, the following statements hold: (i) $\hat{s}_i = N \left(\frac{\ell}{N+\ell} \right)^i (1 - o(1))$.   In particular, $\hat{s}_\tau \leq \sqrt{N}$.*
(ii) $\hat{s}_i \geq \hat{s}_\tau \geq N^{1-\tau\delta}(1 - o(1)) \geq \frac{N^{1/10}}{2}$, and $\frac{\hat{s}_i}{\hat{s}_{i+1}} \geq \frac{N^\delta}{2}$. (iii) Furthermore,

$$\mathbb{P}_{m \sim \mathcal{U}(\mathcal{M}_N^\ell)}[s(m) \text{ } s \text{ a good signature }] = 1 - o(1). \tag{1}$$

Remark 2. By Lemma 2, for any good signature $(s_1, \ldots, s_\tau)$, we have $\frac{s_i}{s_{i+1}} = \Omega(N^\delta)$, $s_\tau = \Omega(N^{1/10})$, and also $\frac{|\bigcup_{(s_1, \ldots, s_\tau) \text{ good}} \mathcal{M}_N^\ell(s_1, \ldots, s_\tau)|}{|\mathcal{M}_N^\ell|} = 1 - o(1)$.

Given a signature $(s_1, \ldots, s_\tau)$, let $\mathcal{P}(s_1, \ldots, s_\tau)$ denote the set of polynomials $\{m \cdot p_T \mid T \subseteq [N], |T| = k, m \in \mathcal{M}_N^\ell(s_1, \ldots, s_\tau), \text{supp}(m) \cap T = \emptyset\}$. Note that all polynomials in $\mathcal{P}(s_1, \ldots, s_\tau)$ are homogeneous of degree at most $\ell + d - k$.

Definition 3. *For any signature $s = (s_1, \ldots, s_\tau)$, let $r_i(s) = s_i$ for $1 \leq i \leq \tau - 1$ and $r_\tau(s) = s_\tau + k$; also, let $r(s) = \sum_i r_i(s) = \sum_i s_i + k$. Usually the signature s will be clear from context, and we use r_i and r instead of $r_i(s)$ and $r(s)$ respectively. The matrix $M(s_1, \ldots, s_\tau)$ is the matrix whose columns are indexed by polynomials $m \cdot p_T \in \mathcal{P}(s_1, \ldots, s_\tau)$ and rows by the monomials $w \in \mathcal{M}_N^{\ell+d-k}(r_1, \ldots, r_\tau)$. The coefficient in row w and column $m \cdot p_T$ is the coefficient of the monomial w in the polynomial $m \cdot p_T$.*

Note that the columns of $M(s_1, \ldots, s_\tau)$ are simply the polynomials in $\mathcal{P}(s_1, \ldots, s_\tau)$ projected to the monomials that label the rows. In particular, a lower bound on the rank of $M(s_1, \ldots, s_\tau)$ implies a lower bound on the rank of the vector space spanned by $\mathcal{P}(s_1, \ldots, s_\tau)$.

It is not too hard to see that $M(s_1, \ldots, s_\tau)$ has $|\mathcal{P}(s_1, \ldots, s_\tau)|$ columns but only $\frac{|\mathcal{P}(s_1, \ldots, s_\tau)|}{\binom{s_\tau + k}{k}}$ rows. Hence, the rank of the matrix is no more than the number of rows in the matrix. The following lemma, proved in Sect. 3.3, shows a lower bound that is quite close to this trivial upper bound.

Lemma 3. *With parameters as above, for any good signature* $s = (s_1, \ldots, s_\tau)$, $\mathrm{rank}(M(s_1, \ldots, s_\tau)) \geqslant \frac{|\mathcal{P}(s_1, \ldots, s_\tau)|}{\binom{s_\tau + k}{k}} (1 - o(1))$.

Since $\mathcal{P}(s_1, \ldots, s_\tau) \subseteq \mathcal{P} \subseteq \langle \partial_k f \rangle_{\leq \ell}$, the above immediately yields a lower bound on $\dim(\langle \partial_k f \rangle_{\leq \ell})$. Our final lower bound, which further improves this, is proved by considering polynomials corresponding to a *set* of signatures.

Definition 4. *Given a set of signatures* $\mathcal{S}$, *define* $\mathcal{M}_N^\ell(\mathcal{S}) = \bigcup_{s \in \mathcal{S}} \mathcal{M}_N^\ell(s)$ *and* $\mathcal{P}(\mathcal{S}) = \bigcup_{s \in \mathcal{S}} \mathcal{P}(s)$. *Also define the matrix* $M(\mathcal{S})$ *as follows: the columns of* $M(\mathcal{S})$ *are labelled by polynomials* $q \in \mathcal{P}(\mathcal{S})$ *and the rows by monomials* $w \in \bigcup_{s \in \mathcal{S}} \mathcal{M}_N^{\ell+d-k}(r_1(s), \ldots, r_\tau(s))$. *The* (w, q)th *entry is the coefficient of* w *in* q.

Note that a lower bound on the rank of $M(\mathcal{S})$ immediately lower bounds the dimension of the space spanned by $\mathcal{P}(\mathcal{S})$ and hence also $\dim \langle \partial_k S_N^d \rangle_{\leq \ell}$.

Definition 5. *A set of signatures is* $\mathcal{S}$ *well-separated if given any distinct signatures* $s = (s_1, \ldots, s_\tau)$ *and* $s' = (s_1', \ldots, s_\tau')$ *from* $\mathcal{S}$, $\max_{i \in [\tau]} |s_i - s_i'| \geqslant 2d$.

To analyze the rank of $M(\mathcal{S})$, we observe that for a well-separated set of signatures $\mathcal{S}$, the matrix $M(\mathcal{S})$ is block-diagonalizable with $|\mathcal{S}|$ blocks, where the blocks are the matrices $M(s)$ for $s \in \mathcal{S}$. Since we already have a lower bound on the ranks of $M(s)$ (for good s), this will allow us to obtain a lower bound on the rank of $M(\mathcal{S})$ as well.

Lemma 4. *Let* $\mathcal{S}$ *be a well-separated set of signatures. Then, the matrix* $M(\mathcal{S})$ *is block-diagonalizable with blocks* $M(s)$ *for* $s \in \mathcal{S}$.

3.3 Bounding the Rank of M: Step 3 of the Proof

We now prove the lower bound on the rank of the matrix $M(s_1, \ldots, s_\tau)$ as claimed in Lemma 3. We block diagonalize it with matrices that have a simple combinatorial structure (their entries are 0 or 1 depending on intersection patterns of the sets that label the rows and columns). We then lower bound the ranks of these matrices: this is the main technical step in the proof.

Lemma 5. *Fix any signature* $(s_1, \ldots, s_\tau)$. *The entry of* $M(s_1, \ldots, s_\tau)$ *in row* $w \equiv [\tilde{w}, R_1, \ldots, R_\tau]$ *and column* $m \cdot p_T$ *with* $m \equiv [\tilde{m}, S_1, \ldots, S_\tau]$ *belongs to* $\{0, 1\}$ *and is not zero if and only if* $\tilde{w} = \tilde{m}$ *and the following system is satisfied:* $T \subseteq R_1, S_1 \subseteq R_1 \cup R_2, S_2 \subseteq R_2 \cup R_3 \ldots S_{\tau-1} \subseteq R_{\tau-1} \cup R_\tau$, *and* $S_\tau \subseteq R_\tau$. *Moreover, the system above implies that* $T \cup S_1 \cup \ldots \cup S_\tau = R_1 \cup \ldots \cup R_\tau$.

Proof. The entry in row w and column $m \cdot p_T$ belongs to $\{0, 1\}$ and is not zero if and only if there exists $A \subseteq [N]$ such that $T \subseteq A$, $|A| = d - k$ and $X_A \cdot m = w$. Assume there is such an A.

Say $w \equiv [\tilde{w}, R_1, \ldots, R_\tau]$ and $m \equiv [\tilde{m}, S_1, \ldots, S_\tau]$. Let $\overline{m} = \prod_{i=1}^{\tau}(X_{S_i})^i$ and $\overline{w} = \prod_{i=1}^{\tau}(X_{R_i})^i$ be the degree at most τ parts of m and w respectively.

Note that $\deg(\overline{w}) - \deg(\overline{m}) = \sum_{i=1}^{\tau} i r_i - \sum_{i=1}^{\tau} i s_i = \tau k = d - k$ by our choice of parameters r_τ and k. Putting this together with the fact that $w = X_A \cdot m$ for $|A| = d - k$, we see that X_A can only 'contribute' to the "degree at most τ" part of m: formally, $\overline{w} = X_A \cdot \overline{m}$ and hence, $\tilde{w} = \tilde{m}$.

Further, since $X_A \cdot \overline{m} = X_{A \setminus T} X_T \prod_{i=1}^{\tau}(X_{S_i})^i = \prod_{i=1}^{\tau}(X_{R_i})^i = \overline{w}$, and $T \cap (S_1 \cup \ldots \cup S_\tau) = \emptyset$, we have $T \subseteq R_1$. Since X_A is multilinear, $S_i \subseteq R_i \cup R_{i+1}$ for all $i \in [\tau - 1]$; $S_\tau \subseteq R_\tau$ is obvious.

Conversely, assume that $\tilde{w} = \tilde{m}$ and the inclusions $T \subseteq R_1$, $S_i \subseteq R_i \cup R_{i+1}$ for all $i \in [\tau - 1]$ and $S_\tau \subseteq R_\tau$ are satisfied. Then $T \cup S_1 \cup \ldots \cup S_\tau \subseteq R_1 \cup \ldots \cup R_\tau$. Since $|T \cup S_1 \cup \ldots \cup S_\tau| = k + \sum_{i=1}^{\tau} s_i = \sum_{i=1}^{\tau} r_i = |R_1 \cup \ldots \cup R_\tau|$, we get $T \cup S_1 \cup \ldots \cup S_\tau = R_1 \cup \ldots \cup R_\tau$. Let $A_i = R_i \setminus S_i$ for $i \in [\tau]$ and $A = A_1 \cup \ldots \cup A_\tau$. The sets A_i are disjoint (because the R_i are disjoint). Moreover, $|A_\tau| = |R_\tau \setminus S_\tau| = r_\tau - s_\tau = k$; and by induction, $|A_i| = |R_i \setminus S_i| = |(R_i \cup \ldots \cup R_\tau) \setminus (S_i \cup \ldots \cup S_\tau)| = \sum_{j=i}^{\tau} r_j - \sum_{j=i}^{\tau} s_j = k$. Hence $|A_i| = k$ for all $i \in [\tau]$. Then $|A| = \tau k = d - k$. Moreover, $T = A_1 \subseteq A$. And it holds that $X_A \prod_{i=1}^{i}(X_{S_i})^i = \prod_{i=1}^{i}(X_{R_i})^i$. Since $\tilde{w} = \tilde{m}$, it follows that $X_A \cdot m = w$. Hence the entry in row w and column $m \cdot p_T$ is 1. ∎

Lemma 6. *Let $(s_1, \ldots, s_\tau)$ be any signature. The matrix $M(s_1, \ldots, s_\tau)$ is block diagonalizable with blocks of size $\binom{r}{r_1 \ r_2 \ \ldots \ r_\tau} \times \binom{r}{s_1 \ s_2 \ \ldots \ s_\tau \ k}$.*

We now lower bound the rank of each block in the block diagonalization.

Lemma 7. *For a good signature $s = (s_1, \ldots, s_\tau)$, and the corresponding $(r_1, \ldots, r_\tau)$ as in Definition 3, $\sum_{s_1' \geq s_1, \ldots, s_{\tau-1}' \geq s_{\tau-1}} \binom{r}{s_1' \ s_2' \ \ldots \ s_{\tau-1}' \ r - \sum_{i=1}^{\tau-1} s_i'} = \binom{r}{s_1 \ s_2 \ \ldots \ s_{\tau-1} \ r - \sum_{i=1}^{\tau-1} s_i}(1 + o(1))$.*

Lemma 8 (Main Technical Lemma). *Fix any good signature $(s_1, \ldots, s_\tau)$. The rank of any diagonal block of $M(s_1, \ldots, s_\tau)$ is $\binom{r}{s_1 \ s_2 \ \ldots \ s_\tau + k}(1 - o(1))$.*

Proof Sketch. Let M' be a diagonal block of the matrix $M(s_1, \ldots, s_\tau)$. Recall from Lemma 6 that such a diagonal block is defined by a monomial $\tilde{w}$ and a subset $R \subseteq [N]$. Rows of this block are labelled with all monomials $w \equiv [\tilde{w}, R_1, \ldots, R_\tau]$ such that $R_1 \cup \ldots \cup R_\tau = R$ and columns of this block are labelled with all polynomials $m \cdot p_T$ where $m \equiv [\tilde{w}, S_1, \ldots, S_\tau]$ is such that $T \cup S_1 \cup \ldots \cup S_\tau = R$. First, we set up some notation.

For a partition $\tilde{B} = (B_1, \ldots, B_p)$ of R, let $\tilde{b} = (b_1, \ldots, b_p)$ be the tuple of part sizes, $b_i = |B_i|$. We say that $\tilde{b}$ is the signature of $\tilde{B}$.

We say $(a_1, \ldots, a_p) \preceq (b_1, \ldots, b_p)$ if $a_i \leq b_i$ for all $i \in [p]$, and $(a_1, \ldots, a_p) \prec (b_1, \ldots, b_p)$ if $(a_1, \ldots, a_p) \preceq (b_1, \ldots, b_p)$ but $(a_1, \ldots, a_p) \neq (b_1, \ldots, b_p)$.

Define the following collections of partitions of R: $X = \{\tilde{R} = (R_1, \ldots, R_\tau) \mid \text{signature}(\tilde{R}) = (r_1, \ldots, r_\tau)\}$, $Y = \{\tilde{S} = (S_1, \ldots, S_\tau, T) \mid$

signature$(\tilde{S}) = (s_1, \ldots, s_\tau, k)\}$, $Z' = \{\tilde{Q} = (Q_1, \ldots, Q_\tau) \mid$ signature$(\tilde{Q}) = (q_1, \ldots, q_\tau); (s_1, \ldots, s_{\tau-1}) \preceq (q_1, \ldots, q_{\tau-1})\}$, and $Z = \{\tilde{Q} = (Q_1, \ldots, Q_\tau) \mid$ signature$(\tilde{Q}) = (q_1, \ldots, q_\tau); (s_1, \ldots, s_{\tau-1}) \prec (q_1, \ldots, q_{\tau-1})\}$.

Note that $|X| = \binom{r}{s_1 \ s_2 \ \cdots \ s_\tau + k}$. Also, $Z' \setminus Z$ is precisely X. By Lemma 7, $|Z'| = |X|(1 + o(1))$. Hence $|Z| = |X| \cdot o(1)$.

The rows and columns of M' are indexed by elements of X and Y respectively (Lemma 6). Let I be the identity matrix with rows/columns indexed by elements of X. We define two auxiliary matrices M_1 and M_2 as follows. The rows and columns of M_1 are indexed by elements of X. The entries of M_1 are in $\{0, 1\}$ and are defined as follows:

$$M_1(\tilde{R}, \tilde{R}') = \begin{cases} 1 & \text{if } R_i' \subseteq R_i \cup R_{i+1} \text{ for each } i \in [\tau - 1] \\ 0 & \text{otherwise.} \end{cases}$$

The rows and columns of M_2 are indexed by elements of X and Z respectively. The entries of M_2 are in $\{0, 1\}$ and are defined as follows:

$$M_2(\tilde{R}, \tilde{Q}) = \begin{cases} 1 & \text{if } Q_i \subseteq R_i \cup R_{i+1} \text{ for each } i \in [\tau - 1] \\ 0 & \text{otherwise.} \end{cases}$$

Our proof proceeds as follows:

1. Show that the columns of M' and M_2 together span the columns of M_1.
2. Show that the columns of M_1 and M_2 together span the columns of I.

It then follows that

$$\text{rank}(M') \geq \text{rank}(M_1) - \text{rank}(M_2) \geq \text{rank}(I) - 2\text{rank}(M_2) \geq |X|(1 - o(1))$$

which is what we had set out to prove.

To prove steps 1 and 2, we describe columns of M', M_1, M_2, I using functions that express whether two partitions are related in a certain way. In particular, we express the inclusion relations described in Lemma 5, which characterise the non-zeroes in M'. The functions we use are mutlivariate polynomials, whose evaluations at the characteristic vectors of row indices give the entries in the rows. A careful choice of a small basis for these functions yields the result (details are omitted in this short version). □

Lemma 3 can now be proved using the block-diagonal decomposition (Lemma 6) and the rank lower bound (Lemma 8).

3.4   Putting it Together

We now have all the ingredients to establish that the shifted partial derivative measure of S_N^d is large.

Proof of Theorem 1. By Lemma 1, $\dim\langle \partial_k S_N^d \rangle_{\leq \ell} \geq \dim(\text{span}(\mathcal{P}))$. This in turn is at least as large as $\text{rank}(M(\mathcal{S}))$, since $M(\mathcal{S})$ is a submatrix of the matrix that describes a basis for $\mathcal{P}$. We now choose a well-separated set of good signatures

$\mathcal{S}$ and apply Lemmas 2, 3, and 4 to lower bound the rank of $M(\mathcal{S})$. This will allow give us our lower bound on $\dim\langle \partial_k S_N^d \rangle_{\leq \ell}$.

Let us see how to choose $\mathcal{S}$. Let $\mathcal{S}_0$ denote the set of all good signatures. For integers $d_1, \ldots, d_\tau \in [2d]$, denote by $\mathcal{S}(d_1, \ldots, d_\tau)$ the signatures $(s_1, \ldots, s_\tau) \in \mathcal{S}_0$ such that $s_i \equiv d_i \pmod{2d}$ for each $i \in [\tau]$. It is easily checked that for any choice of $d_1, \ldots, d_\tau \in [2d]$, the set of signatures $\mathcal{S}(d_1, \ldots, d_\tau)$ is well separated. Since there are $(2d)^\tau$ choices for $d_1, \ldots, d_\tau$, there must be one such that

$$|\mathcal{M}_N^\ell(\mathcal{S}(d_1, \ldots, d_\tau))| \geq \frac{|\mathcal{M}_N^\ell(\mathcal{S}_0)|}{(2d)^\tau}. \tag{2}$$

We fix $d_1, \ldots, d_\tau$ so that the above holds and let $\mathcal{S} = \mathcal{S}(d_1, \ldots, d_\tau)$. This is the set of signatures we will consider.

By Lemma 4, we know that the rank of $M(\mathcal{S})$ is equal to the sum of the ranks of $M(s)$ for each $s \in \mathcal{S}$. Hence, by Lemma 3, we have

$$\mathrm{rank}(M(\mathcal{S})) \geq (1 - o(1)) \cdot \sum_{s \in \mathcal{S}} \frac{|\mathcal{P}(s)|}{\binom{s_\tau + k}{k}} \tag{3}$$

Recall that $\mathcal{P}(s) = \{m \cdot p_T \mid m \in \mathcal{M}_N^\ell(s), |T| = k, \mathrm{supp}(m) \cap T = \emptyset\}$. Hence, for each choice of $m \in \mathcal{M}_N^\ell(s)$, the number of choices of T such that $m \cdot p_T \in \mathcal{P}(s)$ is given by $\binom{N - |\mathrm{supp}(m)|}{k} \geq \binom{N - \ell}{k}$. Adding over all $m \in \mathcal{M}_N^\ell(s)$, we see that $|\mathcal{P}(s)| \geq |\mathcal{M}_N^\ell(s)| \cdot \binom{N - \ell}{k}$.

Plugging this bound into (3), we see that

$$\mathrm{rank}(M(\mathcal{S})) \geq (1 - o(1)) \cdot \sum_{s \in \mathcal{S}} \frac{|\mathcal{M}_N^\ell(s)| \cdot \binom{N-\ell}{k}}{\binom{s_\tau + k}{k}}$$

$$\geq \frac{(1 - o(1)) \cdot \binom{N-\ell}{k}}{(3\sqrt{N}/2)^k} \cdot \sum_{s \in \mathcal{S}} |\mathcal{M}_N^\ell(s)|$$

$$= \frac{(1 - o(1)) \cdot \binom{N-\ell}{k}}{(3\sqrt{N}/2)^k} \cdot |\mathcal{M}_N^\ell(\mathcal{S})| \geq \frac{(1 - o(1)) \cdot \binom{N-\ell}{k}}{(3\sqrt{N}/2)^k} \cdot \frac{|\mathcal{M}_N^\ell(\mathcal{S}_0)|}{(2d)^\tau}$$

where the second inequality follows from the fact that all signatures $s \in \mathcal{S}$ are good, so $s_\tau \leq 3\hat{s}_\tau/2$ and hence $\binom{s_\tau + k}{k} \leq s_\tau^k \leq (3\hat{s}_\tau/2)^k \leq (3\sqrt{N}/2)^k$ (using Lemma 2); the final inequality is a consequence of Eq. (2).

Finally, by Lemma 2 (see also Remark 2), $|\mathcal{M}_N^\ell(\mathcal{S}_0)| \geq |\mathcal{M}_N^\ell|(1 - o(1)) \geq (1 - o(1)) \cdot \binom{N+\ell}{\ell}$, which along with the above computation yields the claimed lower bound on $\mathrm{rank}(M(\mathcal{S}))$. $\qquad\square$

Remark 3. For any multilinear polynomial $F(X)$ on N variables, the quantity $\dim\langle \partial_k F \rangle_{\leq \ell}$ is at most the number of monomial shifts — which is $\binom{N+\ell}{\ell}$ — times the number of possible partial derivatives of order k, which is at most $\binom{N}{k}$. Our result says that this trivial upper bound is (in some sense) close to optimal for the polynomial S_N^d (the $(\sqrt{N})^k$ factor in the denominator can be made $N^{\varepsilon k}$ for any constant $\varepsilon > 0$, see the discussion at the end of the proof of Theorem 2). All previous lower bound results using the shifted partial derivative method also obtain similar statements [4,5,12,13].

4 Lower Bound on the Size of Depth Four Formulas

In this section, we establish the lower bounds claimed in Theorem 2. As in [5], we say that a $\Sigma\Pi\Sigma\Pi$ formula C is a $\Sigma\Pi^{[D]}\Sigma\Pi^{[t]}$ formula if the product gates at level 1 (just above the input variables) have fan-in at most t and the product gates at level 3 have fan-in bounded by D.

The following is implicit in [5] and is stated explicitly in [9].

Lemma 9 ([9], Lemma 4). *Let P be a polynomial on N variables computed by a $\Sigma\Pi^{[D]}\Sigma\Pi^{[t]}$ circuit of top fan-in s. Then, we have*

$$\dim(\langle \partial_k P \rangle_{\leq \ell}) \leq s \cdot \binom{D}{k} \cdot \binom{N + \ell + (t-1)k}{\ell + (t-1)k}.$$

Proof of Theorem 2. For notational simplicity, we present the proof only for $\varepsilon = 1/4$. Let $\tau = 4t + 1$. Let $k = d/(\tau+1)$ and $\delta = 1/(2\tau)$. Assume there exists a $\Sigma^s\Pi^{[D]}\Sigma\Pi^{[t]}$ circuit computing S_N^d. When N is large enough, from Theorem 1 and Lemma 9, it holds that

$$s \geq \frac{\binom{N-\ell}{k}}{\binom{D}{k}} \frac{\binom{N+\ell}{\ell}}{\binom{N+\ell+(t-1)k}{\ell+(t-1)k}} \frac{1 - o(1)}{(3\sqrt{N}/2)^k (2d)^\tau}.$$

For large enough N, we have $\frac{\binom{N-\ell}{k}}{\binom{D}{k}} \geq \left(\frac{N-\ell-k}{D}\right)^k \geq \left(\frac{N}{2D}\right)^k$. Since $kt < d = o(\lg N)$, for large enough N we have $\frac{\binom{N+\ell}{\ell}}{\binom{N+\ell+(t-1)k}{\ell+(t-1)k}} \geq \left(\frac{\ell}{N+\ell}\right)^{kt} \geq \left(\frac{1}{2N^\delta}\right)^{kt} \geq N^{-\delta tk - o(k)}$. Note that $(2d)^\tau \leq (\lg N)^\tau \leq N^{1/10}$.

Putting it all together, we have obtained that asymptotically,

$$s \geq \left(\frac{N}{2D \cdot N^{\delta t + o(1)} \cdot (3\sqrt{N}/2)}\right)^k \cdot \frac{1 - o(1)}{N^{1/10}} \geq \frac{1}{N^{1/10}} \cdot \left(\frac{N^{1/2 - \delta t}}{3D \cdot N^{o(1)}}\right)^k. \quad (4)$$

By our choice of parameters, $t \leq \tau/4$ and hence $1/2 - \delta t \geq 1/2 - 1/8 = 3/8$. Since $D \leq N^{1/4}$, we see that (4) yields a lower bound of $N^{\Omega(k)} = N^{\Omega(d/t)}$.

It is not hard to see that the above proof idea (with some changes in parameters) can be made to give lower bounds of $N^{\Omega(d/t)}$ for $D \leq N^{1-\varepsilon}$ for any constant $\varepsilon > 0$. Specifically, choose $\tau = \Theta\left(\frac{t}{\varepsilon}\right)$ (instead of $4t+1$) and δ such that $\delta\tau = 1 - \Theta(\varepsilon)$ (instead of $\frac{1}{2}$) in the entire proof. We omit the details. □

References

1. Agrawal, M., Vinay, V.: Arithmetic circuits: a chasm at depth four. In: FOCS, pp. 67–75 (2008)
2. Alon, N.: Perturbed identity matrices have high rank: proof and applications. Comb. Probab. Comput. **18**(1–2), 3–15 (2009)
3. Baur, W., Strassen, V.: The complexity of partial derivatives. Theor. Comput. Sci. **22**, 317–330 (1983)

4. Fournier, H., Limaye, N., Malod, G., Srinivasan, S.: Lower bounds for depth 4 formulas computing iterated matrix multiplication. In: Symposium on Theory of Computing, STOC, pp. 128–135 (2014)
5. Gupta, A., Kamath, P., Kayal, N., Saptharishi, R.: Approaching the chasm at depth four. In: Conference on Computational Complexity (CCC) (2013)
6. Hrubes, P., Yehudayoff, A.: Homogeneous formulas and symmetric polynomials. Comput. Complexity **20**(3), 559–578 (2011)
7. Kayal, N.: An exponential lower bound for the sum of powers of bounded degree polynomials. Electronic Colloquium on Computational Complexity (ECCC) **19**, 81 (2012)
8. Kayal, N., Limaye, N., Saha, C., Srinivasan, S.: An exponential lower bound for homogeneous depth four arithmetic formulas. In: Foundations of Computer Science (FOCS) (2014)
9. Kayal, N., Saha, C., Saptharishi, R.: A super-polynomial lower bound for regular arithmetic formulas. In: STOC, pp. 146–153 (2014)
10. Keevash, P., Sudakov, B.: Set systems with restricted cross-intersections and the minimum rank ofinclusion matrices. SIAM J. Discrete Math. **18**(4), 713–727 (2005)
11. Koiran, P.: Arithmetic circuits: the chasm at depth four gets wider. Theor. Comput. Sci. **448**, 56–65 (2012)
12. Kumar, M., Saraf, S.: The limits of depth reduction for arithmetic formulas: it's all about the top fan-in. In: STOC, pp. 136–145 (2014)
13. Kumar, M., Saraf, S.: On the power of homogeneous depth 4 arithmetic circuits. In: FOCS, pp. 364–373 (2014)
14. Kushilevitz, E., Nisan, N.: Communication Complexity. Cambridge University Press, New York (1997)
15. Nisan, N., Wigderson, A.: Lower bounds on arithmetic circuits via partial derivatives. Comput. Complex. **6**(3), 217–234 (1997)
16. Raz, R.: Separation of multilinear circuit and formula size. Theor. Comput. **2**(1), 121–135 (2006)
17. Raz, R.: Multi-linear formulas for permanent and determinant are of super-polynomial size. J. ACM, **56**(2) (2009)
18. Razborov, A.: Lower bounds on the size of bounded depth circuits over a complete basis with logical addition. Math. Notes Acad. Sci. USSR **41**(4), 333–338 (1987)
19. Saptharishi, R.: Unified Approaches to Polynomial Identity Testing and Lower Bounds. Ph.D thesis, Chennai Mathematical Institute (2013)
20. Shpilka, A.: Affine projections of symmetric polynomials. J. Comput. Syst. Sci. **65**(4), 639–659 (2002)
21. Shpilka, A., Wigderson, A.: Depth-3 arithmetic circuits over fields of characteristic zero. Comput. Complex. **10**(1), 1–27 (2001)
22. Tavenas, S.: Improved bounds for reduction to depth 4 and 3. In: Mathematical Foundations of Computer Science (MFCS) (2013)
23. Valiant, L.G.: Completeness classes in algebra. In: 11th ACM Symposium on Theory of Computing (STOC), pp. 249–261. New York, NY, USA (1979)
24. Valiant, L.G., Skyum, S., Berkowitz, S., Rackoff, C.: Fast parallel computation of polynomials using few processors. SIAM J. Comput. **12**(4), 641–644 (1983)
25. Wilson, R.M.: A diagonal form for the incidence matrices of t-subsets vs. k-subsets. Eur. J. Comb. **11**(6), 609–615 (1990)

Parameterized Algorithms for Parity Games

Jakub Gajarský[1], Michael Lampis[2], Kazuhisa Makino[3],
Valia Mitsou[4], and Sebastian Ordyniak[5(✉)]

[1] Faculty of Informatics, Masaryk University, Brno, Czech Republic
[2] LAMSADE, Université Paris Dauphine, Paris, France
[3] RIMS, Kyoto University, Kyoto, Japan
[4] SZTAKI, Hungarian Academy of Sciences, Budapest, Hungary
[5] Institute for Computergraphics and Algorithms, TU Wien, Vienna, Austria
sordyniak@gmail.com

Abstract. Determining the winner of a Parity Game is a major problem in computational complexity with a number of applications in verification. In a parameterized complexity setting, the problem has often been considered with parameters such as (directed versions of) treewidth or clique-width, by applying dynamic programming techniques.

In this paper we adopt a parameterized approach which is more inspired by well-known (non-parameterized) algorithms for this problem. We consider a number of natural parameterizations, such as by Directed Feedback Vertex Set, Distance to Tournament, and Modular Width. We show that, for these parameters, it is possible to obtain recursive parameterized algorithms which are simpler, faster and only require polynomial space. We complement these results with some algorithmic lower bounds which, among others, rule out a possible avenue for improving the best-known sub-exponential time algorithm for parity games.

1 Introduction

A *Parity Game* is an infinite two-player game played on a parity game arena by the two players Even and Odd. A *parity game arena* is a sinkless directed graph, where every vertex is controlled by exactly one of the two players and is addionally assigned a priority (an integer value). Initially, a token is placed on some vertex of the graph (the starting vertex) and at each step of the game the player that controls the vertex that currently contains the token has to move the token along a directed edge to any outneighbor of that vertex. This leads to an infinite path, which is won by the player that corresponds to the parity of the highest priority occuring infinitely often on that path. Given such a parity game arena and a starting vertex v, the problem is to decide which of the two players has a winning strategy on the given arena if the token is initially placed on v.

J. Gajarský—Supported by the research centre Institute for Theoretical Computer Science (CE-ITI), project P202/12/G061.
V. Mitsou—Supported by ERC Starting Grant PARAMTIGHT (No. 280152).
S. Ordyniak—Supported by Employment of Newly Graduated Doctors of Science for Scientific Excellence (CZ.1.07/2.3.00/30.0009).

© Springer-Verlag Berlin Heidelberg 2015
G.F. Italiano et al. (Eds.): MFCS 2015, Part II, LNCS 9235, pp. 336–347, 2015.
DOI: 10.1007/978-3-662-48054-0_28

Although a parity game may, at first glance, seem like an odd kind of game to play, the problem of deciding the winner of a parity game is extremely well-studied. One of the reasons is that solving parity games captures the complexity of model-checking for the modal μ-calculus, which has a large number of applications in software verification (see e.g. [8,12,15]). Despite extensive efforts by the community, it is still a major open question whether there exists a polynomial-time algorithm for this problem [5,11,18,19].

In addition to their wealth of practical applications, parity games are also especially interesting from a complexity-theoretic point of view because of their intriguing complexity status. Despite not being known to be in P, deciding parity games in known to be in NP∩coNP [12], and in fact even in UP∩coUP [17]. The former of these two inclusions follows directly from the (non-trivial) fact that optimal strategies are known to be positional (or memoryless) [20] and the fact that single-player parity games are in P. Parity games are also known to be reducible to (and therefore "easier" than) a number of other natural classes of games, including mean payoff games (which admit a pseudo-polynomial time algorithm) [26] and simple stochastic games [9]. Thus, in a sense, deciding the winner of a parity game is a problem that seems to lie only slightly out of reach of current algorithmic techniques, and could perhaps be solvable in polynomial time. However, the best currently known time upper bound is (roughly) $n^{\sqrt{n}}$ [19].

Since, despite all this effort, it is still not clear if a polynomial-time algorithm for parity games is possible, several more recent works have attempted to tackle this problem from a parameterized complexity perspective. Perhaps the most natural parameter for this problem is the maximum priority p. Although this problem is known to be in XP, i.e., solvable in polynomial-time for fixed p, (by a classical $O(n^p)$ algorithm due to McNaughton [21], later improved to $O(n^{p/3})$ [7,18,24]), it is unclear if this parameterization could be in FPT, i.e., solvable in time $O^*(f(p))^1$ for some function f of p, though this question has been explicitly considered [3].

A direction which has been much more fruitful is to add a second parameter to the problem, usually some structural graph parameter of the input graph. Here, several non-trivial algorithmic results are known. For digraphs whose underlying graph has treewidth k the problem is solvable in time (roughly) $p^{O(k)}$ [22], while a similar running time is achievable if k is the (directed) clique-width of the input digraph [23]. For several directed variations of the notion of treewidth, such as entanglement, DAG-width and kelly-width, algorithms running in $n^{O(k)}$ time are known [1,2,16]. More recently, an algorithm running in $f(p,k)n^{O(1)}$ time for these measures was given in [6], for some computable function f.

Thanks to the above results we now know that parity games are FPT for the most standard structural graph parameters, if parameterized by *both* k and the maximum priority p. It is however, worthy of note that, to the best of our knowledge, this problem is not known to be FPT when the number of priorities

---

[1] As usual in parameterized complexity the $O^*(f(k))$-notation, for some function f of the parameter k, means that there is an algorithm running in time $O(f(k)n^{O(1)})$, where n is the input size of the problem.

is not part of the parameter for any non-trivial graph width. Furthermore, all the above algorithms, which use somewhat complicated dynamic programming techniques, also require space exponential in k. Can this be improved?

Our Contribution: This paper provides a number of algorithms for various natural structural parameterizations of parity games. More specifically, we show the following results:

- There exists an algorithm running in time $O^*(4^k p^{\log k})$ when k is any of the following parameters: the size of the graph's directed feedback vertex set, the number of vertices controlled by one of the two players, or the number of vertices whose deletion makes the graph a tournament.
- There exists an algorithm running in time $O^*\left(k^{O(\sqrt{k})}\right)$ when k is the modular width of the input graph.

Conceptually, the main contribution of this paper is in repurposing the ideas of the classical algorithm of McNaughton [21] (and its sub-exponential time variation by Jurdzinski, Paterson and Zwick [19]) in the parameterized complexity setting. By avoiding the dynamic programming paradigm, we are able to give a number of very simple to implement, recursive algorithms. Notably, all the algorithms of this paper run in polynomial space.

Our first set of algorithmic results introduces the notion of a *dominion hitter*. Informally, a dominion is a confined area of the graph where one player has a winning strategy. The algorithmic importance of dominions was recognized in [19], where the algorithm begins by exhaustively looking for a small dominion in the graph, before running the simple recursive algorithm of [21]. We give a parameterized counterpart of this idea which can be applied whenever we can find a small set of vertices that intersects all dominions. We then also provide three natural example parameterizations where this is the case, showing that this may be an idea of further interest.

We then consider graphs of modular width k. Modular width is a graph parameter that has recently attracted attention in the parameterized complexity community as a more restricted alternative to clique-width [13,14]. We show that, in this more restricted case, the $O^*(p^k)$ complexity of the algorithm for clique-width can be improved to a running time that is FPT parameterized only by k, with a *sub-exponential* parameter dependence. The core algorithmic idea again combines the recursive algorithm of McNaughton, with a judicious search for dominions.

We complement these algorithms with a couple of hardness results. First, as mentioned, one of the key steps in the sub-exponential algorithm of [19] (and in most of our algorithms) consists of exhaustively searching the graph for a dominion. In [19] searching for a dominion of size k is done by checking all sets of k vertices. We give a reduction from MULTI-COLORED CLIQUE showing that this is likely optimal, thus ruling out one possible avenue for improving the algorithm of [19]. Furthermore, in order to demonstrate that the parameterizations we consider are non-trivial, we give a reduction showing that Rabin games (a more general class of infinite games) remain hard for many of the cases of this paper.

2   Definitions and Preliminaries

2.1   Parity Games, Dominions and Attractors

In a parity game the input is a sinkless directed graph $G(V, E)$, where V is partitioned into two sets V_0, V_1, a priority function $\Pr : V \to \mathbb{N}$, and a starting vertex $v \in V$. Throughout the paper we use n to denote $|V|$ and p to denote the maximum priority given to any vertex of G $\max\{\Pr(u) : u \in V\}$. We also use $N^+(u)$ (respectively $N^-(u)$) to denote the set of out-neighbors (in-neighbors) of the vertex u.

In this game there are two players, player 0 (the even player) and player 1 (the odd player), controlling the vertices of V_0 and V_1 respectively. A token is initially placed on the starting vertex v and, in each turn, the player who controls the vertex that currently contains the token selects one of its out-neighbors and moves the token there (an out-neighbor always exists, since the graph has no sinks). The play then goes on to the next round and continues infinitely. Player 0 is the winner of a play if and only if the vertex that has maximum priority out of all the vertices that appear infinitely often in the play has even priority; otherwise, player 1 is the winner. A player has a winning strategy from a starting vertex v, if she has a way to construct a winning play when the game starts from v no matter how the opponent plays. For more information on parity games see for example [15].

We use $W_i(G)$, for $i \in \{0, 1\}$ to denote the winning region of player i in the game graph G, that is, the set of vertices from which player i has a winning strategy (and we will simply write W_i if G is clear from the context). It is a well-known (but non-trivial) fact that each player has a memory-less *positional* strategy, i.e. a strategy where decisions depend on the current position of the token and not on the history, that allows her to win all the vertices of her winning region [4,12,20]. We will make use of the following basic fact:

Fact 1. *For all $i \in \{0, 1\}$ and all vertices $v \in V_i$ we have $v \in W_i$ if and only if $N^+(v) \cap W_i \neq \emptyset$.*

Let us now formally define two notions that will be crucial throughout this paper: dominions and attractors.

Definition 1. *A set of vertices $D \subseteq V$ is an i-dominion, for some $i \in \{0, 1\}$ if the following hold:*

1. *There are no arcs from $D \cap V_{1-i}$ to $V \setminus D$ and every vertex of $D \cap V_i$ has an out-neighbor in D.*
2. *Player i has a winning strategy for all starting vertices in the parity game on the induced subgraph $G[D]$.*

Informally, an i-dominion is a region of the graph where player i can force the token to remain, and by doing so she manages to win the game. It is not hard to see that $W_i(G)$ is always an i-dominion, but smaller i-dominions could also exist.

Definition 2. *An i-attractor of a set of vertices S, for some $i \in \{0,1\}$, denoted* $\mathrm{attr}_i(S)$, *is the smallest superset of S that satisfies the following:*

1. *There are no arcs from $V_i \setminus \mathrm{attr}_i(S)$ to $\mathrm{attr}_i(S)$.*
2. *Every vertex of $V_{1-i} \setminus \mathrm{attr}_i(S)$ has an out-neighbor outside of $\mathrm{attr}_i(S)$.*

Intuitively, an i-attractor of a set S is a region inside which player i can force the token to eventually enter S (this should obviously include S itself).

It is not hard to see that an i-attractor can be calculated in polynomial time by iteratively adding vertices to S, as dictated by the above specifications. One of the reasons that we are interested in attractors is that they allow us to simplify the game once we identify some part of one player's winning region, thanks to the following fact, which is e.g. shown in [19, Lemma 4.5].

Fact 2. *If, for some $i \in \{0,1\}$ and $S \subseteq V$, we have $S \subseteq W_i$, then $W_i(G) = \mathrm{attr}_i(S) \cup W_i(G \setminus \mathrm{attr}_i(S))$.*

We will often make use of the fact that an attractor for one player cannot expand into the region of a dominion controlled by the opponent.

Lemma 1. *For $i \in \{0,1\}$ let D be an i-dominion and $S \subseteq V$ be a set of vertices such that $S \cap D = \emptyset$. Then $\mathrm{attr}_{1-i}(S) \cap D = \emptyset$.*

Proof. Consider the iterative process that adds vertices to S to build $\mathrm{attr}_{1-i}(S)$. No vertex of D can be the first to be added to the attractor in this process: vertices of $D \cap V_i$ have an out-neighbor in D and vertices of $D \cap V_{1-i}$ have all their out-neighbors in D. ☐

Finally, let us point out that a dominion that is completely controlled by a single player can be found (and by Fact 2 removed from the graph) in polynomial time. We will therefore always assume that in all the graphs we consider, no player can win without at some point passing the token to her opponent.

Fact 3. *If there exists a dominion D such that $D \subseteq V_i$ for some $i \in \{0,1\}$, then such a dominion can be found in polynomial time.*

2.2 Attractor-Based Algorithms

One of the main approaches for solving parity games is the attractor-based approach often referred to as McNaughton's algorithm [21,25] as well as its subsequent improvement from [19]. Since these algorithms form the basis of much of the work of this paper, familiarity with them is assumed.

3 Strong Dominion Hitters

In this section we introduce the notion of a *dominion hitter* and show several (parameterized) algorithmic applications. The ideas presented here are heavily inspired by the sub-exponential algorithm of [19], which looks for small dominions in the input graph. Here we will be interested in the intersection of the located dominion with a special set that intersects all dominions. We give two straight-forward applications of this idea (for parameters directed feedback vertex set and number of vertices of one player) and a slightly more involved one (for parameter distance from tournament).

Before moving forward, let us give some useful definitions.

Definition 3. *A set $S \subseteq V$ is a* dominion hitter *if for every (non self-controlled) dominion D of G we have $S \cap D \neq \emptyset$. A set $S \subseteq V$ is a* strong dominion hitter *if, for every $V' \subseteq V$ for which $G[V']$ is a game, $S \cap V'$ is a dominion hitter in $G[V']$.*

Let us explain the intuition behind the definition of strong dominion hitters. First, by Fact 3 we can remove self-controlled dominions in polynomial time from G and all of its subgames. Now, if G contains a strong dominion hitter S with $|S| = k$ then McNaughton's algorithm will run in time $O^*(n^k)$, since the second recursive call will always delete at least one vertex of S (in fact the running time can also be upper-bounded by $O^*(p^k)$, since the first recursive call decreases p).

Our strategy in this section will be to improve upon McNaughton's algorithm significantly by following a strategy similar to that of [19]. Suppose we are given a strong dominion hitter S. First, we look for a dominion that has a small intersection with S. As we will argue, this can be done by solving parity games (recursively) on a much simpler graph where a large part of S has been deleted. If such a dominion is found, it can be removed from the graph. If that fails, then every dominion must have a large intersection with S, therefore we know that McNaughton's algorithm's second recursive call will be on a graph where much of S has been deleted. Proper balancing allows us to obtain a running time of $O^*(p^{\log k} 4^k)$ (Lemma 2).

Let us now describe exactly how dominion hitters can be exploited.

Lemma 2. *There is an algorithm which, given a parity game instance $G(V, E)$ and a strong dominion hitter S with $|S| = k$ decides the problem in $O^*(p^{\log k} 4^k)$ time.*

3.1 Direct Applications

Let us first consider two direct applications of the idea of strong dominion hitters. One is the parameterization of the problem where the parameter is the directed feedback vertex set S of the graph, i.e. a set of vertices whose removal makes the graph a DAG. Since every dominion should contain a directed cycle, S is clearly a strong dominion hitter. Application of Lemma 2 is straightforward.

Theorem 1. *Given an instance of parity games $G(V, E)$ and a directed feedback vertex set $S, |S| \leq k$ of G, there exists an algorithm which decides the problem in time $O^*(p^{\log k} 4^k)$.*

Another example is the parameterization of the problem where the parameter k is equal to $|V_1|$, the number of vertices controlled by one of the players. Because of Fact 3, V_1 is a strong dominion hitter. Once again Lemma 2 can be applied directly.

Theorem 2. *There exists an algorithm which given an instance of parity games $G(V, E)$ such that $|V_1| = k$ decides the problem in time $O^*(4^k p^{\log k})$.*

3.2 More Involved Case

Now we will see how we can apply the idea of strong dominion hitters to a less straightforward case, where the graph is almost a tournament (a tournament is a directed graph having at least one arc between every pair of vertices).

Once again, due to Fact 3, G is free of i-controlled i-dominions (we call these dominions *happy*). In fact, this implies something even stronger: that no happy dominion exists in any subgame of G. Then all i-dominions in G shall have an *unhappy* vertex, i.e. a vertex controlled by player $(1 - i)$. The next fact shall prove useful.

Fact 4. *Let $v_0 \in V_0 \cap W_1$ and $v_1 \in V_1 \cap W_0$ be unhappy vertices. Then v_0 and v_1 are not neighbors.*

Proof. $(v_0, v_1) \notin E$, otherwise the even player would have had a way to escape from v_0 to W_0. With similar reasoning $(v_1, v_0) \notin E$.   □

We first need to establish that parity games on tournaments can be solved in polynomial time, using the above observation. In fact, we can show that in tournaments, after we remove happy dominions, one player wins all the vertices. The fact that parity games are polynomially decidable on tournaments was already shown in [10], but a proof is included in the full version of the paper for the sake of completeness.

We then study the parameterization where the parameter is the vertex-deletion distance of the graph from being a tournament. In this case, the graph is basically a tournament plus a set S of at most k additional vertices which might be missing arcs (this case is more general than the case of a tournament missing a set of k edges).

The algorithm is similar to Lemma 2. If S happens to be a dominion hitter the procedure reduces S by at least $\frac{k}{2}$ either during the preprocessing or during the second recursive call of McNaughton's algorithm. If S is not a dominion hitter, then the dominion found after removing $\mathrm{attr}_i(p)$ might not intersect S. In this case however, we argue that the $(1 - i)$-attractor of the winning region $W_{1-i}(G \setminus \mathrm{attr}_i(p))$ of the smaller graph should absorb all vertices of $V_{1-i} \setminus S$, leaving an instance where player $(1 - i)$ has vertices only in S (up to k vertices). We can then use Theorem 2.

The algorithmic results of this section are thus summarized in the following two theorems.

Theorem 3. *Parity games on tournaments can be solved in polynomial time.*

Theorem 4. *Given a graph $G(V, E)$ on n vertices and a set $S \subseteq V$ with $|S| \leq k$ such that $G \setminus S$ is a tournament, parity games can be solved in time $O^*(p^{\log k} 4^k)$.*

4 Modular Width and Graphs with k Modules

The main result of this section is an FPT algorithm which solves parity games on graphs with small modular width. In fact, the algorithm we present is able to handle the more general case of strongly connected graphs whose vertex set can be partitioned into $k > 1$ modules. For such graphs, we are able to obtain a *sub-exponential FPT* algorithm, even for unbounded p (that is, parameterized just by k), and from this we can then obtain the same result for modular width.

Recall that a set of vertices M of a directed graph $G(V, E)$ is a *module* if for any two vertices $u, v \in M$ we have $N^+(u) \setminus M = N^+(v) \setminus M$ and $N^-(u) \setminus M = N^-(v) \setminus M$, that is, all vertices of M have the same in- and out-neighbors outside of M.

Let us begin this section by presenting an algorithm that decides parity games on G in time $O^*(4^k)$. We then give an improved version with sub-exponential running time.

4.1 Exponential FPT Algorithm

Suppose that we are given a strongly connected graph $G(V, E)$, along with a partition of V into $k > 1$ non-empty modules $M_1, \ldots M_k$. We will call these the graph's *basic* modules.

On a high level, the strategy we will follow is again a variation of McNaughton's algorithm. This algorithm makes two recursive calls, each on a graph obtained from G by removing an attractor of some set. If the removed set contains all of $M_j \cap V_i$ for some $j \in \{1, \ldots, k\}$, $i \in \{0, 1\}$ then we can intuitively think that we are making a lot of progress: such recursive calls cannot have a depth of more than $2k$ before we are able to solve the problem (this is where the $O^*(4^k)$ running time comes from). Thus, our high-level strategy is to avoid branching in cases where the removed set does not simplify the graph in this way.

Let us begin by arguing that the algorithm's second recursive call always makes significant progress in the sense described above. In the remainder we assume, as in the previous section, that the graph has been simplified using Fact 3, so each player must at some point pass the token to her opponent to avoid losing. We now need a helpful lemma.

Lemma 3. *Let M be a non-sink module of G and suppose that $W_i \cap M \neq \emptyset$. Then, $V_i \cap M \subseteq W_i$. Furthermore, there exists a vertex $v \in W_i \setminus M$ such that there is an edge from M to v.*

An informal way to interpret Lemma 3 is that, if a player wins some vertex of a module, then she must be winning all the vertices she controls in that module. We now have the following:

Corollary 1. *Let D be an i-dominion, for some $i \in \{0,1\}$ and M a non-sink module such that $D \cap M \neq \emptyset$. Then $\mathrm{attr}_i(D) \supseteq M \cap V_i$.*

We now know that the second recursive call of McNaughton's algorithm will always remove all the vertices owned by one of the two players in one of the basic modules. What remains is to deal with the first recursive call, where we remove an attractor of the maximum priority vertices. In this case we cannot guarantee that the removal of the attractor will necessarily produce a much simpler graph. However, we will argue that, if the graph is not simplified, the removed vertices were "irrelevant": their presence does not change the winning status of any other vertex. We will then be able to calculate their winning status *without* making the second recursive call using Lemma 3.

The key idea is contained in the following lemma.

Lemma 4. *Let p be the maximum priority in G, v be a vertex with priority p and $i := p \bmod 2$. Let $A := \mathrm{attr}_i(\{v\})$. Suppose that $A \subseteq M_j$ for some basic non-sink module M_j and that if A contains a vertex controlled by some player, then $M_j \setminus A$ also contains a vertex controlled by that player. Then we have $W_i(G) \setminus A = W_i(G \setminus A)$.*

We are now ready to put everything together to obtain the promised algorithm.

Theorem 5. *Consider a parity game on a strongly connected graph $G(V, E)$ where V is partitioned into $k > 1$ non-empty modules. There exists an algorithm that decides the winning regions of the two players and their winning strategies from these regions in time $O^*(4^k)$.*

It's now easy to apply the above algorithm to the case of modular width.

Corollary 2. *There exists a $O^*(4^k)$ algorithm that decides parity games on graphs with modular width k.*

4.2   Sub-exponential FPT Algorithm

Let us now present an improved version of the algorithm of Theorem 5. The idea is inspired by the improved version of McNaughton's algorithm from [19]: we will first look for a dominion that is confined inside at most $\sqrt{k}$ of the k basic modules. If such a dominion is found we can simplify the graph. Otherwise, we know that the second recursive call (when made) will touch at least $\sqrt{k}$ dominions, thus making significant progress.

Theorem 6. *Consider a parity game on a strongly connected graph $G(V, E)$ where V is partitioned into $k > 1$ non-empty modules. There exists an algorithm that decides the winning regions of the two players and their winning strategies from these regions in time $k^{O(\sqrt{k})} n^{O(1)}$.*

Corollary 3. *There exists a $k^{O(\sqrt{k})} n^{O(1)}$ algorithm that decides parity games on graphs with modular width k.*

5   Hardness Results

5.1   Finding Dominions

In this section we consider the parameterized complexity of the following problem: given an instance of parity games, does there exist a dominion consisting of at most k vertices? This problem arises very naturally in the course of examining the sub-exponential time algorithm given in [19]. The first step of this algorithm is to look for a "small" dominion, that is, a dominion that has size k, where k is a parameter to be optimized. The approach proposed in [19] is simply to try out all $\binom{n}{k}$ sets of vertices of size k and solve parity games on the resulting subgraph. Since solving parity games in a graph with k vertices can be done in time sub-exponential in k, the $\binom{n}{k}$ factor dominates the running time of this process. It is therefore natural to ask whether this can be sped up, which in turn would imply that a different value should be chosen for k to get the best worst-case bound on the algorithm. One could plausibly hope to try improving the running time to something like $2^{\sqrt{n}}$ (from $n^{\sqrt{n}}$) using such an approach.

Unfortunately, Theorem 7 establishes that improving significantly upon this brute force method is likely to be very hard. In particular, we give a parameterized reduction from MULTI-COLORED CLIQUE, showing that finding a dominion of size at most k cannot be done in time $n^{o(\sqrt{k})}$ (under standard assumptions).

Theorem 7. *There is no algorithm which, given an instance of parity games decides if there exists a dominion on at most k vertices in time $n^{o(\sqrt{k})}$, unless the ETH fails.*

5.2   Rabin Games

A Rabin game is another type of infinite game played on a graph, which in some sense generalizes parity games. In this section we present a simple reduction which shows that the algorithmic results we have obtained for parity games are unlikely to be extendible to Rabin games. Viewed another way, this reduction shows that the restricted graph classes we have considered are still quite non-trivial. More background on Rabin games is given in the full version.

Our reduction is from MULTI-COLORED CLIQUE. It is essentially a simple tweak of known reductions for parity games (see e.g. the reduction in [3]). The new feature is that we are careful that the graph produced has some very restricted structure. Below, we will say that a vertex is non-trivial if its out-degree is more than 1.

Theorem 8. *There is a polynomial-time reduction from MULTI-COLORED CLIQUE to Rabin games which produces an instance $G(V, E)$ with the following properties:*

- *G can be turned into a DAG by deleting one vertex.*
- *G has modular width $2k + 1$.*
- *Player 0 controls k non-trivial vertices.*
- *Player 1 controls 1 non-trivial vertex.*

Acknowledgements. We would like to thank Danupon Nanongkai for suggesting this problem and for our useful discussions.

References

1. Berwanger, D., Dawar, A., Hunter, P., Kreutzer, S., Obdržálek, J.: The dag-width of directed graphs. J. Comb. Theo. Ser. B **102**(4), 900–923 (2012)
2. Berwanger, D., Grädel, E., Kaiser, L., Rabinovich, R.: Entanglement and the complexity of directed graphs. Theor. Comput. Sci. **463**, 2–25 (2012)
3. Bjorklund, H., Sandberg, S., Vorobyov, S.: On fixed-parameter complexity of infinite games. In: Sere, K., Waldén, M. (eds.) The Nordic Workshop on Programming Theory (NWPT 2003), number 34 in Åbo Akademi, Reports on Computer Science and Mathematics, pp. 29–31. Citeseer (2003)
4. Björklund, H., Sandberg, S., Vorobyov, S.G.: Memoryless determinacy of parity and mean payoff games: a simple proof. Theor. Comput. Sci. **310**(1–3), 365–378 (2004)
5. Björklund, H., Vorobyov, S.G.: A combinatorial strongly subexponential strategy improvement algorithm for mean payoff games. Discrete Appl. Math. **155**(2), 210–229 (2007)
6. Bojanczyk, M., Dittmann, C., Kreutzer, S.: Decomposition theorems and model-checking for the modal μ-calculus. In: Henzinger, T.A., Miller, D. (eds.) Joint Meeting of the Twenty-Third EACSL Annual Conference on Computer Science Logic (CSL) and the Twenty-Ninth Annual ACM/IEEE Symposium on Logic in Computer Science (LICS), CSL-LICS 2014, Vienna, Austria, July 14–18, 2014, p. 17. ACM (2014)
7. Browne, A., Clarke, E.M., Jha, S., Long, D.E., Marrero, W.R.: An improved algorithm for the evaluation of fixpoint expressions. Theor. Comput. Sci. **178**(1–2), 237–255 (1997)
8. Chatterjee, K., Henzinger, T.A.: A survey of stochastic ω-regular games. J. Comput. Syst. Sci. **78**(2), 394–413 (2012)
9. Condon, A.: The complexity of stochastic games. Inf. Comput. **96**(2), 203–224 (1992)
10. Dittmann, C., Kreutzer, S., Tomescu, A.I.: Graph operations on parity games and polynomial-time algorithms. arXiv preprint (2012). arXiv:1208.1640
11. Emerson, E.A., Jutla, C.S.: The complexity of tree automata and logics of programs. SIAM J. Comput. **29**(1), 132–158 (1999)
12. Emerson, E.A., Jutla, C.S., Sistla, A.P.: On model checking for the μ-calculus and its fragments. Theor. Comput. Sci. **258**(1–2), 491–522 (2001)
13. Fomin, F.V., Liedloff, M., Montealegre, P., Todinca, I.: Algorithms parameterized by vertex cover and modular width, through potential maximal cliques. In: Ravi, R., Gørtz, I.L. (eds.) SWAT 2014. LNCS, vol. 8503, pp. 182–193. Springer, Heidelberg (2014)
14. Gajarský, J., Lampis, M., Ordyniak, S.: Parameterized algorithms for modular-width. In: Gutin, G., Szeider, S. (eds.) IPEC 2013. LNCS, vol. 8246, pp. 163–176. Springer, Heidelberg (2013)
15. Grädel, E., Thomas, W., Wilke, T. (eds.): Automata, Logics, and Infinite Games: A Guide to Current Research. LNCS, vol. 2500. Springer, Heidelberg (2002)
16. Hunter, P., Kreutzer, S.: Digraph measures: kelly decompositions, games, and orderings. Theor. Comput. Sci. **399**(3), 206–219 (2008)

17. Jurdzinski, M.: Deciding the winner in parity games is in UP cap co-up. Inf. Process. Lett. **68**(3), 119–124 (1998)
18. Jurdziński, M.: Small progress measures for solving parity games. In: Reichel, H., Tison, S. (eds.) STACS 2000. LNCS, vol. 1770, pp. 290–301. Springer, Heidelberg (2000)
19. Jurdzinski, M., Paterson, M., Zwick, U.: A deterministic subexponential algorithm for solving parity games. SIAM J. Comput. **38**(4), 1519–1532 (2008)
20. Küsters, R.: Memoryless determinacy of parity games. In: Grädel, E., Thomas, W., Wilke, T. (eds.) Automata Logics, and Infinite Games. LNCS, vol. 2500, pp. 95–106. Springer, Heidelberg (2002)
21. McNaughton, R.: Infinite games played on finite graphs. Ann. Pure Appl. Logic **65**(2), 149–184 (1993)
22. Obdržálek, J.: Fast mu-calculus model checking when tree-width is bounded. In: Hunt Jr., W.A., Somenzi, F. (eds.) CAV 2003. LNCS, vol. 2725, pp. 80–92. Springer, Heidelberg (2003)
23. Obdržálek, J.: Clique-width and parity games. In: Duparc, J., Henzinger, T.A. (eds.) CSL 2007. LNCS, vol. 4646, pp. 54–68. Springer, Heidelberg (2007)
24. Schewe, S.: Solving parity games in big steps. In: Arvind, V., Prasad, S. (eds.) FSTTCS 2007. LNCS, vol. 4855, pp. 449–460. Springer, Heidelberg (2007)
25. Zielonka, W.: Infinite games on finitely coloured graphs with applications to automata on infinite trees. Theor. Comput. Sci. **200**(1–2), 135–183 (1998)
26. Zwick, U., Paterson, M.: The complexity of mean payoff games on graphs. Theor. Comput. Sci. **158**(1&2), 343–359 (1996)

Algorithmic Applications of Tree-Cut Width

Robert Ganian[1]([⊠]), Eun Jung Kim[2], and Stefan Szeider[1]

[1] Algorithms and Complexity Group, TU Wien, Vienna, Austria
rganian@gmail.com
[2] CNRS, Université Paris-Dauphine, Paris, France

Abstract. The recently introduced graph parameter *tree-cut width* plays a similar role with respect to immersions as the graph parameter *treewidth* plays with respect to minors. In this paper we provide the first algorithmic applications of tree-cut width to hard combinatorial problems. Tree-cut width is known to be lower-bounded by a function of treewidth, but it can be much larger and hence has the potential to facilitate the efficient solution of problems which are not known to be fixed-parameter tractable (FPT) when parameterized by treewidth. We introduce the notion of nice tree-cut decompositions and provide FPT algorithms for the showcase problems CAPACITATED VERTEX COVER, CAPACITATED DOMINATING SET and IMBALANCE parameterized by the tree-cut width of an input graph G. On the other hand, we show that LIST COLORING, PRECOLORING EXTENSION and BOOLEAN CSP (the latter parameterized by the tree-cut width of the incidence graph) are W[1]-hard and hence unlikely to be fixed-parameter tractable when parameterized by tree-cut width.

1 Introduction

In their seminal work on graph minors, Robertson and Seymour have shown that all finite graphs are not only well-quasi ordered by the *minor* relation, but also by the *immersion* relation[1], the Graph Immersion Theorem [19]. This verified another conjecture by Nash-Williams [17]. As a consequence of this theorem, each graph class that is closed under taking immersions can be characterized by a finite set of forbidden immersions, in analogy to a graph class closed under taking minors being characterized by a finite set of forbidden minors.

In a recent paper [21], Wollan introduced the graph parameter *tree-cut width*, which plays a similar role with respect to immersions as the graph parameter *treewidth* plays with respect to minors. Wollan obtained an analogue to the Excluded Grid Theorem for these notions: if a graph has bounded tree-cut width, then it does not admit an immersion of the r-wall for arbitrarily large r

---

R. Ganian and S. Szeider — Research supported by the FWF Austrian Science Fund (X-TRACT, P26696).

[1] A graph H is an immersion of a graph G if H can be obtained from G by applications of vertex deletion, edge deletion, and edge lifting, i.e., replacing two incident edges by a single edge which joins the two vertices not shared by the two edges.

© Springer-Verlag Berlin Heidelberg 2015
G.F. Italiano et al. (Eds.): MFCS 2015, Part II, LNCS 9235, pp. 348–360, 2015.
DOI: 10.1007/978-3-662-48054-0_29

[21, Theorem 15]. Marx and Wollan [16] proved that for all n-vertex graphs H with maximum degree k and all k-edge-connected graphs G, either H is an immersion of G, or G has tree-cut width bounded by a function of k and n.

In this paper we provide the first algorithmic applications of tree-cut width to hard combinatorial problems. Tree-cut width is known to be lower-bounded by a function of treewidth, but it can be much larger than treewidth if the maximum degree is unbounded (see Subsect. 2.5 for an comparison of tree-cut width to other parameters). Hence tree-cut width has the potential to facilitate the efficient solution of problems which are not known or not believed to be fixed-parameter tractable (FPT) when parameterized by treewidth. For other problems it might allow the strengthening of parameterized hardness results.

We provide results for both possible outcomes: in Sect. 4 we provide FPT algorithms for the showcase problems Capacitated Vertex Cover, Capacitated Dominating Set and Imbalance parameterized by the tree-cut width of an input graph G, while in Sect. 5 we show that List Coloring, Precoloring Extension and Boolean CSP parameterized by tree-cut width (or, for the third problem, by the tree-cut width of the incidence graph) are not likely to be FPT. Table 1 provides an overview of our results. The table shows how tree-cut width provides an intermediate measurement that allows us to push the frontier for fixed-parameter tractability in some cases, and to strengthen W[1]-hardness results in some other cases.

Table 1. Overview of results (*tw* stands for treewidth).

Problem	Parameter		
	tw	tree-cut width	max-degree and tw
Capacitated Vertex Cover	W[1]-hard [4]	FPT[Thm 3]	FPT
Capacitated Dominating Set	W[1]-hard [4]	FPT[Thm 5]	FPT
Imbalance	Open [15]	FPT[Thm 4]	FPT[15]
List Coloring	W[1]-hard [7]	W[1]-hard[Thm 6]	FPT[Obs 4]
Precoloring Extension	W[1]-hard [7]	W[1]-hard[Thm 6]	FPT[Obs 4]
Boolean CSP	W[1]-hard [20]	W[1]-hard[Thm 7]	FPT [20]

Our FPT algorithms assume that a suitable decomposition, specifically a so-called *tree-cut decomposition*, is given as part of the input. Since the class of graphs of tree-cut width at most k is closed under taking immersions [21, Lemma 10], the Graph Immersion Theorem together with the fact that immersions testing is fixed-parameter tractable [10] gives rise to a non-uniform FPT algorithm for testing whether a graph has tree-cut width at most k. In a recent unpublished manuscript, Kim et al. [12] provide a uniform FPT algorithm which constructs a tree-cut decomposition whose width is at most twice the optimal one. Effectively, this result allows us to remove the condition that a tree-cut decomposition is supplied as part of the input from our uniform FPT algorithms.

We briefly outline the methodology used to obtain our algorithmic results. As a first step, in Sect. 3 we develop the notion of *nice tree-cut decompositions*[2] and show that every tree-cut decomposition can be transformed into a nice one in polynomial time. These nice tree-cut decompositions are of independent interest, since they provide a means of simplifying the complex structure of tree-cut decompositions. In Sect. 4 we introduce a general three-stage framework for the design of FPT algorithms on nice tree-cut decompositions and apply it to our problems. The crucial part of this framework is the computation of the "joins." We show that the children of any node in a nice tree-cut decomposition can be partitioned into (i) a bounded number of children with complex connections to the remainder of the graph, and (ii) a potentially large set of children with only simple connections to the remainder of the graph. We then process these by a combination of branching techniques applied to (i) and integer linear programming applied to (ii). The specifics of these procedures differ from problem to problem.

2 Preliminaries

2.1 Basic Notation

We use standard terminology for graph theory, see for instance [3]. All graphs except for those used to compute the torso-size in Subsect. 2.4 are simple; the multigraphs used in Subsect. 2.4 have loops, and each loop increases the degree of the vertex by 2.

Given a graph G, we let $V(G)$ denote its vertex set and $E(G)$ its edge set. The (open) neighborhood of a vertex $x \in V(G)$ is the set $\{y \in V(G) : xy \in E(G)\}$ and is denoted by $N_G(x)$. The closed neighborhood $N_G[v]$ of x is defined as $N_G(v) \cup \{v\}$. For a vertex subset X, the (open) neighborhood of X is defined as $\bigcup_{x \in X} N(x) \setminus X$ and denoted by $N_G(X)$. The set $N_G[X]$ refers to the closed neighborhood of X defined as $N_G(X) \cup X$. We refer to the set $N_G(V(G) \setminus X)$ as $\partial_G(X)$; this is the set of vertices in X which have a neighbor in $V(G) \setminus X$. The degree of a vertex v in G is denoted $deg_G(v)$. When the graph we refer to is clear, we drop the lower index G from the notation. We use $[i]$ to denote the set $\{0, 1, \ldots, i\}$.

2.2 Parameterized Complexity

We refer the reader to [5] for standard notions in parameterized complexity, such as the complexity classes FPT and W[1], FPT Turing reductions and the MULTI-COLORED CLIQUE (MCC) problem.

2.3 Integer Linear Programming

Our algorithms use an Integer Linear Programming (ILP) subroutine. ILP is a well-known framework for formulating problems and a powerful tool for the development of fpt-algorithms for optimization problems.

---

[2] We call them "nice" as they serve a similar purpose as the nice tree decompositions [13], although the definitions are completely unrelated.

Definition 1 (p-Variable Integer Linear Programming Optimization).
Let $A \in \mathbb{Z}^{q \times p}, b \in \mathbb{Z}^{q \times 1}$ and $c \in \mathbb{Z}^{1 \times p}$. The task is to find a vector $x \in \mathbb{Z}^{p \times 1}$ which minimizes the objective function $c \times \bar{x}$ and satisfies all q inequalities given by A and b, specifically satisfies $A \cdot \bar{x} \geq b$. The number of variables p is the parameter.

Theorem 1 ([8,9,11,14]). *p-OPT-ILP can be solved using $O(p^{2.5p+o(p)} \cdot L)$ arithmetic operations in space polynomial in L, L being the number of bits in the input.*

2.4   Tree-Cut Width

The notion of tree-cut decompositions was first proposed by Wollan [21], see also [16]. A family of subsets $X_1, \ldots, X_k$ of X is a *near-partition* of X if they are pairwise disjoint and $\bigcup_{i=1}^{k} X_i = X$, allowing the possibility of $X_i = \emptyset$.

Definition 2. *A tree-cut decomposition of G is a pair $(T, \mathcal{X})$ which consists of a tree T and a near-partition $\mathcal{X} = \{X_t \subseteq V(G) : t \in V(T)\}$ of $V(G)$. A set in the family $\mathcal{X}$ is called a bag of the tree-cut decomposition.*

For any edge $e = (u, v)$ of T, let T_u and T_v be the two connected components in $T - e$ which contain u and v respectively. Note that $(\bigcup_{t \in T_u} X_t, \bigcup_{t \in T_v} X_t)$ is a partition of $V(G)$, and we use $cut(e)$ to denote the set of edges with one endpoint in each partition. A tree-cut decomposition is *rooted* if one of its nodes is called the root, denoted by r. For any node $t \in V(T) \setminus \{r\}$, let $e(t)$ be the unique edge incident to t on the path between r and t. We define the *adhesion* of t ($adh_T(t)$ or $adh(t)$ in brief) as $|cut(e(t))|$; if t is the root, we set $adh_T(t) = 0$.

The *torso* of a tree-cut decomposition $(T, \mathcal{X})$ at a node t, written as H_t, is the graph obtained from G as follows. If T consists of a single node t, then the torso of $(T, \mathcal{X})$ at t is G. Otherwise let $T_1, \ldots, T_\ell$ be the connected components of $T - t$. For each $i = 1, \ldots, \ell$, the vertex set Z_i of $V(G)$ is defined as the set $\bigcup_{b \in V(T_i)} X_b$. The torso H_t at t is obtained from G by *consolidating* each vertex set Z_i into a single vertex z_i. Here, the operation of consolidating a vertex set Z into z is to substitute Z by z in G, and for each edge e between Z and $v \in V(G) \setminus Z$, adding an edge zv in the new graph. We note that this may create parallel edges.

The operation of *suppressing* a vertex v of degree at most 2 consists of deleting v, and when the degree is two, adding an edge between the neighbors of v. Given a graph G and $X \subseteq V(G)$, let the *3-center* of (G, X) be the unique graph obtained from G by exhaustively suppressing vertices in $V(G) \setminus X$ of degree at most two. Finally, for a node t of T, we denote by $\tilde{H}_t$ the 3-center of (H_t, X_t), where H_t is the torso of $(T, \mathcal{X})$ at t. Let the *torso-size* $tor(t)$ denote $|\tilde{H}_t|$.

Definition 3. *The width of a tree-cut decomposition $(T, \mathcal{X})$ of G is $\max_{t \in V(T)} \{adh(t), tor(t)\}$. The tree-cut width of G, or $\mathbf{tcw}(G)$ in short, is the minimum width of $(T, \mathcal{X})$ over all tree-cut decompositions $(T, \mathcal{X})$ of G.*

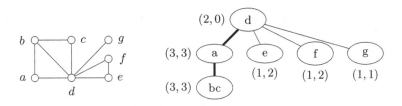

Fig. 1. A graph G and a width-3 tree-cut decomposition of G, including the torso-size (left value) and adhesion (right value) of each node.

We conclude this subsection with some notation related to tree-cut decompositions. For $t \in V(T) \setminus \{r\}$, we let $p_T(t)$ (or $p(t)$ in brief) denote the parent of t in T. For two distinct nodes $t, t' \in V(T)$, we say that t and t' are *siblings* if $p(t) = p(t')$. Given a tree node t, let T_t be the subtree of T rooted at t. Let $Y_t = \bigcup_{b \in V(T_t)} X_b$, and let G_t denote the induced subgraph $G[Y_t]$. The *depth* of a node t in T is the distance of t from the root r. The vertices of $\partial(Y_t)$ are called the *borders* at node t. A node $t \neq r$ in a rooted tree-cut decomposition is *thin* if $adh(t) \leq 2$ and *bold* otherwise (see Fig. 1).

2.5 Relations to Other Width Parameters

Here we review the relations between the tree-cut width and other width parameters, specifically *treewidth* (**tw**), *pathwidth* (**pw**), and *treedepth* (**td**) [18]. We also compare to the maximum over treewidth and maximum degree, which we refer to as *degree treewidth* (**degtw**).

Proposition 1 ([16,21]). *There exists a function h such that $\mathbf{tw}(G) \leq h(\mathbf{tcw}(G))$ and $\mathbf{tcw}(G) \leq 4\mathbf{degtw}(G)^2$ for any graph G.*

Below, we provide an explicit bound on the relationship between treewidth and tree-cut width, and show that it is incomparable with pathwidth and treedepth (see Fig. 2).

Proposition 2. *For any graph G it holds that $\mathbf{tw}(G) \leq 2\mathbf{tcw}(G)^2 + 3\mathbf{tcw}(G)$.*

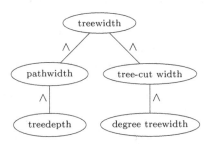

Fig. 2. Relationships between selected graph parameters ($A > B$ means that every graph class of bounded A is also of bounded B, but there are graph classes of bounded B which are not of bounded A).

Proposition 3. *There exists a graph class $\mathcal{H}_1$ of bounded pathwidth and treedepth but unbounded tree-cut width, and there exists a graph class $\mathcal{H}_2$ of bounded tree-cut width but unbounded pathwidth and treedepth.*

3   Nice Tree-Cut Decompositions

In this section we introduce the notion of a *nice tree-cut decomposition* and present an algorithm to transform any tree-cut decomposition into a nice one without increasing the width. A tree-cut decomposition $(T, \mathcal{X})$ rooted at r is *nice* if it satisfies the following condition for every thin node $t \in V(T)$: $N(Y_t) \cap \bigcup_{b \text{ is a sibling of } t} Y_b = \emptyset$.

The intuition behind nice tree-cut decompositions is that we restrict the neighborhood of thin nodes in a way which facilitates dynamic programming.

Lemma 1. *There exists a cubic-time algorithm which transforms any rooted tree-cut decomposition $(T, \mathcal{X})$ of G into a nice tree-cut decomposition of the same graph, without increasing its width or number of nodes.*

The proof of Lemma 1 is based on the following considerations. Let $(T, \mathcal{X})$ be a rooted tree-cut decomposition of G whose width is at most w. We say that a node t, $t \neq r$, is *bad* if it violates the above condition, i.e., $adh(t) \leq 2$ and there is a sibling b of t such that $N(Y_t) \cap Y_b \neq \emptyset$. For a bad node t, we say that b is a *bad neighbor* of t if $N(Y_t) \cap X_b \neq \emptyset$ and b is either a sibling of t or a descendant of a sibling of t.

> REROUTING(t): let t be a bad node and let b be a bad neighbor of t of maximum depth (resolve ties arbitrarily). Then remove the tree edge $e(t)$ from T and add a new tree edge $\{b, t\}$.

> TOP-DOWN REROUTING: as long as $(T, \mathcal{X})$ is not a nice tree-cut decomposition, pick any bad node t of minimum depth. Perform REROUTING(t).

The proof of Lemma 1 then follows by showing that REROUTING does not increase the width of the decomposition and that TOP-DOWN REROUTING terminates after performing at most a quadratic number of REROUTING calls.

The following Theorem 2 builds upon Lemma 1 by additionally giving a bound on the size of the decomposition.

Theorem 2. *If $\mathrm{tcw}(G) = k$, then there exists a nice tree-cut decomposition $(T, \mathcal{X})$ of G of width k with at most $2|V(G)|$ nodes. Furthermore, $(T, \mathcal{X})$ can be computed from any width-k tree-cut decomposition of G in cubic time.*

4   FPT Algorithms

In this section we will introduce a general dynamic programming framework for the design of FPT algorithms on nice tree-cut decompositions. The framework is based on leaf-to-root processing of the decompositions and can be divided into three basic steps:

- **Data Table**: definition of a data table $\mathcal{D}_T(t)$ ($\mathcal{D}(t)$ in brief) for a problem $\mathcal{P}$ associated with each node t of a nice tree-cut decomposition $(T, \mathcal{X})$.
- **Initialization and Termination**: computation of $\mathcal{D}(t)$ in FPT time for any leaf t, and solution of $\mathcal{P}$ in FPT time if $\mathcal{D}(r)$ is known for the root r.
- **Inductive Step**: an FPT algorithm for computing $\mathcal{D}(t)$ for any node t when $\mathcal{D}(t')$ is known for every child t' of t.

Let t be a node in a nice tree-cut decomposition. We use B_t to denote the set of thin children t' of t such that $N(Y_{t'}) \subseteq X_t$, and we let A_t contain every child of t not in B_t. The following lemma is a crucial component of our algorithms, since it bounds the number of children with nontrivial edge connections to other parts of the decomposition.

Lemma 2. *Let t be a node in a nice tree-cut decomposition of width k. Then $|A_t| \leq 2k + 1$.*

In the remainder of this section we employ this high-level framework on the design of FPT algorithms for the following problems: CAPACITATED VERTEX COVER, IMBALANCE, and CAPACITATED DOMINATING SET.

4.1 Capacitated Vertex Cover

The CAPACITATED VERTEX COVER is a generalization of the classical VERTEX COVER problem. Unlike its uncapacitated variant, CAPACITATED VERTEX COVER is known to be W[1]-hard when parameterized by treewidth [4].

A capacitated graph is a graph $G = (V, E)$ together with a capacity function $c : V \to \mathbb{N}_0$. Then we call $C \subseteq V$ a *capacitated vertex cover* of G if there exists a mapping $f : E \to C$ which maps every edge to one of its endpoints so that the total number of edges mapped by f to any $v \in C$ does not exceed $c(v)$. We say that f *witnesses* C.

---

tcw-CAPACITATED VERTEX COVER (tcw-CVC)

Instance: A capacitated graph G on n vertices together with a width-k tree-cut decomposition $(T, \mathcal{X})$ of G, and an integer d.

Parameter: k.

Task: Decide whether there exists a capacitated vertex cover C of G containing at most d vertices.

Data Table, Initialization and Termination. Informally, we store for each node t two pieces of information: the "cost" of covering all edges inside $G[Y_t]$, and how much more it would cost to additionally cover edges incident to Y_t. We formalize below.

For any graph $G = (V, E)$ and $U \subseteq V$, we let $cvc(G, U)$ denote the minimum cardinality of a capacitated vertex cover $C \subseteq U$ of G; if no such capacitated vertex cover exists, we instead let $cvc(G, U) = \infty$. For any node t in a nice tree-cut decomposition of a capacitated graph $G = (V, E)$, we then use a_t to denote $cvc(G[Y_t], Y_t)$.

Let E_t denote the set of all edges with both endpoints in Y_t and let K_t denote the set of edges with exactly one endpoint in Y_t. Then $Q_t = \{ H = (Y_t \cup N(Y_t), E_t \cup E') \mid E' \subseteq K_t \}$. Finally, we define $\beta_t : Q_t \to \mathbb{N}_0$ such that $\beta_t(H) = cvc(H, Y_t) - a_t$ (whereas ∞ acts as an absorbing element).

Definition 4. $\mathcal{D}(t) = (a_t, \beta_t)$.

Next, we show that the number of possible functions β_t is bounded.

Lemma 3. *Let k be the width of a nice tree-cut decomposition $(T, \mathcal{X})$ of G and let t be any node of T. Then $\beta_t(H) \in [k] \cup \{\infty\}$ for every $H \in Q_t$.*

Lemma 4. *Let t be a leaf in a nice tree-cut decomposition $(T, \mathcal{X})$ of a capacitated graph G, and let k be the width of $(T, \mathcal{X})$. Then $\mathcal{D}(t)$ can be computed in time $2^{O(k \cdot \log k)}$.*

Observation 1. *Let (G, d) be an instance of tcw-CVC and let r be the root of a nice tree-cut decomposition of G. Then (G, d) is a yes-instance if and only if $a_r \leq d$.*

Inductive Step. Our next and final goal is to show how to compute $\mathcal{D}(t)$ of a node t once we have $\mathcal{D}(t')$ for each child t' of t. We call this problem CVC JOIN, and we use a two-step approach to solve it. First, we reduce the problem to a simplified version, which we call REDUCED CVC JOIN and which has the following properties: A_t is empty, $adh(t) = 0$, and $G[X_t]$ is edgeless.

Lemma 5. *There is an FPT Turing reduction from CVC JOIN to $2^{O(k^2)}$ instances of REDUCED CVC JOIN which runs in time $2^{O(k^2)} \cdot (|B_t| + 1)$.*

Lemma 6. *There exists an algorithm which solves REDUCED CVC JOIN in time $k^{O(k^2)} \cdot (|B_t| + 1)$.*

Proof (Sketch). We develop an ILP formulation by partitioning B_t into types, which contain nodes with the same β_t and the same neighborhood in X_t. We use variables to express how edges are assigned between types and X_t. □

Corollary 1. *There exists an algorithm which solves CVC JOIN in time $k^{O(k^2)} \cdot (|B_t| + 1)$.*

Theorem 3. *tcw-CVC can be solved in time $k^{O(k^2)} \cdot n + |T|^3$.*

Proof. We use Theorem 2 to transform $(T, \mathcal{X})$ into a nice tree-cut decomposition with at most $2n$ nodes. We then use Lemma 4 to compute $\mathcal{D}(t)$ for each leaf t of T, and proceed by computing $\mathcal{D}(t)$ for nodes in a bottom-to-top fashion by Corollary 1. The total running time is dominated by $\sum_{t \in T}(k^{O(k^2)} \cdot (|B_t| + 1))$, which is bounded by $n \cdot k^{O(k^2)}$ since each node t' appears in at most a single set B_t. Once we obtain $\mathcal{D}(r)$, we can correctly output by Observation 1. □

4.2 Imbalance

The IMBALANCE problem was introduced by Biedl et al. [1]. The problem is FPT when parameterized by **degtw** [15]. In this subsection we prove that IMBALANCE remains FPT even when parameterized by the more general tree-cut width.

Given a linear order R of vertices in a graph G, let $\lhd_R(v)$ and $\rhd_R(v)$ denote the number of neighbors of v which occur, respectively, before ("to the left of") v in R and after ("to the right of") v in R. The *imbalance* of a vertex v, denoted $imb_R(v)$, is then defined as the absolute value of $\rhd_R(v) - \lhd_R(v)$, and the imbalance of R, denoted imb_R, is equal to $\sum_{v \in V(G)} imb_R(v)$.

---

tcw-IMBALANCE (tcw-IMB)
Instance: A graph $G = (V, E)$ with $|V| = n$, and a width-k tree-cut decomposition $(T, \mathcal{X})$ of G, and an integer d.
Parameter: k.
Task: Decide whether there exists a linear order R of V such that $imb_R \leq d$.

Data Table, Initialization and Termination. Let $A \subseteq B$ be sets and let f_A, f_B be linear orders of A, B respectively. We say that f_A is a *linear suborder* of f_B if the elements of A occur in the same order in f_A as in f_B (similarly, f_B is a *linear superorder* of f_A). The information we remember in our data tables can be informally summarized as follows. First, we remember the minimum imbalance which can be achieved by any linear order in Y_t. Second, for each linear order f of vertices which have neighbors outside of Y_t *and* for a restriction on the imbalance on these vertices, we remember how much the imbalance grows when considering only linear superorders of f which satisfy these restrictions. The crucial ingredient is that the restrictions mentioned above are "weak" and we only care about linear superorders of f which do not increase over the optimum "too much"; this allows the second, crucial part of our data tables to remain bounded in k.

For brevity, for $v \in Y_t$ we let $n_t(v)$ denote $|N_{V \setminus Y_t}(v)|$, i.e., the number of neighbors of v outside Y_t. Let f be a linear order of $\partial(Y_t)$ and let τ be a mapping such that $\tau(v \in \partial(Y_t)) \in \{-\infty, -n_t(v), -n_t(v)+1, \ldots, n_t(v), \infty\}$. We then call a tuple of the form (f, τ) an *extract* (of Y_t), and let $\mathcal{L}$ denote the set of all extracts (for nodes with adhesion at most k). The extract $\alpha = (f, \tau)$ is realized in Y_t (by R) if there exists a linear order R of Y_t such that

1. R is a linear superorder of f, and
2. for each $v \in \partial(Y_t)$:
   - if $\tau(v) \in \mathbb{Z}$ then $imb_R(v) = \tau(v)$,
   - if $\tau(v) = -\infty$ then $\rhd_R(v) - \lhd_R(v) < -n_t(v) - 1$,
   - if $\tau(v) = \infty$ then $\rhd_R(v) - \lhd_R(v) > n_t(v) + 1$.

The cost of a realized extract α, denoted $c(\alpha)$, is the minimum value of $\sum_{v \in Y_t} imb_R(v)$ over all R which realize α (notice that edges with only one endpoint in Y_t do not contribute to $c(\alpha)$). If α is not realized in Y_t, we let $c(\alpha) = \infty$.

We store the following information in our data table: the cost of a minimum extract realized in Y_t, and the cost of every extract whose cost is not much larger than the minimum cost. We formalize below; let e_t denote the number of edges with one endpoint in Y_t.

Definition 5. $\mathcal{D}(t) = (a_t, \beta_t)$ where $a_t = \min_{\alpha \in \mathcal{L}} c(\alpha)$ and $\beta_t : \mathcal{L} \to \mathbb{N}_0 \cup \{\infty\}$ such that $\beta_t(\alpha) = c(\alpha) - a_t$ if $c(\alpha) - a_t \leq 4e_t$ and $\beta_t(\alpha) = \infty$ otherwise.

Notice that we are deliberately discarding information about the cost of extracts whose cost exceeds the optimum by over $4e_t$.

Observation 2. *The cardinality of $\mathcal{L}$ is bounded by $k^{O(k)}$, and hence the number of possible functions β_t is bounded by $k^{O(k^2)}$. Additionally, these may be enumerated in the same time.*

Lemma 7. *Let t be a leaf in a nice tree-cut decomposition $(T, \mathcal{X})$ of a graph G, and let k be the width of $(T, \mathcal{X})$. Then $\mathcal{D}(t)$ can be computed in time $k^{O(k)}$.*

Observation 3. *Let (G, d) be an instance of tcw-IMB and let r be the root of a nice tree-cut decomposition of G. Then (G, d) is a yes-instance if and only if $a_r \leq d$.*

Inductive Step. What remains is to show how to compute $\mathcal{D}(t)$ for a node t once $\mathcal{D}(t')$ is known for each child t' of t. Once again, we use the two-step approach of first reducing to a "simpler" problem and then applying a suitable ILP encoding. We call the problem we reduce to REDUCED IMB JOIN.

Lemma 8. *There is an FPT Turing reduction from IMB JOIN to $k^{O(k^2)}$ instances of REDUCED IMB JOIN which runs in time $k^{O(k^2)} \cdot (|B_t| + 1)$.*

Proof (Sketch). We branch over all linear orders of $f(k)$-many "important vertices" in A_t, over all possible extracts in $\mathcal{D}(t)$, and over all extracts in A_t which are compatible with the above. This gives sufficient information to compute the imbalance of vertices in X_t along with constraints on the placement of β_t, which yield an instance of REDUCED IMB JOIN. □

Lemma 9. *There exists an algorithm which solves REDUCED IMB JOIN in time $k^{O(k^4)} \cdot (|B_t| + 1)$.*

Proof (Sketch). The k absolute values can be translated into 2^k-many ILP instances by branching on whether they end up being positive or negative. Vertices in X_t separate the linear order into $k + 1$ "regions", and we can again partition B_t into types. Variables express how many children of type i have a border vertex placed in region j. □

The proof of the theorem below is then analogous to the proof of Theorem 3.

Theorem 4. *tcw-IMB can be solved in time $k^{O(k^4)} \cdot n + |T|^3$.*

4.3 Capacitated Dominating Set

CAPACITATED DOMINATING SET is a generalization of the classical DOMINATING SET problem by the addition of vertex capacities. It is known to be W[1]-hard when parameterized by treewidth [4].

Let $G = (V, E)$ be a capacitated graph with a capacity function $c : V(G) \to \mathbb{N}_0$. We say that $D \subseteq V(G)$ is a *capacitated dominating set* of G if there exists a mapping $f : V \setminus D \to D$ which maps every vertex to one of its neighbors so that the total number of vertices mapped by f to any $v \in D$ does not exceed $c(v)$.

> tcw-CAPACITATED DOMINATING SET (tcw-CDS)
> *Instance*: A capacitated graph G on n vertices together with a width-k tree-cut decomposition $(T, \mathcal{X})$ of G, and an integer d.
> *Parameter*: k.
> *Task*: Decide whether there exists a capacitated dominating set D of G containing at most d vertices.

The methods used to solve tcw-CDS are similar to those used to prove Theorems 3 and 4, and hence we only provide a high-level description of our approach here. For tcw-CDS, the table $\mathcal{D}(t)$ stores information about whether vertices in X_t occur in a dominating set, the residual capacities in $\partial(Y_t)$, and the size of a minimum capacitated dominating set which has these properties. The following steps are then analogous to those above.

Theorem 5. *tcw-CDS can be solved in time $k^{O(k^2)} \cdot n + |T|^3$.*

5 Lower Bounds

We show that LIST COLORING [6] and PRECOLORING EXTENSION [2] are W[1]-hard parameterized by tree-cut width, strengthening the known W[1]-hardness results with respect to treewidth [7].

> tcw-LIST COLORING
> *Instance*: A graph $G = (V, E)$, a width-k tree-cut decomposition $(T, \mathcal{X})$ of G, and for each vertex $v \in V$ a list $L(v)$ of permitted colors.
> *Parameter*: k.
> *Task*: Decide whether there exists a proper vertex coloring c such that $c(v) \in L(v)$ for each $v \in V$.

The tcw-PRECOLORING EXTENSION problem may be defined analogously as LIST COLORING; the only difference is that in PRECOLORING EXTENSION lists are restricted to either contain a single color or all possible colors.

Observation 4. LIST COLORING *and* PRECOLORING EXTENSION *parameterized by* **degtw** *are FPT.*

Theorem 6. *tcw-LIST COLORING and tcw-PRECOLORING EXTENSION are W[1]-hard.*

We also show that the CONSTRAINT SATISFACTION PROBLEM (CSP) is W[1]-hard when parameterized by the tree-cut width of the incidence graph, even when restricted to the Boolean domain; this is not the case for **degtw** [20].

---

tcw-CSP

Instance: A CSP instance $I = (X, D, C)$ together with a width-k tree-cut decomposition $(T, \mathcal{X})$ of the incidence graph G_I of I.

Parameter: k.

Task: Decide whether I is satisfiable.

Theorem 7. *tcw*-BOOLEAN CSP *is* W[1]-*hard.*

The proofs of Theorems 6 and 7 are based on a reduction from MCC.

6 Concluding Notes

We have provided the first algorithmic applications of the new graph parameter tree-cut width, considering a variety of hard combinatorial problems. In some cases we could establish fixed-parameter tractability, in some cases we could establish W[1]-hardness, staking off the potentials and limits of this parameter (see Table 1). The FPT algorithms make use of our new notion of nice tree-cut decompositions, which we believe to be of independent interest. We hope that our results and methods stimulate further work on problems parameterized by tree-cut width, which will result in a more refined parameterized complexity landscape; natural candidate problems include further graph layout problems or the General Factor Problem.

References

1. Biedl, T., Chan, T., Ganjali, Y., Hajiaghayi, M.T., Wood, D.R.: Balanced vertex-orderings of graphs. DAM **148**(1), 27–48 (2005)
2. Biró, M., Hujter, M., Tuza, Z.: Precoloring extension. i. Interval graphs. Discrete Math. **100**(1–3), 267–279 (1992)
3. Diestel, R.: Graph Theory. Graduate Texts in Mathematics. Springer, New York (2000)
4. Dom, M., Lokshtanov, D., Saurabh, S., Villanger, Y.: Capacitated domination and covering: a parameterized perspective. In: Grohe, M., Niedermeier, R. (eds.) IWPEC 2008. LNCS, vol. 5018, pp. 78–90. Springer, Heidelberg (2008)
5. Downey, R.G., Fellows, M.R.: Fundamentals of Parameterized Complexity. Texts in Computer Science. Springer, London (2013)
6. Erdős, P., Rubin, A.L., Taylor, H.: Choosability in graphs. Congressus Numerantium **26**, 125–157 (1979)
7. Fellows, M.R., Fomin, F.V., Lokshtanov, D., Rosamond, F., Saurabh, S., Szeider, S., Thomassen, C.: On the complexity of some colorful problems parameterized by treewidth. Inf. Comput. **209**(2), 143–153 (2011)

8. Fellows, M.R., Lokshtanov, D., Misra, N., Rosamond, F.A., Saurabh, S.: Graph layout problems parameterized by vertex cover. In: Hong, S.-H., Nagamochi, H., Fukunaga, T. (eds.) ISAAC 2008. LNCS, vol. 5369, pp. 294–305. Springer, Heidelberg (2008)

9. Frank, A., Tardos, É.: An application of simultaneous diophantine approximation in combinatorial optimization. Combinatorica 7(1), 49–65 (1987)

10. Grohe, M., Kawarabayashi, K.-I., Marx, D., Wollan, P.: Finding topological subgraphs is fixed-parameter tractable. In: STOC 2011–Proceedings of the 43rd ACM Symposium on Theory of Computing, pp. 479–488. ACM, New York (2011)

11. Kannan, R.: Minkowski's convex body theorem and integer programming. Math. Oper. Res. 12(3), 415–440 (1987)

12. Kim, E., Oum, S.-I., Paul, C., Sau, I., Thilikos, D.: FPT 2-approximation for constructing tree-cut decomposition (2014, Submitted) Manuscript

13. Kloks, T.: Treewidth: Computations and Approximations. Springer, Heidelberg (1994)

14. Lenstra, H.: Integer programming with a fixed number of variables. Math. Oper. Res. 8, 538–548 (1983)

15. Lokshtanov, D., Misra, N., Saurabh, S.: Imbalance is fixed parameter tractable. Inf. Process. Lett. 113(19–21), 714–718 (2013)

16. Marx, D., Wollan, P.: Immersions in highly edge connected graphs. SIAM J. Discrete Math. 28(1), 503–520 (2014)

17. Nash-Williams, C.S.J.A.: On well-quasi-ordering finite trees. Proc. Cambridge Philos. Soc. 59, 833–835 (1963)

18. Nešetřil, J., de Mendez, P.O.: Tree-depth, subgraph coloring and homomorphism bounds. European J. Combin. 27(6), 1024–1041 (2006)

19. Robertson, N., Seymour, P.D.: Graph minors. II. Algorithmic aspects of tree-width. J. Algorithms 7(3), 309–322 (1986)

20. Samer, M., Szeider, S.: Constraint satisfaction with bounded treewidth revisited. J. Comput. Syst. Sci. 76(2), 103–114 (2010)

21. Wollan, P.: The structure of graphs not admitting a fixed immersion. J. Comb. Theo. Ser. B 110, 47–66 (2015). http://arxiv.org/abs/1302.3867 (2013)

Log-Concavity and Lower Bounds
for Arithmetic Circuits

Ignacio García-Marco[1]($\boxtimes$), Pascal Koiran[1], and Sébastien Tavenas[2]

[1] LIP, ENS, Lyon, France
{ignacio.garcia-marco,pascal.koiran}@ens-lyon.fr
[2] Max-Planck-Insitut Für Informatik, Saarbrücken, Germany
Sebastien.tavenas@ens-lyon.org

Abstract. One question that we investigate in this paper is, how can we build log-concave polynomials using sparse polynomials as building blocks? More precisely, let $f = \sum_{i=0}^{d} a_i X^i \in \mathbb{R}^+[X]$ be a polynomial satisfying the log-concavity condition $a_i^2 > \tau a_{i-1} a_{i+1}$ for every $i \in \{1, \ldots, d-1\}$, where $\tau > 0$. Whenever f can be written under the form $f = \sum_{i=1}^{k} \prod_{j=1}^{m} f_{i,j}$ where the polynomials $f_{i,j}$ have at most t monomials, it is clear that $d \leqslant kt^m$. Assuming that the $f_{i,j}$ have only non-negative coefficients, we improve this degree bound to $d = \mathcal{O}(km^{2/3}t^{2m/3}\log^{2/3}(kt))$ if $\tau > 1$, and to $d \leqslant kmt$ if $\tau = d^{2d}$.

This investigation has a complexity-theoretic motivation: we show that a suitable strengthening of the above results would imply a separation of the algebraic complexity classes VP and VNP. As they currently stand, these results are strong enough to provide a new example of a family of polynomials in VNP which cannot be computed by monotone arithmetic circuits of polynomial size.

1 Introduction

Let $f = \sum_{j=0}^{d} a_j X^j \in \mathbb{R}[X]$ be a univariate polynomial of degree $d \in \mathbb{Z}^+$. It is a classical result due to Newton (see [4], §2.22 and §4.3 for two proofs) that whenever all the roots of f are real, then the coefficients of f satisfy the following log-concavity condition:

$$a_i^2 \geqslant \frac{d-i+1}{d-i} \frac{i+1}{i} a_{i-1} a_{i+1} \text{ for all } i \in \{1, \ldots, d-1\}. \qquad (1)$$

Moreover, if the roots of f are not all equal, these inequalities are strict. When $d = 2$, condition (1) becomes $a_1 \geqslant 4a_0 a_2$, which is well known to be a necessary and sufficient condition for all the roots of f to be real. Nevertheless, for $d \geqslant 3$, the converse of Newton's result does not hold any more [13].

This work was supported by ANR project CompA (project number: ANR-13-BS02-0001-01).

I. García-Marco and P. Koiran — UMR 5668 ENS Lyon - CNRS - UCBL - INRIA, Université de Lyon.

© Springer-Verlag Berlin Heidelberg 2015
G.F. Italiano et al. (Eds.): MFCS 2015, Part II, LNCS 9235, pp. 361–371, 2015.
DOI: 10.1007/978-3-662-48054-0_30

When $f \in \mathbb{R}^+[X]$, i.e., when $f = \sum_{j=0}^{d} a_j X^j$ with $a_j \geqslant 0$ for all $j \in \{0, \ldots, d\}$, a weak converse of Newton's result holds true. Namely, a sufficient condition for f to only have real (and distinct) roots is that

$$a_i^2 > 4a_{i-1}a_{i+1} \text{ for all } i \in \{1, \ldots, d-1\}.$$

Whenever a polynomial fulfills this condition, we say that it satisfies the *Kurtz condition* since this converse result is often attributed to Kurtz [13]. Note however that it was obtained some 70 years earlier by Hutchinson [6].

If f satisfies the Kurtz condition, all of its $d + 1$ coefficients are nonzero except possibly the constant term. Such a polynomial is therefore very far from being sparse (recall that a polynomial is informally called *sparse* if the number of its nonzero coefficients is small compared to its degree). One question that we investigate in this paper is: how can we construct polynomials satisfying the Kurtz condition using sparse polynomials as building blocks? More precisely, consider f a polynomial of the form

$$f = \sum_{i=1}^{k} \prod_{j=1}^{m} f_{i,j} \tag{2}$$

where $f_{i,j}$ are polynomials with at most t monomials each. By expanding the products in (2) we see that f has at most kt^m monomials. As a result, $d \leqslant kt^m$ if f satisfies the Kurtz condition. Our goal is to improve this very coarse bound. For the case of polynomials $f_{i,j}$ with nonnegative coefficients, we obtain the following result.

Theorem 1. *Consider a polynomial $f \in \mathbb{R}^+[X]$ of degree d of the form*

$$f = \sum_{i=1}^{k} \prod_{j=1}^{m} f_{i,j},$$

where $m \geqslant 2$ and the $f_{i,j} \in \mathbb{R}^+[X]$ have at most t monomials. If f satisfies the Kurtz condition, then $d = \mathcal{O}(km^{2/3}t^{2m/3}\log^{2/3}(kt))$.

We prove this result in Sect. 2. After that, in Sect. 3, we study the following stronger log-concavity condition

$$a_i^2 > d^{2d}a_{i-1}a_{i+1} \text{ for all } i \in \{1, \ldots, d-1\}. \tag{3}$$

In this setting we prove the following improved analogue of Theorem 1.

Theorem 2. *Consider a polynomial $f \in \mathbb{R}^+[X]$ of degree d of the form*

$$f = \sum_{i=1}^{k} \prod_{j=1}^{m} f_{i,j},$$

where $m \geqslant 2$ and the $f_{i,j} \in \mathbb{R}^+[X]$ have at most t monomials. If f satisfies (3), then $d \leqslant kmt$.

This investigation has a complexity-theoretic motivation: we show in Sect. 4 that a suitable extension of Theorem 2 (allowing negative coefficients for the polynomials f_{ij}) would imply a separation of the algebraic complexity classes VP and VNP. The classes VP of "easily computable polynomial families" and VNP of "easily definable polynomial families" were proposed by Valiant [15] as algebraic analogues of P and NP. As shown in Theorems 2 and 7 as it now stands is strong enough to provide a new example of a family of polynomials in VNP which cannot be computed by monotone arithmetic circuits of polynomial size.

2 The Kurtz Log-Concavity Condition

Our main tool in this section is a result of convex geometry [3]. To state this result, we need to introduce some definitions and notations. For a pair of planar finite sets $R, S \subset \mathbb{R}^2$, the *Minkowski sum* of R and S is the set $R + S := \{y + z \mid y \in R, z \in S\} \subset \mathbb{R}^2$. A finite set $C \subset \mathbb{R}^2$ is *convexly independent* if and only if its elements are vertices of a convex polygon. The following result provides an upper bound for the number of elements of a convexly independent set contained in the Minkowski sum of two other sets.

Theorem 3 *[3, Theorem 1]. Let R and S be two planar point sets with $|R| = r$ and $|S| = s$. Let C be a subset of the Minkowski sum $R + S$. If C is convexly independent we have that $|C| = \mathcal{O}(r^{2/3}s^{2/3} + r + s)$.*

From this result the following corollary follows easily.

Corollary 1. *Let $R_1, \ldots, R_k, S_1, \ldots, S_k, Q_1, Q_2$ be planar point sets with $|R_i| = r$, $|S_i| = s$ for all $i \in \{1, \ldots, k\}$, $|Q_1| = q_1$ and $|Q_2| = q_2$. Let C be a subset of $\cup_{i=1}^{k}(R_i + S_i) + Q_1 + Q_2$. If C is convexly independent, then $|C| = \mathcal{O}(kr^{2/3}s^{2/3}q_1^{2/3}q_2^{2/3} + krq_1 + ksq_2)$.*

Proof. We observe that $\cup_{i=1}^{k}(R_i + S_i) + Q_1 + Q_2 = \cup_{i=1}^{k}((R_i + Q_1) + (S_i + Q_2))$. Therefore, we partition C into k convexly independent disjoint sets $C_1, \ldots, C_k$ such that $C_i \subset (R_i + Q_1) + (S_i + Q_2)$ for all $i \in \{1, \ldots, k\}$. Since $|R_i + Q_1| = rq_1$ and $|S_i + Q_2| \leqslant sq_2$, by Theorem 3, we get that $|C_i| = \mathcal{O}(r^{2/3}s^{2/3}q_1^{2/3}q_2^{2/3} + rq_1 + sq_2)$ and the result follows. $\square$

Theorem 4. *Consider a polynomial $f \in \mathbb{R}^+[X]$ of degree d of the form*

$$f = \sum_{i=1}^{k} g_i h_i,$$

where $g_i, h_i \in \mathbb{R}^+[X]$, the g_i have at most r monomials and the h_i have at most s monomials. If f satisfies the Kurtz condition, then $d = \mathcal{O}(kr^{2/3}s^{2/3}\log^{2/3}(kr) + k(r + s)\log^{1/2}(kr))$.

Proof. We write $f = \sum_{i=0}^{d} c_i X^i$, where $c_i > 0$ for all $i \in \{1, \ldots, d\}$ and $c_0 \geq 0$. Since f satisfies the Kurtz condition, setting $\epsilon := \log(4)/2$ we get that

$$2\log(c_i) > \log(c_{i-1}) + \log(c_{i+1}) + 2\epsilon. \tag{4}$$

for every $i \geq 2$. For every $\delta_1, \ldots, \delta_d \in \mathbb{R}$, we set $C_{(\delta_1, \ldots, \delta_d)} := \{(i, \log(c_i) + \delta_i) \mid 1 \leq i \leq d\}$. We observe that (4) implies that $C_{(\delta_1, \ldots, \delta_d)}$ is convexly independent whenever $0 \leq \delta_i < \epsilon$ for all $i \in \{1, \ldots, d\}$.

We write $g_i = \sum_{j=1}^{r_i} a_{i,j} X^{\alpha_{i,j}}$ and $h_i = \sum_{j=1}^{s_i} b_{i,j} X^{\beta_{i,j}}$, with $r_i \leq r$, $s_i \leq s$ and $a_{i,j}, b_{i,j} > 0$ for all i, j. Then, $c_l = \sum_{i=1}^{k}(\sum_{\alpha_{i,j_1} + \beta_{i,j_2} = l} a_{i,j_1} b_{i,j_2})$. So, setting $M_l := \max\{a_{i,j_1} b_{i,j_2} \mid i \in \{1, \ldots, k\}, \alpha_{i,j_1} + \beta_{i,j_2} = l\}$ for all $l \in \{1, \ldots, d\}$, we have that $M_l \leq c_l \leq kr M_l$, so $\log(M_l) \leq \log(c_l) \leq \log(M_l) + \log(kr)$.

For every $l \in \{1, \ldots, d\}$, we set

$$\lambda_l := \left\lceil \frac{\log(c_l) - \log(M_l)}{\epsilon} \right\rceil \text{ and } \delta_l := \log(M_l) + \lambda_l \epsilon - \log(c_l), \tag{5}$$

and have that $0 \leq \lambda_l \leq \lceil (\log(kr))/\epsilon \rceil$ and that $0 \leq \delta_l < \epsilon$.

Now, we consider the sets

- $R_i := \{(\alpha_{i,j}, \log(a_{i,j})) \mid 1 \leq j \leq r_i\}$ for $i = 1, \ldots, k$,
- $S_i := \{(\beta_{i,j}, \log(b_{i,j})) \mid 1 \leq j \leq s_i\}$ for $i = 1, \ldots, k$,
- $Q := \{(0, \lambda\epsilon) \mid 0 \leq \lambda \leq \lceil \log(kr)/\epsilon \rceil\}$,
- $Q_1 := \{(0, \mu\epsilon) \mid 0 \leq \mu \leq \lceil \sqrt{\log(kr)/\epsilon} \rceil\}$, and
- $Q_2 := \{(0, \nu\lceil \sqrt{\log(kr)/\epsilon} \rceil\epsilon) \mid 0 \leq \nu \leq \lceil \sqrt{\log(kr)/\epsilon} \rceil\}$.

If $(0, \lambda\epsilon) \in Q$, then there exist μ and ν such that $\lambda = \nu\lceil \sqrt{\log(kr)/\epsilon} \rceil + \mu$ where $\mu, \nu \leq \lceil \sqrt{\log(kr)/\epsilon} \rceil$. We have,

$$(0, \lambda\epsilon) = (0, \nu\lceil \sqrt{\log(kr)/\epsilon} \rceil\epsilon) + (0, \mu\epsilon) \in Q_1 + Q_2,$$

so $Q \subset Q_1 + Q_2$. Then, we claim that $C_{(\delta_1, \ldots, \delta_d)} \subset \cup_{i=1}^{k}(R_i + S_i) + Q$. Indeed, for all $l \in \{1, \ldots, d\}$, by (5),

$$\log(c_l) + \delta_l = \log(M_l) + \lambda_l \epsilon = \log(a_{i,j_1}) + \log(b_{i,j_2}) + \lambda_l \epsilon$$

for some $i \in \{1, \ldots, k\}$ and some j_1, j_2 such that $\alpha_{i,j_1} + \beta_{i,j_2} = l$; thus

$$(l, \log(c_l) + \delta_l) = (\alpha_{i,j_1}, \log(a_{i,j_1})) + (\beta_{i,j_2}, \log(b_{i,j_1})) + (0, \lambda_l \epsilon) \in \cup_{i=1}^{k}(R_i + S_i) + Q.$$

Since $C_{(\delta_1, \ldots, \delta_d)}$ is a convexly independent set of d elements contained in $\cup_{i=1}^{k}(R_i + S_i) + Q_1 + Q_2$, a direct application of Corollary 1 yields the result. $\square$

From this result it is easy to derive an upper bound for the general case, where we have the products of $m \geq 2$ polynomials. If suffices to divide the m factors into two groups of approximately $m/2$ factors, and in each group we expand the product by brute force.

Proof of Theorem 1. We write each of the k products as a product of two polyno-
mials $G_i := \prod_{j=1}^{\lfloor m/2 \rfloor} f_{i,j}$ and $H_i := \prod_{j=\lfloor m/2 \rfloor+1}^{m} f_{i,j}$. We can now apply Theorem 4
to $f = \sum_{i=1}^{k} G_i H_i$ with $r = t^{\lfloor m/2 \rfloor}$ and $s = t^{m-\lfloor m/2 \rfloor}$ and we get the result.   □

Remark 1. We observe that the role of the constant 4 in the Kurtz condition
can be played by any other constant $\tau > 1$ in order to obtain the conclusion
of Theorem 1, i.e., we obtain the same result for $f = \sum_{i=0}^{d} a_i X^i$ satisfying that
$a_i^2 > \tau a_{i-1} a_{i+1}$ for all $i \in \{1, \ldots, d-1\}$. For proving this it suffices to replace
the value $\epsilon = \log(4)/2$ by $\epsilon = \log(\tau)/2$ in the proof of Theorem 4 to conclude
this more general result.

For $f = gh$ with $g, h \in \mathbb{R}^+[X]$ with at most t monomials, whenever f satisfies
the Kurtz condition, then f has only real (and distinct) roots and so do g and h.
As a consequence, both g and h satisfy (1) with strict inequalities and we derive
that $d \leqslant 2t$. Nevertheless, in the similar setting where $f = gh + x^i$ for some
$i > 0$, the same argument does not apply and a direct application of Theorem 1
yields $d = \mathcal{O}(t^{4/3} \log^{2/3}(t))$, a bound which seems to be very far from optimal.

Comparison with the Setting of Newton Polygons. A result similar to
Theorem 1 was obtained in [12] for the Newton polygons of bivariate polynomials.
Recall that the Newton polygon of a polynomial $f(X, Y)$ is the convex hull of
the points (i, j) such that the monomial $X^i Y^j$ appears in f with a nonzero
coefficient.

Theorem 5 (Koiran-Portier-Tavenas-Thomassé). *Consider a bivariate polynomial of the form*

$$f(X, Y) = \sum_{i=1}^{k} \prod_{j=1}^{m} f_{i,j}(X, Y) \tag{6}$$

*where $m \geqslant 2$ and the $f_{i,j}$ have at most t monomials. The Newton polygon of f
has $O(kt^{2m/3})$ edges.*

In the setting of Newton polygons, the main issue is how to deal with the cancellations arising from the addition of the k products in (6). Two monomials of
the form $cX^i Y^j$ with the same pair (i, j) of exponents but oppositive values of
the coefficient c will cancel, thereby deleting the point (i, j) from the Newton
polygon.

In the present paper we associate to the monomial cX^i with $c > 0$ the
point $(i, \log c)$. There are no cancellations since we only consider polynomials
$f_{i,j}$ with nonnegative coefficients in Theorems 1 and 4. However, the addition of
two monomials $cX^i, c'X^i$ with the same exponent will "move" the corresponding
point along the coefficient axis. By contrast, in the setting of Newton polygons
points can be deleted but cannot move. In the proof of Theorem 4 we deal with
the issue of "movable points" by an approximation argument, using the fact that
the constant $\epsilon = \log(4)/2 > 0$ gives us a little bit of slack.

3   A Stronger Log-Concavity Condition

The objective of this section is to improve the bound provided in Theorem 1 when $f = \sum_{i=0}^{d} a_i X^i \in \mathbb{R}^+[x]$ satisfies a stronger log-concavity condition, namely, when $a_i^2 > d^{2d} a_{i-1} a_{i+1}$ for all $i \in \{1, \ldots, d-1\}$.

To prove this bound, we make use of the following well-known lemma (a reference and similar results for polytopes in higher dimension can be found in [8]). For completeness, we provide a short proof.

Lemma 1. *If $R_1, \ldots, R_s$ are planar sets and $|R_i| = r_i$ for all $i \in \{1, \ldots, s\}$, then the convex hull of $R_1 + \cdots + R_s$ has at most $r_1 + \cdots + r_s$ vertices.*

Proof. We denote by k_i the number of vertices of the convex hull of R_i. Clearly $k_i \leqslant r_i$. Let us prove that the convex hull of $R_1 + \cdots + R_s$ has at most $k_1 + \cdots + k_s$ vertices. Assume that $s = 2$. We write $R_1 = \{a_1, \ldots, a_{r_1}\}$, then $a_i \in R_1$ is a vertex of the convex hull of R_1 if and only if there exists $w \in S^1$ (the unit Euclidean sphere) such that $w \cdot a_i > w \cdot a_j$ for all $j \in \{1, \ldots, r_1\} \setminus \{i\}$. Thus, R_1 induces a partition of S^1 into k_1 half-closed intervals. Similarly, R_2 induces a partition of S^1 into k_2 half-closed intervals. Moreover, these two partitions induce a new one on S^1 with at most $k_1 + k_2$ half-closed intervals; these intervals correspond to the vertices of $R_1 + R_2$ and; thus, there are at most $k_1 + k_2$. By induction we get the result for any value of s. $\square$

Proposition 1. *Consider a polynomial $f = \sum_{i=0}^{d} a_i X^i \in \mathbb{R}^+[X]$ of the form*

$$f = \sum_{i=1}^{k} \prod_{j=1}^{m} f_{i,j}$$

where the $f_{i,j} \in \mathbb{R}^+[x]$. If f satisfies the condition

$$a_i^2 > k^2 d^{2m} a_{i-1} a_{i+1},$$

then there exists a polynomial $f_{i,j}$ with at least d/km monomials.

Proof. Every polynomial $f_{i,j} := \sum_{l=0}^{d_{i,j}} c_{i,j,l} X^l$, where $d_{i,j}$ is the degree of $f_{i,j}$, corresponds to a planar set

$$R_{i,j} := \{(l, \log(c_{i,j,l})) \mid c_{i,j,l} > 0\} \subset \mathbb{R}^2.$$

We set, $C_{i,l} := \max\{0, \prod_{r=1}^{m} c_{i,r,l_r} \mid l_1 + \cdots + l_m = l\}$, for all $i \in \{1, \ldots, k\}$, $l \in \{0, \ldots, d\}$, and $C_l := \max\{C_{i,l} \mid 1 \leqslant i \leqslant k\}$ for all $l \in \{0, \ldots, d\}$. Since the polynomials $f_{i,j} \in \mathbb{R}^+[X]$ and

$$a_l = \sum_{i=1}^{k} \left(\sum_{l_1 + \cdots + l_m = l} \prod_{r=1}^{m} c_{i,r,l_r} \right)$$

for all $l \in \{0, \ldots, d\}$, we derive the following two properties:

- $C_l \leqslant a_l \leqslant k d^m C_l$ for all $l \in \{0, \ldots, d\}$,
- either $C_{i,l} = 0$ or $(l, \log(C_{i,l})) \in R_{i,1} + \cdots + R_{i,m}$ for all $i \in \{1, \ldots, k\}$, $l \in \{0, \ldots, d\}$. Since $a_l > 0$ for all $l \in \{1, \ldots, d\}$, we have that $C_l > 0$ and $(l, \log(C_l)) \in \bigcup_{i=1}^{k} (R_{i,1} + \cdots + R_{i,m})$

We claim that the points in the set $\{(l, \log(C_l)) \mid 1 \leqslant l \leqslant d\}$ belong to the upper convex envelope of $\bigcup_{i=1}^{k} (R_{i,1} + \cdots + R_{i,m})$. Indeed, if $(a, \log(b)) \in \bigcup_{i=1}^{k} (R_{i,1} + \cdots + R_{i,m})$, then $a \in \{0, \ldots, d\}$ and $b \leqslant C_a$; moreover, for all $l \in \{1, \ldots, d-1\}$, we have that

$$C_l^2 \geqslant a_l^2 / (k^2 d^{2m}) > a_{l-1} a_{l+1} \geqslant C_{l-1} C_{l+1}.$$

Hence, there exist $i_0 \in \{1, \ldots, k\}$ and $L \subset \{1, \ldots, d\}$ such that $|L| \geqslant d/k$ and $C_l = C_{i_0,l}$ for all $l \in L$. Since the points in $\{(l, \log(C_l)) \mid 1 \leqslant l \leqslant d\}$ belong to the upper convex envelope of $\bigcup_{i=1}^{k} (R_{i,1} + \cdots + R_{i,m})$ we easily get that the set $\{(l, \log(C_{i_0,l})) \mid l \in L\}$ is a subset of the vertices in the convex hull of $R_{i_0,1} + \cdots + R_{i_0,m}$. By Lemma 1, we get that there exists j_0 such that $|R_{i_0,j_0}| \geqslant |L|/m \geqslant d/km$ points. Finally, we conclude that f_{i_0,j_0} involves at least d/km monomials. □

Proof of Theorem 2. If $d \leqslant k$ or $d \leqslant m$, then $d \leqslant kmt$. Otherwise, $d^{2d} > k^2 d^{2(d-1)} \geqslant k^2 d^{2m}$ and, thus, f satisfies (3). A direct application of Proposition 1 yields the result. □

4 Applications to Complexity Theory

We first recall some standard definitions from algebraic complexity theory (see e.g. [2] or [15] for more details). Fix a field K. The elements of the complexity class VP are sequences (f_n) of multivariate polynomials with coefficients from K. By definition, such a sequence belongs to VP if the degree of f_n is bounded by a polynomial function of n and if f_n can be evaluated in a polynomial number of arithmetic operations (additions and multiplications) starting from variables and from constants in K. This can be formalized with the familiar model of *arithmetic circuits*. In such a circuit, input gates are labeled by a constant or a variable and the other gates are labeled by an arithmetic operation (addition or multiplication). In this paper we take $K = \mathbb{R}$ since there is a focus on polynomials with nonnegative coefficients. An arithmetic circuit is *monotone* if input gates are labeled by nonnegative constants only.

A family of polynomials belongs to the complexity class VNP if it can be obtained by summation from a family in VP. More precisely, $f_n(\overline{x})$ belongs to VNP if there exists a family $(g_n(\overline{x}, \overline{y}))$ in VP and a polynomial p such that the tuple of variables $\overline{y}$ is of length $l(n) \leqslant p(n)$ and

$$f_n(\overline{x}) = \sum_{\overline{y} \in \{0,1\}^{l(n)}} g_n(\overline{x}, \overline{y}).$$

Note that this summation over all boolean values of $\overline{y}$ may be of exponential size. Whether the inclusion $\mathsf{VP} \subseteq \mathsf{VNP}$ is strict is a major open problem in algebraic complexity.

Valiant's criterion [2, 15] shows that "explicit" polynomial families belong to VNP. One version of it is as follows.

Lemma 2. *Suppose that the function* $\phi : \{0, 1\}^* \rightarrow \{0, 1\}$ *is computable in polynomial time. Then the family* (f_n) *of multilinear polynomials defined by*

$$f_n = \sum_{e \in \{0,1\}^n} \phi(e) x_1^{e_1} \cdots x_n^{e_n}$$

belongs to VNP.

Note that more general versions of Valiant's criterion are known. One may allow polynomials with integer rather than $0/1$ coefficients [2], but in Theorem 7 below we will only have to deal with $0/1$ coefficients. Also, one may allow f_n to depend on any (polynomially bounded) number of variables rather than exactly n variables and in this case, one may allow the algorithm for computing the coefficients of f_n to take as input the index n in addition to the tuple e of exponents (see [9], Theorem 2.3).

Reduction of arithmetic circuits to depth 4 is an important ingredient in the proof of the forthcoming results. This phenomenon was discovered by Agrawal and Vinay [1]. Here we will use it under the form of [14], which is an improvement of [11]. We will also need the fact that if the original circuit is monotone, then the resulting depth 4 circuit is also monotone (this is clear by inspection of the proof in [14]). Recall that a depth 4 circuit is a sum of products of sums of products of inputs; sum gates appear on layers 2 and 4 and product gates on layers 1 and 3. All gates may have arbitrary fan-in.

Lemma 3. *Let C be an arithmetic circuit of size $s > 1$ computing a v-variate polynomial of degree d. Then, there is an equivalent depth 4 circuit Γ of size* $2^{\mathcal{O}\left(\sqrt{d \log(ds) \log(v)}\right)}$ *with multiplication gates at layer 3 of fan-in $\mathcal{O}(\sqrt{d})$. Moreover, if C is monotone, then Γ can also be chosen to be monotone.*

We will use this result under the additional hypothesis that d is polynomially bounded by the number of variables v. In this setting, since $v \leqslant s$, we get that the resulting depth 4 circuit Γ provided by Lemma 3 has size $s^{\mathcal{O}(\sqrt{d})}$.

Before stating the main results of this section, we construct an explicit family of log-concave polynomials.

Lemma 4. *Let $n, s \in \mathbb{Z}^+$ and consider $g_{n,s}(X) := \sum_{i=0}^{2^n - 1} a_i X^i$, with*

$$a_i := 2^{si(2^n - i - 1)} \text{ for all } i \in \{0, \ldots, 2^n - 1\}.$$

Then, $a_i^2 > 2^s \, a_{i-1} a_{i+1}$.

Proof. Take $i \in \{1, \ldots, 2^n - 2\}$, we have that

$$
\begin{aligned}
\log\left(2^s a_{i-1} a_{i+1}\right) &= s + s2^n(i-1) - s(i-1)i + s2^n(i+1) - s(i+1)(i+2) \\
&= 2s2^n i - 2si(i+1) - s \\
&< 2s2^n i - 2si(i+1) \\
&= \log(a_i^2).
\end{aligned}
$$

$\square$

In the next theorem we start from the family $g_{n,s}$ of Lemma 4 and we set $s = n2^{n+1}$.

Theorem 6. *Let $(f_n) \in \mathbb{N}[X]$ be the family of polynomials $f_n(x) = g_{n,n2^{n+1}}(x)$.*

(i) f_n has degree $2^n - 1$ and satisfies the log-concavity condition (3).
(ii) If $\mathsf{VP} = \mathsf{VNP}$, f_n can be written under form (2) with $k = n^{O(\sqrt{n})}$, $m = O(\sqrt{n})$ and $t = n^{O(\sqrt{n})}$.

Proof. It is clear that $f_n \in \mathbb{N}[X]$ has degree $2^n - 1$ and, by Lemma 4, f_n satisfies (3).

Consider now the related family of bivariate polynomials $g_n(X,Y) = \sum_{i=0}^{2^n-1} X^i Y^{e(n,i)}$, where $e(n,i) = si(2^n - i - 1)$. One can check in time polynomial in n whether a given monomial $X^i Y^j$ occurs in g_n: we just need to check that $i < 2^n$ and that $j = e(n,i)$. By mimicking the proof of Theorem 1 in [12] and taking into account Lemma 3 we get that, if $\mathsf{VP} = \mathsf{VNP}$, one can write

$$
g_n(X,Y) = \sum_{i=1}^{k} \prod_{j=1}^{m} g_{i,j,n}(X,Y) \tag{7}
$$

where the bivariate polynomials $g_{i,j,n}$ have $n^{O(\sqrt{n})}$ monomials, $k = n^{O(\sqrt{n})}$ and $m = O(\sqrt{n})$. Performing the substitution $Y = 2$ in (7) yields the required expression for f_n. $\square$

We believe that there is in fact no way to write f_n under form (2) so that the parameters k, m, t satisfy the constraints $k = n^{O(\sqrt{n})}$, $m = O(\sqrt{n})$ and $t = n^{O(\sqrt{n})}$. By part (ii) of Theorem 6, a proof of this would separate VP from VNP. The proof of Theorem 7 below shows that our belief is actually correct in the special case where the polynomials $f_{i,j}$ in (2) have nonnegative coefficients.

The main point of Theorem 7 is to present an unconditional lower bound for a polynomial family (h_n) in VNP derived from (f_n). Note that (f_n) itself is not in VNP since its degree is too high. Recall that

$$
f_n(X) := \sum_{i=0}^{2^n-1} 2^{2n2^n i(2^n-i-1)} X^i. \tag{8}
$$

To construct h_n we write down in base 2 the exponents of "2" and "X" in (8). More precisely, we take h_n of the form:

$$
h_n := \sum_{\substack{\alpha \in \{0,1\}^n \\ \beta \in \{0,1\}^{4n}}} \lambda(n, \alpha, \beta)\, X_0^{\alpha_0} \cdots X_{n-1}^{\alpha_{n-1}} Y_0^{\beta_0} \cdots Y_{4n-1}^{\beta_{4n-1}}, \tag{9}
$$

where $\alpha = (\alpha_0, \ldots, \alpha_{n-1})$, $\beta = (\beta_0, \ldots, \beta_{4n-1})$ and $\lambda(n, \alpha, \beta) \in \{0, 1\}$; we set $\lambda(n, \alpha, \beta) = 1$ if and only if $\sum_{j=0}^{4n-1} \beta_j 2^j = 2n2^n i(2^n - i - 1) < 2^{4n}$, where $i := \sum_{k=0}^{n-1} \alpha_{i,k} 2^k$. By construction, we have:

$$f_n(X) = h_n(X^{2^0}, X^{2^1}, \ldots, X^{2^{n-1}}, 2^{2^0}, 2^{2^1}, \ldots, 2^{2^{4n-1}}). \tag{10}$$

This relation will be useful in the proof of the following lower bound theorem.

Theorem 7. *The family (h_n) in (9) is in* VNP. *If (h_n) is computed by depth 4 monotone arithmetic circuits of size $s(n)$, then $s(n) = 2^{\Omega(n)}$. If (h_n) is computed by monotone arithmetic circuits of size $s(n)$, then $s(n) = 2^{\Omega(\sqrt{n})}$. In particular, (h_n) cannot be computed by monotone arithmetic circuits of polynomial size.*

Proof. Note that h_n is a polynomial in $5n$ variables, of degree at most $5n$, and its coefficients $\lambda(n, \alpha, \beta)$ can be computed in polynomial time. Thus, by Valiant's criterion we conclude that $(h_n) \in$ VNP.

Assume that (h_n) can be computed by depth 4 monotone arithmetic circuits of size $s(n)$. Using (10), we get that $f_n = \sum_{i=1}^{k} \prod_{j=1}^{m} f_{i,j}$ where $f_{i,j} \in \mathbb{R}^+[X]$ have at most t monomials and k, m, t are $\mathcal{O}(s(n))$. Since the degree of f_n is $2^n - 1$, by Theorem 2, we get that $2^n - 1 \leqslant kmt$. We conclude that $s(n) = 2^{\Omega(n)}$.

To complete the proof of the theorem, assume that (h_n) can be computed by monotone arithmetic circuits of size $s(n)$. By Lemma 3, it follows that the polynomials h_n are computable by depth 4 monotone circuits of size $s'(n) := s(n)^{\mathcal{O}(\sqrt{n})}$. Therefore $s'(n) = 2^{\Omega(n)}$ and we finally get that $s(n) = 2^{\Omega(\sqrt{n})}$. □

Lower bounds for monotone arithmetic circuits have been known for a long time (see for instance [7,16]). Theorem 7 provides yet another example of a polynomial family which is hard for monotone arithmetic circuits, with an apparently new proof method.

5   Discussion

As explained in the introduction, log-concavity plays a role in the study of real roots of polynomials. In [10] bounding the number of real roots of sums of products of sparse polynomials was suggested as an approach for separating VP from VNP. Hrubeš [5] suggested to bound the multiplicities of roots, and [12] to bound the number of edges of Newton polygons of bivariate polynomials.

Theorem 6 provides another plausible approach to VP $\neq$ VNP: it suffices to show that if a polynomial $f \in \mathbb{R}^+[X]$ under form (2) satisfies the Kurtz condition or the stronger log-concavity condition (3) then its degree is bounded by a "small" function of the parameters k, m, t. A degree bound which is polynomial bound in k, t and 2^m would be good enough to separate VP from VNP. Theorem 1 improves on the trivial kt^m upper bound when f satisfies the Kurtz condition, but certainly falls short of this goal: not only is the bound on $\deg(f)$ too coarse, but we would also need to allow negative coefficients in the polynomials $f_{i,j}$. Theorem 2 provides a polynomial bound on k, m and t under a stronger log-concavity condition, but still needs the extra assumption that the coefficients in the polynomials $f_{i,j}$ are nonnegative. The unconditional lower bound in Theorem 7 provides a "proof of concept" of this approach for the easier setting of monotone arithmetic circuits.

References

1. Agrawal, M., Vinay, V.: Arithmetic circuits: a chasm at depth four. In: 49th Annual IEEE Symposium on Foundations of Computer Science, FOCS 2008, October 25–28, 2008, Philadelphia, PA, USA, pp. 67–75 (2008)
2. Bürgisser, P.: Completeness and Reduction in Algebraic Complexity Theory. Algorithms and Computation in Mathematics. Springer, Heidelberg (2000)
3. Eisenbrand, F., Pach, J., Rothvoß, T., Sopher, N.B.: Convexly independent subsets of the Minkowski sum of planar point sets. Electron. J. Combin., 15(1): Note 8, 4 (2008)
4. Hardy, G.H., Littlewood, J.E., Pólya, G.: Inequalities. Cambridge Mathematical Library. Cambridge University Press, Cambridge (1988). Reprint of the 1952 edition
5. Hrubes, P.: A note on the real τ-conjecture and the distribution of complex roots. Theo. Comput. 9(10), 403–411 (2013). http://eccc.hpi-web.de/report/2012/121/
6. Hutchinson, J.I.: On a remarkable class of entire functions. Trans. Amer. Math. Soc. 25(3), 325–332 (1923)
7. Jerrum, M., Snir, M.: Some exact complexity results for straight-line computations over semirings. J. ACM (JACM) 29(3), 874–897 (1982)
8. Karavelas, M.I., Tzanaki, E.: The maximum number of faces of the Minkowski sum of two convex polytopes. In: Proceedings of the Twenty-Third Annual ACM-SIAM Symposium on Discrete Algorithms, pp. 11–28 (2012)
9. Koiran, P.: Valiant's model and the cost of computing integers. Comput. Complex. 13, 131–146 (2004)
10. Koiran, P.: Shallow circuits with high-powered inputs. In: Proceedings of the Second Symposium on Innovations in Computer Science (ICS 2011) (2011). http://arxiv.org/abs/1004.4960
11. Koiran, P.: Arithmetic circuits: the chasm at depth four gets wider. Theoret. Comput. Sci. 448, 56–65 (2012)
12. Koiran, P., Portier, N., Tavenas, S., Thomassé, S.: A τ-conjecture for Newton polygons. Found. Comput. Math. 15(1), 185–197 (2014)
13. Kurtz, D.C.: A sufficient condition for all the roots of a polynomial to be real. Amer. Math. Monthly 99(3), 259–263 (1992)
14. Tavenas, S.: Improved bounds for reduction to depth 4 and depth 3. In: Chatterjee, K., Sgall, J. (eds.) MFCS 2013. LNCS, vol. 8087, pp. 813–824. Springer, Heidelberg (2013)
15. Valiant, L.G.: Completeness classes in algebra. In: Proceedings of the Eleventh Annual ACM Symposium on Theory of Computing, STOC 1979, pp. 249–261. ACM, New York, NY, USA (1979)
16. Valiant, L.G.: Negation can be exponentially powerful. In: Proceedings of the Eleventh Annual ACM Symposium on Theory of Computing, pp. 189–196. ACM (1979). Journal version in Theo. Comput. Sci. 12(3), 303–314 (1980)

Easy Multiple-Precision Divisors and Word-RAM Constants

Torben Hagerup$^{(\boxtimes)}$

Institut für Informatik, Universität Augsburg, 86135 Augsburg, Germany
hagerup@informatik.uni-augsburg.de

Abstract. For integers $b \geq 2$ and $w \geq 1$, define the (b, w) *cover size* of an integer A as the smallest nonnegative integer k such that A can be written in the form $A = \sum_{i=1}^{k}(-1)^{\sigma_i}b^{\ell_i}d_i$, where σ_i, ℓ_i and d_i are nonnegative integers and $0 \leq d_i < b^w$, for $i = 1, \ldots, k$. We study the efficient execution of arithmetic operations on (multiple-precision) integers of small (b, w) cover size on a word RAM with words of w b-ary digits and constant-time multiplication and division. In particular, it is shown that if A is an n-digit integer and B is a nonzero m-digit integer of (b, w) cover size k, then $\lfloor A/B \rfloor$ can be computed in $O(1 + (kn + m)/w)$ time. Our results facilitate a unified description of word-RAM algorithms operating on integers that may occupy a fraction of a word or several words.

As an application, we consider the fast generation of integers of a special form for use in word-RAM computation. Many published word-RAM algorithms divide a w-bit word conceptually into equal-sized fields and employ full-word constants whose field values depend in simple ways on the field positions. The constants are either simply postulated or computed with ad-hoc methods. We describe a procedure for obtaining constants of the following form in constant time: The ith field, counted either from the right or from the left, contains $g(i)$, where g is a constant-degree polynomial with integer coefficients that, disregarding mild restrictions, can be arbitrary. This general form covers almost all cases known to the author of word-RAM constants used in published algorithms.

1 Introduction

Let $b \geq 2$ be an integer. If integers are expressed in the usual b-ary positional representation, multiplication and division (with truncation) of nonnegative integers by powers of b are trivial, as they simply amount to a shift and possibly the dropping of some digits. It is also easy to see that multiplication of large integers by the sum or the difference of two powers of b can be carried out in linear time, whereas division by such numbers is a bit trickier.

We generalize this theme in two directions: First, we consider multiplicands and divisors that are composed of possibly more than one or two powers of b. Second, since the arithmetic instructions of a computer usually work on w-digit operands for an integer $w \geq 1$ called the *word size* of the computer, we replace the powers of b by blocks of w consecutive b-ary digits.

© Springer-Verlag Berlin Heidelberg 2015
G.F. Italiano et al. (Eds.): MFCS 2015, Part II, LNCS 9235, pp. 372–383, 2015.
DOI: 10.1007/978-3-662-48054-0_31

Practically all computers are based on the binary number system, and $b = 2$ is clearly the most important special case of the results reported here. Still, the generalization to arbitrary values of b comes largely for free, and some of the results may be interesting also for other values of b, such as $b = 10$ or $b = 2^{64}$.

The considerations above lead us to define a *word cover* of size k of an integer A roughly as a sequence $((\sigma_1, \ell_1, d_1), \ldots, (\sigma_k, \ell_k, d_k))$ of k triples of nonnegative integers with $0 \leq d_i < b^w$ for $i = 1, \ldots, k$ such that $A = \sum_{i=1}^{k} (-1)^{\sigma_i} b^{\ell_i} d_i$, and the *cover size* of A as the smallest size of a word cover of A. We begin by showing that a word cover of minimal size of a given n-digit integer A can be found in $O(1 + n/w)$ time. Note that because w digits can be handled together in constant time, $O(1 + n/w)$ time is what should be considered "linear time"—it is within a constant factor of the number of w-digit words occupied by the input A.

The main result of Sect. 2 is that if A is an n-digit integer and B is an m-digit integer with cover size k, then AB and, if $B \neq 0$, $\lfloor A/B \rfloor$ can be computed in $O(1 + (kn + m)/w)$ time, which may be thought of as "k times linear time". Schönhage and Strassen [12] used the fast Fourier transformation to show that for $l \geq \log n$, the multiplication of two n-bit integers reduces to $O((n/l) \log(n/l))$ multiplications, additions and other linear-time operations on integers of $O(l)$ bits each plus the time needed to convert the input and output to and from a positional representation with a basis of $\Theta(l)$ bits. With a related reduction, they managed to multiply two n-bit integers on a multitape Turing machine in $O(n \log n \log \log n)$ time, a bound that has since been strengthened to $n \log n \, 2^{O(\log^* n)}$ [6]. Used with $l = \Theta(\log n)$, the first reduction mentioned above yields running times of $O(n)$ on the storage modification or pointer machine [11] and of $O(n \log n)$ on the logarithmic-cost RAM [9,11]. For the word RAM, the reduction, now used with $l = w = \Omega(\log n)$, shows that an n-digit integer and an m-digit integer can be multiplied in $O(1 + (n + m) \log(2 + \min\{n, m\}/w)/w)$ time (to multiply an integer x of $n \geq w$ digits by an integer y of $m \geq n$ digits, divide y into $\Theta(m/n)$ "pieces" of $\Theta(n)$ digits each, multiply x by each piece in $O((n/w) \log(2 + n/w))$ time per piece and sum the resulting products in $O(m/w)$ time). Using Newton-Raphson iteration to compute approximate reciprocals, Aho, Hopcroft and Ullman [1] demonstrated that the time needed to divide a $2n$-bit integer by an n-bit integer (with truncation) is within a constant factor of the time needed to multiply two n-bit integers. Their model of computation is different, but the proof is valid also for the word RAM. Generalized slightly, it shows that the integer part of the quotient of an n-digit integer and a nonzero m-digit integer can be computed on a word RAM in $O(1 + n \log(2 + n/w)/w)$ time. A comparison of the bounds reveals that the algorithms developed here are likely to be competitive for small values of k. We shall be particularly interested in integers of *bounded cover size*, i.e., integers whose cover sizes are bounded by a constant.

As a consequence of the findings discussed above, an arbitrary arithmetic expression with a constant number of operators drawn from $\{+, -, \cdot, /\}$ and integer operands of bounded cover size can be evaluated in time linear in the number of words occupied by the operands, provided that every division in the evaluation yields an integer result. To see this, note that if x and y are integers of cover sizes k_x and k_y, respectively, then $x + y$ and $x - y$ have cover sizes bounded by $k_x + k_y$, while xy has cover size at most $2k_x k_y$. Used a constant number of times, therefore, these operations preserve the property of being of constant cover size. The same is not true of division, but provided that the value of every subexpression of the expression to be evaluated is an integer (thus division is true division and not division with truncation), the expression has an equivalent expression in which divisions are "pushed to the outside" so that the only division, if there is one, is carried out at the very end. To wit, $a/b \pm c/d = (ad \pm bc)/bd$, $(a/b) \cdot (c/d) = ac/bd$ and $(a/b)/(c/d) = ad/bc$. During the evaluation of the equivalent expression, the inputs to every arithmetic operation have bounded cover size, so that the operation can be executed in time linear in the number of words occupied by its inputs, which is in turn linear in the number of words occupied by the original operands, as claimed. This observation facilitates the description and analysis of certain word-RAM algorithms by eliminating a need to distinguish between operands that fit in one word and ones that occupy several words. As an example, many—if not most— multiplications in published word-RAM algorithms have at least one operand of the form p/q, where p and q are integers with cover sizes bounded by (small) constants. If the operands are m-bit integers, there is no problem if $m \leq w$—the multiplication takes constant time by definition, independently of cover sizes. In a generalization to operands of arbitrary size, however, one needs results such as those developed here to argue that the time bound generalizes from $O(1)$ to $O(1 + m/w)$. The investigation reported here was begun in an attempt to reduce the complexity of the description of subroutines used in [8], where this issue arises repeatedly.

Many published word-RAM algorithms logically divide a w-digit word into $s = w/f$ fields of f digits each, for some f, and employ full-word constants whose field values depend in simple ways on the field positions. If we denote by $(a_{s-1}, \ldots, a_0)_{b^f}$ the constant whose fields, in the order from left to right, have the values $a_{s-1}, \ldots, a_0$ (i.e., we essentially follow usual notation for positional number systems), the most heavily used constant is $(1, \ldots, 1)_{b^f}$ (which is sometimes denoted as $1_{s,f}$). Another popular constant is $(s, \ldots, 2, 1)_{b^f}$. Both of these are used, e.g., in [2,3]. Constants of this type are either postulated or computed by ad-hoc means. In Sect. 3 we aim for a more systematic treatment.

From a slightly altered perspective, a constant $(a_{s-1}, \ldots, a_0)_{b^f}$ can be viewed as a table that maps each $k \in \{0, \ldots, s-1\}$ to a_k, i.e., as a table of a certain function. The two constants mentioned above correspond to the two functions g_1 and g_2 with $g_1(k) = 1$ and $g_2(k) = k + 1$ for all $k \in \{0, \ldots, s-1\}$. Both g_1 and g_2 are polynomials, of course. We study the efficient computation of a table of an arbitrary polynomial g. If g is of constant degree—the case of greatest practical relevance—we show, under mild restrictions, that a table of g can be computed in $O(1 + sf/w)$ time, the best that one could hope for. Conventions for the representation and handling of negative numbers are left to the reader's discretion.

Although with a more theoretical slant, the present paper can be viewed as continuing a long tradition of "bit trickery" with representatives such as [10, Sect. 7.1.3] and [13].

2   Arithmetic Operations on Easy Numbers

We assume as the model of computation a word RAM with w-digit b-ary words, where b and w are integers with $b \geq 2$ and $w \geq 1$. We consider b, but not w, to be a constant. The value stored in a word is interpreted as an integer in the range $\{0, \ldots, b^w - 1\}$. The word RAM has constant-time instructions for the following operations: Addition, subtraction and multiplication modulo b^w, division with truncation, left and right shifts by an arbitrary number of digit positions (with zeros shifted in), and arbitrary digit-wise operations (such as digit-wise minimum, which generalizes bitwise AND). We assume that w is large enough to allow all words used by the algorithms under consideration to be addressed with w bits. Let $\mathbb{N} = \{1, 2, \ldots\}$ and $\mathbb{N}_0 = \mathbb{N} \cup \{0\}$.

2.1   Word Covers

Definition 2.1. *For $k \in \mathbb{N}_0$, a (b, w) word cover of size k of an integer A is a sequence $((\sigma_1, \ell_1, d_1), \ldots, (\sigma_k, \ell_k, d_k))$ of k triples of integers with $\sigma_i \in \{0, 1\}$ and $0 \leq d_i < b^w$ for $i = 1, \ldots, k$ and $0 \leq \ell_1 \leq \ell_2 \leq \cdots \leq \ell_k$ such that $A = \sum_{i=1}^{k} (-1)^{\sigma_i} b^{\ell_i} d_i$. The (b, w) cover size of A is the smallest nonnegative integer k such that A has a (b, w) word cover of size k.*

When b and w are clear from the context, as will usually be the case, we omit the prefix "(b, w)".

Lemma 2.2. *Given positive integers n and A with $1 \leq A < b^n$, the positions of the leftmost and rightmost nonzero digits in the b-ary representation of A can be computed in $O(1 + n/w)$ time.*

Proof. For $b = 2$ and $n = w$, the claim concerning the leftmost nonzero digit was shown by Fredman and Willard [5, pp. 431–432], and a remark in [4, p. 36] reduces the case of the rightmost nonzero digit to that of the leftmost nonzero digit. The corresponding algorithms easily extend to general values of b and n.   □

Lemma 2.3. *Given nonnegative integers n and A with $0 \leq A < b^n$, a smallest word cover of A can be computed in $O(1 + n/w)$ time.*

Proof. We give a proof only for the most interesting case $b = 2$. A proof for general values of b along the same lines is possible, but less elegant.
     Write $A = (a_{n-1}, \ldots, a_0)_2$ (i.e., the bits of A, in the order from most significant to least significant, are $a_{n-1}, \ldots, a_0$) and take $a_{-1} = a_n = a_{n+1} = \cdots = 0$. Define a *change cover* of A of size $k \in \mathbb{N}_0$ to be a sequence $(\ell_1, \ldots, \ell_k)$ of k nonnegative integers with $\ell_{i+1} \geq \ell_i + w + 1$ for $i = 1, \ldots, k - 1$ such that

- for $i = 1, \ldots, k$, $a_{\ell_i} \neq a_{\ell_i - 1}$;
- for $j = 0, \ldots, n$, if $a_j \neq a_{j-1}$, then $\ell_i \leq j \leq \ell_i + w$ for some $i \in \{1, \ldots, k\}$.

Informally, the definition stipulates that all changes in the binary representation of A be covered by k disjoint intervals of $w + 1$ consecutive bit positions and that each such interval begin with a change.

Assume that $k \in \mathbb{N}_0$ and that a change cover $(\ell_1, \ldots, \ell_k)$ of A of size k is given and define $K^+ = \{i \in \mathbb{N} \mid i \leq k \text{ and } a_{\ell_i + w} = 0\}$ and $K^- = \{1, \ldots, k\} \setminus K^+$. Since $a_j = 0$ for $j < \ell_1$ and for $j \geq \ell_k + w$, $a_{\ell_1} = 1$, and $a_{\ell_i + w} = a_{\ell_i + w + 1} = \cdots = a_{\ell_{i+1} - 1} \neq a_{\ell_{i+1}}$ for $i = 1, \ldots, k - 1$,

$$
\begin{aligned}
A &= \sum_{i=1}^{k} \sum_{j=\ell_i}^{\ell_i + w - 1} 2^j a_j + \sum_{i=1}^{k-1} a_{\ell_i + w} \sum_{j=\ell_i + w}^{\ell_{i+1} - 1} 2^j \\
&= \sum_{i=1}^{k} 2^{\ell_i} \sum_{j=0}^{w-1} 2^j a_{\ell_i + j} + \sum_{i=1}^{k-1} a_{\ell_i + w} (2^{\ell_{i+1}} - 2^{\ell_i + w}) \\
&= \sum_{i=1}^{k} 2^{\ell_i} \left(a_{\ell_i} + \sum_{j=1}^{w-1} 2^j a_{\ell_i + j} \right) + \sum_{i=2}^{k} (1 - a_{\ell_i}) \cdot 2^{\ell_i} - \sum_{i=1}^{k} a_{\ell_i + w} \cdot 2^{\ell_i + w} \\
&= \sum_{i=1}^{k} 2^{\ell_i} \left(1 + \sum_{j=1}^{w-1} 2^j a_{\ell_i + j} \right) - \sum_{i \in K^-} 2^{\ell_i + w} \\
&= \sum_{i \in K^+} 2^{\ell_i} \left(1 + \sum_{j=1}^{w-1} 2^j a_{\ell_i + j} \right) - \sum_{i \in K^-} 2^{\ell_i} \left(2^w - 1 - \sum_{j=1}^{w-1} 2^j a_{\ell_i + j} \right).
\end{aligned}
$$

The formula shows not only that a word cover of A of size k exists, but also how to compute one in $O(1 + k) = O(1 + n/w)$ time.

A change cover $(\ell_1, \ldots, \ell_k)$ of A of minimal size is computed in $O(1 + n/w)$ time by a straightforward greedy algorithm: Start with $k = p = 0$. Then, as long as $a_q \neq a_{q-1}$ for some $q \geq p$, increment k by 1, compute ℓ_k as the smallest such q (with digit-wise operations and Lemma 2.2), and increase p to $\ell_k + w + 1$.

Assume that A has a word cover of size $k \in \mathbb{N}_0$. We complete the proof by demonstrating that then A has a change cover of size at most k. Let us say that two triples (σ, ℓ, d) and (σ', ℓ', d') in a word cover of A overlap if $\max\{\ell, \ell'\} < \min\{\ell, \ell'\} + w$. If this is the case and $\ell \leq \ell'$, then $|(-1)^\sigma 2^\ell d + (-1)^{\sigma'} 2^{\ell'} d'| < 2^{\ell + w}(1 + 2^{\ell' - \ell}) \leq 2^{\ell + 2w}$. Therefore a new word cover of A can be obtained by replacing the triples (σ, ℓ, d) and (σ', ℓ', d') by two suitable nonoverlapping triples with middle components ℓ and $\ell + w$ (inserted in the appropriate positions in the word cover). Starting with a word cover of A of size k and applying this transformation at most $\binom{k}{2}$ times, always to two overlapping triples with minimal second components, we obtain a word cover $(\sigma_1, \ell_1, d_1), \ldots, (\sigma_k, \ell_k, d_k)$ of A of size k in which no two triples overlap, i.e.,

$\ell_{i+1} \geq \ell_i + w$ for $i = 1, \ldots, k-1$. Informally, this is because the transformation moves the leftmost of two overlapping triples further to the left. The absence of overlaps implies that for $j = 0, \ldots, n$, if $a_j \neq a_{j-1}$, then $\ell_i \leq j \leq \ell_i + w$ for some $i \in \{1, \ldots, k\}$. It is not necessarily the case that $\ell_{i+1} \geq \ell_i + w + 1$ for $i = 1, \ldots, k-1$ and that $a_{\ell_i} \neq a_{\ell_i - 1}$ for $i = 1, \ldots, k$. Nonetheless, it is easy to see that the greedy algorithm discussed earlier computes a change cover of A of size at most k. $\square$

2.2 Multiplication and Division by Easy Numbers

At a first reading of the following technical lemma, used in the proof of Theorem 2.5, the reader may find it helpful to disregard the distinction between u and $\tilde{u}$. The lemma essentially says that the quotient of the leading $2w + 4$ b-ary digits of a positive integer u and the leading $w + 2$ b-ary digits of a positive integer v, scaled by the appropriate power of b, is an approximation to u/v with a relative error of at most b^{-w}. The reader may be able to see without the need for a proof that such a statement is true, perhaps with other constant terms substituted for 4 and 2 (arbitrary constants will do for our purposes). The lemma is similar to lemmas developed by others in the context of multiple-precision long division (cf., e.g., [9, Sect. 4.3.1, Theorem B]).

Lemma 2.4. *Let b, w, u, v and $\tilde{u}$ be positive integers with $b \geq 2$ and $u(1 - b^{-w-3}) \leq \tilde{u} \leq u$. Define $p = \max\{\lfloor \log_b \tilde{u} \rfloor - 2w - 3, 0\}$, $q = \max\{\lfloor \log_b v \rfloor - w - 1, 0\}$ and $\hat{u} = \lfloor \tilde{u} \cdot b^{-p} \rfloor$. Take $\hat{v} = \lfloor v \cdot b^{-q} \rfloor + 1$ if $q > 0$, and otherwise take $\hat{v} = v$. Then $(u/v)(1 - b^{-w}) - 1 \leq \lfloor \hat{u}/\hat{v} \rfloor \cdot b^{p-q} \leq u/v$.*

Proof. $\lfloor \hat{u}/\hat{v} \rfloor \leq \tilde{u} \cdot b^{-p}/(v \cdot b^{-q}) \leq (u/v) \cdot b^{q-p}$.
If $p > 0$, $\tilde{u} \cdot b^{-p} = b^{\log_b \tilde{u} - \lfloor \log_b \tilde{u} \rfloor + 2w + 3} \geq b^{2w+3}$ and therefore

$$\hat{u} \geq \tilde{u} \cdot b^{-p}(1 - b^p/\tilde{u}) \geq (1 - b^{-w-3})(1 - b^{-2w-3})u \cdot b^{-p}.$$

Similarly, if $q > 0$, $v \cdot b^{-q} \geq b^{w+1}$. Moreover, we always have $\hat{v} \leq b^{w+2}$. Recall that for all $\epsilon, \epsilon' > 0$, $1/(1 + \epsilon) \geq 1 - \epsilon$ and $(1 - \epsilon)(1 - \epsilon') \geq 1 - \epsilon - \epsilon'$. Consider four cases:

If $p = q = 0$,

$$\lfloor \hat{u}/\hat{v} \rfloor \cdot b^{p-q} = \lfloor \tilde{u}/v \rfloor \geq (u/v)(1 - b^{-w-3}) - 1 \geq (u/v)(1 - b^{-w}) - 1.$$

If $p > 0$ and $q = 0$,

$$\lfloor \hat{u}/\hat{v} \rfloor \cdot b^{p-q} = \lfloor \hat{u}/v \rfloor \cdot b^p \geq (\hat{u}/v)(1 - v/\hat{u})b^p$$
$$\geq (u/v)(1 - b^{-w-3})(1 - b^{-2w-3})(1 - b^{(w+2)-(2w+3)})$$
$$\geq (u/v)(1 - b^{-w-3} - b^{-2w-3} - b^{-w-1}) \geq (u/v)(1 - b^{-w}).$$

If $p = 0$ and $q > 0$,

$$\lfloor \hat{u}/\hat{v} \rfloor \cdot b^{p-q} = \lfloor \tilde{u}/\hat{v} \rfloor \cdot b^{-q} \geq \frac{\tilde{u} \cdot b^{-q}}{v \cdot b^{-q} + 1} - 1 \geq \frac{u(1 - b^{-w-3})}{v(1 + b^q/v)} - 1$$
$$\geq (u/v)(1 - b^{-w-3} - b^{-w-1}) - 1 \geq (u/v)(1 - b^{-w}) - 1.$$

Finally, if $p > 0$ and $q > 0$,

$$\lfloor \widehat{u}/\widehat{v} \rfloor \cdot b^{p-q} \geq \frac{\widehat{u}}{\widehat{v}} \left(1 - \frac{\widehat{v}}{\widehat{u}} \right) b^{p-q}$$

$$\geq \frac{(1 - b^{-w-3})(1 - b^{-2w-3})u}{v(1 + b^q/v)} \left(1 - \frac{v \cdot b^{-q} + 1}{\widehat{u}} \right)$$

$$\geq (u/v)(1 - b^{-w-3} - b^{-2w-3})(1 - b^q/v)(1 - v \cdot b^{-q}/\widehat{u} - 1/\widehat{u}).$$

Let $x = v \cdot b^{-q}$. Then

$$(1 - b^q/v)(1 - v \cdot b^{-q}/\widehat{u} - 1/\widehat{u}) = (1 - 1/x)(1 - x/\widehat{u} - 1/\widehat{u})$$

$$\geq 1 - 1/x - x/\widehat{u} \geq 1 - 1/x - x \cdot b^{-2w-3}.$$

We have $b^{w+1} \leq x \leq b^{w+2}$. The function $x \mapsto 1 - 1/x - x \cdot b^{-2w-3}$ is concave on the interval $b^{w+1} \leq x \leq b^{w+2}$ and therefore attains its minimum at an endpoint of the interval. Thus

$$\lfloor \widehat{u}/\widehat{v} \rfloor \cdot b^{p-q} \geq (u/v)(1 - b^{-w-3} - b^{-2w-3} - b^{-w-1} - b^{-w-2})$$

$$\geq (u/v)(1 - b^{-w}). \qquad \square$$

Theorem 2.5. *Let $n, m, k \in \mathbb{N}_0$ and suppose that A and B are given integers with $0 \leq A < b^n$ and $0 \leq B < b^m$ and that the cover size of B is k. Then AB and, if $B \neq 0$, $\lfloor A/B \rfloor$ can be computed in $O(1 + (kn + m)/w)$ time.*

Proof. Compute a smallest word cover $((\sigma_1, \ell_1, d_1), \ldots, (\sigma_k, \ell_k, d_k))$ of B in $O(1 + m/w)$ time (Lemma 2.3). Let $B^+ = \sum_{i=1}^{k} (1 - \sigma_i) \cdot b^{\ell_i} d_i$ and $B^- = \sum_{i=1}^{k} \sigma_i \cdot b^{\ell_i} d_i$ and consider first the computation of AB. Since $AB = AB^+ - AB^-$, it suffices to consider the special case $\sigma_1 = \cdots = \sigma_k = 0$ (compute AB^+ and AB^- separately with an algorithm for this case and obtain AB with one final subtraction). Assume therefore that $\sigma_1 = \cdots = \sigma_k = 0$.

To compute $AB = \sum_{i=1}^{k} b^{\ell_i} A d_i$, initialize an *accumulator* of $n + m$ digits to the value 0 and, for $i = 1, \ldots, k$, compute $A d_i$ in $O(1 + n/w)$ time and increment the value of the accumulator by $b^{\ell_i} A d_i$. Do this with the usual school method for addition, but process w digit positions together in constant time, begin the addition in digit position ℓ_i, and end it as soon as a digit position to the left of the most significant digit of $b^{\ell_i} A d_i$ does not have an incoming carry. In other words, skip those rightmost and leftmost digit positions of the accumulator that manifestly will not change; denote the number of digit positions actually processed by c_i.

The total time to compute AB is $O(k + (kn + m + \sum_{i=1}^{k} c_i)/w)$. To see that this is $O(1 + (kn + m)/w)$, note that $k = O(1 + m/w)$ and consider a potential Φ equal to the number of digits in the accumulator with a value of $b - 1$. $\Phi = 0$ after the initialization of the accumulator and $\Phi \geq 0$ at the end of the computation, so the net increase in Φ is nonnegative. For $i = 1, \ldots, k$, since $A d_i < b^{n+w}$, the addition of $b^{\ell_i} A d_i$ to the accumulator increases Φ by at most $2(n + w) - c_i + 1$ (every carry to the left of the most significant digit of $b^{\ell_i} A d_i$, except for the last one, changes an accumulator digit from $b-1$ to 0). Thus $\sum_{i=1}^{k}(2(n+w)-c_i+1) \geq$ 0 and $\sum_{i=1}^{k} c_i \leq k(2(n+w)+1) \leq 2kn+(1+m/w)(2w+1) = O(w+kn+m)$.

Assume now that $B > 0$ and consider the computation of $\lfloor A/B \rfloor$. We use the following variant of usual long division: Initialize two variables Q and R to 0 and A, respectively. The quotient $\lfloor A/B \rfloor$ will accumulate in Q, while R at all times contains what remains of A after the subtraction of QB. As long as $R \geq B$, carry out a *round*, consisting of these steps: Use the constant-time algorithm implicit in Lemma 2.4 to compute an approximation $q = s \cdot b^j$ to R/B with $0 \leq s < b^{2w+4}$ and $(R/B)(1 - b^{-w}) - 1 \leq q \leq R/B$, add q to Q and subtract qB from R. Then, if $R \geq B$ still holds, add 1 to Q and subtract B from R. R never increases, and the invariant $A = QB + R$ holds at the beginning and at the end of every round. By the upper bound on q, R remains nonnegative. Since every round subtracts at least B from R, the process clearly terminates. When it does, we have $0 \leq R < B$ and $A = QB + R$, which implies that $Q = \lfloor A/B \rfloor$ (and $R = A \bmod B$). All that remains is to return Q.

Let t be the number of rounds. Take $R_0 = A$ and, for $i = 1,\ldots,t$, denote the value of q used in the ith round by q_i and the value of R at the end of the ith round by R_i. For $i = 1,\ldots,t-1$,

$$R_i = R_{i-1} - q_i B - B \leq R_{i-1} - (R_{i-1}(1 - b^{-w}) - B) - B = R_{i-1} \cdot b^{-w}.$$

Since $R_0 < b^n$, it follows that $t - 1 \leq n/w$.

We still need to describe how to carry out the operations on Q and R. Q is realized as an accumulator of n digits. To add q and possibly subsequently 1 to Q, proceed similarly as for the addition of $b^{\ell_i} A d_i$ to the accumulator in the computation of AB. Since $0 \leq q < b^{2w+4}$ for each relevant q, the same potential Φ as above, now defined in terms of Q, shows that the at most $2t$ additions to Q can be carried out in a total time of $O(t + n/w) = O(1 + n/w)$.

The implementation of R and the operations on R are more complicated. We will assume without loss of generality that $B^+ > 2B^-$—indeed, for $b = 2$ the algorithm of Lemma 2.3 produces word covers with this property. R is realized via two accumulators R^+ and R^- of $n+1$ digits each and initialized to the values $2A$ and A, respectively, with R derived as $R = R^+ - R^-$. Let $K^+ = \{i \in \mathbb{N} \mid i \leq k$ and $\sigma_i = 0\}$ and $K^- = \{1,\ldots,k\} \setminus K^+$. To subtract a value yB from R (y is q or 1), subtract $y \cdot b^{\ell_i} d_i$ from R^+ for all $i \in K^+$ and from R^- for all $i \in K^-$. By assumption, every round subtracts at least twice as much from R^+ as from R^-, and therefore at most twice as much from R^+ as from R. Since R remains nonnegative, these facts imply that R^+ and R^- also remain nonnegative.

Each subtraction of $y \cdot b^{\ell_i} d_i$ is carried out with the usual school method, but processing w digits together in constant time, beginning the subtraction in digit position ℓ_i, and ending it as soon as a digit position to the left of the most significant digit of $y \cdot b^{\ell_i} d_i$ does not have an incoming borrow (i.e., the equivalent of the carry of addition). Consider a potential Φ' equal to the total number of digits with a value of zero in R^+ and R^-. $\Phi' \leq 2(n+1)$ after the initialization of R^+ and R^- and $\Phi' \geq 0$ at the end of the computation, so the net increase in Φ' is at least $-2(n+1)$. The school method for subtraction is invoked at most $2kt =$

$O(k(1 + n/w)) = O(1 + (kn + m)/w)$ times. In each invocation, if the number of digit positions processed is c, Φ' increases by at most $2(2w + 4 + w) - c + 1$ (every borrow to the left of the most significant digit of $y \cdot b^{\ell_i}d_i$, except for the last one, changes a digit in R^+ or R^- from 0 to $b - 1$). Denoting the total number of digit positions processed by C, we therefore obtain the inequalities $2tk(6w + 9) - C \geq -2(n + 1)$ and $C \leq 2n + 30tkw + 2 = O(w + m + (k + 1)n)$. Thus the time spent in school-method subtraction is $O(1 + (kn + m)/w)$.

Subtracting R^- from R^+ whenever we need the value of R to compute the approximate quotient q or to test whether $R \geq B$ would be too expensive. For this reason we instead use an approximation $\widetilde{R}$ to R with $R(1 - b^{-w-3}) \leq \widetilde{R} \leq R$ whose computation will be described below. Lemma 2.4 already allows for the use of such an approximation, so we need not modify our use of the lemma. To carry out the test $R \geq B$, we first test in constant time whether $\widetilde{R} \geq B$. If this is the case, then $R \geq \widetilde{R} \geq B$. If not, we spend $O(1 + n/w)$ time computing R exactly and testing whether $R \geq B$. Since this can happen only in the last two rounds, the total time needed for tests of $R \geq B$ is $O(1 + t + n/w) = O(1 + n/w)$.

Assume that, at a particular point in time, $R^+ = (r_n^+, \ldots, r_0^+)_b$ and $R^- = (r_n^-, \ldots, r_0^-)_b$. We always have $R^+ \geq R^- \geq 0$. To support the efficient computation of $\widetilde{R}$, we maintain throughout the computation the smallest integer h with $0 \leq h \leq n + 1$, called the *discriminating position*, such that $\Delta(h) \leq 1$, where $\Delta(h) = \sum_{i=h}^{n} b^{i-h}(r_i^+ - r_i^-)$. Informally, R^+ and R^- have approached each other to the point where all digits to the left of the discriminating position can be ignored. It is easy to see that if $h > 0$, then $b^{h-1} < R < b^{h+1}$. Since R never grows, this relation shows that if the discriminating position at some point in time is h, it can never later be larger than $h + 1$. Moreover, for every j with $h \leq j \leq n$, $\Delta(j) = 0$ if and only if $r_j^+ = r_j^-$. This in turn implies that the discriminating position h can be updated as follows after every change to R^+ and R^-: Add 1 to h and compute $\Delta(h)$ (take $\Delta(n + 2) = 0$). Then, as long as $\Delta(h) \leq 1$ and $h > 0$, compute $\Delta(h - 1) = b \cdot \Delta(h) + r_{h-1}^+ - r_{h-1}^-$ and, if $\Delta(h - 1) \leq 1$, subtract 1 from h. Processing w digits at a time with digit-wise operations and one of the algorithms of Lemma 2.2, we can maintain the discriminating position in $O(1 + n/w)$ overall time plus $O(1)$ time per update. In more detail, the update of the discriminating position, say from h_1 to h_2, can happen as follows (see Fig. 1): If $\Delta(h_1 + 1) = 0$, find the largest $j \leq h_1$, if any, with $r_j^+ \neq r_j^-$. If such a j exists, call it j^*; otherwise $h_2 = 0$ and we are done. If $\Delta(h_1 + 1) \neq 0$ (then $\Delta(h_1 + 1) = 1$), instead enter the procedure at this point with $j^* = h_1 + 1$. In either case, proceed as follows: If $\Delta(j^*) = r_{j^*}^+ - r_{j^*}^- \neq 1$, $h_2 = j^* + 1$. Otherwise find the largest $j < j^*$, if any, with $r_j^+ \neq 0$ or $r_j^- \neq b - 1$. If such a j exists, adding 1 to it yields h_2; otherwise $h_2 = 0$. With $z = \max\{h - w - 5, 0\}$, take $\widetilde{R} = R = R^+ - R^- = b^h \cdot \Delta(h) + \sum_{i=z}^{h-1} b^i(r_i^+ - r_i^-)$ if $z = 0$, and $\widetilde{R} = b^h \cdot \Delta(h) + \sum_{i=z}^{h-1} b^i(r_i^+ - r_i^-) - b^z$ if $z > 0$. Then $0 \leq R - \widetilde{R} \leq b^{z+1}$. Since $R > b^{h-1}$, we have $R(1 - b^{-w-3}) \leq \widetilde{R} \leq R$, as required. As anticipated above, we can both obtain $\widetilde{R}$ and test the condition $\widetilde{R} \geq B$ in constant time. Altogether, $\lfloor A/B \rfloor$ is computed in $O(1 + (kn + m)/w)$ time.   □

$$
\begin{array}{ll}
 & \quad\ \ h \\
R^+: & 7\,3\,6\,4\,8\,2\,6\,3\,0\ \ \cdots \\
R^-: & 7\,3\,6\,3\,1\,4\,1\,5\,9\ \ \cdots
\end{array}
$$

$$
\begin{array}{ll}
 & \quad\ \ \ h \\
R^+: & 7\,3\,6\,4\,1\,0\,7\,4\,6\ \ \cdots \\
R^-: & 7\,3\,6\,2\,9\,7\,5\,8\,3\ \ \cdots
\end{array}
$$

$$
\begin{array}{ll}
 & \quad h_1 \qquad j^* \qquad\quad h_2 \\
R^+: & 7\,3\,6\,2\,7\,1\,8\,2\,8\,0\,0\,0\,0\,0\,0\,0\,2\,3\,4\ \ \cdots \\
R^-: & 7\,3\,6\,2\,7\,1\,8\,2\,7\,9\,9\,9\,9\,9\,9\,9\,6\,1\,5\ \ \cdots
\end{array}
$$

Fig. 1. Snapshots of the accumulators R^+ and R^- at three points in time in an example with $b = 10$. The first two snapshots indicate the discriminating position h. Together they show that h may increase even though $R^+ - R^-$ decreases. The last snapshot illustrates the computation of first j^* and then the new discriminating position h_2 from the old discriminating position h_1.

A simpler and more natural division algorithm represents R with a single accumulator, rather than the two accumulators R^+ and R^-, and both adds to and subtracts from the single accumulator. However, the running time of such an algorithm possibly is not as indicated in Theorem 2.5.

3   Computing Tables of Polynomials

Suppose that s and f are positive integers and that g is an integer function defined on $\{0, \ldots, s - 1\}$ for which $0 \le g(k) < b^f$ for $k = 0, \ldots, s - 1$. Then the integer $\sum_{k=0}^{s-1} b^{kf} g(k) = (g(s-1), \ldots, g(0))_{b^f}$ can be viewed as a table of g that indicates $g(k)$ for $k = 0, \ldots, s - 1$ as the value of an f-digit field. This section is concerned with the efficient computation of such tables for a class of functions central to applications, namely the class of polynomials.

The approach is simple: We first show how to compute tables of binomial coefficients with constant lower index, i.e., of functions of the form $k \mapsto \binom{k}{i}$. Then, with the aid of Stirling numbers of the second kind, the tables of binomial coefficients are combined to yield tables of powers, i.e., of functions of the form $k \mapsto k^j$. And finally, tables of full polynomials are obtained as the obvious linear combinations of tables of powers.

We first describe the steps in the calculation, arguing only afterwards that the field size f is large enough and that the running time is as indicated.

Theorem 3.1. *Given integers $s, m \ge 1$, $d \ge 0$ and $f \ge m + d \log_b s$ as well as $d + 1$ integers $a_0, \ldots, a_d$ with $0 \le a_j < b^m$ for $j = 0, \ldots, d$, the integer*

$$
\sum_{k=0}^{s-1} b^{kf} \sum_{j=0}^{d} a_j k^j
$$

can be computed in $O((d+1)^2(1 + sf/w)(1 + (m + d\log_b(d+1))/w))$ time.

Proof. For all integers $i \geq 0$ and $x \geq 1$,

$$\sum_{k=0}^{s} \binom{k}{i} x^{k+1} \equiv \left(\sum_{k=1}^{s} x^k\right)^{i+1} \pmod{x^{s+1}}.$$

To see this, observe that for $k = 1, \ldots, s$, the coefficient of x^k is the same on both sides of the $\equiv$ sign: On the left-hand side it is $\binom{k-1}{i}$, and on the right-hand side it is the number of tuples $(j_1, \ldots, j_{i+1}) \in \mathbb{N}^{i+1}$ with $j_1 + \cdots + j_{i+1} = k$, which is also $\binom{k-1}{i}$ (think of marking the integers 0 and k and i additional distinct integers strictly between 0 and k and obtaining the numbers $j_1, \ldots, j_{i+1}$ as the distances between successive marks).

Taking $x = b^f$, we see that all of the tables $\left(\binom{s-1}{i}, \ldots, \binom{0}{i}\right)_{b^f}$, for $i = 0, \ldots, d$, can be computed with d multiplications of

$$(1, 1, \ldots, 1, 0)_{b^f} = \frac{b^{(s+1)f} - 1}{b^f - 1} - 1 \tag{1}$$

$$\underbrace{}_{s \text{ fields}}$$

by itself, with retention only of the remainder modulo $b^{(s+1)f}$, followed by a right shift of each of the intermediate products by one field width.

For arbitrary integers k and $j \geq 0$, by definition of the Stirling numbers $\{{}^j_i\}$ of the second kind [7], $k^j = \sum_{i=0}^{j} \{{}^j_i\} i! \binom{k}{i}$. Correspondingly,

$$((s-1)^j, \ldots, 1^j, 0^j)_{b^f} = \sum_{i=0}^{j} \{{}^j_i\} i! \left(\binom{s-1}{i}, \ldots, \binom{0}{i}\right)_{b^f}.$$

All of the Stirling numbers $\{{}^j_i\}$ with $0 \leq i \leq j \leq d$ can be computed with $O((d+1)^2)$ arithmetic operations according to the recursion $\{{}^j_i\} = i\{{}^{j-1}_i\} + \{{}^{j-1}_{i-1}\}$ $(i, j \in \mathbb{N})$ together with the boundary values $\{{}^0_0\} = 1$ and $\{{}^\ell_0\} = \{{}^0_\ell\} = 0$ for $\ell \in \mathbb{N}$. Thus all of the tables $((s-1)^j, \ldots, 0^j)_{b^f}$, for $j = 0, \ldots, d$, can be computed within the same number of arithmetic operations. The final table is obtained simply as

$$\left(\sum_{j=0}^{d} a_j(s-1)^j, \ldots, \sum_{j=0}^{d} a_j 0^j\right)_{b^f} = \sum_{j=0}^{d} a_j \left((s-1)^j, \ldots, 0^j\right)_{b^f}.$$

All values intended to be stored in single fields in the algorithm are strictly smaller than $b^m \sum_{j=0}^{d}(s-1)^j$. We prove by induction on d that $\sum_{j=0}^{d}(s-1)^j \leq s^d$. The basis, $d = 0$, is trivial. As for the inductive step, if we assume for some $d \geq 1$ that $\sum_{j=0}^{d-1}(s-1)^j \leq s^{d-1}$, we find $\sum_{j=0}^{d}(s-1)^j = \sum_{j=0}^{d-1}(s-1)^j + (s-1)^d \leq s^{d-1} + (s-1)^d \leq s^{d-1}(1 + (s-1)) = s^d$. Therefore all field values are strictly smaller than $b^m \cdot s^d = b^{m+d \log_b s}$, which proves that the condition $f \geq m + d \log_b s$ is sufficient to prevent overflows between fields.

In the computation of the tables of binomial coefficients, each of the d multiplications by $(1, 1, \ldots, 1, 0)_{b^f}$ can, according to (1), be replaced by a multiplication by an integer of cover size at most 2, a division by an integer of cover size at most 2, and a subtraction. Therefore the tables can be computed in $O((d+1)(1+sf/w))$ time. One easily proves by induction that $\{^j_i\} \leq (i+1)^j$ for all $i, j \in \mathbb{N}_0$. For $0 \leq i \leq j \leq d$, both $\{^j_i\}$ and $i!$ are therefore bounded by $(d+1)^d$. This shows that each of the remaining $O((d+1)^2)$ multiplications in the algorithm is of an integer of $O(m + d\log_b(d+1))$ b-ary digits and an integer of at most sf b-ary digits. If we carry out these multiplications simply with the school method, the total running time of the algorithm therefore is $O((d+1)^2(1+sf/w)(1+(m+d\log_b(d+1))/w))$. □

If d is a constant and $m = O(w)$, as is typical of applications, the running time of Theorem 3.1 is $O(1 + sf/w)$, i.e., linear in the number of words occupied by the output. If it is desired to compute a table of $g(k_0), \ldots, g(k_0 + s - 1)$ rather than of $g(0), \ldots, g(s - 1)$, for some integer k_0, one can simply substitute for g the function $k \mapsto g(k_0 + k)$, which is a polynomial of the same degree as g and with coefficients that are not much larger. Similarly, if it is desired to number the fields within a word from left to right rather than from right to left, one can consider functions of the form $k \mapsto g(k_0 - k)$ (a more direct construction is also possible).

References

1. Aho, A.V., Hopcroft, J.E., Ullman, J.D.: The Design and Analysis of Computer Algorithms. Addison-Wesley, Reading (1974)
2. Andersson, A., Hagerup, T., Nilsson, S., Raman, R.: Sorting in linear time? J. Comput. Syst. Sci. **57**(1), 74–93 (1998)
3. Brand, M.: Constant-time sorting. Inform. Comput. **237**, 142–150 (2014)
4. Cole, R., Vishkin, U.: Deterministic coin tossing with applications to optimal parallel list ranking. Inform. Control **70**(1), 32–53 (1986)
5. Fredman, M.L., Willard, D.E.: Surpassing the information theoretic bound with fusion trees. J. Comput. Syst. Sci. **47**(3), 424–436 (1993)
6. Fürer, M.: Faster integer multiplication. SIAM J. Comput. **39**(3), 979–1005 (2009)
7. Graham, R.L., Knuth, D.E., Patashnik, O.: Concrete Mathematics: A Foundation for Computer Science. Addison-Wesley, Reading (1989)
8. Hagerup, T., Kammer, F.: Dynamic data structures for the succinct RAM (2015) (in preparation)
9. Knuth, D.E.: The Art of Computer Programming, Volume 2: Seminumerical Algorithms, 3rd edn. Addison-Wesley, Upper Saddle River (1998)
10. Knuth, D.E.: The Art of Computer Programming, Volume 4A: Combinatorial Algorithms, Part 1. Addison-Wesley, Upper Saddle River (2011)
11. Schönhage, A.: Storage modification machines. SIAM J. Comput. **9**(3), 490–508 (1980)
12. Schönhage, A., Strassen, V.: Schnelle Multiplikation großer Zahlen. Computing **7**(3–4), 281–292 (1971)
13. Warren Jr., H.S.: Hacker's Delight, 2nd edn. Addison-Wesley, Upper Saddle River (2013)

Visibly Counter Languages and the Structure of NC¹

Michael Hahn[✉], Andreas Krebs, Klaus-Jörn Lange, and Michael Ludwig

WSI - University of Tübingen, Sand 13, 72076 Tübingen, Germany
{hahnm,krebs,lange,ludwigm}@informatik.uni-tuebingen.de

Abstract. We extend the familiar program of understanding circuit complexity in terms of regular languages to visibly counter languages. Like the regular languages, the visibly counter languages are NC¹- complete. We investigate what the visibly counter languages in certain constant depth circuit complexity classes are. We have initiated this in a previous work for AC⁰. We present characterizations and decidability results for various logics and circuit classes. In particular, our approach yields a way to understand TC⁰, where the regular approach fails.

1 Introduction

The family of the regular languages is among the best studied objects in theoretical computer science. They arise in many contexts, have many natural characterizations and good closure and decidability properties, and have been used as a tool to understand other objects. The nice behavior of regular languages comes at a price, which is a limited expressiveness. The class of context-free languages contains many important nonregular languages, but at the same time is much harder to analyze. In our present line of work, we attempt to generalize results which are known for the regular case to the nonregular case. A very promising way of generalizing regular languages is given by the *visibly pushdown languages*, which were introduced by Mehlhorn [10] in 1980 and popularized by Alur and Madhusudan [1] in 2004. They have been an active field of research ever since. Visibly pushdown automata are pushdown automata with the restriction that the input letter determines the stack operation: Push, pop or no access to the stack. These automata have good closure and decidability properties that are comparable to those of regular languages, and still cover many important context-free languages. The word problem is as hard as for the regular languages, namely NC¹-complete. We will consider a restriction of visibly pushdown languages. *Visibly counter languages* (VCLs) are languages recognized by *visibly counter automata*, which are visibly pushdown automata only having a counter. In some sense, this class appears to be one of the smallest useful nonregular language classes, and hence is our starting point in the program of generalizing results beyond the realm of regular languages.

Already in the 1980s, based in part on the result that PARITY is not in AC⁰ [6,7], and on Barrington's theorem [3], deep connections between regular

© Springer-Verlag Berlin Heidelberg 2015
G.F. Italiano et al. (Eds.): MFCS 2015, Part II, LNCS 9235, pp. 384–394, 2015.
DOI: 10.1007/978-3-662-48054-0_32

languages and circuit complexity were discovered [5]. These results showed that circuit classes have characterizations in terms of algebraic properties of finite monoids, which has led to the characterization of class separations in terms of properties of regular sets. For instance, the classes ACC^0 and NC^1 differ if and only if no syntactic monoid of a regular set in ACC^0 contains a nonsolvable group. Indeed, the circuit classes AC^0, CC^0, $ACC^0[n]$, ACC^0, and NC^1 can all be characterized using finite monoids and regular languages [12].

What stands out at this point is the absence of TC^0 in these results. Indeed, TC^0 does not relate to any family of regular languages: Assuming $NC^1 \neq ACC^0$, then either a regular language is NC^1-hard or it is contained in ACC^0 – depending on whether its syntactic monoid is solvable or not. This is where our contribution comes into play. We lift the correspondence between circuit classes and algebraic properties of regular languages to visibly counter languages. This greatly extends previous work [8], in which we characterized the visibly counter languages in AC^0.

Our Results. We examine the intersection of visibly counter languages with the circuit classes AC^0, $CC^0[q]$, CC^0, $ACC^0[q]$, ACC^0, TC^0, and NC^1. As our main theorem, we unconditionally prove criteria for visibly counter languages to be complete for these classes (Theorem 18). The results are summarized in Fig. 1, which compares the known families of regular languages complete for various circuit classes with corresponding complete classes of visibly counter languages. Furthermore, given a visibly counter automaton, we can effectively compute the smallest of these complexity classes that contains the language recognized by this automaton. This also provides an effective method to decide whether a language, given by a visibly counter automaton, is in one of these complexity classes (Theorem 20). Decidability results for nonregular languages classes corresponding to circuit complexity classes are rather sparse. We expect that our more general method will help to lift various decidabilty results about regular languages to visibly counter languages.

As most separation questions between those languages classes are still open, we prove equivalences to these separations. For example we show that $CC^0 \neq ACC^0$ iff CC^0 contains a nonregular visibly counter language. Assuming that

Circuit Class	Regular Languages Syntactic Monoids	Visibly Counter Languages Stack	Regular Part
CC^0	solvable groups	no stack	& solvable groups
AC^0	quasi aperiodic	well behaved	& quasi aperiodic
ACC^0	solvable monoids	well behaved	& solvable monoids
TC^0	—	any	& solvable monoids
NC^1	finite monoids	any	& finite monoids

Fig. 1. Circuit classes with complete classes of regular languages [12], and complete visibly counter classes.

$\text{ACC}^0 \neq \text{TC}^0 \neq \text{NC}^1$, we can show that the visibly counter languages intersected with these languages are contained in logic classes using only semilinear predicates. This observation has been explored by [9], who introduced the notion of *extensional uniformity*, which is based on the study of the intersection of a complexity class with a family of formal languages. For $\text{CC}^0[p]$, $\text{ACC}^0[p]$, p being prime, we provide similar but unconditional results.

Our Approach. We obtain our results by splitting the computation of visibly counter automata into two parts. Each position in a word over a visibly alphabet can be assigned a height. The first computation step is determining this height which we treat in Sect. 3. Assuming the height computation is already performed, what is left is the computation performed by the states of the automaton. We call this the *regular part* and treat this in Sect. 4. After we have established those two parts, we derive results on logic in Sect. 5 and on circuits in Sect. 6. Due to space constraints we omit most of the proofs.

2 Preliminaries

An *alphabet* is a finite set Σ. A subset of Σ^* is called a *language*. For a word w, $|w|$ is the length of the word, $|w|_\Sigma$ the number of occurrences of symbols from Σ in w, and w_i is the letter in position i in w. For $L \subseteq \Sigma^*$, the set $F(L) \subseteq \Sigma^*$ is the set of all *factors* of words in L, i.e.: $F(L) = \{y \in \Sigma^* \mid \exists x, z \in \Sigma^* : xyz \in L\}$.

Circuit Complexity. The complexity classes we will consider are defined using circuit families. Precisely, we will study the following classes:

- AC^0: polynomial-size circuit families of constant depth with Boolean gates of arbitrary fan-in.
- $\text{ACC}^0[k_1, ..., k_n]$: AC^0 circuit families supplemented with modulo-k_i-gates for each i. ACC^0 is the union of $\text{ACC}^0[k]$ for all k.
- TC^0: polynomial-size circuit families of constant depth with threshold gates of arbitrary fan-in.
- NC^0: polynomial-size circuit families of constant depth with bounded fan-in.
- NC^1: polynomial-size circuit families of logarithmic depth with bounded fan-in.

If for some $n \in \mathbb{N}$ the circuit with input length n is computable in some complexity bound, we speak of uniformity. One prominent example is so called DLOGTIME-uniformity. For further references on circuit complexity, cf. e.g., [13].

If L, L' are languages, then a *quantifier-free reduction* from L to L' is a constant-depth circuit family for L whose only gates are bounded fan-in gates and an oracle gate for L'. A *quantifier-free Turing reduction* allows in addition for the nesting of oracle gates for L'.

Algebra. Every language $L \subseteq \Sigma^*$ induces a congruence on Σ^*, the *syntactic congruence* of L: For $x, y \in \Sigma^*$, $x \sim_L y$ iff for all $u, v \in \Sigma^*$ we have $uxv \in L \Leftrightarrow uyv \in L$. The *syntactic monoid* is the quotient monoid $\Sigma^* / \sim_L$. The *syntactic morphism* of L is the canonical projection $\eta_L \colon \Sigma^* \to \Sigma^* / \sim_L$.

If $\mathcal{G}$ is a group, then $\mathcal{G}^{(1)}$ is the subgroup generated by elements of the form $xyx^{-1}y^{-1}$ $(x, y \in \mathcal{G})$. We recursively define $\mathcal{G}^{(n)} := (\mathcal{G}^{(n-1)})^{(1)}$. A group is *solvable* if $\mathcal{G}^{(n)} = \{1_{\mathcal{G}}\}$ for some $n \in \mathbb{N}_1$. A monoid is called *solvable* if all its subsemigroups that are groups are solvable groups, otherwise it is *nonsolvable*. A nontrivial group $\mathcal{G}$ is *perfect* if $\mathcal{G}^{(1)} = \mathcal{G}$, in which case $\mathcal{G}$ is not solvable.

Let M, N be two monoids. We say M *divides* N iff there is a subset $N' \subseteq N$ and a surjective monoid morphism $N' \to M$. A nontrivial monoid is *simple* if it has no nontrivial monoid congruences. The simple monoids are the aperiodic unit $U_1 = (\{0, 1\}^*, \cdot)$, the Abelian simple groups $\mathbb{Z}_p$ (p prime), and the non-Abelian simple groups, which are the perfect groups, whose smallest one is the permutation group A_5. The notion of solvability is related to the simple divisors a monoid has: A monoid is solvable if and only if all its simple divisors are commutative, and it is nonsolvable if and only if it is divided by a perfect group.

Let M be a finite monoid. By the *word problem of M*, we denote the set

$$\{w \in M^* : \pi(w) = 1_M\}$$

of words over M evaluating to the identity element of M, where $\pi : M^* \to M$ is the canonical morphism. Observe, this language has a neutral letter.

Logic. Related to circuit classes is the framework of logic over finite words. A nonempty word $w \in \Sigma^+$ defines a structure whose domain is $\{1, ..., |w|\}$. For every $a \in \Sigma$, there is a predicate Q_a such that $w \models Q_a(i)$ iff $w_i = a$. Furthermore, any predicate defined on $\mathbb{N}_1$ can be interpreted canonically in such a structure; such predicates are called *numerical predicates*. Important examples are the order predicate $<$ with arity two and an obvious semantics, and the arithmetic predicate $+$ of arity three: $(i, j, k) \in +$ iff $i+j = k$. We allow first-order existential and universal quantification. Furthermore, we will use quantifiers over finite monoids. Let M be a monoid, $M_+ \subseteq M$, $N \in \mathbb{N}$, let $\psi_1(x), ..., \psi_N(x)$ be formulas, let $\delta : \{0, 1\}^N \to M$ be a map, let $w \in \Sigma^*$ be a word, and let v be an assignment of variables to positions in w. Then $Q^{M, \delta, M_+} x \; \psi_1(x)...\psi_N(x)$ is a formula, which is true in w under the assignment v iff

$$\Pi_{i=1}^{|w|} \delta(w \models \psi_1(i), ..., w \models \psi_N(i)) \in M_+$$

where the product symbol denotes the multiplication of the monoid M. Informally, a quantifier over a monoid labels each position with a monoid element depending on which formulas $\psi_i(x)$ are true at the position, and then checks whether the resulting word over the monoid evaluates to an element of M_+. First order quantification is equivalent to quantification using the monoid $U_1 = (\{0, 1\}, \cdot)$, and modulo quantifiers are equivalent to quantifiers over cyclic groups. Furthermore, there is the majority quantifier Maj, where Maj $x \; \psi(x)$ is true if $\psi(i)$ is true for more than half of the positions i in the word. We write Reg for the set of regular predicates, and SL for the set of semilinear predicates, i.e., those definable using regular predicates and $+$.

If Q is a set of finite monoids and quantifiers, and P is a set of predicates, we write $Q[P]$ for the set of formulas built from P and Q, and also for the set of languages recognized by these formulas. We also write FO for U_1 quantification, and MOD for quantification over cyclic groups. We write arb for the set of all numerical predicates. Then we have $AC^0 = FO[arb]$, $CC^0[k] = \mathbb{Z}_k[arb]$, $CC^0 = MOD[arb]$, $ACC^0[k] = FO + \mathbb{Z}_k[arb]$, $ACC^0 = FO + MOD[arb]$, and $TC^0 = Maj[arb]$. Note that $Maj[<] = Maj[SL]$. See e.g. [13].

Visibly Pushdown Languages. Mehlhorn [10] and later independently Alur and Madhusudan [1] introduced *input-driven* or *visibly pushdown automata*, which are pushdown automata in which the input symbol determines whether a symbol is pushed or popped. This leads to a partitioning of Σ into call, return and internal letters: $\Sigma = \Sigma_{call} \cup \Sigma_{ret} \cup \Sigma_{int}$. We will assume that Σ has some fixed partitioning. The *height* of $w \in \Sigma^*$ is given by $\Delta : \Sigma^* \to \mathbb{Z} : w \mapsto |w|_{\Sigma_{call}} - |w|_{\Sigma_{ret}}$. A word w is *well-matched* if $\Delta(w) = 0$ and $\Delta(w_1...w_i) \geq 0$ for all $1 \leq i \leq |w|$.

The concept of *visibly counter automata* (VCA) was introduced by Bárány et al. [2]. Since every VPA can be determinized and this is also true for visibly counter automata, we restrict attention to deterministic automata:

Definition 1 (mVCA). *An* $m-$VCA $\mathcal{A}$ *over* $\hat{\Sigma} = (\Sigma_{call}, \Sigma_{ret}, \Sigma_{int})$ *is a tuple* $\mathcal{A} = (Q, q_0, F, \hat{\Sigma}, \delta_0, \ldots, \delta_m)$, *where* $m \geq 0$ *is the* threshold, Q *is the set of states,* q_0 *the initial state,* F *the set of final states, and* $\delta_i : Q \times \Sigma \to Q$ *are the transition functions.*

A *configuration* of an $m-$VCA is an element of $Q \times \mathbb{N}$. When reading a letter $\Sigma \in \Sigma$, an $m-$VCA $\mathcal{A}$ performs the transition $(q, k) \xrightarrow{\Sigma} (\delta_{\min(m,k)}(q, \Sigma), k + \Delta(\Sigma))$. Then $w \in L(\mathcal{A})$ iff $(q_0, 0) \xrightarrow{w} (f, h)$ for $f \in F$ and $h \geq 0$. Visibly counter automata, and more generally visibly pushdown automata, can only recognize words where the height of all prefixes is non-negative. All other words are rejected.

Definition 2 (VCL). *A language is a* visibly counter language *if it is recognized by an* $m-$VCA *for some* m.

3 Height Computation for Visibly Counter Languages

A first step towards classifying VCLs with concern to their complexity was made in [8], where we gave a decidable characterization of the VCLs in AC^0. The proof is based on considering separately the computation of the height profile and the simulation of the finite-state control.

For the height computations, the previous work considered the problem of defining unary predicates $H_n(\cdot)$ that are true if a position in a given word iff the symbol at that position has height n.

We defined a notion of *simple height behavior* [8] which guarantees that these height predicates can, in some sense, be approximated in AC^0. Informally, a VCL has simple height behavior if whenever a recognizing automaton loops through an active state q reading a subword x, the height $\Delta(x)$ is determined by q and the length $|w|$. More formally, the following notions were considered:

Definition 3 (simple height behavior).

fixed slope. *We say that a state q has a* fixed slope *if there are numbers $\alpha \in \mathbb{Q}$ and $\gamma \in \mathbb{N}$ so that if for all words $w \in \Sigma^*$ with $(q, h_1) \xrightarrow{w} (q, h_2)$ and $h_1 + \Delta(w') \geq m$ for all prefixes w' of w it holds that:*

- $h_2 = \alpha|w| + h_1$
- $\alpha|w'| - \gamma \leq \Delta(w') \leq \alpha|w'| + \gamma$ *for all prefixes w' of w*

active. *A state q is* active *if there is a word $w \in L(\mathcal{A})$ with positions i and j, $i < j$, such that after reading $w_1 \cdots w_i$, $\mathcal{A}$ is in q, $\Delta(w_1 \cdots w_i) > m + |Q|$ and $\Delta(w_1 \cdots w_i) - \Delta(w_1 \cdots w_j) > |Q|$.*

simple height behavior. *If in a VCA $\mathcal{A}$ all active states have a fixed slope, we say that the recognized language $L(\mathcal{A})$ has simple height behavior.*

The notion of simple height behavior is a property of the accepted language, but uses automata. The next theorem [8] shows that the property is indeed independent of the automaton. Also it shows how hard height computation can be.

Lemma 4. *If a VCL does not have simple height behavior, it is TC^0-hard by quantifier-free Turing reductions.*

In the case of simple height behavior we know that the height is computable in FO[+] which is shown by constructing height predicates [8].

Lemma 5. *Given a VCL L with simple height behavior, we can define for every $k < m$ a monadic predicate $H_k(x)$ in FO[+] such that:*

- $w_{x=i} \models H_k(x)$ *then $\Delta(w_1 \ldots w_{i-1}) = k$ for arbitrary $w \in \Sigma^*$.*
- $w_{x=i} \models H_k(x)$ *iff $\Delta(w_1 \ldots w_{i-1}) = k$ for all $w \in L$.*

We let $H_{\geq m} = \neg \bigvee_{k=0}^{m-1} H_k$ be the negation of these predicates. Hence for $w \notin L$ the predicate might have false-positives, i.e., the predicate might suggest a stack-height greater or equal to m while in fact it is less than m. In later proofs we have take care of this.

So in conclusion we get a dichotomy: Either the height is computable in AC^0, or the language TC^0- hard.

After languages not having a simple stack behavior and languages having a simple stack behavior, we are left with the case that we do not have a height behavior at all. That is the case when the language is regular. But if we really need the counter, then we again get a hardness statement:

Lemma 6. *If L is a nonregular VCL, then it is AC^0-hard by quantifier-free Turing reductions.*

4   The Regular Part of Visibly Counter Languages

In the previous section we ended up with an FO[+] definable height predicate in the case of languages with simple height behavior. In this section we want to look at what is left if we presuppose a completed height computation. We will do so in coding the height up to the threshold m into the input. This is performed by a height transduction [8]:

Definition 7 (Height transduction). *For* $m \in \mathbb{N}$ *we let* $\Sigma_m = \Sigma \times \{0, \ldots, m\}$ *and set* $\Delta_m(w) = \min(\Delta(w), m)$. *Then define*

$\tau_m : \{w : \Delta(w) \geq 0\} \to (\Sigma_m)^*$ *by*

$\tau_m(w_1 w_2 \cdots w_n) =$
$$(w_1, \Delta_m(\epsilon))(w_2, \Delta_m(w_1)) \cdots (w_i, \Delta_m(w_1 \cdots w_{i-1})) \cdots (w_n, \Delta_m(w_1 \cdots w_{n-1}))$$

A word $w \in (\Sigma_m)^*$ *is called* valid *if* $w \in F(\tau_m(\Sigma^*))$. *We call* i *the* label *of the letter* $(a, i) \in \Sigma_m$.

Informally, τ_m labels symbols with their height up to the threshold m. For instance, if a and b are push and pop letters respectively, then $\tau_2(aaaab) = (a, 0)(a, 1)(a, 2)(a, 2)(b, 2)$.

The action of the finite state control is modeled by a regular language, the *regular part* [8]:

Definition 8 (The regular part $R_\mathcal{A}$). *Let* $\mathcal{A} = (Q, q_0, F, \hat{\Sigma}, \delta_0, \ldots, \delta_m)$ *be an* m−VCA. *Let* M *be the finite automaton* $M = (Q, q_0, F, \Sigma_m, \delta)$, *where* $\delta(q, (a, i)) = \delta_i(q, a)$. *Then set* $R_\mathcal{A} = L(M)$.

$R_\mathcal{A}$ is a regular language over the alphabet Σ_m. For $\Delta(w) > 0$, we have $w \in L(\mathcal{A})$ iff $\tau_m(w) \in R_\mathcal{A}$. In general, $R_\mathcal{A}$ also contains words that are not in the image of τ_m, and the choice of $\mathcal{A}$ for L determines what these words are.

Unfortunately we cannot always draw a conclusions regarding $L(\mathcal{A})$, based on properties of $R_\mathcal{A}$. To avoid this problem we defined a normal-form on visibly counter automata [8]:

Definition 9 (Loop-Normal). *An* m−VCA $\mathcal{A}$ *is called* loop-normal *if for all* $x, y \in \Sigma^*$ *with* $\Delta(xy_1 \cdots y_k) \geq m$ *for* $0 \leq k \leq |y|$ *and* $(q_0, 0) \xrightarrow{x} (q, h_1) \xrightarrow{y} (q, h_2)$, $q \in Q$ *then either* q *is a dead state, or one of the following is true, depending on* $\Delta(y)$:

- *If* $\Delta(y) > 0$, *then for each* $z \in \Sigma^*$ *such that* $\Delta(xyz) \geq 0$, *there is a partition of* z *into* $z = z_1 z_3$ *and a word* $z_2 \in \Sigma^*$ *so that for all* $i \geq 0$ *we have that* $xyy^i z_1 z_2^i z_3 \in L(\mathcal{A})$ *iff* $xyz \in L(\mathcal{A})$.
- *If* $\Delta(y) < 0$, *then there is a partition of* x *into* $x = x_1 x_3$ *and a word* $x_2 \in \Sigma^*$ *so that for all* $i \geq 0$ *and for all for each* $z \in \Sigma^*$ *such that* $\Delta(xyz) \geq 0$, *we have that* $x_1 x_2^i x_3 yy^i z \in L(\mathcal{A})$ *iff* $xyz \in L(\mathcal{A})$.
- *If* $\Delta(y) = 0$, *then for all* $i \geq 0$ *and for each* $z \in \Sigma^*$ *such that* $\Delta(xyz) \geq 0$, $xy^i z \in L$ *iff* $xyz \in L(\mathcal{A})$, *and*

if $\delta_i = \delta_m$ *for* $m - |Q| < i < m$.

This normal form has the property that, if the automaton goes through a loop in reading a prefix that can be completed to an accepted word, then it can still be completed when more loops through the same state are appended. In [8] we showed:

Lemma 10. *For every VCL L, there is a loop-normal $m-$VCA $\mathcal{A}$ recognizing L.*

We are interested in which groups are contained in the regular part of a language. The regular languages in some constant-depth circuit class $\mathcal{C}$ are known to be determined already by the class of simple monoids whose word problem is in $\mathcal{C}$ [12]. Broadly speaking, a regular language L is in a circuit class if and only if, for all simple monoids 'contained in' L, their word problems are also in $\mathcal{C}$. In the regular case, the appropriate notion of a monoid M 'being contained' in a language L with syntactic morphism η_L is that there a $t > 0$ such that $\eta_L(\Sigma^t)$ contains a monoid divided by M [4]. In the case of VCLs, the precise meaning of 'being contained' in the language is captured by Q_L:

Definition 11 (Simple monoids of the regular part: Q_L). *Let L be a VCL and let $\mathcal{A}$ be some loop-normal $m-$VCA for L. Then Q_L is the set of simple monoids $\mathcal{N}$ for which there is a number $t > 0$ and a set $N \subseteq (\Sigma_m)^t$ with $N^* \subseteq F(\tau_m(\Sigma^*))$ so that $\eta_{R_{\mathcal{A}}}(N)$ is a monoid divided by $\mathcal{N}$.*

Informally, we take those simple monoids which can be simulated by words over a set N such that all words over N are valid. Our results will not depend on the choice of $\mathcal{A}$ used for constructing Q_L, as long as $\mathcal{A}$ is loop-normal which justifies the definition of Q_L as a property of the language. Then the adequate notion of '$R_{\mathcal{A}}$ contains $\mathcal{G}$' is that $\mathcal{G}$ is an element of Q_L, as shown by the following lem:

Lemma 12. *Let L be a VCL and $\mathcal{G} \in Q_L$ a simple monoid. Then the word problem of $\mathcal{G}$ is reducible to L by quantifier-free Turing reductions.*

Thus, a VCL is at least as hard as the word problems of the monoids contained in Q_L. Putting these hardness results together, we have:

Proposition 13. *Let L be a VCL.*

- *If Q_L contains U_1, then L is AC^0-hard.*
- *If Q_L contains cyclic groups $\mathbb{Z}_{p_1}, \ldots, \mathbb{Z}_{p_k}$, then L is $\mathrm{CC}^0[p_1 \cdots p_k]$-hard*
- *If Q_L contains cyclic groups $\mathbb{Z}_{p_1}, \ldots, \mathbb{Z}_{p_k}$ and U_1, then L is $\mathrm{ACC}^0[p_1 \cdots p_k]$-hard*
- *If Q_L contains a non-Abelian simple group, then L is NC^1-hard.*

Our notion of hardness is via quantifier-free Turing reductions.

We want to define the regular part in logic so that in the next step we can combine it with the height computation and then get a formula for the whole language. The following lemma constructs a formula defining $R_{\mathcal{A}}$ when the input is known to be valid.

Lemma 14. *Let $\mathcal{A}$ be an $m-$VCA recognizing L. There is an $\mathrm{FO} + Q_L[\mathrm{Reg}]$ formula φ with*

$$L(\varphi) \cap \tau_m(\Sigma^*) = R_{\mathcal{A}} \cap \tau_m(\Sigma^*).$$

5 Results on Logic

We can use our findings on height behavior and the regular part to precisely assign logic classes to VCLs:

Theorem 15. *Let L be a VCL. Then:*

- *If L does not have simple height behavior, then it is in* $\mathrm{Maj} + Q_L[<]$.
- *If L has simple height behavior and is not regular, then it is in* $\mathrm{FO} + Q_L[SL]$.
- *If L is regular and* $Q_L \neq \emptyset$, *then L is in* $Q_L[Reg]$.

Remark 16. In the case that L is regular and Q_L is empty, membership of a word w in L only depends on a constant-size prefix and suffix of w.

Indeed, given a VCL, the minimal such logic class that can define it is computable in the following sense. Let S be the set consisting of Maj and the quantifiers over simple finite monoids. For each subset $Q \subseteq S$, we have logic class $Q[+, *]$, so we obtain a lattice $\mathcal{L} = \{Q[+, *] \mid Q \subseteq S\}$ of logic classes. Some classes will be equal, in particular, when A is a set of commutative monoids, $\mathrm{Maj}[+, *] = \mathrm{Maj} + A[+, *]$. We get the following decidability result, assuming a VCL is encoded by an arbitrary $m-$VCA recognizing it.

Corollary 17. *Given a VCL L, one can effectively compute a minimal class* $Q[+, *] \in \mathcal{L}$ *such that L is contained in* $Q[+, *]$.

6 Results on Circuits

The logic classes considered in the last section can be seen as uniform versions of circuit classes. Combining the previous results, we unconditionally get:

Theorem 18. *Let L be a VCL.*

- *L is* AC^0-*complete if L is nonregular and has simple height behavior and* Q_L *does not contain a simple group.*
- *L is* $\mathrm{CC}^0[p_1 \cdots p_k]$-*complete if L is regular and* Q_L *contains exactly the simple cyclic groups* $\mathbb{Z}_{p_1}, \ldots, \mathbb{Z}_{p_k}$.
- *L is* $\mathrm{ACC}^0[p_1 \cdots p_k]$-*complete if L is nonregular and does has simple height behavior and* Q_L *contains* U_1 *and the simple cyclic groups* $\mathbb{Z}_{p_1}, \ldots, \mathbb{Z}_{p_k}$.
- *L is* TC^0-*complete if L does not have simple height behavior and* Q_L *does not contain a non-Abelian simple group.*
- *L is* NC^1-*complete if* Q_L *contains a non-Abelian simple group.*

Our notion of completeness is via quantifier-free Turing reductions.

Proof. Membership in the respective classes follow from Theorem 15 by considering that $\mathrm{AC}^0 = \mathrm{FO}[arb]$, $\mathrm{CC}^0[p_1 \cdots p_k] = (\mathbb{Z}_{p_1} + \ldots \mathbb{Z}_{p_k})[arb]$, and $\mathrm{ACC}^0[p_1 \cdots p_k] = (\mathrm{FO} + \mathbb{Z}_{p_1} + \ldots \mathbb{Z}_{p_k})[arb]$, furthermore, $\mathrm{TC}^0 = \mathrm{Maj}[arb] = \mathrm{Maj} + A[arb]$ for A being a set of commutative monoids, and $\mathrm{NC}^1 = G[arb]$ for G any nonsolvable group by Barrington's theorem. Hardness follows from Lemma 6 in the first case and from Theorem 4 in the fourth case. In the other cases, it follows from Proposition 13. $\qquad\square$

We also obtain decidability results. We can show this rather abstractly for arbitrary circuit classes. Precisely, by a (constant-depth) circuit class C over a set of gates, we mean the class of languages recognized by constant-depth, polynomial size circuit families over this set of gates. We assume that any class is closed under quantifier-free reductions, i.e., has bounded fan-in Boolean gates. By results of [12], the regular languages in any such class are already determined by the simple monoids whose word problems are in the class. For all such classes, we have:

Lemma 19. *Let C be a circuit class. There is a computable reduction from the membership problem of* $\mathrm{VCL} \cap C$ *to the membership problem of* $\mathrm{Reg} \cap C$.

Proof. Compute the logic class $Q[+, *]$ from Corollary 17. If Q contains Maj and $C \not\supseteq \mathrm{TC}^0$, we know $L(\mathcal{A}) \notin C$. Otherwise, check whether the word problems of the simple finite monoids in Q are in C. □

Since membership in $\mathrm{TC}^0 \cap \mathrm{Reg}$ and $\mathrm{ACC}^0 \cap \mathrm{Reg}$ at most depends on the presence of nonsolvable groups and thus is decidable, we get:

Corollary 20. *Given a VCL L, it is decidable whether L is in AC^0, CC^0, ACC^0, or TC^0, respectively.*

However, it has to be noted that for most classes, the concrete decision algorithm depends on open questions about the separation of classes, as applying it to suitable languages would immediately settle these questions. The algorithm is unconditionally known in the case of AC^0 due to its separation from ACC^0, and also for $\mathrm{CC}^0[p]$ and $\mathrm{ACC}^0[p]$ for p a prime number by a result of [11]. We also get:

Corollary 21. *Given a VCL L and a prime p, it is decidable whether L is complete for AC^0, $\mathrm{CC}^0[q]$, $\mathrm{ACC}^0[q]$, TC^0, or NC^1 respectively.*

Under the separation conjectures, we also obtain uniformity results by using Theorem 15:

Corollary 22. *If $\mathrm{TC}^0 \neq \mathrm{NC}^1$, then $\mathrm{TC}^0 \cap \mathrm{VCL} \subseteq \mathrm{Maj}[<] \subseteq \mathrm{FO}[<]$-uniform TC^0.*

If $\mathrm{ACC}^0 \neq \mathrm{TC}^0$, then $\mathrm{ACC}^0 \cap \mathrm{VCL} \subseteq \mathrm{FO} + \mathrm{MOD}[+] \subseteq \mathrm{FO}[+]$-uniform ACC^0.

We can relate open questions about circuits to visibly counter automata:

Corollary 23. *The following holds:*

- $\mathrm{ACC}^0 = \mathrm{TC}^0$ *if and only if $\mathrm{ACC}^0 \cap \mathrm{VCL} = \mathrm{TC}^0 \cap \mathrm{VCL}$.*
- $\mathrm{ACC}^0 = \mathrm{CC}^0$ *if and only if $\mathrm{CC}^0 \cap \mathrm{VCL} = \mathrm{ACC}^0 \cap \mathrm{VCL}$.*
- $\mathrm{ACC}^0 \neq \mathrm{CC}^0$ *if and only if $\mathrm{CC}^0 \cap \mathrm{VCL} = \mathrm{CC}^0 \cap \mathrm{Reg}$.*

Hence, while the separation between ACC^0 and TC^0 cannot be reduced to a question about regular languages, it can indeed be viewed as a question about VCLs.

7 Discussion

This work considered relations between families of formal languages on the one side and logic and complexity on the other side. This line of research has hitherto mainly been pursued for regular languages, since their syntactic monoids are finite and thus open to algebraic methods. We made a step forward to extend these results towards nonregular languages. The most natural candidate to choose were the visibly pushdown languages, where we focused on VCLs. We established a one-to-one correspondence between VCLs and logic. Using this, we derived decidable characterizations for ACC^0, CC^0, and TC^0, and related our results to the open questions of the relationships of these classes.

As a next step we naturally want to extend our approach. For instance, we will try to generalize our results to all visibly pushdown languages, but we will probably need additional methods.

References

1. Alur, R., Madhusudan, P.: Visibly pushdown languages. In: Babai, L. (eds.) STOC, pp. 202–211. ACM (2004)
2. Bárány, V., Löding, C., Serre, O.: Regularity problems for visibly pushdown languages. In: Durand, B., Thomas, W. (eds.) STACS 2006. LNCS, vol. 3884, pp. 420–431. Springer, Heidelberg (2006)
3. David, A., Barrington, M.: Bounded-Width Polynomial-Size Branching Programs Recognize Exactly Those Languages in NC^1. J. Comput. Syst. Sci. **38**(1), 150–164 (1989)
4. David, A., Thérien, D., Straubing, H., Compton, K.J., Barrington, M.: Regular Languages in NC^1. J. Comput. Syst. Sci. **44**(3), 478–499 (1992)
5. David, A., Thérien, D., Barrington, M.: Finite monoids and the fine structure of NC^1. J. ACM **35**(4), 941–952 (1988)
6. Furst, M.L., Saxe, J.B., Sipser, M.: Parity, circuits, and the polynomial-time hierarchy. In: FOCS, pp. 260–270 (1981)
7. Håstad, J.: Almost optimal lower bounds for small depth circuits. In: STOC, pp. 6–20. ACM (1986)
8. Krebs, A., Lange, K., Ludwig, M.: Visibly Counter Languages and Constant Depth Circuits. In: Mayr, E.W., Ollinger, N. (eds.) 32nd International Symposium on Theoretical Aspects of Computer Science (STACS 2015), Garching, Germany, vol. 30 of LIPIcs, pp. 594–607. Schloss Dagstuhl - Leibniz-Zentrum fuer Informatik, 4–7 March 2015
9. McKenzie, P., Thomas, M., Vollmer, H.: Extensional uniformity for boolean circuits. SIAM J. Comput. **39**(7), 3186–3206 (2010)
10. Mehlhorn, K.: Pebbling mountain ranges and its application to DCFL-recognition. In: de Bakker, J., van Leeuwen, J. (eds.) Automata Languages and Programming. LNCS, pp. 422–435. Springer, Berlin Heidelberg (1980)
11. Smolensky, R.: Algebraic methods in the theory of lower bounds for boolean circuit complexity. In: STOC, pp. 77–82 (1987)
12. Straubing, H.: Finite Automata, Formal Logic, and Circuit Complexity. Birkhäuser, Boston (1994)
13. Vollmer, H.: Introduction to Circuit Complexity - A Uniform Approach. Texts in theoretical computer science. Springer, Heidelberg (1999)

The Price of Connectivity
for Cycle Transversals

Tatiana R. Hartinger[1], Matthew Johnson[2]([✉]),
Martin Milanič[1], and Daniël Paulusma[2]

[1] UP IAM and UP FAMNIT, University of Primorska, Koper, Slovenia
`tatiana.hartinger@iam.upr.si, martin.milanic@upr.si`
[2] School of Engineering and Computing Sciences,
Durham University, Durham, UK
{`matthew.johnson2,daniel.paulusma`}`@durham.ac.uk`

Abstract. For a family of graphs $\mathcal{F}$, an $\mathcal{F}$-transversal of a graph G is a subset $S \subseteq V(G)$ that intersects every subset of $V(G)$ that induces a subgraph isomorphic to a graph in $\mathcal{F}$. Let $t_{\mathcal{F}}(G)$ be the minimum size of an $\mathcal{F}$-transversal of G, and $ct_{\mathcal{F}}(G)$ be the minimum size of an $\mathcal{F}$-transversal of G that induces a connected graph. For a class of connected graphs $\mathcal{G}$, the price of connectivity for $\mathcal{F}$-transversals is the supremum of the ratios $ct_{\mathcal{F}}(G)/t_{\mathcal{F}}(G)$ over all $G \in \mathcal{G}$. We perform an in-depth study into the price of connectivity for various well-known graph families $\mathcal{F}$ that contain an infinite number of cycles and that, in addition, may contain one or more anticycles or short paths. For each of these families we study the price of connectivity for classes of graphs characterized by one forbidden induced subgraph H. We determine exactly those classes of H-free graphs for which this graph parameter is bounded by a multiplicative constant, bounded by an additive constant, or equal to 1. In particular, our tetrachotomies extend known results of Belmonte et al. (EuroComb 2012, MFCS 2013) for the case when $\mathcal{F}$ is the family of all cycles.

1 Introduction

Let $\mathcal{F}$ be a family of graphs. An $\mathcal{F}$-*transversal* of a graph $G = (V, E)$ is a subset $S \subseteq V$ that intersects every subset of V that induces a subgraph isomorphic to a graph in $\mathcal{F}$. Equivalently, S is an $\mathcal{F}$-transversal of G if $G - S$ is $\mathcal{F}$-*free*; that is, it does not contain an induced subgraph isomorphic to any graph in $\mathcal{F}$ (if $\mathcal{F} = \{H\}$ then we write H-free instead).

In certain cases, $\mathcal{F}$-transversals are well-studied. For example, a *vertex cover* is an $\mathcal{F}$-transversal for any family $\mathcal{F}$ that contains P_2 but not P_1 (here, P_k is the path on k vertices). Note that, for any $\{P_2\}$-transversal S of a graph G, the graph $G - S$ is an independent set. To give another example, a *feedback vertex set* is

---

This work was supported by a London Mathematical Society Scheme 4 Grant and by EPSRC Grant EP/K025090/1, and, in part, by the Slovenian Research Agency (I0-0035, research program P1-0285, research projects N1-0032, J1-5433, J1-6720, and J1-6743, and a Young Researchers Grant).

G.F. Italiano et al. (Eds.): MFCS 2015, Part II, LNCS 9235, pp. 395–406, 2015.
DOI: 10.1007/978-3-662-48054-0_33

an $\mathcal{F}$-transversal for $\mathcal{F} = \{C_3, C_4, C_5, \ldots\}$. In this case, for any $\mathcal{F}$-transversal S of a graph G, the graph $G - S$ is a forest. As the examples suggest, it is natural to study minimum size $\mathcal{F}$-transversals.

We can put an additional constraint on an $\mathcal{F}$-transversal S of a graph G by requiring that the subgraph of G induced by S is connected. Minimum size *connected* $\mathcal{F}$-transversals of a graph have also been considered. Minimum size connected vertex covers are well-studied, and minimum size connected feedback vertex sets have also received attention (see, for example, [7,8]). We study the following question:

What is the effect of adding the connectivity constraint on the minimum size of a $\mathcal{F}$-transversal for a graph family $\mathcal{F}$?

To address this question we use the *price of connectivity* as our comparison measure. The price of connectivity was introduced by Cardinal and Levy [5] for vertex cover. We define it as follows. For a graph G, let $t_{\mathcal{F}}(G)$ denote the minimum size of an $\mathcal{F}$-transversal S of G, and $ct_{\mathcal{F}}(G)$ the minimum size of a connected $\mathcal{F}$-transversal S' of G. Then, for a class of connected graphs $\mathcal{G}$, the price of connectivity of $\mathcal{F}$-transversals is the supremum of the ratios $ct_{\mathcal{F}}(G)/t_{\mathcal{F}}(G)$ over all $G \in \mathcal{G}$.

We briefly survey existing work starting with a result of Cardinal and Levy [5], who proved that the price of connectivity for $\{P_2\}$-transversal (vertex cover) is at most $2/(1 + \epsilon)$ for connected graphs with average degree ϵn. Camby et al. [3] proved that the price of connectivity for $\{P_2\}$-transversal is at most 2 for the class of all connected graphs and that this bound is asymptotically sharp for paths and cycles. They also gave forbidden induced subgraph characterizations of classes of graphs such that the price of connectivity for $\{P_2\}$-transversal for every connected induced subgraph is at most t, for each $t \in \{1, 4/3, 3/2\}$.

Belmonte et al. [1,2] studied the price of connectivity for feedback vertex set, that is, for $\mathcal{F}$-transversals where $\mathcal{F} = \{C_3, C_4, C_5 \ldots\}$. They characterized exactly those finite families $\mathcal{H}$ for which the price of connectivity for feedback vertex set is bounded by a constant [2]. If $|\mathcal{H}| = 1$ they also considered additive bounds: they determined exactly those graphs classes $\mathcal{G}$ of H-free graphs for which, for all $G \in \mathcal{G}$, $ct_{\mathcal{F}}(G) - t_{\mathcal{F}}(G)$ is bounded by a constant (and they found exactly when that constant is zero) [1].

The price of connectivity can also be defined for other graph measures that are defined as the size of a smallest subset of vertices that satisfies a prescribed constraint. We give two further examples. A result of Duchet and Meyniel [6] implies that the price of connectivity for dominating set is at most 3 for all connected graphs. A result of Zverovich [9] implies that the price of connectivity for dominating set is exactly 1 for connected (P_5, C_5)-free graphs. Camby and Schaudt [4] gave an additive bound of 1 for every connected (P_6, C_6)-free graph G. The same authors proved that the price of connectivity for dominating set is at most 2 for connected (P_8, C_8)-free graphs and at most 3 for connected (P_9, C_9)-free graphs; both bounds were shown to be sharp. Camby and Schaudt [4] proved that the problem of deciding whether the price of connectivity for dominating set is at most r is $P^{\mathrm{NP}[\log]}$-complete for fixed r, $1 < r < 3$. Grigoriev and Sitters [7] proved that the price of connectivity for face hitting set is at most 11 for connected planar graphs of minimum degree at least 3. Schweitzer and Schweitzer [8] reduced this bound to 5 and proved tightness.

We consider a number of families $\mathcal{F}$ that contain cycles, paths and complements of cycles. We study the price of connectivity of $\mathcal{F}$-transversals for graph classes characterized by one forbidden induced subgraph. Before we can present our results we need to introduce the following terminology and notation.

Definition 1. *Let H be a graph and let $\mathcal{G}$ be the class of connected H-free graphs. Let $\mathcal{F}$ be a family of graphs. We say that $\mathcal{G}$ is:*

(a) *$\mathcal{F}$-unbounded if for every function $f : \mathbb{N} \to \mathbb{N}$ there exists a graph $G \in \mathcal{G}$ such that $ct_{\mathcal{F}}(G) > f(t_{\mathcal{F}}(G))$;*

(b) *$\mathcal{F}$-multiplicative if $ct_{\mathcal{F}}(G) \leq c_H t_{\mathcal{F}}(G)$ for some constant c_H and for every $G \in \mathcal{G}$;*

(c) *$\mathcal{F}$-additive if $ct_{\mathcal{F}}(G) \leq t_{\mathcal{F}}(G) + d_H$ for some constant d_H and for every $G \in \mathcal{G}$; and*

(d) *$\mathcal{F}$-identical if $ct_{\mathcal{F}}(G) = t_{\mathcal{F}}(G)$ for every $G \in \mathcal{G}$.*

For graphs F and G, we write $F \subseteq_i G$ to denote that F is an induced subgraph of G. We let C_n, K_n and P_n denote the cycle, complete graph, and path on n vertices, respectively. The *disjoint union* of two vertex-disjoint graphs G and H is the graph $G + H$ that has vertex set $V(G) \cup V(H)$ and edge set $E(G) \cup E(H)$ where $V(G) \cap V(H) = \emptyset$. We denote the disjoint union of r copies of G by rG. A graph is a *linear forest* if it is the disjoint union of a set of paths.

The *complement* $\overline{G}$ of a graph G has the same vertex set as G and an edge between two distinct vertices if and only if these vertices are not adjacent in G. A *hole* is a cycle of length at least 4. An *antihole* is the complement of a hole. A cycle, hole or antihole is *even* if it contains an even number of vertices; otherwise it is *odd*. A hole is *long* if it is of length at least 5, and a *long antihole* is the complement of a long hole. A graph is *odd-hole-free* or *odd-antihole-free* if it contains no induced odd holes or no induced odd antiholes, respectively. An *even-hole-free* graph is defined similarly. A graph is *chordal* if it has no induced hole, that is, if it has no induced cycles of length at least 4. A graph is *weakly chordal* if it has no induced long hole and no induced long antihole. A graph is *perfect* if the chromatic number of every induced subgraph equals the size of a largest clique in that subgraph. By the Strong Perfect Graph Theorem, a graph is perfect if and only if it is odd-hole-free and odd-antihole-free. A graph is a *split graph* if its vertex set can be partitioned into a clique and an independent set. Split graphs coincide with the $(2K_2, C_4, C_5)$-free graphs. A graph is *threshold* if it is $(2P_2, P_4)$-free, *trivially perfect* if it is (C_4, P_4)-free, *cotrivially perfect* if it is $(C_4, 2P_2)$-free and a *cograph* if it is P_4-free.

Our Results. Table 1 summarizes our results together with related past work. Results can be seen both according to the family $\mathcal{F}$ and the corresponding property of the graph $G - S$, where S is an $\mathcal{F}$-transversal of G. We note that when $\mathcal{F}$ is the family of even cycles or of holes there is an open case. In all

Table 1. The price of connectivity of $\mathcal{F}$-transversals for various families of graphs $\mathcal{F}$ on graph classes defined by one forbidden induced subgraph H. The results on cycles in the first row are due to Belmonte et al. [1] and the multiplicativity result on cycles and P_2 in the ninth row is due to Camby et al. [3]. All other results are new and presented in this paper. †For even cycles and holes the condition is not complete as in these cases we do not know if H-free graphs are $\mathcal{F}$-additive for $H \subseteq_i P_3 + P_2 + sP_1$.

$\mathcal{F}$	Property of $G - S$	Condition for $\mathcal{F}$-multiplicativity (for $\mathcal{F}$-boundedness)	Condition for $\mathcal{F}$-additivity	Condition for $\mathcal{F}$-identity
cycles	forest	H is a linear forest [1]	$H \subseteq_i P_5 + sP_1$ or $H \subseteq_i sP_3$ [1]	$H \subseteq_i P_3$ [1]
odd cycles	bipartite	H is a linear forest	$H \subseteq_i P_5 + sP_1$ or $H \subseteq_i sP_3$	$H \subseteq_i P_3$
even cycles† (equiv.: even holes)	even-hole-free	H is a linear forest	$H \subseteq_i P_4 + sP_1$ †	$H \subseteq_i P_3$
holes†	chordal	H is a linear forest	$H \subseteq_i P_4 + sP_1$ †	$H \subseteq_i P_3$
odd holes	odd-hole-free	H is a linear forest	$H \subseteq_i P_4 + sP_1$	$H \subseteq_i P_4$
odd holes and odd antiholes	perfect	H is a linear forest	$H \subseteq_i P_4 + sP_1$	$H \subseteq_i P_4$
long holes	long-hole-free	H is a linear forest	$H \subseteq_i P_4 + sP_1$	$H \subseteq_i P_4$
long holes and long antiholes	weakly chordal	H is a linear forest	$H \subseteq_i P_4 + sP_1$	$H \subseteq_i P_4$
cycles and P_2 (equiv.: $\{P_2\}$)	edgeless	no restriction [3]	$H \subseteq_i P_5 + sP_1$ or $H \subseteq_i sP_3$	$H \subseteq_i P_3$
holes and $2P_2$ (equiv.: $\{C_4, C_5, 2P_2\}$)	split	no restriction	$H \subseteq_i P_4 + sP_1$ or $H \subseteq_i P_3 + sP_2$	$H \subseteq_i P_3$
holes and $2P_2, P_4$ (equiv.: $\{C_4, 2P_2, P_4\}$)	threshold	no restriction	$H \subseteq_i P_4 + sP_1$	$H \subseteq_i P_3$
holes and P_4 (equiv.: $\{C_4, P_4\}$)	trivially perfect	no restriction	$H \subseteq_i P_4 + sP_1$	$H \subseteq_i P_3$
long holes and $2P_2$ (equiv.: $\{C_5, 2P_2\}$)	$(C_5, 2P_2)$-free	no restriction	$H \subseteq_i P_4 + sP_1$	$H \subseteq_i P_3$ $H \subseteq_i P_2 + P_1$
long holes and $2P_2, P_4$ (equiv.: $\{2P_2, P_4\}$)	cotrivially perfect	no restriction	$H \subseteq_i P_4 + sP_1$	$H \subseteq_i P_3$ or $H \subseteq_i P_2 + P_1$
long holes and P_4 (equiv.: $\{P_4\}$)	cograph	no restriction	$H \subseteq_i P_4 + sP_1$	$H \subseteq_i P_4$

other cases, the stated conditions in Table 1 are both necessary and sufficient for $\mathcal{F}$-multiplicativity ($\mathcal{F}$-boundedness), $\mathcal{F}$-additivity, and $\mathcal{F}$-identity, respectively, in the class of connected H-free graphs.

From Table 1 we can draw a number of conclusions. If a transversal that intersects (small) paths is wanted, we obtain multiplicative bounds for any class of H-free graphs. In all other cases, H may not contain a cycle or a claw (so is a linear forest). We also see that when we add a requirement that all triangles are intersected, there is always a jump from $H = P_4 + sP_1$ to $H = P_5 + sP_1$ for the additive bound. In general, it can be noticed that adding small graphs to $\mathcal{F}$ has differing effects. We say that a family of graphs $\mathcal{F}$ or a graph F *positively (negatively) influences* a family of graphs $\mathcal{F}'$ if the row in the table for their union contains more (fewer) bounded cases than the row for $\mathcal{F}'$. So, for example, $2P_2$ does not influence $\{C_4, C_5, C_6, \ldots\} \cup \{P_4\}$, and P_4 does not influence the family of long holes. Moreover, odd holes do not influence even holes, whereas even holes influence odd holes positively.

In the remainder of our paper, after presenting some known and new basic results in Sect. 2, we present a number of general theorems, from which the results in Table 1 directly follow. We emphasize that all proofs of these theorems are algorithmic in nature, that is, they can be translated directly into polynomial-time algorithms that modify an $\mathcal{F}$-transversal into a connected $\mathcal{F}$-transversal of appropriate cardinality.

We provide a brief guide to Table 1. Theorem 2 implies the second row. Theorem 3 implies the third and fourth row, and Theorem 4 implies the next four rows. Proofs for the remaining rows have been omitted due to space restrictions.

2   Preliminaries and Some Basic Results

We consider finite undirected graphs with no multiple edges and no self-loops. Let $G = (V, E)$ be a connected graph. The *distance* between two vertices u and v is the length of a shortest path between them. The maximum distance in G is called the *diameter* of G. A set $D \subseteq V$ *dominates* G if every vertex $u \in V \setminus D$ is adjacent to at least one vertex in D. We also say that $G[D]$ dominates G. If $D = \{u, v\}$ for two adjacent vertices u, v, then uv is called a *dominating edge* of G. A set $D \subseteq V$ *dominates* a set $S \subseteq V \setminus D$ if every vertex in S is adjacent to at least one vertex in D. We say that two vertex-disjoint subgraphs of a graph G are *adjacent* if there is at least one edge between a vertex in one of them and a vertex in the other one. Similarly, we say that a vertex of G is adjacent to a subgraph of $G - v$ if G has an edge joining v with a vertex of the subgraph. The *join* of two vertex-disjoint graphs G and H is the graph obtained from the disjoint union $G + H$ by adding to it all possible edges of the form xy with $x \in V(G)$ and $y \in V(H)$.

For $r \geq 1$, $s \geq 1$, the *complete bipartite graph* $K_{r,s}$ is a bipartite graph whose vertex set can be partitioned into two sets of sizes r and s such that there is an edge joining each pair of vertices from distinct sets. The graph $K_{1,3}$ is also called a *claw*.

We now give a number of new results that we use as lemmas in our other results. Some proofs are omitted for reasons of space.

Lemma 1. *For every family $\mathcal{F}$ of graphs, the class of connected P_4-free graphs is $\mathcal{F}$-additive.*

We also need to generalize a result that was proved by Belmonte et al. [1] for the graph $H = P_5$.

Lemma 2. *For a family of graphs $\mathcal{F}$ and a graph H, if the class of connected H-free graphs is $\mathcal{F}$-additive then so is the class of connected $(H + sP_1)$-free graphs for all $s \geq 1$.*

Lemma 3. *Let G be a connected graph with diameter d. Let A be a subgraph of G consisting of r components. Then G has a connected subgraph A' that contains A and that has less than $|V(A)| + (r - 1)d$ vertices.*

The following theorem is used in all our tetrachotomies. The third part was shown by Belmonte et al. [1] for the case when $\mathcal{F}$ is the family of all cycles, and our proof is a modification of theirs.

Theorem 1. *Let $\mathcal{F}$ be a family of graphs and let H be a graph. Then, the following statements hold:*

  (i) *If $\mathcal{F}$ contains a linear forest, then the class of all graphs is $\mathcal{F}$-multiplicative.*
  (ii) *If H is a linear forest, then the class of H-free graphs is $\mathcal{F}$-multiplicative.*
  (iii) *If $\mathcal{F}$ contains an infinite number of cycles and no linear forests and H is not a linear forest, then the class of H-free graphs is $\mathcal{F}$-unbounded.*

3   Transversals of Families of Odd Cycles

In this section we assume we are given a family $\mathcal{F}$ that contains all odd cycles, although we will show more general results whenever possible. To prove our results we need a number of lemmas, the first of which has been proven by Belmonte et al. [1] for the special case when the family $\mathcal{F}$ consists of all cycles.

Lemma 4. *For any family of graphs $\mathcal{F}$ with $K_r \in \mathcal{F}$ for some integer $r \geq 1$, the class of connected P_5-free graphs is $\mathcal{F}$-additive.*

We now give a technical lemma (which we also apply in some other proofs).

Lemma 5. *Let $s \geq 1$ be an integer and let G be an sP_3-free connected graph with a subset $S \subseteq V(G)$ and an independent set $U \subseteq V(G) \setminus S$. Suppose that some component of $G[S]$, say Z, contains an induced copy of $(s-1)P_3$. Then there exists a set S' with $S \subseteq S'$ of size at most $|S| + 8s^2 + 2s$ such that*

  • $G[S']$ *has a component Z' that contains all vertices of $V(Z) \cup (S' \setminus S)$;*
  • *every vertex of $U' = U \setminus S'$ is adjacent to at most one component of $G[S']$ not equal to Z';*
  • *every component of $G[S']$ not equal to Z' is adjacent to at most one vertex of U'.*

The following lemma generalizes the corresponding result of Belmonte et al. [1] when $\mathcal{F}$ is the family of all cycles. We use a similar approach as used in their proof but our arguments (which are based on bipartiteness instead of cycle-freeness) are different and this proof demonstrates some techniques used several times in obtaining our results.

Lemma 6. *For any family of graphs $\mathcal{F}$ containing either all odd cycles or P_2 and for any fixed $s \geq 1$, the class of connected sP_3-free graphs is $\mathcal{F}$-additive.*

Proof. The proof is by induction on s. Let $s = 1$. Then every connected sP_3-free graph G is complete. Hence, every minimum $\mathcal{F}$-transversal of G is connected.

Now let $s \geq 2$. Let G be a connected sP_3-free graph. We may assume by induction that G contains an induced copy Γ_0 of an $(s-1)P_3$. Let S be a minimum $\mathcal{F}$-transversal of G. Let Γ be a minimum connected induced subgraph of G that contains Γ_0. Because G is sP_3-free, G has diameter less than $4s$. Then, by Lemma 3, we find that Γ has size less than $3(s-1)+(s-2)4s = 4s^2 - 5s - 3$. Let $S' = S \cup V(\Gamma)$. Then we have that $|S'| \leq |S| + 4s^2 - 5s - 3$.

If S' is connected then we take $d_{sP_3} = 4s^2 - 5s - 3$ as our desired constant and we are done. Suppose S' is not connected. Below we describe how to refine S'. During this process, we always use Z to denote the component of S' containing Γ, and we will never remove a vertex of Z from S'; in fact, one can think of the proof as "growing" Z and connecting it to the other vertices of S' until $Z = S'$.

Observe that the sP_3-freeness of G implies that every component of S' other than Z is complete. Throughout the proof, we let A denote the union of clique components of S', so $V(A) = S' \setminus V(Z) = S \setminus V(Z)$. We also note that the graph $G - S'$ is bipartite, as even its supergraph $G - S$ contains no odd cycles by the definition of S. Hence we can partition $G - S'$ into two (possibly empty) sets U_1 and U_2 so that U_1 and U_2 are independent sets.

We start with the following two claims, both of which follow from Lemma 5, which we apply twice, namely once with respect to U_1 and once with respect to U_2. By Lemma 5 this leads to a total increase of S' by an additive factor of at most $2(8s^2 + 2s) = 16s^2 + 4s$.

Claim 1: Without loss of generality, we may assume that every vertex of $U_1 \cup U_2$ is adjacent to at most one component of A.

Claim 2: Without loss of generality, we may assume that every component of A is adjacent to at most one vertex of U_1 and to at most one vertex of U_2.

Using Claims 1 and 2 we prove the following crucial claim.

Claim 3: Without loss of generality, every vertex of every component of A has exactly one neighbour in U_1 and exactly one neighbour in U_2.

We prove Claim 3 as follows. Let A^* be the union of components for which the statement of Claim 3 does not hold. Let D be a component of A^*. By Claim 2, D is adjacent to at most one vertex of U_1 and to at most one vertex of U_2. First suppose that D is non-adjacent to U_1 or to U_2, say D is not adjacent to U_1. Because G is connected, this means that D is adjacent to (exactly one) vertex $z \in U_2$, say $v \in D$ is adjacent to z. As D belongs to A^*, we find that D contains a vertex v' not adjacent to z. Hence, $vv'z$ is an induced P_3. Now suppose that D is adjacent to U_1 and to U_2, say D has vertices u, v (possibly $u = v$) so that u is adjacent to $x \in U_1$ and v is adjacent to $z \in U_2$. Then, as D is in A^*, there exists a vertex v' that is non-adjacent to at least one of x, z, say to z. Again, $vv'z$ is an induced P_3. As G is sP_3-free and no vertex in $U_1 \cup U_2$ is adjacent to more than one component of A by Claim 1, we deduce that A^* contains at most $s - 1$ components. Moreover, each vertex $z \in U_1 \cup U_2$ involved in an induced P_3 as described above must be adjacent to Z (due to sP_3-freeness of G and the fact that Z contains an induced $(s-1)P_3$). Hence, we can add these vertices to Z increasing the size of Z, and thus the size of S', by at most $s - 1$. The remaining components of A have the desired property. Moreover, Claims 1 and 2 are still valid. This completes the proof of Claim 3.

Due to Claim 3 we may assume without loss of generality that each vertex v in each component D of A has exactly two neighbours in $G - S'$, namely one neighbour in U_1 and one neighbour in U_2. By Claim 2, these neighbours are the same for all vertices in D. Hence, we may denote these two neighbours by s_D and t_D, respectively.

Consider a component D of A. If one of its neighbours in $U_1 \cup U_2$, say s_D, is adjacent to Z, then replacing S' with $(S' \cup \{s_D\}) \setminus \{v\}$ and Z with the connected component of S' containing $Z \cup \{s_D\}$ does not result in an odd cycle in $G - S'$. Moreover, such a swap does not increase the size of S' either. It does, however, reduce the number of vertices of S' that are not in Z (which is our goal). Consequently, we perform these swaps until, in the end, both the neighbours s_D and t_D of each component of A are not adjacent to Z. In particular this implies that s_D and t_D are adjacent, so $V_D \cup \{s_D, t_D\}$ is a clique. Then, due to Claims 1–3, the components in A together with their neighbours in $U_1 \cup U_2$ induce a union of complete graphs. This union is a disjoint union, as otherwise G would contain an induced P_3 not adjacent to Z and, as Z has an induced $(s-1)P_3$, we would obtain an induced sP_3 in G. Note that the swaps did not change the size of S'.

Let U_1' and U_2' denote the subsets of U_1 and U_2, respectively, that consist of vertices adjacent to no components of A. Let W_1 consist of all vertices s_D adjacent to U_2' and let W_2 consist of all vertices t_D adjacent to U_1'. Note that $W_1 \subseteq U_1 \setminus U_1'$ and that $W_2 \subseteq U_2 \setminus U_2'$. Because G is connected and no s_D or t_D is adjacent to Z or to some other component of A not equal to D, we find that $W_1 \cup W_2$ contain at least one of s_D, t_D for each component D of A.

We choose smallest sets U_1'' and U_2'' in U_1' and U_2', respectively, that dominate W_2 and W_1, respectively. By minimality, each vertex $u \in U_1''$ must have a "private" neighbour t_D in W_2, and hence together with t_D and s_D, corresponds to a "private" P_3. Consequently, as G is sP_3-free and $U_1'' \subseteq U_1$ is an independent set, U_1'' has size at most $s - 1$. Similarly, U_2'' has size at most $s - 1$. Moreover, each vertex in $U_1'' \cup U_2''$ is adjacent to Z (again due to the sP_3-freeness of G).

Figure 1 shows an example in which the components of A consist on three cliques (the first two of size two and the last one of size one) to illustrate the situation.

We now do as follows. First, for each component D of A we pick one of its vertices v and swap v with s_D if $s_D \in W_1$ and otherwise we swap v with t_D (note that $t_D \in W_2$ in that case). We also add all vertices of $U_1'' \cup U_2''$ to Z and thus to S'. The results of these swaps are as follows. First, $G[S']$ has become connected. Second, S' has increased in size at most by $2(s-1)$, which is allowed. Third, $G - S'$ is still bipartite (as swapping a vertex of a component D of A with s_D or t_D does not create any odd cycles). Consequently, we have found a connected $\mathcal{F}$-transversal of size at most $|S| + 4s^2 - 5s - 3 + 16s^2 + 4s + (s-1) + 2(s-1) = |S| + 20s^2 + 2s - 6$, so we can take $d_{sP_3} = 20s^2 + 2s - 6$. $\square$

Belmonte et al. [1] proved that the class of connected $(P_2 + P_4, P_6)$-free graphs is not $\mathcal{F}$-additive if $\mathcal{F}$ is the class of all cycles. We have the following more general result.

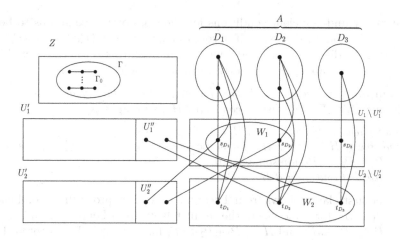

Fig. 1. The situation in the proof of Lemma 6.

Lemma 7. *For any family of cycles $\mathcal{F}$ with $C_3 \in \mathcal{F}$, the class of connected $(P_2 + P_4, P_6)$-free graphs is not $\mathcal{F}$-additive.*

We now state the following result. To prove it we use the previous lemmas for the first two claims. For the third claim we observe that connected P_3-free graphs are complete (proving the case $H \subseteq_i P_3$) and that $K_{2,2,2}$ is a counterexample for the case $H \not\subseteq_i P_3$.

Theorem 2. *For any graph H and for any family of cycles $\mathcal{F}$ containing all odd cycles, the class of connected H-free graphs is*

- *$\mathcal{F}$-multiplicative if and only if H is a linear forest;*
- *$\mathcal{F}$-additive if and only if $H \subseteq_i P_5 + sP_1$ or $H \subseteq_i sP_3$ for some $s \geq 0$;*
- *$\mathcal{F}$-identical if and only if $H \subseteq_i P_3$.*

4 Cycle Families with 4-Cycles but No 3-Cycles

In this section we consider families of cycles $\mathcal{F}$ such that $C_3 \notin \mathcal{F}$ but $C_4 \in \mathcal{F}$. To prove our results we need a lemma.

Lemma 8. *For any family $\mathcal{F}$ of cycles with $C_3 \notin \mathcal{F}$ and $C_4 \in \mathcal{F}$:*

- *the class of connected P_5-free graphs is not $\mathcal{F}$-additive.*
- *the class of connected $P_2 + P_4$-free graphs is not $\mathcal{F}$-additive.*
- *the class of connected $2P_3$-free graphs is not $\mathcal{F}$-additive.*
- *the class of connected $3P_2$-free graphs is not $\mathcal{F}$-additive.*

We now state our result for infinite families of cycles $\mathcal{F}$ with $C_3 \notin \mathcal{F}$ and $C_4 \in \mathcal{F}$. It does not provide a complete characterization as we are unable to give necessary and sufficient conditions for the class of H-free graphs to be $\mathcal{F}$-additive. This would be possible if it could be shown that $(P_3 + P_2 + sP_1)$-free graphs are $\mathcal{F}$-additive for all $s \geq 0$. Due to Lemma 2, this is the case if and only if $(P_3 + P_2)$-free graphs are $\mathcal{F}$-additive, which we conjecture to be true.

Theorem 3. *For any graph H and for any infinite family of cycles $\mathcal{F}$ with $C_3 \notin \mathcal{F}$ and $C_4 \in \mathcal{F}$, the class of connected H-free graphs is*

- *$\mathcal{F}$-multiplicative if and only if H is a linear forest;*
- *$\mathcal{F}$-additive if $H \subseteq_i P_4 + sP_1$ for some $s \geq 0$, but not if $H \not\subseteq_i P_4 + sP_1$ nor $H \not\subseteq_i P_3 + P_2 + sP_1$ for some $s \geq 0$;*
- *$\mathcal{F}$-identical if and only if $H \subseteq_i P_3$.*

Proof. The first claim follows from Theorem 1. We now prove the second claim. If $H \subseteq_i P_4 + sP_1$ for some $s \geq 0$, the result follows from Lemmas 1 and 2. Now suppose $H \not\subseteq_i P_4 + sP_1$ and $H \not\subseteq_i P_3 + P_2 + sP_1$ for any $s \geq 0$. By Theorem 1, we may assume that H is a linear forest. Then $P_5 \subseteq_i H$, $P_2 + P_4 \subseteq_i H$, $2P_3 \subseteq_i H$ or $3P_2 \subseteq_i H$ and we can use Lemma 8.

We now prove the third claim. If $H \subseteq_i P_3$ then any connected H-free graph is complete, so the result follows directly. If $H \not\subseteq_i P_3$ then, by Theorem 1, we may assume that H is a linear forest. Hence, $3P_1 \subseteq_i H$ or $P_1 + P_2 \subseteq_i H$.

If $P_1 + P_2 \subseteq_i H$, then we have that the complete bipartite graph $G = K_{3,3}$ is a connected H-free graph (since it is $P_1 + P_2$-free). And $t_{\mathcal{F}}(G) = 2 < 3 = ct_{\mathcal{F}}(G)$ so the class of connected H-free graphs is not $\mathcal{F}$-identical.

Finally, suppose that $3P_1 \subseteq_i H$, and let G be the complement of the graph shown in Fig. 2. Since $\overline{G}$ is triangle-free and every two vertices of $\overline{G}$ have a common non-neighbour, G is a connected $3P_1$-free graph. As every $\mathcal{F}$-transversal of G must intersect every induced $2P_2$ in $\overline{G}$, the minimum $\mathcal{F}$-transversals of G are in bijective correspondence with the four edges of the 4-cycle in $\overline{G}$. So $t_{\mathcal{F}}(G) = 2 < 3 = ct_{\mathcal{F}}(G)$, and the class of connected H-free graphs is also not $\mathcal{F}$-identical in this case. □

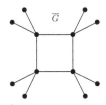

Fig. 2. The complement of a graph G with $t_{\mathcal{F}}(G) < ct_{\mathcal{F}}(G)$ whenever $C_3 \notin \mathcal{F}$ and $C_4 \in \mathcal{F}$.

5   Cycle Families with 5-Cycles but No 3- or 4-Cycles

In this section we consider families of cycles $\mathcal{F}$ such that $C_3, C_4 \notin \mathcal{F}$ but $C_5 \in \mathcal{F}$.

We first give the following lemma; note that C_3 and C_4 are both induced subgraphs of $\overline{2P_4}$.

Lemma 9. *Let $\mathcal{F}$ be a graph family with $C_5 \in \mathcal{F}$ that contains no induced subgraphs of $\overline{sP_4}$ for any $s \geq 1$. Then the class of connected $2P_2$-free graphs is not $\mathcal{F}$-additive.*

Proof. We describe a family of connected $2P_2$-free graphs that is not $\mathcal{F}$-additive, where $\mathcal{F}$ is any family of cycles as in the statement of the lemma. The graphs in the family consist of $k \geq 2$ copies of the join of two P_4s, say $H_1, \ldots, H_k$. For each of them there is a new vertex v_i adjacent to both endpoints of the two P_4s, and in addition there are all possible edges between vertices in different H_i's.

We first show that every graph G in this family is $2P_2$-free. Every edge e of G has at least one endpoint in some H_i, say in H_1. Deleting the closed neighbourhood of e results in the subgraph induced by a subset of $\{v_1, \ldots, v_k\}$ (if $e \in E(H_1)$), or in the subgraph induced by $\{u, v_2, \ldots, v_k\}$ for some $u \in V(H_1)$ (otherwise). In either case, the resulting graph is edgeless. Therefore, G is $2P_2$-free. Let G be a graph in this family, and let k be the number of H_i's. We have $t_{\mathcal{F}}(G) \leq k$ since deleting the vertices $v_1, \ldots, v_k$ results in a graph that is isomorphic to $\overline{2kP_4}$ and thus $\mathcal{F}$-free. On the other hand, every connected $\mathcal{F}$-transversal S of G must contain at least two vertices from each subgraph induced by $V(H_i) \cup \{v_i\}$, for every i (otherwise it either misses an induced C_5 or contains only v_i, making it isolated in $G[S]$). Therefore, $ct_{\mathcal{F}}(G) \geq 2k$, which establishes the non-$\mathcal{F}$-additivity of the family.   □

Lemma 10. *Let $\mathcal{F}$ be a family of graphs that contains C_5 but no induced subgraph of $\overline{4P_4}$. Then the class of connected $3P_1$-free graphs is not $\mathcal{F}$-identical.*

Theorem 4. *For any graph H and for any graph family $\mathcal{F}$ which only contains graphs with an induced P_4, including C_5 and an infinite number of other cycles but no linear forests and no induced subgraphs of $\overline{sP_4}$ for any $s \geq 1$, the class of connected H-free graphs is*

- *$\mathcal{F}$-multiplicative if and only if H is a linear forest;*
- *$\mathcal{F}$-additive if and only if $H \subseteq_i P_4 + sP_1$ for some $s \geq 0$;*
- *$\mathcal{F}$-identical if and only if $H \subseteq_i P_4$.*

Proof. The first claim follows from Theorem 1. We now prove the second claim. First suppose that $H \subseteq_i P_4 + sP_1$ for some $s \geq 0$. Then the class of connected H-free graphs is $\mathcal{F}$-additive due to Lemmas 1 and 2. Now suppose that $H \not\subseteq_i P_4 + sP_1$ for any $s \geq 0$. By Theorem 1, we may assume that H is a linear forest. Hence, $2P_2 \subseteq_i H$ and we use Lemma 9. Finally, we show the third claim. Recall that if $H \subseteq_i P_4$ then any H-free graph is already $\mathcal{F}$-free. Suppose that $H \not\subseteq_i P_4$. If $2P_2 \subseteq_i H$ we use Lemma 9 again. Hence $3P_1 \subseteq_i H$. In that case we use Lemma 10. This completes the proof of Theorem 4.   □

6   Conclusions

We extended the tetrachotomy result of Belmonte et al. [1] for the family $\mathcal{F}$ of all cycles by giving tetrachotomy results for a number of natural families $\mathcal{F}$

containing cycles and anticycles (see Table 1). Let us recall that a tetrachotomy for the price of connectivity of $\mathcal{F}$-transversals when $\mathcal{F}$ is the family of even cycles or of all holes is still an open case. To settle it, it would suffice to show that the set of connected $(P_3 + P_2)$-free graphs is $\mathcal{F}$-additive which we conjecture to be true. We also have no tetrachotomy for families $\mathcal{F}$ that contain C_3 but that miss some other odd cycle. The partial results below show that a more refined analysis is needed to obtain complete results in this direction.

We first summarize our current knowledge. By Theorem 1 we know that the class of H-free graphs is $\mathcal{F}$-multiplicative if and only if H is a linear forest. We also know, due to Lemma 7, that the class of connected $(P_2 + P_4, P_6)$-free graphs is not $\mathcal{F}$-additive. Moreover, the class of connected H-free graphs is $\mathcal{F}$-identical if and only if $H \subseteq_i P_3$, as we can use the example of $G = K_{2,2,2}$ from Theorem 2. Hence, using Lemmas 1, 2, and 4, we see that what remains is to check, for every $s \geq 2$, whether the class of H-free graphs is $\mathcal{F}$-additive if $H = sP_3$. We can show that already for $s = 2$ this is true for some families $\mathcal{F}$ and false for others.

Proposition 1. *For any family of cycles $\mathcal{F}$ containing C_3 and C_5, the class of connected $2P_3$-free graphs is $\mathcal{F}$-additive.*

Proposition 2. *For any family $\mathcal{F}$ of cycles with $C_3 \in \mathcal{F}$ and $C_5 \notin \mathcal{F}$, the class of connected $2P_3$-free graphs is not $\mathcal{F}$-additive.*

References

1. Belmonte, R., van 't Hof, P., Kamiński, M., Paulusma, D.: The price of connectivity for feedback vertex set. In: Proceedings of the EuroComb 2013 CRMS 16, pp. 123–128 (2013)
2. Belmonte, R., van 't Hof, P., Kamiński, M., Paulusma, D.: Forbidden induced subgraphs and the price of connectivity for feedback vertex set. In: Csuhaj-Varjú, E., Dietzfelbinger, M., Ésik, Z. (eds.) MFCS 2014, Part II. LNCS, vol. 8635, pp. 57–68. Springer, Heidelberg (2014)
3. Camby, E., Cardinal, J., Fiorini, S., Schaudt, O.: The Price of Connectivity for Vertex Cover. Discrete Math. Theor. Comput. Sci. **16**, 207–224 (2014)
4. Camby, E., Schaudt, O.: The price of connectivity for dominating set: Upper bounds and complexity. Discrete Appl. Math. **17**, 53–59 (2014)
5. Cardinal, J., Levy, E.: Connected vertex covers in dense graphs. Theoret. Comput. Sci. **411**, 2581–2590 (2010)
6. Duchet, P., Meyniel, H.: On Hadwiger's number and the stability number. Ann. Discret. Math. **13**, 71–74 (1982)
7. Grigoriev, A., Sitters, R.: Connected Feedback Vertex Set in Planar Graphs. In: Paul, C., Habib, M. (eds.) WG 2009. LNCS, vol. 5911, pp. 143–153. Springer, Heidelberg (2010)
8. Schweitzer, P., Schweitzer, P.: Connecting face hitting sets in planar graphs. Inf. Process. Lett. **111**, 11–15 (2010)
9. Zverovich, I.E.: Perfect connected-dominant graphs. Discussiones Math. Graph Theor. **23**, 159–162 (2003)

Upper and Lower Bounds on Long Dual Paths in Line Arrangements

Udo Hoffmann[1], Linda Kleist[1], and Tillmann Miltzow[2]([✉])

[1] Technische Universität Berlin, Berlin, Germany
{uhoffmann,kleist}@math.tu-berlin.de
[2] Freie Universität Berlin, Berlin, Germany
t.miltzow@gmail.com

Abstract. Given a line arrangement $\mathcal{A}$ with n lines, we show that there exists a path of length $n^2/3 - O(n)$ in the dual graph of $\mathcal{A}$ formed by its faces. This bound is tight up to lower order terms. For the bicolored version, we describe an example of a line arrangement with $3k$ blue and $2k$ red lines with no alternating path longer than $14k$. Further, we show that any line arrangement with n lines has a coloring such that it has an alternating path of length $\Omega(n^2/\log n)$. Our results also hold for pseudoline arrangements.

1 Introduction

A *line arrangement* $\mathcal{A}$ is a set of lines in the Euclidean plane. The set of lines partitions the plane into faces, edges and vertices and thus defines a plane graph $G(\mathcal{A})$. A line arrangement is called *simple* if no two lines are parallel and every point of the Euclidean plane is covered by at most 2 lines. That is, every two lines intersect exactly once and every vertex of $G(\mathcal{A})$ has degree 4. A *(dual) path* in a line arrangement is a sequence of faces such that consecutive faces share an edge and no face appears more than once. Alternatively, such a path can be seen as a simple path in the dual graph $G^*(\mathcal{A})$ of the plane graph $G(\mathcal{A})$, where the faces of the arrangement are the vertices and two faces of the arrangement are adjacent when they share an edge.

We are interested in the length of a longest path of a simple arrangement. In particular, we want to bound the length of the longest path in $\mathbf{A}_n$, the set of arrangements with n lines: We are interested in bounds on the function

$$f(n) := \min_{\mathcal{A} \in \mathbf{A}_n} \max_{P \in \mathcal{A}} |P|$$

where P is a dual path in $\mathcal{A}$ and $|P|$ its length.

U. Hoffmann—Supported by the Deutsche Forschungsgemeinschaft within the research training group 'Methods for Discrete Structures' (GRK 1408).

T. Miltzow—Supported by the ERC grant PARAMTIGHT: Parameterized complexity and the search for tight complexity results", no. 280152.

G.F. Italiano et al. (Eds.): MFCS 2015, Part II, LNCS 9235, pp. 407–419, 2015.
DOI: 10.1007/978-3-662-48054-0_34

This question has a colored counterpart. If we color the lines of a line arrangement with either red or blue (such that each color appears at least once), we talk about a *bicolored line arrangement*. A dual path in a bicolored arrangement is *alternating* if no two consecutive edges which certify the contact of the faces are of the same color. For the bicolored version, we are interested in the length of a longest alternating path of a simple arrangement. We are particularly interested in the case when the number of red lines and the number of blue lines are roughly the same.

1.1   State of the Art

Aichholzer *et al.* [1] introduced the problems. To the best of our knowledge, no one else studied these questions previously.

We briefly repeat the current state of art, that is results in [1]: For the uncolored case, they show that any line arrangement has a dual path of length $\frac{1}{4}n^2 - O(n)$. They present two ideas. The first is based on Tutte's famous result that every 4-connected planar graph admits a Hamilton cycle [7]. By some local transformations, they manage to transform the plane dual graph $G^*(\mathcal{A})$ to a plane 4-connected graph G' such that the Hamilton path of G' can be translated to one of $G^*(\mathcal{A})$. Their second idea is to use the levels of the line arrangement in order to construct a long path in a straightforward manner. This is an idea which we also pursue and therefore come back to in more detail. For the upper bound, they argue that a path of length $\frac{1}{3}n^2 + O(n)$ is the best possible lower bound for all arrangements with n lines. Note that this is roughly 2/3 of the faces.

In a bicolored line arrangement, a face is called *bicolored* if the lines bounding it are not all of the same color. They show that the graph induced by the bicolored faces is (alternating-)connected, i.e. there exists an alternating dual path between any two bicolored faces. This result implies the existence of an alternating path of length $\Omega(n)$ in any bicolored arrangement with n lines. This is due to the fact that each bicolored arrangement has two unbounded bicolored faces of distance at least n. Thus, any alternating path connecting them has length $\Omega(n)$. They also give arrangements with a linear upper bound: Consider a bicolored line arrangement with $n-1$ red lines and 1 blue line. Then, only n blue edges exist and any alternating path has length in $O(n)$.

1.2   Results and Outline

Our results hold not only for *line arrangements*, but also in the more general setting of *pseudoline arrangements*. Definitions and further properties can be found in the preliminaries.

Our main result is a (up to lower order terms) tight lower bound on the length of a longest path. Section 2 is devoted to its proof.

Theorem 1 (Uncolored Arrangement). *In every simple arrangement of n pseudolines in the Euclidean plane, there exists a path of length $\frac{1}{3}n^2 - O(n)$.*

For the bicolored case, we were able to improve the upper bound, to a more balanced scenario.

Theorem 2 (Bicolored Arrangement). *There exists a simple arrangement of* $3k$ *red and* $2k$ *blue lines where any alternating dual path has length of at most* $14k$, *for every odd* k.

The example for $k = 3$ is given in Fig. 1. The idea of this construction is to use monochromatic faces in order to separate the bicolored ones in a way that the graph induced by the bicolored faces is similar to a star with a cycle instead of a center vertex. A detailed proof can be found in the full version.

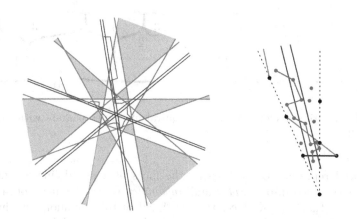

Fig. 1. Left: An arrangement with alternating path of linear length. The red lines extend the sides of a regular $3k$-gon. A longest path is given in green. Right: A path gets stuck in one "wedge" (Color figure online).

This result leads us to the question of whether every pseudoline arrangement can be colored such that there exists a long alternating path. Theorem 3 states that a random coloring does the job with high probability. This implies that there is a coloring for any pseudoline arrangement such that there exists an alternating path of this length.

Theorem 3 (Random Bicolorings). *In a random bicoloring of an arrangement of* n *pseudolines, there exists an alternating path of length* $\Omega\left(\frac{n^2}{\log n}\right)$ *with high probability.*

The idea of the proof is to find paths in tunnels with logarithmic width and glue them together to one long path. The proof can be found in the full version and in [6]. Concise proof ideas can be found in a preliminary version [5].

1.3   Preliminaries

A pseudoline arrangement is *simple* if no three pseudolines have a common point of intersection. Any simple line arrangement is also a simple pseudoline arrangement. A pseudoline arrangement partitions the plane into cells of dimension 0, 1, and 2 – the *vertices*, *edges* and *faces*. We consider sequences of faces of the arrangement such that consecutive faces share an edge and no face appears more than once. We refer to such sequences as *dual paths* and define the *length* of a dual path to be the number of involved faces. We drop the dual, when its clear from the context. We now summarize some useful facts about arrangements and their

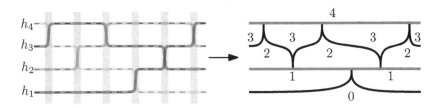

Fig. 2. Left: A wiring diagram. Right: An equivalent tunnel diagram with tunnels of width $w = 2$.

combinatorial representations. Clearly, the faces of a pseudoline arrangement $\mathcal{A}$ can be partitioned into *bounded* and *unbounded* faces. The faces of $\mathcal{A}$ can be bicolored, such that any two faces that are incident to a common edge have different colors, i.e., the dual graph is bipartite. A simple pseudoline arrangement in the Euclidean plane has $\binom{n}{2}$ vertices, n^2 edges and $\binom{n}{2} + n + 1$ faces. The number of unbounded faces is $2n$.

It is standard to represent line arrangements by *wiring diagrams*. For an introduction to wiring diagrams, we refer to [2] and [4]. For an example consider the left of Fig. 2.

Orienting the lines in the wiring diagram from left to right, induces an acyclic orientation of the edges. We call the unbounded face above h_n *top face* and the one below h_1 *bottom face* of W. All the other unbounded faces are either *left* or *right* unbounded. For every oriented pseudoline ℓ, the top face of W is to the left of ℓ and the bottom face to the right of ℓ.

The *i-th level* of a pseudoline arrangement $\mathcal{A}$ is the set of edges with exactly i lines strictly below. The edges of level i form an x-monotone curve in every pseudoline arrangement. We define the *(face)-level* of a face f as the number of lines strictly below f. It is easy to check that this is well defined. The bottom face f_{bottom} of a wiring diagram W is the unique face of level 0. Alternatively, one could define the level of a face f using the length of the shortest dual path from f to f_{bottom}.

There are exactly two unbounded faces in each level; in L_i, we denote the left one by l_i, and the right one by r_i, see left of Fig. 4.

We denote by L_i the set of faces with level i. Let w be some integer that divides $n + 1$. We define the i-th tunnel of width 2 as

$$\mathcal{T}_i = \bigcup_{j=2i}^{2i+1} L_j \quad , \text{ for } i = 0, \ldots, \tfrac{n+1}{2} - 1.$$

We denote by $l = \tfrac{n+1}{2} - 1$ the index of the last tunnel. To handle the case that n is even, note that only the last tunnel will be smaller and nothing else changes. Edges bounding faces of different tunnels are called *wall edges*. All other edges are called *tunnel edges*. It will be convenient to represent a pseudoline arrangement by a *tunnel diagram*. In a *tunnel diagram* each tunnel $\mathcal{T}_i$ is drawn in a horizontal stripe, except for the first and last tunnel, which are drawn in an half plane. This implies that all the wall edges separating the same tunnel are on one line. Consider the right of Fig. 2 for an example. Note that the tunnel diagram of a wiring diagram has the same level structure as long as the bottom face is the same.

2   Long Paths in Pseudoline Arrangements

First, we note that there are line arrangement with n lines, introduced by Füredi and Palásti [3], in which the longest path is of length $\tfrac{1}{3}n^2 + O(n)$. Hence, up to lower order terms Theorem 1 is tight. For more details we refer to [1].

We start with a brief overview of the proof. In the first step of the path construction, we find a path in each tunnel and connect all paths in a consistent manner at their ends. This gives a path P of length $\tfrac{1}{4}n^2 - O(n)$. To strengthen this result we prolong the current path P by incorporating sufficiently many faces, which are not yet used by P. We call these unused faces *bad*. (For a precise definition, see bellow.) The set of bad faces has strong structural properties. Most importantly, the set of bad faces induces a set of paths. These paths can be incorporated until only isolated bad faces remain. This has to be done carefully, due to two reasons: Firstly, some unbounded faces cannot be incorporated. Secondly, after one rerouting, the structure of our path changes and a second rerouting may not be possible. After eliminating adjacent bad faces, every remaining unused bad face is given two units of charge. The charged faces distribute the charge to traversed faces. It is possible that some traversed faces obtain two units of charge. In these cases, we reroute again or redistribute the charge. For this second round of rerouting, we identified the specific configurations that can appear and prove that no other situation occurs. At last, we will conclude using the charging scheme that roughly two third of all the faces are traversed and this will conclude the proof.

Recall a tunnel $\mathcal{T}_i$ is the union of the two face-levels L_{2i-1} and L_{2i} for $1 \leq i \leq \lfloor \tfrac{n}{2} \rfloor$. The set of edges separating tunnels $\mathcal{T}_{i-1}$ and $\mathcal{T}_i$ is called *top wall* of tunnel $\mathcal{T}_i$; the one separating $\mathcal{T}_i$ and $\mathcal{T}_{i+1}$ *bottom wall* of tunnel $\mathcal{T}_i$. An edge e of a face f is a *wall edge* if e is part of a wall, otherwise it is a *tunnel edge*, see also Fig. 3. Depending on the type of shared edge, two adjacent faces are *tunnel neighbors* or *wall neighbors*.

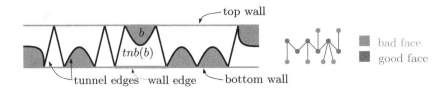

Fig. 3. Left: Schematic tunnel structure in primal view. Right: Schematic tunnel structure in dual view.

Observe that all vertices (which are of degree 4) lie on some wall and are adjacent to exactly two wall edges and two tunnel edges which do not alternate. A face is called *bad* if all its vertices lie on the same wall, otherwise it is called *good*. A path is called *bad path* if all involved faces are bad. Further, every face has either wall edges to the top or to the bottom, but never both. For every wall, its edges can be oriented from the left unbounded to the right unbounded edge. This orientation induces a linear order of the edges of a wall (and on the vertices). We call this the *wall order*. For two faces f and f' adjacent to the same wall, f is left of f' if its leftmost wall edge is left of the one of f'. (Since each face has wall edges on exactly one wall, this is well defined.)

The set of tunnel edges belonging to a tunnel can be interpreted as a curve, also *tunnel curve*, that runs from left to right. If the curve touches the same wall twice in a row, the face, which is bounded by the wall and the curve, is bad. Hence, a bad face b has exactly one tunnel edge, yielding a unique *tunnel neighbor* $tnb(b)$. If the curve connects the top and bottom wall, the adjacent faces of this edge are both good. The dual graph of a tunnel is a caterpillar where the backbone is formed by good faces which alternatingly have top and bottom wall edges. The leaves are bad faces. These tunnel properties are summarized in Observation 1.

Lemma 1. *Structural tunnel properties*

(a) Every face f has a wall edge; all wall edges of f are either top or bottom wall edges of f's tunnel.

(b) If a good and bad face share a tunnel edge, then their wall edges belong to different walls.

(c) Every bad face b has exactly one tunnel neighbor $tnb(b)$ and $tnb(b)$ is good.

(d) Bad faces within the same tunnel are not adjacent.

(e) There are two unbounded faces at the beginning and at the end of the tunnel, one of which is good and the other is bad.

(f) There exist a unique path between each two faces of a tunnel.

(g) The graph $G[\mathcal{T}_i]$ is a caterpillar. Its backbone is formed by good faces which alternatingly have top and bottom wall edges. The leaves are bad faces.

Besides the tunnel properties, we need a property which is based on simplicity of the arrangement. Given two faces f and f', we say f is *nested* inside f' if all wall edges of f are also wall edges of f'.

Lemma 2. *Bounded bad faces are not nested within another face.*

This observation follows from the fact that a nested bad face has only two adjacent faces, which is a contradiction to the fact that bounded faces are at least of degree 3.

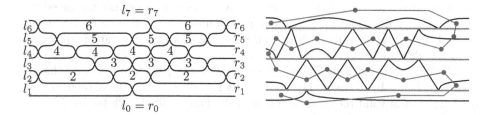

Fig. 4. Left: A Wiring diagram labeled with the face-level. Right: Tunnel diagram together with the glueable paths family $\mathcal{P}$.

We now define an ordered set of tunnel paths $\mathcal{P} := \{P_i \mid i \in [[\lfloor \frac{n}{2} \rfloor]]\}$ where

$$P_i := \text{path in tunnel } \mathcal{T}_i \begin{cases} \text{from } l_{2i-1} \text{ to } r_{2i}, & i \text{ odd,} \\ \text{from } l_{2i} \text{ to } r_{2i-1}, & i \text{ even.} \end{cases}$$

By Lemma 1 (g) the paths P_i are essentially the backbones of $G[\mathcal{T}_i]$ connecting some left unbounded face with some right unbounded face. We think of path P_i as oriented from the left to the right unbounded face. The path family $\mathcal{P}$ has the property of being *glueable*, that is the paths are pairwise disjoint and for odd i, the end faces of P_i and P_{i+1} are adjacent, and for even i, the start faces of P_i and P_{i+1} are adjacent. Hence, the paths in $\mathcal{P}$ can be combined (glued) to one path by attaching them alternating in a forward and reverse orientation: $P := P_1, \overleftarrow{P_2}, P_3, \overleftarrow{P_4}, \dots$ (Here $\overleftarrow{P}$ denotes the path that traverses the vertices of P in reverse order.)

Proposition 3. *Let Q be a path from l_i to r_j in $\mathcal{A}$.*
If $i, j \leq n/2$, then $|Q| \geq i + j + 1$. If $i, j \geq n/2$, then $|Q| \geq 2n - i - j + 1$.

Proof. The path Q must cross at least the i lines starting above l_i and the j lines ending above r_j. Crossing any of these lines yields one new face. Hence, the path is at least of length $i + j + 1$. For $i, j \geq n/2$, consider the lines starting and ending below l_i and r_j. □

Proposition 4. *P is at least of length $n^2/4 - O(n)$.*

Proof. Summing the length of the paths in $\mathcal{P}$, we immediately obtain that P is of quadratic length, by Proposition 3.

$$|P| = \sum_{i=1}^{\lfloor \frac{n}{2} \rfloor} |P_i| \geq 2 \sum_{i=1}^{\lfloor \frac{n}{4} \rfloor} 4i \geq n^2/4 - n \qquad \square$$

In the following, we prolong the path P in order to improve the leading coefficient of the length bound from $1/4$ to $1/3$. A path (or a glueable path family) partitions the set of faces into *traversed* and *not traversed* faces. Since unbounded faces turn out to be special cases, we need to treat them separately. Let F denote the set of all faces of $\mathcal{A}$ and $U \subset F$ the unbounded faces. Given a glueable path family $\mathcal{Q}$, we partition the bounded faces of F into the set T of bounded faces traversed by $\mathcal{Q}$, and the set N of bounded faces not traversed by $\mathcal{Q}$. To simplify notation: for a glueable path family $\mathcal{Q}^x$, we denote these sets as T^x and N^x. A path family $\mathcal{Q}'$ is a *valid rerouting* of a glueable path family $\mathcal{Q}$ if $\mathcal{Q}'$ is glueable and $T \subseteq T'$.

The proof concept is as follows: We start with the path family $\mathcal{P}$ and charge each face in N by 2 units, i.e. $ch(f) = 2$ for all $f \in N$. By a sequence of discharging and valid rerouting steps, we will obtain a final path family $\mathcal{P}^*$ and final charge function ch^* such that:

(1) $ch^*(f) = 0$ for all $f \in N^*$.
(2) $ch^*(f) \leq 1$ for all $f \in T^*$.
(3) $ch^*(f) \leq 2$ for all $f \in U$.
(4) $\sum_f ch^*(f) = 2|N^*|$.

These conditions imply the wished result as follows: By definition, $|F| = |T^*| + |N^*| + |U|$, with $|F| = \frac{n(n+1)}{2} + 1$ and $|U| = 2n$. The conditions give $2|N^*| \leq |T^*| + 2|U|$. Clearly, the final path P^* contains all faces in T^*:

$$|T^*| \geq 2|N^*| - 2|U| = 2(|F| - |T^*| - 2|U|) \implies |T^*| \geq \frac{2|F|}{3} - \frac{4|U|}{3} = \frac{n^2}{3} - O(n)$$

In other words, these conditions imply that $2/3$ of the bounded faces are traversed. Since there are only $O(n)$ unbounded faces, the wished result follows.

Initial Charge. We give some initial charge $ch : F \to \mathbb{N}$ to the faces in the following way:

$$ch(f) = \begin{cases} 2 & f \in N, \\ 0 & \text{else.} \end{cases}$$

Hence condition (4) is fulfilled by definition. This property will be maintained in the entire process, because we delete charge of faces that become traversed. Note that charged faces are, by definition, bad and bounded.

HeadRule. A charged face (bad and bounded) sends its charge to two different faces, one unit through its leftmost wall edge and the other through the right vertex of its leftmost wall edge, see Fig. 5.

Lemma 5. *The charge of a face f is sent to faces of an adjacent tunnel. Moreover, charge through vertices is sent to good faces.*

Its important to note that we may divert some of this charge. This is why we think of it as "being on its way" that is, we distinguish between charge of a face which is *sent* and *obtained*. In particular, we partially *overwrite* this HeadRule in the following two phases!

Fig. 5. Discharging of bounded not traversed faces through leftmost wall edge and second leftmost vertex.

Rerouting and Discharging Step 1. First, we guarantee that no face in N obtains charge. In particular, we eliminate adjacent charged faces by incorporating them in some path. If this is not possible, we discharge. Consider the set of (all not only bounded) not traversed faces N_{All} in $\mathcal{P}$. We exploit some structural properties of the graph $G[N_{All}]$ induced by the set N_{All}.

Lemma 6. *Structural properties of $G[N_{All}]$*

(a) *If face f is good, then $f \notin N_{All}$.*
(b) *If face f is bad and $f \notin N_{All}$, then there exists i such that P_i starts or ends in f.*
(c) *The graph $G[N_{All}]$ is a collection of paths. The wall orders induce a left-right orientation on the paths.*
(d) *Let $Q = b_1, b_2, \ldots$ be a maximal path in $G[N_{All}]$ oriented from left to right. If $\deg_{\mathcal{A}}(b_1) \geq 3$, then the penultimate wall neighbor of b_1 is traversed.*
(e) *Let $Q = b_1, b_2, \ldots$ be a maximal path in $G[N_{All}]$ oriented from left to right. If $|Q| \geq 5$, then there exist $j \leq 4$ such that b_j has a traversed wall neighbor.*

Step 1. *We construct a valid rerouting $\mathcal{P}^{(1)}$ of $\mathcal{P}$ and discharge such that*

(P1) $ch(f) = 0$ for all $f \in N^{(1)}$,
(P2) $ch(f) \leq 2$ for all $f \in F$, and
(P3) $\sum_f ch(f) = 2|N^{(1)}|$.

Let Q be a maximal path in $G[N_{All}]$ with $|Q| \geq 2$, where b_1, b_2 denote the first two (bad) faces of Q. The aim is to replace single edges of a current path in $\mathcal{P}$ by a path through not yet traversed faces. For this replacement, we enter the bad path by a traversed wall neighbor of a bad face. We use the sufficient conditions for the existence of such neighbor by Lemma 6(d) and (e) and distinguish three cases, see Fig. 6. We initialize $\mathcal{P}_{curr} := \mathcal{P}$ and redefine some of the paths during the phases of step 1.

Step (1a). *If $\deg_{\mathcal{A}}(b_1) \geq 3$ and $|Q| \geq 2$:*
We set f_{enter} to the penultimate wall neighbor of b_1 and f_{exit} to the tunnel neighbor of b_2. Let P_i^{curr} be the path of $\mathcal{P}_{curr}$ containing f_{enter}. (We will show that f_{enter} and f_{exit} are consecutive in P_i^{curr}.) The idea is to insert b_1 and b_2 in between f_{enter} and f_{exit} in P_i^{curr}. For the formal definition, let P'

denote the prefix of P_i^{curr} ending right before f_{enter} and P'' denote the suffix of P_i^{curr} starting after f_{exit}. (If f_{enter} and f_{exit} are adjacent, it holds that $P_i^{curr} = P', f_{enter}, f_{exit}, P''$.) We alter P_i^{curr} by inserting b_1, b_2:

$$P_i^{curr} := P', f_{enter}, b_1, b_2, f_{exit}, P''$$

Since b_1, b_2 are now traversed, their status switches from N to T and (if they are charged,) their charge is deleted.
If $(|Q| - 2) \geq 2$, apply Step 1a) to $Q' := Q - \{b_1, b_2\}$.

Step (1b). If $\deg_{\mathcal{A}}(b_1) = 2$ and $|Q| \geq 5$:
Determine $j_{\min}$, the smallest j such that b_j has a traversed left wall neighbor. f_{enter} is set to the rightmost traversed wall neighbor of $b_{j_{\min}}$ and f_{exit} as tunnel neighbor of $b_{(j_{\min}+1)}$. Let P_i^{curr} denote the path of $\mathcal{P}_{curr}$ containing f_{enter}. Let P' denote the prefix of P_i^{curr} ending right before f_{enter} and P'' denote the suffix of P_i^{curr} starting after f_{exit}. We reroute as in a):

$$P_i^{curr} := P', f_{enter}, b_{j_{\min}}, b_{(j_{\min}+1)}, f_{exit}, P''$$

Since $b_{(j_{\min})}, b_{(j_{\min}+1)}$ are now traversed, their status switches from N to T and (if they are charged) their charge is deleted.
   If $j_{\min} = 4$, the total charge of b_3 is terminatory sent to the unbounded face b_1.
   If $(|Q| - j_{min} - 1) \geq 2$, apply Step 1a) to $Q' := Q - \{b_1, \ldots, b_{(j_{\min}+1)}\}$.

Step (1c). If $\deg_{\mathcal{A}}(b_1) = 2$ and $|Q| \leq 4$:
The total charge of b_3 and b_4 (if they exist and are charged) is terminatory sent to the unbounded faces b_1 and b_2, respectively.

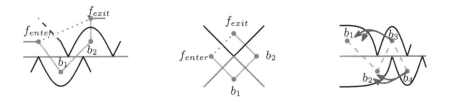

Fig. 6. Left/middle: Rerouting in Step (1a) and (1b). Right: Discharging in Step (1c) – charge from b_3 and b_4 is sent to unbounded faces.

We denote the path family obtained after handling all path in $G[N_{\text{All}}]$ by $\mathcal{P}^{(1)}$, and the corresponding set of traversed and not traversed bounded faces by $T^{(1)}$ and $N^{(1)}$, respectively.

Proposition 7. $\mathcal{P}^{(1)}$ is a valid rerouting of $\mathcal{P}$.

Next, we state a crucial property which we apply afterwards in order to prove that the claimed conditions are fulfilled.

Lemma 8. *Bad faces still sending charge were rightmost of a bad path in the graph* $G[N_{All}]$.

Lemma 9. *A good face obtains its charge through at most one vertex and one edge. (See Fig. 7.)*

Proposition 10. *After Step 1, (P1)–(P3) hold.*

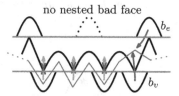
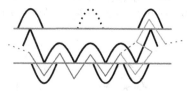

Fig. 7. After Step 1, good faces obtain at most two units of charge. Only last bad faces may send charge. Left: A good face with two units of charge and the path is displayed. Right: The path after rerouting is displayed.

Rerouting and Discharging Step 2. It remains to guarantee that faces in T obtain at most one unit of charge.

Step 2. *We construct a valid rerouting* $\mathcal{P}^{(2)}$ *of* $\mathcal{P}^{(1)}$ *and discharge such that*

(Q1) $ch(f) = 0$ *for all* $f \in N^{(2)}$,
(Q2) $ch(f) \leq 1$ *for all* $f \in T^{(2)}$,
(Q3) $ch(f) \leq 2$ *for all* $f \in U$, *and*
(Q4) $\sum_f ch(f) = 2|N^{(2)}|$.

In order to obtain the wished properties, it only remains to consider bounded traversed faces obtaining charge of more than 1 unit. By Lemmas 5 and 9, a (bounded) traversed face f with charge of more than one unit, is good and obtains its charge through exactly one vertex from a face b_v and one edge from a face b_e, which are in different tunnels, see Fig. 7. In this case, we either redistribute the charge or incorporate the two not yet traversed faces b_v and b_e.

Let f be a face in tunnel T_i then without loss of generality b_e is in tunnel T_{i-1} and b_v is in T_{i+1}. Denote the first and second good face after f within its tunnel (in the left right order of the original P_i) by f_1 and f_2. The predecessor, successor and second successor of f in the possibly modified P_i, we denote by pr, s_1, and s_2. Since b_e is a face of degree ≥ 3, these faces exist. In particular, the existence of f_2 implies the existence of all other faces as well.

By Lemma 8, b_e and b_v were the rightmost bad faces of a bad path. This implies that f_1 is a good face with exactly one vertex on the opposite tunnel wall from its wall edges, otherwise b_e would not have been last of a bad path. Consequently, the edge (f, f_1) was not replaced in Step 1. Hence, $f_1 = s_1$. We make a case distinction based on whether or not (f_1, f_2) is an edge in some path, see Fig. 8.

Step (2a). *If* (f_1, f_2) *is an edge in* P_i, *then* $f_2 = s_2$. *Let* P' *be the prefix of* P_i *until* pr *and* P'' *be the suffix of* P_i *starting after* f_2. *We redefine* P_i *as follows:*

$$P_i := P', pr, b_v, f_1, f, b_e, f_2, P''$$

Since b_v, b_e *are now traversed, their status switches from* N *to* T *and their charge is deleted.*

Step (2b). *If* (f_1, f_2) *was replaced in Step 1 (when some* b_1, b_2 *where incorporated), we send one unit of charge from* f *to* b_1.

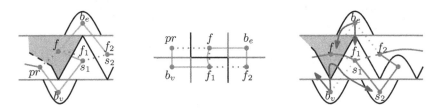

Fig. 8. Left: Rerouting in Step 2. Right: Discharging in Step 2 – One unit of f (from b_v) is sent to s_2.

Proposition 11. $\mathcal{P}^{(2)}$ *is a valid rerouting of* $\mathcal{P}^{(1)}$.

Proposition 12. *(Q1)-(Q4) of Step 2 hold.*

This establishes the claimed properties, and Theorem 1 follows.

3 Open Problem

We have shown that the linear upper bound on the longest alternating path also holds if the number of red and blue lines are in a ratio two to three. It is completely unclear, what happens if the number of red lines *equals* the number of blue lines.

Does every pseudoline arrangement with n red and n blue lines have an alternating dual path of length $\Omega(n^2)$?

Acknowledgment. We thank Nieke Aerts, Stefan Felsner, Heuna Kim, and Piotr Micek for interesting and helpful discussions on the topic. We thank the anonymous reviewers for helpful comments.

References

1. Aichholzer, O., Cardinal, J., Hackl, T., Hurtado, F., Korman, M., Pilz, A., Silveira, R., Uehara, R., Valtr, P., Vogtenhuber, B., Welzl, E.: Cell-paths in mono- and bichromatic line arrangements in the plane. Discrete Math. Theor. Comput. Sci. **16**(3), 317–332 (2014)
2. Felsner, S.: Geometric Graphs and Arrangements. Vieweg, Verlag (2004)
3. Füredi, Z., Palásti, I.: Arrangements of lines with a large number of triangles. Proc. Am. Math. Soc. **92**(4), 561–566 (1984)
4. Goodman, J.E., O'Rourke, J. (eds.): Handbook of Discrete and Computational Geometry. CRC Press, Boca Raton (2010)
5. Hoffmann, U., Kleist, L., Miltzow, T.: Long paths in line arrangements. Eur. Workshop Comput. Geom. **2015**, 157–160 (2015)
6. Hoffmann, U., Kleist, L., Miltzow, T.: Upper and lower bounds on long dual-paths in line arrangements. ArXiv e-prints 1506.03728 (2015)
7. Tutte, W.T.: A theorem on planar graphs. Trans. Am. Math. Soc. **82**, 99–116 (1956)

A Numbers-on-Foreheads Game

Sune K. Jakobsen[(✉)]

School of Mathematical Sciences and School of Electronic Engineering and Computer
Science, Queen Mary University of London, Mile End Road, London E1 4NS, UK
S.K.Jakobsen@qmul.ac.uk

Abstract. Is there a joint distribution of n random variables over the
natural numbers, such that they always form an increasing sequence and
whenever you take two subsets of the set of random variables of the same
cardinality, their distribution is almost the same?
We show that the answer is yes, but that the random variables will
have to take values as large as $2^{2^{\cdot^{\cdot^{\cdot^{2^{\Theta\left(\frac{1}{\epsilon}\right)}}}}}}$, where $\epsilon \leq \epsilon_n$ measures how
different the two distributions can be, the tower contains $n-2$ 2's and the
constants in the Θ notation are allowed to depend on n. This result has
an important consequence in game theory: It shows that even though you
can define extensive form games that cannot be implemented on players
who can tell the time, you can have implementations that approximate
the game arbitrarily well.

1 Introduction

A group of gamblers are standing in a circle so that each gambler can see all the
other gamblers' foreheads but not their own, and the gamblers are not allowed
to communicate. A dealer then sticks one natural number on each gamblers'
forehead, and ask each gambler to choose two numbers i and j (the other gambler
do not learn the numbers). If the gambler had the i'th smallest number he wins
1 dollar from the dealer, if he has the j'th smallest he loses 1 dollar to the dealer.
Does she have a randomised strategy that ensures that in expectation she does
not lose money?

A randomised dealer strategy is just a distribution on $(X_1, \ldots, X_n)$ where
n is the number of gamblers and $X_1 < X_2 < \cdots < X_n$ are random variables
taking integers. We show that if such a strategy ensures that the dealer does not
lose in expectation, then for any k any two k-subsets of $\{X_1, \ldots, X_n\}$ will have
the same distribution. This has an important consequence in game theory. It
is well known that some extensive form games, for example the absent minded
driver game [10], cannot be implemented on agents with perfect memory. In
order to avoid such games, we often require games to have *perfect recall*, that
is, the players remember which information sets they have previously been in,
and what choices they made. However, there are also games with perfect recall
that cannot be implemented on agents with a sense of time, because the information sets do not respect any ordering [5]. If we can find a distribution on

© Springer-Verlag Berlin Heidelberg 2015
G.F. Italiano et al. (Eds.): MFCS 2015, Part II, LNCS 9235, pp. 420–431, 2015.
DOI: 10.1007/978-3-662-48054-0_35

$(X_1, \ldots, X_n)$ such that learning the values of some subset $\{X_{i_1}, \ldots, X_{i_k}\}$ only tells you the *cardinality* k of the subset, but does not give you any information about $i_1, \ldots, i_k$, then we can use this to time any game with perfect recall and at most n nodes in each history. We simply play the root at time X_1, then the next node at time X_2 and so on. The agents would learn some times X_{i_j}, but the only information an agent would get from this, is the number of nodes he has had, and agents alway know this in any game with perfect recall.

However, we show that no such distribution of $(X_1, \ldots, X_n)$ exists. In the other direction, we show that the dealer can ensure that she only lose ϵ dollars in expectation for any $\epsilon > 0$. Unfortunately, to do that, she will need to use numbers as large as $2^{2^{\cdot^{\cdot^{2^{\Theta(\frac{1}{\epsilon})}}}}}$, when ϵ is sufficiently small. Here the tower contains $n-2$ 2's, and the constants in the Θ notation are allowed to depend on n. This result implies (see [5]) that any extensive form game can be approximated arbitrarily well by games where the players know the time at any node.

1.1 Related Work

The setup where each person in a group has some information on their forehead or their hat, arises naturally in many different areas such as communication complexity [2], network coding [11], coding theory [8] and derandomization [1], and numbers-on-foreheads problems can illustrate the difference between common knowledge and mutual knowledge [3]. See [7] or [6] for an overview of many hat problems and numbers-on-foreheads problems. The only one of these that concerns the rank of the numbers on the foreheads is the problem studied in [1]. In this problem n people each have a real number on their forehead and they are asked to (simultaneously) pick either a red or a blue hat, in such a way that if you order the players according to their numbers, the hats would be alternating blue and red. There is a strategy that achieves this with certainty, and this strategy can be used to derandomize auctions [1].

1.2 Notation and Preliminaries

For a real number x we define $[x] = \{i \in \mathbb{N} | i \leq x\}$. We let log denote the base 2 logarithm, and let $\exp_2$ denote the function $x \mapsto 2^x$. Hence, $\exp_2^n(x)$ denotes iteration of $\exp_2$, so $\exp_2^n(x) = 2^{2^{\cdot^{\cdot^{2^x}}}}$ where the tower contains n 2's. Random variables are denoted by capital letter, and the values they take by small letters. The *total variation distance* (also called *statistical distance*) between two discrete random variables X_1 and X_2 is given by $\delta(X_1, X_2) = \sum_x \left| \frac{\Pr(X_1 = x) - \Pr(X_2 = x)}{2} \right| = \sum_x \max(\Pr(X_2 = x) - \Pr(X_1 = x), 0)$ where the sum is over all possible values of X_1 and X_2. This measure is symmetric in X_1 and X_2. For an event E we use $\delta(X_1, X_2)|_E$ to denote $\delta(X_1|_E, X_2|_E)$.

We say that X_1 and X_2 are ϵ-*indistinguishable* if $\delta(X_1, X_2) \leq \epsilon$. Given a tuple $X = (X_1, \ldots, X_n)$ we let X_{-i} denote $(X_1, \ldots, X_{i-1}, X_{i+1}, \ldots, X_n)$ and for $I = \{i_1, \ldots, i_m\} \subset [n]$ with $i_1 < \ldots i_k$ we define $X_I = (X_{i_1}, \ldots, X_{i_m})$. Similar for x_{-i} and x_i. We say that $(X_1, \ldots, X_n)$ *has ϵ-indistinguishable m-subsets* if for any two subsets $I, J \subset [n]$ of size m, the two random tuples X_I and X_J are ϵ-indistingushable. We slightly abuse notation and say that $(X_1, \ldots, X_n)$ *has ϵ-indistinguishable subsets* if for all $m < n$ it has ϵ-indistinguishable m-subsets. Finally, we say that X is *ϵ-indistinguishable neighbouring $n-1$-subsets* if $\delta(X_{-i}, X_{-(i+1)}) \leq \epsilon$ for all $i \in [n-1]$. We will need the following propositions about total variation distance. All omitted and sketched proofs can be found in the full version on this paper [4].

Proposition 1 (Triangle inequality). *For random variables X_1, X_2, X_3 we have $\delta(X_1, X_3) \leq \delta(X_1, X_2) + \delta(X_2, X_3)$.*

Proposition 2. *If X_1, X_2 and Y are random variables, Y is independent from X_1 and X_2 and f is a function, then $\delta(f(X_1, Y), f(X_2, Y)) \leq \delta(X_1, X_2)$.*

Proposition 3. *Let $X_1, \ldots, X_n, Y_1, \ldots, Y_n, I$ be independent random variables with X_i and Y_i distributed on $\mathcal{X}_i$, and I distributed on $[n]$. Let $X = X_I$ and $Y = Y_I$. We have $\delta(X, Y) \leq \sum_{i=1}^{n} \Pr(I = i)\delta(X_i, Y_i)$, with equality if all the $\mathcal{X}_i$'s are pairwise disjoint.*

We will assume that the reader knows some game theory, and knows the minimax theorem. For an introduction to game theoretical concepts see [9].

2 Relation Between Gambling Games and Total Variation Distance

In this section we will show that several similar problems are the same up to a constant factor (depending on n) on ϵ. First we show that the total variation distance can be seen as a measure of the advantage in a betting game.

Proposition 4 (Total variation as betting advantage). *For random variables Y_1 and Y_2 we define a one-player game: First y_1 and y_2 is chosen according to the distribution of Y_1 and Y_2. Then independently i is chosen uniformly on $\{1, 2\}$. The player learns y_i and makes a guess about i. If correct he gets utility 1 if wrong he gets utility -1.*

The expected utility the player gets using the optimal strategy is $\delta(Y_1, Y_2)$.

Proof. As this is a one-player game, the optimal strategy is deterministic. A deterministic strategy is a function g that for each possible value y of Y_1 or Y_2 gives the value in $\{1, 2\}$ that the player should guess. If $g(y) = 1$ the contribution of y to the expected output when gambler use strategy g is $\Pr(Y_1 = y)\Pr(I = 1) - \Pr(Y_2 = y)\Pr(I = 2) = \frac{\Pr(Y_1=y)-\Pr(Y_2=y)}{2}$. Similarly, if $g(y) = 2$ then y's contribution to the expected outcome is $\frac{\Pr(Y_2=y)-\Pr(Y_1=y)}{2}$. Clearly, the best strategy is to choose the positive one of these two, in which case the contribution of y is $\left| \frac{\Pr(Y_1=y)-\Pr(Y_2=y)}{2} \right|$. Summing over all y's gives $\delta(Y_1, Y_2)$.

We will now define two games between Dealer and Gambler, and show that they are related. Given n and N we define Game 1: First Dealer chooses some natural numbers $1 \leq X_1 < \cdots < X_n \leq N$. Then a number i_0 is chosen uniformly at random from $[n]$. Then Gambler learns X_{-i_0} and chooses two numbers i and j. If $i = i_0$ Gambler wins 1 dollar from Dealer, if $j = i_0$ he loses 1 dollar to Dealer. This is just the game from the introduction seen from the perspective of a single gambler.

Game 1 is a two-player zero-sum games, and because we only allow Dealer to choose natural numbers between 1 and N, each of these players only have finitely many pure strategies. By the minimax theorem such a game has a value v_1 such that: Dealer has a probabilistic strategy, that is a distribution on $(X_1, \ldots, X_n)$, such that no matter what strategy Gambler uses he cannot earn more than v_1 dollars in expectation and Gambler has a probabilistic strategy, such that no matter which numbers Dealer chooses, Gambler will win at least v_1 dollars in expectation.

We want to figure out how this value changes with n and N. In order to do this we define Game 2, which is less natural but easier to analyse: First Dealer chooses some natural numbers $1 \leq X_1 < \cdots < X_n \leq N$ and Gambler chooses two numbers i_1 and i_2. A fair coin is flipped to decide if $K = 1$ or 2. Gambler learns X_{-i_K} and guesses if $K = 1$ or 2. If Gambler is correct he wins 1 dollar from Dealer, if he is wrong he loses 1 dollar from Dealer. This is also a zero-sum game with finitely many pure strategies, so the minimax theorem says that Game 2 also have a value v_2.

Proposition 5. $\frac{2}{n} v_2 \leq v_1$

Proof. Suppose that Gambler has a strategy that ensures an expected outcome of v_2 in Game 2. To show the statement we construct a strategy that ensures expected outcome of $\frac{2}{n} v_2$ in Game 1.

Before Game 1 starts, Gambler chooses two numbers i_1 and i_2 using his strategy for Game 2. We then play Game 1: Dealer chooses numbers $x_1 < \cdots < x_n$ and i_0 is chosen uniformly from $[n]$. Gambler sees x_{i_0}. Gambler still plays as if he was playing Game 2 and had chosen i_1 and i_2. If he would have guessed $K = 1$ he sets $i = i_1, j = i_2$ and if he would have guessed 2 he sets $i = i_2, j = i_1$.

As Gambler choice of i_1 and i_2 cannot affect i_0 and X, there is probability $\frac{2}{n}$ that $i_0 \in \{i_1, i_2\}$. Given that this happens the expected outcome is exactly the same as the expected outcome in Game 2. Given that $i_0 \notin \{i_1, i_2\}$ Gambler will neither lose nor win money. Thus, the expected outcome of the strategy is $\frac{2}{n} v_2$.

Similarly, we can show the following Proposition.

Proposition 6. $\frac{v_1}{n-1} \leq v_2$.

Proposition 7. $v_2 \leq \epsilon$ if and only if there exists $(X_1, \ldots, X_n)$ with $1 \leq X_1 < \ldots X_n \leq N$ with ϵ-indistinguishable $n-1$-subsets.

Proof. Follows from Proposition 4.

Theorem 1. *Fix parameters $n \geq 2$ and N. Let v_1 be the value of Game 1 and let ϵ_0 be the infimum over all values ϵ such that there exists a distribution of $X = (X_1, \ldots, X_n)$ where $1 \leq X_1 < X_2 < \cdots < X_n \leq N$ are all integers and X has ϵ-indistinguishable $n-1$-subsets. Then $\frac{2}{n}\epsilon_0 \leq v_1 \leq (n-1)\epsilon_0$.*

Proof. Follows from Propositions 5, 6 and 7.

Thus, to get upper and lower bounds for v_1 it is enought to get bounds for ϵ_0. All the above problems have only involved $n-1$-subsets on the n numbers. The following proposition show that if any two neighbouring $n-1$-subsets look the same, then any two subsets of the same size looks the same, so we only need to consider neighbouring $n-1$-subsets.

Proposition 8. *Fix $n \in \mathbb{N}$ and $\epsilon > 0$. If $(X_1, \ldots, X_n)$ has ϵ-indistinguishable neighbouring $n-1$-subsets, is has $n^2\epsilon$-indistinguishable subsets.*

Proof (Sketch). This follows from repeated use of the triangle inequality and the fact that if $Y = (Y_1, \ldots, Y_k)$ and $Z = (Z_1, \ldots, Z_k)$ then $\delta(Y_{-i}, Z_{-i}) \leq \delta(Y, Z)$.

3   Upper Bound

In this section we will construct random variables $(X_1, \ldots, X_n)$ such that X_{-i} and X_{-j} are ϵ-indistinguishable for all i and j. First we consider the case $n = 2$.

Proposition 9. *Given ϵ, there exists random variables (X_1, X_2) such that $1 \leq X_1 < X_2 \leq \lceil \frac{1}{\epsilon} \rceil + 1$ are integers and (X_1, X_2) has ϵ-indistinguishable 1-subsets.*

Proof. Let X_1 be uniformly distributed on $\left[\lceil \frac{1}{\epsilon} \rceil\right]$ and let $X_2 = X_1 + 1$. Then $\delta(X_1, X_2) = \lceil \frac{1}{\epsilon} \rceil^{-1} \leq \left(\frac{1}{\epsilon}\right)^{-1} = \epsilon$ so (X_1, X_2) has ϵ-indistinguishable 1-subsets.

Proposition 10. *Let $n_1 > n_2$ and let U_{n_1}, U_{n_2} be independent random variables uniformly distributed on $[n_1]$ respectively $[n_2]$. Then $\delta(U_{n_1}, U_{n_2} + U_{n_1}) = \frac{n_2+1}{2n_1}$.*

This follows from a simple computation. We are now ready for the construction of a distribution of $X = (X_1, \ldots, X_n)$ for $n \geq 3$. First we give the construction for $n = 3$ and then we construct a distribution for n given a distribution for $n - 1$. For this recursive construction to work we need to assume more than just having ϵ-indistinguishable $n-2$-subsets about the distribution for $n-1$ so we cannot use the distribution for $n = 2$ as the start of the recursive definition.

Lemma 1. *For all $n \geq 3$ and all $\epsilon \in (0,1)$ there exists $X = (X_1, \ldots, X_n)$ where the X_i takes values in $\mathbb{N}$, with a joint distribution such that $X_1 < \cdots < X_n$ and*

1. *X has $\epsilon(12^{-n})$-indistinguishable neighbouring $n-1$-subsets,*
2. *$\forall i \in [n-1] : \Pr(X_{i+1} - X_i < n + 4 - \log(\epsilon)) \leq \epsilon 2^{-n-3}$,*
3. *X_n never takes values above $\exp_2^{n-2}\left(4\lceil \frac{1}{\epsilon} \rceil + 6\right) - 4n - 2 + 2\log(\epsilon)$.*

Proof. We fix ϵ and prove the statement by induction in n, so first we show the statement for $n = 3$. Let X_1 be uniformly distributed on $\left[2^{4\lceil\frac{1}{\epsilon}\rceil+4}\right]$ and let K be uniformly distributed on $\left[\left\lceil\frac{1}{\epsilon}\right\rceil + 3, \left\lceil\frac{1}{\epsilon}\right\rceil + 4, \ldots, 4\left\lceil\frac{1}{\epsilon}\right\rceil + 3\right]$. We now define $X_2 = X_1 + 2^K$ and $X_3 = X_2 + 2^K = X_1 + 2^{K+1}$. It is straightforward to check that (X_1, X_2, X_3) satisfy the 3 requirements.

For the induction step, assume that $X = (X_1, \ldots, X_n)$ satisfy the statement for n. We want to construct $Y = (Y_1, \ldots, Y_{n+1})$ that shows that the statement holds for $n + 1$. To do this we construct a joint distribution of (X, D, Y), where $D = (D_1, \ldots, D_n)$. We choose X so that it satisfy the requirements for n, and given these, we let D_i be uniformly distributed on $[2^{X_i + 4n - 2\log(\epsilon)}]$ and let Y_1 be uniformly distributed on $\left[\exp_2^{n-1}\left(4\left\lceil\frac{1}{\epsilon}\right\rceil + 6\right)/2 - 4n - 6 + 2\log(\epsilon)\right]$. All these are independent given X. We define $Y_{i+1} = Y_i + D_i$ for $i \in [n]$.

We now check that Y satisfy the requirements. First we want to show that if we are given the tuple (Y_1, D_{-i}) then it will not make much of a difference if we add D_i to D_{i+1}. That is, we want to bound

$$\delta((Y_1, (D_1, \ldots, D_{i-1}, D_i + D_{i+1}, D_{i+2}, \ldots, D_n)), (Y_1, D_{-i})).$$

To do this, we first get from Proposition 10 that

$$\delta(D_i + D_{i+1}, D_{i+1})|_{(X_i, X_{i+1}=(x_i, x_{i+1})} = \frac{\lfloor 2^{x_i + 4n - 2\log(\epsilon)}\rfloor + 1}{2 \cdot \lfloor 2^{x_{i+1} + 4n - 2\log(\epsilon)}\rfloor} \leq 2^{x_i - x_{i+1}}.$$

Now Proposition 3 gives us

$$\delta((D_i + D_{i+1}, X_i, X_{i+1}), (D_{i+1}, X_i, X_{i+1}))$$
$$= \sum_{(x_i, x_{i+1})} \Pr((X_i, X_{i+1}) = (x_i, x_{i+1}))\delta(D_i + D_{i+1}, D_{i+1})|_{(X_i, X_{i+1})=(x_i, x_{i+1})}.$$

From requirement (2), we have $\Pr(X_{i+1} - X_i < n + 4 - \log(\epsilon)) \leq \epsilon 2^{-n-3}$. When $x_{i+1} - x_i < n + 4 - \log(\epsilon)$ we have

$$\delta(D_i + D_{i+1}, D_i)|_{(X_i, X_{i+1})=(x_i, x_{i+1})} \leq 1$$

as δ only takes values in $[0, 1]$. In all other cases, we have

$$\delta(D_i + D_{i+1}, D_i)|_{(X_i, X_{i+1})=(x_i, x_{i+1})} \leq 2^{-(n+4)+\log(\epsilon)} = \epsilon 2^{-(n+4)}.$$

Summing up gives

$$\delta((D_i + D_{i+1}, X_i, X_{i+1}), (D_{i+1}, X_i, X_{i+1})) \leq \epsilon 2^{-n-3} + \epsilon 2^{-n-4} \leq \epsilon 2^{-n-2}.$$

Given X_i and X_{i+1} and either $D_i + D_{i+1}$ or D_{i+1}, there is a random function giving $(Y_1, (D_1, \ldots, D_{i-1}, D_i + D_{i+1}, D_{i+2}, \ldots, D_n))$ respectively (Y_1, D_{-i}). Thus by Proposition 2 we have

$$\delta((Y_1, (D_1, \ldots, D_{i-1}, D_i + D_{i+1}, D_{i+2}, \ldots, D_n)), (Y_1, D_{-i}))$$
$$\leq \delta((D_i + D_{i+1}, X_i, X_{i+1}), (D_{i+1}, X_i, X_{i+1})) \leq \epsilon 2^{-n-2}.$$

This is the upper bound we wanted. Clearly there is a random function, not depending on i, that given X_{-i} returns (Y_1, D_{-i}) such that when input have the correct distribution, then the output have the correct distribution. Thus,

$$\delta((Y_1, D_{-i}), (Y_1, D_{-(i+1)}) \le \delta(X_{-i}, X_{-(i+1)}) \le \epsilon(1 - 2^{-n}).$$

For $i \ge 2$ we use the fact that the Y_j's can be computed from the D_j's and Y_1 and then use the triangle inequality to get

$$
\begin{aligned}
&\delta(Y_{-i}, Y_{-(i+1)}) \\
&= \delta((Y_1, (D_1, \ldots D_{i-2}, D_{i-1} + D_i, D_{i+1}, \ldots, D_n)), \\
&\quad (Y_1, (D_1, \ldots, D_{i-1}, D_i + D_{i+1}, \ldots, D_n))) \\
&\le \delta((Y_1, (D_1, \ldots D_{i-2}, D_{i-1} + D_i, D_{i+1}, \ldots, D_n)), (Y_1, D_{-(i-1)})) \\
&\quad + \delta((Y_1, D_{-(i-1)}), (Y_1, D_{-i}) \\
&\quad + \delta((Y_1, D_{-i}), (Y_1, D_1, \ldots, D_{i-1}, D_i + D_{i+1}, \ldots, D_n)) \\
&\le 2 \cdot \epsilon 2^{-n-2} + \epsilon(1 - 2^{-n}) = \epsilon(1 - 2^{-(n+1)}).
\end{aligned}
$$

This shows requirement (1) in all cases except $\delta(Y_{-1}, Y_{-2})$. The proof of that case is similar. Requirements (2) and (3) are straightforward to check.

Corollary 1. *For fixed $n \ge 2$ there exists a distribution of $X = (X_1, \ldots, X_n)$ where $1 \le X_1 < X_2 < \cdots < X_n \le N(\epsilon)$ are all integers and X has ϵ-indistinguishable subsets and $N(\epsilon) = \exp_2^{n-2}(O(\frac{1}{\epsilon}))$.*

Proof. Follows from Propositions 8 and 9 and Lemma 1.

4   Lower Bound

In this section we will show lower bounds on how large values X_n needs to take if $X = (X_1, \ldots, X_n)$ has ϵ-indistinguishable $n - 1$-subsets and we always have $X_1 \ge 0$ and $X_{i+1} \ge X_i + 1$. We no longer require that the X_i are integers, only that there are at least one apart. This weaker requirement makes the induction argument easier. Clearly, any lower bound we show under the assumption that the X_i's are at least one apart will also be a lower bound in the case where the X_i have to take integer values. Conversely, if you have a distribution of X with ϵ-indistinguishable $n - 1$-subsets and $X_1 \ge 0, X_{i+1} \ge X_i + 1$ you can define X' by $X'_i = 1 + \lfloor X_i \rfloor$. Then $X_1 < X_2 < \cdots < X_n$ will be natural numbers and by Proposition 2 X' will have ϵ-indistinguishable $n - 1$-subsets.

Proposition 11. *If X_1 and X_2 are discrete random variables taking real values in an interval $[a, b]$ and $\mathbb{E}X_2 \ge \mathbb{E}X_1 + 1$ then $\delta(X_1, X_2) \ge \frac{1}{b-a}$.*

Proof. If $a \ne 0$ we can subtract a from X_1 and X_2, and set the new b to be $b - a$ and a to be 0. We will still have $\mathbb{E}X_2 \ge \mathbb{E}X_1 + 1$ and the distance $\delta(X_1, X_2)$ and $b - a$ are not affected by this. So in the following we will assume $a = 0$. Now

$$\mathbb{E}X_2 - \mathbb{E}X_1 = \sum_x x(\Pr(X_2 = x) - \Pr(X_1 = x))$$

$$\leq \sum_x x \max(\Pr(X_2 = x) - \Pr(X_1 = x), 0)$$

$$\leq b \sum_x \max(\Pr(X_2 = x) - \Pr(X_1 = x), 0) = b\delta(X_2, X_1).$$

So $\delta(X_1, X_2) = \delta(X_2, X_1) \geq \frac{1}{b} = \frac{1}{b-a}$.

Proposition 12 (Lower bound, $n = 2$). *If X_1, X_2 are random variables over the non-negative real numbers such that $X_2 \geq X_1 + 1$ and (X_1, X_2) has ϵ-indistinguishable 1-subsets, then X_2 takes values of at least $\frac{1}{\epsilon}$.*

Proof. We only need to show $\delta(X_2, X_1) \leq \epsilon$. As $X_2 \geq X_1 + 1$ we have $\mathbb{E}X_2 \geq \mathbb{E}X_1 + 1$ and the statement follows from Proposition 11. ∎

Here we allow X_1 to be 0. If we required X_1 and X_2 to be natural numbers, the lower bound would be $\lceil \frac{1}{\epsilon} \rceil + 1$, which exactly matches our construction in Proposition 9. Combining Proposition 11 with Proposition 2 we get:

Proposition 13. *If X_1 and X_2 are random variables with domain $\mathcal{X}$ and $f : \mathcal{X} \to [a, b]$ is a function such that $\mathbb{E}f(X_2) \geq \mathbb{E}f(X_1) + 1$ then $\delta(X_1, X_2) \geq \frac{1}{b-a}$.*

Proposition 14 (Lower bound, $n = 3$). *Let X_1, X_2, X_3 be random variables taking non-negative real numbers such that $X_2 \geq X_1 + 1$ and $X_3 \geq X_2 + 1$. If (X_1, X_2, X_3) has ϵ-indistinguishable 2-subsets, then X_3 must take values of at least $2^{\frac{1}{\epsilon}}$.*

Proof. We have $X_3 - X_1 = (X_3 - X_2) + (X_2 - X_1)$ so by using Jensen's inequality on log we get $\log(X_3 - X_1) - 1 = \log\left(\frac{X_3 - X_1}{2}\right) \geq \frac{\log(X_3 - X_2) + \log(X_2 - X_1)}{2}$. Thus $\mathbb{E}(2\log(X_3 - X_1)) - 2 \geq \mathbb{E}\log(X_3 - X_2) + \mathbb{E}\log(X_2 - X_1)$, so we must have at least one of $\mathbb{E}\log(X_3 - X_1) \geq \mathbb{E}\log(X_3 - X_2) + 1$ and $\mathbb{E}\log(X_3 - X_1) \geq \mathbb{E}\log(X_2 - X_1) + 1$. Assume without loss of generality that the first one is the case. As (X_1, X_2, X_3) has ϵ-indistinguishable 2-subsets we have $\delta((X_1, X_3), (X_2, X_3)) \leq \epsilon$ so Proposition 11 tell us that the log's must take values in an interval of length $\frac{1}{\epsilon}$. As the X_i's always differ by at least one, the log's only take non-negative values. Hence, $\log(X_3 - X_1) \geq \frac{1}{\epsilon}$ with positive probability, so $X_3 \geq X_3 - X_1 \geq 2^{\frac{1}{\epsilon}}$ with positive probability. ∎

In later proofs, we would like to ignore events that only happen with small probability, and argue that this does not increase the total variation distance between two random variables too much. We can do that using the next proposition.

Proposition 15. *Let X_1, X_2 and T be random variables with some joint distribution, where T only takes values 0 and 1 and $\Pr(T = 0) = \epsilon < 1$ and $\delta(X_1, X_2) = \delta$. Define $(X_1', X_2') = (X_1, X_2)|_{T=1}$. Then $\delta(X_1', X_2') \leq \frac{\delta + \epsilon}{1 - \epsilon}$.*

We will now consider the case $n = 4$. Before we show the lower bound, we will show that if X has ϵ-indistinguishable 3-subsets then (X_1, X_2, X_3, X_4) will with high probability be in one of two cases. Intuitively, one of these cases corresponds to the gaps $X_2 - X_1, X_3 - X_2, X_4 - X_3$ increasing and the other corresponds to the gaps decreasing.

Proposition 16. *Let X_1, X_2, X_3, X_4 be discrete random variables taking real values such that $X_1 < X_2 < X_3 < X_4$. Assume that $X = (X_1, X_2, X_3, X_4)$ has ϵ-indistinguishable 3-subsets. Then with probability at least $1 - 9\epsilon$ we have $\frac{X_3 - X_2}{X_4 - X_1} < \frac{1}{4}$ and one of the following:*

1. $X_3 < \frac{X_1 + X_4}{2}$ and $X_2 \le \frac{X_1 + X_3}{2}$, or
2. $X_2 > \frac{X_1 + X_4}{2}$ and $X_3 > \frac{X_2 + X_4}{2}$.

Proof. Define $f(a, b, c) = \frac{b-a}{c-a}\epsilon^{-1}$. Then $0 < f(X_1, X_2, X_4) < f(X_1, X_3, X_4) < \epsilon^{-1}$. From the assumption about X we have $\delta(X_{-3}, X_{-2}) \le \epsilon$, so Proposition 11 implies that $\mathbb{E}\left(f(X_1, X_3, X_4) - f(X_1, X_2, X_4)\right) < 1$, so $\mathbb{E}\frac{X_3 - X_2}{X_4 - X_1} < \epsilon$. In particular $\Pr\left(\frac{X_3 - X_2}{X_4 - X_1} \ge \frac{1}{4}\right) < 4\epsilon$. Define T to be the random variable that is $T = 1$ when $\frac{X_3 - X_2}{X_4 - X_1} < \frac{1}{4}$ and otherwise is 0. Let $X' = X|_{T=1}$. Now X' has $\frac{\Pr(T=0)+\epsilon}{\Pr(T=1)}$-indistinguishable 3-subset: to see e.g. that $\delta(X'_{-4}, X'_{-3}) \le \frac{\Pr(T=0)+\epsilon}{\Pr(T=1)}$ we use Proposition 15 on (X_{-4}, X_{-3}, T), and similar for all other pairs of 3-subsets.

Now define $g(a, b, c) = 1$ if $b > \frac{a+c}{2}$ and otherwise $g(a, b, c) = 0$. As $4(X'_3 - X'_2) \le X'_4 - X'_1$ we have $g(X'_{-4}) \ge g(X'_{-2}) \ge g(X'_{-3}) \ge g(X'_{-1})$. Here the middle inequality follows from $X'_3 > X'_2$. To show the first inequality, assume for contradiction that it is wrong for some particular value x of X'. Then we must have $g(x_{-2}) = 1$, so $x_3 > \frac{x_1 + x_4}{2}$. But that implies $\frac{x_3 - x_1}{x_4 - x_1} > \frac{1}{2}$ and as $\frac{x_3 - x_2}{x_4 - x_1} < \frac{1}{4}$ this implies $x_2 > \frac{x_1 + x_3}{2}$ and the first inequality is true. The last inequality is similar.

By Proposition 2 we know that $\delta(g(X'_{-4}), g(X'_{-1})) \le \frac{\Pr(T=0)+\epsilon}{\Pr(T=1)}$, so $\mathbb{E}(g(X'_{-4}) - g(X'_{-1})) \le \frac{\Pr(T=0)+\epsilon}{\Pr(T=1)}$. Because g only takes the values 0 and 1 and $g(X'_{-4}) \ge g(X'_{-1})$ we have $\Pr(g(X'_{-4}) \ne g(X'_{-1})) \le \frac{\Pr(T=0)+\epsilon}{\Pr(T=1)}$. Let T' be the random variable that is 0 when $T = 0$ or $g(X_{-4}) \ne g(X_{-1})$. We have

$$\Pr(T' = 0) = \Pr(T = 0) + \Pr(T = 1, g(X_{-4}) \ne g(X_{-1}))$$
$$\le \Pr(T = 0) + \Pr(T = 1)\frac{\Pr(T = 0) + \epsilon}{\Pr(T = 1)} \le 4\epsilon + 4\epsilon + \epsilon = 9\epsilon.$$

If $g(X_{-4}) = g(X_{-1}) = 0$ we have $X_3 \le \frac{X_2 + X_4}{2} < \frac{X_1 + X_4}{2}$ and $X_2 \le \frac{X_1 + X_3}{2}$ and we are in the first case of the conclusion of the proposition. Similarly, if $g(X_{-4}) = g(X_{-1}) = 1$ we have $X_2 > \frac{X_1 + X_3}{2} > \frac{X_1 + X_4}{2}$ and $X_3 > \frac{X_2 + X_4}{2}$.

Proposition 17 (Lower bound, $n = 4$). *Let X_1, X_2, X_3, X_4 be random variables over the non-negative real numbers such that $X_{i+1} \ge X_i + 1$ for $i \in \{1, 2, 3\}$ and let $\epsilon < \frac{1}{9}$. If $X = (X_1, X_2, X_3, X_4)$ has ϵ-indistinguishable 3-subsets, then X_4 must take values of at least $\exp_2^2\left(\frac{1 - 9\epsilon}{20\epsilon}\right)$.*

Proof. First we consider the case where we always have $X_3 \leq \frac{X_1+X_4}{2}$ and $X_2 \leq \frac{X_1+X_3}{2}$. From this we conclude that $X_4 - X_1 \geq 2(X_3 - X_1) \geq 4(X_2 - X_1)$. In other words $\log(X_4 - X_1) \geq \log(X_3 - X_1) + 1$ and $\log(X_3 - X_1) \geq \log(X_2 - X_1) + 1$. We claim that if X has ϵ-indistinguishable 3-subsets, then $(\log(X_2 - X_1), \log(X_3 - X_1), \log(X_4 - X_1))$ has ϵ-indistinguishable 2-subsets. To show e.g. that $\delta((\log(X_3 - X_1), \log(X_4 - X_1)), (\log(X_2 - X_1), \log(X_3 - X_1))) \leq \epsilon$ we define $f(x, y, z) = (\log(y - x), \log(z - x))$ and use Proposition 2 together with the assumption that $\delta((X_{-2}), (X_{-4})) \leq \epsilon$. Similar for all other pair of 2-subsets of $\{\log(X_2 - X_1), \log(X_3 - X_1), \log(X_4 - X_1)\}$. As the X_i's differ by one, the log's are always non-negative, and we have shown that they differ by one. Hence, by Proposition 14 $\log(X_4 - X_1)$ most take values of at least $2^{\frac{1}{\epsilon}}$. Thus, X_4 must take values of at least $\exp_2^2(\frac{1}{\epsilon})$.

This was assuming $X_3 \leq \frac{X_1+X_4}{2}$ and $X_2 \leq \frac{X_1+X_3}{2}$. If we instead assume $X_2 \geq \frac{X_1+X_4}{2}$ and $X_3 \geq \frac{X_2+X_4}{2}$ we can look at $\log(X_4 - X_3), \log(X_4 - X_2)$ and $\log(X_4 - X_1)$, and get the same result.

Next, suppose that we are only promised that for each value of X we are in one of those cases, but that it is not always the same of the two cases. Let I be a random variable that is 1 when we are in the case where the gaps increase and 0 in the case where the gaps decrease. Given X_{-i} we can see which case we are in, even if we do not know the value of i: we simply plug the three numbers into the function g from the proof of Proposition 16. Proposition 3 gives us $\delta(X_{-4}, X_{-1}) = \sum_{i=0}^{1} \Pr(I = i)\delta(X_{-4}|_{I=i}, X_{-1}|_{I=i})$, and similar for all other pairs for 3-subsets. There must be an i_0 such that $\Pr(I = i_0) \geq \frac{1}{2}$, so if X has ϵ-indistinguishable 3-subsets, then $X|_{I=i_0}$ must have 2ϵ-indistinguishable 3-subsets, and hence X_4 must takes some value of at least $\exp_2^2(\frac{1}{2\epsilon})$.

Finally, without any promises on X we know from Proposition 16 that with probability $1 - 9\epsilon$ one of the two requirement holds. Let T be a random variable that is 1 when one of these holds and 0 otherwise. Define $X' = X|_{T=1}$. Using Proposition 15 we can show that X' has $\frac{10\epsilon}{1-9\epsilon}$-indistinguishable 3-subsets. As X' always satisfy one of the two requirements, X'_4 (and hence X_4) must take values of at least $\exp_2^2\left(\left(2\frac{10\epsilon}{1-9\epsilon}\right)^{-1}\right) = \exp_2^2\left(\frac{1-9\epsilon}{20\epsilon}\right)$.

In the proof of a lower bound for general n, we can use Proposition 16 to argue that any four consecutive X_i will either have increasing or decreasing gaps. We can then use the following proposition to argue that all the gaps must be either increasing or decreasing.

Proposition 18. *Let $x_1 < \cdots < x_n$ be a sequence such that for all $i \in [n - 3]$ we have $\frac{x_{i+2}-x_{i+1}}{x_{i+3}-x_i} < \frac{1}{4}$ and one of the following two conditions holds:*

1. $x_{i+2} < \frac{x_i+x_{i+3}}{2}$ and $x_{i+1} \leq \frac{x_i+x_{i+2}}{2}$, or
2. $x_{i+1} > \frac{x_i+x_{i+3}}{2}$ and $x_{i+2} > \frac{x_{i+1}+x_{i+3}}{2}$.

Then it must be the same of the two conditions that holds for every i. If it is the first then $x_{i+1} - x_i \geq x_i - x_1$ for all $i \in \{2, \ldots, n-1\}$. If it is the second then $x_i - x_{i-1} \geq x_n - x_i$ for all $i \in \{2, \ldots n-1\}$.

Proof (Sketch). In the first case, gaps between consecutive x_js are increasing for $x_i, x_{i+1}, x_{i+2}, x_{i+3}$, and in the second they are decreasing. As $x_i, \ldots, x_{i+3}$ and $x_{i+1}, \ldots, x_{i+4}$ has 3 x_js, and hence two gaps between consecutive x_js, in common, it must be the same case that hold for i and $i + 1$, and by induction the same that hold for all i. The second part of the proposition follows from an easy computation.

Theorem 2 (Lower bound, $n \geq 4$). *Let $n \geq 4, \epsilon < \left(18^{n-3}(n-2)!\right)^{-1}$ and let $X_1, \ldots, X_n$ be random variables over the non-negative real numbers such that $X_{i+1} \geq X_i + 1$ for $i \in [n-1]$. If $X = (X_1, \ldots, X_n)$ has ϵ-indistinguishable $n-1$-subsets, then X_n must take values of at least $\exp_2^{n-2}\left(\left(18^{n-3}(n-2)!\epsilon\right)^{-1}\right)$.*

Proof. We show this by induction on n. The case $n = 4$ we know from Proposition 17 that X_4 must take values of at least $\exp_2^2\left(\frac{1-9\epsilon}{20\epsilon}\right)$. For $\epsilon < \frac{1}{18^{n-3}(n-2)!} = \frac{1}{36}$ we clearly have $\exp_2^2\left(\frac{1-9\epsilon}{20\epsilon}\right) \geq \exp_2^{n-2}\left(\left(18^{n-3}(n-2)!\epsilon\right)^{-1}\right)$.

Assume for induction that the theorem is true for $n - 1$. For each $i \in [n-3]$ consider $X_i, X_{i+1}, X_{i+2}, X_{i+3}$. If X has ϵ-indistinguishable $n-1$ subsets, then $(X_i, X_{i+1}, X_{i+2}, X_{i+3})$ has ϵ-indistinguishable 3-subsets: to show e.g. that $\delta((X_i, X_{i+1}, X_{i+2}), (X_{i+1}, X_{i+2}, X_{i+3})) < \epsilon$ we define f to be the function that given an $n - 1$-tuple returns the i'th, $i + 1$'th and $i + 2$'th element and use Proposition 2 together with the assumption that $\delta(X_{-(i+3)}, X_{-i}) \leq \epsilon$.

Define T_i to be 1 if $\frac{X_{i+2}-X_{i+1}}{X_{i+3}-X_i} < \frac{1}{4}$ and $X_{i+2} < \frac{X_i+X_{i+3}}{2}$ and $X_{i+1} \leq \frac{X_i+X_{i+2}}{2}$ and define T_i to be 2 if $\frac{X_{i+2}-X_{i+1}}{X_{i+3}-X_i} < \frac{1}{4}$ and $X_{i+1} > \frac{X_i+X_{i+3}}{2}$ and $X_{i+2} > \frac{X_{i+1}+X_{i+3}}{2}$ and define $T_i = 0$ otherwise. By Proposition 16, $\Pr(T_i = 0) \leq 9\epsilon$. We define T to be 1 if all the T_i's are 1, we define it to be 2 if all the T_i's are 2 and we define $T = 0$ otherwise. We know from Proposition 18 that "otherwise" only happens if one of the T_i are 0. So by Proposition 16 and the union bound $\Pr(T = 0) = \Pr(\exists i : T_i = 0) \leq (n-3)9\epsilon < 1$. Define $X' = X|_{T \neq 0}$. From Proposition 15 we conclude that X' has $\frac{\epsilon+(n-3)9\epsilon}{1-(n-3)9\epsilon}$-indistinguishable $n-1$-subsets. We must have $\Pr(T = t|T \neq 0) \geq \frac{1}{2}$ for some $t \in \{1, 2\}$. In the following we will assume that this is the case for $t = 1$, the proof for $t = 2$ is very similar. By the same argument as in the proof of Proposition 17 we see that, $X'|_{T=1}$ has $2\frac{\epsilon+(n-3)9\epsilon}{1-(n-3)9\epsilon}$-indistinguishable $n-1$-subsets, and as $\epsilon < \left(18^{n-3}(n-2)!\right)^{-1} < \frac{8}{81(n-2)(n-3)}$ a computation shows that we have $2\frac{\epsilon+(n-3)9\epsilon}{1-(n-3)9\epsilon} \leq 18(n-2)\epsilon$ so $X'|_{T=1}$ has $(18(n-2)\epsilon)$-indistinguishable $n-1$-subsets. As consecutive X_i''s always differ by at least 1, the random variables $\log(X_2' - X_1'), \log(X_3' - X_1'), \ldots, \log(X_n' - X_1')$ take non-negative values, and as $T = 1$ we know from Proposition 18 that consecutive $\log(X_i' - X_1')$'s always differ by at least 1. We have $18(n-2)\epsilon < \frac{18(n-2)}{18^{n-3}(n-2)!} = \frac{1}{18^{(n-1)-3}((n-1)-2)!}$, so by the induction hypothesis $\log(X_n' - X_1')$

must take values of at least $\exp_2^{n-3}\left(\left(18^{(n-1)-3}((n-1)-2)!\,(18(n-2)\epsilon)\right)^{-1}\right) = \exp_2^{n-3}\left(\left(18^{n-3}(n-2)!\epsilon\right)^{-1}\right)$ so X_n', and hence X_n, must take values of at least $\exp_2^{n-2}\left(\left(18^{n-3}(n-2)!\epsilon\right)^{-1}\right)$.

In the case $\Pr(T = 2|T \neq 0) > \frac{1}{2}$ we consider $\log(X_n - X_{n-1}), \log(X_n - X_{n-2}), \ldots, \log(X_n - X_1)$ instead of $\log(X_2 - X_1), \log(X_3 - X_1), \ldots, \log(X_n - X_1)$ but otherwise the proof is the same.

Putting the above together, we get our lower bound.

Theorem 3. *For fixed n there exists a distribution of $X = (X_1, \ldots, X_n)$ where $1 \le X_1 < X_2 < \cdots < X_n \le N(\epsilon)$ are all integers and X has ϵ-indistinguishable subsets and $N(\epsilon) = \exp_2^{n-2}(\Theta(\frac{1}{\epsilon}))$.*

5   Conclusion

We have shown that for any n and $\epsilon > 0$ there is a distribution of $(X_1, \ldots, X_n)$ with $1 \le X_1 < \cdots < X_n$ integers such that any two subsets of $\{X_1, \ldots, X_n\}$ of the same size are ϵ-indistinguishable. This could in theory be used to approximately time games that cannot be exactly timed. However, we have also proved a lower bound that shows that even for reasonable values of n and ϵ this is impossible in practice. For example for $n = 4$ and $\epsilon = \frac{1}{300}$ we would need to use values much larger than the universe's age in Planck times.

References

1. Aggarwal, G., Fiat, A., Goldberg, A.V., Hartline, J.D., Immorlica, N., Sudan, M.: Derandomization of auctions. In: Proceedings of the 37th ACM Symposium on Theory of Computing, pp. 619–625. ACM (2005)
2. Chandra, A.K., Furst, M.L., Lipton, R.J.: Multi-party protocols. In: Proceedings of the 15th ACM Symposium on Theory of Computing, pp. 94–99. ACM (1983)
3. Conway, J.H., Paterson, M.S., Moscow, U.S.S.R.: A headache-causing problem. In Een pak met een korte broek (1977). http://www.tanyakhovanova.com/BlogStuff/Conway/Headache.pdf
4. Jakobsen, S.K.: A numbers-on-foreheads game. arXiv, abs/1502.02849 (2015)
5. Jakobsen, S.K., Sørensen, T.B., Conitzer, V.: Timeability of extensive-form games. arxiv:1502.03430 (2015)
6. Krzywkowski, M.: On the hat problem, its variations, and their applications. Ann. Univ. Paedagog. Crac. Stud. Math. 9(1), 55–67 (2010)
7. Krzywkowski, M.: Hat problem on a graph. Ph.D. thesis, University of Exeter, April 2012
8. Lenstra Jr., H.W., Seroussi, G.: On hats and other covers. In: Proceedings IEEE International Symposium on Information Theory, p. 342. IEEE (2002)
9. Nisan, N., Roughgarden, T., Tardos, E., Vazirani, V.V.: Algorithmic Game Theory. Cambridge University Press, New York, NY, USA (2007)
10. Piccione, M., Rubinstein, A.: On the interpretation of decision problems with imperfect recall. Games Econ. Beha 20, 3–24 (1997)
11. Riis, S.: Information flows, graphs and their guessing numbers. Electr. J. Comb., 14(1), 17 p. (2007)

Faster Lightweight Lempel-Ziv Parsing

Dmitry Kosolobov[(✉)]

Ural Federal University, Ekaterinburg, Russia
dkosolobov@mail.ru

Abstract. We present an algorithm that computes the Lempel-Ziv decomposition in $O(n(\log \sigma + \log\log n))$ time and $n\log \sigma + \epsilon n$ bits of space, where ϵ is a constant rational parameter, n is the length of the input string, and σ is the alphabet size. The $n\log \sigma$ bits in the space bound are for the input string itself which is treated as read-only.

1 Introduction

The Lempel-Ziv decomposition [13] is a basic technique for data compression and plays an important role in string processing. It has several modifications used in various compression schemes. The decomposition considered in this paper is used in LZ77-based compression methods and in several compressed text indexes designed to efficiently store and search massive highly-repetitive data sets.

The standard algorithms computing the Lempel-Ziv decomposition work in $O(n\log \sigma)^1$ time and $O(n\log n)$ bits of space, where n is the length of the input string and σ is the alphabet size. It is known that this is the best possible time for the general alphabets [12]. However, for the most important case of integer alphabet, there exist algorithms working in $O(n)$ time and $O(n\log n)$ bits (see [7] for references). When σ is small, this number of bits is too big compared to the $n\log \sigma$ bits of the input string and can be prohibitive. To address this issue, several algorithms using $O(n\log \sigma)$ bits were designed.

Time	Bits of space	Note	Author(s)
$O(n\log \sigma)$	$O(n\log \sigma)$		Ohlebusch and Gog [15]
$O(n\log^3 n)$	$n\log \sigma + O(n)$	online	Okanohara and Sadakane [16]
$O(n\log^2 n)$	$O(n\log \sigma)$	online	Starikovskaya [18]
$O(n\log n)$	$O(n\log \sigma)$	online	Yamamoto et al. [19]
$O(n\log n\log\log \sigma)$	$n\log \sigma + \epsilon n$		Kärkkäinen et al. [10]
$O(n(\log \sigma + \log\log n))$	$n\log \sigma + \epsilon n$		this paper

The main contribution of this paper is a new algorithm computing the Lempel-Ziv decomposition in $O(n(\log \sigma + \log\log n))$ time and $n\log \sigma + \epsilon n$ bits

---

[1] Throughout the paper, log denotes the logarithm with the base 2.

© Springer-Verlag Berlin Heidelberg 2015
G.F. Italiano et al. (Eds.): MFCS 2015, Part II, LNCS 9235, pp. 432–444, 2015.
DOI: 10.1007/978-3-662-48054-0_36

of space, where ϵ is a constant rational parameter. The $n \log \sigma$ bits in the space bound are for the input string itself which is treated as read-only. The following table lists the time and space required by existing approaches to the Lempel-Ziv parsing in $O(n \log \sigma)$ bits of space.

By a more careful analysis, one can show that when ϵ is not a constant, the running time of our algorithm is $O(\frac{n}{\epsilon}(\log \sigma + \log \frac{\log n}{\epsilon}))$; we omit the details here.

Preliminaries. Let w be a string of length n. Denote $|w| = n$. We write $w[0], w[1], \ldots, w[n-1]$ for the letters of w and $w[i..j]$ for $w[i]w[i+1]\cdots w[j]$. A string can be *reversed* to get $\overleftarrow{w} = w[n-1]\cdots w[1]w[0]$ called the *reversed* w. A string u is a *substring* (or *factor*) of w if $u = w[i..j]$ for some i and j. The pair (i, j) is not necessarily unique; we say that i specifies an *occurrence* of u in w. A string can have many occurrences in another string. For $i, j \in \mathbb{Z}$, the set $\{k \in \mathbb{Z} \colon i \le k \le j\}$ is denoted by $[i..j]$; $[i..j)$ denotes $[i..j-1]$.

Throughout the paper, s denotes the input string of length n over the integer alphabet $[0..\sigma)$. Without loss of generality, we assume that $\sigma \le n$ and σ is a power of two. Thus, s occupies $n \log \sigma$ bits. Simplifying the presentation, we suppose that $s[0]$ is a special letter that is smaller than any letter in $s[1..n-1]$.

Our model of computation is the unit cost word RAM with the machine word size at least $\log n$ bits. Denote $r = \log_\sigma n = \frac{\log n}{\log \sigma}$. For simplicity, we assume that $\log n$ is divisible by $\log \sigma$. Thus, one machine word can contain a string of length $\le r$; we say that it is a *packed string*. Any substring of s of length r can be packed in a machine word in constant time by standard bitwise operations. Therefore, one can compare any two substrings of s of length k in $O(k/r + 1)$ time.

The *Lempel-Ziv decomposition of* s is the decomposition $s = z_1 z_2 \cdots z_l$ such that each z_i is either a letter that does not occur in $z_1 z_2 \cdots z_{i-1}$ or the longest substring that occurs at least twice in $z_1 z_2 \cdots z_i$ (e.g., $s = a \cdot b \cdot b \cdot abbabb \cdot c \cdot ab \cdot ab$). The substrings $z_1, z_2, \ldots, z_l$ are called the *Lempel-Ziv factors*. Our algorithm consecutively reports the factors in the form of pairs $(|z_i|, p_i)$, where p_i is either the position of a nontrivial occurrence of z_i in $z_1 z_2 \cdots z_i$ (it is called an *earlier occurrence of* z_i) or z_i itself if z_i is a letter that does not occur in $z_1 z_2 \cdots z_{i-1}$. The reported pairs are not stored in main memory.

Fix a rational constant $\epsilon > 0$. It suffices to prove that our algorithm works in $O(n(\log \sigma + \log \log n))$ time and $n \log \sigma + O(\epsilon n)$ bits: the substitution $\epsilon' = c\epsilon$, where c is the constant under the bit-O, gives the required $n \log \sigma + \epsilon' n$ bits with the same working time. We use different approaches to process the Lempel-Ziv factors of different lengths. In Sect. 2 we show how to process "short" factors of length $<r/2$. In Sect. 3 we describe new compact data structures that allow us to find all "medium" factors of length $<(\log n/\epsilon)^2$. In Sect. 4 we apply the clever technique of [5] for the analysis of all other "long" factors.

2   Short Factors

In this section we consider the Lempel-Ziv factors of length $< r/2$, so we assume $r \ge 2$. Suppose the algorithm has reported the factors $z_1, z_2, \ldots, z_{k-1}$ and now we process z_k. Denote $p = |z_1 z_2 \cdots z_{k-1}|$. We maintain arrays $H_1, H_2, \ldots, H_{\lceil r/2 \rceil}$ defined as follows: for $i \in [1..\lceil \frac{r}{2} \rceil]$, the array H_i contains σ^i integers such that for any $x \in [0..\sigma^i)$, either $H_i[x]$ equals the position from $[0..p)$ of an occurrence in s of the packed string x of length i or $H_i[x] = -1$ if there are no such positions.

For each $i \in [1..r]$ and $j \in [0..n]$, denote by x_i^j the packed string $s[j..j+i-1]$. We have $H_1[x_1^p] = -1$ iff z_k is a letter that does not appear in $s[0..p-1]$; in this case the algorithm reports z_k immediately. Further, we have $H_{\lceil r/2 \rceil}[x_{\lceil r/2 \rceil}^p] \neq -1$ iff $|z_k| \geq \frac{r}{2}$; this case is considered in Sects. 3 and 4. Suppose $H_1[x_1^p] \neq -1$ and $H_{\lceil r/2 \rceil}[x_{\lceil r/2 \rceil}^p] = -1$. Our algorithm finds the minimal $q \in [2..\lceil \frac{r}{2} \rceil]$ such that $H_q[x_q^p] = -1$. Then we obviously have $|z_k| = q-1$ and $H_{|z_k|}[x_{|z_k|}^p]$ is the position of an earlier occurrence of z_k. Clearly, the algorithm works in $O(|z_k|)$ time.

The inequality $r = \log n / \log \sigma \geq 2$ implies $\sigma \leq \sqrt{n}$. Thus, $H_1, H_2, \ldots, H_{\lceil r/2 \rceil}$ altogether occupy at most $\sigma^{\lceil r/2 \rceil} r \log n \leq \sigma^{\frac{r}{2}} \sigma^{\frac{1}{2}} r \log n \leq n^{\frac{3}{4}} r \log n = o(n)$ bits.

To maintain $H_1, \ldots, H_{\lceil r/2 \rceil}$, we consecutively examine the positions $j = 0, 1, \ldots, p-1$ and for those positions, for which $H_{\lceil r/2 \rceil}[x_{\lceil r/2 \rceil}^j] = -1$, we perform the assignments $H_1[x_1^j] \leftarrow j, H_2[x_2^j] \leftarrow j, \ldots, H_{\lceil r/2 \rceil}[x_{\lceil r/2 \rceil}^j] \leftarrow j$. Hence, we execute these assignments for at most $\sigma^{\lceil r/2 \rceil}$ positions and the overall time required for the maintenance of $H_1, \ldots, H_{\lceil r/2 \rceil}$ is $O(n + r\sigma^{\lceil r/2 \rceil}) = O(n)$.

3   Medium Factors

Suppose the algorithm has reported the Lempel-Ziv factors $z_1, z_2, \ldots, z_{k-1}$ and already decided that $|z_k| \geq \frac{r}{2}$ applying the procedure of Sect. 2. Denote $p = |z_1 z_2 \cdots z_{k-1}|$, $\tau = \lceil \frac{\log n}{\epsilon} \rceil$, and $b = \lceil \epsilon n / (\log \sigma + \log \log n) \rceil$. We assume $p+b+\tau^2 < n$; the case $p+b+\tau^2 \geq n$ is analogous. Our algorithm processes $s[0..p+b]$ and reports not only z_k but also all Lempel-Ziv factors starting in positions $[p..p+b]$.

The algorithm consists of three phases: the first one builds for other phases an indexing data structure on the string $s[p..p+b]$ in $O(b \log \sigma)$ time and $O(b(\log \sigma + \log \log n)) = O(\epsilon n)$ bits; the second phase scans $s[0..p+b]$ in $O(n)$ time and fills a bit array $lz[0..b]$ so that for any $i \in [0..b]$, $lz[i] = 1$ iff there is a Lempel-Ziv factor starting in the position $p+i$; finally, the last phase scans $s[0..p+b]$ in $O(n)$ time and reports earlier occurrences of the found Lempel-Ziv factors. Thus, the overall time required by this algorithm is $O((n + b \log \sigma) \frac{n}{b}) = O(n(\log \sigma + \log \log n))$.

The data structures we use can search only the Lempel-Ziv factors of length $< \tau^2$; we delegate the longer factors to the procedure of Sect. 4. This restriction allows us to make our structures fast and compact. More precisely, our algorithm consecutively computes the lengths of the Lempel-Ziv factors starting in $[p..p+b]$ and once we have found a factor of length $\geq \tau^2$, we invoke the procedure of Sect. 4 to compute the length and an earlier occurrence of this factor.

3.1   Main Tools

Let x be a string of length $d+1$. Denote $\overleftarrow{x}_i = \overleftarrow{x[0..i]}$. The *suffix array of* $\overleftarrow{x}$ is the permutation $SA[0..d]$ of the integers $[0..d]$ such that $\overleftarrow{x}_{SA[0]} < \overleftarrow{x}_{SA[1]} <$

$\ldots < \overline{x}_{SA[d]}$ in the lexicographical order. The *Burrows-Wheeler transform* [6] of $\overline{x}$ is the string $BWT[0..d]$ such that $BWT[i] = x[SA[i]+1]$ if $SA[i] < d$ and $BWT[i] = x[0]$ otherwise. We equip BWT with the function Ψ defined as follows: $\Psi(i) = SA^{-1}[SA[i] + 1]$ if $SA[i] < d$ and $\Psi(i) = 0$ otherwise.

Lemma 1 (see [9]). *The string BWT and the function Ψ for a string $\overline{x}$ of length $d+1$ over the alphabet $[0..\sigma)$ can be constructed in $O(d \log \log \sigma)$ time and $O(d \log \sigma)$ bits of space; Ψ is encoded in $O(d \log \sigma)$ bits with $O(1)$ access time.*

In the *dynamic weighted ancestor (WA for short)* problem one has (1) a weighted tree, where the weight of each vertex is greater than the weight of parent, (2) the queries finding for a vertex v and number i the ancestor of v with the minimal weight $\geq i$, (3) the updates inserting new vertices. Let v be a vertex of a trie T ($v \in T$ for short). Denote by $lab(v)$ the string written on the path from the root to v. We treat tries as weighted trees: $|lab(v)|$ is the weight of v.

Lemma 2 (see [11]). *For a weighted tree with at most k vertices, the dynamic WA problem can be solved in $O(k \log k)$ bits of space with queries and updates working in $O(\log k)$ amortized time.*

One can easily modify the proof of [11] for a special case of this problem when the weights are integers $[0..\tau^2]$ and the height of the tree is bounded by τ^2.

Lemma 3 (see [11]). *Let T be a weighted tree with at most $m \leq n$ vertices, the weights $[0..\tau^2]$, and the height $\leq \tau^2$. The dynamic WA problem for T can be solved in $O(m(\log m + \log \log n))$ bits of space with queries and updates working in $O(1)$ amortized time using a shared table of size $o(n)$ bits.*

Denote by $lcp(t_1, t_2)$ the length of the longest common prefix of the strings t_1 and t_2. Denote $rlcp(i, j) = \min\{\tau^2, lcp(x'_{SA[i]}, x'_{SA[j]})\}$.

Lemma 4 (see [2]). *For a string x of length $d+1$, using BWT of $\overline{x}$, one can compute an array $rlcp[0..d-1]$ such that $rlcp[i] = rlcp(i, i+1)$, for $i \in [0..d)$, in $O(d \log \sigma)$ time and $O(d \log \sigma)$ bits; the array occupies $O(d \log \log n)$ bits.*

3.2 Indexing Data Structure

Trie. Denote $d = 1+b+\tau^2$. The algorithm creates a string x of length $d+1$ and copies the string $s[p..p+b+\tau^2]$ in $x[1..d]$; $x[0]$ is set to a special letter less than any letter in $x[1..d]$. Let SA be the suffix array of $\overline{x}$ (we use it only conceptually). Denote $x'_i = x[i-\tau^2+1..i]$ (we assume that $x[-1], x[-2], \ldots$ are equal to $x[0]$). Here we discuss the design of our indexing data structure, a carefully packed in $O(d(\log \sigma + \log \log n))$ bits augmented compact trie of the strings $x'_0, x'_1, \ldots, x'_d$.

For simplicity, suppose d is a multiple of r. The skeleton of our structure is a compact trie Q_0 of the strings $\{x'_{SA[jr]} : j \in [0..d/r]\}$. We augment Q_0 with the WA structure of Lemma 3. Each vertex $v \in Q_0$ contains the following fields:

(1) the pointer to the parent of v (if any); (2) the pointers to the children of v in the lexicographical order; (3) the length of $lab(v)$; (4) the length of the string written on the edge connecting v to its parent (if any).

Notice that the fields (3)–(4) fit in $O(\log\log n)$ bits. Clearly, Q_0 occupies $O((d/r)\log n) = O(d\log\sigma)$ bits of space. The pointers to the substrings of x written on the edges of Q_0 are not stored, so, one cannot use Q_0 for searching.

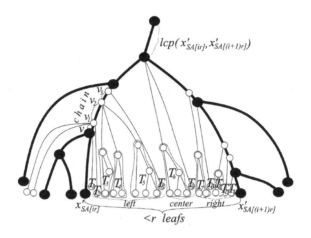

Fig. 1. Solid vertices and edges are from Q_0.

We create an array $L[0..d/r]$ such that for $i \in [0..d/r]$, $L[i]$ is the pointer to the leaf of Q_0 corresponding to $x'_{SA[ir]}$. Now we build a compact trie Q inserting the strings $\{x'_{SA[ir+j]}\}_{j=1}^{r-1}$ in Q_0 for each $i \in [0..d/r)$ as follows. For a fixed i, these strings add to Q_0 trees $T_1, \ldots, T_l$ attached to the branches $x'_{SA[ir]}$ and $x'_{SA[(i+1)r]}$ in Q_0 (see Fig. 1). We store $T_1, \ldots, T_l$ in a contiguous memory block F_i. The pointer to F_i is stored in the leaf of Q_0 corresponding to $x'_{SA[ir]}$, so, one can find F_i in $O(1)$ time using L. Since $T_1, \ldots, T_l$ have at most $2r$ vertices in total, $O(\log\log n)$ bits per vertex suffice for the fields 1)–4). Now we discuss how $T_1, \ldots, T_l$ are attached to Q_0. Consider $v \in Q_0$ and the vertices $v_1, \ldots, v_h$ splitting the edge connecting v to its parent in Q_0. Let $T_{i_1}, \ldots, T_{i_g}$ be the trees that must be attached to $v, v_1, \ldots, v_h$ (see Fig. 1). We add to v a memory block N_v containing the WA structure of Lemma 3 for the chain $v, v_1, \ldots, v_h$ with the weights $|lab(v)|, |lab(v_1)|, \ldots, |lab(v_h)|$. Each of the vertices $v, v_1, \ldots, v_h$ in this chain contains the $O(\log\log n)$-bit pointers (inside F_i) to the roots of $T_{i_1}, \ldots, T_{i_g}$ attached to this vertex. Hence, N_v occupies $O((h+g)\log\log n)$ bits. One can find the children for each of the vertices $v, v_1, \ldots, v_h$ in $O(1)$ time using Q_0 and the chain in the block N_v. Further, one can find, for any $j \in [1..g]$, the parent of the root of T_{i_j} in $O(1)$ time by a WA query on Q_0 to find a suitable v and a WA query on the chain in N_v. Finally, we augment each T_i with the WA structure of Lemma 3. Thus, by Lemma 3, $T_1, \ldots, T_l$ add at most $O(r\log\log n)$ bits to Q.

For each $i \in [0..d/r)$, we augment the leaf referred by $L[i]$ with an array $L_i[0..r-2]$ such that for $j \in [0..r-2]$, $L_i[j]$ is the $O(\log\log n)$-bit pointer (inside F_i)

to the leaf of Q corresponding to $x'_{SA[ir+1+j]}$. So, for any $j \in [0..d]$, one can easily find the leaf of Q corresponding to $x'_{SA[j]}$ in $O(1)$ time via L and $L_{\lfloor j/r \rfloor}$. Finally, the whole described structure Q occupies $O(d(\log \sigma + \log \log n))$ bits.

Prefix Links. Consider $v \in Q$. Denote by $[i_v..j_v]$ the longest segment such that for each $i \in [i_v..j_v]$, $x'_{SA[i]}$ starts with $lab(v)$ (see Fig. 2). Let BWT be the Burrows-Wheeler transform of $\tilde{x}$. Denote the set of the letters of $BWT[i_v..j_v]$ by P_v. We associate with v the *prefix links* mapping each $c \in P_v$ to an integer $p_v(c) \in [i_v..j_v]$ such that $x[SA[p_v(c)]+1] = c$ (there might be many such $p_v(c)$; we choose any). The prefix links correspond to the well-known *Weiner-links*. Hence, Q has at most $O(d)$ prefix links. Observe that $P_u \supset P_v$ for any ancestor u of v. The problem is to store the prefix links in $O(d(\log \sigma + \log \log n))$ bits.

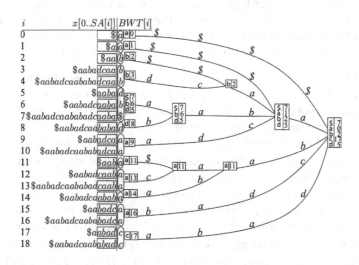

Fig. 2. $\tau^2 = 4$, the prefix links associated with vertices are in squares.

Fix $i \in [0..d)$. Denote by V_i the set of the vertices $v \notin Q_0$ such that v does not have descendants from Q_0 and lies between branches $x'_{SA[ir]}$ and $x'_{SA[(i+1)r]}$. We associate with each $v \in V_i$ a dictionary D_v mapping each $c \in P_v$ to $p_v(c) - ir$ and store all D_v, for $v \in V_i$, in a contiguous memory block H_i. Since $|V_i| < r$ and P_v is a subset of $BWT[ir..(i+1)r]$, we have $p_v(c) - ir \in [1..r)$ and all D_v, for $v \in V_i$, occupy overall $O(\sum_{v \in V_i} |P_v|(\log \sigma + \log \log n)) = O(r^2(\log \sigma + \log \log n))$ bits of space. Therefore, we can store in each $v \in V_i$ the $O(\log \log n)$-bit pointer to D_v (inside H_i). The pointer to H_i itself is stored in the leaf referred by $L[i]$.

Consider $v \notin Q_0$ such that v lies on an edge connecting a vertex $w \in Q_0$ to its parent in Q_0. Let $x'_{SA[j_1r]}$ and $x'_{SA[j_2r]}$ be the strings corresponding to the leftmost and rightmost descendant leaves of w contained in Q_0. We split P_v on three subsets: $P_1 = \{c \in P_v : p_v(c) < j_1r\}$, $P_2 = \{c \in P_v : p_v(c) > j_2r\}$, $P_3 = P_v \setminus (P_1 \cup P_2)$. Clearly $P_3 \subset P_w \subset P_v$. Hence, we can use P_w instead of P_3 and store only the sets P_1 and P_2 in a way similar to that discussed above.

Suppose $v \in Q_0$. Let for $c \in P_v$, $j_c \in [i_v..j_v]$ be the position of the first occurrence of c in $BWT[i_v..j_v]$. Clearly, we can set $p_v(c) = j_c$. We add to v a dictionary mapping each $c \in P_v$ to $h_c = |\{c' \in P_v : j_{c'} < j_c\}|$. Denote $q = |P_v|$. Since $q \le \sigma$, the dictionary occupies $O(q \log \sigma)$ bits. Now it suffices to map h_c to j_c. Let $j'_0, \ldots, j'_{q-1}$ denote all j_c, for $c \in P_v$, in increasing order. Obviously $j'_{h_c} = j_c$. The idea is to sample each $(\tau^2 \log n)$th position in BWT. We add to v a bit array $A_v[0..q-1]$ indicating the sampled $j'_0, \ldots, j'_{q-1}$: $A_v[0] = 1$ and for $h \in [1..q)$, $A_v[h] = 1$ iff $j'_{h-1} < l\tau^2 \log n \le j'_h$ for an integer l; A_v is equipped with the structure of [17] supporting the queries $\mathrm{r}_{A_v}(h) = \sum_{i=0}^{h} A_v[i]$ in $O(1)$ time and $o(q)$ additional bits. The sampled sequence $\{j'_h : A_v[h] = 1\}$ is stored in an array B_v. Finally, we add an array $C_v[0..q-1]$ such that $C_v[h] = j'_h - B_v[\mathrm{r}_{A_v}(h)-1]$. Now we map h to j'_h as follows: $j'_h = B_v[\mathrm{r}_{A_v}(h)-1] + C_v[h]$. Clearly, each value of C_v is in the range $[0..\tau^2 \log n]$ and hence, C_v occupies $O(q \log(\tau^2 \log n)) = O(q \log \log n)$ bits. It suffices to estimate the space consumed by B_v. Since the number of the vertices in Q_0 is $O(d/r)$ and the height of Q is at most τ^2, all B_v arrays occupy at most $O((d/r) \log n + \frac{d}{\tau^2 \log n} \tau^2 \log n) = O(d \log \sigma)$ bits in total.

Construction of Q. Initially, Q contains one leaf corresponding to $x'_{SA[0]}$. We consecutively insert $x'_{SA[1]}, \ldots, x'_{SA[d]}$ in Q in groups of r elements. During the construction, we maintain on Q a set of the dynamic WA structures of Lemma 3 in such a way that one can answer any WA query on Q in $O(1)$ time.

Suppose we have inserted $x'_{SA[0]}, \ldots, x'_{SA[ir]}$ in Q and now we are to insert $x'_{SA[ir+1]}, \ldots, x'_{SA[(i+1)r]}$. We first allocate the memory block F_i required for new vertices. Using Lemma 4, we compute $rlcp(j-1, j)$ for all $j \in [ir+1..(i+1)r]$. Since $rlcp(j_1, j_2) = \min\{rlcp(j_1, j_2-1), rlcp(j_2-1, j_2)\}$, the algorithm can compute $rlcp(ir, ir+j)$ for all $j \in [1..r]$ in $O(r)$ time. Using the WA query on the leaf $x'_{SA[ir]}$ and the value $rlcp(ir, (i+1)r)$, we find the position where we insert a new leaf $x'_{SA[(i+1)r]}$. Similarly, using the WA queries, we consecutively insert $x'_{SA[ir+j]}$ for $j = 1, 2, \ldots$ as long as $rlcp(ir, ir+j) > rlcp(ir, (i+1)r)$ and then all other $x'_{SA[(i+1)r-j]}$ for $j = 1, 2, \ldots$ (Fig. 1). All related WA structures, the arrays L, L_i, the pointers, and the fields for the vertices are built in an obvious way.

One can construct the prefix links of a vertex from those of its children in $O(q \log \sigma)$ time, where q is the number of the links in the children. As there are at most $O(d)$ prefix links, one DFS traverse of Q builds them in $O(d \log \sigma)$ time.

Finally, using the result of [8], the algorithm converts in $O(d \log \sigma)$ time all dictionaries in the prefix links of the resulting trie Q in the perfect hashes with $O(1)$ access time. So, one can access any prefix link in $O(1)$ time.

3.3 Algorithm for Medium Factors

In the *dynamic marked descendant problem* one has a tree, a set of marked vertices, the queries asking whether there is a marked descendant of a given vertex, and the updates marking a given vertex. We assume that each vertex is a descendant of itself. We solve this problem on Q as follows (see arXiv:1504.06712).

Lemma 5. *In $O(d(\log \sigma + \log \log n))$ bits one can solve the dynamic marked descendant problem on Q so that any k queries and updates take $O(k+d)$ time.*

Filling lz. Denote $s_i = s[0..i]$. Let for $i \in [0..p{+}d)$, t_i denotes the longest prefix of $\overleftarrow{s}_i$ presented in Q. We add to each $v \in Q$ an $O(\log \log n)$-bit field $v.mlen$ initialized to τ^2. Also, we use an integer variable f that initially equals 0.

The algorithm increases f computing $|t_f|$ in each step and augments Q as follows. Suppose $v \in Q$ is such that t_{f-1} is a prefix of $lab(v)$ and other vertices with this property are descendants of v. We say that v *corresponds to* t_{f-1}. We are to find the vertex of Q corresponding to t_f. Suppose $p_v(s[f])$ is defined. By Lemma 1, one can compute $i = \Psi(p_v(s[f]))$ in $O(1)$ time. Obviously, $x'_{SA[i]}$ starts with $s[f]t_{f-1}$. We obtain the leaf corresponding to $x'_{SA[i]}$ in $O(1)$ time via L and $L_{\lfloor i/r \rfloor}$ and then find $w \in Q$ corresponding to t_f by the WA query on the obtained leaf and the number $\min\{\tau^2, |t_{f-1}|{+}1\}$. Suppose $p_v(s[f])$ is undefined. If v is the root of Q, then we have $|t_f| = 0$. Otherwise, we recursively process the parent u of v in the same way as v assuming $t_{f-1} = lab(u)$. Finally, once we have found $w \in Q$ corresponding to t_f, we mark the parent of w using the structure of Lemma 5 and assign $w.mlen \leftarrow \min\{w.mlen, |lab(w)|{-}|t_f|\}$.

Let $i \in [p..f{+}1]$ such that $|s[i..f{+}1]| \leq \tau^2$. Suppose all positions $[0..f]$ are processed as described above. It is easy to verify that the string $s[i..f{+}1]$ has an occurrence in $s[0..f]$ iff either the vertex $v \in Q$ corresponding to $\overleftarrow{s[i..f{+}1]}$ has a marked descendant or the parent of v is marked and $|lab(v)| - v.mlen \geq |s[i..f{+}1]|$. Based on this observation, the algorithm computes lz as follows.

```
1: for (t ← p; t ≤ p + b; t ← t + max{1, z}) do
2:     for (z ← 0, v ← the root of Q; true; v ← w, z ← z + 1) do
3:         increase f processing Q accordingly until f = t + z − 1
4:         if z ≥ τ² then invoke the procedure of Sect. 4 to find z and break;
5:         find w ∈ Q corresp. to s[t..t+z] using v, prefix links, WA queries
6:         if w is undefined then break;
7:         if w do not have marked descendants then
8:             if parent(w) is not marked or |lab(w)| − w.mlen ≤ z then break;
9:     lz[t−p] ← 1;
```

The lengths of the Lempel-Ziv factors are accumulated in z. The above observation implies the correctness. Line 5 is similar to the procedure described above. Since $O(n)$ queries to the prefix links and $O(n)$ markings of vertices take $O(n)$ time, by standard arguments, one can show that the algorithm takes $O(n)$ time.

Searching of Occurrences. Denote by Z the set of all Lempel-Ziv factors of lengths $[r/2..\tau^2)$ starting in $[p..p{+}b)$. Obviously $|Z| = O(d/r)$. Using lz, we build in $O(d \log \sigma)$ time a compact trie R of the strings $\{\overleftarrow{z} : z \in Z\}$. We add to each $v \in R$ such that $z_v = \overleftarrow{lab(v)} \in Z$ the list of all starting positions of the Lempel-Ziv factors z_v in $[p..p{+}b)$. Obviously, R occupies $O((d/r) \log n) = O(d \log \sigma)$ bits. We construct for the strings Z a succinct Aho-Corasick automaton of [1] occupying $O((d/r) \log n) = O(d \log \sigma)$ bits. In [1] it is shown that the reporting states of the automaton can be associated with vertices of R, so that we can scan $s[0..p+d-1]$ in $O(n)$ time and store the found positions of the first occurrences of the strings Z in R. Finally, by a DFS traverse on R, we obtain for each string of Z the position of its first occurrence in $s[0..p+d-1]$. To find earlier occurrences of other Lempel-Ziv factors starting in $[p..p+b]$, we use the algorithms of Sects. 2 and 4.

4 Long Factors

4.1 Main Tools

Let $k \in \mathbb{N}$. A set $D \subset [0..k)$ is called a *difference cover* of $[0..k)$ if for any $x \in [0..k)$, there exist $y, z \in D$ such that $y - z \equiv x \pmod{k}$. Obviously $|D| \geq \sqrt{k}$. Conversely, for any $k \in \mathbb{N}$, there is a difference cover of $[0..k)$ with $O(\sqrt{k})$ elements and it can be constructed in $O(k)$ time (see [5]).

Lemma 6 (see [5]). *Let D be a difference cover of $[0..k)$. For any integers i, j, there exists $d \in [0..k)$ such that $(i - d) \bmod k \in D$ and $(j - d) \bmod k \in D$.*

An *ordered tree* is a tree whose leaves are totally ordered (e.g., a trie).

Lemma 7 (see [14]). *In $O(k \log k)$ bits of space we can maintain an ordered tree with at most k vertices under the following operations:*

1. *insertion of a new leaf (possibly splitting an edge) in $O(\log k)$ time;*
2. *searching of the leftmost/rightmost descendant leaf of a vertex in $O(\log k)$ time.*

Lemma 8 (see [3]). *A linked list can be designed to support the following operations:*

1. *insertion of a new element in $O(1)$ amortized time;*
2. *determine whether x precedes y for given elements x and y in $O(1)$ time.*

The ordinary searching in tries is too slow for our purposes, so, using packed strings and fast string dictionaries (see "*ternary trees*"), we improve our tries with the operations provided in the following lemma.

Lemma 9 (see arXiv:1504.06712). *In $O(k \log n)$ bits of space we can maintain a compact trie for at most k substrings of s under the following operations:*

1. *insertion of a string w in $O(|w|/r + \log n)$ amortized time;*
2. *searching of a string w in $O(|u|/r + \log n)$ time, where u is the longest prefix of w present in the trie; we scan w from left to right r letters at a time and report the vertices of the trie corresponding to the prefixes of lengths $r, 2r, \ldots, \lfloor |u|/r \rfloor r$, and $|u|$ immediately after reading these prefixes.*

In the *dynamic tree range reporting problem* one has ordered trees T_1 and T_2 and a set of pairs $Z = \{(x_1^i, x_2^i)\}$, where x_1^i and x_2^i are leaves of T_1 and T_2, respectively (see Fig. 3); the query asks, for given vertices $v_1 \in T_1$ and $v_2 \in T_2$, to find a pair $(x_1, x_2) \in Z$ such that x_1 and x_2 are descendants of v_1 and v_2, respectively; the update inserts new pairs in Z or new vertices in T_1 and T_2. To solve this problem, we apply the structure of [4] and Lemmas 7 and 8.

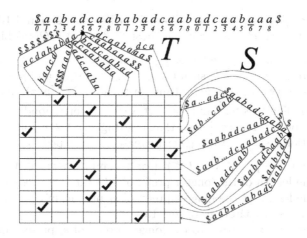

Fig. 3. $\tau = 3$, $D = \{0, 1, 3, 6\}$ is a diff. cover of $[0..\tau^2)$, positions in M are underlined.

Lemma 10 (see arXiv:1504.06712). *The dynamic tree range reporting problem with $|Z| \leq k$ can be solved in $O(k \log k)$ bits of space with updates and queries working in $O(\log k)$ amortized time.*

4.2 Algorithm for Long Factors

Data Structures. At the beginning, using the algorithm of [5], our algorithm constructs a difference cover D of $[0..\tau^2)$ such that $|D| = \Theta(\tau)$. Denote $M = \{i \in [0..n): i \bmod \tau^2 \in D\}$. The set M is the basic component in our constructions.

Suppose the algorithm has reported the Lempel-Ziv factors $z_1, z_2, \ldots, z_{k-1}$ and already decided that $|z_k| \geq \tau^2$ applying the procedure of Sect. 3. Denote $p = |z_1 z_2 \cdots z_{k-1}|$. We use an integer variable z to compute the length of $|z_k|$ and z is initially equal to τ^2. Let us first discuss the related data structures.

We use an auxiliary variable t such that $p \leq t < p + z$ at any time of the work; initially $t = p$. Denote $s_i = s[0..i]$. Our main data structures are compact tries S and T: S contains the strings $\overleftarrow{s_i}$ and T contains the strings $s[i+1..i+\tau^2]$ for all $i \in [0..t) \cap M$ (we append τ^2 letters $s[0]$ to the right of s so that $s[i+1..i+\tau^2]$ is always defined). Both S and T are augmented with the structures supporting the searching of Lemma 9 and the tree range queries of Lemma 10 on pairs of leaves of S and T. Since $s[0]$ is a sentinel letter, each $\overleftarrow{s_i}$, for $i \in [0..t) \cap M$, is represented in S by a leaf. The set of pairs for our tree range reporting structure contains the pairs of leaves corresponding to $\overleftarrow{s_i}$ in S and $s[i+1..i+\tau^2]$ in T for all $i \in [0..t) \cap M$ (see Fig. 3). Also, we add to S the WA structure of Lemma 2.

Let us consider vertices $v \in S$ and $v' \in T$ corresponding to strings $\overleftarrow{t}_v$ and $t_{v'}$, respectively. Denote by treeRng(v, v') the tree range query that returns either **nil** or a suitable pair of descendant leaves of v and v'. We have treeRng$(v, v') \neq$ **nil** iff there is $i \in [0..t) \cap M$ such that $s[i-|t_v|+1..i]s[i+1..i+|t_{v'}|] = \overleftarrow{t}_v t_{v'}$.

Since $|M| \le \frac{n}{\tau^2}|D| = O(\frac{n}{\tau})$, it follows from Lemmas 2, 9 and 10 that S and T with all related structures occupy at most $O(\frac{n}{\tau}\log n) = O(\epsilon n)$ bits.

The Algorithm. Suppose the factor z_k occurs in a position $x \in [0..p)$; then, by Lemma 6, there is a $d \in [0..\tau^2)$ such that $x + |z_k| - d \in M$ and $p + |z_k| - d \in M$. Based on this observation, our algorithm, for each $t \in M \cap [p..z)$, finds the vertex $v \in S$ corresponding to $\overleftarrow{s[p..t]}$ and the vertex $v' \in T$ corresponding to as long as possible prefix of $s[t+1..n+\tau^2]$ such that $\text{treeRng}(v, v') \ne \textbf{nil}$ and with the aid of this bidirectional search, we further increase z if it is possible.

1: **for** $(t \leftarrow \min\{i \ge p: i \in M\};\ t < p + z;\ t \leftarrow \min\{i > t: i \in M\})$ **do**
2:     $x \leftarrow$ the length of the longest prefix of $s[t+1..t+\tau^2]$ present in T
3:     $y \leftarrow$ the length of the longest prefix of $\overleftarrow{s}_t$ present in S
4:     **if** $y < t - p + 1$ **then** go to line 13
5:     $v \leftarrow$ the vertex corresp. to the longest prefix of $\overleftarrow{s}_t$ present in S
6:     $v \leftarrow \text{weiAnc}(v, t - p + 1)$;
7:     **for** $j = t, t+r, t+2r, \ldots, t+\lfloor x/r \rfloor r, x$ and $v' \in T$ corresp. to $s[t+1..j]$ **do**
8:         **if** $j \ge p + z$ **then**   ▷ $|s[p..j]| > |s[p..p+z-1]|$
9:             **if** $\text{treeRng}(v, v') = \textbf{nil}$ **then**
10:                 $j \leftarrow \max\{j': \text{treeRng}(v, u) \ne \textbf{nil}$ for $u \in T$ corresp. $s[t+1..j']\}$;
11:             $z \leftarrow \max\{z, j - p + 1\}$;
12:             **if** $\text{treeRng}(v, v') = \textbf{nil}$ **then break**;
13:     insert $s[t+1..t+\tau^2]$ in T, $\overleftarrow{s}_t$ in S; process the pair of the corresp. leaves

Some lines need further clarification. Here $\text{weiAnc}(v, i)$ denotes the WA query that returns either the ancestor of v with the minimal weight $\ge i$ or **nil** if there is no such ancestor; we assume that any vertex is an ancestor of itself. Since M has period τ^2, one can compute, for any t, $\min\{i > t: i \in M\}$ in $O(1)$ time using an array of length τ^2 for example. The operations on T in lines 2 and 13 take, by Lemma 9, $O(\tau^2/r + \log n)$ time. To perform the similar operations on S in lines 3, 5 and 13, we use other techniques (discussed below) working in the same time. The loop in line 7 executes exactly the procedure described in Lemma 9. To compute j in line 10, we perform the binary search on at most r ancestors of the vertex v'; thus, we invoke $\text{treeRng}\ O(\log r)$ times in line 10.

Let us prove the correctness. Suppose we have $\tau^2 \le z < |z_k|$ in some iteration. It suffices to show that the algorithm cannot terminate with this value of z. Let z_k occur in a position $x \in [0..p)$. By Lemma 6, there is a $d \in [0..\tau^2)$ such that $x + z - d \in M$ and $p + z - d \in M$. Thus, the string $s[p..p+z-d]$ is presented in S when $t = p + z - d$ and we find the corresponding vertex v in line 6. Moreover, the string $s[p+z-d+1..p+z]$ is presented in T and we find the vertex corresponding to this or a longer string in the loop 7–12. Denote this vertex by w; w is either v' or u in line 10. Obviously, $\text{treeRng}(v, w) \ne \textbf{nil}$, so, we increase z in line 11.

Let us estimate the running time. The main loop performs $O(|z_k|/\tau)$ iterations. The operations in lines 2, 3, 5 and 13 require, as mentioned above, $O(\tau^2/r + \log n)$ time (some of them will be discussed in the sequel). One WA query and one modification of the tree range reporting structure take, by Lemmas 2 and 10, $O(\log n)$ time. By Lemma 9, the traverse of T in line 7 requires

$O(\tau^2/r + \log n)$ time. For each fixed t, every time we perform treeRng query in line 9, except probably for the first and last queries, we increase z by r. Hence, the algorithm executes at most $O(|z_k|/\tau + |z_k|/r)$ such queries in total. Finally, in line 10 we invoke treeRng at most $O(\log r)$ times for every fixed t. Putting everything together, we obtain $O(\frac{|z_k|}{\tau}(\tau^2/r + \log n) + \frac{|z_k|}{r}\log n + \frac{|z_k|}{\tau}\log r \log n) = O(|z_k|\log \sigma + |z_k|\log r) = O(|z_k|(\log \sigma + \log \log n))$ overall time.

One can find the position of an early occurrence of z_k from the pairs of leaves reported in lines 9 and 10. Now let us discuss how to insert and search strings in S.

Operations on S. The operations on S are based on the fact that for any $i \in [\tau^2..n) \cap M$, $i - \tau^2 \in M$. Let u and v be leaves of S corresponding to some $\bar{s}_j$ and $\bar{s}_k$. To compare $\bar{s}_j$ and $\bar{s}_k$ in $O(1)$ time via u and v, we store all leaves of S in a linked list K of Lemma 8 in the lexicographical order. To calculate $lcp(\bar{s}_j, \bar{s}_k)$ in $O(\log n)$ time via u and v, we put all leaves of S in an augmented search tree B. Finally, we augment S with the ordered tree structure of Lemma 7.

Denote $s'_i = \overleftarrow{s[i-\tau^2+1..i]}$. We add to S a compact trie S' containing s'_i for all $i \in [0..t) \cap M$ (we assume $s[0]=s[-1]=\ldots$, so, S' is well-defined). The vertices of S' are linked to the respective vertices of S. Let w be a leaf of S' corresponding to a string s'_i. We add to w the set $H_w = \{(p_1^j, p_2^j): j \in [0..t) \cap M \text{ and } s'_j = s'_i\}$, where p_1^j and p_2^j are the pointers to the leaves of S corresponding to $\overleftarrow{s}_{j-\tau^2}$ and $\bar{s}_j$, respectively; H_w is stored in a search tree in the lexicographical order of the strings $\overleftarrow{s}_{j-\tau^2}$ referred by p_1^j, so, one can find, for any $k \in [0..t+\tau^2) \cap M$, the predecessor or successor of the string $\overleftarrow{s}_{k-\tau^2}$ in H_w in $O(\log n)$ time. It is straightforward that all these structures occupy $O(\frac{n}{\tau}\log n) = O(\epsilon n)$ bits.

Suppose S contains $\bar{s}_i$ for all $i \in [0..t) \cap M$ and we insert $\bar{s}_t$. We first search s'_t in S'. Suppose S' does not contain s'_t. We insert s'_t in S' in $O(\tau^2/r + \log n)$ time, by Lemma 9, then add to S the vertices corresponding to the new vertices of S' and link them to each other. Using the structure of Lemma 7 on S, we find the position of $\bar{s}_t$ in K in $O(\log n)$ time. All other structures are easily modified in $O(\log n)$ time. Now suppose S' has a vertex w corresponding to s'_t. In $O(\log n)$ time we find in H_w the pairs (p_1^j, p_2^k) and (p_1^j, p_2^k) such that p_1^j points to the predecessor $\overleftarrow{s}_{j-\tau^2}$ of $\overleftarrow{s}_{t-\tau^2}$ in H_w and p_1^k points to the successor $\overleftarrow{s}_{k-\tau^2}$. So, the leaf corresponding to $\bar{s}_t$ must be between $\bar{s}_j$ and $\bar{s}_k$. Using B, we calculate $lcp(\bar{s}_j, \bar{s}_t) = lcp(\bar{s}_{j-\tau^2}, \bar{s}_{t-\tau^2}) + \tau^2$ and, similarly, $lcp(\bar{s}_k, \bar{s}_t)$ in $O(\log n)$ time and then find the position where to insert the new leaf by WA queries on S. All other structures are simply modified in $O(\log n)$ time. Thus, the insertion takes $O(\tau^2/r + \log n)$ time. One can use a similar algorithm for the searching of $\bar{s}_t$.

References

1. Belazzougui, D.: Succinct dictionary matching with no slowdown. In: Amir, A., Parida, L. (eds.) CPM 2010. LNCS, vol. 6129, pp. 88–100. Springer, Heidelberg (2010)

2. Beller, T., Gog, S., Ohlebusch, E., Schnattinger, T.: Computing the longest common prefix array based on the Burrows-Wheeler transform. J. Discrete Algorithms **18**, 22–31 (2013)
3. Bender, M.A., Cole, R., Demaine, E.D., Farach-Colton, M., Zito, J.: Two simplified algorithms for maintaining order in a list. In: Möhring, R.H., Raman, R. (eds.) ESA 2002. LNCS, vol. 2461, pp. 152–164. Springer, Heidelberg (2002)
4. Blelloch, G.E.: Space-efficient dynamic orthogonal point location, segment intersection, and range reporting. In: SODA 2008. pp. 894–903. SIAM (2008)
5. Burkhardt, S., Kärkkäinen, J.: Fast lightweight suffix array construction and checking. In: Baeza-Yates, R., Chávez, E., Crochemore, M. (eds.) CPM 2003. LNCS, vol. 2676, pp. 55–69. Springer, Heidelberg (2003)
6. Burrows, M., Wheeler, D.J.: A block-sorting lossless data compression algorithm. Technical report 124 (1994)
7. Fischer, J., I, T., Köppl, D.: Lempel Ziv Computation In Small Space (LZ-CISS). In: Cicalese, F., Porat, E., Vaccaro, U. (eds.) CPM 2015. LNCS, vol. 9133, pp. 172–184. Springer, Heidelberg (2015)
8. Hagerup, T., Miltersen, P.B., Pagh, R.: Deterministic dictionaries. J. Algorithms **41**(1), 69–85 (2001)
9. Hon, W.K., Sadakane, K., Sung, W.K.: Breaking a time-and-space barrier in constructing full-text indices. In: FOCS 2003. pp. 251–260. IEEE (2003)
10. Kärkkäinen, J., Kempa, D., Puglisi, S.J.: Lightweight lempel-ziv parsing. In: Demetrescu, C., Marchetti-Spaccamela, A., Bonifaci, V. (eds.) SEA 2013. LNCS, vol. 7933, pp. 139–150. Springer, Heidelberg (2013)
11. Kopelowitz, T., Lewenstein, M.: Dynamic weighted ancestors. In: SODA 2007. pp. 565–574. SIAM (2007)
12. Kosolobov, D.: Lempel-Ziv factorization may be harder than computing all runs. In: STACS 2015. LIPICS, vol. 30, pp. 582–593 (2015)
13. Lempel, A., Ziv, J.: On the complexity of finite sequences. IEEE Trans. Inf. Theor. **22**(1), 75–81 (1976)
14. Navarro, G., Sadakane, K.: Fully functional static and dynamic succinct trees. ACM Trans. Algorithms (TALG) **10**(3), 16 (2014)
15. Ohlebusch, E., Gog, S.: Lempel-ziv factorization revisited. In: Giancarlo, R., Manzini, G. (eds.) CPM 2011. LNCS, vol. 6661, pp. 15–26. Springer, Heidelberg (2011)
16. Okanohara, D., Sadakane, K.: An online algorithm for finding the longest previous factors. In: Halperin, D., Mehlhorn, K. (eds.) ESA 2008. LNCS, vol. 5193, pp. 696–707. Springer, Heidelberg (2008)
17. Raman, R., Raman, V., Rao, S.S.: Succinct indexable dictionaries with applications to encoding k-ary trees and multisets. In: SODA 2002. pp. 233–242. SIAM (2002)
18. Starikovskaya, T.: Computing lempel-ziv factorization online. In: Rovan, B., Sassone, V., Widmayer, P. (eds.) MFCS 2012. LNCS, vol. 7464, pp. 789–799. Springer, Heidelberg (2012)
19. Yamamoto, J., I, T., Bannai, H., Inenaga, S., Takeda, M.: Faster compact on-line Lempel-Ziv factorization. In: STACS 2014. LIPICS, vol. 25, pp. 675–686 (2014)

Parallel Identity Testing for Skew Circuits with Big Powers and Applications

Daniel König[(✉)] and Markus Lohrey

Universität Siegen, Siegen, Germany
{koenig,lohrey}@eti.uni-siegen.de

Abstract. Powerful skew arithmetic circuits are introduced. These are skew arithmetic circuits with variables, where input gates can be labelled with powers x^n for binary encoded numbers n. It is shown that polynomial identity testing for powerful skew arithmetic circuits belongs to coRNC^2, which generalizes a corresponding result for (standard) skew circuits. Two applications of this result are presented: (i) Equivalence of higher-dimensional straight-line programs can be tested in coRNC^2; this result is even new in the one-dimensional case, where the straight-line programs produce strings. (ii) The compressed word problem (or circuit evaluation problem) for certain wreath products belongs to coRNC^2. Full proofs can be found in the long version [13].

1 Introduction

Polynomial Identity Testing (PIT) is the following computational problem: The input is an arithmetic circuit, built up from addition gates, multiplication gates, and input gates that are labelled with variables $(x_1, x_2, \ldots)$ or constants $(-1, 0, 1)$, and it is asked whether the output gate evaluates to the zero polynomial (in this paper, we always work in the polynomial ring over the coefficient ring $\mathbb{Z}$ or $\mathbb{Z}_n$ for $n \geq 2$). Based on the Schwartz-Zippel-DeMillo-Lipton Lemma, Ibarra and Moran [9] proved that PIT over $\mathbb{Z}$ or $\mathbb{Z}_p$ belongs to the class coRP (the complements of problems in randomized polynomial time). Whether there is a deterministic polynomial time algorithm for PIT is an important problem. In [10] it is shown that if there exists a language in $\mathsf{DTIME}(2^{\mathcal{O}(n)})$ that has circuit complexity $2^{\Omega(n)}$, then $\mathsf{P} = \mathsf{BPP}$ (and hence $\mathsf{P} = \mathsf{RP} = \mathsf{coRP}$). There is also an implication that goes the other way round: Kabanets and Impagliazzo [11] have shown that if $\mathsf{PIT} \in \mathsf{P}$, then (i) there is a language in $\mathsf{NEXPTIME}$ that does not have polynomial size circuits, or (ii) the permanent is not computable by polynomial size arithmetic circuits. Both conclusions represent major open problems in complexity theory. This indicates that derandomizing PIT will be very difficult. On the other hand, for arithmetic formulas (where the circuit is a tree) and skew arithmetic circuits (where for every multiplication gate, one of the two input gates is a constant or a variable), PIT belongs to coRNC (but it is still not known to be in P), see [11, Corollary 2.1]. This holds, since arithmetic formulas and skew arithmetic circuits can be evaluated in NC if the variables are

© Springer-Verlag Berlin Heidelberg 2015
G.F. Italiano et al. (Eds.): MFCS 2015, Part II, LNCS 9235, pp. 445–458, 2015.
DOI: 10.1007/978-3-662-48054-0_37

substituted by concrete (binary coded) numbers. Then, as for general PIT, the Schwartz-Zippel-DeMillo-Lipton Lemma yields a coRNC-algorithm.

In this paper, we identify a larger class of arithmetic circuits, for which polynomial identity testing is still in coRNC; we call these circuits *powerful skew circuits*. In such a circuit, we require that for every multiplication gate, one of the two input gates is either a constant or a power x^N of a variable x, where the exponent N is given in binary notation. One can replace this power x^N by a subcircuit of size $\log N$ using iterated squaring, but the resulting circuit is no longer skew. The main result of this paper states that PIT for powerful skew circuits over the rings $\mathbb{Z}[x]$ and $\mathbb{Z}_p[x]$ (p prime) is still in coRNC (in fact, coRNC^2). For this, we use an identity testing algorithm of Agrawal and Biswas, [1] which computes the output polynomial of the circuit modulo a polynomial $q(x)$ of polynomially bounded degree, which is randomly chosen from a certain sample space. Moreover, in our application, all computations can be done in the ring $\mathbb{F}_p[x]$ for a prime number p of polynomial size. This allows us to compute the big powers x^N modulo $q(x)$ in NC^2 using an algorithm of Fich and Tompa [5]. It should be noted that the application of the Agrawal-Biswas algorithm is crucial. If, instead we would use the Schwartz-Zippel-DeMillo-Lipton Lemma, then we would be forced to compute $a^N \bmod m$ for randomly chosen numbers a and m with polynomially many bits. Whether this problem (modular powering) belongs to NC is an open problem [6, Problem B.5.6].

We present two applications of our coRNC identity testing algorithm. The first one concerns the equivalence problem for straight-line programs. Here, a straight-line program (SLP) is a context-free grammar G that computes a single word val(G). In this context, SLPs are extensively used in data compression and algorithmics on compressed data, see [15] for an overview. Equivalence for SLPs, i.e., the question whether val(G) = val(H) for given SLPs G, H, can be decided in polynomial time. This result was independently discovered in [8,18,20]. All known algorithms for SLP-equivalence are sequential and it is not clear how to parallelize them. Here, we exhibit an NC^2-reduction from SLP-equivalence to PIT for skew powerful circuits. Hence, equivalence for SLPs belongs to coRNC. Moreover, our reduction immediately generalizes to higher dimensional pictures for which SLPs can be defined in a fashion similar to the one-dimensional (string) case, using one concatenation operation in each dimension. For two-dimensional SLPs, Berman et al. [3] proved that equivalence belongs to coRP using a reduction to PIT. We improve this result to coRNC. Whether equivalence of two-dimensional (resp., one-dimensional) SLPs belongs to P (resp., NC) is open.

Our second application concerns the compressed word problem for groups. Let G be a finitely generated (f.g.) group, and let Σ be a finite generating set for G. For the compressed word problem for G, briefly CWP(G), the input is an SLP (as defined in the preceding paragraph) over the alphabet $\Sigma \cup \Sigma^{-1}$, and it is asked whether val(G) evaluates to the group identity. The compressed word problem is a succinct version of the classical word problem (Does a given word over $\Sigma \cup \Sigma^{-1}$ evaluate to the group identity?). One of the main motivations for the compressed word problem is the fact that the classical word problem for certain groups (automorphism groups, group extensions) can be reduced to the compressed word problem for simpler groups [16, Sect. 4.2]. For finite groups (and monoids) the compressed word problem was studied in Beaudry et al. [2], and for infinite groups the problem was studied for the first time in [14]. Subsequently, several important classes of f.g. groups with polynomial time compressed word problems were found: f.g. nilpotent groups, graph groups (also known as right-angled Artin groups), and virtually special groups. The latter include all Coxeter groups, one-relator groups with torsion, fully residually free groups, and fundamental groups of hyperbolic 3-manifolds; see [16]. For f.g. linear groups, i.e., f.g. groups of matrices over a field, the compressed word problem reduces to PIT (over $\mathbb{Z}$ or $\mathbb{Z}_p$, depending on the characteristic of the field) and hence belongs to coRP [16, Theorem 4.15]. Recently it was shown in [12] that the compressed word problem for a f.g. nilpotent group belongs to NC^2 (the result can be extended to so called f.g. nilpotent by finite solvable groups). Here, we prove that the compressed word problem for the wreath product $\mathbb{Z} \wr \mathbb{Z}$ is equivalent w.r.t. NC^2-reductions to PIT for powerful skew circuits. In particular, $\mathsf{CWP}(\mathbb{Z} \wr \mathbb{Z})$ belongs to coRNC. This result generalizes to wreath products $G \wr \mathbb{Z}^n$, where G is a direct product of copies of $\mathbb{Z}$ and $\mathbb{Z}_p$ for primes p. In contrast, it was shown in [16, Theorem 4.21] that $\mathsf{CWP}(G \wr \mathbb{Z})$ is coNP-hard for every non-abelian group G.

2   Background from Complexity Theory

Recall that RP is the set of all problems A for which there exists a polynomial time bounded randomized Turing machine R such that: (i) if $x \in A$ then R accepts x with probability at least $1/2$, and (ii) if $x \notin A$ then R accepts x with probability 0. The class coRP is the class of all complements of problems from RP.

We use standard definitions concerning circuit complexity, see e.g. [21] for more details. In particular we will consider the class NC^i of all problems that can be solved by a polynomial size circuit family of depth $O(\log^i n)$ that uses NOT-gates and AND-gates and OR-gates of fan-in two. The class NC is the union of all classes NC^i. All circuit families in this paper will be logspace-uniform, which means that the n-th circuit in the family can be computed in logspace from the unary encoding of n.

To define a randomized version of NC^i, one uses circuit families with additional inputs. So, let the n-th circuit $\mathcal{C}_n$ in the family have n normal input gates plus m random input gates, where m is polynomially bounded in n. For an input $x \in \{0,1\}^n$, its acceptance probability is $\mathsf{Prob}[\mathcal{C}_n \text{ accepts } x] = 2^{-m} \cdot |\{y \in \{0,1\}^m \mid \mathcal{C}_n(x,y) = 1\}|$. Here, $\mathcal{C}_n(x,y) = 1$ means that the circuit $\mathcal{C}_n$ evaluates to 1 if the i-th normal input gate gets the i-th bit of the input string x, and the i-th random input gate gets the i-th bit of the random string y. Then, the class RNC^i is the class of all problems A for which there exists a polynomial size circuit family $(\mathcal{C}_n)_{n \geq 0}$ of depth $O(\log^i n)$ with random input gates that uses NOT-gates and AND-gates and OR-gates of fan-in two, such that for all inputs $x \in \{0,1\}^*$ of length n: (i) if $x \in A$, then $\mathsf{Prob}[\mathcal{C}_n \text{ accepts } x] \geq 1/2$, and (ii) if $x \notin A$, then $\mathsf{Prob}[\mathcal{C}_n \text{ accepts } x] = 0$. As usual, coRNC^i is the class of all complements of problems from RNC^i. Section B.9 in [6] contains several problems that are known to be in RNC, but not known to be in NC; the most prominent example is the existence of a perfect matching in a graph.

3 Polynomials and Circuits

We deal with multivariate polynomial rings $R[x_1, \ldots, x_k]$, where R is the ring of integers $\mathbb{Z}$ or the ring $\mathbb{Z}_n$ of integers modulo $n \geq 2$. For computational problems, we distinguish between two representations of polynomials. Consider the polynomial

$$p(x_1, \ldots, x_k) = \sum_{1 \leq i \leq l} a_i x_1^{e_{i,1}} \cdots x_k^{e_{i,k}}$$

The *standard representation* of $p(x)$ is the sequence of tuples $(a_i, e_{i,1}, \ldots, e_{i,k})$, where the coefficient a_i is represented in binary notation (of course this is only important for the coefficient ring $\mathbb{Z}$) and the exponents $e_{i,j}$ are represented in unary notation. Let $|p| = \sum_{i=1}^{n}(\lceil \log |a_i| \rceil + e_{i,1} + \cdots + e_{i,k})$. The *succinct representation* of $p(x)$ coincides with the standard version, except that the exponents $e_{i,j}$ are represented in binary notation. Let $\|p\| = \sum_{i=1}^{n}(\lceil \log |a_i| \rceil + \lceil \log e_{i,1} \rceil + \cdots + \lceil \log e_{i,k} \rceil)$. We use the following result of Eberly [4] (see also [7]).

Proposition 1. *Iterated addition, iterated multiplication, and division with remainder of polynomials from $\mathbb{Z}[x]$ or $\mathbb{F}_p[x]$ (p is a prime that can be part of the input in binary encoding) that are given in standard representation belong to* NC^1.

Consider a commutative semiring $\mathcal{S} = (S, \oplus, \otimes)$. An arithmetic circuit (or just circuit) over $\mathcal{S}$ is a triple $\mathcal{C} = (V, \mathsf{rhs}, A_0)$, where V is a finite set of *gates* or *variables*, $A_0 \in V$ is the output gate, and rhs (for *right-hand side*) maps every $A \in V$ to an expression (the right-hand side of A) of one of the following three forms: (i) a semiring element $s \in S$ (such a gate is an *input gate*), (ii) $B \oplus C$ with $B, C \in V$ (such a gate is an *addition gate*), (iii) $B \otimes C$ with $B, C \in V$ (such a gate is a *multiplication gate*). Moreover, the directed graph $(V, \{(A, B) \in V \times V \mid B$ occurs in $\mathsf{rhs}(A)\})$ (the graph of $\mathcal{C}$) has to be acyclic. Every gate $A \in V$ evaluates to an element $\mathsf{val}_{\mathcal{C}}(A) \in S$ in the natural way and we set $\mathsf{val}(\mathcal{C}) = \mathsf{val}_{\mathcal{C}}(A_0)$. A circuit over $\mathcal{S}$ is called skew if for every multiplication gate A one of the two gates in $\mathsf{rhs}(A)$ is an input gate.

A branching program over $\mathcal{S}$ is a tuple $\mathcal{A} = (V, E, \lambda, s, t)$, where (V, E) is a directed acyclic graph, $\lambda : E \to S$ assigns to each edge a semiring element, and $s, t \in V$. Let $\mathcal{P}$ be the set of all paths from s to t. For a path $p = (v_0, v_1, \ldots, v_n) \in \mathcal{P}$ ($v_0 = s$, $v_n = t$) we define $\lambda(p) = \prod_{i=1}^{n} \lambda(v_{i-1}, v_i)$ as the product (w.r.t. $\otimes$) of all edge labels along the path. Finally, the value defined by $\mathcal{A}$ is $\mathsf{val}(\mathcal{A}) = \sum_{p \in \mathcal{P}} \lambda(p)$. Skew circuits and branching programs are basically the same objects (the edge labels of the branching program correspond to the constant inputs of multiplication gates in the skew circuit).

It is well known that the value defined by a branching program $\mathcal{A}$ can be computed using matrix powers. W.l.o.g. assume that $\mathcal{A} = (\{1, \ldots, n\}, E, \lambda, 1, n)$

and consider the adjacency matrix M of $\mathcal{A}$, i.e., the $(n \times n)$-matrix M with $M[i, j] = \lambda(i, j)$. Then val$(\mathcal{A})$ is the $(1, n)$-entry of the matrix $\sum_{i=0}^{n} M^i$. For many semirings $\mathcal{S}$, this fact can be used to get an NC^2-algorithm for computing val$(\mathcal{A})$. The $n+1$ matrix powers M^i $(0 \le i \le n)$ can be computed in parallel, and every power can be computed by a balanced tree of height $\log i \le \log n$, where every tree node computes a matrix product. Hence, we obtain an NC^2-algorithm, if (i) the number of bits needed to represent a matrix entry in M^n is polynomially bounded in n and the number of bits of the entries in M, and (ii) the product of two matrices over the semiring $\mathcal{S}$ can be computed in NC^1. Point (ii) holds if products of two elements and iterated sums in $\mathcal{S}$ can be computed in NC^1. For the following important semirings these facts are well known, see e.g. [21]: $(\mathbb{Z}[x], +, \cdot)$, $(\mathbb{Z}_n[x], +, \cdot)$ for $n \ge 2$, $(\mathbb{Z} \cup \{\infty\}, \min, +)$, and $(\mathbb{Z} \cup \{-\infty\}, \max, +)$. Here, we assume that polynomials are given in the *standard representation*. For the polynomial ring $\mathbb{Z}[x]$ also note that every entry $p(x)$ of the matrix power M^n is a polynomial of degree $n \cdot m$, where m is the maximal degree of a polynomial in M, and all coefficients are bounded by a^n (and hence need at most $n \log a$ bits), where a is the maximal absolute value of a coefficient in M. Hence point (i) above holds, and we get:

Lemma 1. *The output value of a given skew circuit (or branching program) over one of the following semirings can be computed in* NC^2:

(a) $(\mathbb{Z}[x], +, \cdot)$ *and* $(\mathbb{Z}_n[x], +, \cdot)$ *for* $n \ge 2$ *(polynomials are given in the standard representation, and n can be part of the input in binary representation)*
(b) $(\mathbb{Z} \cup \{\infty\}, \min, +)$ *and* $(\mathbb{Z} \cup \{-\infty\}, \max, +)$ *(integers are given in binary representation)*

Point (a) of Lemma 1 also holds for the polynomial rings $(\mathbb{Z}[x_1, \ldots, x_k], +, \cdot)$ and $(\mathbb{Z}_n[x_1, \ldots, x_k], +, \cdot)$ as long as the number k of variables is not part of the input: The polynomial $\prod_{i=1}^{k}(x_i + 1)$ can be defined by a branching program with $O(k)$ edges labeled by the polynomials $x_i + 1$, but the product of these polynomials has 2^k monomials. Also note that it is important that we use the standard representation for polynomials in (a): The polynomial $\prod_{i=1}^{n}(x^{2^i} + 1)$ has 2^n monomials but can be represented by a branching program with $O(n)$ edges labeled by the polynomials $x^{2^i} + 1$.

In this paper, we will deal with circuits over a polynomial ring $R[x_1, \ldots, x_k]$, where R is $(\mathbb{Z}, +, \cdot)$ or $(\mathbb{Z}_n, +, \cdot)$. By definition, in such a circuit every input gate is labelled with a polynomial from $R[x_1, \ldots, x_k]$. Usually, one considers circuits where the right-hand side of an input gate is a polynomial given in *standard representation* (or, equivalently, a constant $a \in R$ or variable x_i); we will also use the term "standard circuits" in this case. For succinctness reasons, we will also consider circuits over $R[x_1, \ldots, x_k]$, where the right-hand sides of input gates are polynomials given in *succinct representation*. For general circuits this makes no real difference (since a big power $x_i^{n_i}$ can be defined by a subcircuit of size $O(\log n_i)$ using iterated squaring), but for skew circuits we will gain additional succinctness. We will use the term "powerful skew circuits". Formally, a *powerful skew circuit* over the polynomial ring $R[x_1, \ldots, x_k]$ is a skew circuit over the ring

$R[x_1, \ldots, x_k]$ as defined above, where the right-hand side of every input gate is a polynomial that is given in succinct representation (equivalently, we could require that the right-hand side is a constant $a \in R$ or a power x_i^n with n given in binary notation). We define the size of a powerful skew circuit $\mathcal{C}$ as follows: First, define the size $\mathsf{size}_{\mathcal{C}}(A)$ of a gate $A \in V$ as follows: If A is an addition gate or a multiplication gate, then $\mathsf{size}_{\mathcal{C}}(A) = 1$, and if A is an input gate with $\mathsf{rhs}(A) = p(x_1, \ldots, x_k)$, then $\mathsf{size}_{\mathcal{C}}(A) = \|p(x_1, \ldots, x_k)\|$. Finally, we define the size of $\mathcal{C}$ as $\sum_{A \in V} \mathsf{size}_{\mathcal{C}}(A)$.

A *powerful branching program* is a branching program (V, E, λ, s, t) over a polynomial ring $R[x_1, \ldots, x_k]$, where every edge label $\lambda(e)$ is a polynomial that is given in succinct representation. The size of a powerful branching program is $\sum_{e \in E} \|\lambda(e)\|$. From a given powerful skew circuit one can compute in logspace an equivalent powerful branching program and vice versa.

Note that the transformation of a powerful skew circuit over $R[x_1, \ldots, x_k]$ into an equivalent standard skew circuit (where every input gate is labelled by a polynomial given in standard representation) requires an exponential blow-up. For instance, the smallest standard skew circuit for the polynomial x^n has size n, whereas x^n can be trivially obtained by a powerful skew circuit of size $\lceil \log n \rceil$.

A central computational problem in computational algebra is *polynomial identity testing*, briefly PIT. Let R be a ring that is effective in the sense that elements of R can be encoded by natural numbers in such a way that addition and multiplication in R become computable operations. Then, PIT for the ring R is the following problem: Given a number $k \geq 1$ and a circuit $\mathcal{C}$ over the ring $R[x_1, \ldots, x_k]$, is $\mathsf{val}(\mathcal{C})$ the zero-polynomial? For the rings $\mathbb{Z}$ and $\mathbb{Z}_p$ (p prime) the following result was shown in [9]; for $\mathbb{Z}_n$ with n composite, it was shown in [1].

Theorem 1. *For each of the rings $\mathbb{Z}$ and $\mathbb{Z}_n$ ($n \geq 2$), PIT belongs to the class* coRP.

Note that the number k of variables is part of the input in PIT. On the other hand, there is a well-known reduction from PIT to PIT restricted to univariate polynomials (polynomials with a single variable) [1]. For a multivariate polynomial $p(x_1, \ldots, x_k) \in R[x_1, \ldots, x_k]$ let $\deg(p, x_i)$ be the degree of p in the variable x_i. It is the largest number d such that x_i^d appears in a monomial of p. Let $p(x_1, \ldots, x_k)$ be a polynomial and let $d \in \mathbb{N}$ such that $\deg(p, x_i) < d$ for all $1 \leq i \leq k$. We define the univariate polynomial $U(p, d) = p(y^1, y^d, \ldots, y^{d^{k-1}})$. It is obtained from $p(x_1, \ldots, x_k)$ by replacing every monomial $a \cdot x_1^{n_1} \cdots x_k^{n_k}$ by $a \cdot y^N$, where $N = n_1 + n_2 d + \cdots n_k d^{k-1}$ is the number with base-d representation $(n_1, n_2, \ldots, n_k)$. The polynomial p is the zero-polynomial if and only if $U(p, d)$ is the zero-polynomial. The following lemma can be shown for arbitrary circuits, but we will only need it for powerful skew circuits.

Lemma 2. *Given a powerful skew circuit $\mathcal{C}$ for the polynomial $p(x_1, \ldots, x_k)$, one can compute in* NC^2 *(i) the binary encoding of a number d with $\deg(p, x_i) < d$ for all $1 \leq i \leq k$ and (ii) a powerful skew circuit $\mathcal{C}'$ for $U(p, d)$.*

Note that the above reduction from multivariate to univariate circuits does not work for standard skew circuits: the output circuit will be powerful skew even if the input circuit is standard skew. For instance, the polynomial $\prod_{i=1}^{k} x_i$ (which can be produced by a standard skew circuit of size k) is transformed into the polynomial y^{2^k-1}, for which the smallest standard skew circuit has size $\Omega(2^k)$.

4   PIT for Powerful Skew Circuits

The main result of this paper is:

Theorem 2. *For each of the rings $\mathbb{Z}$ and $\mathbb{F}_p$ (p is a prime that can be part of the input in unary encoding), PIT for powerful skew circuits belongs to the class* coRNC^2.

The proof of Theorem 2 has two main ingredients: The randomized PIT algorithm of Agrawal and Biswas [1] and the modular polynomial powering algorithm of Fich and Tompa [5]. Let us start with the Agrawal-Biswas algorithm. We only need the version for the polynomial ring $\mathbb{F}_p[x]$, where p is a prime number. Consider a polynomial $P(x) \in \mathbb{F}_p[x]$ of degree d. The algorithm of Agrawal and Biswas consists of the following steps (later we will apply this algorithm to the polynomial defined by a powerful skew circuit), where $0 < \epsilon < 1$ is an error parameter:

1. Let ℓ be a number with $\ell \geq \log d$ and $t = \max\{\ell, \frac{1}{\epsilon}\}$
2. Find the smallest prime number r such that $r \neq p$ and r does not divide any of $p - 1, p^2 - 1, \ldots, p^{\ell-1} - 1$. It is argued in [1] that $r \in O(\ell^2 \log p)$.
3. Randomly choose a tuple $b = (b_0, \ldots, b_{\ell-1}) \in \{0, 1\}^\ell$ and compute the polynomial $T_{r,b,t}(x) = Q_r(A_{b,t}(x))$, where $Q_r(x) = \sum_{i=0}^{r-1} x^i$ is the rth cyclotomic polynomial and $A_{b,t} = x^t + \sum_{i=0}^{\ell-1} b_i \cdot x^i$.
4. Accept, if $P(x) \bmod T_{r,b,t} = 0$, otherwise reject.

Clearly, if $P(x) = 0$, then the above algorithm accepts with probability 1. For a non-zero polynomial $P(x)$, Agrawal and Biswas proved:

Theorem 3 ([1]). *Let $P(x) \in \mathbb{F}_p[x]$ be a non-zero polynomial of degree d. The above algorithm rejects $P(x)$ with probability at least $1 - \varepsilon$.*

The second result we are using was shown by Fich and Tompa:

Theorem 4 ([5]). *The following computation can be done in* NC^2:

Input: A unary encoded prime number p, polynomials $a(x), q(x) \in \mathbb{F}_p[x]$ such that $\deg(a(x)) < \deg(q(x)) = d$, and a binary encoded number m.
Output: The polynomial $a(x)^m \bmod q(x)$.

Remark 1. In [5], it is stated that the problem can be solved using circuits of depth $(\log n)^2 \log \log n$ for the more general case that the underlying field is $\mathbb{F}_{p^\ell}$, where p and ℓ are given in unary representation. The main bottleneck is the computation of an iterated matrix product $A_1 A_2 \cdots A_k$ (for k polynomial in n) of $(d \times d)$-matrices over the field $\mathbb{F}_{p^\ell}$. In our situation (where the field is $\mathbb{F}_p$) we easily obtain an NC^2-algorithm for this step: Two $(d \times d)$-matrices over $\mathbb{F}_p$ can be multiplied in NC^1. Then we compute the product $A_1 A_2 \cdots A_k$ by a balanced binary tree of depth $\log k$. Also logspace-uniformity of the circuits is not stated explicitly in [5], but follows easily since only standard arithmetical operations on binary coded numbers are used.

Proof of Theorem 2. By Lemma 2 we can restrict to univariate polynomials. We first prove the theorem for the case of a powerful skew circuit $\mathcal{C}$ over the field $\mathbb{F}_p$, where the prime number p is part of the input but specified in unary notation.

Let p be a unary encoded prime number and $\mathcal{A} = (\{1,\dots,n\},1,n,\lambda)$ a powerful branching program with n nodes that is equivalent to $\mathcal{C}$. Let $P(x) = \mathsf{val}(\mathcal{A}) \in \mathbb{F}_p[x]$. Fix an error probability $0 < \varepsilon < 1$. Our randomized NC^2-algorithm is based on the Agrawal-Biswas identity test. It accepts with probability 1 if $\mathsf{val}(\mathcal{A}) = 0$ and accepts with probability at most ϵ if $P(x) \neq 0$. Let us go through the four steps of the Agrawal-Biswas algorithm to see that they can be implemented in NC^2.

Step 1. An upper bound on the degree of $P(x)$ can be computed in NC^2 as in the proof of Lemma 2. For the number ℓ we can take the number of bits of this degree bound, which is polynomial in the input size.

Step 2. For the prime number r we know that $r \in O(\ell^2 \log p)$, which is a polynomial bound. Hence, we can test in parallel all possible candidates for r. For a certain candidate r, we check in parallel whether it is prime (recall that r is of polynomial size) and whether it divides any of the numbers $p - 1$, $p^2 - 1, \dots, p^{\ell-1} - 1$. The whole computation is possible in NC^1.

Step 3. Let $b = (b_0, \dots, b_{\ell-1}) \in \{0,1\}^\ell$ be the chosen tuple. Computing the polynomial $T_{r,b,t}(x) = Q_r(A_{b,t}(x))$, where $Q_r(x) = \sum_{i=0}^{r-1} x^i$ and $A_{b,t} = x^t + \sum_{i=0}^{\ell-1} b_i \cdot x^i$, is an instance of iterated multiplication (for the powers $A_{b,t}(x)^i$) and iterated addition of polynomials. Hence, by Proposition 1 also this step can be carried out in NC^1. Note that the degree of $T_{r,b,t}(x)$ is $\ell \cdot (r-1) \in O(\ell^3 \log p)$, i.e., polynomial in the input size.

Step 4. For the last step, we have to compute $P(x) \bmod T_{r,b,t}(x)$. For this, we consider in parallel all monomials $a \cdot x^k$ that occur in an edge label of our powerful branching program $\mathcal{A}$. Recall that $a \in \mathbb{F}_p$ and k is given in binary notation. Using the Fich-Tompa algorithm we compute $x^k \bmod T_{r,b,t}(x)$ in NC^2. We then replace the edge label $a \cdot x^k$ by $a \cdot (x^k \bmod T_{r,b,t}(x))$. Let $\mathcal{B}$ the resulting branching program. Every polynomial that appears as an edge label in $\mathcal{B}$ is now given in standard form. Hence, by Lemma 1 we can compute in NC^2 the output polynomial $\mathsf{val}(\mathcal{B})$. Clearly, $P(x) \bmod T_{r,b,t}(x) = \mathsf{val}(\mathcal{B}) \bmod T_{r,b,t}(x)$, which can be computed in NC^1 by Proposition 1.

Let us now prove Theorem 2 for the ring $\mathbb{Z}$. Let $\mathcal{A} = (\{1,\dots,n\},1,n,\lambda)$ be a powerful branching program over $\mathbb{Z}$ with n nodes and let $P(x) = \mathsf{val}(\mathcal{A})$. Let us first look at the coefficients of $P(x)$. Let m be the maximum absolute value $|a|$, where $a \cdot x^k$ is an edge label of $\mathcal{A}$. Since there are at most 2^n many paths from s to t in $\mathcal{A}$, every coefficient of $P(x)$ belongs to the interval $[-(2m)^n, (2m)^n]$.

Let $k = (\lceil \log(m) \rceil + 1) \cdot (n + 1)$ and $p_1, \ldots, p_k$ be the first k prime numbers. Each prime p_i is polynomially bounded in k (and hence the input size) and the list of primes can be computed in logspace by doing all necessary divisibility checks on binary encoded numbers having only $O(\log k)$ bits. The Chinese remainder theorem implies that $P(x) = 0$ if and only if $P(x) \equiv 0 \bmod p_i$ for all $1 \le i \le k$. We can carry out the latter tests in parallel using the above algorithm for a unary encoded prime number. The overall algorithm accepts if we accept for every prime p_i. If $P(x) = 0$, then we will accept for every $1 \le i \le k$ with probability 1, hence the overall algorithm accepts with probability 1. On the other hand, if $P(x) \ne 0$, then there exists a prime p_i $(1 \le i \le k)$ such that the algorithm rejects with probability at least $1 - \varepsilon$. Hence, the overall algorithm will reject with probability at least $1 - \varepsilon$ as well.   □

Our coRNC^2 identity testing algorithm for powerful skew circuits only works for the coefficient rings $\mathbb{Z}$ and $\mathbb{Z}_p$ with p prime. It is not clear how to extend it to $\mathbb{Z}_n$ with n composite. The Agrawal-Biswas identity testing algorithm also works for $\mathbb{Z}_n$ with n composite. But the problem is that the Fich-Tompa algorithm only works for polynomial rings over finite fields.

5   Multi-dimensional Straight-Line Programs

Let Γ be a finite alphabet. For $l \in \mathbb{N}$ let $[0, l] = \{0, 1, \ldots, l\}$. An n-*dimensional picture* over Γ is a mapping $p : \prod_{j=1}^{n}[0, l_j - 1] \to \Gamma$ for some $l_j \in \mathbb{N}$. Let $\mathrm{dom}(p) = \prod_{j=1}^{n}[0, l_j - 1]$. For $1 \le j \le n$ we define $|p|_j = l_j$ as the length of p in the j-th dimension. Note that one-dimensional pictures are simply finite words. Let Γ_n^* denote the set of n-dimensional pictures over Γ. On this set we can define partially defined concatenation operations $\circ_i$ $(1 \le i \le n)$ as follows: For pictures $p, q \in \Gamma_n^*$, the picture $p \circ_i q$ is defined if and only if $|p|_j = |q|_j$ for all $1 \le j \le n$ with $i \ne j$. In this case, we have $|p \circ_i q|_j = |p|_j \, (= |q|_j)$ for $j \ne i$ and $|p \circ_i q|_i = |p|_i + |q|_i$. Let $l_j = |p \circ_i q|_j$. For a tuple $(k_1, \ldots, k_n) \in \prod_{j=1}^{n}[0, l_j - 1]$ we finally set $(p \circ_i q)(k_1, \ldots, k_n) = p(k_1, \ldots, k_n)$ if $k_i < |p|_i$ and $(p \circ_i q)(k_1, \ldots, k_n) = q(k_1, \ldots, k_{i-1}, k_i - |p|_i, k_{i+1}, \ldots, k_n)$ if $k_i \ge |p|_i$. These operations generalize the concatenation of finite words.

An n-*dimensional straight-line program (SLP)* over the terminal alphabet Γ is a triple $\mathbb{A} = (V, \mathrm{rhs}, S)$, where V is a finite set of variables, $S \in V$ is the start variable, and rhs maps each variable A to its right-hand side $\mathrm{rhs}(A)$, which is either a terminal symbol $a \in \Gamma$ or an expression of the form $B \circ_i C$, where $B, C \in V$ and $1 \le i \le n$ such that (i) the relation $\{(A, B) \in V \times V \mid B \text{ occurs in } \mathrm{rhs}(A)\}$ is acyclic, and (ii) one can assign to each $A \in V$ and $1 \le i \le n$ a number $|A|_i$ with the following properties: If $\mathrm{rhs}(A) \in \Gamma$ then $|A|_i = 1$ for all i. If $\mathrm{rhs}(A) = B \circ_i C$ then $|A|_i = |B|_i + |C|_i$ and $|A|_j = |B|_j = |C|_j$ for all $j \ne i$. These conditions ensure that every variable A evaluates to a unique n-dimensional picture $\mathrm{val}_{\mathbb{A}}(A)$ such that $|\mathrm{val}_{\mathbb{A}}(A)|_i = |A|_i$ for all $1 \le i \le n$. Finally, $\mathrm{val}(\mathbb{A}) = \mathrm{val}_{\mathbb{A}}(S)$ is the picture defined by $\mathbb{A}$. We define the size of the SLP $\mathbb{A} = (V, \Gamma, S, P)$ as $|\mathbb{A}| = |V|$. Note that the length of the picture $\mathrm{val}(\mathbb{A})$ (in each dimension) can be exponential in $|\mathbb{A}|$. A one-dimensional SLP is a context-free grammar that generates a single word [15]. Two-dimensional SLPs were studied in [3].

Given two n-dimensional SLPs we want to know whether they evaluate to the same picture. In [3] it was shown that this problem belongs to coRP by translating it to PIT. For an n-dimensional picture $p : \mathrm{dom}(p) \to \{0,1\}$ we define the polynomial

$$f_p(x_1, ..., x_n) = \sum_{(k_1,...,k_n) \in \mathrm{dom}(p)} p(k_1, ..., k_n) \prod_{i=1}^{n} x_i^{k_i}.$$

We consider f_p as a polynomial from $\mathbb{Z}_2[x_1, \ldots, x_n]$. For two n-dimensional pictures p and q such that $|p|_i = |q|_i$ for all $1 \le i \le n$ we clearly have $p = q$ if and only if $f_p + f_q = 0$ (recall that coefficients are from $\mathbb{Z}_2$). In [3], it was observed that from an SLP $\mathbb{A}$ for a picture P, one can easily construct an arithmetic circuit $\mathcal{C}$ for the polynomial f_p, which leads to a coRP-algorithm for equality testing. For instance, $\mathrm{rhs}_{\mathbb{A}}(A) = B \circ_k C$ is translated into $\mathrm{rhs}_{\mathcal{C}}(A) = B + x_k^N \cdot C$, where $N = |B|_k$ can be precomputed in NC^2 (using Lemma 1). This shows that the circuit $\mathcal{C}$ is actually powerful skew and can be constructed in NC^2 (see the appendix for details). Hence, we get:

Theorem 5. *The question whether two n-dimensional SLPs $\mathcal{A}$ and $\mathcal{B}$ evaluate to the same n-dimensional picture is in coRNC^2 (here, n is part of the input).*

It should be noted that even in the one-dimensional case (where SLP-equality can be tested in polynomial time [8,18,20]), no randomized NC-algorithm was known before. For equality testing for SLPs of dimension $n \ge 2$ it remains open whether a polynomial time algorithm exists. For the one-dimensional case, a polynomial time algorithm exists [8,18,20], but no NC-algorithm is known.

6   Circuits Over Wreath Products

As a second application of PIT for powerful skew circuits we consider the circuit evaluation problem, also known as the compressed word problem, for wreath products of finitely generated abelian groups. We assume some basic familiarity with group theory.

6.1   Compressed Word Problems

Let G be a finitely generated (f.g.) group and let Σ be a finite generating set for G, i.e., every element of G can be written as a finite product of elements from Σ and inverses of elements from Σ. Let $\Gamma = \Sigma \cup \{a^{-1} \mid a \in \Sigma\}$. For a word $w \in \Gamma^*$ we write $w = 1$ in G if and only if the word w evaluates to the identity of G. The *word problem* for G asks, whether $w = 1$ in G for a given input word. There exist finitely generated groups and in fact finitely presented groups (groups that are defined by finitely many defining relations) with an undecidable word problem [19]. Here, we are interested in the *compressed word problem* for a f.g. group. For this, the input word w is given in compressed form by a one-dimensional SLP as defined in Sect. 5. Recall that a one-dimensional picture over an alphabet Γ is simply a finite word over Γ. In the following we always mean one-dimensional SLPs when using the term SLP. The compressed word problem for G asks, whether $\mathrm{val}(\mathbb{A}) = 1$ in G for a given SLP $\mathbb{A}$.

As mentioned in the introduction, there are important classes of groups with a polynomial time compressed word problem. Moreover, for f.g. linear groups the compressed word problem belongs to coRP. Concerning the parallel complexity, it was recently shown in [12] (using results from [2]) that $\mathsf{CWP}(G) \in \mathsf{NC}^2$, whenever G has a f.g. nilpotent normal subgroup H such that the quotient G/H is finite and solvable (a so called f.g. nilpotent by finite solvable group). To the knowledge of the authors, there are no other known examples of groups with a compressed word problem in NC.

6.2 Wreath Products

Let G and H be groups. The restricted wreath product $H \wr G$ is defined as follows: Elements of $H \wr G$ are pairs (f, g), where $g \in G$ and $f : G \to H$ is a mapping such that $f(a) \neq 1_H$ for only finitely many $a \in G$ (1_H is the identity element of H). Multiplication in $H \wr G$ is defined as follows: Let $(f_1, g_1), (f_2, g_2) \in H \wr G$. Then $(f_1, g_1)(f_2, g_2) = (f, g_1 g_2)$, where $f(a) = f_1(a) f_2(g_1^{-1} a)$.

For readers, who have not seen this definition before, the following intuition might be helpful: An element $(f, g) \in H \wr G$ can be thought as a finite collection of elements of H that are sitting in certain elements of G (the mapping f) together with a distinguished element of G (the element g), which can be thought as a cursor moving around G. If we want to compute the product $(f_1, g_1)(f_2, g_2)$, we do this as follows: First, we shift the finite collection of H-elements that corresponds to the mapping f_2 by g_1: If the element $h \in H \setminus \{1_H\}$ is sitting at $a \in G$ (i.e., $f_2(a) = h$), then we remove h from a and put it to the new location $g_1 a \in G$. This new collection corresponds to the mapping $f_2' : a \mapsto f_2(g_1^{-1} a)$. After this shift, we multiply the two collections of H-elements pointwise: If the elements h_1 and h_2 are sitting at $a \in G$ (i.e., $f_1(a) = h_1$ and $f_2'(a) = h_2$), then we put the product $h_1 h_2$ into the G-location a. Finally, the new distinguished G-element (the new cursor position) becomes $g_1 g_2$.

Proofs of the following lemmas can be found in the long version [13].

Lemma 3. *The group $(A \times B) \wr G$ embeds into $(A \wr G) \times (B \wr G)$.*

Lemma 4. *For every $k \geq 1$ and every finitely generated group G, $\mathsf{CWP}(G \wr \mathbb{Z}^k)$ is NC^2-reducible to $\mathsf{CWP}(G \wr \mathbb{Z})$.*

6.3 $\mathsf{CWP}(\mathbb{Z} \wr \mathbb{Z})$ and Identity Testing for Powerful Skew Circuits

In this section, we show that $\mathsf{CWP}(\mathbb{Z} \wr \mathbb{Z})$ (resp., $\mathbb{Z}_n \wr \mathbb{Z}$) and PIT for powerful skew circuits over $\mathbb{Z}[x]$ (resp., $\mathbb{Z}_n[x]$) are equivalent w.r.t. NC^2-reductions.

We consider the generators a and t of $\mathbb{Z} \wr \mathbb{Z}$, where $a = (0, 1)$ and $t = (f, 0)$ with $f(0) = 1$ and $f(x) = 0$ for $x \neq 0$. So, multiplying with a (resp., a^{-1}) on the right corresponds to moving the cursor to the left (resp., right) and multiplying with t (resp., t^{-1}) on the right corresponds to adding (resp., subtracting) one from the value at the current cursor position. Let $\Gamma = \{a, t, a^{-1}, t^{-1}\}$. The main result of this section is:

Theorem 6. *The compressed word problem for $\mathbb{Z} \wr \mathbb{Z}$ (resp., $\mathbb{Z}_n \wr \mathbb{Z}$) is equivalent w.r.t. NC^2-reductions to PIT for powerful skew circuits over the ring $\mathbb{Z}[x]$ (resp., $\mathbb{Z}_n[x]$).*

Going from PIT for powerful skew circuits over the ring $\mathbb{Z}[x]$ to $\mathsf{CWP}(\mathbb{Z} \wr \mathbb{Z})$ is relatively easy (and the same reduction works for $\mathbb{Z}_n$ instead of $\mathbb{Z}$): We encode a polynomial $p(x) = a_n x^n + a_{n-1} x^{n-1} + \cdots + a_1 x + a_0$ by the group element $g(p(x)) = (f, 0) \in \mathbb{Z} \wr \mathbb{Z}$, where $f(k) = a_k$ for all $0 \le k \le n$ and $f(z) = 0$ for all other integers z. The idea now is to construct from a given powerful skew circuit $\mathcal{C}$ over the ring $\mathbb{Z}[x]$ an SLP $\mathbb{A}$ over the alphabet Γ such that $\mathrm{val}(\mathbb{A})$ evaluates (in $\mathbb{Z} \wr \mathbb{Z}$) to the group element $g(\mathrm{val}(\mathcal{C}))$. Note that $g(p_1(x) + p_2(x)) = g(p_1(x))g(p_2(x))$ in the group $\mathbb{Z} \wr \mathbb{Z}$. This allows to deal with addition gates in $\mathcal{C}$. Multiplication gates in general circuits cannot be handled in this way. But fortunately, $\mathcal{C}$ is powerful skew, and we can make use of the following identity, where $m, n \in \mathbb{N}$: $g(m \cdot x^n \cdot p(x)) = a^n g(p(x))^m a^{-n}$ (conjugation by a^n corresponds to multiplication with the monomial x^n).

Going from $\mathsf{CWP}(\mathbb{Z} \wr \mathbb{Z})$ back to PIT for powerful skew circuits over the ring $\mathbb{Z}[x]$ is based on the same correspondence between polynomials and elements of $\mathbb{Z} \wr \mathbb{Z}$, but is slightly more technical. The problem is that for a group element $(f, z) \in \mathbb{Z} \wr \mathbb{Z}$ there might be a negative $a \in \mathbb{Z}$ with $f(a) \ne 0$. Hence, encoding f by a polynomial would in fact lead to a Laurent polynomial. But we can avoid this problem by conjugating all elements of $\mathbb{Z} \wr \mathbb{Z}$ with a large enough power a^n such that the domain of the above function f is contained in the non-negative integers; see [13] for details.

By Lemmas 3 and 4, $\mathsf{CWP}((G \times H) \wr \mathbb{Z}^n)$ is NC^2-reducible to $\mathsf{CWP}(G \wr \mathbb{Z})$ and $\mathsf{CWP}(H \wr \mathbb{Z})$. Together with Theorems 2 and 6 we obtain the following result:

Corollary 1. *Let G be a finite direct product of copies of $\mathbb{Z}$ and $\mathbb{Z}_p$ for primes p. Then, for every $n \ge 1$, $\mathsf{CWP}(G \wr \mathbb{Z}^n)$ belongs to coRNC^2.*

It is not clear, whether in Corollary 1 we can replace G by an arbitrary finitely generated abelian group. On the other hand, if we apply Theorem 1 instead of Theorem 2 we obtain:

Corollary 2. *Let G be f.g. abelian and let H be f.g. virtually abelian (i.e., H has a f.g. abelian subgroup of finite index). Then $\mathsf{CWP}(G \wr H)$ belongs to coRP.*

Proof. Let $K \le H$ be a f.g. abelian subgroup of finite index m in H. Since K has the form $A \times \mathbb{Z}^k$ for some $k \ge 0$ with A finite abelian, we can (by increasing the finite index m) assume that $K = \mathbb{Z}^k$ for some $k \ge 0$. It is shown in [17] that $G^m \wr \mathbb{Z}^k \cong G^m \wr K$ is isomorphic to a subgroup of index m in $G \wr H$. If the group A is a finite index subgroup of the group B, then $\mathsf{CWP}(B)$ is polynomial-time many-one reducible to $\mathsf{CWP}(A)$ [16, Theorem 4.4]. Since $G^m \wr \mathbb{Z}^k$ is a finite index subgroup of $G \wr H$, it suffices to show that $\mathsf{CWP}(G^m \wr \mathbb{Z}^k) \in \mathsf{coRP}$. Since G^m is finitely generated abelian, it suffices to consider $\mathsf{CWP}(\mathbb{Z}_n \wr \mathbb{Z}^k)$ ($n \geq 2$) and $\mathsf{CWP}(\mathbb{Z} \wr \mathbb{Z}^k)$. The case $k = 1$ is clear, so assume that $k \geq 1$. By Corollary 1, $\mathsf{CWP}(\mathbb{Z} \wr \mathbb{Z}^k) \in \mathsf{coRNC}$ and by Theorems 1 and 6, $\mathsf{CWP}(\mathbb{Z}_n \wr \mathbb{Z}^k) \in \mathsf{coRP}$. $\square$

In the full version [13] of this paper, Corollary 1 is further applied to the compressed word problem for quotients of free groups with respect to commutator subgroups.

References

1. Agrawal, M., Biswas, S.: Primality and identity testing via Chinese remaindering. J. Assoc. Comput. Mach. **50**(4), 429–443 (2003)
2. Beaudry, M., McKenzie, P., Péladeau, P., Thérien, D.: Finite monoids: from word to circuit evaluation. SIAM J. Comput. **26**(1), 138–152 (1997)
3. Berman, P., Karpinski, M., Larmore, L.L., Plandowski, W., Rytter, W.: On the complexity of pattern matching for highly compressed two-dimensional texts. J. Comput. Syst. Sci. **65**(2), 332–350 (2002)
4. Eberly, W.: Very fast parallel polynomial arithmetic. SIAM J. Comput. **18**(5), 955–976 (1989)
5. Fich, F.E., Tompa, M.: The parallel complexity of exponentiating polynomials over finite fields. J. Assoc. Comput. Mach. **35**(3), 651–667 (1988)
6. Greenlaw, R., Hoover, H.J., Ruzzo, W.L.: Limits to Parallel Computation: P-Completeness Theory. Oxford University Press, Oxford (1995)
7. Hesse, W., Allender, E., Barrington, D.A.M.: Uniform constant-depth threshold circuits for division and iterated multiplication. J. Comput. Syst. Sci. **65**, 695–716 (2002)
8. Hirshfeld, Y., Jerrum, M., Moller, F.: A polynomial algorithm for deciding bisimilarity of normed context-free processes. Theor. Comput. Sci. **158**(1&2), 143–159 (1996)
9. Ibarra, O.H., Moran, S.: Probabilistic algorithms for deciding equivalence of straight-line programs. J. Assoc. Comput. Mach. **30**(1), 217–228 (1983)
10. Impagliazzo, R., Wigderson, A.: P = BPP if E requires exponential circuits: Derandomizing the XOR lemma. In: Proceedings of the STOC 1997, pp. 220–229. ACM Press (1997)
11. Kabanets, V., Impagliazzo, R.: Derandomizing polynomial identity tests means proving circuit lower bounds. Comput. Complex. **13**(1–2), 1–46 (2004)
12. König, D., Lohrey, M.: Evaluating matrix circuits. In: Xu, D., Du, D., Du, D. (eds.) COCOON 2015. LNCS, vol. 9198, pp. 235–248. Springer, Heidelberg (2015)
13. König, D., Lohrey, M.: Parallel identity testing for algebraic branching programs with big powers and applications. arXiv.org (2015). http://arxiv.org/abs/1502.04545
14. Lohrey, M.: Word problems and membership problems on compressed words. SIAM J. Comput. **35**(5), 1210–1240 (2006)
15. Lohrey, M.: Algorithmics on SLP-compressed strings: asurvey. Groups Complex. Cryptol. **4**(2), 241–299 (2012)
16. Lohrey, M.: The Compressed Word Problem for Groups. SpringerBriefs in Mathematics. Springer, New York (2014)

17. Lohrey, M., Steinberg, B., Zetzsche, G.: Rational subsets and submonoids of wreath products. Inf. Comput. **243**, 191–204 (2015)
18. Mehlhorn, K., Sundar, R., Uhrig, C.: Maintaining dynamic sequences under equality tests in polylogarithmic time. Algorithmica **17**(2), 183–198 (1997)
19. Novikov, P.S.: On the algorithmic unsolvability of the word problem in group theory. Amer. Math. Soc. Transl. Ser. **2**(9), 1–122 (1958)
20. Plandowski, W.: Testing equivalence of morphisms on context-free languages. In: van Leeuwen, J. (ed.) ESA 1994. LNCS, vol. 855, pp. 460–470. Springer, Heidelberg (1994)
21. Vollmer, H.: Introduction to Circuit Complexity. Springer, New York (1999)

On Probabilistic Space-Bounded Machines with Multiple Access to Random Tape

Debasis Mandal[1]($\boxtimes$), A. Pavan[1], and N.V. Vinodchandran[2]

[1] Department of Computer Science, Iowa State University, Ames, USA
{debasis,pavan}@cs.iastate.edu
[2] Department of Computer Science and Engineering,
University of Nebraska-Lincoln, Lincoln, USA
vinod@cse.unl.edu

Abstract. We investigate probabilistic space-bounded Turing machines that are allowed to make multiple passes over the random tape. As our main contribution, we establish a connection between derandomization of such probabilistic space-bounded classes to the derandomization of probabilistic time-bounded classes. Our main result is the following.

– For some integer $k > 0$, if all the languages accepted by bounded-error randomized log-space machines that use $O(\log n \log^{(k+3)} n)$ random bits and make $O(\log^{(k)} n)$ passes over the random tape is in deterministic polynomial-time, then $\mathrm{BPTIME}(n) \subseteq \mathrm{DTIME}(2^{o(n)})$. Here $\log^{(k)} n$ denotes log function applied k times iteratively.

This result can be interpreted as follows: If we restrict the number of random bits to $O(\log n)$ for the above randomized machines, then the corresponding set of languages is trivially known to be in P. Further, it can be shown that (proof is given in the main body of the paper) if we instead restrict the number of passes to only $O(1)$ for the above randomized machines, then the set of languages accepted is in P. Thus our result implies that any non-trivial extension of these simulations will lead to a non-trivial and unknown derandomization of $\mathrm{BPTIME}(n)$. Motivated by this result, we further investigate the power of multi-pass, probabilistic space-bounded machines and establish additional results.

1 Introduction

In this paper we investigate probabilistic space-bounded Turing machines that are allowed to access their random bits multiple times. In the traditional definition of probabilistic space-bounded computations, a probabilistic machine can access its random tape in a *one-way*, read-only manner and the random tape does not count towards the space complexity of the probabilistic machine. In particular, the machine cannot reread the random bits unless they are stored in its work tapes. This access mechanism is the most natural one as it corresponds

D. Mandal and A. Pavan—Research Supported in part by NSF grants 0916797, 1421163.

N.V. Vinodchandran—Research Supported in part by NSF grants 0916525, 1422668.

© Springer-Verlag Berlin Heidelberg 2015
G.F. Italiano et al. (Eds.): MFCS 2015, Part II, LNCS 9235, pp. 459–471, 2015.
DOI: 10.1007/978-3-662-48054-0_38

to modeling probabilistic machines as coin-tossing machines, originally defined by Gill [8]. The complexity class BPL is the class of languages accepted by such bounded-error probabilistic machines that use logarithmic space and halt in polynomial time[1]. The class RL is its one-sided error counterpart. Whether BPL or even RL can be derandomized to deterministic log-space is one of the central questions in the study of space-bounded computation. In spite of clear and steady progress in space-bounded derandomization, this question is far from settled [4,11,15,17–19,21].

Even though one-way access to the random tape is the standard in investigating probabilistic space-bounded computations, researchers have considered space-bounded models where the base probabilistic machines are allowed to read contents of the random tape multiple times [3,5,6,16]. However, our understanding of such multiple-access models is limited. For example, consider the class of languages that are accepted by bounded-error probabilistic log-space machines that have an unrestricted *two-way* access to the random tape (we denote this class by 2-*way*BPL). While we know that BPL is in deterministic polynomial time, it is not known whether 2-*way*BPL is even in deterministic sub-exponential time (note that it is in BPP). This is because, while one-way access machines can be characterized using certain graph reachability problem, we do not have such a nice combinatorial characterization for two-way access machines. It is also interesting to note that allowing two way access to the random tape for a space-bounded machine makes the corresponding nonuniform classes more closer to randomized circuit complexity classes. It is known that log-space uniform NC_1 is in deterministic log-space, where NC_1 is the class of languages accepted by polynomial-size, bounded fan-in $O(\log n)$-depth circuits. However, a randomized version of this inclusion is not known to hold. That is, we do not know whether (uniform) $BPNC_1$ is contained in BPL. However, it is a folklore that (uniform) $BPNC_1$ is in 2-*way*BPL (for example, see [16]).

This paper revisits probabilistic space-bounded machines that are allowed to access their random bits multiple times. In particular, we study a model where the probabilistic space-bounded machines are allowed to make *multiple passes* over the random tape, where in each pass the machines access their random tapes in a traditional one-way manner. This model was first considered by David, Papakonstantinou, and Sidiropoulos [6]. Clearly, the *multi-pass* model is intermediate between the standard model and the two-way access model. Our focus is to investigate the consequences of *derandomizing* such machines. Our main conceptual contribution is that derandomizing such probabilistic space-bounded machines leads to derandomization of probabilistic *time-bounded* classes.

Our Results. As our main result, we show a connection between derandomization of multi-pass probabilistic space-bounded machines and derandomization of probabilistic (linear) time-bounded machines. In particular, we prove the following theorem.

---

[1] We only consider probabilistic machines that halt on all inputs on all settings of the random tape. If we do not require the machines to halt, then we get a potentially larger class of languages [13,20].

Theorem 1. *For some constant $k > 0$, if every language decided by a bounded-error probabilistic log-space machine that uses $O(\log n \log^{(k+3)} n)$ random bits and makes $O(\log^{(k)} n)$ passes over its random tape is in P, then $\text{BPTIME}(n) \subseteq \text{DTIME}(2^{o(n)})$.*

Here $\log^{(k)} n$ denotes log function applied k times iteratively. Showing that $\text{BPTIME}(n)$ is a subset of $\text{DTIME}(2^{o(n)})$ is a significant open problem. The best unconditional derandomization of $\text{BPTIME}(n)$ till date is due to Santhanam and van Melkebeek [22] who showed that any bounded-error linear time probabilistic machine can be simulated in $\text{DTIME}(2^{\epsilon n})$, where $\epsilon > 0$ is a constant that depends on the number of tapes and the alphabet size of the probabilistic machine. Here we show that derandomizing a slightly non-constant pass probabilistic space-bounded machine that uses slightly larger than $O(\log n)$ random bits yields a non-trivial derandomization of $\text{BPTIME}(n)$. Notice that if we restrict the number of random bits from $O(\log n \log^{(k+3)}(n))$ to $O(\log n)$, then the corresponding set of languages is trivially in P. If we restrict the number of passes from $O(\log^{(k)} n)$ to $O(1)$, we can still show that the set of languages accepted is in P (refer to Sect. 3.2 for a proof). Thus, the above theorem states that any extension of these simulations will lead to a non-trivial and unknown derandomization of $\text{BPTIME}(n)$.

We also present some upper bounds on the class of languages accepted by multi-pass probabilistic space-bounded machines. Even though we are unable to prove that the hypothesis of Theorem 1 holds, we show that for every constants $k \geq 3$ and $\epsilon > 0$, languages accepted by probabilistic log-space machines that use $O(\log n \log^{(k+3)} n)$ random bits and make $O(\log^{(k)} n)$ passes over its random tapes are in $\text{DSPACE}(\log n (\log \log n)^{1/2+\epsilon})$, which in turn is contained in $\text{DTIME}(n^{(\log \log n)^{1/2+\epsilon}})$. We also show that any $k(n)$-pass, $s(n)$-space, probabilistic machine can be simulated by traditional $k(n)s(n)$-space bounded probabilistic machines. Thus, in particular, a constant number of passes do not add power to the traditional one-way random tape machines.

Finally, we extend some well-known results regarding standard probabilistic log-space classes to multi-pass, probabilistic log-space classes.

Prior Work on Multiple Access Models. As mentioned earlier, the literature on probabilistic space-bounded Turing machines with multiple access to random bits is limited compared to the standard model. Borodin, Cook, and Pippenger [3], while investigating deterministic simulations of probabilistic space-bounded classes, raised the question whether two-way probabilistic $s(n)$-space-bounded machines can be simulated deterministically in $O(s^2(n))$ space. Karpinsky and Verbeek [13] showed that the answer is negative in general. They showed that two-way log-space probabilistic machines that are allowed to run for time $2^{n^{O(1)}}$ time can simulate PSPACE with zero error probability. Another relevant result is due to Nisan [16] who showed BPL can be simulated by zero-error, probabilistic, space-bounded machines that has a two-way access to the random tape. Nisan also showed that 2-*way*BPL is same as almost-logspace, where almost-logspace is the class of languages accepted by deterministic log-space machines relative to a random oracle.

Probabilistic space-bounded machines that can make multiple passes over the random tape was first considered by David, Papakonstantinou, and Sidiropoulos [6]. They showed that any pseudo-random generator that fools traditional $k(n)s(n)$-space bounded machines can also fool $k(n)$-pass $s(n)$-space bounded machines. As a corollary, they obtain that polylog-pass, randomized log-space is contained in deterministic polylog-space. David, Nguyen, Papakonstantinou, and Sidiropoulos [5] considered probabilistic space-bounded machine that have access to a stack and a two-way/multiple pass access to the random tape and related them to traditional classes.

2 Preliminaries

We assume familiarity with the standard complexity classes [2]. We are interested in probabilistic space-bounded machines that can access random bits multiple times. For such machines, the random bits appear on a special read-only tape called the *random tape*. In addition to the random tape, these machines have one read-only input tape and few read-write work tapes, as standard in space-bounded machine models. The total space used by the work tapes signify the space bound of such a machine. We also assume that all the probabilistic space-bounded machines halt in time at most exponential in their space bounds. Thus, the number of random bits used by them is at most exponential in their space-bounds. More formally, our multi-pass machines are defined as follows:

Definition 1. *A language L is in $k(n)$-pass BPSPACE$[s(n), r(n)]$ if there exists an $O(s(n))$-space bounded probabilistic Turing machine M such that on any input x of length n:*

- *M makes $k(n)$ passes over the random tape, during each pass it accesses the contents of random tape in a one-way manner,*
- *M uses at most $O(r(n))$ random bits, and*
- *the probability that M incorrectly decides x is at most $1/3$.*

In our notation, BPL = 1-*pass* BPSPACE$[\log n, n^{O(1)}]$. In Sect. 3, we observe that a constant factor in the number of passes does not add power to the model and hence $O(1)$-*pass* BPSPACE$[\log n, n^{O(1)}]$ is also same as BPL.

Since we assume that every space-bounded probabilistic machine halts on all settings of random tape on all inputs, the running time of the multi-pass machine is bounded by $2^{O(s(n))}$ where $s(n)$ is the space bound. Thus, this machine can only access random tape of length $2^{O(s(n))}$. Indeed, when the number of random bits is exponential in space, i.e., $r(n) = 2^{O(s(n))}$, we simply write the above class as $k(n)$-*pass* BPSPACE$(s(n))$. Further, when the space of the $k(n)$-pass BPSPACE machine is bounded by $O(\log n)$, we simply write the class as $k(n)$-*pass* BPL$[r(n)]$.

$\log^{(k)}(\cdot)$ is the iterated logarithmic function applied k times with itself.

3 Space-Bounded Machines with Multiple Passes Over the Random Tape

In this section, we consider probabilistic space-bounded machines that are allowed to make multiple passes over the random tape. First, we establish results that connect derandomization of such machines to derandomization of probabilistic time classes. Next, we consider the problem of simulating multi-pass, probabilistic, space-bounded machines with the traditional one-pass machines. Finally, we provide a space-efficient deterministic nonuniform simulation of the multi-pass machines.

3.1 Derandomization of Probabilistic Time

As our main result of this section, we show that a time-efficient derandomization of probabilistic log-space machines that use very few random bits and make very few passes over their random tape, yields a non-trivial derandomization of probabilistic time. In particular, we show the following theorem.

Theorem 2. *If for some constant $k > 0$, $O(\log^{(k)} n)$-pass* BPL $\left[\log n \log^{(k+3)} n\right]$ *is in* P, *then* BPTIME$(n) \subseteq$ DTIME$(2^{o(n)})$.

Remark. We use the iterated logarithmic function for simplicity, but the above theorem can be proved with any "nice" slowly growing function $f(n) \in \omega(1)$.

We establish the theorem by first proving that every BPTIME(n) machine can be simulated by a bounded-error probabilistic space-bounded machine that makes $O(\log^{(k)} n)$ passes over the random tape and uses $o(n)$ space, on inputs of size n. There is a trade-off between the number of passes and space used by the simulating machine and this trade-off is essential in the proof. More formally, we prove the following theorem.

Theorem 3. *For every constant $k > 0$,*

$$\text{BPTIME}(n) \subseteq O(\log^{(k)} n)\text{-pass BPSPACE}\left[n/\log^{(k+3)} n, n\right].$$

Remark. The above theorem may be of independent interest. Hopcroft, Paul, and Valiant [10] showed that DTIME$(n) \subseteq$ DSPACE$(o(n))$. The analogous inclusion relationship for probabilistic classes is not known. That is, we do not know unconditionally if BPTIME(n) is a subset of BPSPACE$(o(n))$ (see [12,14] for conditional results in this direction). The above theorem can be viewed as a partial solution; if we allow the space-bounded machine to have a slightly non-constant number of passes, then BPTIME(n) can be simulated in $o(n)$ probabilistic space.

We first give the proof of Theorem 2 assuming Theorem 3. The proof uses a simple padding argument.

Proof of Theorem 2. Let $k > 0$ be a constant for which the hypothesis in the statement of Theorem 2 holds. Let L be a language in BPTIME(n). By Theorem 3, L is in $O(\log^{(k-1)} n)$-*pass* BPSPACE $\left[n / \log^{(k+2)} n, n \right]$. Let

$$L' = \left\{ \langle x, 0^{2^{n / \log^{(k+2)} n} - n} \rangle \mid x \in L, |x| = n \right\}.$$

It is easy to see that L' is in $O(\log^{(k)} n)$-*pass* BPL $\left[\log n \log^{(k+3)} n \right]$. By our hypothesis, L' is in P. So $L' \in$ DTIME(n^ℓ) for some $\ell > 0$. From this it follows that for some $\ell > 0$, L is in DTIME $\left(2^{\frac{\ell n}{\log^{(k+2)} n}} \right)$ and thus in DTIME($2^{o(n)}$). □

Now we move on to proving Theorem 3. The proof relies on the classical result of Hopcroft, Paul, and Valiant [10] who showed that every deterministic machine running in time $O(n)$ can be simulated by a deterministic machine that uses $O(n / \log n)$ space. If we adopt their proof to the case of probabilistic machines, we obtain that every bounded-error probabilistic machine running in time $O(n)$ can be simulated by a bounded-error, probabilistic machine that uses space $O(n / \log n)$; however the simulating machine makes an exponential number of passes over the random tape. We observe that the number of passes can be greatly reduced at the expense of a little increase in space. This is essentially achieved by using a careful choice of parameters than those used in [10].

To proceed with the proof we need the notions of *block-respecting Turing machines* and *pebbling games* [10].

Definition 2. *Let M be a multi-tape Turing machine running in time $t(n)$. Let $b(n)$ be a function such that $1 \leq b(n) \leq t(n)/2$. Divide the computation of M into $a(n)$ time segments so that each segment has $b(n) = t(n)/a(n)$ steps. Also divide each tape of M into $a(n)$ blocks so that each block has exactly $b(n)$ cells. We say that the machine M is $b(n)$-block respecting if during each time segment every tape head of the machine M visits cells of exactly one block. I.e, a tape head can cross a block boundary only at time step $c \cdot b(n)$ for some integer $c > 0$.*

Hopcroft, Paul, and Valiant showed that every ℓ-tape Turing machine running in time $t(n)$ can be simulated by a $(\ell + 1)$ tape $b(n)$-block respecting Turing machine running in time $O(t(n))$ for any $b(n)$ such that $1 \leq b(n) \leq t(n)/2$.

Pebbling Game. Let G be a directed acyclic graph and w be a special vertex of the graph. We say that a vertex u is a predecessor of vertex v if there is an edge from u to v. The goal is to place a pebble on the special vertex w using as little pebbles as possible, subject to the following constraints: we can place a pebble on a vertex v only if all predecessors of v have pebbles on them, a pebble can be removed from a vertex at any time.

Hopcroft, Paul, and Valiant showed (by means of a clever divide-and-conquer algorithm) that every bounded-degree graph with n vertices can be pebbled using $O(n / \log n)$ pebbles, and there is a deterministic algorithm S that does this pebbling in time $O(2^{n^2})$. Now we are ready to prove Theorem 3.

Proof of Theorem 3. Fix $k > 0$, and set $b(n) = n/\log^{(k+2)} n$. Let L be a language in BPTIME(n) and let M be an ℓ-tape, $b(n)$-block respecting probabilistic machine that accepts L in time $t(n) = O(n)$. Set $a(n) = t(n)/b(n)$. Without loss of generality, we can assume that M reads the contents of the random-tape in a one-way manner. The computation graph G_M of M is an edge-labeled graph defined as follows. The vertex set is $V = \{1, \cdots, a(n)\}$. For $1 \leq i < a(n)$, $\langle i, i+1 \rangle \in E$ with label 0, implying the computation at time segment $i+1$ requires the computation of the time segment i. Assume that the tape heads are numbered $1, \cdots, \ell$. We place an edge from i to j with label $h \in [1, \ell]$ if the following holds: Suppose that during the time segment i the tape head h is in some block b and the next time segment that the tape head h revisits block b is j (*i.e.*, the computation at time segment j requires the content of the block b from the time-segment i). This process defines a multi-graph.

Given $1 \leq i \leq a(n)$, let $B_1(i), \cdots, B_\ell(i)$ be the blocks that each of the ℓ tape heads are visiting during time segment i. Let $C(i)$ be a string that describes the contents of blocks $B_1(i), \cdots, B_\ell(i)$ and the state q at the end of time segment i. The following observation is crucial.

Observation 1. *Suppose a vertex j has predecessors $i_1, \cdots, i_r$ $(1 \leq r \leq \ell)$. Then we can simulate the computation of M during time segment j by knowing $C(i_1), \cdots, C(i_r)$. Thus we can compute $C(j)$.*

Using this observation, as in [10], we simulate M by a machine M' as follows. We describe a simulation algorithm that gets a bounded degree graph (degree $\leq \ell + 1$) G as input and attempts to simulate M.

> Call the pebbling algorithm S on graph G. If S places pebble on vertex i, then compute $C(i)$ and store $C(i)$. If S removes pebble from a vertex i, then erase $C(i)$.

Note that a priori, we do not know the correct computation graph G_M. We will first assume that the correct computation graph G_M is known to us and thus can be given as input to the above algorithm. Latter we will remove this restriction. By Observation 1, it follows that the above algorithm correctly simulates M (under the assumption that G_M is known). Now we bound the space, time, number of passes, and number of random bits required by the above simulation algorithm (under the assumption that G_M is known). We start with the following two claims.

Claim 1. The total space used by the above simulation is $O\left(\frac{a(n)b(n)}{\log a(n)} + 2^{a^2(n)}\right)$.

Claim 2. The above simulation algorithm makes at most $O(2^{a^2(n)})$ passes over the random tape.

Now we address the assumption that the computation graph G_M is known.

Observation 2. *Suppose that G is not the correct computation graph. If the above simulation algorithm gets G as input, then it will discover that G is not the correct computation graph, using $O(\frac{a(n)b(n)}{\log a(n)} + 2^{a^2(n)})$ space and by making $O(2^{a^2(n)})$ passes over the random tape.*

So our final simulation of M proceeds as follows: Iterate through all possible computation graphs; for each graph G, attempt to simulate M using the above algorithm. If it discovers that G is not a correct computation graph, then proceed to next graph.

By Claim 1 and Observation 2, each iteration needs $O(\frac{a(n)b(n)}{\log b(n)} + 2^{a^2(n)})$ space. Since we can reuse space from one iteration to the next iteration, total space is bounded by $O(\frac{a(n)b(n)}{\log b(n)} + 2^{a^2(n)})$. By Claim 2 and Observation 2, each iteration can be done making $2^{a^2(n)}$ passes. Since there are at most $2^{a^2(n)}$ possible computation graphs, the total number of passes is $2^{2a^2(n)}$.

By plugging in the values of $a(n)$ and $b(n)$ from above, we obtain that the space used by the simulating machine is $O(n/\log^{(k+3)} n)$ and the number of passes is $O(\log^{(k)} n)$. Finally, note that the number of random bits used by the simulating machine remains same as the number of random bits used by M, i.e., $O(n)$. This completes the proof of the Theorem.   □

The proofs of Claims 1, 2, and Observation 2 will appear in the full version of the paper.

3.2 Simulating Multiple Passes with Single Pass

An obvious question at this point is the following: Can we simulate multi-pass probabilistic machines with traditional one-pass probabilistic space bounded machines? The main result of this subsection shows that passes can be traded for space. This helps us to obtain an upper bound on the deterministic space to simulate a multi-pass probabilistic space-bounded machine. We first start with the following lemma whose proof appears in the full version of the paper.

Lemma 1. *If a language L is in $k(n)$-pass BPSPACE$[s(n), r(n)]$, then there is a probabilistic $O(k(n)s(n))$-space bounded machine N that has one-way access to the random tape and for every $x \in \Sigma^n$,*

$$\Pr[N(x) = L(x)] \geq \frac{1}{2} + \frac{1}{2^{O(k(n)s(n))}}.$$

Moreover, N uses $O(r(n) + k(n)s(n))$ random bits.

Using above Lemma, we obtain the following.

Theorem 4. $k(n)$-pass BPSPACE$(s(n)) \subseteq$ BPSPACE$(k(n)s(n))$.

Proof. Let L be a language that is accepted by a $k(n)$-pass BPSPACE$(s(n))$ machine M. By definition, this machine uses $2^{O(s(n))}$ random bits. Thus by Lemma 1, there is an $O(k(n)s(n))$-space bounded, one-pass, probabilistic machine N that uses $O(2^{O(s(n))} + k(n)s(n))$ random bits, and

$$Pr[L(x) = N(x)] \geq \frac{1}{2} + \frac{1}{2^{O(k(n)s(n))}}.$$

We can amplify the success probability of N to 2/3 by simulating it $2^{O(k(n)s(n))}$ times and taking the majority vote. This will use $2^{O(k(n)s(n))}$ random bits. Thus we obtain a $O(k(n)s(n))$-space bounded machine that uses $2^{O(k(n)s(n))}$ random bits. Thus L is in BPSPACE($k(n)s(n)$).   □

Corollary 1. $O(1)$-*pass* BPL = BPL.

We relate the above lemma to the result of [6].

Theorem 5. *Let M be a $k(n)$-pass, $s(n)$-space bounded machine that uses $r(n)$ random bits. Any pseudo-random generator that fools $k(n)s(n)$-space bounded machines (that read their input in a one-way manner) running on $r(n)$-bit input strings also fools M.*

Their result states that to deterministically simulate $k(n)$-pass, $s(n)$-space bounded probabilistic machines, a pseudo-random generator against standard $O(k(n)s(n))$-space bounded probabilistic machine suffices. Lemma 1 can be interpreted as an explanation of their result, as it shows that any $k(n)$-pass, $s(n)$-space bounded machine can indeed be simulated by a standard $O(k(n)s(n))$ space bounded machine.

Next, we consider the main result of this subsection: deterministic simulation of the class $k(n)$-*pass* BPSPACE($s(n)$). By above theorem, this is a subclass of BPSPACE($k(n)s(n)$). By the celebrated results of Nisan [15] and Saks and Zhou [21], it follows that this class is a subset of DSPACE($k^{3/2}(n)s^{3/2}(n)$). Observe below that we can get rid of the polynomial factor off the number of passes, more formally,

Theorem 6. $k(n)$-*pass* BPSPACE($s(n)$) $\subseteq$ DSPACE($k(n)s^{3/2}(n)$).

Proof. Let L be a language that is accepted by a $k(n)$-*pass* BPSPACE($s(n)$) machine M. Since M halts in $2^{O(s(n))}$ time, $k(n)$ is bounded by $2^{O(s(n))}$ and M uses $2^{O(s(n))}$ random bits. Thus, by Lemma 1, there is an $O(k(n)s(n))$-space bounded, one-pass, probabilistic machine N that uses $2^{O(s(n))}$ random bits and

$$Pr[L(x) = N(x)] \geq \frac{1}{2} + \frac{1}{2^{O(k(n)s(n))}}.$$

Saks and Zhou [21], building on Nisan's [15] work showed that any language accepted by a probabilistic machine using $O(s(n))$-space, $2^{O(r(n))}$ random bits, with success probability as low as $1/2 + 1/2^{O(s(n))}$, is in deterministic space $O(s(n)r^{1/2}(n))$. Applying this to our case, we obtain that N can be simulated by a deterministic space-bounded machine that uses $O(k(n)s^{3/2}(n))$.   □

Recall that our hypothesis in Theorem 2 states that for some $k > 0$ if the complexity class $\log^{(k)} n\text{-}pass$ $\mathrm{BPL}(\log n \log^{(k+2)} n)$ is in P. We obtain the following upper bound for this class, by applying Lemma 1 and the techniques of Saks and Zhou [21]. The proof is similar to the proof of Theorem 6.

Corollary 2. *For any constants $k \geq 3$ and $\epsilon > 0$,*

$$\log^{(k)} n\text{-}pass\ \mathrm{BPL}[\log n \log^{(k+3)} n] \subseteq \mathrm{DSPACE}(\log n (\log \log n)^{\frac{1}{2}+\epsilon})$$
$$\subseteq \mathrm{DTIME}(n^{(\log \log n)^{1/2+\epsilon}}).$$

3.3 Deterministic Simulation of Multi-pass Machines with Linear Advice

Fortnow and Klivans [7] showed that standard (one-pass) randomized log-space machines (BPL) can be simulated by deterministic log-space machines that have access to a linear amount of advice. I.e., they showed that BPL is a subset of $L/O(n)$. On the other hand, using Adleman's technique [1], it can be shown that randomized log-space machines with two-way access to the random tape can be simulated in deterministic log-space using a polynomial amount of advice [16]. Thus, any multi-pass, randomized log-space machine can be simulated in deterministic log-space with polynomial amount of advice. Can we bring down the advice to linear? We show that this is indeed possible with a small increase in space.

Let M be a $O(\log n)$-pass, randomized log-space machine. By Theorem 4, M can be simulated by a one-pass randomized machine that uses $O(\log^2 n)$ space, and by applying the techniques of Fortnow and Klivans [7], it follows that M can be simulated in deterministic space $O(\log^2 n)$ with linear advice. Below we show that we can improve the space bound of the deterministic machine to $O(\log n \log \log n)$. More formally, we prove the following.

Theorem 7. *For any $k(n) \in \omega(1)$, a $k(n)$-pass $s(n)$-space bounded randomized machine using $R(n) = 2^{r(n)}$ random bits can be simulated by a deterministic machine using $O(s(n) + r(n) \log(k(n)s(n)))$ space that uses an advice of size $O(r(n)k(n)s(n) + n)$. I.e.,*

$$k(n)-pass\ \mathrm{BPSPACE}[s(n), 2^{r(n)}] \subseteq$$
$$\mathrm{DSPACE}(s(n) + r(n) \log k(n)s(n))/O(r(n)k(n)s(n) + n).$$

Before we sketch a proof of the theorem, we note the following corollaries.

Corollary 3. *For every constant $k > 0$,*

$$O(\log^k n)\text{-}pass\mathrm{BPL} \subseteq \mathrm{DSPACE}(\log n \log \log n)/O(n).$$

Corollary 4. *For every $0 < \epsilon < 1$, $n^\epsilon\text{-}pass$ $\mathrm{BPL} \subseteq \mathrm{DSPACE}(\log^2 n)/O(n)$.*

Proof Sketch of Theorem 7. Consider a $k(n)$-pass $s(n)$-space bounded one-sided randomized machine M that uses $R(n) = 2^{r(n)}$ random bits. By Theorem 5, any pseudorandom generator that fools standard (one-pass) $O(k(n)s(n))$ space-bounded machine using $2^{r(n)}$ random bits also fools M. For our proof, we use Nisan's generator [15]. Note that Nisan's generator, for our choice of parameters of space and random bits, stretches a seed of length $\ell(n) = O(r(n)k(n)s(n))$ to $2^{r(n)}$. The seed for this pseudorandom generator is a tuple consisting of $r(n)$ hash functions from $\{0,1\}^{O(s(n)k(n))}$ to $\{0,1\}^{O(s(n)k(n))}$ and one string of length $O(s(n)k(n))$. Let us denote the hash functions by $h_1, \cdots, h_{r(n)}$ and the string by y. Consider the following simulation M' of M: M' has a seed of length $\ell(n)$ on its random tape. On any input of length n, it will apply Nisan's generator on the contents of random tape and simulate M on the output of the generator. Note that M' accesses the contents of the random tape in a 2-way manner.

By applying Adleman's technique to M', we can fix n seeds which will act as good random string for all strings of length n. We can hardwire the n seeds (each of length $\ell(n)$) and obtain a deterministic simulation. Note that the space used by this simulation is $O(s(n)k(n))$ as we need $O(s(n)k(n))$ space to simulate Nisan's generator and the length of the advice is $O(n\ell(n))$. However, we observe that for any i, the ith bit of the output of the generator can in fact be computed in space $O(r(n) \times \log(k(n)s(n)))$. This is because, the output of Nisan's generator is a concatenation of blocks of strings where each block is of length $O(k(n)s(n))$. Each block is obtained by composing $r(n)$ hash functions (in some predetermined order based on the index of the block) on the input y. Note that the output of each individual hash function can be computed in $O(\log(s(n)k(n)))$ space. Thus, composition of $r(n)$ hash functions can be computed in $O(r(n)\log(s(n)k(n)))$ space. So we get that a $k(n)$-pass $s(n)$-space-bounded randomized machine using $R(n) = 2^{r(n)}$ random bits can be simulated deterministically in space $O(s(n) + r(n)\log(k(n)s(n)))$ with $O(n \times \ell(n))$ length advice. Instead of using n independent seeds of length $\ell(n)$, we do a random walk on an expander graph of size $O(2^{\ell(n)})$ and reduce the advice size to $O(n + \ell(n))$, as done by Fortnow and Klivans [7] (using the work of Gutfreund and Viola [9]).

The proof can be extended to two-sided error multi-pass Turing machines as well, as mentioned in [7]. We omit the details.   □

4   Conclusions

This paper establishes that time efficient derandomization of probabilistic, log-space machines that make a non-constant passes over the random tape yields a new non-trivial derandomization of probabilistic time. This result suggests that it is fruitful to further study multi-pass, probabilistic, space-bounded machines. One interesting question that arises is on error reduction. Let M be a $k(n)$-pass, $s(n)$-space bounded, bounded-error probabilistic space-bounded machine with error probability less than $1/3$. Can we reduce the error probability to $1/2^{e(n)}$ for a polynomial e without substantial increase in passes and space used? Note that by increasing the number of passes to $O(k(n)e(n))$ this is indeed possible.

On the other hand, if we were not to increase the number of passes, then by increasing the space to $O(e(n)s(n))$ we can achieve the same reduction in error probability. Can we do better? Can we reduce the error probability to $1/2^{e(n)}$ while keeping the number of passes to $O(k(n))$ and the space bound to $O(s(n))$?

References

1. Adleman, L.M.: Two theorems on random polynomial time. In: Proceedings of the 19th IEEE Symposium on Foundations of Computer Science (FOCS), pp. 75–83 (1978)
2. Arora, S., Barak, B.: Computational Complexity: A Modern Approach. Cambridge University Press, Cambridge (2009)
3. Borodin, A., Cook, S., Pippenger, N.: Parallel computation for well-endowed rings and space-bounded probabilistic machines. Inf. Control 58(1–3), 113–136 (1983)
4. Chung, K.M., Reingold, O., Vadhan, S.: S-T connectivity on digraphs with a known stationary distribution. ACM Trans. Algorithms 7(3), 30 (2011)
5. David, M., Nguyen, P., Papakonstantinou, P.A., Sidiropoulos, A.: Computationally Limited Randomness. In: Proceedings of the Innovations in Theoretical Computer Science (ITCS) (2011)
6. David, M., Papakonstantinou, P.A., Sidiropoulos, A.: How strong is Nisan's pseudo-random generator? Inf. Process. Lett. 111(16), 804–808 (2011)
7. Fortnow, L., Klivans, A.R.: Linear advice for randomized logarithmic space. In: Durand, B., Thomas, W. (eds.) STACS 2006. LNCS, vol. 3884, pp. 469–476. Springer, Heidelberg (2006)
8. Gill, J.: Computational complexity of probabilistic Turing machines. SIAM J. Comput. 6(4), 675–695 (1977)
9. Gutfreund, D., Viola, E.: Fooling parity tests with parity gates. In: Jansen, K., Khanna, S., Rolim, J.D.P., Ron, D. (eds.) RANDOM 2004 and APPROX 2004. LNCS, vol. 3122, pp. 381–392. Springer, Heidelberg (2004)
10. Hopcroft, J.H., Paul, W.J., Valiant, L.G.: On Time Versus Space. J. ACM 24(2), 332–337 (1977)
11. Impagliazzo, R., Nisan, N., Wigderson, A.: Pseudorandomness for network algorithms. In: Proceedings of the 26th ACM Symposium on Theory of Computing (STOC), pp. 356–364 (1994)
12. Karakostas, G., Lipton, R.J., Viglas, A.: On the complexity of intersecting finite state automata and NL versus NP. Theor. Comput. Sci. 302, 257–274 (2003)
13. Karpinski, M., Verbeek, R.: There is no polynomial deterministic space simulation of probabilistic space with a two-way random-tape generator. Inf. Control 67(1985), 158–162 (1985)
14. Lipton, R.J., Viglas, A.: Non-uniform depth of polynomial time and space simulations. In: Lingas, A., Nilsson, B.J. (eds.) FCT 2003. LNCS, vol. 2751, pp. 311–320. Springer, Heidelberg (2003)
15. Nisan, N.: Pseudorandom generators for space-bounded computation. Combinatorica 12(4), 449–461 (1992)
16. Nisan, N.: On read once vs. multiple access to randomness in logspace. Theor. Comput. Sci. 107(1), 135–144 (1993)
17. Raz, R., Reingold, O.: On recycling the randomness of states in space bounded computation. In: Proceedings of the 31st ACM Symposium on Theory of Computing (STOC), pp. 159–168 (1999)

18. Reingold, O.: Undirected connectivity in log-space. J. ACM **55**(4), 1–24 (2008)
19. Reingold, O., Trevisan, L., Vadhan, S.: Pseudorandom walks on regular digraphs and the RL vs. L problem. In: Proceedings of the 38th ACM Symposium on Theory of Computing (STOC), p. 457 (2006)
20. Saks, M.: Randomization and derandomization in space-bounded computation. In: Proceedings of the 11th IEEE Conference on Computational Complexity (1996)
21. Saks, M., Zhou, S.: BPSPACE(S) $\subseteq$ DSPACE($S^{3/2}$). J. Comput. Syst. Sci. **403**, 376–403 (1999)
22. Santhanam, R., van Melkebeek, D.: Holographic proofs and derandomization. SIAM J. Comput. **35**(1), 59–90 (2005)

Densest Subgraph in Dynamic Graph Streams

Andrew McGregor$^{(\boxtimes)}$, David Tench, Sofya Vorotnikova, and Hoa T. Vu

University of Massachusetts, Amherst, USA
{mcgregor,dtench,svorotni,hvu}@cs.umass.edu

Abstract. In this paper, we consider the problem of approximating the densest subgraph in the dynamic graph stream model. In this model of computation, the input graph is defined by an arbitrary sequence of edge insertions and deletions and the goal is to analyze properties of the resulting graph given memory that is sub-linear in the size of the stream. We present a single-pass algorithm that returns a $(1+\epsilon)$ approximation of the maximum density with high probability; the algorithm uses $O(\epsilon^{-2} n \operatorname{polylog} n)$ space, processes each stream update in $\operatorname{polylog}(n)$ time, and uses $\operatorname{poly}(n)$ post-processing time where n is the number of nodes. The space used by our algorithm matches the lower bound of Bahmani et al. (PVLDB 2012) up to a poly-logarithmic factor for constant ϵ. The best existing results for this problem were established recently by Bhattacharya et al. (STOC 2015). They presented a $(2 + \epsilon)$ approximation algorithm using similar space and another algorithm that both processed each update and maintained a $(4 + \epsilon)$ approximation of the current maximum density in $\operatorname{polylog}(n)$ time per-update.

1 Introduction

In the dynamic graph stream model of computation, a sequence of edge insertions and deletions defines an input graph and the goal is to solve a specific problem on the resulting graph given only one-way access to the input sequence and limited working memory. Motivated by the need to design efficient algorithms for processing massive graphs, over the last four years there has been a considerable amount of work designing algorithms in this model [1–5,8,9,11,17,19,21,22,24,25]. Specific results include testing edge connectivity [3] and node connectivity [19], constructing spectral sparsifiers [21], approximating the densest subgraph [8], maximum matching [5,9,11,24], correlation clustering [1], and estimating the number of triangles [25]. For a recent survey of the area, see [27].

In this paper, we consider the densest subgraph problem. Let G_U be the induced subgraph of graph $G = (V, E)$ on nodes U. Then the *density* of G_U is defined as

$$d(G_U) = |E(G_U)|/|U|,$$

where $E(G_U)$ is the set of edges in the induced subgraph. We define the *maximum density* as

This work was supported by NSF Awards CCF-0953754, IIS-1251110, CCF-1320719, and a Google Research Award.

G.F. Italiano et al. (Eds.): MFCS 2015, Part II, LNCS 9235, pp. 472–482, 2015.
DOI: 10.1007/978-3-662-48054-0_39

$$d^* = \max_{U \subseteq V} d(G_U).$$

and say that the corresponding subgraph is the *densest subgraph*. The densest subgraph can be found in polynomial time [10,15,18,23] and more efficient
approximation algorithms have been designed [10]. Finding dense subgraphs is an
important primitive when analyzing massive graphs; applications include community detection in social networks and identifying link spam on the web, in
addition to applications on financial and biological data. See [26] for a survey of
applications and existing algorithms for the problem.

1.1 Our Results and Previous Work

We present a single-pass algorithm that returns a $(1+\epsilon)$ approximation with high
probability[1]. For a graph on n nodes, the algorithm uses the following resources:

- *Space:* $O(\epsilon^{-2} n \operatorname{polylog} n)$. The space used by our algorithm matches the lower
  bound of Bahmani et al. [7] up to a poly-logarithmic factor for constant ϵ.
- *Per-update time:* $\operatorname{polylog}(n)$. We note that this is the worst-case update time
  rather than amortized over all the edge insertions and deletions.
- *Post-processing time:* $\operatorname{poly}(n)$. This will follow by using any exact algorithm
  for densest subgraph [10,15,18] on the subgraph generated by our algorithm.

The most relevant previous results for the problem were established recently
by Bhattacharya et al. [8]. They presented two algorithms that use similar space
to our algorithm and process updates in $\operatorname{polylog}(n)$ amortized time. The first
algorithm returns a $(2 + \epsilon)$ approximation of the maximum density of the final
graph while the second (the more technically challenging result) outputs a $(4+\epsilon)$
approximation of the current maximum density after every update while still
using only $\operatorname{polylog}(n)$ time per-update. Our algorithm improves the approximation factor to $(1+\epsilon)$ while keeping the same space and update time. It is possible
to modify our algorithm to output a $(1 + \epsilon)$ approximation to the current maximum density after each update but the simplest approach would require the
post-processing step to be run after every edge update and this would not be
efficient.

Bhattacharya et al. were one of the first to combine the space restriction
of graph streaming with the fast update and query time requirements of fully-
dynamic algorithms from the dynamic graph algorithms community. Epasto,
Lattanzi, and Sozio [14] present a fully-dynamic algorithm that returns a $(2 + \epsilon)$
approximation of the current maximum density. Other relevant work includes
papers by Bahmani, Kumar, and Vassilvitskii [7] and Bahmani, Goel, and
Munagala [6]. The focus of these papers is on designing algorithms in the MapReduce model but the resulting algorithms can also be implemented in the data
stream model if we allow multiple passes over the data.

---

[1] Throughout this paper, we say an event holds with high probability if the probability
is at least $1 - n^{-c}$ for some constant $c > 0$.

1.2   Our Approach and Paper Outline

The approach we take in this paper is as follows. In Sect. 2, we show that if we sample every edge of a graph independently with a specific probability then we generate a graph that is a) sparse and b) can be used to estimate the maximum density of the original graph. This is not difficult to show but requires care since there are an exponential number of subgraphs in the subsampled graph that we will need to consider.

In Sect. 3, we show how to perform this sampling in the dynamic graph stream model. This can be done using the ℓ_0 sampling primitive [12,20] that enables edges to be sampled uniformly from the set of edges that have been inserted but not deleted. However, a naive application of this primitive would necessitate $\Omega(n)$ per-update processing. To reduce this to $O(\text{polylog } n)$ we reformulate the sampling procedure in such a way that it can be performed more efficiently. This reformulation is based on creating multiple partitions of the set of edges using pairwise independent hash functions and then sampling edges within each group in the partition. The use of multiple partitions is somewhat reminiscent of that used in the Count-Min sketch [13].

2   Subsampling Approximately Preserves Maximum Density

In the section, we consider properties of a random subgraph of the input graph G. Specifically, let G' be the graph formed by sampling each edge in G independently with probability p where

$$p = c\epsilon^{-2} \log n \cdot \frac{n}{m}$$

for some sufficiently large constant $c > 0$ and $0 < \epsilon < 1/2$. We may assume that m is sufficiently large such that $p < 1$ because otherwise we can reconstruct the entire graph in the allotted space using standard results from the sparse recovery literature [16].

We will prove that, with high probability, the maximum density of G can be estimated up to factor $(1+\epsilon)$ given G'. While it is easy to analyze how the density of a specific subgraph changes after the edge sampling, we will need to consider all 2^n possible induced subgraphs and prove properties of the subsampling for all of them.

The next lemma shows that $d(G'_U)$ is roughly proportional to $d(G_U)$ if $d(G_U)$ is "large" whereas if $d(G_U)$ is "small" then $d(G'_U)$ will also be relatively small.

Lemma 1. *Let U be an arbitrary set of k nodes. Then,*

$$\mathbb{P}\left[d(G'_U) \geq pd^*/10\right] \leq n^{-10k} \qquad \textit{if } d(G_U) \leq d^*/60$$

$$\mathbb{P}\left[|d(G'_U) - pd(G_U)| \geq \epsilon pd(G_U)\right] \leq 2n^{-10k} \qquad \textit{if } d(G_U) > d^*/60 \ .$$

Proof. We start by considering the density of the entire graph $d(G) = m/n$ and therefore conclude that the maximum density, d^*, is at least m/n. Hence, $p \geq (c\epsilon^{-2} \log n)/d^*$.

Let X be the number of edges in G'_U and note that $\mathbb{E}[X] = pkd(G_U)$. First assume $d(G_U) \leq d^*/60$. Then, by an application of the Chernoff Bound (e.g., [28, Theorem 4.4]), we observe that

$$\mathbb{P}[d(G'_U) \geq pd^*/10] = \mathbb{P}[X \geq pkd^*/10] \leq 2^{-pkd^*/10} < 2^{-ck(\log n)/10}$$

and this is at most n^{-10k} for sufficiently large constant c.

Next assume $d(G_U) > d^*/60$. Hence, by an application of an alternative form of the Chernoff Bound (e.g., [28, Theorems 4.4 and 4.5]), we observe that

$$\begin{aligned}
\mathbb{P}[|d(G'_U) - pd(G_U)| \geq \epsilon pd(G_U)] &= \mathbb{P}[|X - pkd(G_U)| \geq \epsilon pkd(G_U)] \\
&\leq 2\exp(-\epsilon^2 pkd(G_U)/3) \\
&\leq 2\exp(-\epsilon^2 pkd^*/180) \\
&\leq 2\exp(-ck(\log n)/180).
\end{aligned}$$

and this is at most $2n^{-10k}$ for sufficiently large constant c.   □

Corollary 1. *With high probability, for all $U \subseteq V$:*

$$d(G'_U) \geq (1 - \epsilon)pd^* \quad \Rightarrow \quad d(G_U) \geq \frac{1 - \epsilon}{1 + \epsilon} \cdot d^*.$$

Proof. There are $\binom{n}{k} \leq n^k$ subsets of V that have size k. Hence, by appealing to Lemma 1 and the union bound, with probability at least $1 - 2n^{-9k}$, the following two equations hold,

$$d(G'_U) \geq pd^*/10 \quad \Rightarrow \quad d(G_U) > d^*/60$$

$$d(G_U) > d^*/60 \quad \Rightarrow \quad d(G_U) \geq \frac{d(G'_U)}{p(1 + \epsilon)}$$

for all $U \subseteq V$ such that $|U| = k$. Since $(1 - \epsilon)pd^* \geq pd^*/10$, together these two equations imply

$$d(G'_U) \geq (1 - \epsilon)pd^* \quad \Rightarrow \quad d(G_U) \geq \frac{d(G'_U)}{p(1 + \epsilon)} \geq \frac{1 - \epsilon}{1 + \epsilon} \cdot d^*$$

for all sets U of size k. Taking the union bound over all values of k establishes the corollary.   □

We next show that the densest subgraph in G' corresponds to a subgraph in G that is almost as dense as the densest subgraph in G.

Theorem 1. *Let $U' = \mathrm{argmax}_U \, d(G'_U)$. Then with high probability,*

$$\frac{1 - \epsilon}{1 + \epsilon} \cdot d^* \leq d(G_{U'}) \leq d^*.$$

Proof. Let $U^* = \mathrm{argmax}_U \, d(G_U)$. By appealing to Lemma 1, we know that $d(G'_{U^*}) \geq (1 - \epsilon)pd^*$ with high probability. Therefore

$$d(G'_{U'}) \geq d(G'_{U^*}) \geq (1 - \epsilon)pd^*,$$

and the result follows by appealing to Corollary 1.   □

3   Implementing in the Dynamic Data Stream Model

In this section, we show how to sample each edge independently with the pre-scribed probability in the dynamic data stream model. The resulting algorithm uses $O(\epsilon^{-2}n\operatorname{polylog} n)$ space. The near-linear dependence on n almost matches the $\Omega(n)$ lower bound proved by Bahmani et al. [7]. The main theorem we prove is:

Theorem 2. *There exists a randomized algorithm in the dynamic graph stream model that returns a $(1 + \epsilon)$-approximation for the density of the densest sub-graph with high probability. The algorithm uses $O(\epsilon^{-2}n\operatorname{polylog} n)$ space and $O(\operatorname{polylog} n)$ update time. The post-processing time of the algorithm is polyno-mial in n.*

To sample the edges with probability p in the dynamic data stream model there are two main challenges:

1. Any edge we sample during the stream may subsequently be deleted.
2. Since p depends on m, we do not know the value of p until the end of the stream.

To address the first challenge, we appeal to an existing result on the ℓ_0 sampling technique [20]: there exists an algorithm using $\operatorname{polylog}(n)$ space and update time that returns an edge chosen uniformly at random from the final set of edges in the graph. Consequently we may sample r edges uniformly at random using $O(r\operatorname{polylog} n)$ update time and space. To address the fact we do not know p apriori, we could set $r \gg pm = c\epsilon^{-2}n\log n$, and then, at the end of the stream when p and m are known a) choose $X \sim \mathbf{Bin}(m, p)$ where $\mathbf{Bin}(\cdot, \cdot)$ denotes the binomial distribution and b) randomly pick X distinct random edges amongst the set of r edges sampled (ignoring duplicates). This approach will work with high probability if r is sufficiently large since X is tightly concentrated around $\mathbb{E}[X] = pm$. However, a naive implementation of this algorithm would require $\omega(n)$ update time. The main contribution of this section is to demonstrate how to ensure $O(\operatorname{polylog} n)$ update time.

3.1   Reformulating the Sampling Procedure

We first describe an alternative sampling process that, with high probability, returns a set of edges S where each edge in S has been sampled independently with probability p as required. The purpose of this alternative formulation is that it will allow us to argue that it can be emulated in the dynamic graph stream model efficiently.

Basic Approach. The basic idea is to partition the set of edges into different groups and then sample edges within groups that do not contain too many edges. We refer to such groups as "small". We determine which of the edges in a small group are to be sampled in two steps:

– *Fix the number X of edges to sample:* Let $X \sim \mathbf{Bin}(g, p)$ where g is the number of edges in the relevant group.
– *Fix which X edges to sample:* We then randomly pick X edges without replacement from the relevant group.

It is not hard to show that this two-step process ensures that each edge in the group is sampled independently with probability p. At this point, the fate of all edges in small groups has been decided: they will either be returned in the final sample or definitely not returned in the final sample.

We next consider another partition of the edges and again consider groups that do not contain many edges. We then determine the fate of the edges in such groups whose fate has not hitherto been determined. We keep on considering different partitions until every edge has been included in a small group and has had its fate determined.

Lemma 2. *Assume for every edge there exists a partition such that the edge is in a small group. Then the distribution over sets of sampled edges is the same as the distribution had each edge been sampled independently with probability p.*

Proof. The proof does not depend on the exact definition of "small" and the only property of the partitions that we require is that every edge is in a small group of some partition. We henceforth consider a fixed set of partitions with this property.

We first consider the jth group in the ith partition. Let g be the number of edges in this group. For any subset Q of ℓ edges in this group, we show that the probability that Q is picked by the two-step process above is indeed p^ℓ.

$$
\begin{aligned}
\mathbb{P}\left[\forall e \in Q, e \text{ is picked}\right] &= \sum_{t=\ell}^{g} \mathbb{P}\left[\forall e \in Q, e \text{ is picked} \mid X = t\right] \mathbb{P}\left[X = t\right] \\
&= \sum_{t=\ell}^{g} \frac{\binom{g-\ell}{t-\ell}}{\binom{g}{t}} \cdot \binom{g}{t} \cdot p^t (1-p)^{g-t} \\
&= p^\ell \sum_{t=\ell}^{g} \binom{g-\ell}{t-\ell} \cdot p^{t-\ell}(1-p)^{g-t} = p^\ell.
\end{aligned}
$$

and hence edges within the same group are sampled independently with probability p. Furthermore, the edges in different groups of the same partition are sampled independently from each other.

Let $f(e)$ be the first partition in which e is placed in a group that is small and let $W_i = \{e : f(e) = i\}$. Restricting Q to edges in W_i in the above analysis establishes that edges in each W_i are sampled independently. Since $f(e)$ is determined by the fixed set of partitions rather than the randomness of the sampling procedure, we also conclude that edges in different W_i are sampled independently. As we assume that every edge belongs to at least one small group in some partition, if we let r be the total number of partitions, then $\{W_i\}_{i \in [r]}$ partition the set of edges E. Hence, all edges in E are sampled independently with probability p. $\square$

Details of Alternative Sampling Procedure. The partitions considered will be determined by pairwise independent hash functions and we will later argue that it is sufficient to consider only $O(\log n)$ partitions. Each hash function will partition the m edges into $n\epsilon^{-2}$ groups. In expectation the number of edges in a group will be $\epsilon^2 m/n$ and we define a group to be small if it contains at most $t = 4\epsilon^2 m/n$ edges. We therefore expect to sample less than $4p\epsilon^2 m/n = 4c \log n$ edges from a small group. We will abort the algorithm if we attempt to sample significantly more edges than this from some small group. The procedure is as follows:

- Let $h_1, \ldots, h_r : \binom{n}{2} \to [n\epsilon^{-2}]$ be pairwise independent hash functions where $r = 10 \log n$.
- Each h_i defines a partition of E comprising of sets of the form

$$E_{i,j} = \{e \in E : h_i(e) = j\}.$$

Say $E_{i,j}$ is *small* if it is of size at most $t = 4\epsilon^2 m/n$. Let D_i be the set of all edges in the small sets determined by h_i.
- For each small $E_{i,j}$, let

$$X_{i,j} = \mathbf{Bin}(|E_{i,j}|, p)$$

and abort if

$$X_{i,j} \geq \tau \quad \text{where} \quad \tau = 24c \log n.$$

Let $S_{i,j}$ be a set of $X_{i,j}$ edges sampled without replacement from $E_{i,j}$.
- Let S be set of edges that were sampled among some D_i that are not in $D_1 \cup D_2 \cup \ldots \cup D_{i-1}$, i.e., edges whose fate had not already been determined.

$$S = \bigcup_{i=1}^r \{e \in D_i : e \in \cup_j S_{i,j} \text{ and } e \notin D_1 \cup D_2 \cup \ldots \cup D_{i-1}\}$$

Analysis. There are two main things that we need to show to establish that the above process emulates our basic sampling approach with high probability. First, we will show that with high probability for every edge e there exists i and j such that $e \in E_{i,j}$ and $E_{i,j}$ is small. This ensures that we will make a decision on whether e is included in the final sample. Second, we will show that it is very unlikely we abort because some $X_{i,j}$ is too large.

Lemma 3. *With probability at least $1 - n^{-8}$, for every edge e there exists i such that $e \in E_{i,j}$ and $E_{i,j}$ is small.*

Proof. Fix $i \in [r]$ and let $j = h_i(e)$. Then $\mathbb{E}\left[|E_{i,j}|\right] \leq 1 + \epsilon^2(m-1)/n \leq 2\epsilon^2 m/n$ assuming $m \geq \epsilon^{-2}n$. By an application of the Markov bound:

$$\mathbb{P}\left[|E_{i,j}| \geq 4m\epsilon^2/n\right] \leq 1/2.$$

Since each h_i is independent,

$$\mathbb{P}\left[|E_{i,h_i(e)}| \geq 4m\epsilon^2/n \text{ for all } i\right] \leq 1/2^r = 1/n^{10}.$$

Therefore by the union bound over all $m \leq n^2$ edges there exists a good partition for each e with probability at least $1 - n^{-8}$. $\qquad\square$

Lemma 4. *With high probability, all $X_{i,j}$ are less than $\tau = 24c \log n$.*

Proof. Since $E_{i,j}$ is small then $\mathbb{E}[X_{i,j}] = |E_{i,j}|p \leq 4\epsilon^2 pm/n = 4c \log n$. Hence, by an application of the Chernoff bound,

$$\mathbb{P}[X_{i,j} \geq 24c \log n] \leq 2^{-24c \log n} \leq n^{-10}.$$

Taking the union bound over all $10 \log n$ values of i and $\epsilon^{-2}n$ values of j establishes the lemma. $\square$

3.2 The Dynamic Graph Stream Algorithm

We are now ready to present the dynamic graph stream algorithm. To emulate the above sampling process in the dynamic graph stream model, we proceed as follows:

1. *Pre-Processing:* Pick the hash functions $h_1, h_2, \ldots, h_r$. These define the sets $E_{i,j}$.
2. *During One Pass:*
   - Compute the size of each $E_{i,j}$ and m. Note that m is necessary to define p.
   - Sample τ edges $S'_{i,j}$ uniformly without replacement from each $E_{i,j}$.
3. *Post-Processing:*
   - Randomly determine the values $X_{i,j}$ based on the exact values of $|E_{i,j}|$ and m for each $E_{i,j}$ that is small. If $X_{i,j}$ exceeds τ then abort.
   - Let $S_{i,j}$ be a random subset of $S'_{i,j}$ of size $X_{i,j}$.
   - Return $p^{-1} \max_U d(G'_U)$ where G' is the graph with edges:

$$S = \bigcup_{i=1}^{r} \{e \in D_i : e \in \cup_j S_{i,j} \text{ and } e \notin D_1 \cup D_2 \cup \ldots \cup D_{i-1}\}$$

Note that is possible to compute $|E_{i,j}|$ using a counter that is incremented or decremented whenever an edge e is added or removed respectively that satisfies $h_i(e) = j$. We may evaluate pairwise independent hash functions in $O(\text{polylog } n)$ time. The exact value of $\max_U d(G'_U)$ can be determined in polynomial time using the result of Charikar [10]. To prove Theorem 2, it remains to describe how to sample τ edges *without replacement* from each $E_{i,j}$.

Sampling Edges Without Replacement Via ℓ_0-Sampling. To do this, we use the ℓ_0-sampling algorithm of Jowhari et al. [20]. Their algorithm returns, with high probability, a random edge from $E_{i,j}$ and the space and update time of the algorithm are both $O(\text{polylog } n)$. Running τ independent instantiations of this algorithm immediately enables us to sample τ edges uniformly from $E_{i,j}$ *with replacement*.

However, since their algorithm is based on linear sketches, there is an elegant way (at least, more elegant than simply over sampling and removing duplicates) to ensure that all samples are distinct. Specifically, let $\mathbf{x}$ be the characteristic vector of the set $E_{i,j}$. Then, τ instantiations of the algorithm of Jowhari et al. [20] generate random projections

$$\mathcal{A}_1(\mathbf{x}) \, , \, \mathcal{A}_2(\mathbf{x}) \, , \, \ldots \, , \, \mathcal{A}_\tau(\mathbf{x})$$

of $\mathbf{x}$ such that a random non-zero entry of $\mathbf{x}$ (which corresponds to an edge from $E_{i,j}$) can be identified by processing each $\mathcal{A}_i(\mathbf{x})$. Let e_1 be the edge reconstructed from $\mathcal{A}_1(\mathbf{x})$. Rather than reconstructing an edge from $\mathcal{A}_2(\mathbf{x})$, which could be the same as e_1, we instead reconstruct an edge e_2 from

$$\mathcal{A}_2(\mathbf{x}) - \mathcal{A}_2(\mathbf{i}_{e_1}) = \mathcal{A}_2(\mathbf{x} - \mathbf{i}_{e_1})$$

where $\mathbf{i}_{e_1}$ is the characteristic vector of the set $\{e_1\}$. Note that e_2 is necessarily different from e_1 since $\mathbf{x} - \mathbf{i}_e$ is the characteristic vector of the set $E_{i,j} \setminus \{e_1\}$. Similarly we reconstruct e_j from

$$\mathcal{A}_j(\mathbf{x}) - \mathcal{A}_j(\mathbf{i}_{e_1}) - \mathcal{A}_j(\mathbf{i}_{e_2}) - \ldots - \mathcal{A}_j(\mathbf{i}_{e_{j-1}}) = \mathcal{A}_2(\mathbf{x} - \mathbf{i}_{e_1} - \ldots - \mathbf{i}_{e_{j-1}})$$

and note that e_j is necessarily distinct from $\{e_1, e_2, \ldots, e_{j-1}\}$.

4 Conclusion

We presented the first algorithm for estimating the density of the densest subgraph up to a $(1 + \epsilon)$ factor in the dynamic graph stream model. Our algorithm used $O(\epsilon^{-2} n \operatorname{polylog} n)$ space, $\operatorname{polylog}(n)$ per-update processing time, and $\operatorname{poly}(n)$ post-processing to return the estimate. The most relevant previous results, by Bhattacharya et al. [8], were a $(2 + \epsilon)$ approximation in similar space and a $(4 + \epsilon)$ approximation with $\operatorname{polylog}(n)$ per-update processing time that also outputs an estimate of the maximum density after each edge insertion or deletion. A natural open question is whether it is possible to use ideas contained in this paper to improve the approximation factor for the problem of maintaining a running estimate of the maximum density.

References

1. Ahn, K.J., Cormode, G., Guha, S., McGregor, A., Wirth, A.: Correlation clustering in data streams. In: Proceedings of the 32nd International Conference on Machine Learning, ICML 2015, Lille, France, July 6–11, 2015 (2015)
2. Ahn, K.J., Guha, S., McGregor, A.: Analyzing graph structure via linear measurements. In: Twenty-Third Annual ACM-SIAM Symposium on Discrete Algorithms, SODA 2012, pp. 459–467 (2012)
3. Ahn, K.J., Guha, S., McGregor, A.: Graph sketches: sparsification, spanners, and subgraphs. In: 31st ACM SIGMOD-SIGACT-SIGART Symposium on Principles of Database Systems, pp. 5–14 (2012)
4. Ahn, K.J., Guha, S., McGregor, A.: Spectral sparsification in dynamic graph streams. In: Raghavendra, P., Raskhodnikova, S., Jansen, K., Rolim, J.D.P. (eds.) RANDOM 2013 and APPROX 2013. LNCS, vol. 8096, pp. 1–10. Springer, Heidelberg (2013)

5. Assadi, S., Khanna, S., Li, Y., Yaroslavtsev, G.: Tight bounds for linear sketches of approximate matchings. CoRR, abs/1505.01467 (2015)
6. Bahmani, B., Goel, A., Munagala, K.: Efficient primal-dual graph algorithms for mapreduce. In: Bonato, A., Graham, F.C., Prałat, P. (eds.) WAW 2014. LNCS, vol. 8882, pp. 59–78. Springer, Heidelberg (2014)
7. Bahmani, B., Kumar, R., Vassilvitskii, S.: Densest subgraph in streaming and mapreduce. PVLDB **5**(5), 454–465 (2012)
8. Bhattacharya, S., Henzinger, M., Nanongkai, D., Tsourakakis, C.E.: Space- and time-efficient algorithm for maintaining dense subgraphs on one-pass dynamic streams. In: STOC (2015)
9. Bury, M., Schwiegelshohn, C.: Sublinear estimation of weighted matchings in dynamic data streams. CoRR, abs/1505.02019 (2015)
10. Charikar, M.: Greedy approximation algorithms for finding dense components in a graph. In: Jansen, K., Khuller, S. (eds.) APPROX 2000. LNCS, vol. 1913, pp. 84–95. Springer, Heidelberg (2000)
11. Chitnis, R.H., Cormode, G., Esfandiari, H., Hajiaghayi, M., McGregor, A., Monemizadeh, M., Vorotnikova, S.: Kernelization via sampling with applications to dynamic graph streams. CoRR, abs/1505.01731 (2015)
12. Cormode, G., Firmani, D.: A unifying framework for ℓ_0-sampling algorithms. Distrib. Parallel Databases **32**(3), 315–335 (2014)
13. Cormode, G., Muthukrishnan, S.: An improved data stream summary: the count-min sketch and its applications. J. Algorithms **55**(1), 58–75 (2005)
14. Epasto, A., Lattanzi, S., Sozio, M.: Efficient densest subgraph computation in evolving graphs. In: WWW (2015)
15. Gallo, G., Grigoriadis, M.D., Tarjan, R.E.: A fast parametric maximum flow algorithm and applications. SIAM J. Comput. **18**(1), 30–55 (1989)
16. Gilbert, A.C., Indyk, P.: Sparse recovery using sparse matrices. Proc. IEEE **98**(6), 937–947 (2010)
17. Goel, A., Kapralov, M., Post, I.: Single pass sparsification in the streaming model with edge deletions. CoRR, abs/1203.4900 (2012)
18. Goldberg, A.V.: Finding a maximum density subgraph. Technical report, Berkeley, CA, USA (1984)
19. Guha, S., McGregor, A., Tench, D.: Vertex and hypergraph connectivity in dynamic graph streams. In: PODS (2015)
20. Jowhari, H., Saglam, M., Tardos, G.: Tight bounds for lp samplers, finding duplicates in streams, and related problems. In: PODS, pp. 49–58 (2011)
21. Kapralov, M., Lee, Y.T., Musco, C., Musco, C., Sidford, A.: Single pass spectral sparsification in dynamic streams. In: FOCS (2014)
22. Kapralov, M., Woodruff, D.P.: Spanners and sparsifiers in dynamic streams. In: ACM Symposium on Principles of Distributed Computing, PODC 2014, Paris, France, July 15–18, 2014, pp. 272–281 (2014)
23. Khuller, S., Saha, B.: On finding dense subgraphs. In: Albers, S., Marchetti-Spaccamela, A., Matias, Y., Nikoletseas, S., Thomas, W. (eds.) ICALP 2009, Part I. LNCS, vol. 5555, pp. 597–608. Springer, Heidelberg (2009)
24. Konrad, C.: Maximum matching in turnstile streams. CoRR, abs/1505.01460 (2015)
25. Kutzkov, K., Pagh, R.: Triangle counting in dynamic graph streams. In: Ravi, R., Gørtz, I.L. (eds.) SWAT 2014. LNCS, vol. 8503, pp. 306–318. Springer, Heidelberg (2014)

26. Lee, V., Ruan, N., Jin, R., Aggarwal, C.: A survey of algorithms for dense subgraph discovery. In: Aggarwal, C.C., Wang, H. (eds.) Managing and Mining Graph Data. Advances in Database Systems, vol. 40, pp. 303–336. Springer, US (2010)
27. McGregor, A.: Graph stream algorithms: a survey. SIGMOD Rec. **43**(1), 9–20 (2014)
28. Mitzenmacher, M., Upfal, E.: Probability and Computing: Randomized Algorithms and Probabilistic Analysis. Cambridge University Press, New York (2005)

The Offline Carpool Problem Revisited

Saad Mneimneh$^{(\boxtimes)}$ and Saman Farhat

Hunter College and The Graduate Center,
City University of New York (CUNY), New York, USA
saad@hunter.cuny.edu, sfarhat@gc.cuny.edu

Abstract. The carpool problem is to schedule for every time $t \in \mathbb{N}$ l tasks taken from the set $[n]$ ($n \geq 2$). Each task i has a weight $w_i(t) \geq 0$, where $\sum_{i=1}^{n} w_i(t) = l$. We let $c_i(t) \in \{0,1\}$ be 1 iff task i is scheduled at time t, where (carpool condition) $w_i(t) = 0 \Rightarrow c_i(t) = 0$.

The carpool problem exists in the literature for $l = 1$, with a goal to make the schedule fair, by bounding the absolute value of $E_i(t) = \sum_{s=1}^{t} [w_i(s) - c_i(s)]$. In the typical online setting, $w_i(t)$ is unknown prior to time t; therefore, the only sensible approach is to bound $|E_i(t)|$ at all times. The optimal online algorithm for $l = 1$ can guarantee $|E_i(t)| = O(n)$. We show that the same guarantee can be maintained for a general l. However, it remains far from an ideal $|E_i(T)| < 1$ when all tasks have reached completion at some future time $t = T$.

The main contribution of this paper is the offline version of the carpool problem, where $w_i(t)$ is known in advance for all times $t \leq T$, and the fairness requirement is strengthened to the ideal $|E_i(T)| < 1$ while keeping $E_i(t)$ bounded at all intermediate times $t < T$. This problem has been mistakenly considered solved for $l = 1$ using Tijdeman's algorithm, so it remains open for $l \geq 1$. We show that achieving the ideal fairness with an intermediate $O(n^2)$ bound is possible for a general l.

Keywords: Carpool problem · Fair scheduling · Graphs · Flows · Online and offline algorithms

1 Introduction

The carpool problem was first systematically studied by Fagin and Williams [1] to resolve issues of fairness related to the following scenario: There are n people. Each day a subset of these people will participate in a carpool and only one of them must be designated to drive. Fairness dictates that each person should drive a number of times (approximately) equal to the sum of the inverses of the number of people who showed up on the days that person participated in the carpool.

The problem has been generalized (see for instance [2,3]) by introducing weights: We schedule for every time $t \in \mathbb{N}$ one task taken from the set $[n]$

S. Mneimneh—Partially supported by the CoSSMO institute at CUNY.
S. Farhat—Supported by a CUNY Graduate Center Fellowship.

G.F. Italiano et al. (Eds.): MFCS 2015, Part II, LNCS 9235, pp. 483–492, 2015.
DOI: 10.1007/978-3-662-48054-0_40

$(n \geq 2)$. Each task i has a weight $w_i(t) \geq 0$, where $\sum_i w_i(t) = 1$. We say that $c_i(t) \in \{0, 1\}$ is 1 iff i is the scheduled task at time t. Since exactly one task is scheduled at any time, $\sum_i w_i(t) = \sum_i c_i(t) = 1$, and $\sum_i E_i(t) = 0$. In addition, the following **carpool condition** is enforced:

Definition 1 (Carpool condition). $w_i(t) = 0 \Rightarrow c_i(t) = 0$ *for every* $i \in [n]$.

In effect, $w_i(t) > 0$ indicates the presence of the task at time t. Translating back to the original scenario, this is just to say a person must show up on the day he is the designated driver. When $w_i(t)$ is the same for every $i \in \{j | w_j(t) > 0\}$, we retrieve the special case introduced in [1].

We generalize the carpool problem further. Assume $l \in \mathbb{N}$ ($l \leq n$).

The Carpool Problem: Our version of the carpool problem is to schedule l tasks for every time $t \in \mathbb{N}$, where $\sum_{i=1}^{n} w_i(t) = \sum_{i=1}^{n} c_i(t) = l$ and $w_i(t) \leq 1$, and $c_i(t) \in \{0, 1\}$ is 1 iff task i is scheduled at time t, subject to the carpool condition.

The typical setting is online, i.e. $w_i(t)$ is unknown prior to time t, and one has to make immediate choices to construct a fair schedule by bounding the absolute value of $E_i(t) = \sum_{s=1}^{t} [w_i(s) - c_i(s)]$ at all times. It was shown in [3] that the online *greedy* algorithm that schedules at time t one (so $l = 1$) task $i \in \{j \in [n] | w_j(t) > 0\}$ such that $\theta_i(t) = \sum_{s=1}^{t-1} [w_i(s) - c_i(s)] + w_i(t)$ is maximized will achieve $|E_i(t)| = O(n)$ at all times. In the online setting, this (deterministic) bound is asymptotically optimal.

We consider the offline version of the carpool problem, where $w_i(t)$ is known in advance for all times $t < T$ for some fixed T. In this setting, the notion of fairness is strengthened to $|E_i(T)| < 1$. In other words, if T is the completion time of the schedule, each task i will have been served as closely as possible to its share $\sum_{s=1}^{T} w_i(s)$.

Definition 2 (Fair schedule). *A schedule is fair iff* $|E_i(T)| < 1$ *for every* $i \in [n]$, *where* $E_i(t) = \sum_{s=1}^{t} [w_i(s) - c_i(s)]$ *and*

$$c_i(t) = \begin{cases} 1 & i \text{ is a scheduled task at time } t \\ 0 & \text{otherwise} \end{cases}$$

The above fairness property of carpool makes it attractive to many load balancing problems; however, in many applications, the correctness/guarantee of the scheduling algorithm will also require $|E_i(t)|$ to be bounded for all $t < T$ (see for instance [2]). We refer to this property in the offline context as *non-bursty*. While an online algorithm is naturally non-bursty, fairness as in Definition 2 alone may not be enough when moving to the offline version. Therefore, in addition to fairness, we will require the following:

Definition 3 (Non-bursty schedule). *A schedule is non-bursty iff* $|E_i(t)|$ *is bounded (independent of t) at all times $t < T$ for every $i \in [n]$.*

The problem of constructing an offline schedule that is fair and non-bursty was thought to be solved for $l = 1$. Section 2 covers the detail behind that misconception. For now, to further motivate our work, observe that the offline setting is important not only when the instance is known in advance, but also for recovery. For example, consider a scenario in which a backlog of T time steps has been created as a result of an interruption in the scheduler at time t_0. The scheduler looses all information and, therefore, if $|E_i(t_0)| < B$ for every $i \in [n]$, B will be added to whatever bound is achieved after recovery. When scheduling is resumed at time t, the scheduler works on the backlog starting with tasks presented at $t_0, t_0 + 1, \dots$. It is then only logical to make use of the information in the backlog. As such, an offline algorithm offers a chance to maintain the same guarantee prior to the failure: $|E_i(t + T - 1)| < B + 1$, as opposed to the $B + O(n)$ bound of an online algorithm, which adds another $O(n)$.

2   Related Work

The carpool problem exists in the literature only for $l = 1$. The work in [1,3, 4,6,9] (and their references) provide extensive treatment of the online setting. But the offline version of the carpool problem has not been explicitly mentioned in the literature. An exception lies in a few instances, e.g. [5,7], that model the offline carpool problem for $l = 1$ as a flow problem, as in Fig. 1, where $m_i = \lceil \sum_{t=1}^{T} w_i(t) \rceil$, and an edge from vertex i to vertex t exists iff $w_i(t) > 0$.

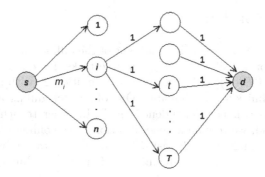

Fig. 1. The carpool problem for $l = 1$ as a flow.

The theory of flows, integer programs, and their linear program relaxations suggest that an integer maximum flow of value T exists and corresponds to a valid schedule in which $|E_i(T)| < 1$ for every $i \in [n]$, see for instance [5,8]. In addition, it is not hard to generalize the flow of Fig. 1 to handle any l. But this does not immediately yield an obvious bound on $|E_i(t)|$ for every $t < T$.

A short note in [4] (in the context of a problem called vector rounding) mistakenly claims that Tijdeman's algorithm [10] solves the offline carpool problem (for $l = 1$). Indeed, the carpool problem is closely related to the chairman assignment problem popularized by Tijdeman in [10], where tasks are persistent (a task can be scheduled at any time) so the carpool condition $w_i(t) = 0 \Rightarrow c_i(t) = 0$ is **dropped**.

For the chairman assignment problem, Tijdeman's algorithm guarantees $|E_i(t)| < 1$ at all times. To schedule a task at time t, the algorithm considers the set of tasks that satisfy $\theta_i(t) = \sum_{s=1}^{t-1}[w_i(s) - c_i(s)] + w_i(t) \geq 1/(2n-2)$ (this set must be non-empty), finds the smallest $t_0 \leq T$ such that $\theta_i(t) + \sum_{s=t+1}^{t_0} w_i(s) \geq 1 - 1/(2n-2)$ for some i in the set, and makes $c_i(t) = 1$ (if no such t_0 exists, the algorithm schedules any task i). There is no guarantee that $w_i(t) > 0$ for the scheduled task i. Figure 2 shows a counterexample with $n = 4$ and $T = 5$.

task	1	2	3	4	5
1	0	1/12	1/6	**0**	4/5
2	1/12	0	1/6	1/100	0
3	1/2	**1/2**	1/4	1/100	0
4	**5/12**	5/12	**5/12**	98/100	1/5

Fig. 2. Tijdeman's algorithm fails for carpool: weights in bold represent the schedule of tasks over time. Task 1 is scheduled at time $t = 4$ so $c_1(4) = 1$, but $w_1(4) = 0$.

Following [4], we see no subsequent attempt in the literature to construct an offline schedule that is both fair and non-bursty. The generalization to l tasks is absent as well.

3   Our Contribution

We believe this is the first explicit treatment of the offline carpool problem (also generally in the context of scheduling multiple tasks, i.e. $l > 1$). We show that achieving fairness ($|E_i(T)| < 1$ according to Definition 2) with an intermediate $O(n^2)$ bound (Definition 3) is possible. Our offline scheduling algorithm has a running time with a linear dependence on T. In order to achieve the $O(n^2)$ intermediate bound, we rely on a generalization of the online *greedy* algorithm described in [3] to handle multiple tasks. The schedule obtained by this algorithm is then used to transform a fair schedule to a fair and non-bursty one.

4   Generalizing the Online Algorithm

The online *greedy* algorithm described in [3] for $l = 1$ schedules at time t one task $i \in \{j \in [n] | w_j(t) > 0\}$ such that $\theta_i(t) = \sum_{s=1}^{t-1}[w_i(s) - c_i(s)] + w_i(t)$ is maximized, and achieves $|E_i(t)| < (n-1)/2$ at all times. We will describe an online algorithm with the same guarantee for a general l, called *l-greedy*.

The *l-greedy* algorithm schedules at time t the l tasks corresponding to the l largest elements in $\{\theta_i(t)|w_i(t) > 0\}$. This online algorithm can run offline in $O(lnT)$ time. We will show below that this algorithm has the same guarantee as *greedy*.

Theorem 1. *The l-greedy algorithm has at worst the same guarantee as greedy.*

Proof. We construct an imaginary instance of *greedy* that schedules the same tasks. Without loss of generality, assume that $X = \{1, 2, \ldots, l\}$ is the set of tasks scheduled at time t by the *l-greedy* algorithm. Starting from $t = 1$, divide each t into l times $t_1, \ldots, t_l$, and define $(1 \leq i, j \leq l)$

$$w_i'(t_k) = \begin{cases} w_i(t) & i = k \\ 0 & \text{otherwise} \end{cases}$$

For all tasks $i \notin X$, define $w_i'(t_k)$ such that $\sum_{k=1}^{l} w_i'(t_k) = w_i(t)$ by dividing $\sum_{i \notin X} w_i(t)$ arbitrarily among $t_1, \ldots, t_l$ to make the sum of the weights for each time t_k equal to 1. Figure 3 illustrates this construction.

t		t_1	t_2	$\ldots$	t_l
$w_1(t)$		$w_1(t)$	0		0
$w_2(t)$		0	$w_2(t)$		0
$\vdots$		$\vdots$	$\vdots$		$\vdots$
$w_l(t)$		0	0		$w_l(t)$
$w_{l+1}(t)$		$w_{l+1}'(t_1)$	$w_{l+1}'(t_2)$		$w_{l+1}'(t_l)$
$\vdots$		$\vdots$	$\vdots$		$\vdots$
$w_n(t)$		$w_n'(t_1)$	$w_n'(t_2)$		$w_n'(t_l)$

Fig. 3. Constructing the *greedy* instance. The weights as seen by the *l-greedy* algorithm are shown on the left, where $X = \{1, 2, \ldots, l\}$ are the scheduled tasks at time t. The weights as seen by *greedy* at times $t_1, \ldots, t_l$ are shown on the right, $w_i'(t_i) = w_i(t)$ for $i \in X$, $\sum_{k=1}^{l} w_i'(t_k) = w_i(t)$, and $\sum_{i=1}^{n} w_i'(t_k) = 1$.

Consider $\theta_i'(t_k)$ as seen by a *greedy* algorithm now acting on $t_1, \ldots, t_l$. Suppose that *greedy* has been scheduling the same tasks as *l-greedy*. If $i \in X$, then $\theta_i'(t_i) = \theta_i(t)$; if $i \notin X$, then $\theta_i'(t_k) \leq \theta_i(t)$. Therefore, $\theta_i'(t_i)$ is the largest for t_i among all tasks j with $w_j'(t_i) > 0$. So *greedy* will schedule for $t_1, \ldots, t_l$ exactly the tasks in X. $\qquad\square$

We conclude that the *l-greedy* algorithm achieves the same guarantee as *greedy*, namely $|E_i(t)| = O(n)$ for every $i \in [n]$ and $\sum_{i=1}^{n} |E_i(t)| = O(n^2)$.

5   A Fair Schedule

As we mentioned in Sect. 2, it is possible to use flows to prove the existence of (and obtain) a fair schedule. But for completeness, we describe in this section a canonical algorithm to transform any arbitrary schedule, given by $c_i(t)$ for $i \in [n]$ and $t \leq T$, to a fair schedule (which will also serve as a proof for the existence of such a schedule). We construct a directed multigraph $G = (\mathbb{V} = [n], \mathbb{E})$ such that $(i, j)_t \in \mathbb{E}$ is an edge from i to j iff $c_i(t) = 1$ and $c_j(t) = 0$ and $w_j(t) > 0$. Such an edge means that task i is scheduled at time t but task j could be scheduled instead.

A path $\{(i_1, i_2)_{t_1}, (i_2, i_3)_{t_2}, \ldots, (i_r, i_{r+1})_{t_r}\}$ with $E_{i_1}(T) < 0$ and $E_{i_r}(T) > 0$ represents a way to modify the schedule: make $c_{i_1}(t_1) = 0$ and $c_{i_2}(t_1) = 1$, $c_{i_2}(t_2) = 0$ and $c_{i_3}(t_2) = 1$, $\ldots$, $c_{i_r}(t_r) = 0$ and $c_{i_{r+1}}(t_r) = 1$. After making the changes, update the edges of the multigraph accordingly to reflect the new schedule. This will increase $E_{i_1}(T)$ by 1 and decrease $E_{i_{r+1}}(T)$ by 1. Therefore, if $E_{i_1}(T) \leq -1$ and/or $E_{i_{r+1}}(T) \geq 1$, the schedule is improved (by decreasing $|E_{i_1}(T)|$ and/or $|E_{i_{r+1}}(T)|$). Below we show that we can always make such improvements until the schedule becomes fair.

Lemma 1. *If $E_i(T) < 0$ ($E_i(T) > 0$), there is a path in G from i to some j (from some j to i) with $E_j(T) > 0$ ($E_j(T) < 0$). This path may be used to improve the schedule as stated above.*

Proof. We prove the case when $E_i(T) < 0$, the second case is symmetric. So assume $E_i(T) < 0$ and let the set A consist of all vertices j reachable from i such that $E_j(T) \leq 0$. The proof is by contradiction, so we can assume that there are no outgoing edges from A to the rest of the multigraph. Since $i \in A$, we know that $E = \sum_{j \in A} E_j(T) = E_i(T) + \sum_{j \in A, j \neq i} E_j(T) < 0$. Now consider the set $B = \{t | c_j(t) = 1 \text{ for some } j \in A\}$. Since A has no outgoing edges, if $j \notin A$ and $t \in B$, then $c_j(t) = 0 \Rightarrow w_j(t) = 0$ (so $w_j(t) - c_j(t) \leq 0$). Therefore, it should be clear that $e = \sum_{j \in A} \sum_{t \in B} [w_j(t) - c_j(t)] \geq \sum_{j \in A} \sum_{t \in B} [w_j(t) - c_j(t)] + \sum_{j \notin A} \sum_{t \in B} [w_j(t) - c_j(t)] = \sum_j \sum_{t \in B} [w_j(t) - c_j(t)] = \sum_{t \in B} \sum_j [w_j(t) - c_j(t)] = 0$. But $E = e + \sum_{j \in A} \sum_{t \notin B} [w_j(t) - 0] \geq 0$, a contradiction.   □

Theorem 2. *There exists a fair schedule for every instance of the offline carpool problem.*

Proof. If a schedule is not fair, then there exists a task i such that $E_i(T) \geq 1$ or $E_i(T) \leq -1$. Therefore, one could apply Lemma 1. When we consider the sum

$$S = \sum_{i \in [n]} \lfloor |E_i(T)| \rfloor$$

we observe that it will decrease by either 1 or 2 after each iteration of Lemma 1: $E_i(T) \geq 1$ decreases by 1 for some i, or $E_j(T) \leq -1$ increases by 1 for some j, or both. Therefore, the sum will eventually reach 0. When this happens, $|E_i(T)| < 1$ for every $i \in [n]$ and the schedule is fair.   □

Since we can write the above sum in two parts

$$S = \sum_{i \in \{j | E_j(T) > 0\}} \lfloor E_i(T) \rfloor - \sum_{i \in \{j | E_j(T) < 0\}} \lceil E_i(T) \rceil$$

and the first part is at most $\sum_{i,t} w_i(t) = lT$ and the second part is at least $-\sum_{i,t} c_i(t) = -lT$, we conclude that $S = O(lT)$. This is an upper bound on the number of iterations needed to modify any arbitrary schedule. But given a specific initial schedule, one could refine this bound. For the schedule obtained by the l-greedy algorithm of Sect. 4, $S = O(n^2)$. Therefore, we have $O(n^2)$ iterations, each will process the multigraph starting from some vertex in $O(lnT)$ time (the size of the multigraph) leading to an $O(ln^3T)$ time algorithm.

6   A Non-bursty Schedule

We now describe a canonical algorithm to transform a fair schedule, given by $c_i(t)$ for $i \in [n]$ and $t \leq T$, to a non-bursty schedule without affecting its fairness. For this, we assume the existence of an auxiliary non-bursty schedule (but not necessarily fair). Let S_t^{orig} and S_t^{aux} be the set of tasks scheduled at time t by those schedules respectively. We construct a directed multigraph $G = (\mathbb{V} = [n], \mathbb{E})$ such that $(i, j)_t \in \mathbb{E}$ is an edge from i to j if $i \in S_t^{orig} - S_t^{aux}$ and $j \in S_t^{aux} - S_t^{orig}$. Such an edge represents a discrepancy between the two schedules at time t; the original schedules task i but the auxiliary schedules task j. We make all such edges $(i, j)_t$ for a given t form a maximal matching in $[n]$ (so there are exactly $|S_t^{orig} - S_t^{aux}| \leq l$ of them). Consequently, the out-degree of vertex i is the number of times task i is scheduled by the original schedule but not by the auxiliary and, similarly, the in-degree of vertex i is the number of times task i is scheduled by the auxiliary schedule but not by the original.

A cycle $\{(i_1, i_2)_{t_1}, (i_2, i_3)_{t_2}, \dots, (i_r, i_1)_{t_r}\}$ represents a way to bring closer the two schedules by modifying the original schedule as follows: make $c_{i_1}(t_1) = 0$ and $c_{i_2}(t_1) = 1$, $c_{i_2}(t_2) = 0$ and $c_{i_3}(t_2) = 1$, $\dots$, $c_{i_r}(t_r) = 0$ and $c_{i_1}(t_r) = 1$. After making these changes, eliminate the cycle from the multigraph. The time needed to eliminate all cycles is $O(n + lT)$ (the size of the multigraph), and that's when we obtain the modified schedule.

Given the modification to the original schedule as described above, we will use E^{mod} and E^{aux} to refer to these quantities in the modified and the auxiliary schedules, respectively. Observe that $E_i^{mod}(T) = E_i(T)$ is kept unchanged for every $i \in [n]$. Since the auxiliary schedule is non-bursty, one would expect that the original schedule will be transformed as such. Below we quantify this intuition.

Let $H_1 = (\mathbb{V} = [n], \mathbb{E}_1)$ be the subgraph of G obtained by the elimination of all the cycles in G. This multigraph can be converted into a simple weighted directed acyclic graph (DAG) $H_2 = (\mathbb{V} = [n], \mathbb{E}_2)$ such that $e = (i, j) \in \mathbb{E}_2$ is an edge from i to j iff $(i, j)_t \in \mathbb{E}_1$ for some t with weight $w(e) = |\{t | (i, j)_t \in \mathbb{E}_1\}|$. Define $w(i, j) = w(e)$ if $e = (i, j) \in \mathbb{E}_2$ and $w(i, j) = 0$ otherwise. Observe that $\sum_j w(i, j)$ is now the number of times task i is scheduled by the modified schedule but not by the auxiliary and, similarly, $\sum_j w(j, i)$ is the number of times task i is scheduled by the auxiliary schedule but not by the modified.

Lemma 2. *If the auxiliary schedule (non-bursty) guarantees an intermediate bound $|E_i^{aux}(t)| < f(n)$, then*

$$\left| \sum_{j \in [n]} w(i, j) - \sum_{j \in [n]} w(j, i) \right| = E_i^{mod}(T) - E_i^{aux}(T) < 1 + f(n)$$

for every $i \in [n]$.

490   S. Mneimneh and S. Farhat

Proof. Given the above interpretation of $\sum_j w(i,j)$ and $\sum_j w(j,i)$, observe that $\sum_j [w(i,j) - w(j,i)] = E_i^{mod}(T) - E_i^{aux}(T)$. Therefore, the result is immediate because the modified schedule is fair (so $|E_i^{mod}(T)| < 1$) and the auxiliary schedule satisfies $|E_i^{aux}(T)| < f(n)$.   □

Lemma 3. *If the auxiliary schedule (non-bursty) guarantees an intermediate bound $|E_i^{aux}(t)| < f(n)$, then*

$$\sum_{j \in [n]} w(i,j) < \frac{n}{2}[1 + f(n)]$$

$$\sum_{j \in [n]} w(j,i) < \frac{n}{2}[1 + f(n)]$$

for every $i \in [n]$.

Proof. Given vertex i, divide the DAG into m sets of vertices $V_1, \ldots, V_r, \ldots, V_m$ such that $V_r = \{i\}$ and all edges are from V_{k-1} to V_k for $k = 1, \ldots, m$ ($V_0 = \emptyset$), as shown in Fig. 4.

Let

$$I_k = \sum_{i \in V_{k-1}, j \in V_k} w(i,j)$$

and observe that

$$I_{k+1} - I_k < |V_k|[1 + f(n)]$$

by Lemma 2. Therefore, by summing up these inequalities over $k = 1, \ldots, r-1$, we get

$$I_r < (|V_1| + \ldots + |V_{r-1}|)[1 + f(n)]$$

By a symmetric argument, we also have

$$I_{r+1} < (|V_{r+1}| + \ldots + |V_m|)[1 + f(n)]$$

and invoking Lemma 2 on $I_r - I_{r+1}$ gives

$$I_r < (|V_{r+1}| + \ldots + |V_m| + 1)[1 + f(n)]$$

Finally,

$$2I_r < (|V_1| + \ldots + |V_{r-1}| + |V_{r+1}| + \ldots + |V_m| + 1)[1 + f(n)] = n[1 + f(n)]$$

and the same is true for I_{r+1} by symmetry. This concludes the proof because

$$\sum_{j \in [n]} w(i,j) = I_{r+1} \text{ and } \sum_{j \in [n]} w(j,i) = I_r \qquad □$$

Lemmas 2 and 3 can be trivially generalized by changing $1 + f(n)$ to $B + f(n)$ if the fair schedule is replaced by a weaker schedule that guarantees $|E_i(T)| < B$. Given Lemmas 2 and 3, we just proved the following theorem.

Theorem 3. *If the auxiliary schedule (non-bursty) guarantees an intermediate bound $|E_i^{aux}(t)| < f(n) = \Omega(1)$, then the modified schedule (fair) satisfies $|E_i^{mod}(t)| = O(nf(n))$ at all times for every $i \in [n]$.*

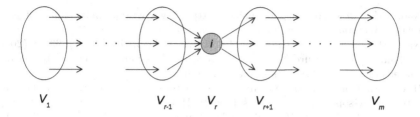

Fig. 4. Schematic illustration for the proof of Lemma 3.

7   The Final Algorithm

We now describe an offline algorithm for the carpool problem that is fair ($|E_i(T)| < 1$) and non-bursty with an $O(n^2)$ intermediate bound ($|E_i(t)| = O(n^2)$ for all $t < T$). The algorithm has a running time with a linear dependence of T, which is a nice feature.

1. Obtain an initial schedule using the *l-greedy* algorithm of Sect. 4.
2. Modify the schedule to be fair using the canonical algorithm of Sect. 5.
3. Modify the schedule to be non-bursty (and fair) using the canonical algorithm of Sect. 6, and the schedule of the *l-greedy* algorithm as the auxiliary schedule.

Since the *l-greedy* algorithm guarantees $|E_i(t)| = O(n)$ at all times, Theorem 3 implies that the above algorithm is non-bursty with an intermediate bound of $O(n^2)$.

8   Conclusion

The carpool problem is a scheduling problem where every task must receive its fair share, but may not be available at all times. We believe this is the first explicit treatment of the offline version of the carpool problem, which has been only studied in the online setting. In a typical offline setting, the goal would be to guarantee fairness by the schedule completion time T while avoiding an unbounded deviation from fairness at all times $t < T$. We achieved this goal by combining offline and online algorithms.

References

1. Fagin, R., Williams, J.H.: A fair carpool scheduling algorithm. IBM J. Res. Dev. **27**(2), 133–139 (1983)
2. Mneimneh, S.: Load balancing in a switch without buffers. In: IEEE Workshop on High Performance Switching and Routing, Poznan (2006)
3. Coppersmith, D., Nowicki, T., Paleologo, G., Tresser, C., Wu, C.W.: The optimality of the online greedy algorithm in carpool and chairman assignment problems. ACM Trans. Algorithms, **7**(3), Article 37, July 2011
4. Ajtai, M., Aspnes, J., Naor, M., Rabini, Y., Schulman, L.J., Waarts, O.: Fairness in Scheduling. J. Algorithms **29**(2), 306–357 (1988)

5. Naor, M.: How to Carpool Fairly. http://www.wisdom.weizmann.ac.il/naor/PAPERS/carpool_fair.pps
6. Naor, M.: On fairness in the carpool problem. J. Algorithms **55**(1), 93–98 (2005)
7. Williamson, D.: Lecture Notes on Network Flows, Chapter 3. http://people.orie.cornell.edu/dpw/techreports/cornell-flow.pdf
8. Havet, F.: Combinatorial Optimization, Chapter 11 on Fractional Relaxation. http://www-sop.inria.fr/members/Frederic.Havet/
9. Boavida, J.B., Kamat, V., Nakum, D., Nong, R., Wu, C.W., Zhang, X.: Algorithms for the Carpool Problem. http://www.ima.umn.edu/2005-2006/MM8.9-18.06/activities/Wu-Chai/team6_rep.pdf
10. Tijdeman, R.: The chairman assignment problem. Discrete Math. **32**, 323–330 (1980)

On Sampling Simple Paths in Planar Graphs According to Their Lengths

Sandro Montanari[1][(✉)] and Paolo Penna[2]

[1] Department of Computer Science, ETH Zurich, Zurich, Switzerland
sandro.montanari@inf.ethz.ch
[2] LIAFA, Université Paris Diderot, Paris, France
penna@liafa.univ-paris-diderot.fr

Abstract. We consider the problem of sampling simple paths between two given vertices in a planar graph and propose a natural Markov chain exploring such paths by means of "local" modifications. This chain can be tuned so that the probability of sampling a path depends on its *length* (for instance, output shorter paths with higher probability than longer ones). We show that this chain is always ergodic and thus it converges to the desired sampling distribution for any planar graph. While this chain is not rapidly mixing in general, we prove that a simple restricted variant is. The restricted chain samples paths on a 2D lattice which are monotone in the vertical direction. To the best of our knowledge, this is the first example of a rapidly mixing Markov chain for sampling simple paths with a probability that depends on their lengths.

1 Introduction

Sampling (or generating) a "random" object from a large set of combinatorial objects is a fundamental problem arising in Statistical Physics, Mathematics, and Computer Science. Because the number of such objects is typically huge (i.e., exponential in the size of the input), direct enumeration is unfeasible. An efficient sampling procedure is thus an important tool for studying statistical properties of "typical" instances. For many problems, sampling with uniform distribution and counting the number of objects are two computationally equivalent tasks [13], which in most cases are #*P*-hard. This includes counting simple paths in graphs [23], even when restricting to planar graphs [19]. For these cases, sampling is instead considered according to distributions that are "close" to the desired one [22]. A most relevant technique for the design of this kind of sampling procedures is the *Markov chain Monte Carlo* method [3].

We consider the task of sampling simple paths between two given vertices in planar graphs according to a fixed probability distribution using Markov chain Monte Carlo. Since in several applications it is natural to ask for paths of some fixed length, we thus consider the *weighted* version of the sampling problem in which the distribution depends on the length of the paths.

The Markov chain Monte Carlo method involves the design of a Markov chain whose states are the objects we wish to sample, and whose stationary distribution

© Springer-Verlag Berlin Heidelberg 2015
G.F. Italiano et al. (Eds.): MFCS 2015, Part II, LNCS 9235, pp. 493–504, 2015.
DOI: 10.1007/978-3-662-48054-0_41

is the probability we want to use to sample them. The sampling procedure is then a "random walk" on the chain for a fixed number of steps, until we are certain that the probability of sampling one object is (approximately) the stationary distribution of the chain. A most critical part of this method is proving that the Markov chain is *rapidly mixing*, i.e., the number of steps needed to reach the stationary distribution is polynomially bounded by the size of the input.

While rapidly mixing Markov chains are known for several hard problems, like graph coloring [6–8,11], knapsack [18], perfect matchings [12], independent sets [2,9,16], there is essentially no positive result for the case of simple *st*-paths on general graphs. The only chain proposed for this setting is by Roberts and Kroese [21], but it is however *not* rapidly mixing.

Most of the positive results consider restricted paths over a lattice structure. In the simplest instance of these restrictions we have paths using only downward and rightward edges of the two-dimensional grid. This case has been analyzed by Luby et al. [15] also for *multiple* source-destination paths, where the chain can "get stuck" because all paths must be disjoint and thus some non-local moves are introduced. If we further impose the paths to stay above the main diagonal of a square grid, the so-called *staircase walks*, the number of such paths is given by the famous Catalan numbers [5]. Martin and Randall [17] consider a Markov chain for sampling such paths where the weight of a path is the number of times it hits the diagonal. In the sampler by Greenberg et al. [10], the weight of a path is instead the number of faces below it (the authors also consider the more general case of higher dimensional lattices). Finally, Randall and Sinclair [20] consider the case where only one end of the path is fixed, and provide an efficient sampler for all such paths of a given length in the *infinite* d-dimensional lattice.

1.1   Our Results

We study a natural Markov chain in which a current *st*-path in a given (undirected, unweighted) planar graph is modified according to a simple local rerouting operation (see Sect. 3). Roughly speaking, rerouting operations resulting in longer paths are "accepted" only with small probability, while those resulting in shorter paths are always accepted. We show that the chain always converges to the Gibbs distribution on the paths weighted according to their lengths. In other words, the probability of sampling a specific path x depends only on its length $\ell(x)$ and it is of the form

$$\pi(x) \propto \lambda^{\ell(x)}, \tag{1}$$

where $\lambda > 0$ is a parameter that can be used to "tune" the chain. Setting $\lambda = 1$ yields a uniform sampler over all *st*-paths, while smaller/larger values provide samples biased towards shorter/longer paths.

Despite this chain being *not* rapidly mixing in general planar graphs, we obtain an efficient sampler with Gibbs distribution (1) for the following setting, depicted in Fig. 1. The paths from s to t are monotone in the vertical direction and the graph is any sub-grid of the 2D lattice without internal holes.

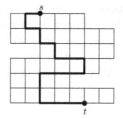

Fig. 1. An example of a vertical-monotone path in a sub-grid.

The new Markov chain is a restriction of the original one maintaining the "vertical-monotonicity" of the paths (Sect. 5). Our main technical contribution is a rigorous proof that this chain is rapidly mixing for all $\lambda \in (0,1]$. In the proof we combine the technique of path coupling without contraction [1] with the idea of modifying the chain by making some transitions "more lazy". Note that in this restricted setting, an efficient sampler can also be obtained using dynamic programming (see the full version of the paper for the details). However, our main interest is in the analysis of the mixing time of the proposed Markov chain, which is a variant of a well-known "mountain/valley" chain [17].

We show that our results are tight in the following sense (Sect. 4). First, the original "unrestricted" chain is *not* rapidly mixing for $\lambda = 1$ in some planar graphs. This is true even for sub-graphs of the 2D lattice, and thus for the chain sampling all paths in a grid, without restricting to vertical-monotone ones. Both for the restricted and the unrestricted chains, we show that the mixing time is exponential in the number of vertices for *every* $\lambda > 1$. The latter result is in part expected because determining if a planar graph has an Hamiltonian path is NP-hard (and for sufficiently large λ a sampler can be used to solve this problem). However, our negative results on the Markov chain are stronger in the sense that the chain remains slowly mixing even for very simple graphs where this problem can be easily solved. Thus, these results give a certain indication of the limitation of "local" chains. As for the case $\lambda = 1$, the existence of an efficient (uniform) sampler for planar graphs remains an interesting open problem.

2   Preliminaries

Planar Graphs. Given a planar graph $G = (V, E)$, we use n and m to denote respectively the number of vertices $|V|$ and of undirected edges $|E|$. By planarity, the vertices in V can be drawn as points in the plane in a way such that the edges in E are non-crossing curves; we denote such a drawing as a *plane embedding* of G. In a plane embedding, any maximal region of the plane enclosed by edges of E is called a *face*; the infinite region not enclosed by any edge is called the *outer face*. We use f to denote the overall number of faces, including the outer face. According to Euler's formula, the number of faces satisfies $f = m - n + 2$. Note also that in any planar graph $f \in O(n)$. Namely, $f \leq 2n - 4$ with equality achieved by triangulated graphs, i.e., planar graphs in which every face is a triangle.

A *path* is a sequence of vertices $x = (v_1, \ldots, v_l)$ such that $(v_i, v_{i+1}) \in E$, for all $i = 1, \ldots, l - 1$; we denote the number of edges along x as $|x| = l - 1$. In a *simple path* no vertex appears more than once. For any two vertices s and t, an *st-path* is a simple path starting at s and ending at t. Without loss of generality, we assume the graph to be 2-connected (we can otherwise easily reduce to this case).

Markov Chains and Mixing Time. We consider Markov chains $\mathcal{M}$ whose state space Ω is finite. In our application, Ω is the set of all simple st-paths in a given planar graph. The transition matrix $P \in \mathbb{R}^{|\Omega| \times |\Omega|}$ defines the transitions of $\mathcal{M}$. That is, $P(x, y)$ is the probability that the chain moves from state x to state y in one step. Thus, $P^t(x, y)$ is the probability of moving in t steps from state x to state y, where P^t is the tth power of matrix P.

A Markov chain as above is *irreducible* if, for all $x, y \in \Omega$, there exists a $t \in \mathbb{N}$ such that $P^t(x, y) > 0$. In other words, every state can be reached with non-zero probability regardless of the starting state. A Markov chain is *aperiodic* if, for all $x \in \Omega$, $\gcd\{t \in \mathbb{N} \mid P^t(x, x) > 0\} = 1$. It is well known [14] that an irreducible and aperiodic Markov chain converges to its unique *stationary distribution* π. That is, there exists a unique vector $\pi \in \mathbb{R}^{|\Omega|}$ such that $\pi P = \pi$ and, for all $x, y \in \Omega$, it holds

$$\lim_{t \to \infty} P^t(x, y) = \pi(y).$$

An aperiodic and irreducible Markov chain is called *ergodic*.

The *mixing time* of a Markov chain is the time needed for the distribution $P^t(x, \cdot)$ to get "sufficiently close" to the stationary distribution for any starting state x. Formally, the *mixing time* is defined as

$$t_{mix}(\epsilon) := \min_{t \in \mathbb{N}} \max_{x \in \Omega} \{||P^t(x, \cdot) - \pi||_{TV} \le \epsilon\}, \tag{2}$$

where $||P^t(x, \cdot) - \pi||_{TV} = \frac{1}{2} \sum_{y \in \Omega} |P^t(x, y) - \pi(y)|$ is the total variation distance. It is common [14] to define $t_{mix} := t_{mix}(1/4)$ since $t_{mix}(\epsilon) \le \lceil \log_2(1/\epsilon) \rceil t_{mix}$. A Markov chain is *rapidly mixing* if $t_{mix}(\epsilon)$ is bounded from above by a polynomial in $\log(|\Omega|)$ and in $\log(1/\epsilon)$.

A rapidly mixing Markov chain can be used to efficiently sample elements from Ω with probability arbitrarily close to π. Simply simulate a random walk on the chain from an arbitrary initial state x for $t = t_{mix}(\epsilon)$ time steps and return the state of the chain at time t. According to (2), the probability $P^t(x, y)$ of the returned state y is approximately $\pi(y)$.

3   A Markov Chain for Planar Graphs

We now define a Markov chain $\mathcal{M}_{paths}$ whose state space Ω is the set of all simple st-paths of a given planar graph. The transitions of $\mathcal{M}_{paths}$ are defined by rerouting a current path along one of its adjacent faces (see Fig. 2).

Fig. 2. Rerouting of x along face a.

Fig. 3. A graph admitting multiple embeddings with different sets of faces.

Definition 1. *Let x be an st-path and a be a face adjacent to at least one edge of x. We say that x can be rerouted along a if the edges in x and a form a single sub-path of x of length at least one, and the path y obtained by replacing all edges common to x and a with the edges in a that do not belong to x is simple. In this case, the rerouting operation consists of replacing x with y.*

Note that we forbid rerouting operations that reduce the length of the current path by introducing a cycle and short-cutting it afterwards. The reason is that these operations are not "reversible" and would prevent us from using well-established methods to determine the stationary distribution[1].

Markov Chain $\mathcal{M}_{paths}$. Given a parameter $\lambda > 0$, the transition from the current state x to the next one are defined according to the following rule:

1. With probability $\frac{1}{2}$ do nothing. Otherwise,
2. Select a face a uniformly at random. If x cannot be rerouted along a then do nothing. Otherwise,
3. Move to the path y obtained by rerouting x along a with probability

$$A(x,y) := \min\left\{1, \frac{\lambda^{|y|}}{\lambda^{|x|}}\right\},$$

and do nothing with remaining probability $1 - A(x,y)$.

For $\lambda < 1$, rerouting operations increasing the length of the current path by ℓ are accepted with probability λ^ℓ, while those reducing it are always accepted. The converse happens for $\lambda > 1$.

Remark 2. We stress that a planar graph can have different embeddings, and the transitions of $\mathcal{M}_{paths}$ depend on the particular given one. Figure 3 shows an

---

[1] The analysis of non-reversible Markov chains is in general rather difficult and it is considered an interesting problem also for simple chains [4].

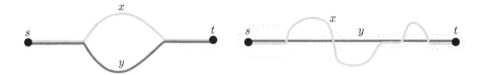

Fig. 4. Paths with $\Delta_{xy} = 2$ (left) and $\Delta_{xy} = 4$ (right).

example of a graph admitting an embedding in which there is a face with three vertices. In the next section we show that the chain is ergodic for any embedding.

Remark 3. For the sake of simplicity we forbid rerouting along the outer face. Doing so would not change the presented results significantly except in making the proofs less readable.

4   Analysis of $\mathcal{M}_{paths}$

Ergodicity. For all states $x \in \Omega$ it holds that $P(x, x) \geq 1/2$, and thus $\mathcal{M}_{paths}$ is aperiodic. To show ergodicity, it then suffices to prove that any two states $x, y \in \Omega$ are connected by a path with non-zero probability. Before stating the main theorem, we first introduce the following notion of "distance" between st-paths.

Definition 4. *Given two st-paths x and y, a maximal sub-path common to x and y is an ordered sequence of vertices appearing in both paths and not contained in a longer sequence with the same property. We let Δ_{xy} denote the number of maximal sub-paths common to x and y.*

Note that the definition also allows "degenerate" sub-paths of just one vertex. For instance, if x and y have only the starting and the ending vertices in common, then $\Delta_{xy} = 2$. Figure 4 shows examples of paths with different values of Δ_{xy}.

Theorem 5. *For any plane embedding of a given graph and any pair of vertices s and t, the Markov chain $\mathcal{M}_{paths}$ is ergodic with diameter at most $2n^2$.*

Proof (Sketch). Any two paths x and y such that $\Delta_{xy} = 2$ are connected by a sequence of at most f paths. Moreover, for $\Delta_{xy} > 2$, there is an intermediate path x' such that $\Delta_{xx'} = 2$ and $\Delta_{x'y} < \Delta_{xy}$. Since $\Delta_{xy} < n$ and $f < 2n$ in planar graphs, the theorem follows.   □

Stationary Distribution. To characterize the stationary distribution of $\mathcal{M}_{paths}$, we show that, for some probability distribution π, it holds

$$\pi(x) \cdot P(x, y) = \pi(y) \cdot P(y, x), \qquad \text{for all } x, y \in \Omega. \tag{3}$$

It is well-known that π is then the stationary distribution of $\mathcal{M}_{paths}$; this property is known as the *detailed balance condition* [14, Proposition 1.19].

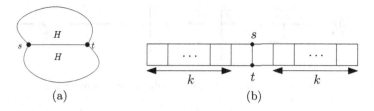

Fig. 5. Graphs with exponential bottleneck ratio.

Theorem 6. *For any planar graph and any two vertices s and t, the stationary distribution of the Markov chain $\mathcal{M}_{paths}$ is*

$$\pi(x) = \frac{\lambda^{|x|}}{Z(\lambda)}, \qquad \text{where } Z(\lambda) = \sum_{z \in \Omega} \lambda^{|z|}. \tag{4}$$

The parameter λ can be used to tune the stationary distribution of $\mathcal{M}_{paths}$. For example, by setting $\lambda = 1$ the stationary distribution is the uniform distribution over all simple st-paths.

Mixing Time. We conclude this section by providing some negative results concerning the mixing time of $\mathcal{M}_{paths}$.

Theorem 7. *There exist planar graphs G and vertices s, t such that $\mathcal{M}_{paths}$ with $\lambda = 1$ is not rapidly mixing.*

Proof. We apply the well-known bottleneck theorem [14] which says that, for any subset of states $R \subset \Omega$ such that $\pi(R) \leq 1/2$, the following bound on the mixing time of the chain holds:

$$t_{mix} \geq \frac{\pi(R)}{4Q(R, \bar{R})} \qquad \text{where } Q(R, \bar{R}) = \sum_{x \in R, y \in \bar{R}} \pi(x) P(x, y). \tag{5}$$

Let G be the planar graph obtained by combining two copies of a planar graph H as follows. The two copies share only a single edge connecting s and t, and all other vertices and edges in H are duplicated (see Fig. 5a). The set R consists of the subset of st-paths of G that use only edges in one of the two copies, say the upper one. Note that R contains the common single-edge path $x^* = (s, t)$, that this is the only path in R with transitions to some $y \in \bar{R}$, and that there are at most two transitions from x^* to some other state (each edge is adjacent to at most two faces). Therefore

$$Q(R, \bar{R}) = \sum_{x \in R, y \in \bar{R}} \pi(x) P(x, y) \leq 2\pi(x^*) = 2/|\Omega|.$$

In order to apply the bottleneck theorem we need $\pi(R) \leq 1/2$. This can be easily achieved by adding two more paths to the bottom copy of H (recall that R consists of all paths in the upper copy of H). We then get

$$t_{mix} \geq \frac{\pi(R)}{4Q(R, \bar{R})} \geq \frac{|R|/|\Omega|}{8/|\Omega|} = |R|/8. \qquad \square$$

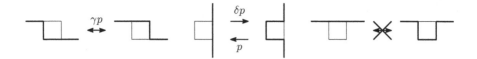

Fig. 6. The transitions of $\mathcal{M}_{mon}$; p is the probability of selecting a face u.a.r.

Corollary 8. *There exists an infinite family of planar graphs such that the mixing time of $\mathcal{M}_{paths}$ satisfies $t_{mix} \in \Omega(\lambda^{n/2})$ for all $\lambda > 1$.*

Proof (Sketch). The claim follows by considering the bottleneck ratio of the graph in Fig. 5b.   □

We note that Theorem 7 holds also for graphs that have a very simple structure, like two square grids sharing only edge (s, t), and for outerplanar graphs.

5   A Rapidly Mixing Chain for Vertical-Monotone Paths

We now present the rapidly mixing Markov chain $\mathcal{M}_{mon}$, which is a natural modification of $\mathcal{M}_{paths}$ for the case where the graph is a sub-graph of the two-dimensional lattice (grid) with no holes. That is, every face is either a cell of the grid or it is the outer face. The chain $\mathcal{M}_{mon}$ samples paths that are *vertical-monotone*, that is, that are only monotone in the vertical direction (if we follow the path from s to t, it never goes up). We thus assume that s lies above or at the same y-coordinate of t. Though it is straightforward to generate such paths *uniformly* at random, our goal is a weighted sampler with probability biased towards shorter paths according to the parameter λ, i.e., with distribution of the form (1).

Markov Chain $\mathcal{M}_{mon}$. The chain is a modification of $\mathcal{M}_{paths}$ in which some transitions are disallowed and others are "more lazy" (see Fig. 6). Specifically, the chain does not allow to replace an horizontal edge with three edges, and transitions swapping two consecutive edges of a face are only performed with probability

$$\gamma := \frac{1 + \delta}{2} \quad \text{where } \delta = \lambda^2 \text{ and } \lambda \in (0, 1].$$

This choice of γ will be useful for the analysis of the mixing time. Note that we restrict to the case $\lambda \leq 1$, because the lower bound for $\lambda > 1$ of Corollary 8 holds also for $\mathcal{M}_{mon}$. Note further that $\mathcal{M}_{mon}$ is ergodic with diameter $\leq 2f$, and its stationary distribution is the same as for $\mathcal{M}_{paths}$.

5.1   Mixing Time of $\mathcal{M}_{mon}$

To bound the mixing time we use the method of *path coupling without contraction* [1]. A *path coupling* for a chain $\mathcal{M}$ can be specified by providing distributions

$$\mathbb{P}_{x,y}[X = x', Y = y'], \qquad \text{for all } x, y \in \Omega \text{ such that } P(x, y) > 0, \qquad (6)$$

satisfying, for all $x, y \in \Omega$ such that $P(x, y) > 0$,

$$\mathbb{P}_{x,y}[X = x'] = P(x, x') \qquad \text{for all } x' \in \Omega, \tag{7}$$
$$\mathbb{P}_{x,y}[Y = y'] = P(y, y') \qquad \text{for all } y' \in \Omega. \tag{8}$$

We use ρ to denote the shortest-path distance in the Markov chain, i.e., $\rho(x, y)$ is the minimum number of transitions to go from x to y.

Lemma 9 (Theorem 2 in [1]). *Suppose we have a path coupling for a Markov chain $\mathcal{M}$ such that, for all x, y with $P(x, y) > 0$, it holds*

$$\mathbb{E}_{x,y}[\rho(X, Y)] \leq 1. \tag{9}$$

Then, the Markov chain $\mathcal{M}^*$ with transition matrix $P^* = (P + p_{min}I)/(1 + p_{min})$ has mixing time $t^*_{mix} \in O\left(D^2/p_{min}\right)$, where p_{min} and D denote respectively the smallest non-zero transition probability and the diameter of $\mathcal{M}$.

Note that $\mathcal{M}^*$ is the chain with transition probabilities

$$P^*(x, y) := \begin{cases} \frac{P(x,x) + p_{min}}{1 + p_{min}} & \text{if } y = x, \\ \frac{P(x,y)}{1 + p_{min}} & \text{otherwise.} \end{cases}$$

Therefore $\mathcal{M}^*$ and $\mathcal{M}$ have the same stationary distribution. This suggests naturally to run the chain $\mathcal{M}^*_{mon}$ for efficiently sample vertical-monotone paths.

Path Coupling for $\mathcal{M}_{mon}$. For the sake of clarity, for every face a we define the following shorthand:

$$p_a(x) := P(x, x \oplus a),$$

where $x \oplus a$ denotes the path obtained by rerouting x along a. We define a path coupling by specifying, for every pair (x, y) such that x and y differ in one face d, the probabilities in (6) to move to a pair (x', y'):

$$(x, y) \mapsto (x \oplus d, y) \qquad \text{with probability } p_d(x), \tag{10}$$
$$(x, y) \mapsto (x, y \oplus d) \qquad \text{with probability } p_d(y), \tag{11}$$

and for every other face a $\neq$ d

$$(x, y) \mapsto (x \oplus a, y \oplus a) \qquad \text{with probability } \min\{p_a(x), p_a(y)\}, \tag{12}$$
$$(x, y) \mapsto (x \oplus a, y) \qquad \text{with probability } \max\{0, p_a(x) - p_a(y)\}, \tag{13}$$
$$(x, y) \mapsto (x, y \oplus a) \qquad \text{with probability } \max\{0, p_a(y) - p_a(x)\}. \tag{14}$$

Finally, with all remaining probability

$$(x, y) \mapsto (x, y). \tag{15}$$

One can easily check that this is indeed a path coupling, that is, (7)–(8) are satisfied. The difficulty is in proving the condition necessary to apply Lemma 9.

Lemma 10. *The path coupling defined above satisfies condition (9).*

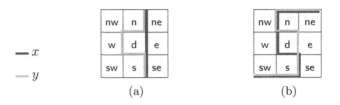

Fig. 7. Analysis of path coupling for grids (main idea).

Proof (Sketch). In the coupling, the distance between two paths x and y can either increase by 1 or decrease by 1. Since the initial distance is 1, we can thus write the expected distance after one coupling step as

$$\mathbb{E}_{x,y}[\rho(X,Y)] = 0 \cdot p_0 + 1 \cdot p_1 + 2 \cdot p_2 = 1 + p_2 - p_0,$$

where p_0 is the probability that the distance decreases, p_2 is the probability that it increases, and $p_1 = 1 - p_0 - p_2$.

The transitions reducing the distance are those corresponding to (10)–(11), that always happen with probability

$$p_0 = p_{\mathsf{d}}(x) + p_{\mathsf{d}}(y) = p(1 + \delta) = 2p\gamma,$$

where $p = \frac{1}{2f}$ is the probability that one particular face is chosen.

The distance increases instead to 2 if, for instance, the coupling uses a face for rerouting y while x stays the same. Since the coupling attempts to reroute both paths whenever possible (12), the probability that the distance becomes 2 is due to (13) and (14) only. We thus have to consider only faces for which the probability of rerouting is *different* for the two paths, that is, $p_{\mathsf{a}}(x) \neq p_{\mathsf{a}}(y)$. We illustrate the proof only for the cases in Fig. 7; the remaining ones can be found in the full version of this paper.

For the case of Fig. 7a, the transitions increasing the distance correspond to the four faces around d (namely, $\mathsf{n}, \mathsf{s}, \mathsf{e}, \mathsf{w}$). We thus get that

$$p_2 = p\delta + p\delta + p(\gamma - \delta) + p(\gamma - \delta) = 2p\gamma,$$

and conclude then that (9) holds for this case.

In the scenario of Fig. 7b we can apply the same analysis. Here it is crucial to observe that faces ne and sw do not contribute to p_2 because in $\mathcal{M}_{mon}$ these transitions are not allowed. This is one of the cases where the monotonicity of the paths plays a crucial role. We thus get that

$$p_2 = p(1 - \gamma) + p(1 - \gamma) + p\delta + p\delta = 2\gamma p.$$

Therefore, (9) holds also for this case. □

Since the diameter of $\mathcal{M}_{mon}$ is $O(n)$ and the minimum non-zero transition probability is $p_{min} = \frac{\lambda^2}{2f}$, we can establish the following bound on the mixing time.

Theorem 11. *The mixing time of $\mathcal{M}^*_{mon}$ is $O\left(n^3/\lambda^2\right)$ for every $\lambda \in (0, 1]$.*

6   Conclusion

We have studied a natural Markov chain $\mathcal{M}_{paths}$ for sampling st-paths in any given planar graph. We have shown that this chain is always ergodic and its stationary distribution is the Gibbs distribution on the paths weighted according to the number of edges. The chain is, in general, not rapidly mixing, but it might be possible that, in some graphs, modifications of this "basic" chain yield a rapidly mixing one. We have shown that this is indeed the case when restricting the sampling to vertical-monotone paths on sub-graphs of the 2D lattice. Another possible direction might be to introduce *non-local* transitions. In this case, the chain should probably be designed "ad-hoc" for a specific class of graphs. It would also be interesting to find graph classes for which $\mathcal{M}_{paths}$ is rapidly mixing. We conjecture that this might be the case for regular square grids.

An interesting related question is how to *count* the number of paths of a certain length. Note that the case of vertical-monotone paths provides an excellent example to also show the limitations of our local chain. Indeed, in this case the dynamic programming algorithm for counting all paths (even non simple) in general graphs would work here. Moreover, the procedure can be easily adapted to sample paths with uniform distribution among those with a fixed length. This procedure in the end leads to an exact Gibbs sampler (1) for all values of $\lambda > 0$.

Acknowledgments. We are grateful to Francesco Pasquale for comments on an earlier version of this work. We also wish to thank the anonymous referees for pointing out relations to the class of outerplanar graphs, and the use of dynamic programming for the vertical-monotone paths. Part of this work has been done while the second author was at ETH Zurich. This work is supported by the EU FP7/2007-2013 (DG CONNECT.H5-Smart Cities and Sustainability), under grant agreement no. 288094 (project eCOMPASS), and by the French ANR Project DISPLEXITY.

References

1. Bordewich, M., Dyer, M.E.: Path coupling without contraction. J. Discrete Algorithms **5**(2), 280–292 (2007)
2. Bordewich, M., Dyer, M.E., Karpinski, M.: Path coupling using stopping times and counting independent sets and colorings in hypergraphs. Random Struct. Algorithms **32**(3), 375–399 (2008)
3. Bubley, R.: Randomized Algorithms: Approximation, Generation, and Counting. Springer, Heidelberg (2011)
4. Diaconis, P., Holmes, S., Neal, R.M.: Analysis of a nonreversible Markov chain sampler. Ann. Appl. Probab. **10**, 726–752 (2000)
5. Došlić, T.: Seven (lattice) paths to log-convexity. Acta Applicandae Math. **110**(3), 1373–1392 (2010)
6. Dyer, M., Flaxman, A.D., Frieze, A.M., Vigoda, E.: Randomly coloring sparse random graphs with fewer colors than the maximum degree. Random Struct. Algorithms **29**(4), 450–465 (2006)
7. Dyer, M., Frieze, A., Hayes, T.P., Vigoda, E.: Randomly coloring constant degree graphs. Random Struct. Algorithms **43**(2), 181–200 (2013)

8. Dyer, M.E., Frieze, A.M.: Randomly coloring random graphs. Random Struct. Algorithms **36**(3), 251–272 (2010)
9. Dyer, M.E., Greenhill, C.S.: On markov chains for independent sets. J. Algorithms **35**(1), 17–49 (2000)
10. Greenberg, S., Pascoe, A., Randall, D.: Sampling biased lattice configurations using exponential metrics. In: SODA, pp. 76–85 (2009)
11. Hayes, T.P., Vera, J.C., Vigoda, E.: Randomly coloring planar graphs with fewer colors than the maximum degree. Random Struct. Algorithms **46**(2), 29–44 (2014)
12. Jerrum, M., Sinclair, A., Vigoda, E.: A polynomial-time approximation algorithm for the permanent of a matrix with nonnegative entries. J. ACM **51**(4), 671–697 (2004)
13. Jerrum, M., Valiant, L.G., Vazirani, V.V.: Random generation of combinatorial structures from a uniform distribution. Theor. Comput. Sci. **43**, 169–188 (1986)
14. Levin, D.A., Peres, Y., Wilmer, E.L.: Markov chains and mixing times. Am. Math. Soc. **30** (2009)
15. Luby, M., Randall, D., Sinclair, A.: Markov chain algorithms for planar lattice structures. SIAM J. Comput. **31**(1), 167–192 (2001)
16. Luby, M., Vigoda, E.: Approximately counting up to four. In: STOC, pp. 682–687 (1997)
17. Martin, R.A., Randall, D.: Sampling adsorbing staircase walks using a new Markov chain decomposition method. In: FOCS, pp. 492–502 (2000)
18. Morris, B., Sinclair, A.: Random walks on truncated cubes and sampling 0–1 knapsack solutions. SIAM J. Comput. **34**(1), 195–226 (2004)
19. Provan, J.S.: The complexity of reliability computations in planar and acyclic graphs. SIAM J. Comput. **15**(3), 694–702 (1986)
20. Randall, D., Sinclair, A.: Self-testing algorithms for self-avoiding walks. J. Math. Phy. **41**(3), 1570–1584 (2000)
21. Roberts, B., Kroese, D.P.: Estimating the number of s-t paths in a graph. J. Graph Algorithms Appl. **11**(1), 195–214 (2007)
22. Sinclair, A., Jerrum, M.: Approximate counting, uniform generation and rapidly mixing Markov chains. Inf. Comput. **82**(1), 93–133 (1989)
23. Valiant, L.G.: The complexity of enumeration and reliability problems. SIAM J. Comput. **8**(3), 410–421 (1979)

Degree-Constrained Subgraph Reconfiguration is in P

Moritz Mühlenthaler[(✉)]

Department of Computer Science, University of Erlangen-Nuremberg,
Erlangen, Germany
moritz.muehlenthaler@cs.fau.de

Abstract. The degree-constrained subgraph problem asks for a subgraph of a given graph such that the degree of each vertex is within some specified bounds. We study the following reconfiguration variant of this problem: Given two solutions to a degree-constrained subgraph instance, can we transform one solution into the other by adding and removing individual edges, such that each intermediate subgraph satisfies the degree constraints and contains at least a certain minimum number of edges? This problem is a generalization of the matching reconfiguration problem, which is known to be in P. We show that even in the more general setting the reconfiguration problem is in P.

1 Introduction

A reconfiguration problem asks whether a given solution to a combinatorial problem can be transformed into another given solution in a step-by-step fashion such that each intermediate solution is "proper", where the definition of proper depends on the problem at hand. For instance, in the context of vertex coloring reconfiguration, "step-by-step" typically means that the color of a single vertex is changed at a time, and "proper" has the usual meaning in the graph coloring context. An issue of particular interest is the relation between the complexity of the underlying combinatorial problem and its reconfiguration variant. This complexity relation has been studied for classical combinatorial problems including for example graph coloring, satisfiability, matching, and the shortest path problem [1,2,4]. Surprisingly, the reconfiguration variants of some tractable problems turn out to be intractable [4], and vice versa [7]. An overview of recent results on reconfiguration problems can be found in [6].

In this work we investigate the complexity of the reconfiguration problem associated with the (a,b)-degree-constrained subgraph (ab-DCS) problem. Let $G = (V,E)$ be a graph and let $a,b : V \to \mathbb{N}$ be two functions called *degree bounds* such that for each vertex v of G we have $0 \leq a(v) \leq b(v) \leq \delta_G(v)$, where $\delta_G(v)$ denotes the degree of v in G. The task is to decide if there is a subgraph S of G that satisfies the degree constraints, that is, for each vertex

M. Mühlenthaler—Research funded in parts by the School of Engineering of the University of Erlangen-Nuremberg.

© Springer-Verlag Berlin Heidelberg 2015
G.F. Italiano et al. (Eds.): MFCS 2015, Part II, LNCS 9235, pp. 505–516, 2015.
DOI: 10.1007/978-3-662-48054-0_42

v of S, $\delta_S(v)$ is required to be at least $a(v)$ and at most $b(v)$. Typically, the intention is to find among all subgraphs of G that satisfy the degree constraints one with the greatest number of edges. This problem is a generalization of the classical maximum matching problem and can be solved in polynomial time by a combinatorial algorithm [3,5]. The ab-DCS reconfiguration problem is defined as follows:

Definition 1. *(st-DCSCONN)*

INSTANCE: *An ab-DCS instance, source and target solutions M, N, an integer $k \geq 1$.*

QUESTION: *Is it possible to transform M into N by adding/removing a single edge in each step such that each intermediate subgraph satisfies the degree constraints and contains at least $\min\{|E(M)|, |E(N)|\} - k$ edges?*

Our main result is the following

Theorem. *st-DCSCONN can be solved in polynomial time.*

It was shown by Ito et al. in [4, Proposition 2] that the analogous matching reconfiguration problem can be solved in polynomial time. According to our result, the reconfiguration problem remains tractable even in the more general ab-DCS reconfiguration setting. The proof of the main result essentially contains an algorithm that determines a suitable sequence of edge additions/removals if one exists. The number of reconfiguration steps is bounded by $O(|E|^2)$. The algorithm also provides a certificate for No-instances.

2   Notation

In this paper we deal with subgraphs of some simple graph $G = (V, E)$, which is provided by ab-DCS a problem instance. The subgraphs of concern are induced by subsets of E. For notational convenience, we identify these subgraphs with the subsets of E and can therefore use standard set-theoretic notation ($\cap$, $\cup$, $-$) for binary operations on the subgraphs. Let H and K be two subgraphs of G, denoted by $H, K \subseteq G$. By $E(H)$ we refer explicitly to the set of edges of the graph H. We write $H + K$ for the union of H and K if they are disjoint. By $H \triangle K$ we denote the *symmetric difference* of H and K, that is $H \triangle K := (H - K) + (K - H)$. If e is an edge of G we may write $H + e$ and $H - e$ as shorthands for $H + \{e\}$ and $H - \{e\}$, respectively. We denote the degree of a vertex v of H by $\delta_H(v)$. A walk $v_0 \xrightarrow{e_0} \ldots \xrightarrow{e_{t-1}} v_t$ in G is a *trail* if $e_0, \ldots, e_{t-1}$ are distinct. The vertices v_0 and v_t are called *end vertices*, all other vertices are called *interior*. A trail without an edge is called *empty*. A non-empty trail is *closed* if its end vertices agree, otherwise it is *open*. A closed trail is also called a *cycle*. In a slight abuse of notation we will sometimes consider trails in G simply as subgraphs of G and combine them with other subgraphs using the notation introduced above. A trail is called (K, H)-*alternating* if its edges, in the order given by the trail, are alternatingly chosen from $K - H$ and $H - K$. An odd-length (K, H)-alternating trail T is called K-*augmenting* if $|E(K \triangle T)| > |E(K)|$.

Let G be the graph and $a, b : V \rightarrow \mathbb{N}$ be the degree bounds of an ab-DCS instance. A subgraph $M \subseteq G$ that satisfies the degree constraints is called ab-constrained. A vertex v of M is called a-tight (b-tight) in M if $\delta_M(v) = a(v)$ ($\delta_M(v) = b(v)$). A vertex is called ab-fixed in M if it is both a-tight and b-tight in M. We say that M is a-tight (b-tight) if each vertex of M is a-tight (b-tight). A closed (M, N)-alternating trail $T = v_0 - \ldots - v_t$ of even length is called alternatingly ab-tight in M if for each i, $0 \le i \le t$, v_i is a-tight iff i is even and b-tight iff i is odd, or vice versa.

3 ab-Constrained Subgraph Reconfiguration

Throughout this section, we assume that we are given some st-DCSCONN instance (G, M, N, a, b, k), where $G = (V, E)$ is a graph, $a, b : V \rightarrow \mathbb{N}$ are degree bounds, $M, N \subseteq G$ are ab-constrained, and $k \ge 1$. A reconfiguration step, or move, adds/removes an edge to/from a subgraph. Given an M-edge e and an N-edge e', an elementary move on M yields $M - e + e'$, either by adding e' after removing e or vice versa. M is k-reconfigurable to N if there is a sequence of reconfiguration moves that transforms M into N such that each intermediate subgraph respects the degree constraints and contains at least $\min\{|E(M)|, |E(N)|\} - k$ edges. Clearly, if M is k-reconfigurable to N then M is also k'-reconfigurable to N for any $k' > k$. M is internally k-reconfigurable to N if it is k-reconfigurable under the additional restriction that each intermediate subgraph is contained in $M \triangle N$. If M is not internally k-reconfigurable to N but still k-reconfigurable to N then we say that M is externally k-reconfigurable to N.

The general procedure for deciding if M is k-reconfigurable to N is the following: First, we check for the presence of obstructions that render a reconfiguration impossible. If it turns out that reconfiguration is still possible we reconfigure (M, N)-alternating trails in $M \triangle N$, one by one, until we either finish successfully or we obtain a certificate for M not being k-reconfigurable to N. Curiously, it turns out that if M is not 2-reconfigurable to N then M is not k-reconfigurable to N for any $k \ge 2$.

3.1   Obstructions

When transforming M into N, certain parts of M may be "fixed" and therefore make a proper reconfiguration impossible. Similar obstructions occur for example in vertex coloring reconfiguration, where certain vertices are fixed in the sense that their color cannot be changed [2]. In our case we identify a certain subgraph of G which depends on M and N and cannot be changed at all.

Let v be an ab-fixed vertex of G. When reconfiguring M to N, no M-edge incident to an ab-fixed vertex can be removed and no $(G - M)$-edge incident to an ab-fixed vertex can be added during a reconfiguration process without violating the degree constraints. Hence an edge is fixed if it is incident to an ab-fixed vertex. However, we may identify larger parts of G that are fixed due to the given subgraph M and the degree bounds. If we consider each ab-fixed edge to be M-fixed, then we can identify further M-fixed edges based on the following observations: First if a vertex v is incident to exactly $b(v)$ M-edges and each of them is M-fixed then all edges incident to v are M-fixed. Similarly, if v is incident to exactly $a(v)$ M-edges and each $(G - M)$-edge incident to v is M-fixed then each edge incident to v is M-fixed. Algorithm 1 shows how to identify an M-fixed subgraph of G based on these observations. By $\mathrm{I}_M(v)$ we denote the set of M-edges incident to the vertex v. Some bookkeeping could be employed to speed things up, but it is not necessary for our argument.

Algorithm 1. M-FIXEDSUBGRAPH

---

    input : ab-DCS instance (G, a, b), ab-constrained $M \subseteq G$
    output: M-fixed subgraph $F \subseteq G$
    $F \longleftarrow \emptyset;$ $F' \longleftarrow \{e \in G \mid e \text{ is fixed}\}$
    while $|F'| > |F|$ **do**
        $F \longleftarrow F'$
        if $\delta_M(v) = b(v)$ *and* $\mathrm{I}_M(v) \subseteq F$ **then**
            $F' \longleftarrow F' \cup \mathrm{I}_G(v)$
        if $\delta_M(v) = a(v)$ *and* $\mathrm{I}_{G-M}(v) \subseteq F$ **then**
            $F' \longleftarrow F' \cup \mathrm{I}_G(v)$
    return F

Proposition 1. *Let M, N be ab-constrained subgraphs of G and let $F \subseteq G$ be M-fixed. If $(M \bigtriangleup N) \cap F$ is non-empty then M is not k-reconfigurable to N for any $k \geq 1$.* □

That is, any M-fixed edge in $(M \bigtriangleup N)$ is a NO-certificate. As a consequence, we can check if M and N agree on the subgraph $F \subseteq G$ found by Algorithm 1 as a preprocessing step. At this point in particular, but also later on it will be convenient to consider *subinstances* of a given st-DCSCONN instance $\mathcal{I} = (G, M, N, a, b, k)$. If $H \subseteq G$ then the corresponding subinstance $\mathcal{I}_H$ is the instance $(H, M \cap H, N \cap H, a_H, b_H, k)$, where

$$a_H(v) = \max\{0, a(v) - \delta_{(G-H) \cap M}(v)\}$$
$$b_H(v) = b(v) - \delta_{(G-H) \cap M}(v) \ .$$

Proposition 2. *If $(M \bigtriangleup N) \cap F = \emptyset$ then $\mathcal{I}$ is a YES-instance if and only if $\mathcal{I}_{G-F}$ is a YES instance.* □

The graph $G - F$ does not have any fixed vertices and hence the $(M - F)$-fixed subgraph produced by Algorithm 1 is empty. Removing the M-fixed subgraph of G in a preprocessing step will considerably simplify our arguments later on. It should be immediate that no fixed edges can be introduced by reconfiguring an ab-constrained subgraph.

3.2 Internal Alternating Trail Reconfiguration

The next Lemma is our fundamental tool for reconfiguring alternating trails in $M \bigtriangleup N$. For any such trail T, it provides necessary and sufficient conditions for

$T \cap M$ being internally 1-reconfigurable to $T \cap N$ by performing only elementary moves. Behind the scenes, T is recursively divided into subtrails which need to be reconfigured in a certain order. If successful, the reconfiguration procedure performs exactly $|E(T)|$ edge additions/removals.

Lemma 1. *Let* $T = v_0 \xrightarrow{e_0} \dots \xrightarrow{e_{t-1}} v_t$ *be a* (M, N)-*alternating trail of even length in* $M \triangle N$. *Then* $T \cap M$ *is internally 1-reconfigurable to* $T \cap N$ *using only elementary moves if and only if* T *it satisfies each of the following conditions:*

1. *T contains no ab-fixed vertex.*
2. *If T is open then v_0 is not a-tight and v_t is not b-tight in M.*
3. *If T is closed then each of the following is true:*
   (a) T is not b-tight in M
   (b) T is not a-tight in M
   (c) T is not alternatingly ab-tight in M

Proof. Without loss of generality, let e_0 be an M-edge. We first show the necessity of conditions 1–3. By Proposition 1, if T contains an M-fixed vertex then $T \cap M$ is not k-reconfigurable to $T \cap N$ for any $k \geq 1$. If T is open and v_0 is a-tight in M, then e_0 cannot be removed from $T \cap M$ without violating the degree constraints. Likewise, if T is open and v_t is b-tight then e_t cannot be added to $T \cap M$. If T is closed and b-tight in M, i.e., 3a is violated, then no N-edge can be added after removing any M-edge. Similarly, if T is closed and a-tight in M, i.e., 3b is violated, then no M-edge can be removed after adding a single N-edge. If T is closed and alternatingly ab-tight then no edge can be added to or removed from $T \cap M$, so it cannot be internally 1-reconfigurable to $T \cap N$. In summary, if any of the conditions 1–3 is violated then $T \cap M$ is not internally 1-reconfigurable to $T \cap N$ using elementary moves.

In order to show the sufficiency of conditions 1–3 we employ the following general strategy: We partition T into (M, N)-alternating subtrails R, Q, S, each of even length and at least two of them non-empty. We show that for an appropriate choice of these subtrails there is an ordering, say Q, R, S, such that the first two subtrails are non-empty and Q satisfies conditions 1–3 in M, R satisfies the same conditions in $M - (Q \cap M) + (Q \cap N)$, and S, if non-empty, in turn satisfies the conditions in $M - (Q \cap M) + (Q \cap N) - (R \cap M) + (R \cap N)$. Therefore, each non-empty subtrail can be dealt with individually in a recursive fashion as long as the ordering is respected. The base case of the recursion consists of an (M, N)-alternating trail $B = u - v - w$ of length two. We show that if B satisfies 1–3 then $B \cap M$ is internally 1-reconfigurable to $B \cap N$ by an elementary move. Since B satisfies conditions 1 and 2, u is not a-tight and w is not b-tight. If v is b-tight we can remove $u - v$ from M and add $v - w$ to $M - (u - v)$ without violating the degree constraints in any step. Similarly, if v is not b-tight we can add $v - w$ to M and afterwards remove $u - v$ from $M + (v - w)$ without violating the degree constraints. So $B \cap M$ is internally 1-reconfigurable to $B \cap N$ by an elementary move as required. For the general recursion, we consider two main cases: T is either open or closed.

We first assume that T is open. Since T satisfies conditions 1 and 2, there is some i, $0 \le i < t - 1$, i even, such that v_i is not a-tight and v_{i+2} is not b-tight in M. To see this, assume that there is no such i. Then, by induction, for each $0 \le i < t - 1$, i even, v_{i+2} must b-tight because v_i is not a-tight. However, by condition 2, v_t is not b-tight in M, a contradiction. We pick R, Q, and S as follows

$$\underbrace{v_0 \xrightarrow{e_0} v_1 \xrightarrow{e_1} \dots \xrightarrow{e_{i-1}} v_i}_{R}, \quad \underbrace{v_i \xrightarrow{e_i} v_{i+1} \xrightarrow{e_{i+1}} v_{i+2}}_{Q}, \quad \underbrace{v_{i+2} \xrightarrow{e_{i+2}} v_{i+3} \xrightarrow{e_{i+3}} \dots \xrightarrow{e_{t-1}} v_t}_{S} .$$

Note that Q is an open trail satisfying conditions 1–3 and R, S, if non-empty, can be open or closed. At this point $Q \cap M$ is internally 1-reconfigurable to $Q \cap N$ as described above, and the result is $M - e_i + e_{i+1}$. It is readily verified that if R and S are open then they satisfy conditions 1–3 in $M - e_i + e_{i+1}$ and can therefore be treated independently after reconfiguring Q. However, at least one of R, S being closed leads to a slight complication.

Let us assume that S is closed. Note that, by assumption, v_{i+2} is not b-tight in M and therefore it cannot be a-tight in $M - e_i + e_{i+1}$. Thus, if any of the conditions 1–3 is violated in M then it cannot be violated in $M - e_i + e_{i+1}$. Since 3a–3c cannot be violated simultaneously we conclude that S satisfies conditions 1–3 either in M or in $M - e_i + e_{i+1}$. An analogous argument shows that R satisfies conditions 1–3 either in M or in $M - e_i + e_{i+1}$. Therefore, there is an ordering of R, Q, S that is consistent with the general strategy outlined above and depends on the tightness of the vertices v_i and v_{i+2} in M.

It remains to be shown that if T is closed and T satisfies properties 1–3 then $T \cap M$ is internally 1-reconfigurable to $T \cap N$ using only elementary moves. For this purpose we find a partition of T into subtrails that is compatible with our general strategy above. In particular, we show that if T is closed and satisfies properties 1–3 then T can be partitioned into two non-empty open trails R and Q, both of even length, such that Q satisfies 1–3. Furthermore, we show that R satisfies conditions 1–3 in $M - (Q \cap M) + (Q \cap N)$. That is, R can be dealt with after reconfiguring $Q \cap M$ to $Q \cap N$. We pick any two vertices of T that are connected by an M-edge, say v_0 and v_1, and consider the following cases:

(i) v_0 is neither a-tight nor b-tight, or v_1 is neither a-tight nor b-tight
(ii) v_0 is a-tight and v_1 is b-tight, or v_0 is b-tight and v_1 is a-tight
(iii) v_0 and v_1 are both b-tight
(iv) v_0 and v_1 are both a-tight

Case (i) We assume without loss of generality, that v_0 is neither a-tight nor b-tight, since if v_0 is a-tight or b-tight, then v_1 must be neither a-tight nor b-tight and if this is the case we can rearrange the vertices of T in the following way

$$T = v_1 - v_0 - v_{t-1} - \dots - v_2 - v_1 \tag{1}$$

and establish that v_0 is neither a-tight nor b-tight. Now we choose R and Q as follows:

$$R = v_2 - v_3 - \dots - v_t, \quad Q = v_0 - v_1 - v_2$$

If v_2 is not b-tight, then Q satisfies 1–3 and $Q \cap M$ can be reconfigured instantly to $Q \cap N$ by an elementary move. That is, we obtain $M - e_0 + e_1$ without violating the degree constraints. Now v_0 cannot be b-tight and v_2 cannot be a-tight in $M - e_0 + e_1$. Therefore, R now satisfies 1–3 and can be reconfigured as shown in the first main case of the proof. If v_2 is b-tight in M, then, by analogous considerations, R satisfies 1–3 and Q satisfies conditions 1–3 in $M - (R \cap M) + (R \cap N)$.

Case (ii) Without loss of generality, we assume that v_0 is a-tight and v_1 is b-tight, since if not, we can rearrange the vertices of T according to Eq. (1). Due to property 3c, T is not alternatingly ab-tight, so there is some i, $0 \leq i < t$, i even, such that v_i is not a-tight or v_{i+1} is not b-tight. If v_i is not a-tight, we choose R and Q to be

$$Q = v_i - v_{i+1} - \ldots - v_t (= v_0), \quad R = v_0 - v_1 - \ldots - v_i$$

Otherwise, v_{i+1} is not b-tight and we pick R and Q as follows

$$Q = v_1 - v_0 - \ldots - v_{i+1}, \quad R = v_{i+1} - v_i - \ldots - v_1$$

Either way, Q satisfies satisfies conditions 1–3 in M and R satisfies the same conditions in $M - (Q \cap M) + (Q \cap N)$.

Case (iii) If v_0 and v_1 are both b-tight then, by property 3a, there is some i, $0 \leq i < t$, such that v_i is not b-tight. Without loss of generality, we assume that i is even. We pick Q and R as follows:

$$Q = v_0 - v_1 - \ldots - v_i, \quad R = v_i - v_{i-1} - \ldots - v_0$$

Then Q satisfies 1–3 and R satisfies the same conditions in $M - (Q \cap M) + (Q \cap N)$.

Case (iv) This case is analogous to case (iii). By property 3b, there is some i, $0 \leq i < t$, such that v_i is not a-tight. Again, without loss of generality, we assume that i is even. We pick Q and R as follows

$$Q = v_0 - v_1 - \ldots - v_i, \quad R = v_i - v_{i-1} - \ldots - v_0$$

and conclude that Q satisfies 1–3 in M and R satisfies 1–3 in $M - (Q \cap M) + (Q \cap N)$.

From our consideration of the various cases we conclude that a given (M, N)-alternating trail T that satisfies conditions 1–3 can be recursively partitioned into subtrails as outlined in the general strategy above. Since the single base case of the recursion employs only elementary moves on edges of T, $T \cap M$ is internally 1-reconfigurable to $T \cap N$ using only elementary moves.   □

An (M, N)-alternating trail is *maximal* if there is no suitable edge in $M \triangle N$ to extend the trail at one of its end nodes. The subsequent lemmas establish sufficient conditions for maximal alternating trails to be internally 1- or

2-reconfigurable. Such trails are important in the proof of Theorem 1. Note however, that in contrast to Lemma 1 we are not restricted to elementary moves. The proofs are quite technical and have been omitted. In the following, let $T = v_0 \xrightarrow{e_0} \ldots \xrightarrow{e_{t-1}} v_t$ be a maximal (M, N)-alternating trail in $M \triangle N$ such that no vertex of T is ab-fixed.

Lemma 2. *If T is open and has even length then $T \cap M$ is internally 1-reconfigurable to $T \cap N$.*   □

Lemma 3. *If T has odd length and e_0 is an N-edge then*

1. *$T \cap M$ is internally 1-reconfigurable to $T \cap N$, and*
2. *$T \cap N$ is internally 2-reconfigurable to $T \cap M$.*   □

Lemma 4. *If T is closed and has even length then*

1. *$T \cap M$ is internally 2-reconfigurable to $T \cap N$ if T is not alternatingly ab-tight, and*
2. *$T \cap M$ is internally 1-reconfigurable to $T \cap N$ if T is neither b-tight nor alternatingly ab-tight.*   □

3.3   External Alternating Trail Reconfiguration

In the following we deal with even-length alternating cycles that are either alternatingly ab-tight or b-tight. The two cases are somewhat special since we will need to consider edges that are not part of the cycles themselves. Let $C = u_0 \xrightarrow{e_0} \ldots \xrightarrow{e_{t-1}} u_t$ be an (M, N)-alternating cycle of even length in $M \triangle N$. We will first consider the case that C is b-tight. 2-reconfigurability of C is established by Lemma 4. We generalize the approach from [4, Lemma 1] to ab-constrained subgraphs to obtain a characterization of 1-reconfigurable b-tight even cycles in the case that M and N are maximum and no vertex of G is ab-tight. We denote by $\mathrm{NOTA}(M) := \{v \in V(G) \mid\ \ v$ is not a-$tight$ in $M\}$ and $\mathrm{NOTB}(M) := \{v \in V(G) \mid\ \ v$ is not b-$tight$ in $M\}$ the sets of vertices that are not a-tight in M and not b-tight in M, respectively. Let

$$\mathrm{EVEN}(M) := \{\ v \in \mathrm{NOTA}(M)\ |\text{There is some even-length } M\text{-alternating}$$
$$vw\text{-trail starting with an } M\text{-edge}$$
$$\text{s.t. } w \in \mathrm{NOTB}(M)\}$$

$$\mathrm{NOTB}(G) := \{\qquad v \in V(G) \mid \text{There is some maximum } M \subseteq G \text{ satisfying}$$
$$\text{the degree constraints s.t. } v \in \mathrm{NOTB}(M)\}$$

The proof of the following lemma is analogous to that in [4, Lemma 2] and has been omitted due to space restrictions.

Lemma 5. *If M is maximum then $\mathrm{EVEN}(M) = \mathrm{NOTB}(G)$.*   □

Lemma 6. *If G contains no ab-fixed vertices, M and N are maximum, and C is b-tight then $C \cap M$ is 1-reconfigurable to $C \cap N$ if and only if there is a vertex v of C such that $v \in \mathrm{EVEN}(M)$.*

Proof. We first show that if $v \in C \cap \text{EVEN}(M)$ then $C \cap M$ is 1-reconfigurable to $C \cap N$. If $v \in C \cap \text{EVEN}(M)$ then there is a M-alternating vw-trail T of even length starting with an M-edge such that w is not b-tight. In the order given by the trail there is an earliest edge e such that none of the successors of e in T are C-edges. We distinguish two cases: First, assume that e is an M-edge. Without loss of generality, let u_0 be the C-vertex incident to e. We obtain the M-alternating subtrail $T' = u_0 - \ldots - w$ and $M' = M \triangle T$ is maximum and satisfies the degree constraints. Further, since u_0 is b-tight in M and w is not b-tight M, $T' \cap M$ is 1-reconfigurable to $T' \cap M'$ by Lemma 1. Then u_0 is not b-tight in M' and $C \cap M$ is 1-reconfigurable to $C \cap M'$ by Lemma 1. Since u_0 is not b-tight and w is not a-tight in $C \cap M'$, $T' \cap M'$ is 1-reconfigurable to $T' \cap M$ and thus we can undo the changes to M caused by the reconfiguration on T'.

In the second case we assume that e is a $(G-M)$-edge. Then there is a latest C-edge on T such that none the successors of e in T are C-edges. Without loss of generality, let $e = u_0 - u_1$ such that we obtain an M-alternating subtrail $T' = u_0 - u_1 - \ldots - w$. Then u_0 is not a-tight in $T' \cap M$, w is not b-tight in $T' \cap M$, and therefore $T' \cap M$ is internally 1-reconfigurable to $T' - M$ by Lemma 1. In the resulting subgraph, u_0 is not b-tight and w is b-tight. By using Lemma 1 again, we can reconfigure the remaining parts of C and undo the modifications to $M - C$ caused by the previous step, by considering the trail $w - \ldots - u_1 - u_2 - \ldots - u_t(= u_0)$. Hence, in both cases $C \cap M$ is 1-reconfigurable to $C \cap N$. Note that each time we invoke Lemma 1, we use the assumption that no vertex of G is ab-fixed.

We now show that $v \in C \cap \text{EVEN}(M)$ is a necessary condition for $C \cap M$ to be 1-reconfigurable to $C \cap N$. If no vertex of C is in $\text{EVEN}(M) = \text{NOTB}(G)$ (Lemma 5), then each vertex v of C is essentially ab-fixed in the sense that no maximum ab-constrained subgraph exists such that v is not b-tight. Therefore $C \cap M$ cannot be 1-reconfigurable to $C \cap N$. $\qquad\square$

We now characterize k-reconfigurable alternatingly ab-tight cycles in $M \triangle N$ assuming that G contains no ab-fixed vertices. Such cycles cannot occur in the matching reconfiguration setting since no vertex of a cycle is a-tight in this case. There is some conceptual similarity to the proof of Lemma 4, but for the purpose of proving Theorem 1 we cannot assume that M and N are both maximum. Therefore, we cannot rely on Lemma 5, which simplifies the problem of finding a maximum ab-constrained subgraph M' such that a certain vertex is not b-tight to checking for the existence of an alternating trail. Instead, we now check if there is some ab-constrained subgraph M' such that the tightness of the C-vertices in M' differs from their tightness in M. The existence of a suitable M' can be checked in polynomial time by constructing and solving suitable ab-DCS instances.

Lemma 7. *If C is alternatingly ab-tight and G contains no ab-fixed vertices then $C \cap M$ is k-reconfigurable to $C \cap N$ for any $k \geq 1$ if and only if there is some ab-constrained $M' \subseteq G$ such that $C \cap M = C \cap M'$ and C is not alternatingly ab-tight in M'.*

Proof. Let us first assume that there is some ab-constrained $M' \subseteq G$ such that $C \cap M = C \cap M'$ and C is not alternatingly ab-tight in M'. We show that $C \cap M$ is 1-reconfigurable to $C \cap N$. Since C is not alternatingly ab-tight in M' there is some vertex v of C that is b-tight in M but not in M', or there is some vertex u of C that is a-tight in M but not in M'. We will consider in detail the case that there is some v of C that is b-tight in M but not in M'. The other case is analogous. Since $\delta_M(v) < \delta_{M'}(v)$ there is an open (M, M')-alternating vw-trail T in $M \triangle M'$ starting at v such that T cannot be extended at w. Without loss of generality, we assume that $v = u_0$ and $u_0 \;—\; u_1$ is a $(C - M)$-edge.

We consider two subcases: T has either even or odd length. First assume that T is even. Let $R = w \;—\; \ldots \;—\; u = u_0 \;—\; u_1 \;—\; \ldots \;—\; u_t$. Since T is even, $v = u_t$ is not a-tight in M and w is not b-tight in M. Furthermore, $R \triangle M$ satisfies the degree constraints. By Lemma 1, $R \cap M$ is 1-reconfigurable to $R \cap (R \triangle M) = ((C \cap N) + (T \cap M'))$. As a result, $C \cap M$ has been reconfigured to $C \cap N$. We now need to undo the changes in $T = R - C$ caused by the previous reconfiguration. Observe that v is not b-tight and w is not a-tight in $T \cap M'$. Therefore, we can invoke Lemma 1 again to reconfigure $T \cap M'$ to $T \cap M$. In the second subcase we assume that T is odd and let $R = w \;—\; \ldots \;—\; u = u_0 \;—\; u_1 \;—\; \ldots \;—\; u_{t-1}$ and let $S = u_{t-1} \;—\; u_0 \;—\; \ldots \;—\; w$. Then R and S are open and have even length, w is not a-tight and u_{t-1} is not b-tight in M. Therefore, by Lemma 1, $R \cap M$ is 1-reconfigurable to $R \cap (R \triangle M)$. Now u_{t-1} is not a-tight and w is not b-tight in $R \triangle M$. Therefore we can use Lemma 1 again in order to reconfigure $S \cap (R \triangle M)$ to $S \cap (S \triangle (R \triangle M)) = S \cap (M - e)$. As a result $C \cap M$ has been reconfigured to $C \cap N$.

In order to prove the converse statement, assume that there is no ab-constrained $M' \subseteq G$ such that $C \cap M = C \cap M'$ and C is not alternatingly ab-tight in M'. Then each vertex of C is essentially ab-fixed. Therefore, $C \cap M$ is not k-reconfigurable to $C \cap N$ for any $k \geq 1$. □

3.4 Reconfiguring ab-constrained Subgraphs

For the overall task of deciding if M is reconfigurable to N we will iteratively partition $M \triangle N$ into alternating trails as shown in Algorithm 2. Given M and N such that $M \triangle N$ is non-empty, the algorithm outputs a decomposition of $M \triangle N$ into trails $T_0, \ldots, T_{i-1}$ and a list of ab-constrained subgraphs $M_0, \ldots, M_{i-1}$ for some $i \geq 0$ such $M = M_0$ and $N = M_{i-1}$ such that for each j, $1 \leq j < i$, $M_{j+1} = M_j \triangle T_j$. In each iteration i we need to find an M-augmenting uv-trail in $M_i \triangle N$ such that u and v are not b-tight in M_i. Since M_i is ab-constrained we can use for example the technique from [5, Sect. 2] to reduce the problem to obtaining an alternating path in an auxiliary graph that is constructed from $M \triangle N$. This approach produces a suitable M-augmenting trail if it exists in polynomial time. Since the number of iterations performed by the algorithm is bounded by $|E(M \triangle N)|$, the overall running time is polynomial in the size of the input graph. We are now ready to prove the main theorem. The structure of the proof is somewhat similar to the proof that matching reconfiguration can be solved in polynomial time, see [4, Proposition 2].

Algorithm 2. ALTERNATINGTRAILDECOMPOSITION

---

input : ab-constrained $M, N \subseteq G$ s.t. $M \triangle N$ non-empty
output: Lists of alternating trails and ab-constrained subgraphs
$i \longleftarrow 0;\quad M_0 \longleftarrow M$
while $M_i \neq N$ **do**
    Find M-augmenting uv-trail T in $M_i \triangle N$ s.t. u and v are not b-tight in M_i
    if *such T does not exist* **then**
        Let T be any maximal (M_i, N)-alternating trail in $M_i \triangle N$
    $T_i \longleftarrow T$
    $M_{i+1} \longleftarrow M_i \triangle T$
    $i \longleftarrow i+1$
return $[M_0, \ldots, M_{i-1}], [T_0, \ldots, T_{i-1}]$

Theorem 1. *st*-DCSCONN *can be solved in polynomial time.*

Proof. Let $\mathcal{I} = (G', M', N', a', b', k)$ be a st-DCSCONN instance. Let $F \subseteq G'$ be the M'-fixed subgraph produced by Algorithm 1. If $(M' \triangle N') \cap F$ is non-empty then some M'-fixed edge needs to be reconfigured, which is impossible by Proposition 1. Otherwise, we consider the subinstance $\mathcal{I}_{G'-F} = (G, M, N, a, b, k)$ (see Proposition 2).

Without loss of generality we assume that $|E(M)| \leq |E(N)|$. We process the alternating trails in $M \triangle N$ output by Algorithm 2 one by one in the given order. During the process, we observe the following types of (M, N)-alternating trails: (i) even-length trails that are not b-tight or alternatingly ab-tight cycles, (ii) M-augmenting trails, (iii) N-augmenting trails, (iv) b-tight even-length cycles, and (v) alternatingly ab-tight even-length cycles. Since $|E(M)| \leq |E(N)|$ there are at least as many type (ii)-trails as type (iii)-trails. Note that each condition in the Lemmas 1–7 can be checked in polynomial time. Cycles that are not reconfigurable according to Lemma 6 or 7 are NO-certificates. We distinguish the following cases:

Case $k \geq 2$. By construction, in each step i, if T_i is in categories (i)–(iv) then $T \cap M_i$ is 2-reconfigurable to $T \cap M_{i+1}$ by Lemmas 2, 3 and 4. If T_i is an even-length alternatingly ab-tight cycle (type v)) then Lemma 7 gives necessary and sufficient conditions under which $T_i \cap M_i$ is k-reconfigurable to $T_i \cap M_{i+1}$ for any $k \geq 1$. These conditions can be checked in polynomial time and do not depend on what edges are present in M_i outside of T_i The reconfigurability of T_i is a property solely of T_i, G, and the degree bounds.

Case $k = 1$ and $|E(M)| < |E(N)|$. Trails of types (i) and (ii) are 1-reconfigurable by Lemmas 2 and 3. Due to the preference given to M-augmenting trails in Algorithm 2, if T_i is an N-augmenting trail or a b-tight cycle then $|E(M_i)| \geq |E(N)|$. Therefore by Lemma 3 or 6 we have that $T_i \cap M_i$ is 2-reconfigurable to M_{i+1}, but no intermediate subgraph is of size less than $|E(M)| - 1$. The alternatingly ab-tight cycles can be dealt with just as in the previous case.

Case $k = 1$ and $|E(M)| = |E(N)|$, both not maximum. In this case we increase the size of N by one, using an N-augmenting T, to obtain N'. We first reconfigure

M to N' as in the case before. If successful, the result N' is 2-reconfigurable to N by Lemma 3. No intermediate subgraph is of size less than $|E(M)| - 1$.

Case $k = 1, |E(M)| = |E(N)|, both\,maximum$. Since M and N are maximum, each trail T_i is of type (i), (iv), or (v). Therefore, each open trail is 1-reconfigurable by Lemma 2. We need to check the 1-reconfigurability of each cycle according to Lemmas 6 (type (iv)) and 7 (type (v)). $\square$

The running time of the decision procedure is dominated by the time needed to check the conditions of Lemmas 6 and 7. Overall, this amounts to solving $O(|V(G)| \cdot |E(G)|^2)$ ab-DCS instances, which takes time $O(|E(G)|^{\frac{3}{2}})$ per instance using the algorithm from [3].

Acknowledgements. We would like to thank the anonymous referees for their constructive comments and valuable remarks on this paper.

References

1. Bonsma, P.S.: The complexity of rerouting shortest paths. Theor. Comput. Sci. **510**, 1–12 (2013)
2. Cereceda, L., van den Heuvel, J., Johnson, M.: Finding paths between 3-colorings. J. Graph Theory **67**(1), 69–82 (2011)
3. Gabow, H.N.: An efficient reduction technique for degree-constrained subgraph and bidirected network flow problems. In: Proceedings of the Fifteenth Annual ACM Symposium on Theory of Computing, STOC 1983, pp. 448–456. ACM, New York (1983)
4. Ito, T., Demaine, E.D., Harvey, N.J.A., Papadimitriou, C.H., Sideri, M., Uehara, R., Uno, Y.: On the complexity of reconfiguration problems. Theor. Comput. Sci. **412**(1214), 1054–1065 (2011)
5. Shiloach, Y.: Another look at the degree constrained subgraph problem. Inf. Process. Lett. **12**(2), 89–92 (1981)
6. van den Heuvel, J.: The complexity of change. In: Blackburn, S.R., Gerke, S., Wildon, M.: (eds.) Surveys in Combinatorics 2013. London Mathematical Society Lectures Note Series, vol. 409 (2013)
7. Wrochna, M.: Homomorphism reconfiguration via homotopy. In: 32nd International Symposium on Theoretical Aspects of Computer Science, STACS 2015, 4–7 March 2015, Garching, Germany, pp. 730–742 (2015)

Generalized Pseudoforest Deletion: Algorithms and Uniform Kernel

Geevarghese Philip[1], Ashutosh Rai[2(✉)], and Saket Saurabh[2]

[1] Max-Planck-Institut für Informatik (MPII), Saarbrücken, Germany
gphilip@mpi-inf.mpg.de
[2] The Institute of Mathematical Sciences, Chennai, India
{ashutosh,saket}@imsc.res.in

Abstract. FEEDBACK VERTEX SET (FVS) is one of the most well studied problems in the realm of parameterized complexity. In this problem we are given a graph G and a positive integer k and the objective is to test whether there exists $S \subseteq V(G)$ of size at most k such that $G - S$ is a forest. Thus, FVS is about deleting as few vertices as possible to get a forest. The main goal of this paper is to study the following interesting problem: How can we generalize the family of forests such that the nice structural properties of forests and the interesting algorithmic properties of FVS can be extended to problems on this class? Towards this we define a graph class, $\mathcal{F}_l$, that contains all graphs where each connected component can transformed into forest by deleting at most l edges. The class $\mathcal{F}_1$ is known as pseudoforest in the literature and we call $\mathcal{F}_l$ as *l-pseudoforest*. We study the problem of deleting k-vertices to get into $\mathcal{F}_l$, *l*-PSEUDOFOREST DELETION, in the realm of parameterized complexity. We show that *l*-PSEUDOFOREST DELETION admits an algorithm with running time $c_l^k n^{\mathcal{O}(1)}$ and admits a kernel of size $f(l)k^2$. Thus, for every fixed l we have a kernel of size $\mathcal{O}(k^2)$. That is, we get a uniform polynomial kernel for *l*-PSEUDOFOREST DELETION. For the special case of $l = 1$, we design an algorithm with running time $7.5618^k n^{\mathcal{O}(1)}$. Our algorithms and uniform kernels combine iterative compression, expansion lemma and protrusion machinery.

1 Introduction

In the field of graph algorithms, vertex deletion problems constitute a considerable fraction. In these problems we need to delete a small number of vertices such that the resulting graph satisfies certain properties. Many well known problems like VERTEX COVER and FEEDBACK VERTEX SET (given a graph G and a positive integer k, does there exists $S \subseteq V(G)$ of size at most k such that $G - S$ is a forest) fall under this category. Most of these problems are NP-complete due to a classic result by Lewis and Yannakakis [9]. These problems are one of the most well studied problems in all those algorithmic paradigms that are meant for coping with NP-hardness, such as approximation algorithms and parameterized complexity. The topic of this paper is a generalization of FEEDBACK VERTEX SET in the realm of parameterized complexity.

© Springer-Verlag Berlin Heidelberg 2015
G.F. Italiano et al. (Eds.): MFCS 2015, Part II, LNCS 9235, pp. 517–528, 2015.
DOI: 10.1007/978-3-662-48054-0_43

The field of parameterized complexity tries to provide efficient algorithms for NP-complete problems by going from the classical view of single-variate measure of the running time to a multi-variate one. It aims at getting algorithms of running time $f(k)n^{\mathcal{O}(1)}$, where k is an integer measuring some aspect of the problem. These algorithms are called fixed parameter tractable (FPT) algorithms and the integer k is called the *parameter*. In most of the cases, the solution size is taken to be the parameter, which means that this approach gives faster algorithms when the solution is of small size. It is known that a decidable problem is FPT if and only if it is kernelizable: a kernelization algorithm for a problem Q takes an instance (x, k) and in time polynomial in $|x| + k$ produces an equivalent instance (x', k') (i.e., $(x, k) \in Q$ iff $(x', k') \in Q$) such that $|x'| + k' \le g(k)$ for some computable function g. The function g is the *size of the kernel*, and if it is polynomial, we say that Q admits a polynomial kernel. The study of kernelization is a major research frontier of parameterized complexity and many important recent advances in the area are on kernelization. For more background, the reader is referred to the monographs [2,10].

The FEEDBACK VERTEX SET problem has been widely studied in the field of parameterized algorithms. A series of results have improved the running times to $\mathcal{O}^*(3.619^k)$ in deterministic setting [8] and $\mathcal{O}^*(3^k)$ in randomized setting [1], where the $\mathcal{O}^*$ notation hides the polynomial factors. The main goal of this paper is to study the following interesting problem: How can we generalize the family of forests such that the nice structural properties of forests and the interesting algorithmic properties of FVS can be extended to problems on this class? There are two ways of quantitatively generalizing forests: given a positive integer l we define graph classes $\mathcal{G}_l$ and $\mathcal{F}_l$. The graph class $\mathcal{G}_l$ is defined as those graphs that can be made forest by deleting at most l edges. On the other hand the graph class, $\mathcal{F}_l$ contains all graphs where each connected component can be made forest by deleting at most l edges. Graphs in $\mathcal{G}_l$ are called *almost l-forest*. The class $\mathcal{F}_1$ is known as pseudoforest in the literature and we call $\mathcal{F}_l$ as *l-pseudoforest*. In this paper we study the problem of deleting k-vertices to get into $\mathcal{F}_l$, l-PSEUDOFOREST DELETION, in the realm of parameterized complexity.

Recently, a subset of authors [11] looked at a generalization of the FEEDBACK VERTEX SET problem in terms of $\mathcal{G}_l$. In particular they studied the problem of deleting k-vertices to get into $\mathcal{G}_l$, ALMOST FOREST DELETION, parameterized by $k + l$. They obtained a $2^{\mathcal{O}(l+k)}n^{\mathcal{O}(1)}$ algorithm and a kernel of size $\mathcal{O}(kl(k + l))$. One property of almost-l-forests which is crucial in the design of FPT and kernelization algorithms of [11] is that any almost-l-forests on n vertices can have at most $n + l - 1$ edges. The same can not be said about l-pseudoforests and they can turn out to be significantly more dense. So while the techniques used for arriving at FPT and kernelization results for FEEDBACK VERTEX SET give similar results for ALMOST FOREST DELETION, they break down when applied directly to l-PSEUDOFOREST DELETION. So we had to get into the theory of *protrusions*. Protrusions of a graph are subgraphs which have small boundary and a small treewidth. A protrusion replacer is an algorithm which identifies large protrusions and replaces them with smaller ones. Fomin et al. [5] use protrusion-replacer to arrive at FPT and kernelization results for PLANAR-$\mathcal{F}$ DELETION.

We first apply the techniques used in [5] to get an FPT algorithm for l-PSEUDOFOREST DELETION. To that end, we have to show that l-PSEUDOFOREST DELETION has a protrusion replacer, which we do by showing that the property of being an l-pseudoforest is strongly monotone and minor-closed. We arrive at a running time of $\mathcal{O}^*(c_l^k)$ for l-PSEUDOFOREST DELETION where c_l is a function of l alone. If we try to apply the machinery of [5] to get a kernelization algorithm for l-PSEUDOFOREST DELETION, it only gives a kernel of size k^c where the constant c depends on l. We use the similarity of this problem with FEEDBACK VERTEX set and apply Gallai's theorem and Expansion Lemma to decrease the maximum degree of the graph. This, when combined with techniques used in [5], gives us a kernel of size ck^2, where the constant c depends on l. These kind of kernels are more desired as it gives $\mathcal{O}(k^2)$ kernel for every fixed l, while the non-uniform kernelization does give a polynomial kernel for every fixed l, but the exponent's dependency on l makes the size of the kernel grow very quickly when compared to uniform-kernelization case. This result is one of the main contributions of the paper and should be viewed as another result similar to the one obtained recently by Giannopoulou et al. [7].

We also looked at a special case for of l-PSEUDOFOREST DELETION, namely PSEUDOFOREST DELETION, where we ask whether we can delete at most k vertices to get to a *pseudoforest*. A pseudoforest is special case of l-pseudoforest for $l = 1$, i.e. in a pseudoforest, each connected component is just one edge away from being a tree. In other words, it is a class of graphs where every connected component has at most one cycle. We apply the well known technique of iterative compression along with a non-trivial measure and an interesting base case to arrive at an $\mathcal{O}^*(7.5618^k)$ algorithm for this problem. We also give an explicit kernel with $\mathcal{O}(k^2)$ vertices for the problem.

2   Preliminaries

In this section, we first give the notations and definitions which are used in the paper. Then we state some basic properties about l-pseudoforests and some known results which will be used.

Notations and Definitions: For a graph G, we denote the set of vertices of the graph by $V(G)$ and the set of edges of the graph by $E(G)$. We denote $|V(G)|$ and $|E(G)|$ by n and m respectively, where the graph is clear from context. For a set $S \subseteq V(G)$, the *subgraph of G induced by S* is denoted by $G[S]$ and it is defined as the subgraph of G with vertex set S and edge set $\{(u, v) \in E(G) : u, v \in S\}$ and the subgraph obtained after deleting S is denoted as $G - S$. For a partition (X, Y) of $V(G)$, by $E(X, Y)$ we denote the edges which have one end point in X and the other in Y. A *flower* in a graph is set $\mathcal{C}$ of cycles which are vertex disjoint except for one vertex v, which is shared by all cycles in the set. The vertex v is called *center* of the flower while the cycles in $\mathcal{C}$ are called *petals* of the flower. All vertices adjacent to a vertex v are called neighbours of v and the set of all such vertices is called *open* neighbourhood of v, denoted by $N_G(v)$. The *closed* neighbourhood of a vertex v is denoted by $N_G[v]$ and defined as $N_G[v] = N_G(v) \cup \{v\}$. For a set of vertices $S \subseteq V(G)$, we define $N_G(S) = (\cup_{v \in S} N(v)) \setminus S$ and $N_G[S] = \cup_{v \in S} N[v]$. We drop the subscript G when the graph is clear from the context. For $0 < \alpha \leq 1$, we say that a vertex subset $S \subseteq V(G)$ is an α-cover, if the sum of vertex degrees $\sum_{v \in S} d(v)$ is at least $2\alpha |E(G)|$. A *forest* is a graph which does not contain any cycles. An *l-pseudoforest* is a graph which every component is at most l edges away from being a tree, i.e. the graph can be transformed into a forest by deleting at most l edges from each of its connected components. When l is equal to 1, we call the graph a *pseudoforest* instead of a 1-pseudoforest. For a connected component C of a graph, we call the quantity $|E(G[C])| - |C| + 1$ the *excess of* C and denote it by $\mathsf{ex}(C)$. It can also be equivalently defined as the minimum number of edges we need to delete from a connected component to get to a tree. For a graph G, let $\mathcal{C}$ be the set of its connected components. We define the *excess of a graph* G, denoted by $\mathsf{ex}(G)$ as follows.

$$\mathsf{ex}(G) = \max_{C \in \mathcal{C}} \mathsf{ex}(C)$$

It is easy to see that a graph G is an l-pseudoforest if and only if $\mathsf{ex}(G) \leq l$. We define the l-PSEUDOFOREST DELETION problem as follows.

---

l-PSEUDOFOREST DELETION
Input: A graph G, integers l and k.
Parameter: k
Question: Does there exist $X \subseteq V(G)$ such that $G - X$ is an l-pseudoforest?

The set X is called an *l-pseudoforest deletion set* of G. Similarly we can define PSEUDOFOREST DELETION to be the problem where we ask whether we can delete $S \subseteq V(G)$ such that $|S| \leq k$ and $G - S$ is a pseudoforest.

Treewidth. Let G be a graph. A *tree-decomposition* of a graph G is a pair $(\mathbb{T}, \mathcal{X} = \{X_t\}_{t \in V(\mathbb{T})})$ such that

- $\cup_{t \in V(\mathbb{T})} X_t = V(G)$,
- for every edge $xy \in E(G)$ there is a $t \in V(\mathbb{T})$ such that $\{x, y\} \subseteq X_t$, and
- for every vertex $v \in V(G)$ the subgraph of $\mathbb{T}$ induced by the set $\{t \mid v \in X_t\}$ is connected.

The *width* of a tree decomposition is $\max_{t \in V(\mathbb{T})} |X_t| - 1$ and the *treewidth* of G is the minimum width over all tree decompositions of G and is denoted by $\mathbf{tw}(G)$.

t-**Boundaried Graphs and Gluing.** A t-boundaried graph is a graph G and a set $B \subset V(G)$ of size at most t with each vertex $v \in B$ having a label $l_G(v) \in \{1, \ldots, t\}$. Each vertex in B has a unique label. We refer to B as the boundary of G. For a t-boundaried G the function $\delta(G)$ returns the boundary of G. Observe that a t-boundaried graph may have no boundary at all.

Two t-boundaried graphs G_1 and G_2 can be glued together to form a graph $G = G_1 \oplus G_2$. The gluing operation takes the disjoint union of G_1 and G_2

and identifies the vertices of $\delta(G_1)$ and $\delta(G_2)$ with the same label. If there are vertices $u_1, v_1 \in \delta(G_1)$ and $u_2, v_2 \in \delta(G_2)$ such that $l_{G_1}(u_1) = l_{G_2}(u_2)$ and $l_{G_1}(v_1) = l_{G_2}(v_2)$ then G has vertices u formed by unifying u_1 and u_2 and v formed by unifying v_1 and v_2. The new vertices u and v are adjacent if $(u_1, v_1) \in E(G_1)$ or $(u_2, v_2) \in E(G_2)$.

Protrusions and Protrusion Replacement. For a graph G and $S \subseteq V(G)$, we define $\partial_G(S)$ as the set of vertices in S that have a neighbour in $V(G) \setminus S$. An r-protrusion in a graph G is a set $X \subseteq V(G)$ such that $|\partial(X)| \leq r$ and $\mathbf{tw}(G[X]) \leq r$. Let G be a graph containing an r-protrusion X and let X' be an r-boundaried graph. Let $\hat{X} = X \setminus \partial(X)$. Then the act of *replacing X by X'* means replacing G by $\hat{G} \oplus X'$, where $\hat{G}$ is the boundaried graph $G - \hat{X}$ with boundary $\partial(X)$.

A protrusion replacer for a parameterized graph problem Π is a family of algorithms, with one algorithm for every constant r. The rth algorithm has the following specifications. There exists a constant r' (which depends on r) such that given an instance (G, k) and an r-protrusion X in G of size at least r', the algorithm runs in time $\mathcal{O}(|X|)$ and outputs an instance (G', k') such that $(G', k') \in \Pi$ if and only if $(G, k) \in \Pi$, $k' \leq k$ and G' is obtained from G by replacing X by an r-boundaried graph X' with less than r' vertices. Observe that since X has at least r' vertices and X' has less than r' vertices this implies that $|V(G')| < |V(G)|$.

Protrusion Decomposition. A graph G has an (α, β)-protrusion decomposition if $V(G)$ has a partition $P = R_0, R_1, \ldots, R_t$ where $\max\{t, |R_0|\} \leq \alpha$, each $N_G[R_i], i \in [t]$ is a β-protrusion of G, and for all $i \geq 1$, $N(R_i) \subseteq R_0$. We call the sets $R_i = N_G[R_i], i \in [t]$ protrusions of P.

2.1   Preliminary Results

Here we present some basic properties of l-pseudoforests as well as some known results which will be used in other sections the paper.

Lemma 1. *Treewidth of an l-pseudoforest is at most $l + 1$.*

Theorem 2. *l-PSEUDOFOREST DELETION has a protrusion replacer.*

Proving the theorem requires some basic concepts of Counting Monadic Second Order Logic and Graph Minors. We defer the details to the full version of the paper. Now we state the final result in the section.

Lemma 3 ([5]). *If an n-vertex graph G has a vertex subset X such that $\mathbf{tw}(G - X) \leq b$, then G admits a $((4|N[X]|)(b + 1), 2(b + 1))$-protrusion decomposition.*

3   A $c_l^k n^{\mathcal{O}(1)}$ Algorithm for l-pseudoforest Deletion

In this section we will present a $c_l^k n^{\mathcal{O}(1)}$ algorithm for l-PSEUDOFOREST DELETION. By Theorem 2, we know that l-PSEUDOFOREST DELETION has a protrusion replacer. We first state the following theorem which will be used while designing the FPT algorithm as well as in the kernelization algorithm.

Theorem 4 (Linear Time Protrusion Replacement Theorem, [4]). *Let Π be a problem that has a protrusion replacer which replaces r protrusions of size at least r' for some fixed r. Let s and β be constants such that $s \geq r' \cdot 2^r$ and $r \geq 3(\beta+1)$. Given an instance (G, k) as input, there is an algorithm that runs in time $\mathcal{O}(m+n)$ and produces an equivalent instance (G', k') with $|V(G')| \leq |V(G)|$ and $k' \leq k$. If additionally G has a (α, β)-protrusion decomposition such that $\alpha \leq \frac{n}{244s}$, then we have that $|V(G')| \leq (1 - \delta)|V(G)|$ for some constant $\delta > 0$.*

We call the algorithm in Theorem 4 Linear Time Protrusion Replacer (LPR). We show that given an instance (G, k) of l-PSEUDOFOREST DELETION, there exists a constant ρ such that LPR can be used to get to an equivalent instance (G', k') where any l-pseudoforest deletion set of G' is also an ρ-cover of G'.

Lemma 5. *There exist constants ρ, r, s and $c < 1$ such that if we run LPR with parameters r, s on an instance (G, k) of l-PSEUDOFOREST DELETION such that G has an l-pseudoforest deletion set S which is not a ρ-cover, then the output instance (G', k') satisfies $|V(G)| - |V(G')| \geq c|V(G)|$.*

Lemma 6. *There is an algorithm that given an instance (G, k) of l-PSEUDOFOREST DELETION, takes $\mathcal{O}((n + m) \log n)$ time and outputs an equivalent instance (G', k') such that $|V(G')| \leq |V(G)|$ and $k' \leq k$. Furthermore there exists a constant $0 < \rho < 1$ such that every l-pseudoforest deletion set S of G' is a ρ-cover of G'.*

This algorithm uses the algorithm of Lemma 5 as a subroutine. Since each iteration decreases the size of the vertex set by constant fraction, we can achieve the desired running time.

Now we define the notion of *Buckets* which will be used by the algorithm crucially. Given graph G, we make a partition $P = \{B_1, B_2, \dots, B_{\lceil \log n \rceil}\}$ of $V(G)$ as follows.

$$B_i = \left\{v \in V(G) \mid \frac{n}{2^i} < d(v) \leq \frac{n}{2^{i-1}}\right\} \text{ for all } i \in [\lceil \log n \rceil]$$

We call the sets B_i *buckets*. We call an instance (G, k) of l-PSEUDOFOREST DELETION *irreducible* if applying the algorithm of Lemma 6 returns (G, k) itself, in which case, every l-pseudoforest deletion set S of G is a ρ-cover of G. Let (G, k) be an irreducible instance of and let X be an l-pseudoforest deletion set of G of size at most k. A bucket B_i is said to be *good* if $|B_i \cap X| \geq d|B_i|$ and *big* if $|B_i| > i\lambda$, where d and λ are constants such that $(2d + 2\lambda) < \rho$.

Lemma 7. *Let (G, k) be an irreducible YES instance of l-PSEUDOFOREST DELETION. Then G has a bucket that is both big and good.*

Theorem 8. *l-PSEUDOFOREST DELETION can be solved in time $\mathcal{O}(c_l^k(m + n) \log n)$ for an instance (G, k) and the constant c_l depends only on l.*

The algorithm first applies Lemma 6 to get to an irreducible graph G' and then branches on all possible big subsets of the big buckets. The correctness follows from Lemma 7. After this, a careful running time analysis yields the desired running time.

4   A $c^k n^{\mathcal{O}(1)}$ Algorithm for Pseudoforest Deletion

In this section we will present a $c^k n^{\mathcal{O}(1)}$ algorithm for PSEUDOFOREST DELE-TION. We use the well known technique of iterative compression, and arrive at the desired running time after defining a non-trivial measure. In the spirit of the iterative compression technique, we start by defining the PSEUDOFOREST DELETION DISJOINT COMPRESSION problem.

PSEUDOFOREST DELETION DISJOINT COMPRESSION
Input: A graph G, a pseudoforest deletion set S of G, integer k
Parameter: k
Question: Does there exist a pseudoforest deletion set of G disjoint from S of size at most k?

To solve PSEUDOFOREST DELETION DISJOINT COMPRESSION, we first state a set of reduction rules.

Reduction Rule 1. *If there exists a vertex v of degree at most 1 in the graph, delete it.*

Reduction Rule 2. *If there exists $v \in V(G) \setminus S$ such that $G[S \cup \{v\}]$ is not a pseudoforest, delete v and decrease k by 1.*

Reduction Rule 3. *If there exists a vertex $v \in V(G) \setminus S$ of degree two, such that at least one of its neighbours are in $V(G) \setminus S$, delete v, and put an edge between its neighbours (even if they were already adjacent). If both of its edges are to the same vertex, delete v and put a self loop on the adjacent vertex (even if it has self loop(s) already).*

We say that an instance (G, S, k) of PSEUDOFOREST DELETION DISJOINT COM-PRESSION is a *good* instance if $G - S$ is a disjoint union of cycles such that for all $v \in V(G) \setminus S$, v has exactly one neighbour in S and $G[S]$ has no connected components which are trees, i.e., all the connected components of $G[S]$ have a cycle.

Lemma 9. PSEUDOFOREST DELETION DISJOINT COMPRESSION *can be solved in polynomial time on good instances.*

Proof of the lemma reduces a good instance (G, S, k) of PSEUDOFOREST DELE-TION DISJOINT COMPRESSION to finding minimum vertex cover of $G - S$, which can be done in polynomial time.

Lemma 10. *After the exhaustive application of reduction Rules 1, 2 and 3, either there exists $v \in V(G) \setminus S$ such that v has at least two neighbours in S or $G - S$ is a disjoint union of cycles.*

Theorem 11. PSEUDOFOREST DELETION *can be solved in $\mathcal{O}^*(7.5618^k)$ time.*

Algorithm 1. PFD(G, S, k)

Input: A graph G, $S \subseteq V(G)$ and positive integer k
Output: YES, if H has a pseudoforest deletion set of size at most k disjoint
　　　　　from S, NO otherwise.
1　If any of $G[S]$ and $G - S$ is not a pseudoforest, return NO .
2　If G is a pseudoforest then return YES . Else if $k \leq 0$ then return NO .
3　Apply Reduction Rule 1 or 2 or 3, whichever applicable, to get an instance
　(G', S', k'). Return PFD(G', S', k').
4　If (G, k) is a good instance, then solve the problem in polynomial time.
5　If $G - S$ has a vertex v which has at least two neighbours in S, return
　PFD($G - \{v\}, S, k - 1$)$\lor$ PFD($G, S \cup \{v\}, k$).
6　Else pick any vertex $v \in V(G) \setminus S$
7　Case 1. v has a self loop, return PFD($G - \{v\}, S, k - 1$)$\lor$ PFD($G, S \cup \{v\}, k$).
8　Case 2. v is part of a cycle of length 2, return PFD($G - \{v\}, S, k - 1$)$\lor$
　PFD($G - \{u\}, S \cup \{v\}, k - 1$)$\lor$ PFD($G, S \cup \{u, v\}, k$), where u is the other
　vertex in the cycle.
9　Case 3. v is part of a cycle of length ≥ 3, return PFD($G - \{v\}, S, k - 1$)$\lor$
　PFD($G - \{u, w\}, S \cup \{v\}, k - 2$)$\lor$ PFD($G - \{w\}, S \cup \{u, v\}, k - 1$)$\lor$
　PFD($G - \{u\}, S \cup \{v, w\}, k - 1$)$\lor$ PFD($G, S \cup \{u, v, w\}, k$), where u and w are
　v's neighbours in the cycle.

Proof. We use Algorithm 1 to solve PSEUDOFOREST DELETION DISJOINT COM-PRESSION. The correctness of the algorithm follows from the correctness of reduction rules and the fact that the branching is exhaustive. To analyze the running time of the algorithm, for an input instance $I = (G, S, k)$, we define a measure $\phi(I) := k + \mathsf{tc}(S)$, where $\mathsf{tc}(S)$ denotes the number of connected components of $G[S]$ which are trees.

A careful running time analysis of Algorithm 1 with respect to the measure shows that PSEUDOFOREST DELETION DISJOINT COMPRESSION can be solved in $\mathcal{O}^*(6.5618^k)$ time. After that, using standard iterative compression techniques, we get a running time of $\mathcal{O}^*(7.5618^k)$ for PSEUDOFOREST DELETION.   □

5   Kernels

In this section, we give kernelization algorithm for l-PSEUDOFOREST DELETION. First, we give a procedure to reduce the maximum degree of the graph. Then we arrive at the kernel by exploiting the small maximum degree using protrusions. In the end, we give an explicit kernel for PSEUDOFOREST DELETION.

5.1   Degree Reduction

For bounding the maximum degree of the graph, we give a set of reduction rules which we apply in polynomial time. If the maximum degree of the input graph is already bounded by $(k + l)(3l + 8)$, then we have already succeeded in bounding the maximum degree of the graph. Hence, for the rest of this subsection, we assume that there exists a vertex with degree greater than $(k + l)(3l + 8)$.

Reduction Rule 4. *If there exists a vertex v of degree at most 1 in the graph, delete it.*

Reduction Rule 5. *If any edge has multiplicity more that $l+2$, then delete all but $l+2$ copies of that edge.*

Lemma 12. *Any connected component in an l-pseudoforest can have at most l edge disjoint cycles.*

Reduction Rule 6. *If there is a vertex v with more than l self loops, delete v and decrease k by 1.*

We now look at a vertex which, after application of reduction Rules 4, 5 and 6, still has high degree. The idea is that either a high degree vertex participates in many cycles (and contributes many excess edges) and hence should be part of the solution, or only a small part of its neighbourhood is relevant for the solution. We formalize these notions and use them to find flowers using Gallai's theorem and apply a set of reduction rules. Given a set $T \subseteq V(G)$, by T-path we mean set of paths of positive length with both endpoints in T.

Theorem 13 (Gallai, [6]). *Given a simple graph G, a set $T \subseteq V(G)$ and an integer s, one can in polynomial time find either*

- *a family of $s+1$ pairwise vertex-disjoint T-paths, or*
- *a set B of at most $2s$ vertices, such that in $G - B$ no connected component contains more than one vertex of T.*

We would want to have the neighborhood of a high degree vertex as the set T for applying Gallai's theorem and for detecting flowers. But we need to be careful, as the graph in its current form contains multiple edges and self loops. Let v be a vertex with high degree. The vertices in $N(v)$ which have at least two parallel edges to v can be greedily picked to form a petal of the flower. Let L be the set of vertices in $N(v)$ which have at least two parallel edges to v.

Reduction Rule 7. *If $|L| > k+l$, delete v and decrease k by 1.*

Let $\widehat{G}$ be the graph $G - L$ with all parallel edges replaced with single edges, and all self loops removed. It is not hard to show that finding an f-flower in G centered at v is equivalent to finding an $f - |L|$ flower in $\widehat{G}$ centered at v for any $f \geq |L|$. Now we apply Gallai's theorem on $\widehat{G}$ with $T = N(v)$ and $s = k+l-|L|$. If the theorem returns a collection of vertex disjoint T-paths, then it is easy to see that they are in one to one correspondence with cycles including v, and hence can be considered petals of the flower centered at v.

Reduction Rule 8. *If the application of Gallai's theorem returns a flower with more than s petals, then delete v and decrease k by 1.*

Now, we deal with the case when the application of Gallai's theorem returns a set B of at most $2(k+l-|L|)$ vertices, such that in $G-B$, no connected component contains more than one vertex of T. Let $Z = B \cup L$. Clearly, $|Z| \leq 2(k+l) - |L|$.

Now we look at the set of connected components of $G - (Z \cup \{v\})$. Let us call this set C. We first state a reduction rule which relies on the fact that if too many connected components in C have a cycle, then v has to be part of any l-pseudoforest deletion set of G of size at most k.

Reduction Rule 9. *If more than $k + l$ components of C contain a cycle, then delete v and reduce k by 1.*

Lemma 14. *After applying reduction Rules 4, 5, 6, 7, 8 and 9 exhaustively, there are at least $2(l + 2)(k + l)$ components in C which are trees and connected to v with exactly one edge.*

Before we proceed further, we state the Expansion Lemma. Let G be a bipartite graph with vertex bipartition (A, B). For a positive integer q, a set of edges $M \subseteq E(G)$ is called a q-expansion of A into B if every vertex of A is incident with exactly q edges of M, and exactly $q|A|$ vertices in B are incident to M.

Lemma 15 (Expansion Lemma, [3]). *Let $q \geq 1$ be a positive integer and G be a bipartite graph with vertex bipartition (A, B) such that $|B| \geq q|A|$ and there are no isolated vertices in B. Then there exist nonempty vertex sets $X \subseteq A$ and $Y \subseteq B$ such that there is a q-expansion of X into Y and no vertex in Y has a neighbor outside X, that is, $N(Y) \subseteq X$. Furthermore, the sets X and Y can be found in time polynomial in the size of G.*

Let D the set of connected components which are trees and connected to v with exactly one edge. We have shown that $|D| \geq 2(l+2)(k+l)$. Now we construct an auxiliary bipartite graph H as follows. In one partition of H, we have a vertex for every connected component in D, and the other partition is Z. We put an edge between $A \in D$ and $w \in Z$ if some vertex of A is adjacent to w. Since every connected component in D is a tree and has only one edge to v, some vertex in it has to have a neighbour in Z, otherwise Reduction Rule 4 would apply. Now we have that $|Z| \leq 2(k + l)$ and every vertex in D is adjacent to some vertex in Z, we may apply expansion lemma with $q = l + 2$. This means, that in polynomial time, we can compute a nonempty set $\widehat{Z} \subseteq Z$ and a set of connected components $\widehat{D} \subseteq D$ such that:

1. $N_G(\bigcup_{D \in \widehat{D}} D) = \widehat{Z} \cup \{v\}$, and
2. Each $z \in \widehat{Z}$ will have $l + 2$ private components $A_z^1, A_z^2, \ldots A_z^{l+2} \in \widehat{D}$ such that $z \in N_G(A_z^i)$ for all $i \in [l + 2]$. By private we mean that the components $A_z^1, A_z^2, \ldots A_z^{l+2}$ are all different for different $z \in \widehat{Z}$.

Lemma 16. *For any l-pseudoforest deletion set of G that does not contain v, there exists an l-pseudoforest deletion set X' in G such that $|X'| \leq |X|$, $X' \cap (\bigcup_{A \in \widehat{D}} A) = \emptyset$ and $\widehat{Z} \subseteq X'$.*

Reduction Rule 10. *Delete all edges between v and $\bigcup_{A \in \widehat{D}} A$ and put $l + 2$ parallel edges between v and z for all $z \in \widehat{Z}$.*

Now we are ready to state the degree bound.

Theorem 17. *Given an instance (G, k) of l-PSEUDOFOREST DELETION, in polynomial time, we can get an equivalent instance (G', k') such that $k' \leq k$, $|V(G')| \leq |V(G)|$ and maximum degrees of G' is at most $(k + l)(3l + 8)$.*

Proof. First we show that either either the degree is already bounded or one of the reduction rules apply. Then, define a measure which is polynomial in the size of the graph and show that each of the reduction rules decrease the measure by a constant. We define the measure of a graph G to be $\phi(G) := 2|V(G)| + |E_{\leq l+2}|$, where $E_{\leq l+2}$ is set of edges with multiplicity at most $l + 2$. Then we show that each of the reduction rules either terminates the algorithm or decreases the measure by a constant.   □

5.2   Kernels Through Protrusions

We show that l-PSEUDOFOREST DELETION admits polynomial kernels using the concept of protrusions once again. We first show the following.

Lemma 18. *If (G, k) is a YES instance for l-PSEUDOFOREST DELETION and the maximum degree of a vertex in G is bounded by d, then G has a $(4kd(l + 2), 2(l + 2))$-protrusion decomposition.*

The proof of lemma is a straight-forward implication of combination of Lemmas 1 and 3.

Theorem 19. *l-PSEUDOFOREST DELETION admits a kernel with ck^2 vertices, where the constant c is a function of l alone.*

Proof. We combine Theorem 17 with Lemma 18 to get a protrusion decomposition of G and then we apply Theorem 4 to get the kernel.   □

5.3   An Explicit Kernel for Pseudoforest Deletion

To arrive at an explicit kernel for PSEUDOFOREST DELETION, in addition to bounding the maximum degree of the graph from above, we also need to show that the minimum degree of the graph is at least 3. We observe that the vertices of degree at most 1 are already reduced by Reduction Rule 4. It is easy to see that reduction rules similar to Reduction Rule 3 can be applied reduce vertices of degree 2.

Lemma 20. *If a graph G has minimum degree at least 3, maximum degree at most d, and pseudoforest deletion set of size at most k, then it has at most $k(d + 1)$ vertices and at most $2kd$ edges.*

Theorem 21. PSEUDOFOREST DELETION *admits a kernel with $\mathcal{O}(k^2)$ edges and $\mathcal{O}(k^2)$ vertices.*

6 Conclusions

In this paper we studied the problem of deleting vertices to get to a graph where each component is l edges away from being a tree. We obtained uniform kernels as well as an FPT algorithm for the problem when parameterized by the solution size. It would be interesting to study other classical problems (such as OCT) from this view-point where we want to delete vertices to get to some graph class plus l edges.

References

1. Cygan, M., Nederlof, J., Pilipczuk, M., Pilipczuk, M., van Rooij, J.M.M., Wojtaszczyk, J.O.: Solving connectivity problems parameterized by treewidth in single exponential time. In: FOCS, pp. 150–159 (2011)
2. Downey, R.G., Fellows, M.R.: Fundamentals of Parameterized Complexity. Texts in Computer Science. Springer, London (2013)
3. Fomin, F.V., Lokshtanov, D., Misra, N., Philip, G., Saurabh, S.: Hitting forbidden minors: approximation and kernelization. In: STACS, pp. 189–200 (2011)
4. Fomin, F.V., Lokshtanov, D., Misra, N., Ramanujan, M.S., Saurabh, S.: Solving d-sat via backdoors to small treewidth. In: SODA, pp. 630–641 (2015)
5. Fomin, F.V., Lokshtanov, D., Misra, N., Saurabh, S.: Planar F-deletion: approximation, kernelization and optimal FPT algorithms. In: FOCS (2012)
6. Gallai, T.: Maximum-minimum stze und verallgemeinerte faktoren von graphen. Acta Mathematica Academiae Scientiarum Hungarica **12**(1–2), 131–173 (1964)
7. Giannopoulou, A.C., Jansen, B.M.P., Lokshtanov, D., Saurabh, S.: Uniformkernelization complexity of hitting forbidden minors. CoRR, abs/1502.03965 (2015). to appear in ICALP 2015
8. Kociumaka, T., Pilipczuk, M.: Faster deterministic feedback vertex set. Inf. Process. Lett. **114**(10), 556–560 (2014)
9. Lewis, J.M., Yannakakis, M.: The node-deletion problem for hereditary properties is NP-complete. J. Comput. Syst. Sci. **20**(2), 219–230 (1980)
10. Niedermeier, R.: Invitation to Fixed Parameter Algorithms. Oxford Lecture Series in Mathematics and its Applications. Oxford University Press, Oxford (2006)
11. Rai, A., Saurabh, S.: Bivariate complexity analysis of almost forest deletion. In: COCOON (2015) (to appear)

Efficient Equilibria in Polymatrix Coordination Games

Mona Rahn[1] and Guido Schäfer[1,2]([✉])

[1] Centrum Wiskunde and Informatica (CWI), Amsterdam, The Netherlands
{M.M.Rahn,g.schaefer}@cwi.nl
[2] VU University Amsterdam, Amsterdam, The Netherlands

Abstract. We consider polymatrix coordination games with individual preferences where every player corresponds to a node in a graph who plays with each neighbor a separate bimatrix game with non-negative symmetric payoffs. In this paper, we study α-*approximate* k-*equilibria* of these games, i.e., outcomes where no group of at most k players can deviate such that each member increases his payoff by at least a factor α. We prove that for $\alpha \geq 2$ these games have the finite coalitional improvement property (and thus α-approximate k-equilibria exist), while for $\alpha < 2$ this property does not hold. Further, we derive an almost tight bound of $2\alpha(n-1)/(k-1)$ on the price of anarchy, where n is the number of players; in particular, it scales from unbounded for pure Nash equilibria ($k = 1$) to 2α for strong equilibria ($k = n$). We also settle the complexity of several problems related to the verification and existence of these equilibria. Finally, we investigate natural means to reduce the inefficiency of Nash equilibria. Most promisingly, we show that by fixing the strategies of k players the price of anarchy can be reduced to n/k (and this bound is tight).

1 Introduction

In this paper, we are interested in strategic games where the players are associated with the nodes of a graph and can benefit from coordinating their choices with their neighbors. More specifically, we consider *polymatrix coordination games with individual preferences*: We are given an undirected graph $G = (N, E)$ on the set of players (nodes) $N := \{1, \ldots, n\}$. Every player $i \in N$ has a finite set of strategies S_i to choose from and an individual preference function $q^i : S_i \to \mathbb{R}^+$. Each player $i \in N$ plays a separate bimatrix game with each of his neighbors in $N_i := \{j \in N \mid \{i, j\} \in E\}$. In particular, every edge $\{i, j\} \in E$ is associated with a payoff function $q^{ij} : S_i \times S_j \to \mathbb{R}^+$, specifying a non-negative payoff $q^{ij}(s_i, s_j)$ that both i and j receive if they choose strategies s_i and s_j, respectively. Given a joint strategy $s = (s_1, \ldots, s_n)$ of all players, the overall payoff of player i is defined as

$$p_i(s) := q^i(s_i) + \sum_{j \in N_i} q^{ij}(s_i, s_j). \tag{1}$$

© Springer-Verlag Berlin Heidelberg 2015
G.F. Italiano et al. (Eds.): MFCS 2015, Part II, LNCS 9235, pp. 529–541, 2015.
DOI: 10.1007/978-3-662-48054-0_44

These games naturally model situations in which each player has individual preferences over the available options (possibly not having access to all options) and may benefit in varying degrees from coordinating with his neighbors. For example, one might think of students deciding which language to learn, co-workers choosing which project to work on, or friends determining which mobile phone provider to use. On the other hand, these games also capture situations where players prefer to anti-coordinate, e.g., competing firms profiting equally by choosing different markets.

A special case of our games are *polymatrix coordination games* (without individual preferences, i.e., $q^i = 0$ for all i) which have previously been investigated by Cai and Daskalakis [9]. Among other results, the authors show that pure Nash equilibria are guaranteed to exist, but that finding one is PLS-complete. Polymatrix coordination games capture several other well-studied games among which are party affiliation games [6], cut games [11] and congestion games with positive externalities [12].

Yet another special case which will be of interest in this paper are *graph coordination games*. Here every edge $\{i, j\} \in E$ is associated with a non-negative edge weight w_{ij} and the payoff function q^{ij} is simply defined as $q^{ij}(s_i, s_j) = w_{ij}$ if $s_i = s_j$ and $q^{ij}(s_i, s_j) = 0$ otherwise. Intuitively, in this game every player (node) $i \in N$ chooses a color s_i from the set of colors S_i available to him and receives a payoff equal to the total weight of all incident edges to neighbors choosing the same color. These games have recently been studied by Apt et al. [2] for the special case of unit edge weights.

This paper is devoted to the study of equilibria in polymatrix coordination games with individual preferences. It is not hard to see that these games always admit pure Nash equilibria. However, in general these equilibria are highly inefficient. One of the most prominent notions to assess the inefficiency of equilibria is the *price of anarchy* [13]. It is defined as the ratio in social welfare of an optimal outcome and a worst-case equilibrium. Here the social welfare of a joint strategy s refers to the sum of the payoffs of all players, i.e., $\mathrm{SW}(s) = \sum_{i \in N} p_i(s)$.

The high inefficiency of our games even arises in the special case of graph coordination games as has recently been shown in [2]. To see this, fix an arbitrary graph $G = (N, E)$ with unit edge weights and suppose each player $i \in N$ can choose between a private color c_i (only available to him) and a common color c. Then each player i choosing his private color c_i constitutes a Nash equilibrium in which every player has a payoff of zero. In contrast, if every player chooses the common color c then each player i obtains his maximum payoff equal to the degree of i. As a consequence, the price of anarchy is unbounded. The example demonstrates that the players might be unable to coordinate on the (obviously better) common choice because they cannot escape from a bad initial configuration by unilateral deviations. In particular, observe that the example breaks if two (or more) players can deviate simultaneously. This suggests that one should consider more refined equilibrium notions where deviations of groups of players are allowed.

In our studies, we focus on a general equilibrium notion which allows us to differentiate between both varying sizes of coalitional deviations and different degrees of player reluctance to deviate. More specifically, in this paper we consider α-*approximate k-equilibria* as the solution concept, i.e., outcomes that are resilient to deviations of at most k players such that each member increases his payoff by at least a factor of $\alpha \geq 1$. Subsequently, we call these equilibria also (α, k)-*equilibria* for short. In light of this refined equilibrium notion, several natural questions arise and will be answered in this paper: Which are the precise values of α and k that guarantee the existence of (α, k)-equilibria? What is the price of anarchy of these equilibria as a function of α and k? How about the complexity of problems related to the verification and existence of such equilibria? And finally, are there efficient coordination mechanisms to reduce the price of anarchy?

Table 1. Complexity of graph coordination games. The parameters α and k are assumed to be part of the input unless they are stated to be fixed. [a] Shown to be efficiently computable for forests.

Problem		Complexity
Verification	(α, k)-equilibrium (k constant)	P
	(α, k)-equilibrium (α fixed)	co-NP-complete
	α-approximate strong equilibrium	P
Existence	k-equilibrium ($k \geq 2$ fixed)	NP-complete
	strong equilibrium	NP-complete[a]

Our Contributions. We study (α, k)-equilibria of graph and polymatrix coordination games. Our main contributions are summarized below.

1. *Existence:* We prove that for $\alpha \geq 2$ polymatrix coordination games have the finite (α, k)-improvement property, i.e., every sequence of α-improving k-deviations is finite (and thus results in an (α, k)-equilibrium). We also exhibit an example showing that for $\alpha < 2$ this property does not hold in general. For graph coordination games we show that if the underlying graph is a tree then (α, k)-equilibria exist for every α and k. On the other hand, if the graph is a pseudotree (i.e., a tree with exactly one cycle) the existence of (α, k)-equilibria cannot be guaranteed for every $\alpha < \varphi$ and $k \geq 2$, where $\varphi = \frac{1}{2}(1 + \sqrt{5})$ is the golden ratio.

2. *Inefficiency:* We show that the price of anarchy of (α, k)-equilibria for polymatrix coordination games is at most $2\alpha(n-1)/(k-1)$. We also provide a lower bound of $2\alpha(n-1)/(k-1) + 1 - 2\alpha$. In particular, the price of anarchy drops from unbounded for pure Nash equilibria ($k = 1$) to 2α for strong equilibria ($k = n$), both of which are tight bounds.

3. *Complexity:* We settle the complexity of several problems related to the verification and existence of (α, k)-equilibria in graph coordination games. Naturally all hardness results extend to the more general class of polymatrix coordination games with individual preferences. A summary of our results is given in Table 1.

4. *Coordination mechanisms:* We investigate two natural mechanisms that a central coordinator might deploy to reduce the price of anarchy of pure Nash equilibria: (i) asymmetric sharing of the common payoffs q^{ij} and (ii) strategy imposition of a limited number of players. Concerning (i), we show that there is no payoff distribution rule that reduces the price of anarchy in general. As to (ii), we prove that by (temporarily) fixing the strategies of k players according to an arbitrarily given joint strategy s, the resulting Nash equilibrium recovers at least a fraction of k/n of the social welfare SW(s) and this is best possible. Exploiting this in combination with a 2-approximation algorithm for the optimal social welfare problem [12], we derive an efficient algorithm to reduce the price of anarchy to at most $2n/k$ for a special class of polymatrix coordination games with individual preferences.

Related Work. Apt et al. [2] study k-equilibria in graph coordination games with unit edge weights, which constitute a special case of our games. They identify several graph structural properties that ensure the existence of such equilibria. Interestingly, most of these results do not carry over to our weighted graph coordination games, therefore demanding for the new approach of considering approximate equilibria.

Many of the mentioned games have been studied from a computational complexity point of view. In particular, Cai and Daskalakis [9] show that the problem of finding a pure Nash equilibrium in a polymatrix coordination game is PLS-complete. Further, they show that finding a mixed Nash equilibrium is in PPAD ∩ PLS. While this suggests that the latter problem is unlikely to be hard, it is not known whether it is in P. It is easy to see that these results also carry over to our polymatrix coordination games with individual preferences.[1]

For the special case of party affiliation games efficient algorithms to compute an approximate Nash equilibrium are known [7,10]. The current best approximation guarantee is $3 + \varepsilon$, where $\varepsilon > 0$, due to Caragiannis, Fanelli and Gavin [10]. The algorithm crucially exploits that party affiliation games admit an exact potential whose relative gap (called *stretch*) between any two Nash equilibria is bounded by 2. The latter property is not satisfied in our games, even for graph coordination games (as the example outlined in the Introduction shows).

A class of games that is closely related to our graph coordination games are *additively separable hedonic games* [8]. As in our games, the players are embedded in a weighted graph. Every player chooses a coalition and receives as payoff the total weight of all edges to neighbors in the same coalition. These games were originally studied in a cooperative game theory setting. More recently, researchers also address computational issues of these games (see, e.g., [4]). It is important to note that in hedonic games every player can choose every coalition, while in our graph coordination games players may only have limited options.

Anshelevich and Sekar [1] study coordination games with individual preferences where the players are nodes in a graph and profit from neighbors

---

[1] In [9] the bimatrix games on the edges may have negative payoffs and this is exploited in the PLS-completeness proof. However, we can accommodate this in our model by adding a sufficiently large constant to each payoff.

choosing the same color. However, in their setting the edge weight between two neighbors can be distributed asymmetrically and all players are assumed to have the same strategy set. Among other results, they give an algorithm to compute a $(2, n)$-equilibrium and show how to efficiently compute an approximate equilibrium that is not too far from the optimal social welfare.

Concerning the social welfare optimization problem, a 2-approximation algorithm is given in [12] for the special case of polymatrix coordination games with individual preferences where the bimatrix game of each edge has positive entries only on the diagonal.

Our Techniques. Most of our existence results use a generalized potential function argument for coalitional deviations. In our proof of the upper bound on the inefficiency of (α, k)-equilibria we first argue locally for a fixed coalition of players and then use a sandwich bound in combination with a counting argument to derive the upper bound. Most of our lower bounds and hardness results follow by exploiting specific properties and deep structural insights of graph coordination games with edge weights.

It is worth mentioning that our algorithm to compute a strong equilibrium for graph coordination games on trees reveals a surprising connection to a sequential-move version of the game. In particular, we show that if we fix an arbitrary root of the tree and consider the induced sequential-move game then every subgame perfect equilibrium corresponds to a strong equilibrium of the original game. As a consequence, strong equilibria exist and can be computed efficiently. Further, this in combination with our strong price of anarchy bound shows that the *sequential price of anarchy* [14] for these induced games is at most 2, which is a significant improvement over the unbounded price of anarchy for the strategic-form version of the game. This result is of independent interest.

We also note that the k/n bound on the social welfare which is guaranteed by our strategy imposition algorithm is proven via a *smoothness argument* [16]. Besides some other consequences, this implies that our bound also holds for more permissive solution concepts such as correlated and coarse correlated equilibria (see [16] for more details).

2   Preliminaries

Let $\mathcal{G} = (G, (S_i)_{i \in N}, (q^i)_{i \in N}, (q^{ij})_{\{i,j\} \in E})$ be a polymatrix coordination game with individual preferences (w.i.p.) where $G = (N, E)$ is the underlying graph. Recall that we identify the player set N with $\{1, \ldots, n\}$. We first introduce some standard game-theoretic concepts.

We call a subset $K := \{i_1, \ldots, i_k\} \subseteq N$ of players a *coalition of size* k. We define the set of joint strategies of players in K as $S_K := S_{i_1} \times \cdots \times S_{i_k}$ and use $S := S_N$ to refer to the set of joint strategies of all players. Given a joint strategy $s \in S$, we use s_K to refer to $(s_{i_1}, \ldots, s_{i_k})$ and s_{-K} to refer to $(s_i)_{i \notin K}$. By slightly abusing notation, we also write (s_K, s_{-K}) instead of s. If there is a strategy x such that $s_i = x$ for every player $i \in K$, we also write $s = (x_K, s_{-K})$.

Given a joint strategy s and a coalition K, we say that $s' = (s'_K, s_{-K})$ is a *deviation of coalition K from s* if $s'_i \neq s_i$ for every player $i \in K$; we also denote this by $s \to_K s'$. If we constrain to deviations of coalitions of size at most k, we call such deviations *k-deviations*. We call a deviation *α-improving* if every player in the coalition improves his payoff by at least a factor of $\alpha \geq 1$, i.e., for every $i \in K$, $p_i(s') > \alpha p_i(s)$; we also call such deviations *(α, k)-improving*. We omit the explicit mentioning of the parameters if $\alpha = 1$ or $k = 1$. A joint strategy s is an *α-approximate k-equilibrium* (also called *(α, k)-equilibrium* for short) if there is no (α, k)-improving deviation from s. If $k = 1$ or $k = n$ then we also refer to the respective equilibrium notion as *α-approximate Nash equilibrium* and *α-approximate strong equilibrium* [3].

We say that a finite strategic game has the finite *(α, k)-improvement property* (or *(α, k)-FIP* for short) if every sequence of (α, k)-improving deviations is finite. This notion generalizes the finite improvement property introduced by Monderer [15] for $\alpha = k = 1$. A function $\Phi : S \to \mathbb{R}$ is called an *(α, k)-generalized potential* if for every joint strategy s, for every (α, k)-improving deviation $s' := (s'_K, s_{-K})$ from s it holds that $\Phi(s') > \Phi(s)$. It is not hard to see that if a finite game admits an (α, k)-generalized potential then it has the (α, k)-FIP.

The *social welfare* of a joint strategy s is defined as $\mathrm{SW}(s) := \sum_{i \in N} p_i(s)$. For $K \subseteq N$, we define $\mathrm{SW}_K(s) := \sum_{i \in K} p_i(s)$. A joint strategy s^* of maximum social welfare is called a *social optimum*. Given a finite game that has an (α, k)-equilibrium, its *(α, k)-price of anarchy (POA)* is the ratio $\mathrm{SW}(s^*)/\mathrm{SW}(s)$, where s^* is a social optimum and s is an (α, k)-equilibrium of smallest social welfare. In the case of division by zero, we interpret the outcome as ∞. Note that if $\alpha' \geq \alpha$ and $k' \leq k$, then every (α, k)-equilibrium is an (α', k')-equilibrium. Hence the (α, k)-PoA lower bounds the (α', k')-PoA.

Due to lack of space, several proofs or parts thereof are omitted from this extended abstract and will be given in the full version of the paper.

3   Existence

We first give a characterization of the values α and k for which our polymatrix coordination games with individual preferences have the (α, k)-FIP.

Theorem 1. *Let $\mathcal{G}$ be a polymatrix coordination game w.i.p. Then:*

1. *$\mathcal{G}$ has the $(\alpha, 1)$-FIP for every α.*
2. *$\mathcal{G}$ has the (α, k)-FIP for every $\alpha \geq 2$ and for every k.*

Proof. Observe that every α-improving deviation is also α'-improving for $\alpha \geq \alpha'$. It is thus sufficient to prove the claims above for $\alpha = 1$ and $\alpha = 2$, respectively.

The proof idea for the first claim is to show that the game admits an exact potential and thus has the FIP.

We prove the second claim for $\alpha = 2$ by showing that $\Phi(s) := \mathrm{SW}(s)$ is a $(2, k)$-generalized potential. Given a joint strategy s and two sets $K, K' \subseteq N$, define

$$Q_s(K, K') := \sum_{i \in K, \, j \in N_i \cap K'} q^{ij}(s) \quad \text{and} \quad Q_s(K) := \sum_{i \in K} q^i(s).$$

Consider a $(2, k)$-improving deviation $s' = (s'_K, s_{-K})$ from s. Let $\bar{K}$ be the complement of K. We have $\mathrm{SW}_K(s) = Q_s(K, K) + Q_s(K, \bar{K}) + Q_s(K)$. Note that $\mathrm{SW}_K(s') > 2\mathrm{SW}_K(s)$ because the deviation is 2-improving. Thus,

$$Q_{s'}(K, K) + Q_{s'}(K, \bar{K}) + Q_{s'}(K) > 2\big(Q_s(K, K) + Q_s(K, \bar{K}) + Q_s(K)\big). \quad (2)$$

The social welfare of s can be written as

$$\mathrm{SW}(s) = Q_s(K, K) + 2Q_s(K, \bar{K}) + Q_s(\bar{K}, \bar{K}) + Q_s(K) + Q_s(\bar{K}).$$

Note that $Q_s(\bar{K}, \bar{K}) = Q_{s'}(\bar{K}, \bar{K})$ and $Q_s(\bar{K}) = Q_{s'}(\bar{K})$. Using (2), we obtain

$$\begin{aligned}
\varPhi(s') - \varPhi(s) &= Q_{s'}(K, K) + 2Q_{s'}(K, \bar{K}) + Q_{s'}(K) \\
&\quad - Q_s(K, K) - 2Q_s(K, \bar{K}) - Q_s(K) \\
&> Q_s(K, K) + Q_{s'}(K, \bar{K}) + Q_s(K) \geq 0.
\end{aligned}$$

Thus $\varPhi(s)$ is a $(2, k)$-generalized potential which concludes the proof. $\square$

The next theorem shows that in general our polymatrix coordination games do not have the (α, k)-FIP for $\alpha < 2$.

Theorem 2. *For all $\alpha < 2$ there is a polymatrix coordination game $\mathcal{G}$ that has a cycle of $(\alpha, n-1)$-improving deviations.*

We derive some more refined insights for the special case of graph coordination games.

Theorem 3. *The following holds for graph coordination games:*

1. *Let $\mathcal{G}$ be a graph coordination game on a tree. Then $\mathcal{G}$ has a strong equilibrium.*
2. *There is a graph coordination game $\mathcal{G}$ on a graph with one cycle such that no (α, k)-equilibrium exists for every $\alpha < \varphi$ and $k \geq 2$, where $\varphi := \frac{1}{2}(1 + \sqrt{5}) \approx 1.62$ is the golden ratio.*

Note that Theorem 3 shows that for $k \geq 2$ a k-equilibrium may not exist. In contrast, Nash equilibria always exist by Theorem 1. Further, the graph used to show the second claim is a pseudoforest[2]. For graph coordination games with unit edge weights, this guarantees the existence of a strong equilibrium [2].

4 Inefficiency

We analyze the price of anarchy of our polymatrix coordination games. The upper bound in the special case of $(\alpha, k) = (1, n)$ follows from a result in [5].

---

[2] A graph is a *pseudoforest* if each of its connected components has at most one cycle.

Theorem 4. *The (α, k)-price of anarchy in polymatrix coordination games w.i.p. is between $2\alpha(n-1)/(k-1) + 1 - 2\alpha$ and $2\alpha(n-1)/(k-1)$. The upper bound of 2α is tight for α-approximate strong equilibria.*

Proof (Upper Bound). Let s be an (α, k)-equilibrium (which we assume to exist) and let s^* be a social optimum. Fix an arbitrary coalition $K = \{i_1, \ldots, i_k\}$ of size k. Then there is a player $i \in K$ such that $p_i(s_K^*, s_{-K}) \leq \alpha p_i(s)$. Denote by $p_i^K(s^*) := q^i(s^*) + \sum_{j \in N_i \cap K} q^{ij}(s^*)$ the total payoff that i gets from players in K under s^* (including himself). Because all payoffs are non-negative, we have

$$p_i^K(s^*) \leq q^i(s_i^*) + \sum_{j \in N_i \cap K} q^{ij}(s_i^*, s_j^*) + \sum_{j \in N_i \cap \bar{K}} q^{ij}(s_i^*, s_j) = p_i(s_K^*, s_{-K}). \quad (3)$$

Thus, $p_i^K(s^*) \leq \alpha p_i(s)$. Rename the nodes in K such that $i_k = i$ and repeat the arguments above with $K \setminus \{i_k\}$ instead of K. Continuing this way, we obtain that for every player $i_x \in K$, $x \in \{1, \ldots, k\}$, $p_{i_x}^{\{i_1, \ldots, i_x\}}(s^*) \leq \alpha p_{i_x}(s)$.

We thus have

$$\sum_{i \in K} \left(q^i(s^*) + \frac{1}{2} \sum_{j \in N_i \cap K} q^{ij}(s^*) \right) = \sum_{x=1}^{k} \left(q^{i_x}(s^*) + \sum_{i_y \in N_{i_x} \cap K : y < x} q^{i_x i_y}(s^*) \right)$$

$$= \sum_{x=1}^{k} p_{i_x}^{\{i_1, \ldots, i_x\}}(s^*) \leq \alpha \sum_{i \in K} p_i(s).$$

Summing over all coalitions K of size k, we obtain

$$\sum_{K : |K| = k} \left(\sum_{i \in K} \left(q^i(s^*) + \frac{1}{2} \sum_{j \in N_i \cap K} q^{ij}(s^*) \right) \right) \leq \alpha \sum_{K : |K| = k} \sum_{i \in K} p_i(s). \quad (4)$$

Consider the right-hand side of (4). Note that every player $i \in N$ occurs in $\binom{n-1}{k-1}$ many coalitions of size k because we can choose $k - 1$ out of $n - 1$ remaining players to form a coalition of size k containing i. Thus

$$\sum_{K : |K| = k} \sum_{i \in K} p_i(s) = \binom{n-1}{k-1} \sum_{i \in N} p_i(s) = \binom{n-1}{k-1} \text{SW}(s). \quad (5)$$

Similarly, the first term of the left-hand side of (4) yields

$$\sum_{K : |K| = k} \sum_{i \in K} q^i(s^*) = \binom{n-1}{k-1} \sum_{i \in N} q^i(s^*) \geq \frac{1}{2} \binom{n-2}{k-2} \sum_{i \in N} q^i(s^*).$$

Now, consider the second term of the left-hand side of (4). Every pair (i, j) with $i \in N$ and $j \in N_i$ occurs in $\binom{n-2}{k-2}$ many coalitions of size k because we can choose $k - 2$ out of $n - 2$ remaining players to complete a coalition of size k containing both i and j. Thus for the left-hand side of (4) we obtain

$$\sum_{K:|K|=k} \left(\sum_{i \in K} \left(q^i(s^*) + \frac{1}{2} \sum_{j \in N_i \cap K} q^{ij}(s^*) \right) \right)$$

$$\geq \frac{1}{2} \binom{n-2}{k-2} \left(\sum_{i \in N} q^i(s^*) + \sum_{i \in N} \sum_{j \in N_i} q^{ij}(s^*) \right) = \frac{1}{2} \binom{n-2}{k-2} \mathrm{SW}(s^*). \quad (6)$$

Combining (5) and (6) with inequality (4), we obtain that the (α, k)-price of anarchy is at most $2\alpha \binom{n-1}{k-1} / \binom{n-2}{k-2} = 2\alpha \frac{n-1}{k-1}$. $\qquad \square$

5   Complexity

In this section, we study the complexity of various computational problems on graph coordination games.

Theorem 5. *Let $\mathcal{G}$ be a graph coordination game. Given a joint strategy s, the problem of deciding whether s is an (α, k)-equilibrium*

1. *is in P, if $k = O(1)$ or $k = n$;*
2. *is co-NP-complete for every fixed α.*

Proof (Sketch). We sketch the proof of the first claim for $k = n$. A crucial insight is that if there is an α-improving deviation from s then there is one which is *simple*, i.e., $s' = (s'_K, s_{-K})$ where the subgraph $G[K]$ induced by K is connected and all nodes in K deviate to the same color $s'_K = x$ for some x.

Fix some color x and let $G_x := (N_x, E_x)$ be the subgraph of G induced by the set of nodes N_x that can choose color x but do not do so in s. For each $u \in N_x$ define $d_u := \alpha p_u(s) - w(\{\{u, v\} \in E \mid s_v = x\}) - q^u(x)$. Now, a deviation of a coalition $K \subseteq N_x$ to (x_K, s_{-K}) is α-improving if and only if for every node $u \in K$ the total weight of all incident edges in the induced subgraph $G_x[K]$ is larger than d_u. We prove that an inclusionwise maximal $K \subseteq N_x$ satisfying this property can be found in polynomial time. This way we can verify for every color x whether an α-improving deviation exists. $\qquad \square$

Deciding whether a graph coordination game admits a k-equilibrium is hard for every $k \geq 2$. Note that for unit edge weights 2-equilibria are guaranteed to exist and can be found efficiently, as shown in [2].

Theorem 6. *Let $\mathcal{G}$ be a graph coordination game. Then the problem of deciding whether there is a k-equilibrium is NP-complete for every fixed $k \geq 2$.*

Proof $(k = 2)$. We give a reduction from MINIMUM MAXIMAL MATCHING which is known to be NP-complete [17]: Given a graph $G = (V, E)$ and a number l, does there exist an inclusionwise maximal matching of size at most l?

Let (G, l) be an instance of this problem with $G = (V, E)$ and $n = |V|$. We add $n - 2l$ gadgets $H_1, \ldots, H_{n-2l}$ to G, where an illustration of gadget H_i is given in Fig. 1. The dashed edge from v_0^i to G indicates that v_0^i is connected to all vertices in G and each of these edges has weight 3. We assign to each node

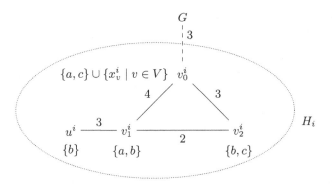

Fig. 1. The gadget H_i.

$v \in V$ the color set $S_v = \{x_v^i \mid i = 1, \ldots, n - 2l\} \cup \{y_e \mid e = \{v, w\} \in E\}$, i.e., v can either choose a 'gadget color' x_v^i or a color corresponding to some adjacent edge in E. Every edge in E has weight 4. Note that for all joint strategies of nodes in V the set of unicolor edges in E constitutes a matching. The idea is that in every 2-equilibrium $n - 2l$ nodes in V are needed to 'stabilize' the gadgets and the $2l$ remaining nodes in V form a maximal matching.

Assume that G has a maximal matching $M \subseteq E$ with $|M| \leq l$. We construct a 2-equilibrium s. For every matched node $v \in V(M)$, choose the color corresponding to the adjacent matching edge. On the unmatched nodes in V and nodes of the form v_0^i, we assign colors in such a way that every gadget has one outgoing edge (indicated by the dashed edge) that is unicolor. This is possible because there are at least $n - 2l$ unmatched nodes in V. If there are uncolored nodes in V left, assign arbitrary colors to them. Finally, let v_1^i and v_2^i choose color b for every i. We claim that s is a 2-equilibrium: The matched nodes obtain a payoff of 4, which is the maximal payoff nodes in V can get; so they are not part of any improving deviation. Let $v \in V$ be unmatched. Then v cannot deviate together with another unmatched node to increase the payoff because M is maximal. Further, all gadget nodes are 'taken': every v_0^i has a payoff of 3, which a joint deviation with v cannot increase. This implies that v cannot be part of any improving deviation. Lastly, it is easy to see that pairs of gadget nodes cannot profitably deviate. This proves that s is a 2-equilibrium.

Conversely, assume that a joint strategy s is a 2-equilibrium. Let M consist of the unicolor edges in G. By the choice of the color assignment, M is a matching. M is maximal because if there were two unmatched adjacent nodes, then they could form a profitable deviating coalition. It remains to show that $|M| \leq l$. It is not hard to see that if there is a gadget without an outgoing unicolor edge, then there is a 2-improving deviation in H_i. So at least $n - 2l$ nodes choose gadget colors, implying that $|V(M)| \leq 2l$ and thus $|M| \leq l$. □

On the positive side, we can compute a strong equilibrium in polynomial time if the underlying graph is a tree.

Theorem 7. *Let $\mathcal{G}$ be a graph coordination game on a tree. Then there is a polynomial-time algorithm to compute a strong equilibrium.*

Proof (Sketch). The idea is as follows: We fix an arbitrary root r of the tree and consider the induced sequential-move game. This game has a subgame perfect equilibrium s which can be computed in polynomial time by backwards induction. Let $\bar{s}$ be the corresponding joint strategy of $\mathcal{G}$ if every player plays his best response according to s. We can prove that $\bar{s}$ is a strong equilibrium of $\mathcal{G}$. □

6 Coordination Mechanisms

In this section, we investigate means that a central designer could use to reduce the inefficiency of Nash equilibria.

In our games the common payoff q^{ij} of the bimatrix game on edge $\{i, j\} \in E$ is distributed equally to both i and j. An idea that arises is to use different *payoff sharing rules* to reduce the inefficiency. Unfortunately, it is not hard to see from the example given in the Introduction that the price of anarchy remains unbounded no matter which payoff sharing rule is used.

We therefore consider another natural approach. Suppose the central designer can impose strategies on a subset of the players to reduce the inefficiency. Let $\mathcal{G}$ be a polymatrix coordination game w.i.p. Further, let $K \subseteq N$ be a subset of the players and fix a joint strategy $f_K \in S_K$ for players in K. We define $\mathcal{G}[f_K]$ as the game with players from $N \setminus K$ that arises from $\mathcal{G}$ if we fix the strategies of all players in K according to f_K. We say that f_K *guarantees* social welfare z if $\mathrm{SW}(f_K, s_{-K}) \geq z$ for all Nash equilibria s_{-K} of $\mathcal{G}[f_K]$. We also call f_K a *joint strategy of size* $|K|$.

Suppose that f_K guarantees social welfare z. Then once all players in $\mathcal{G}[f_K]$ have reached a Nash equilibrium we can release all players in K and let them play their best responses too. By Theorem 1, the social welfare can only increase subsequently. As a result, the final Nash equilibrium has social welfare at least z. So we can view f_K as a 'temporary advice' for the players in K. A similar idea has been put forward in [6].

We first show that determining the minimum number of players to guarantee a certain social welfare is hard, even for graph coordination games.

Theorem 8. *Let $\mathcal{G}$ be a graph coordination game. Given a joint strategy s, the problem of finding a minimal k such that there is a joint strategy f_K of size k that guarantees social welfare $\mathrm{SW}(s)$ is NP-hard. The claim also holds if f_K is restricted to be s_K.*

In light of the above hardness results, we resort to approximation algorithms.

Theorem 9. *Let $\mathcal{G}$ be a polymatrix coordination game w.i.p. Given a joint strategy s and a number k, we can find in polynomial time a coalition K of size k such that s_K guarantees social welfare $\frac{k}{n}\mathrm{SW}(s)$ and this is tight.*

Using the 2-approximation algorithm in [12] for the social welfare optimization problem, we obtain the following result:

Corollary 1. *Let $\mathcal{G}$ be a polymatrix coordination game w.i.p. where the bimatrix game of every edge has positive entries on the diagonal only. Given a number k, we can compute a joint strategy f_K of size K that guarantees social welfare $\frac{k}{2n}SW(s^*)$, where s^* is a social optimum.*

References

1. Anshelevich, E., Sekar, S.: Approximate equilibrium and incentivizing social coordination. In: Proceedings of 28th Conference on Artificial Intelligence (AAAI), pp. 508–514 (2014)
2. Apt, K.R., Rahn, M., Schäfer, G., Simon, S.: Coordination games on graphs (extended abstract). In: Liu, T.-Y., Qi, Q., Ye, Y. (eds.) WINE 2014. LNCS, vol. 8877, pp. 441–446. Springer, Heidelberg (2014)
3. Aumann, R.J.: Acceptable points in general cooperative n-person games. In: Luce, R.D., Tucker, A.W. (eds.) Contribution to the Theory of Game IV. Annals of Mathematical Study, vol. 40, pp. 287–324. Princeton University Press, Princeton (1959)
4. Aziz, H., Brandt, F., Seedig, H.G.: Stable partitions in additively separable hedonic games. In: Proceedings of the 10th International Conference on Autonomous Agents and Multiagent Systems (AAMAS), pp. 183–190 (2011)
5. Bachrach, Y., Syrgkanis, V., Tardos, É., Vojnović, M.: Strong price of anarchy, utility games and coalitional dynamics. In: Lavi, R. (ed.) SAGT 2014. LNCS, vol. 8768, pp. 218–230. Springer, Heidelberg (2014)
6. Balcan, M., Blum, A., Mansour, Y.: Improved equilibria via public service advertising. In: Proceedings of the 20th ACM-SIAM Symposium on Discrete Algorithms (SODA), pp. 728–737 (2009)
7. Bhalgat, A., Chakraborty, T., Khanna., S.: Approximating pure Nash equilibrium in cut, party affiliation, and satisfiability games. In: Proceedings of the 11th ACM Conference on Electronic Commerce (EC), pp. 73–82 (2010)
8. Bogomolnaia, A., Jackson, M.O.: The stability of hedonic coalition structures. Games Econ. Behav. **38**(2), 201–230 (2002)
9. Cai, Y., Daskalakis, C.: On minmax theorems for multiplayer games. In: Proceedings of the 22nd ACM-SIAM Symposium on Discrete Algorithms (SODA), pp. 217–234 (2011)
10. Caragiannis, I., Fanelli, A., Gravin, N.: Short sequences of improvement moves lead to approximate equilibria in constraint satisfaction games. In: Lavi, R. (ed.) SAGT 2014. LNCS, vol. 8768, pp. 49–60. Springer, Heidelberg (2014)
11. Christodoulou, G., Mirrokni, V.S., Sidiropoulos, A.: Convergence and approximation in potential games. Theo. Comput. Sci. **438**, 13–27 (2012)
12. de Keijzer, B., Schäfer, G.: Finding social optima in congestion games with positive externalities. In: Epstein, L., Ferragina, P. (eds.) ESA 2012. LNCS, vol. 7501, pp. 395–406. Springer, Heidelberg (2012)
13. Koutsoupias, E., Papadimitriou, C.: Worst-case equilibria. Comput. Sci. Rev. **3**(2), 65–69 (2009)
14. Leme, R.P., Syrgkanis, V., Tardos, E.: The curse of simultaneity. In: Proceedings of the 3rd Conference on Innovations in Theoretical Computer Science (ITCS), pp. 60–67 (2012)

15. Monderer, D., Shapley, L.S.: Potential games. Games Econ. Behav. **14**(1), 124–143 (1996)
16. Roughgarden, T.: Intrinsic robustness of the price of anarchy. In: Proceedings of the 41st ACM Symposium on Theory of Computing (STOC), pp. 513–522 (2009)
17. Yannakakis, M., Gavril, F.: Edge dominating sets in graphs. SIAM J. Appl. Math. **38**(3), 364–372 (1980)

Finding Consensus Strings with Small Length Difference Between Input and Solution Strings

Markus L. Schmid$^{(\boxtimes)}$

Trier University, Fachbereich IV – Abteilung Informatikwissenschaften,
54286 Trier, Germany
MSchmid@uni-trier.de

Abstract. The parameterised complexity of the CLOSEST SUBSTRING PROBLEM and the CONSENSUS PATTERNS PROBLEM with respect to the parameter $(\ell - m)$ is investigated, where ℓ is the maximum length of the input strings and m is the length of the solution string. We present an exact exponential time algorithm for both problems, which is based on an alphabet reduction. Furthermore, it is shown that for most combinations of $(\ell - m)$ and one of the classical parameters (m, ℓ, number of input strings k, distance d), we obtain fixed-parameter tractability, but even for constant $(\ell - m)$ and constant alphabet size, both problems are NP-hard.

Keywords: Parameterised complexity · Hard string problems

1 Introduction

Consensus string problems consist in finding a (preferably long) string that is sufficiently similar to a given set of strings (or to substrings of these given strings). They are among the most classical hard string problems and have many applications, mostly in computationally biology and coding theory (see [1]). In order to give a mathematically sound definition, we need a measure for the similarity of strings – or rather a distance function – and a classical approach is to use the Hamming distance $\mathsf{d_H}$. In this regard, the central problem considered in this paper is to find, for given strings $s_1, s_2, \ldots, s_k$, a string s of length m that has a Hamming distance of at least d from some length-m substrings $s'_1, s'_2, \ldots, s'_k$ of the input strings.

CLOSEST SUBSTRING (CLOSESUBSTR)

Instance: Strings $s_1, s_2, \ldots, s_k$ over some alphabet Σ with $|s_i| \leq \ell$, $1 \leq i \leq k$, for some $\ell \in \mathbb{N}$, and numbers $m, d \in \mathbb{N}$.
Question: Is there a string s with $|s| = m$ such that, for every i, $1 \leq i \leq k$, s_i has a substring s'_i with $\mathsf{d_H}(s, s'_i) \leq d$?

If we require $\sum_{i=1}^{k} \mathsf{d_H}(s, s'_i) \leq d$ instead of $\mathsf{d_H}(s, s'_i) \leq d$, $1 \leq i \leq k$, then the problem is called CONSENSUS PATTERNS (CONSPAT).

© Springer-Verlag Berlin Heidelberg 2015
G.F. Italiano et al. (Eds.): MFCS 2015, Part II, LNCS 9235, pp. 542–554, 2015.
DOI: 10.1007/978-3-662-48054-0_45

Both CloseSubstr and ConsPat are NP-hard and they have been intensely studied in the multivariate setting (see [2,3,5,8] and, for a survey, [1]). The most commonly considered parameters are k, m, d and $|\Sigma|$. The existing results show that CloseSubstr and ConsPat are fixed-parameter intractable, even for highly parameterised cases. For example, CloseSubstr is W[1]-hard if parameterised by (k, m, d) [3] or $(k, d, |\Sigma|)$ [8]. For ConsPat, the situation looks slightly better: ConsPat parameterised by (k, m, d) or $(k, |\Sigma|)$ is still W[1]-hard [3], but it becomes fixed-parameter tractable if parameterised by $(d, |\Sigma|)$ [8]. By simple enumeration, CloseSubstr and ConsPat are in FPT with respect to $(m, |\Sigma|)$.

In contrast to that, an NP-hard consensus string problem that exhibits a better parameterised complexity (see [6]), is the CLOSEST STRING PROBLEM (CloseStr), which is similar to CloseSubstr (the "computationally harder sister problem of CloseStr", according to [1]), with the strong additional restriction that $|s_1| = |s_2| = \ldots = |s_k| = m$. Analogously, we can also define CloseStr as CloseSubstr with the restriction $(\ell - m) = 0$. The NP-hardness of CloseStr implies that bounding $(\ell - m)$ by a constant does not yield polynomial time solvability of CloseSubstr; thus, the question arises whether some of the positive fixed-parameter tractability results for CloseStr carry over to CloseSubstr and ConsPat if $(\ell - m)$ is considered a parameter. This paper is devoted to an investigation of this parameter $(\ell - m)$, which can also be seen as an attempt towards Challenge 4 formulated in [1], which consists in finding new parameters of CloseSubstr that yield fixed-parameter tractability.

Our Contribution.[1] We first present an exact exponential time algorithm for CloseSubstr and ConsPat, based on an alphabet reduction, that runs in time $\mathcal{O}^*((k(\ell - m + 1))^m)$.[2]

For CloseSubstr, parameter $(\ell - m)$ alone cannot lead to fixed-parameter tractability, since even for $(\ell - m) = 0$ and $|\Sigma| = 2$ (i.e., CloseStr with binary alphabet) the problem remains NP-hard (see [5]). However, it is comparatively easy to show that in fact the tractable cases of CloseStr carry over to CloseSubstr if we additionally take $(\ell - m)$ as parameter (the parameter $((\ell - m), d)$ requires a bit more work, for which we adapt an fpt-algorithm for CloseStr parameterised by d presented in [6]).

For ConsPat, the situation is more complicated. Firstly, setting $(\ell - m) = 0$ makes the problem easily polynomial time solvable. In order to answer the question about its fixed-parameter tractability with respect to $((\ell - m), |\Sigma|)$ in the negative, we conduct a new reduction for which the alphabet reduction of the exact exponential algorithm mentioned above shall play an important role. For parameters $((\ell - m), k)$ and $((\ell - m), d)$ fixed-parameter tractability follows from simple enumeration algorithms, but the case of parameter $((\ell - m), m)$ is open. Obviously, the combined parameter $((\ell - m), m)$ is equivalent to parameter ℓ, which, to the knowledge of the author, has been neglected in the multivariate analysis of ConsPat. In this regard, we can at least note that parameter ℓ leads to fixed-parameter tractability if any of k, d or $|\Sigma|$ is also treated as a parameter.

---

[1] A compact presentation of all results is provided by Tables 1 and 2.

[2] By $\mathcal{O}^*$ we denote the $\mathcal{O}$-notation that suppresses polynomial factors.

Due to space constraints, not all results are formally proven.

Basic Definitions. The set of strings over an alphabet Σ is denoted by Σ^*, by $|v|$ we denote the length of a string v, alph(v) is the smallest Γ with $v \in \Gamma^*$, a string u is called a *substring* of v, if $v = v'uv''$; if $v' = \varepsilon$ or $v'' = \varepsilon$, then u is a *prefix* or *suffix*, respectively, where ε is the empty string. For a position j, $1 \leq j \leq |v|$, we refer to the symbol at position j of v by the expression $v[j]$ and $v[j..j'] = v[j]v[j + 1] \ldots v[j']$, $j < j' \leq |v|$. The *Hamming distance* for strings u and v with $|u| = |v|$ is defined by $d_H(u, v) = |\{j \mid 1 \leq j \leq |u|, u[j] \neq v[j]\}|$.

We assume the reader to be familiar with the basic concepts of (classical) complexity theory. Next, we shall briefly summarise the fundamentals of parameterised complexity (see also [4]). Decision problems are considered as languages over some alphabet Γ. A *parameterisation* (*of* Γ) is a polynomial time computable mapping $\kappa : \Gamma^* \to \mathbb{N}$ and a *parameterised problem* is a pair (Q, κ), where Q is a problem (over Γ) and κ is a parameterisation of Γ. We usually define κ implicitly by describing which part of the input is the parameter. A parameterised problem (Q, κ) is *fixed-parameter tractable* if there is an *fpt-algorithm* for it, i. e., an algorithm that solves Q on input x in time $\mathcal{O}(f(\kappa(x)) \times p(|x|))$ for recursive f and polynomial p. The class of fixed-parameter tractable problems is denoted by FPT. Note that if a parameterised problem becomes NP-hard if the parameter is set to a constant, then it is not in FPT unless P = NP.

For the problems CLOSESUBSTR and CONSPAT, we consider the parameters k, m, d, $|\Sigma|$, ℓ and $(\ell - m)$, which shall always be denoted in this way. The parameterised versions of the problems are denoted by simply listing the considered parameters in parentheses; if a parameter is bounded by a constant, we explicitly state the constant, e. g., CLOSESUBSTR$(d, (\ell - m))$ is the problem CLOSESUBSTR parameterised by d and $(\ell - m)$), and CONSPAT$((\ell - m) = c, |\Sigma| = c')$, $c, c' \in \mathbb{N}$, denotes the variant of CONSPAT, where the parameters $(\ell - m)$ and $|\Sigma|$ are bounded by c and c', respectively.

We conclude this section by introducing some more convenient terminology. Let $s_1, s_2, \ldots, s_k$ and m, d be an instance of CLOSESUBSTR or CONSPAT, and let s be a fixed candidate for a solution string. We say that s is *aligned* with s_i at position j, $1 \leq j \leq |s_i| - m + 1$, in order to denote that s is compared to the substring $s_i[j..j + m - 1]$. Once we have fixed such an alignment, every single position j of s is aligned with (or *corresponds* to) a position of every input string, which can either be a *match* or a *mismatch*. In the case of CONSPAT, it is also convenient to interpret position j to be aligned with the *column* $x_1 x_2 \ldots x_k$, where x_i is the aligned symbol of s_i. Every position j has then between 0 and k mismatches with respect to its corresponding column.

2   Alphabet Reduction and Exact Exponential Algorithm

The problems CLOSESUBSTR and CONSPAT can both be solved by enumerating all length-m strings over the input alphabet Σ and check for each such string whether or not is a solution string, which can be done in time $\mathcal{O}(km(\ell - m + 1))$. This yields an algorithm with running time $\mathcal{O}^*(|\Sigma|^m)$ or, since $|\Sigma| \leq k\ell$, $\mathcal{O}^*((k\ell)^m)$. In this section, we improve this naive algorithm by an alphabet reduction, such that a running time of $\mathcal{O}^*((k(\ell - m + 1))^m)$ is achieved.

In order to illustrate the basic idea of this alphabet reduction, let $s_1, \ldots, s_k \in \Sigma^*$, $m, d \in \mathbb{N}$ be a CLOSESUBSTR or CONSPAT instance. For every i, $1 \leq i \leq k$, we define $\gamma_i = |s_i| - m$. Since a solution string s can be aligned with s_i only at positions $1, 2, \ldots, \gamma_i + 1$, any position j of s can only be aligned with one of the positions $j, j + 1, \ldots, \gamma_i + j$ of s_i. Thus, regardless of the actual alignment, the mismatches caused by position j only depend on the substrings $s_i[j..j + \gamma_i]$. This suggests that renaming the strings s_i, such that, for every j, $1 \leq j \leq m$, the structure of the substrings $s_i[j..j + \gamma_i]$, $1 \leq i \leq k$, is preserved, yields an equivalent instance. We sketch such a renaming procedure.

For every j, $1 \leq j \leq m$, let $\Gamma_j = \bigcup_{i=1}^{k} \text{alph}(s_i[j..j + \gamma_i - 1])$ and $\Delta_j = \{s_i[j + \gamma_i] \mid 1 \leq i \leq k\}$. Let Σ' be some alphabet with $|\Sigma'| = \max\{|\Gamma_j \cup \Delta_j| \mid 1 \leq j \leq m\}$. For all values $j = 1, 2, \ldots, m$, we injectively rename (i. e., different symbols are replaced by different symbols) the substrings $s_i[j..j + \gamma_i]$, $1 \leq i \leq k$, with symbols from Σ'. However, since these substrings overlap, in every step j, $j \geq 2$, the substrings $s_i[j..j + \gamma_i - 1]$ are already renamed and therefore, except for the first step, we only have to rename the length-1 substrings $s_i[j + \gamma_i]$, which can be done as follows. For all i, $1 \leq i \leq k$, if $s_i[j + \gamma_i] \in \Gamma_j$, then we rename $s_i[j + \gamma_i]$ as has been done before in the substrings $s_i[j..j + \gamma_i - 1]$. All the remaining *new* symbols $\Delta_j \setminus \Gamma_j$ are injectively renamed by some symbols from $\Sigma' \setminus \bigcup_{i=1}^{k} \text{alph}(s_i[j..j + \gamma_i - 1])$.

For the strings $s_1 = abcac$, $s_2 = dbef$, $s_3 = ghabcf$ and $m = 3$, we have $\gamma_1 = 2$, $\gamma_2 = 1$, $\gamma_3 = 3$ and $\max\{|\bigcup_{i=1}^{3} \text{alph}(s_i[j..j + \gamma_i])| \mid 1 \leq j \leq 3\} = 6$. For $\Sigma' = \{A, B, \ldots, F\}$, the renaming described above proceeds as follows:

$$
\begin{array}{llll}
s_1 = a\,b\,c\,a\,c & A\,B\,C\,a\,c & A\,B\,C\,A\,c & A\,B\,C\,A\,C & = t_1 \\
s_2 = d\,b\,e\,f & \Rightarrow D\,B\,e\,f & \Rightarrow D\,B\,D\,f & \Rightarrow D\,B\,D\,E & = t_2 \\
s_3 = g\,h\,a\,b\,c\,f & E\,F\,A\,B\,c\,f & E\,F\,A\,B\,C\,f & E\,F\,A\,B\,C\,E & = t_3
\end{array}
$$

This reduces the alphabet by 2 symbols and it can be easily verified that, for every j, $1 \leq j \leq 3$, the substrings $t_1[j..j + \gamma_1], t_2[j..j + \gamma_2]$ and $t_3[j..j + \gamma_3]$ are isomorphic to the substrings $s_1[j..j + \gamma_1], s_2[j..j + \gamma_2]$ and $s_3[j..j + \gamma_3]$. From this property, we can also conclude that the instances s_1, s_2, s_3 and t_1, t_2, t_3 are equivalent; thus, we obtain the following result.

Lemma 1. *Let $P \in \{\text{CLOSESUBSTR}, \text{CONSPAT}\}$ and let $s_1, s_2, \ldots, s_k \in \Sigma^*$, $m, d \in \mathbb{N}$ be an instance of P. Then there exists an equivalent P instance $t_1, t_2, \ldots, t_k \in \Sigma'^*$, m, d, with $|t_i| = |s_i|$, $1 \leq i \leq k$, and $|\Sigma'|$ is of size $\max\{|\bigcup_{i=1}^{k} \text{alph}(s_i[j..j + (|s_i| - m)])| \mid 1 \leq j \leq m\}$. The strings $t_1, t_2, \ldots, t_k$ can be computed in time $\mathcal{O}(|\Sigma| + k\ell)$.*

An instance of CLOSESUBSTR or CONSPAT can now be solved by first reducing the alphabet to Σ' by the renaming procedure and then solve the instance by checking for every length m-string over Σ' whether it is a solution string. Since $|\Sigma'|$ is bounded by $k(\ell - m + 1)$, we obtain the following result:

Theorem 1. *The problems* CLOSESUBSTR *and* CONSPAT *can each be solved in time* $\mathcal{O}(|\Sigma| + k\ell + (k(\ell - m + 1))^m km(\ell - m + 1)) = \mathcal{O}^*((k(\ell - m + 1))^m)$.

This alphabet reduction can also be interpreted as a *kernelisation* with respect to the parameters $k, (\ell - m), m$. However, with respect to these parameters, fixed-parameter tractability can be shown more directly. Therefore, Lemma 1 has no application in proving fixed-parameter tractability results, but it shall be an important tool later on in Sect. 4 for proving a hardness result.

3   Closest Substring

The parameterised complexity of CLOSESUBSTR is well understood with respect to parameters k, m, d, $|\Sigma|$ and ℓ (see left side of Table 1).[3]

Table 1. Old and new results about CLOSESUBSTR.

k	m	d	$\|\Sigma\|$	ℓ	Result	Reference
–	–	–	–	p	FPT	[2]
p	–	–	2	–	W[1]-hard	[3]
p	p	p	–	–	W[1]-hard	[3]
–	p	–	p	–	FPT	Trivial
p	–	p	2	–	W[1]-hard	[8]

k	m	d	$\|\Sigma\|$	$(\ell - m)$	Results	Ref.
–	–	–	2	0	NP-hard	Prop. 1
p	–	–	–	p	FPT	Thm. 2
–	p	–	–	p	FPT	Thm. 2
–	–	p	–	p	FPT	Thm. 3

In the following, we shall take a closer look at the parameter $(\ell - m)$. If we restrict the parameter $(\ell - m)$ in the strongest possible way, i.e., requiring $(\ell - m) = 0$, then the input strings and the solution string have the same length; thus, CLOSESUBSTR collapses to the problem CLOSESTR. Unfortunately, CLOSESTR is NP-hard even if $|\Sigma| = 2$ (see [5]), which shows the fixed-parameter intractability of CLOSESUBSTR with respect to $(\ell - m)$ and $|\Sigma|$:

Proposition 1. CLOSESUBSTR$((\ell - m) = 0, |\Sigma| = 2)$ *is* NP-*hard.*

However, as we shall see next, adding one of k, m or d to the parameter $(\ell - m)$ yields fixed-parameter tractability. For the parameters $(k, (\ell - m))$ and $(m, (\ell - m))$ this can be easily concluded from known results.

Theorem 2. CLOSESUBSTR$(k, (\ell - m))$, CLOSESUBSTR$(m, (\ell - m)) \in$ FPT.

Proof. Every input string s_i has at most $(\ell - m + 1)$ substrings of length m, so the number of possible alignments of a candidate solution string is at most $(\ell - m + 1)^k$. After an alignment is chosen, the problem is equivalent to solving

---

[3] In all tables, **p** means that the label of this column is treated as a parameter and an integer entry means that the result holds even if this parameter is set to the given constant; problems that are hard for W[1] are not in FPT (under complexity theoretical assumptions, see [4]).

CLOSESTR, which is fixed-parameter tractable if parameterised by k (see [6]). This proves the first statement.

If both m and $(\ell - m)$ are parameters, then also ℓ is a parameter. From CLOSESUBSTR(ℓ) $\in$ FPT (see [2]), the second statement follows.   □

The only case left is the one where $(\ell - m)$ and d are parameters. In comparison to the cases discussed above, an fpt-algorithm for this variant of the problem is more difficult to find. It turns out that an fpt-algorithm for CLOSESTR(d) presented in [6] can be adapted to the problem CLOSESUBSTR($d, (\ell - m)$).

Theorem 3. CLOSESUBSTR($d, (\ell - m)$) $\in$ FPT.

Proof. Let $s_1, s_2, \ldots, s_k \in \Sigma^*$ and $m, d \in \mathbb{N}$ be a CLOSESUBSTR instance. If a solution string s exists, then it must be possible to construct s by changing at most d symbols in some length-m substring of some s_i. This yields a search tree approach: we start with a length-m substring of s_1 and then we branch into $m|\Sigma|$ new nodes by considering all possibilities of changing a symbol of s into another one. We repeat this procedure d times and for every such constructed string, we check in polynomial time whether it is a solution string. We shall now improve this procedure such that the branching factor is bounded by $(\ell - m + 1)(d + 1)$, which results in a search tree of size $((\ell - m + 1)(d + 1))^d$.

In a first step, we branch from the root into the at most $(\ell - m + 1)$ substrings of s_1. After that we branch in every node according to the following rule. Let s' be the string at the current node. We first check whether s' is a solution string in time $\mathcal{O}(k(\ell - m + 1)m)$. If s' is a solution string, then we can stop. If, on the other hand, s' is not a solution string, then there exists an input string s_i such that all its length-m substrings have too large a distance from s'. Let s_i' be the length-m substring of s_i that is aligned to s (i.e., the assumed solution string). In order to transform s' into s, we have to change $s'[j]$ into $s_i'[j]$ for a position j, $1 \leq j \leq m$, with $s'[j] \neq s_i'[j]$ and $s_i'[j] = s[j]$ (since otherwise this modification cannot lead to s). Since $d_H(s, s_i') \leq d$, there are at most d positions j with $s_i'[j] \neq s[j]$; thus, if we choose any $d + 1$ positions among all positions j with $s'[j] \neq s_i'[j]$, we will necessarily also select one that satisfies the properties described above (note that there are at least $d + 1$ positions with $s'[j] \neq s_i'[j]$, since $d_H(s', s_i') > d$). Consequently, for some $A \subseteq \{j \mid 1 \leq j \leq m, s'[j] \neq s_i'[j]\}$, $|A| = d + 1$, and every $j \in A$, we construct a new string from s' by changing $s'[j]$ to $s_i'[j]$. This procedure is correct under the assumption from above that s_i' is aligned with s in a solution. Since we have no knowledge of the correct solution alignment, we have to construct $d + 1$ new strings for each of the $(\ell - m + 1)$ substrings of s_i, which results in a branching factor of $(\ell - m + 1)(d + 1)$.

The total running time of this procedure is $\mathcal{O}((\ell - m + 1)((\ell - m + 1)(d + 1))^d k(\ell - m + 1)m) = \mathcal{O}(((\ell - m + 1)(d + 1))^{d+1} k(\ell - m + 1)m)$.   □

We conclude this section by some remarks about Theorem 3. The fundamental idea of the algorithm, i.e., changing only $d + 1$ symbols in every branching, is the same as for the fpt-algorithm for CLOSESTR of [6]. However, to demonstrate that if $(\ell - m)$ is also parameter, then this idea works for the more general problem CLOSESUBSTR, too, it is necessary to present it in a comprehensive way.

In every node, the construction of the successor nodes depends on some s_i, which we are free to choose. Furthermore, the successor nodes can be partitioned into $(\ell - m + 1)$ groups of $(d + 1)$ successors that all correspond to the same choice of the length-m substring of s_i. Thus, in the $d+1$ branches of each group, whenever successors are constructed again with respect to s_i, we can always choose the same substring, which results in a branching factor of only $d + 1$. Moreover, if we can choose between several s_i's, then we could always select one for which the substring has already been chosen in some predecessor. This heuristic can considerably decrease the size of the search tree.

4   Consensus Patterns

Apart from CONSPAT$(d, |\Sigma|) \in$ FPT (see [8]), CONSPAT shows a comparatively unfavourable fixed-parameter behaviour as CLOSESUBSTR (left side of Table 2).

Table 2. Old and new results about CONSPAT.

k	m	d	$(\ell - m)$	$|\Sigma|$	ℓ	Results	Ref.
−	p	−	p	−	p	Open	Open Prob. 4
−	−	−	−	p	p	FPT	Thm. 5
p	−	−	−	−	p	FPT	Thm. 5
−	−	p	−	−	p	FPT	Thm. 5
−	−	−	6	5	−	NP-hard	Thm. 7
p	−	−	p	−	−	FPT	Thm. 6
−	−	p	p	−	−	FPT	Thm. 6

k	m	d	$|\Sigma|$	Results	Ref.
p	−	−	2	W[1]-hard	[3]
p	p	p	−	W[1]-hard	[3]
−	p	−	p	FPT	Trivial
−	−	p	p	FPT	[8]

We note that the parameter ℓ is missing from the left side of Table 2 and, to the knowledge of the author, it seems as though this parameter has been neglected in the multivariate analysis of the problem CONSPAT. Unfortunately, we are not able to answer the most important respective question, i. e., whether or not CONSPAT$(\ell) \in$ FPT. Since ℓ is a trivial upper bound for the parameters m and $(\ell - m)$, we state this open question in the following form:

Open Problem 4. *Is* CONSPAT$(\ell, m, (\ell - m))$ *in* FPT*?*

For all other combinations of parameters including ℓ, fixed-parameter tractability can be easily shown:

Theorem 5. CONSPAT$(|\Sigma|, \ell)$, CONSPAT(k, ℓ), CONSPAT$(d, \ell) \in$ FPT.

Proof. The problem CONSPAT$(|\Sigma|, m)$ is in FPT (see left side of Table 2). Since $m \le \ell$ and $|\Sigma| \le \ell k$, this directly implies the first two statements.

Obviously, ℓ bounds $(\ell - m)$. Furthermore, CONSPAT$((\ell - m), d) \in$ FPT (see Theorem 6 below), which proves the third statement.   □

4.1   The Parameter $(\ell - m)$

We shall now turn to the parameter $(\ell - m)$. Unlike as for CLOSESUBSTR, the NP-hardness of CONSPAT is not preserved if $(\ell - m)$ is bounded by 0. More precisely, if $|s_1| = |s_2| = \ldots = |s_k| = m$, then the length-$m$ string s that minimises $\sum_{i=1}^{k} \mathsf{d_H}(s, s_i)$ is easily constructed by setting $s[j]$, $1 \le j \le m$, to one of the symbols that occur the most often among the symbols $s_1[j], s_2[j], \ldots, s_k[j]$.

Nevertheless, similar to CLOSESUBSTR, CONSPAT$((\ell - m) = c, |\Sigma| = c')$ is NP-hard, too, for small constants $c, c' \in \mathbb{N}$ (see Theorem 7). Before we prove this main result of the paper, we consider the other combinations of parameters.

Theorem 6. CONSPAT$(k, (\ell - m))$, CONSPAT$(d, (\ell - m)) \in$ FPT.

Proof. We can solve CONSPAT by first choosing length-m substrings $s'_1, s'_2, \ldots, s'_k$ of the input strings and then compute in polynomial time a length-m string s that minimises $\sum_{i=1}^{k} \mathsf{d_H}(s, s'_i)$ as described above. Since there are at most $(\ell - m + 1)^k$ possibilities of choosing the substrings, the first statement is implied.

In order to prove the second statement, we observe that if $k \le d$, then we can solve CONSPAT$(d, (\ell - m))$ by the fpt-algorithm for CONSPAT$(k, (\ell - m))$. If, on the other hand, $k > d$, then the possible solution string s must be a substring of some input string s_i, since otherwise $\sum_{i=1}^{k} \mathsf{d_H}(s, s'_i) \ge k > d$. Thus, we only have to check the $(\ell - m + 1)k$ length-m substrings of the input strings.   □

If CONSPAT is parameterised by $(\ell - m)$ and m, then we arrive again at the problem already mentioned in Open Problem 4. Consequently, there are only two cases left open: the parameter $(\ell - m)$ and the combined parameter $((\ell - m), |\Sigma|)$. We answer the question whether for these cases we have fixed-parameter tractability in the negative, by showing that CONSPAT remains NP-hard, even if $(\ell - m)$ and $|\Sigma|$ are small constants.

Theorem 7. CONSPAT$((\ell - m) = 6, |\Sigma| = 5)$ *is* NP-*hard.*

The existing reductions proving hardness results for CONSPAT (see [3]) construct $\binom{k}{2}$ strings representing the same graph $\mathcal{G} = (V, E)$ by listing its edges. A solution string then selects an edge from each string such that the selected edges form a k-clique. Thus, it must be possible to align the solution string with $|E|$ different substrings. This means that $\ell - m$ necessarily depends on $|E|$ and therefore this general idea of reduction is unsuitable for our case.

We choose a different approach, but we also use a graph problem, for which we first need the following definitions. Let $\mathcal{G} = (V, E)$ be a graph with $V = \{v_1, v_2, \ldots, v_n\}$. A vertex s is the *neighbour* of a vertex t if $\{t, s\} \in E$ and $N_{\mathcal{G}}[t] = \{s \mid \{t, s\} \in E\} \cup \{t\}$ is called the *closed neighbourhood* of t (or simply *neighbourhood*, for short). If, for some $k \in \mathbb{N}$, every vertex of $\mathcal{G}$ has exactly k neighbours, then $\mathcal{G}$ is k-*regular*. A *perfect code* for $\mathcal{G}$ is a subset $C \subseteq V$ with $|N_{\mathcal{G}}[t] \cap C| = 1$, $t \in V$. Next, we define the problem to decide whether or not a given 3-regular graph has a perfect code, which is NP-hard (see [7]):

3-Regular Perfect Code (3RPerCode)

Instance: A 3-regular graph $\mathcal{G}$.
Question: Does $\mathcal{G}$ contain a perfect code?

We now define a reduction from 3RPerCode to ConsPat. To this end, let $\mathcal{G} = (V, E)$ be a 3-regular graph with $V = \{v_1, v_2, \ldots, v_n\}$ and, for every i, $1 \leq i \leq n$, N_i is the neighbourhood of v_i. In order to define the neighbourhoods in a more convenient way, we use mappings $\wp_r : \{1, 2 \ldots, n\} \rightarrow \{1, 2 \ldots, n\}$, $1 \leq r \leq 4$, that map an $i \in \{1, 2 \ldots, n\}$ to the index of the rth vertex (with respect to some arbitrary order) of neighbourhood N_i, i.e., for every i, $1 \leq i \leq n$, $N_i = \{v_{\wp_1(i)}, v_{\wp_2(i)}, v_{\wp_3(i)}, v_{\wp_4(i)}\}$.

We now transform $\mathcal{G}$ into strings over $\Sigma = V \cup \{\star\}$. The size of $|\Sigma|$ obviously depends on $|V|$ and therefore is not constant; we shall later show how our construction can be modified in such a way that an alphabet of size 5 is sufficient. For every i, $1 \leq i \leq n$, N_i is transformed into $s_i = \star^6 t_{i,1} t_{i,2} \ldots t_{i,n} \star^6$ with $t_{i,i} = v_{\wp_1(i)} \star v_{\wp_2(i)} \star v_{\wp_3(i)} \star v_{\wp_4(i)} \star$ and, for every j, $1 \leq j \leq n$, $i \neq j$, $t_{i,j} = \alpha_1 \star \alpha_2 \star \alpha_3 \star \alpha_4 \star$, where, for every r, $1 \leq r \leq 4$, $\alpha_r = v_{\wp_r(i)}$ if $v_{\wp_r(i)} \in N_j$ and $\alpha_r = \star$ otherwise. Furthermore, for every r, $1 \leq r \leq 4$, we construct a string $q_r = \star^6 v_{\wp_r(1)} \star^7 v_{\wp_r(2)} \star^7 \ldots v_{\wp_r(n)} \star^7$. Moreover, $m = 8n + 6$ and $d = 4n^2 + 11n$. Note that $\ell = 8n + 12$; thus, $(\ell - m) = 6$. The ConsPat instance consists now of the strings $s_1, s_2, \ldots, s_n$ and $n + 1$ copies of each of the strings q_r, $1 \leq r \leq 4$.

Let us explain this reduction in an intuitive way and illustrate it with an example. For every i, $1 \leq i \leq n$, N_i is completely represented in the string s_i by $t_{i,i} = v_{\wp_1(i)} \star v_{\wp_2(i)} \star v_{\wp_3(i)} \star v_{\wp_4(i)} \star$. Furthermore, every $t_{i,j}$ with $i \neq j$ contains exactly the vertices from $N_i \cap N_j$, but they are listed according to N_i, i.e., every $t_{i,j}$ corresponds to the list $v_{\wp_1(i)} \star v_{\wp_2(i)} \star v_{\wp_3(i)} \star v_{\wp_4(i)} \star$ in which all elements not in N_j have been erased (i.e., replaced by $\star$). In the solution strings, there will be n special positions, each of which is aligned with position $1, 3, 5$ or 7 of $t_{i,j}$; thus, selecting one of the 4 possible vertices of $t_{i,j}$ (or $\star$ if no vertex is present). If $v_{\wp_r(i)}$ is selected from $t_{i,i}$, then also the rth vertex in every $t_{i,j}$, $i \neq j$, must be selected. Due to the order in which the vertices are listed, this is either the exact same vertex $v_{\wp_r(i)}$ in case that $v_{\wp_r(i)} \in N_j$ or $\star$ otherwise.

The 3-regularity of $\mathcal{G}$ allows us to bound $|t_{i,j}|$; thus, bounding $(\ell - m)$ as well. Moreover, it implies that in every s_i there are exactly 16 occurrences of vertices, which we exploit in the definition of the distance bound d. We illustrate these definitions with an example: let $N_1 = (v_1, v_4, v_5, v_8)$, $N_4 = (v_5, v_9, v_4, v_1)$, $N_5 = (v_1, v_5, v_{10}, v_4)$, $N_8 = (v_1, v_8, v_{11}, v_{15})$. Then $s_1 = \star^6 t_{1,1} t_{1,2} \ldots t_{1,n} \star^6$ with

$$t_{1,1} = v_1 \star v_4 \star v_5 \star v_8 \star, \quad t_{1,4} = v_1 \star v_4 \star v_5 \star \star \star, \quad t_{1,5} = v_1 \star v_4 \star v_5 \star \star \star,$$

$$t_{1,8} = v_1 \star \star \star \star \star v_8 \star, \quad t_{1,9} = \star \star v_4 \star \star \star \star \star, \quad t_{1,10} = \star \star \star \star v_5 \star \star \star,$$

$$t_{1,11} = \star \star \star \star \star \star v_8 \star, \quad t_{1,15} = \star \star \star \star \star \star v_8 \star .$$

In addition to these strings s_i, we use $n + 1$ copies of the length-m strings q_r,

$$q_1 = \quad \star^6 \quad v_1 \quad \star^7 \quad v_{\wp_1(2)} \quad \star^7 \quad \ldots \quad v_{\wp_1(n)} \quad \star^7 ,$$

$$q_2 = \quad \star^6 \quad v_4 \quad \star^7 \quad v_{\wp_2(2)} \quad \star^7 \quad \ldots \quad v_{\wp_2(n)} \quad \star^7 ,$$

$$q_3 = \quad \star^6 \quad v_5 \quad \star^7 \quad v_{\wp_3(2)} \quad \star^7 \quad \ldots \quad v_{\wp_3(n)} \quad \star^7 ,$$

$$q_4 = \quad \star^6 \quad v_8 \quad \star^7 \quad v_{\wp_4(2)} \quad \star^7 \quad \ldots \quad v_{\wp_4(n)} \quad \star^7 ,$$

which contain the neighbourhoods in form of columns separated by symbols $\star$ and serve the purpose of enforcing a certain structure of the solution string.

Before moving on, we first introduce more convenient notations in order to facilitate the following technical statements. Since the strings q_r have a length of m, the alignment of a candidate solution string s only concerns the strings s_i. For every j, $1 \leq j \leq m$, the *weight of position j (of s)* is the number of mismatches between $s[j]$ and the corresponding symbols in the input strings. Hence, s is a solution string if its *total weight*, i.e., the sum of the weights of all positions, is at most d. For every i, $1 \leq i \leq n$, we define the position $\delta_i = 8(i-1) + 7$, i.e., the δ_i are the positions of the strings q_r that contain a symbol from V and the positions of the strings s_j, where a substring $t_{j,i}$ starts.

We have to show that $\mathcal{G}$ has a perfect code if and only if there is a solution string for the CONSPAT instance. The next lemma proves the *only if* direction, which is the easier one.

Lemma 2. *If $\mathcal{G}$ has a perfect code, then there exists a solution string.*

How a solution string translates into a perfect code is more difficult to show. To this end, we first observe that the string s with the lowest possible weight necessarily adheres to a certain structure, which, in conjunction with the fact that it has a weight of at most d, allows then to extract a perfect code for $\mathcal{G}$.

Lemma 3. *If there exists a solution string, then $\mathcal{G}$ has a perfect code.*

Proof. Without loss of generality, we assume that s and the way it is aligned results in a total weight of $d' \leq d$ that is minimal among all possible total weights. By a sequence of separate claims, we show that s has a certain structure.

Claim: $s[\delta_i] \in N_i$, for every i, $1 \leq i \leq n$.

Proof of Claim: We assume to the contrary that, for some i, $1 \leq i \leq n$, $s[\delta_i] \notin N_i$. The symbol $s[\delta_i]$ corresponds to a symbol from N_i in every q_r and to a symbol from $N_i \cup \{\star\}$ in every s_j (the latter is due to the fact that position δ_i of s must be aligned with a position of $t_{j,i}$). Thus, position δ_i of s contributes at least $4n + 4$ to the total weight, if $s[\delta_i] = \star$ and at least $5n + 4$, if $s[\delta_i] \in V \setminus N_i$. If we change $s[\delta_i]$ to $v_{\wp_1(i)} \in N_i$, then it matches the corresponding symbol in all $n+1$ copies of q_1. Hence, it contributes at most $(4n+4) - (n+1) + n = 4n + 3$ to the total weight, if $s[\delta_i] = \star$ and at most $(5n+4) - (n+1) = 4n+3$, if $s[\delta_i] \in V \setminus N_i$. This is a contradiction to the minimality of d'. $\square$

Claim: $s[j] = \star$, for every j, $1 \leq j \leq m$, with $j \notin \{\delta_1, \delta_2, \ldots, \delta_n\}$.

Proof of Claim: If, for some j, $1 \leq j \leq m$, with $j \notin \{\delta_1, \delta_2, \ldots, \delta_n\}$, $s[j] \neq \star$, then position j contributes at least $4n+4$ to the total weight, since it constitutes a mismatch with the corresponding symbol in all strings q_r. If we change $s[j]$ to $\star$, then position j contributes a weight of at most n, since it matches with respect to all strings q_r and can only have mismatches with respect to the n strings s_i. This is a contradiction to the minimality of d'.   □

Claim: For every i, $1 \leq i \leq n$, s is aligned with s_i at position $1, 3, 5$ or 7.

Proof of Claim: The only possible positions where s can be aligned with some s_i are $1, 2, \ldots, 7$. If s is aligned with some s_i at position $2, 4$ or 6, then, for every j, $1 \leq j \leq n$, position δ_j corresponds to the $2^{\text{nd}}, 4^{\text{th}}$ or 6th position of $t_{i,j}$, which is $\star$. Since $s[\delta_j] \in N_j$, $1 \leq j \leq n$, all n occurrences of symbols from V in s cause mismatches. Moreover, all 16 occurrences of symbols from V in s_i correspond to occurrences of $\star$ in s. This yields $n + 16$ mismatches between s and s_i. If s is aligned at a position $1, 3, 5$ or 7, then it is still possible that all the n symbols at positions δ_j, $1 \leq j \leq n$, cause mismatches, but since 4 of these positions correspond to occurrences of symbols from V in s_i, there are only at most 12 additional mismatches between occurrences of symbols from V in s_i and $\star$ in s. This is a contradiction to the minimality of d'.   □

Consequently, $s = \star^6 v_{p_1} \star^7 v_{p_2} \star^7 \ldots v_{p_n} \star^7$ with $v_{p_i} \in N_i$, $1 \leq i \leq n$, and s is aligned with position $1, 3, 5$ or 7 of s_i, $1 \leq i \leq n$. Next, we show that $d' = d$.

Claim: $d' = d$.

Proof of Claim: Every position δ_i of s matches with the corresponding symbol of exactly one of the 4 strings q_r, $1 \leq r \leq 4$, and therefore causes 3 mismatches. All other positions $j \notin \{\delta_1, \delta_2, \ldots, \delta_n\}$ are matches with respect to the strings q_r. Thus, the weight caused by the mismatches between s and the strings q_r is exactly $3(n+1)n$. This implies that a weight of at most $d - 3(n+1)n = n^2 + 8n$ is caused by mismatches between s and all strings s_i. However, the minimum number of mismatches between s and any fixed string s_i is $(n-4) + 12 = n + 8$, i.e., at most 4 of the n positions δ_j of s match the corresponding symbol in s_i and the other $(n-4)$ positions δ_j cause mismatches with $\star$, and the remaining 12 occurrences of symbols from V in s_i are mismatches with $\star$ in s). Hence, the minimum weight due to the strings s_i is $n(n+8) = n^2 + 8n$, too; thus, $d' = d$. □

Since $v_{p_i} \in N_i$, $1 \leq i \leq n$, every symbol v_{p_i} can be interpreted as a vertex selected from N_i. In order to conclude that these vertices $\{v_{p_1}, v_{p_2}, \ldots, v_{p_n}\}$ form a perfect code, we have to show that if a vertex is select from N_i, then it must also be selected from all neighbourhoods in which it is contained, i.e., $v_{p_i} \in N_j$ implies $v_{p_i} = v_{p_j}$, which is established by the following claim.

Claim: If s is aligned with s_i at position $u \in \{1, 3, 5, 7\}$, then, for every j, $1 \leq j \leq n$, with $v_j \in N_i$, $v_{p_j} = v_{\wp_r(i)}$, where $r = \frac{u+1}{2}$.

Proof of Claim: We first assume that s is aligned with s_i at position $u = 1$. This means that, for every j, $1 \leq j \leq n$, the position δ_j of s, which carries the symbol v_{p_j}, is aligned with position δ_j of s_i. By the structure of s_i, we know that exactly the 4 positions $\delta_{j'}$ of s_i with $v_{j'} \in N_i$ carry the symbol $v_{\wp_1(i)}$, whereas all other $n - 4$ positions $\delta_{j''}$ with $v_{j''} \notin N_i$ carry the symbol $\star$. In particular, this implies that there are at least $n - 4$ mismatches due to the positions δ_j, $1 \leq j \leq n$, of s. Furthermore, all other 12 occurrences of symbols from V in s_i (i.e., the ones not corresponding to a position δ_j) constitute mismatches with $\star$ in s. Since the total number of mismatches between s and s_i is at most $(n - 4) + 12$, the 4 positions $\delta_{j'}$ of s_i with $v_{j'} \in N_i$ must be matches and therefore $s[\delta_{j'}] = v_{p_j} = v_{\wp_1(i)}$. The cases $u \in \{3, 5, 7\}$ can be handled analogously; the only difference is that positions δ_j of s are aligned with positions $\delta_j + (u - 1)$ of s_i. □

By the claims from above, $C = \{v_{p_1}, v_{p_2}, \ldots, v_{p_n}\} \subseteq V$ with $v_{p_i} \in N_i$, $1 \leq i \leq n$. If, for some j, $1 \leq j \leq n$, $|N_j \cap C| > 1$, then there exists an i, $1 \leq i \leq n$, $i \neq j$, such that $v_{p_i} \in N_j$ and $v_{p_j} \in N_j$ with $v_{p_i} \neq v_{p_j}$; a contradiction to the previous claim. Consequently, C is a perfect code, which concludes the proof. □

In order to complete the proof of Theorem 7, it remains to show how the alphabet size can be bounded by 5. To this end, we first slightly modify the reduction by adding substrings $\star^6$ between substrings $t_{i,j}$ and $t_{i,j+1}$ of s_i and between substrings $v_{\wp_r(j)} \star^7$ and $v_{\wp_r(j+1)} \star^7$ of q_r, i.e., $s_i = \star^6 t_{i,1} \star^6 t_{i,2} \star^6 \ldots t_{i,n} \star^6$, $1 \leq i \leq n$, and $q_r = \star^6 v_{\wp_r(1)} \star^7 \star^6 v_{\wp_r(2)} \star^7 \star^6 \ldots v_{\wp_r(n)} \star^7$, $1 \leq r \leq 4$. Furthermore, we set $m = 8n + 6 + 6(n - 1) = 14n$ and $d = 4n^2 + 11n$. We note that $\ell = 8n + 12 + 6(n - 1) = 14n + 6$ and therefore $(\ell - m) = 6$. This reduction is still correct (in fact, the proofs apply in the same way).

Due to the newly introduced substrings $\star^6$ of the modified reduction, for every j, $1 \leq j \leq m$, $\bigcup_{i=1}^{n} \text{alph}(s_i[j..j + (|s_i| - m)]) = N_{i'} \cup \{\star\}$, for some i', $1 \leq i' \leq n$. Hence, for every j, $1 \leq j \leq m$, $|\bigcup_{i=1}^{n} \text{alph}(s_i[j..j + (|s_i| - m)]) \cup \{q_r[j] \mid 1 \leq r \leq 4\}| = 5$; thus, by Lemma 1, there is an equivalent instance over an alphabet of size 5 (with $(\ell - m) = 6$), which can be computed in polynomial time.

While Theorem 7 proves the fixed-parameter intractability of $\text{ConsPat}((\ell - m), |\Sigma|)$ (assuming $P \neq NP$), it leaves a gap with respect to smaller constant bounds for $(\ell - m)$ and $|\Sigma|$. In order to improve the reduction with respect to $(\ell - m)$, the problem HITTING SET comes to mind, which is still NP-hard if every set has 2 elements. However, for this problem it is not clear how to cater for the size bound of the desired hitting set. We conjecture that an analogous reduction, with slightly lower $(\ell - m)$, from NEGATION FREE 1-IN-3 3SAT is possible, but the proof would be more involved, since we do not have the regularity property. The parameter $|\Sigma|$ is probably more interesting, since for applications in computational biology the alphabet size is typically 4. In this regard, the parameterised complexity of $\text{ConsPat}((\ell - m), |\Sigma| = 4)$ is still open.

References

1. Bulteau, L., Hüffner, F., Komusiewicz, C., Niedermeier, R.: Multivariate algorithmics for np-hard string problems. EATCS Bull. **114**, 31–73 (2014)
2. Evans, P.A., Smith, A.D., Wareham, H.T.: On the complexity of finding common approximate substrings. Theoret. Comput. Sci. **306**, 407–430 (2003)

3. Fellows, M.R., Gramm, J., Niedermeier, R.: On the parameterized intractability of motif search problems. Combinatorica **26**, 141–167 (2006)
4. Flum, J., Grohe, M.: Parameterized Complexity Theory. Springer-Verlag, New York (2006)
5. Frances, M., Litman, A.: On covering problems of codes. Theor. Comput. Sys. **30**, 113–119 (1997)
6. Gramm, J., Niedermeier, R., Rossmanith, P.: Fixed-parameter algorithms for closest string and related problems. Algorithmica **37**, 25–42 (2003)
7. Kratochvíl, J., Křivánek, M.: On the computational complexity of codes in graphs. In: Chytil, M.P., Koubek, V., Janiga, L. (eds.) Mathematical Foundations of Computer Science 1988. LNCS, pp. 396–404. Springer, Heidelberg (1988)
8. Marx, D.: Closest substring problems with small distances. SIAM J. Comput. **38**, 1382–1410 (2008)

Active Linking Attacks

Henning Schnoor$^{(\boxtimes)}$ and Oliver Woizekowski

Institut Für Informatik, Christian-Albrechts-Universität Kiel
Olshausenstraße 40, 24098 Kiel, Germany
{henning.schnoor,oliver.woizekowski}@email.uni-kiel.de

Abstract. We study linking attacks on communication protocols. We observe that an *active* attacker is strictly more powerful in this setting than previously-considered passive attackers. We introduce a formal model to reason about active linking attacks, formally define security against these attacks and give conditions for both security and insecurity of protocols. In addition, we introduce a composition-like technique that allows to obtain security proofs by only studying small components of a protocol.

Introduction

A typical goal of a protocol using web services, or more generally the purpose of a distributed algorithm, is to compute values based on information that is distributed among several parties: A user may have a specific set of input values, a web service then can compute, given these values, a function whose result—possibly combined with further data supplied by the user—is later used as an input to a further web service. Such protocols can be synthesized to respect the privacy of individual values (e.g., address or credit card number) [3,4].

In addition to privacy of values, a crucial aspect in such a setting is *linkability*: If an adversary can connect different values to the *same* user session, this may be a privacy violation, even if exposure of the individual values is not critical. For example, it might be harmless if an adversary learns customer names and products sold by a shop as long as these values cannot be *linked*, i.e., the adversary does not learn who ordered what. Linkability has been studied in the context of eHealth protocols [6] and anonymous internet usage [5], similar questions have been considered in [2,7,9,12]. Work on linkability usually focuses on passive adversaries (see, e.g., [13]) and cryptographic attacks.

In this paper, we initiate a study of *active*, information-theoretic attacks in this context: While there is, of course, extensive literature on active attacks on *cryptographic* aspects of protocols, to the best of our knowledge, active *information-theoretic* attacks have not been considered yet. To focus on information-theoretic aspects, we abstract away from cryptographic properties and study security on an information-theoretic level: Assuming that all channels are secure (using sender-anonymity of all messages built into the model), what can a set of dishonest parties in a protocol gain by answering queries not according to the protocol, but by choosing her answers to maximize the possibility

© Springer-Verlag Berlin Heidelberg 2015
G.F. Italiano et al. (Eds.): MFCS 2015, Part II, LNCS 9235, pp. 555–566, 2015.
DOI: 10.1007/978-3-662-48054-0_46

of linking other participant's data during the protocol run? We stress that the "incorrect" answers chosen by the adversary are not false or forged messages that do not correspond to the protocol, but actually use "factually wrong" answers in the actual payload of the exchanged messages.

Consequently, we do not use a model as usually considered for the study of cryptographic protocols (as, e.g., [10]), but model protocols more abstractly as a sequence of "function calls," where a user can query a webserver (without revealing her identity). In particular, linking attacks in our sense cannot be countered by cryptographic means, since they rely on the actual *function* that a protocol computes. Our contributions are as follows:

- We define a formal model that takes into account anonymous channels and nested web service queries, modeled by anonymous function calls.
- We give a formal definition of active linking attacks, and formalize a class of such attacks, which we call *tracking strategies*.
- For a large, natural class of protocols, we give a complete characterization of secure protocols and possible attacks: Active linking attacks can be mounted if and only if tracking strategies in the above sense exist.
- We demonstrate an *embedding* technique which generalizes composition. This technique can be used to simplify security proofs.

Our security results can be compared to classic results on lossless decomposition of databases, where complete database tuples can be reproduced from partial ones [1,8]. However, the database setting only corresponds to passive attacks, and hence our notion of active attackers requires completely different proof techniques. We consider our security proofs against active attackers (Theorem 3.3) to be the main technical contribution of this paper.

An Example

A customer C with address wants to learn the shipping cost for product ordered from shop S with delivery service company service D. C knows the values address and product, S knows the function parceltype, determining the type of parcel $p = \mathsf{parceltype}(\mathsf{product})$ needed to package product, p is a number between 0 and some n. The company D knows

$$
\begin{aligned}
C \to S &: (\mathsf{product}) \\
S \to C &: p = \mathsf{parceltype}(\mathsf{product}) \\
C \to D &: (p, \mathsf{address}) \\
D \to C &: \mathsf{deliveryprice}(p, \mathsf{address})
\end{aligned}
$$

Fig. 1. Simple Protocol τ_{ex}

the function deliveryprice determining the shipping cost $\mathsf{deliveryprice}(p, \mathsf{address})$ of a parcel of type p to address. This setting yields the straight-forward protocol given in Fig. 1. (We abstract away from cryptographic properties and assume secure channels between all parties.)

C expects that S and D cannot link product and address, even if they work together: S learns product but not address; D learns address but not product. If many users run the protocol in parallel and C cannot be identified by her IP address (e.g., uses an anonymity service), and C waits a while between her two messages to avoid linking due to timing, then ideally S and D should be unable to determine which of their respective queries come from the same customer.

This reasoning is indeed correct for a passive attacker. However, it overlooks that S and D control part of the user's data—namely the value p—and therefore can mount the following *active* attack:

1. S replies to the first received query with a 0, every later query is answered with a 1. S stores the value of product from the first query.
2. D waits for a query of the form $(0, \text{address})$ and sends address to S.
3. S knows that address received from D comes from the same user as the first query and hence can link this user's address and product.

This allows S and D to produce a "matching" pair of address and product, even with many interleaving protocol runs and anonymous connections from C to S and D. After one such run, the value 0 can be used to track another session. Similarly, $n-1$ sessions can be tracked in parallel. The strategy can be refined in order to track a session in which a particular product was ordered. The attack uses the 0-value for p as a "session cookie" that identifies a particular session. We stress that this attack does not violate a cryptographic property, but abuses control over user data to violate a privacy property of the protocol. In particular, the attack cannot be countered by cryptographic means.

This paper is organized as follows: In Sect. 1, we introduce our protocol model and state our security definition. In Sect. 2, we generalize the above strategy to tracking strategies. In Sect. 3, we present techniques to prove security of protocols, including characterizations of secure and insecure protocols for a large natural class of protocols. We then conclude in Sect. 4 with open questions. The proofs of our results can be found in our technical report, [11].

Fig. 2. Model of τ_{ex}

1   Protocol Model and Security Definition

Our model provides anonymous channels between the user and each web service, since linking is trivial if the user can be identified by e.g., an IP address. For simplicity, we assume that all relevant web services are controlled by a single adversary. As discussed in the introduction, since we abstract away from cryptographic properties, we treat the queries a user sends to a web service as function calls—we therefore identify the webservice with the function it is supposed to compute in the protocol.

Since the web services are under the adversary's control, the replies received from the function calls (i.e., web services) are completely controlled by the adversary. In particular, the same function call made twice in the protocol can be answered with different values by the adversary (in Sect. 2, we will see that it is indeed necessary for the adversary to reply inconsistently in this sense for some protocols, also note that such a behavior may be correct even for an honest service). To model interleaving of sessions, a schedule determines the order in which queries are performed.

A protocol in our sense is defined by a structure of user's input values and nested function calls. We therefore model a protocol as a directed acyclic graph, where the nodes represent input values (these can be seen as variables and form the set $Vars(\tau)$ in our formal definition below) or functions (the remaining nodes in the set $\tau \setminus Vars(\tau)$ in the definition below).

An edge $u \to f$ in a protocol models that the value of u (either an input value or a function result) is used as input to f. For simplicity, we assume that all values and query results in τ are Boolean; other values can be modeled by introducing function domains or by encoding values as sequences of Booleans. The representation of a protocol is similar to Boolean circuits (see [14]).

Definition 1.1. *A protocol τ is a directed acyclic graph (V, E) with a subset $\emptyset \neq Vars(\tau) \subseteq V$ such that each node in $Vars(\tau)$ has in-degree 0.*

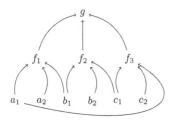

Fig. 3. Protocol τ_{priv}

In Fig. 2, the protocol τ_{ex} from the introduction is formalized in our model, another example (which we will discuss in detail below) is presented in Fig. 3. Our protocols do not fix the order of function calls (except that if $f_1 \to f_2$ is an edge in τ, each user must query f_1 before f_2). However, our results also hold for the case that the protocol fixes an order.

Nodes of τ without outgoing edges are called *output nodes*. If τ only has a single output node, this is the *root* of τ. We often identify τ and its set of nodes, i.e., talk about nodes $f \in \tau$ and subsets $\tau' \subseteq \tau$. $u \rightsquigarrow v$ means that there is a directed path from u to v in τ.

1.1   Protocol Execution

We first informally describe how a protocol τ is executed in our model. We identify a *user* with her *local session* containing her input values: A user or local session is an assignment $I\colon Vars(\tau) \to \{0,1\}$. During a protocol run, users store query results. To model this, local sessions are extended to assignments $I\colon \tau \to \{0,1\}$ during a protocol run. For a non-variable node $f \in \tau$, the value $I(f)$ then contains the result of the f-function call for user I. *Assign* is the set of all such assignments $I\colon V \to \{0,1\}$, where $V \subseteq \tau$.

A *global session* of τ is based on a multiset $S = \{I_1, \ldots, I_m\}$ of users. Each I_i performs a function call for each non-variable node f of τ as follows: Let $u_1, \ldots, u_n$ be the predecessors of f. The query arguments are the function name f and the user values for $u_1, \ldots, u_n$, i.e., $I_i(u_1), \ldots, I_i(u_n)$. Hence the adversary learns which function the user calls and the arguments for this call, but does *not* see the value i identifying the user or values $I_i(f)$ for $f \notin \{u_1, \ldots, u_n\}$. This makes explicit that our model assumes anonymous channels. The adversary can reply to I_i's f-query immediately or first wait for further queries. When she eventually replies with the bit r, the user stores this reply: We model this by extending I_i with the value $I_i(f) = r$.

An adversary *strategy* chooses one of three options in every situation:

1. *reply* to a previously-received function call,
2. *wait* for the next function call (even if there are unanswered function calls),
3. *print* an $I \in Assign$; the adversary wins if $I \in S$, and fails otherwise.

We now define our model more formally.

Schedules and Global Sessions. Function calls can be performed in any order that calls the predecessors of each node before the node itself. Clearly, in a situation where some user completes the entire protocol run before a second user even performs her initial query will always allow the adversary to correctly link all the values from the first user's session. In particular, this is true if there is only one user. Therefore, we only call a protocol insecure if the adversary can be successful for every possible interleaving of user sessions, and is successful in the presence of arbitrarily many users. Session interleavings are modelled with a schedule. Formally, an $n - user\ schedule\ for\ \tau$ is a sequence of pairs (i, f) where $i \in \{1, \ldots, n\}$, $f \in \tau \setminus Vars(\tau)$ where each such pair appears exactly once, and if $f \to g$ is an edge in τ with $f \notin Vars(\tau)$, then (i, f) appears in τ before (i, g). The pair (i, f) represents the f-query of the user I_i.

A *global session* for τ is a pair (S, σ) where S is a multiset of local sessions for τ, and σ is a $|S|$-user schedule for τ.

Protocol State. A *protocol state* contains complete information about a protocol run so far. It is defined as a pair (s, σ), where s is a sequence over $\tau \times Assign \times \mathbb{N} \times \{0, 1, \perp\} \times (\mathbb{N} \cup \{\perp\})$ and σ is a suffix of a schedule encoding the function calls remaining to be performed. An element $s_i = (f, I, i, r, t)$ in s encodes the f-call of user I_i as described above, here I is the assignment defined as $I(u) = I_i(u)$ for all $u \in \tau$ where $u \to f$ is an edge in τ. Therefore, this assignment I is exactly the set of arguments that the adversary receives for this function call. The value r is the adversary's reply to the function call, t records the time of the reply (both r and t are $\perp$ for a yet unanswered call).

The initial state of a global session (S, σ) is $((f, \emptyset, 1, \perp, |S|), \sigma)$ for some $f \in \tau$; this initializes σ and tells the adversary the number of users. Two actions modify the state: Users perform function calls, and the adversary replies. (The adversary's print-action ends the protocol run.) The above-discussed call of a function $f \in \tau$ by user I_i is performed in a state (s, σ) if the first element of σ is (i, f) and the adversary chooses the *wait* action. This action adds the tuple $(f, I, i, \perp, \perp)$ to the sequence s, where I encodes the input values for f (see above), and removes the first element of σ. In a state (s, σ), if s contains an element $s_k = (f, I, i, \perp, \perp)$ (an unanswered function call by the user I_i), the adversary's reply to this call with bit r exchanges s_k in s with (f, I, i, r, t),

where t is a counter that is increased with each reply answered by the adversary.[1]
The assignment I_i is then extended with $I_i(f) = r$ (cp. above). The remaining
schedule is unchanged.

Adversary Knowledge and Strategies. A strategy chooses an action for
each state (s, σ). The action may only depend on information available to the
adversary, which is captured in $view\,(s, \sigma)$, obtained from s by erasing each
tuple's third component and ignoring σ. This models that the adversary has
complete information except for the index of the user session from which a
request comes and the remaining schedule. (Giving the adversary access to either
makes tracking a user session trivial.) An *adversary strategy* for τ is a function Π
with input $view\,(s, \sigma)$ for a state (s, σ), and the output is one of the actions (*wait*,
reply to element s_k with r, *print* assignment I), with the following restrictions:

- *reply* can only be chosen if s contains an unanswered query $(f, I, i, \bot, \bot)$,
- *wait* is only available if the first query in σ can be performed, i.e., the first
  element of σ is (i, f) where $I_i(u)$ is defined for all u with $u \to f$.[2]

For a global session and an adversary strategy, the resulting τ-run is the
resulting sequence of states arising from performing Π stepwise, until the remaining schedule is empty and all queries have been answered, or the adversary's print
action has been performed.

Definition 1.2. *A protocol τ is* insecure *if there is a strategy Π such that for
every global session (S, σ), an action* print(I) *for some $I \in S$ occurs during the
τ-run for (S, σ) with strategy Π. Otherwise, τ is* secure.

2   Insecure Protocols: Tracking Strategies

We now generalize the tracking strategy discussed in the introduction. That
strategy used the value 0 produced by parceltype as a "session cookie" to track
the input values of a designated user session. In our definition below, the node
t_{init} plays the role of parceltype in the example. At this node, tracking a user
session is initialized by first replying with the session cookie: When the first
t_{init}-query in a global session is performed, the adversary replies with the session
cookie's value, 0. The adversary stores the user's values used as arguments to
t_{init}, this gives a partial assignment I_{track} which is later extended by additional
values: When the user later calls a function f with $t_{init} \to f$ with the value
0 for the argument representing t_{init}'s value, this query belongs to the tracked
user session. The arguments for this f-call that contain the values of additional
variables are then used to extend the assignment I_{track}, and the call is answered

---

[1] The purpose of the value t is only to give the adversary complete information about
his actions in the protocol run so far in the function $view\,(.)$ (see below).
[2] Whether *wait* is available does not follow from $view\,(s, \sigma)$. We can extend $view\,(s, \sigma)$
with a flag for the availability of *wait*, for simplification we omit this.

with the session cookie to allow tracking this session in the remainder of the protocol. In general, the f-call will not necessarily have further user values as input, but instead receive return values from different previous function calls. However, since the adversary controls the replies to these calls as well, she can use them to simply "forward" the value of a user variable. If the values of all user variables can be forwarded to a node where tracking in the above sense happens, then by repeating these actions, the adversary eventually extends I_{track} to a complete local session, which constitutes a successful active linking attack. The following definition captures the protocols for which this attack is successful:

Definition 2.1. *A set $T \subseteq \tau$ is called a* tracking strategy *if the following conditions hold:*

1. *Synchronization condition: There is a $\rightsquigarrow$-smallest element t_{init} in T, i.e. for every $u \in T$ we have $t_{init} \rightsquigarrow u$.*
2. *Cover condition: For every $x \in Vars(\tau)$ there is a path p_x such that:*
   (a) $x \rightsquigarrow t$ via path p_x for some $t \in T$,
   (b) if $x \neq y$, then p_x and p_y do not share a node from the set $\tau \setminus T$.

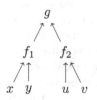

Fig. 4. No synchronization

The set T contains the nodes which perform tracking, i.e., which use the session cookie as reply to track the user session. The remaining nodes are used to simply forward one input value of the user to a later part in the protocol. The cover condition guarantees that all input variables can be forwarded in this fashion.

The synchronization condition requires some node t_{init} that can initialize tracking. The strategy then ensures that the session cookie used by each node in T identifies the same session. This cookie is passed on to tracking nodes appearing later in the protocol run (indirectly via the users, who echo the result of their f-query f for all g with $f \rightarrow g$). Without such a t_{init}, the adversary might use the session cookie for queries that belong to different users. A protocol for which a tracking strategy exists is always insecure:

Theorem 2.2. *Let τ be a protocol such that there exists a tracking strategy for τ. Then τ is insecure.*

We conjecture that the converse of Theorem 2.2 holds as well, i.e., that a protocol is insecure if and only if a tracking strategy exists. For a large class of protocols, we have proved this conjecture, see Theorem 3.3. Also, whether a tracking strategy exists for a protocol can be tested efficiently with a standard application of network flow algorithms.

Consider the example in Fig. 4. We demonstrate that there is no tracking strategy for this protocol. In a tracking strategy, both f_1 and f_2 must be tracking in order to capture all input bits. Both f_1 and f_2 then each collect a partial local session and use the session cookie 0 to identify these.

The adversary then waits for a function call to g with arguments $(0,0)$—which, however, may never occur: For a schedule starting with $(1, f_1)$, $(2, f_2)$, the initial calls of f_1 and f_2 belong to different user sessions, and two g-calls $g(1,0)$ and $g(0,1)$ will occur, belonging to different users. We will show (Example 3.1) that the protocol does not only fail to have a tracking strategy, but is indeed secure: The adversary does not have *any* strategy for a successful linking attack.

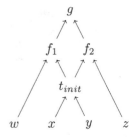

Fig. 5. t_{init} is a synchronizer

The situation is different for the protocol in Fig. 5, where a *synchronizer* t_{init} is placed before f_1 and f_2: $T = \{t_{init}, f_1, f_2\}$ is a tracking strategy for this protocol. The synchronization condition is satisfied, since every node in T can be reached from t_{init} via a directed path, the cover condition is satisfied as well.

	w	x	y	z
I_1	1	0	0	1
I_2	0	0	0	0

Fig. 6. Sessions I_1, I_2

In this strategy, the adversary stores the first values of x and y received as arguments for t_{init} in I_{track}, and replies to this first query with 0. The node f_1 waits for a zero in its second argument and adds the received value for w to I_{track}. Analogously, f_2 waits for a zero in its first argument and stores z. After both f_1 and f_2 have received this zero, I_{track} is a complete assignment. The adversary thus can print the linked local session.

In the protocol from Fig. 5, the adversary's strategy must be *inconsistent* as follows: Even when two local sessions agree on the values x and y, the strategy chooses different replies to their t_{init}-calls. This is necessary, since every consistent adversary strategy fails: Consider a global session comprising only the two local sessions shown in Fig. 6 and the following schedule:

1. First, I_1 calls t_{init}, which yields $I_1(t_{init}) = 0$ because t_{init} is tracking. The adversary stores $I_{track}(x) = 0$ and $I_{track}(y) = 0$.
2. Next, I_2 calls t_{init}, which gives $I_2(t_{init}) = 0$ because of consistency and I_1 and I_2 agree on x and y.
3. Lastly, I_1 calls f_1 with values $I_1(w) = 1$ and $I_1(t_{init}) = 0$, which gives 0 since f_1 is tracking and has not been called before. The adversary stores $I_{track}(w) = 1$. Analogously, I_2 calls f_2 with values $I_2(z) = 0$ and $I_2(z) = 0$, which also yields zero because of tracking. The adversary stores $I_{track}(z) = 0$.

Hence we have that $I_{track}(w) = 1$ and $I_{track}(z) = 0$. However, there is no user session with this property, hence the adversary strategy fails.

3   Secure Protocols: Proof Techniques and a Characterization

We now present criteria implying security of protocols. We start with *flat* protocols in Sect. 3.1, for which we give a complete example security proof, and state two general security results. Using an embedding technique, we generalize these results in Sect. 3.2 from flat protocols to more general ones.

3.1   Secure Flat Protocols

	x	y	u	v	g-query
I_1	0	0	0	0	$(0,\gamma)$
I_2	0	1	0	1	$(0,\delta)$
I_3	1	0	1	0	$(\alpha,0)$
I_4	1	1	1	1	$(\beta,0)$

	x	y	u	v	g-query
I_1	0	0	0	1	$(0,\delta)$
I_2	0	1	0	0	$(0,\gamma)$
I_3	1	0	1	1	$(\alpha,0)$
I_4	1	1	1	0	$(\beta,0)$

Fig. 7. Two Global Sessions for τ

A protocol is *flat* if it has a root and its depth (i.e., length of longest directed path) is 2. A flat protocol can be written[3] as $\tau = g(f_1(\overrightarrow{x_1}), \dots, f_n(\overrightarrow{x_n}))$, where $\overrightarrow{x_i}$ is a sequence of variables ($\overrightarrow{x_i}$ and $\overrightarrow{x_j}$ are not necessarily disjoint). For example, the flat protocol from Fig. 3 can be written as $g(f_1(a_1, a_2, b_1), f_2(b_1, b_2, c_1), f_3(c_1, c_2, a_1))$.

We now present an example and two classes of secure flat protocols. The following example is the simplest non-degenerate secure protocol.

Example 3.1. The protocol $\tau = g(f_1(x,y), f_2(u,v))$ (see Fig. 4) is secure.

As discussed above, there is no tracking strategy for τ. However, this leaves open whether there is some other strategy to perform a successful attack on τ. We now show that this is not the case: There is no successful adversary strategy for τ, tracking or otherwise.

Proof. We only consider global sessions consisting of 4 different local sessions $I_1, \dots, I_4$, where each Boolean combination appears as input to f_1 and f_2: For each $\alpha, \beta \in \{0,1\}$, there is some I_i with $I_i(x) = \alpha$ and $I_i(y) = \beta$, and a session I_j with $I_j(u) = \alpha$ and $I_j(v) = \beta$. We only consider schedules σ that first perform all f_1-calls, then all f_2-calls followed by all g-calls, and perform the queries for each function in lexicographical order. It suffices to show that the adversary does not have a strategy for this case, clearly then a general strategy does not exist either. In sessions like this, the adversary always receives the same set of arguments for f_1 and f_2; therefore these queries contain no information for the adversary. Since no f_i is queried more than once with the same arguments, we only need to consider consistent strategies, i.e., strategies that do not change the answer to a given function call during the protocol run. As a result, each adversary strategy Π consists of functions $f_1, f_2 \colon \{0,1\}^2 \to \{0,1\}$ and a rule for the print action. We construct, depending on f_1 and f_2, a global session where Π fails. Hence let f_1 and f_2 be functions as above.

Since f_1 and f_2 cannot be injective, we assume without loss of generality that $f_1(0,0) = f_1(0,1) = 0$ and $f_2(1,0) = f_2(1,1) = 0$. We further define $\alpha = f_1(1,0)$, $\beta = f_1(1,1)$, $\gamma = f_2(0,0)$, and $\delta = f_2(0,1)$.

Consider the two global sessions in Fig. 7, with a schedule as above. For the local sessions appearing, the tables list the values of the variables and the parameters for the resulting g-function call, which for the session I_i is the pair

---

[3] One can without loss of generality assume that there is no variable x and an edge $x \to g$ for the output node g of a flat protocol.

$(f_1(I_i(x), I_i(y)), f_2(I_i(u), I_i(v)))$. These g-calls are observed by the adversary. For both global sessions, the adversary makes the same observations: At both f_1 and f_2, each Boolean pair appears as an argument once; the arguments for g are $(0, \gamma)$, $(0, \sigma)$, $(\alpha, 0)$, and $(\beta, 0)$—performed in lexicographical order. Therefore, the adversary cannot distinguish the two global sessions and her strategy prints out the same assignment in both. Since the sessions have disjoint sets of local sessions, the adversary fails in at least one of them; the protocol is indeed secure.

The ideas from the above proof can be used to prove the following theorem. Its proof is more involved in part since here, we have to consider inconsistent adversary strategies. The theorem states that if a flat protocol can be partitioned into two variable-disjoint components, neither of which grant the adversary enough "channels" to forward all user inputs to the output node, then it is secure.

Theorem 3.2. *Let τ be a flat protocol of the form $\tau = g(f_1(\overrightarrow{x_1}), \ldots, f_n(\overrightarrow{x_n}))$, such that $\{1, \ldots, n\} = I_1 \cup I_2$ for sets I_1 and I_2 with*

- *if $i \in I_1$ and $j \in I_2$, then $\overrightarrow{x_i} \cap \overrightarrow{x_j} = \emptyset$,*
- *$|\cup_{i \in I_j} \overrightarrow{x_i}| > |I_j| \geq 1$ for $j \in \{1, 2\}$.*

Then τ is secure.

We now consider flat protocols $\tau = g(f_1(\overrightarrow{x_1}), \ldots, f_n(\overrightarrow{x_n}))$ where each f_i has a private variable $x \in \overrightarrow{x_i} \setminus \cup_{j \neq i} \overrightarrow{x_j}$, i.e., a variable that is an input only to f_i. For these protocols, the converse of Theorem 2.2 is true as well:

Theorem 3.3. *Let τ be a flat protocol where each f_i has a private variable. Then τ is insecure if and only if a tracking strategy for τ exists.*

The proof of Theorem 3.3 converts τ to a normal form, and applies a rather complex construction to obtain a global session (S, σ) such that for every local session $I \in S$, there is a global session (S_I, σ_I) which is indistinguishable from (S, σ) for the adversary but does not contain I. Due to this indistinguishability, the adversary has to print the same local session on (S, σ) and each (S_I, σ_I), and hence fails on (S, σ) or on some (S_I, σ_I). The simplest non-trivial example for which Theorem 3.3 implies security is the protocol τ_{priv} (see Fig. 3), which itself already requires a surprisingly complex security proof.

3.2   Security Proofs for General Protocols

Using a technique we call *embedding*, we can "lift" the results obtained for flat protocols in Sect. 3.1 to more general protocols. For simplicity, we only consider *layered* protocols τ, which are protocols of the form $\tau = L_0 \cup \cdots \cup L_n$, where $L_i \cap L_j = \emptyset$ for $i \neq j$, each predecessor of a node in L_i is in L_{i+1}, variable nodes only appear in L_n, and output nodes only in L_0. One can easily rewrite every protocol into a layered one without affecting security. We generalize Theorem 3.2 to protocols of arbitrary depth: As soon as the "disjoint" structure of Theorem 3.2 appears in some level of τ, τ is secure.

For $f \in \tau$, $Vars(f)$ denotes $\{x \in Vars(\tau) \mid x \leadsto f$ is a path in $\tau\}$, i.e., the set of input values that influence the queries made at the node f. For a set $S \subseteq \tau$, with $Vars(S)$ we denote the set $\cup_{u \in S} Vars(u)$.

Corollary 3.4. *Let τ be a layered protocol with levels $L_0, \ldots, L_n$ with some i such that $L_i = I_1 \cup I_2$ with $Vars(I_1) \cap Vars(I_2) = \emptyset$ and $|Vars(I_1)| > |I_1|$, $|Vars(I_2)| > |I_2|$. Then τ is secure.*

We now give a similar generalization of flat protocols $\tau = g(f_1(\overrightarrow{x_1}), \ldots, f_n(\overrightarrow{x_n}))$ where each of the f_i has a private variable. For such a protocol τ, a tracking strategy exists if and only if there is one node f_i that has access to all variables except the one private variable of each other f_j (f_i itself and the other f_j of course each may have more than one private variable). This characterization leads to the following generalization of Theorem 3.3.

Corollary 3.5. *Let τ be a layered protocol with levels $L_0, \ldots, L_n$ with some i such that for each $f \in L_i$, there is a variable $x_f \in Vars(f) \setminus Vars(L_i \setminus \{f\})$ and there is no $f \in L_i$ with $Vars(f) \supseteq Vars(L_i) \setminus \{x_{f'} \mid f' \in L_i\}$. Then τ is secure.*

4 Conclusion

We have initiated the study of active linking attacks which exploit that the adversary has control over user data. We introduced a model, formalized a security definition and gave a sound criterion—tracking strategies—for insecurity of protocols. We also gave sound criteria for security of protocols. For a large class of protocols, these criteria are complete, i.e., they detect all insecure protocols. The question whether this completeness holds in general remains open.

Further interesting open questions concern relaxations of some of the assumptions in our model. On the one hand, our security definition takes a "worst-case" point of view, since the adversary is free to answer all queries only with the goal of successfully complete the linking attack (and does not care that this might expose her). On the other hand, our security definition is "optimistic" since it assumes arbitrary interleaving of protocol messages from different users, realistically, the adversary will be able to make some assumptions about the occurring interleavings.

References

1. Aho, A.V., Beeri, C., Ullman, J.D.: The theory of joins in relational databases. ACM Trans. Database Syst. **4**(3), 297–314 (1979)
2. Arapinis, M., Chothia, T., Ritter, E., Ryan, M.: Analysing unlinkability and anonymity using the applied pi calculus. In: CSF, pp. 107–121. IEEE Computer Society (2010)
3. Bhargavan, K., Corin, R., Fournet, C., Gordon, A.D.: Secure sessions for web services. ACM Trans. Inf. Syst. Secur. **10**(2) (2007)

4. Backes, M., Maffei, M., Pecina, K., Reischuk, R.M.: G2C: cryptographic protocols from goal-driven specifications. In: Mödersheim, S., Palamidessi, C. (eds.) TOSCA 2011. LNCS, vol. 6993, pp. 57–77. Springer, Heidelberg (2012)
5. Biryukov, A., Pustogarov, I., Weinmann, R.-P.: TorScan: tracing long-lived connections and differential scanning attacks. In: Foresti, S., Yung, M., Martinelli, F. (eds.) ESORICS 2012. LNCS, vol. 7459, pp. 469–486. Springer, Heidelberg (2012)
6. Dong, N., Jonker, H., Pang, J.: Formal analysis of privacy in an ehealth protocol. In: Foresti, S., Yung, M., Martinelli, F. (eds.) ESORICS 2012. LNCS, vol. 7459, pp. 325–342. Springer, Heidelberg (2012)
7. Eigner, F., Maffei, M.: Differential privacy by typing in security protocols. In: CSF, pp. 272–286. IEEE (2013)
8. Maier, D., Mendelzon, A.O., Sagiv, Y.: Testing implications of data dependencies. ACM Trans. Database Syst. 4(4), 455–469 (1979)
9. Narayanan, A., Shmatikov, V.: Robust de-anonymization of large sparse datasets. In: IEEE Symposium on Security and Privacy, pp. 111–125. IEEE Computer Society (2008)
10. Rusinowitch, M., Turuani, M.: Protocol insecurity with a finite number of sessions, composed keys is NP-complete. Theoret. Comput. Sci. 1–3(299), 451–475 (2003)
11. Schnoor, H., Woizekowski, O.: Active linkability attacks. CoRR, abs/1311.7236 (2014)
12. Sweeney, L.: Achieving k-anonymity privacy protection using generalization and suppression. Int. J. Fuzziness Knowl. Based Syst. 10(5), 571–588 (2002)
13. Veeningen, M., de Weger, B., Zannone, N.: Symbolic privacy analysis through linkability and detectability. In: Fernández-Gago, C., Martinelli, F., Pearson, S., Agudo, I. (eds.) Trust Management VII. IFIP AICT, vol. 401, pp. 1–16. Springer, Heidelberg (2013)
14. Vollmer, H.: Introduction to Circuit Complexity - A Uniform Approach. Texts in theoretical computer science. Springer, Heidelberg (1999)

On the Complexity of Master Problems

Martijn van Ee[1](✉) and René Sitters[1,2]

[1] VU University Amsterdam, Amsterdam, The Netherlands
{m.van.ee,r.a.sitters}@vu.nl
[2] Centrum voor Wiskunde en Informatica (CWI), Amsterdam, The Netherlands
r.a.sitters@cwi.nl

Abstract. A master solution for an instance of a combinatorial problem is a solution with the property that it is optimal for any sub instance. For example, a master tour for an instance of the TSP problem has the property that restricting the solution to any subset S results in an optimal solution for S. The problem of deciding if a TSP instance has a master tour is known to be polynomially solvable. Here, we show that the master tour problem is Δ_2^p-complete in the scenario setting, that means, the subsets S are restricted to some given sets. We also show that the master versions of Steiner tree and maximum weighted satisfiability are also Δ_2^p-complete, as is deciding whether the optimal solution for these problems is unique. Like for the master tour problem, the special case of the master version of Steiner tree where every subset of vertices is a possible scenario turns out to be polynomially solvable. All the results also hold for metric spaces.

Keywords: Computational complexity · Polynomial hierarchy · Universal optimization · Unique optimal solutions

1 Introduction

In the *master tour problem*, defined in Deineko et al. [1], one is given n vertices with pairwise distances. A master tour is a tour on, i.e., an ordering of, the vertices with the property that restricting the tour to a subset of the vertices results in the optimal tour on that subset. Now, the question is whether a given instance has a master tour. More formally, let τ denote the tour on all vertices, τ_S the tour τ restricted to S, $c(\tau_S)$ the value of τ_S and OPT_S the value of the optimal tour on set S. Given vertices V, with $|V| = n$, and distance matrix C, does there exist a tour τ such that $c(\tau_S) = \text{OPT}_S$ for all $S \subseteq V$? In [1] it is shown that this decision problem is polynomially solvable.

This problem is related to the field of universal and *a priori* optimization. In *universal TSP*, one has to construct a tour τ and the quality of this tour is measured by the worst case ratio of $c(\tau_s)$ and OPT_S over all $S \subseteq V$. In [2] it is shown that you can always find a tour with worst case ratio $O(\log^2 n)$. It was shown in Gorodezky et al. [3] that there are instances such that all tours have a worst case ratio of $\Omega(\log n)$. The master tour problem is the problem of deciding

© Springer-Verlag Berlin Heidelberg 2015
G.F. Italiano et al. (Eds.): MFCS 2015, Part II, LNCS 9235, pp. 567–576, 2015.
DOI: 10.1007/978-3-662-48054-0_47

whether there exists a tour τ with worst case ratio equal to 1. The stochastic alternative of universal TSP is called *a priori* TSP. In this problem, one also has to construct one tour. After that, a subset of V is drawn according to some probability distribution. The goal is now to minimize the expected length of the restricted tour.

There are several models for the probability distribution. One can consider the black-box model (the distribution is unknown but one can sample to learn it), the independent model (each vertex is active independently) or the scenario model (an explicit list of subsets with its probabilities is given). Schalekamp and Shmoys [4] showed an $O(\log n)$-approximation for the black-box model. It was shown in [3] that this is the best we can hope for. In the independent decision model, Shmoys and Talwar [5] obtained a randomized 4-approximation. The approximability of *a priori* TSP in the scenario model is still open [5]. An approach for getting approximation results is using the optimal scenario lower bound, i.e., you compare the value of the solution with $\sum_j p_j T_j$, where p_j and T_j are the probability and the optimal tour length for scenario j respectively. The master tour problem can now be viewed as the question whether the optimal value is tight with this lower bound. This was our main motivation to look at the master tour problem in the scenario model.

In this paper, we extend the master tour problem as defined by Deineko et al. [1] to the master tour problem with scenarios. In this model we are given, additionally, a set $\mathcal{S}$ of scenarios and the question becomes whether there exists a tour τ such that $c(\tau_S) = \mathrm{OPT}_S$ for all $S \in \mathcal{S}$. We also investigate other problems in the "master" context, namely the Steiner tree problem and the maximum weighted satisfiability problem. In the *master tree problem with scenarios*, one is given a weighted graph G on n vertices and a set of scenarios $\mathcal{S}$, where the scenarios are subsets of vertices. The question is whether there exists a spanning tree T of G such that $c(T_S) = \mathrm{OPT}_S$ for all $S \in \mathcal{S}$. Here, restricting a tree on a scenario S means that vertices not in S will be deleted provided that this does not disconnect the tree. In the *master satisfiability problem with scenarios*, one is given a Boolean formula consisting of clauses $C_1, \ldots, C_m$ with weights $w_1, \ldots, w_m$ using variables $x_1, \ldots, x_n$ and a set of scenarios $\mathcal{S}$, where the scenarios are subsets of clauses. The question is whether there exists an assignment to $x_1, \ldots, x_n$ such that this assignment is optimal for every scenario in $\mathcal{S}$. In this paper, we show that the master tree problem with scenarios is polynomially solvable if $\mathcal{S} = 2^V$. This was also the case for the master tour problem. Further, we show that all three problems are Δ_2^p-complete in general.

The complexity class Δ_2^p is defined as the class of problems that are polynomially solvable on a deterministic Turing machine augmented with an oracle for an NP-complete problem. The class is also known as P^{NP} or $\mathrm{P}^{\mathrm{SAT}}$. It contains both NP and coNP and is contained in $\Sigma_2^p = \mathrm{NP}^{\mathrm{NP}}$ and $\Pi_2^p = \mathrm{coNP}^{\mathrm{NP}}$. It was defined by Stockmeyer [6] as part of the polynomial hierarchy. The first natural complete problem was found by Papadimitriou [7]. He showed Δ_2^p-completeness for the *unique optimum traveling salesman problem*, the problem of deciding whether the optimal TSP-tour is unique. Later, Krentel [8] showed that deciding whether the optimal value is equivalent with $0 \mod k$ is Δ_2^p-complete for the maximum weighted satisfiability problem, the traveling salesman problem, integer programming and the knapsack problem. He also showed that the problem of deciding whether $x_n = 1$ in the maximum satisfying assignment, i.e., the lexicographically maximum assignment satisfying all clauses, is also complete in this class. The next lemma shows that the problems we consider fall in the complexity class described above.

Lemma 1. *The following problems are contained in Δ_2^p.*

- *The master tour problem with scenarios*
- *The master tree problem with scenarios*
- *The master satisfiability problem with scenarios*

Proof. We prove the result for the master tour problem with scenarios. The other two results are proven similarly. We start with solving TSP for each scenario using a TSP-oracle. This gives an optimal value k_i for scenario S_i. Now, the problem becomes to decide about the existence of a tour on V such that the tour restricted to S_i has value k_i for all i. This problem is in NP, hence it can be solved by an oracle. Concluding, the master tour problem with scenarios is solvable in polynomial time on a deterministic Turing machine with an NP-oracle. □

Other problems that fall naturally in this class are problems on unique optimal solutions. Papadimitriou showed that the unique optimum traveling salesman problem is contained in this class. A similar proof can be used to show that the *unique optimum Steiner tree problem* , the problem of deciding whether the set of vertices used by the optimal solution is unique, is also in Δ_2^p. The same holds for the *unique optimum maximum weighted satisfiability problem*. In this paper, we also show that these problems are also Δ_2^p-complete.

In the next section, we show Δ_2^p-completeness for the master satisfiability problem with scenarios and the unique optimum maximum weighted satisfiability problem. The former is used to show completeness of the master tour problem with scenarios in Sect. 3. After that, we will discuss the complexity of the master tree problem with scenarios. It will be shown that the problem is polynomially solvable when each subset of V is a scenario. Here, it is also shown that the problem is Δ_2^p-complete in general as is the unique optimum Steiner tree problem. We conclude with a summary and discussion.

2   The Master Satisfiability Problem with Scenarios

Unlike the master tour problem, the master satisfiability problem is not polynomially solvable when all possible subsets are a scenario, unless P=NP. In this case every single clause is a scenario. This means that the assignment should satisfy each clause in order to be optimal for all scenarios. Hence, the problem is equivalent to the Satisfiability problem and thus NP-complete. For general scenarios, we obtain the following result. For this, we need that it is Δ_2^p-complete to decide whether variable x_n is set to true in the maximum satisfying assignment [8].

Note that the maximum satisfying assignment is defined as the lexicographically maximal assignment satisfying all clauses. Here, the ordering on the assignments is defined by the lexicographically ordering of the binary strings $(x_1, x_2, \ldots, x_n)$. We denote this decision problem by *lexicographically maximum SAT*. Note that this problem remains Δ_2^p-complete if we restrict ourselves to satisfiable formulas.

Theorem 1. *The master satisfiability problem with scenarios is Δ_2^p-complete.*

Proof. By Lemma 1, the problem is contained in Δ_2^p. To show hardness, we make a reduction from lexicographically maximum SAT. Given is a Boolean formula φ consisting of m clauses $C_1, \ldots, C_m$ using n variables $x_1, \ldots, x_n$. We create an instance for the master satisfiability problem on the same variables and we create $m + n$ clauses. The first m clauses are C_1 up to C_m. These clauses get weight 2^n. The last n clauses correspond to the variables, i.e., clause $m + i$ is clause (x_i). Clauses $m + i$ will get a weight of 2^{n-i}. We define two scenarios. The first one contains all $m + n$ clauses, the second one only contains clause $m + n$. It is easy to see, by the choice of the weights, that the maximum weight assignment in the new instance corresponds to the lexicographically maximum assignment in the original instance. We conclude with observing that there is a master assignment if and only if $x_n = 1$ in the lexicographically maximum assignment.   □

The theorem above will be needed to show completeness in Δ_2^p for the master tour problem with scenarios. Next, we will show that the unique optimum version of this problem is also Δ_2^p-complete. This problem was never considered before. A problem that was considered in the literature is the *unique satisfiability problem*, the problem of deciding whether there is exactly one assignment satisfying all clauses. It was shown that this problem is in D^p [9], the class of languages that are an intersection of an NP-language with a coNP-language. The class is equivalent to BH_2, the second level of the Boolean hierarchy [10]. In [11], it was shown that the unique satisfiability problem is not D^p-complete under regular reductions. Valiant and Vazirani [12] showed that the problem is D^p-complete under randomized reductions. For the weighted version, we get the following result.

Theorem 2. *The unique optimum maximum weighted satisfiability problem is Δ_2^p-complete.*

Proof. The problem is contained in Δ_2^p, as we already observed in the introduction. We reduce this problem from lexicographically maximum SAT. Given is a Boolean formula φ consisting of m clauses $C_1, \ldots, C_m$ using n variables $x_1, \ldots, x_n$. We create the following instance for the maximum weighted satisfiability problem. We use variables $x_1, \ldots, x_n$ and a new variable d. There will be $m + n + 1$ clauses. Again the first m correspond to the m clauses of φ and the next n correspond to the variables of φ. Clause $m + n + 1$ is clause $(\overline{x_n} \vee d)$. The weight of the first m clauses will be 2^{n+1}, the clause (x_i) gets weight 2^{n+1-i} and clause $m + n + 1$ gets weight 1. It is easy to see, by the choice of the weights, that the maximum weight assignment in the new instance corresponds to the lexicographically maximum assignment in the original instance. If $x_n = 1$ in the lexicographically maximum assignment, the optimal solution for maximum weighted satisfiability will set $x_n = 1$ and $d = 1$ to get the unique optimal solution. On the other hand, if $x_n = 0$, the lexicographically maximum assignment together with either $d = 1$ or $d = 0$ gives an optimal solution. Hence, the optimal solution is not unique in this case. Thus, we have proven that $x_n = 1$ in the lexicographically maximum assignment if and only if the optimal solution for maximum weighted satisfiability is unique.   □

3   The Master Tour Problem with Scenarios

As mentioned in the introduction, it was shown in [1] that the master tour problem is polynomially solvable if $S = 2^V$. The authors showed that an instance has a master tour if and only if the distance matrix is a permuted Kalmansson matrix. The main result in [1] was that permuted Kalmansson matrices can be recognized in polynomial time, hence the master tour problem is polynomially solvable in this case. Before we discuss the general case, we note that the previous result extends beyond the case $S = 2^V$. A closer examination of the proof shows that you need at least all subsets of V of size 4 to get the characterization above. Since [1] also showed how to find a master tour, we get the following result.

Theorem 3 ([1]). *Finding a master tour can be done in polynomial time if the set of scenarios S contains at least each subset of V of size 4.*

We now show that the problem is Δ_2^p-complete for general S.

Theorem 4. *The master tour problem with scenarios is Δ_2^p-complete.*

Proof. By Lemma 1, the problem is contained in Δ_2^p. To show hardness, we reduce the master satisfiability problem to the master tour problem. The reduction is an extension of the reduction of Krentel [8] and uses ideas from Papadimitriou [7]. We are given an instance of the master satisfiability problem. So we have a Boolean formula φ containing clauses $C_1, \ldots, C_m$ with weights $w_1, \ldots, w_m$ using variables $x_1, \ldots, x_n$ and a set of scenarios S with $|S| = k$. Each scenario is a set of clauses. We will show the reduction for the case when all clauses have three literals. The proof naturally extends to the general case.

We construct the graph in Fig. 1. For each variable, we create two vertices connected by two edges. One edge will correspond to setting the variable to true, the other edge will correspond to false. These variable gadgets are connected in a chain. For each clause, we create two vertices and four edges. The first three edges will correspond to the three literals in the clause. The fourth edge will correspond to not satisfying the clause. The clauses will also be connected by a chain, where there is an extra intermediate vertex between two clause gadgets. We also add edges between the intermediate vertices in the chain of clause gadgets. These edges ensure that the shortcutting of the tour is done correctly. Finally, the chain of variable gadgets and the chain of clause gadgets are connected using two edges.

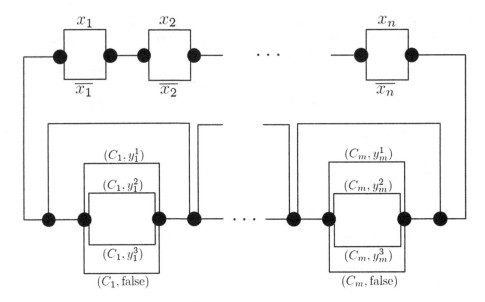

Fig. 1. Created instance for the master tour problem (without NAND-gadgets). Edge (C_j, y_j^i) corresponds to the ith literal of clause j.

The edges in the variable gadgets and the clause gadgets are connected through NAND-gadgets [7]. These gadgets make sure that the two edges it connects are not both used in the solution. There will be a NAND-gadget between variable edge x_i and every clause edge corresponding to literal $\overline{x_i}$. Similarly, there will be a NAND-gadget between variable edge $\overline{x_i}$ and every clause edge corresponding to literal x_i. To illustrate how the NAND-gadget works, we depicted an OR-gadget in Fig. 2. A tour can only get through this gadget by entering it at one endpoint of an edge and leave at the other endpoint. In this way, exactly one of the edges will be used in the solution. The NAND-gadget is an extension of this OR-gadget. It uses more vertices which ensure that at most one of these edges will be used. For a further description of the NAND-gadgets and its proof of correctness, the reader is referred to [7]. Note that the NAND-gadgets also ensure that there are no multiple edges in the graph.

We use the following edges costs. All edges in the graph constructed, except edges corresponding to not satisfying the clause, have length 0, whereas non-edges have length M, where M is a large number. The edge corresponding to not satisfying C_i will get cost w_i. Note that the optimal traveling salesman tour in the created graph corresponds to a maximum weight assignment of φ, i.e., the tour minimizes the weight of the unsatisfied clauses.

We now create the following k scenarios for the master tour problem. Each scenario contains all vertices in the variable chain and all intermediate vertices in the clause chain. Moreover, the scenario corresponding to the jth scenario of the master satisfiability problem contains the vertices corresponding to the clauses in scenario j. It also contains the vertices of the relevant NAND-gadgets.

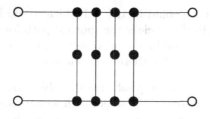

Fig. 2. OR-gadget

Suppose there is an assignment that is optimal for all scenarios. Use this assignment to make the following tour. For each variable gadget, use edge x_i if x_i is set to true and vice versa. If a clause is satisfied, choose one of the edges corresponding to a satisfied literal. If a clause is not satisfied, you have to use the corresponding edge. Note that if a clause is not in the scenario, the tour shortcuts the clause gadget through the edge between the intermediate vertices. Now, the tour restricted to a scenario corresponds to the maximum weight assignment, so it is the optimal tour for that scenario. Hence, we have a master tour. On the other hand, suppose that there is a master tour. Then this tour has to correspond to an assignment being optimal for all scenarios.   □

Note that the master satisfiability problem with scenarios is already Δ_2^p-complete for two scenarios, which implies that this is also the case for the master tour problem with scenarios. The result still holds in the metric case, since adding a large weight to all distances does not change the complexity of the problem.

4   The Master Tree Problem with Scenarios

We start with showing that the master tree problem with scenarios is polynomially solvable when $\mathcal{S} = 2^V$. Note that V is a scenario, which implies that the master tree is a minimum spanning tree. Therefore, no edge (i, k) with $c_{ij} + c_{jk} < c_{ik}$ for some j will be used. So, without loss of generality we assume that our instance is metric, i.e., our distances satisfy the triangle inequality. Further, observe that every pair of vertices is a scenario. This implies that the master tree should contain a shortest path for each pair of vertices. We get the following characterization. Note that a metric space is a tree metric if there exists a tree such that the distances in the metric space correspond to the shortest path distances in the tree.

Theorem 5. *An instance of the master tree problem with $\mathcal{S} = 2^V$ has a master tree if and only if the graph distances form a tree metric.*

Proof. If the graph distances form a tree metric, one can choose the tree as solution. It is easy to see that every subset is connected in the cheapest way. Now, suppose that the graph distances do not form a tree metric. This means that there is a cycle in the underlying graph with the property that each edge in this cycle is cheaper than the sum of edge weights of the other edges in the cycle. Otherwise, this edge could be deleted which would break the cycle. This implies that each edge in this cycle is the shortest path for its endpoints. Hence, every edge in the cycle should be included to get a master tree. But if they are all included, there is a cycle. So, there is no master tree.   □

Hence, the master tree problem is solvable in polynomial time. Note that this result implies that the master tree problem with scenarios is polynomial solvable if S contains all pairs of vertices and V. As stated in the introduction, it turns out that the general problem is Δ_2^p-complete. The reduction is a slight variant of a standard reduction from the satisfiability problem [13].

Theorem 6. *The master tree problem with scenarios is Δ_2^p-complete.*

Proof. By Lemma 1 the problem is in Δ_2^p. To show hardness, we reduce the problem from lexicographically maximum SAT. Given an instance of this problem, we construct the graph in Fig. 3. In this figure, the non-present edge weights are equal to 1. Edge weights of non-present edges get a weight equal to the shortest path distance in the graph. The black vertices correspond to terminals, whereas the white vertices correspond to Steiner vertices. Further, we have that $q \geq 2$. We create two scenarios. The first scenario contains all terminals, the second scenario only contains r and x_n. Note that any Steiner tree contains at least one of x_i^1 and x_i^0 for all i. Given that there is a satisfying assignment, the optimal Steiner tree will use exactly one out of x_i^1 and x_i^0 for all i. Because $q^{-i} > \sum_{j=i+1}^n q^{-j}$, the optimal solution will correspond to the maximum satisfying assignment. If the maximum satisfying assignment has $x_n = 1$, then there is an optimal Steiner tree using edges (r, x_n^1) and (x_n^1, x_n), so there is a master tree. If there is a master tree, it should use x_n^1 and since it only uses the true-vertex or the false-vertex, the optimal Steiner tree contains the true-vertex and $x_n = 1$ in the maximum satisfying assignment.   □

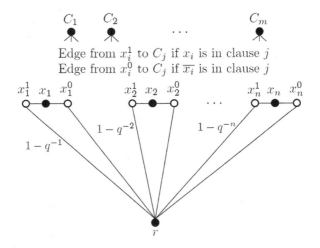

Fig. 3. Created instance for Steiner tree.

Note that this reduction can also be used to show that computing the optimal Steiner tree is OptP-complete [8]. This basically means that computing the optimal value for the Steiner tree problem is as hard as computing the optimal values for TSP, maximum weighted satisfiability, integer programming and the knapsack problem. The reduction can also be used to show that deciding whether a given Steiner node is in the optimal solution is Δ_2^p-complete. Finally, we will show that if one adjust the reduction slightly, it turns out that deciding whether the set of vertices used by the optimal Steiner tree is unique is also Δ_2^p-complete.

Theorem 7. *The unique optimum Steiner tree problem is Δ_2^p-complete.*

Proof. As observed in the introduction, the problem is in Δ_2^p. Given an instance of lexicographically maximum SAT, we construct the graph in Fig. 4. As in Theorem 6, all weights that are not present in the figure are equal to 1. If $x_n = 1$, the optimal solution uses vertex x_n^1 and the set of vertices used by the optimum is unique. If $x_n = 0$, then there are two sets of vertices resulting in an optimal Steiner tree: one using vertex d and one which does not.   □

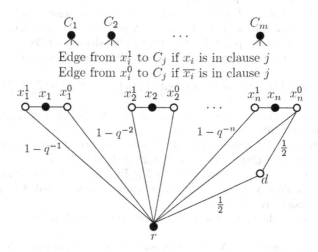

Fig. 4. Created instance for Steiner tree.

5   Conclusion

In this paper, we showed that five new problems are complete for Δ_2^p. This was shown for the master tour problem with scenarios, the master tree problem with scenarios, the master satisfiability problem with scenarios, the unique optimum Steiner tree problem and the unique optimum maximum weighted satisfiability problem. We also showed that the master tree problem is polynomially solvable when all subsets of V are a scenario. In the introduction, it was discussed that master problems have a natural connection to universal optimization. We think that the main open question in the field of universal optimization is the approximability of the *a priori* TSP in the scenario model. Here, we showed that it is Δ_2^p-complete to decide whether the optimal value is tight with the optimal scenario lower bound. Further research should be performed in order to investigate the approximability of this problem.

References

1. Deineko, V.G., Rudolf, R., Woeginger, G.J.: Sometimes travelling is easy: The master tour problem. SIAM J. Discrete Math. **11**(1), 81–93 (1998)
2. Gupta, A., Hajiaghayi, M.T., Räcke, H.: Oblivious network design. In: Proceedings of the Seventeenth Annual ACM-SIAM Symposium on Discrete Algorithms, pp. 970–979 (2006)
3. Gorodezky, I., Kleinberg, R.D., Shmoys, D.B., Spencer, G.: Improved lower bounds for the universal and a priori TSP. In: Serna, M., Shaltiel, R., Jansen, K., Rolim, J. (eds.) APPROX 2010. LNCS, vol. 6302, pp. 178–191. Springer, Heidelberg (2010)
4. Schalekamp, F., Shmoys, D.B.: Algorithms for the universal and *a priori* TSP. Oper. Res. Lett. **36**(1), 1–3 (2008)
5. Shmoys, D.B., Talwar, K.: A constant approximation algorithm for the *a priori* traveling salesman problem. In: Lodi, A., Panconesi, A., Rinaldi, G. (eds.) IPCO 2008. LNCS, vol. 5035, pp. 331–343. Springer, Heidelberg (2008)
6. Stockmeyer, L.J.: The polynomial-time hierarchy. Theor. Comput. Sci. **3**(1), 1–22 (1976)
7. Papadimitriou, C.H.: On the complexity of unique solutions. J. ACM **31**(2), 392–400 (1984)
8. Krentel, M.W.: The complexity of optimization problems. J. Comput. Syst. Sci. **36**(3), 490–509 (1988)
9. Papadimitriou, C.H., Yannakakis, M.: The complexity of facets (and some facets of complexity). J. Comput. Syst. Sci. **28**(2), 244–259 (1984)
10. Wechsung, G.: On the boolean closure of NP. In: Budach, L. (ed.) Fundamentals of Computation Theory. Lecture Notes in Computer Science, vol. 199, pp. 485–493. Springer, Heidelberg (1985)
11. Blass, A., Gurevich, Y.: On the unique satisfiability problem. Inf. Control **55**(1–3), 80–88 (1982)
12. Valiant, L.G., Vazirani, V.V.: NP is as easy as detecting unique solutions. Theor. Comput. Sci. **47**(3), 85–93 (1986)
13. Hauptmann, M.: Approximation complexity of optimization problems: Structural foundations and Steiner tree problems. Ph.D thesis, Rheinischen Friedrich-Wilhelms-Universität Bonn (2004)

Efficient Algorithm for Computing All Low s-t Edge Connectivities in Directed Graphs

Xiaowei Wu and Chenzi Zhang[✉]

The University of Hong Kong, Pok Fu Lam, Hong Kong
{xwwu,czzhang}@cs.hku.hk

Abstract. Given a directed graph with n nodes and m edges, the (strong) edge connectivity $\lambda(u,v)$ between two nodes u and v is the minimum number of edges whose deletion makes u and v not strongly connected. The problem of computing the edge connectivities between all pairs of nodes of a directed graph can be done in $O(m^\omega)$ time by Cheung, Lau and Leung (FOCS 2011), where ω is the matrix multiplication factor (≈ 2.373), or in $\tilde{O}(mn^{1.5})$ time using $O(n)$ computations of max-flows by Cheng and Hu (IPCO 1990).

We consider in this paper the "low edge connectivity" problem, which aims at computing the edge connectivities for the pairs of nodes (u,v) such that $\lambda(u,v) \leq k$. While the undirected version of this problem was considered by Hariharan, Kavitha and Panigrahi (SODA 2007), who presented an algorithm with expected running time $\tilde{O}(m + nk^3)$, no algorithm better than computing all-pairs edge connectivities was proposed for directed graphs. We provide an algorithm that computes all low edge connectivities in $O(kmn)$ time, improving the previous best result of $O(\min(m^\omega, mn^{1.5}))$ when $k \leq \sqrt{n}$. Our algorithm also computes a minimum u-v cut for each pair of nodes (u,v) with $\lambda(u,v) \leq k$.

1 Introduction

Given an undirected graph, the edge connectivity between two nodes is the minimum number of edges whose deletion disconnects those two nodes, which by Menger's Theorem [12] is also the maximum number of edge-disjoint paths between them. The definition of edge connectivity can be naturally generalized to directed graphs [1,6,13] (it is denoted by "strong edge connectivity" in some literatures): given a digraph $G(V,E)$, the edge connectivity $\lambda(u,v)$ between two nodes $u, v \in V$ is the minimum number of edges whose deletion makes u and v not strongly connected. The edge connectivity of a graph is the minimum edge connectivity between any two nodes in the graph. Computing the edge connectivity is a classic and well-studied problem.

Given two nodes u and v in a digraph, the edge connectivity $\lambda(u,v) = \min\{f(u,v), f(v,u)\}$, where $f(u,v)$ is the max-flow from u to v, if we attach unit capacity to each edge. Given a unit capacity network with m edges and n nodes, Even and Tarjan [5] showed that Dinic's algorithm [4] for computing the s-t max-flow terminates in $O(\min\{m^{\frac{3}{2}}, mn^{\frac{2}{3}}\})$ time. The above algorithm was

© Springer-Verlag Berlin Heidelberg 2015
G.F. Italiano et al. (Eds.): MFCS 2015, Part II, LNCS 9235, pp. 577–588, 2015.
DOI: 10.1007/978-3-662-48054-0_48

the fastest algorithm for computing unit capacity max-flow for almost 40 years until very recently Lee and Sidford [11] proposed an $\tilde{O}(m\sqrt{n})$ time algorithm using a new method to solve LP.

The problem of computing the edge connectivities between all pairs of nodes of a digraph was also considered. Note that the problem can be trivially solved by computing $O(n^2)$ max-flows, which yields a total running time of $\tilde{O}(mn^{2.5})$ by Lee and Sidford [11]. Cheung et al. [3] considered the problem and proposed an $O(m^\omega)$ time randomized algorithm, where ω is the matrix multiplication factor (≈ 2.373), using the idea of network coding. We provide in this paper an efficient algorithm that computes the edge connectivities $\lambda(u,v)$ for all pairs of nodes (u,v) such that $\lambda(u,v) \leq k$ in $O(kmn)$ time, for any integer $k \geq 1$. Our algorithm also computes a minimum u-v cut for each such pair of nodes (u,v).

Gomory-Hu Tree. It was observed by Gomory and Hu [7] long ago that the edge connectivities between all pairs of nodes in an undirected graph $G(V,E)$ can be represented by a weighted tree T on all nodes V such that

- the edge connectivity between any two nodes $u,v \in V$ equals the weight of the lightest edge on the unique u-v path in T.
- the partition of the nodes produced by removing this edge from T forms a minimum u-v cut in graph G.

Any tree satisfying both conditions is called a *cut-equivalent tree*, or *Gomory-Hu tree* of G; if a tree satisfies only the first condition, then it is called a *flow-equivalent tree* of G. The computation of a Gomory-Hu tree of any undirected graph can be reduced to the computation of n max-flows [7,8], which yields a total running time of $\tilde{O}(mn^{1.5})$ using the current fastest unit capacity max-flow algorithm [11]. Currently the above running time is the best for any deterministic cut-equivalent tree construction. For randomized Gomory-Hu tree construction, Hariharan et al. [10] proposed an algorithm that with high probability computes a Gomory-Hu tree for any unweighted undirected graph in $\tilde{O}(mn)$ time.

The definition of Gomory-Hu tree and flow-equivalent tree can be naturally generalized to digraphs. Schnorr [13] attempted to construct the Gomory-Hu tree for general weighted digraphs. However, it was later pointed out by Benczur [1] that for general weighted digraphs, Gomory-Hu trees do not exist. We generalize the counter-example of Benczur [1] and show that the Gomory-Hu tree does not exist even in some unweighted digraph.

Fact 1 (Non-existence of Gomory-Hu tree for Digraphs). *There exists an unweighted digraph that does not have any Gomory-Hu tree.*

Contrary to the Gomory-Hu tree, a flow-equivalent tree always exists for any weighted digraph. Cheng and Hu [2] generalized the result of Gomory and Hu [7] to construct a flow-equivalent tree using $O(n)$ computations of max-flows, which yields a total running time of $\tilde{O}(mn^{1.5})$. Actually Cheng and Hu proved something even more powerful: given any set V of n nodes, if we attach arbitrary weight $w(S)$ to each subset $S \subseteq V$ of nodes, the minimum weight cuts that separate all $\binom{n}{2}$ pairs of nodes can be represented by an *ancestor tree*, which is a tree spanning all nodes V.

We show in this paper that in directed unweighted graphs, the problem of computing the all-pairs edge connectivities and the computation of flow-equivalent tree are highly related to each other. By the following theorem, the result of Cheng and Hu [2] and that of Cheung et al. [3] hold for both problems.

Theorem 1 (Reducibility). *For any digraph G with n nodes, the all-pairs edge connectivities problem and the flow-equivalent tree problem are $O(n^2)$-reducible*

- *given the edge connectivities $\lambda(u,v)$ of all pairs of (u,v), a flow-equivalent tree of G can be constructed in $O(n^2)$ time.*
- *given a flow-equivalent tree of G, the edge connectivities $\lambda(u,v)$ of all pairs of (u,v) can be computed in $O(n^2)$ time.*

Low Edge Connectivities. In many applications, computing the edge connectivities of pairs of nodes which are poorly connected in the graph is more important. In particular, we consider the problem of computing the edge connectivities for the pairs of nodes (u,v) whose edge connectivities are at most k in the input graph, for any integer $k \geq 1$. Using the same definition from Hariharan et al. [9], the output should be represented succinctly as a weighted tree T whose nodes are $V_1, V_2, \ldots, V_l$, a partition of V, with the property that (1) for all $i \in [l]$, $\lambda(u,v) > k$ for all $u,v \in V_i$; (2) edge connectivity between $u \in V_i$ and $v \in V_j$, if $i \neq j$, is equal to the weight of the lightest edge in the unique V_i-V_j path in T. We call the above weighted tree a k-*edge-connectivity tree*. Note that for $k \geq \Delta = \max_{u,v \in V} \lambda(u,v)$, the k-edge-connectivity tree is a flow-equivalent tree. The problem for undirected graphs was considered by Hariharan et al. [9], who presented a randomized algorithm with expected running time $\tilde{O}(m + nk^3)$.

We consider in this paper the same problem in digraphs. For the special case when $k = 1$, Georgiadis et al. [6] showed that the 1-edge-connectivity tree can be constructed in linear time. However, for general k, the best algorithm to solve this problem involves computing all-pairs edge connectivities, which requires $\tilde{O}(mn^{1.5})$ time by Cheng and Hu [2]. We improve in this paper the above result to $O(kmn)$. It is easy to verify from our proofs that the following result holds even in directed multigraphs (in which case $m = \omega(n^2)$ is possible).

Theorem 2 (Computing Low Edge Connectivities). *Given a digraph $G(V, E)$ and an integer $k \geq 1$, a k-edge-connectivity tree of G can be computed in $O(kmn)$ time.*

While it is shown by Cheng and Hu that the $\binom{n}{2}$ edge connectivities can be computed using $O(n)$ computations of max-flows, improving their running time in the low edge connectivity case is non-trivial. As we compute $\lambda(u,v)$, we actually obtain a minimum u-v cut, which defines a partition of nodes and this piece of information can be reused in the computation of the edge connectivities of other pairs of nodes. The above observation is crucial for Cheng and Hu's algorithm. However, in the low edge connectivity problem, if $\lambda(u,v) \geq k$, then we can not afford to compute a minimum u-v cut.

Instead, we decompose the computation of edge connectivities such that for each pair of nodes (u, v), the lower bound for $\lambda(u, v)$ is increased by 1 (if possible) in each iteration. We maintain partitions of nodes in our algorithm and attach a *seed* node to each partition. Using the seeds to represent the edge connectivities between nodes in the same partition and a crucial *merge-flow* subroutine, we are able to reduce the total computation time in each iteration to $O(mn)$, which directly yields Theorem 2.

2 Preliminaries

Given a subset $S \subseteq V$ of nodes, let $d_-(S) = |\{(u, v) \in E | u \in S, v \in V\backslash S\}|$ be the *out-degree* of S, $d_+(S) = |\{(u, v) \in E | u \in V\backslash S, v \in S\}|$ be the *in-degree* of S and $d(S) = \min\{d_-(S), d_+(S)\}$ be the *degree* of S.

Definition 1 (Edge-Connectivity). *Given $u, v \in V$, the* edge-connectivity *$\lambda(u, v)$ between u and v is the minimum number of edges whose removal makes u and v not strongly connected. We assume $\lambda(u, u) = \infty$ for all $u \in V$.*

Given two nodes $u, v \in V$, we use $f(u, v)$ to denote the max-flow from u to v (assume that unit capacity is attached to each directed edge in the graph). By the above definition, we have $\lambda(u, v) = \min\{f(u, v), f(v, u)\} = \lambda(v, u)$. Moreover, there must exist at least one $S \subsetneq V$ such that $u \in S$, $v \in V\backslash S$ and $d(S) = d(V\backslash S) = \lambda(u, v)$. By Menger's Theorem, we have the following basic fact.

Fact 2 (i-Edge-Connectivity is an Equivalence Relation) *For any integer $i \geq 1$, given any nodes $a, b, c \in V$ such that $\lambda(a, b) \geq i$ and $\lambda(b, c) \geq i$, we have $\lambda(a, c) \geq i$.*

Throughout this paper, we use n to denote the number of nodes, m for the number of edges and $\Delta = \max_{u, v \in V, u \neq v} \lambda(u, v)$ to denote the maximum edge-connectivity between any two nodes in V. Unless otherwise stated, a graph is always directed and unweighted.

Definition 2 (Blocks, Partition). *Given a graph $G(V, E)$, for any integer $i \geq 0$, an i-edge-connected (i-ec) block is a subset of nodes $B \subseteq V$ such that $\forall u, v \in B, \lambda(u, v) \geq i$ and $\forall u \in B, v \in V\backslash B, \lambda(u, v) < i$. An i-edge-connected partition, denoted by Ω_i, is the collection of all i-edge-connected blocks of G.*

Definition 3 (Seed). *We attach a unique seed $r(B) \in B$ to each block B. For any $(i + 1)$-ec block B' that is a subset of a i-ec block B, if $r(B) \in B'$, then we set $r(B') = r(B)$.*

By definition, we have $\Omega_0 = \{V\}$ and $\Omega_{\Delta+1} = \{\{u\} | u \in V\}$. Note that $\cup_{i=0}^{\Delta+1} \Omega_i$ is a Laminar family and hence we can organize all the blocks by a tree rooted at V such that block B is the *parent block* of block $B' \in \Omega_i$ iff $B' \subseteq B$ and $B \in \Omega_{i-1}$. We call B' a *child block* of B. If $r(B') = r(B)$, then we call B' the *closest* child block of B. Note that if $u \in V$ is the seed of some block in Ω_i, then u will be a seed for exactly one block in Ω_j, for each $j = i + 1, i + 2, \ldots, \Delta + 1$.

Proof of **Fact** 1: The following graph does not have a Gomory-Hu tree, where "↔" stands for two directed edges in two directions. It is easy to see that the given graph has four 5-edge-connected blocks: A, B, C, D.

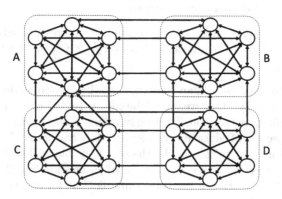

Suppose the given graph has a Gomory-Hu tree T, then if we contract all edges of weight at least 5 in T, then all nodes in the same 5-edge-connected block should become a single super-node. Hence in the contracted tree, there are only four super-nodes: u_A, u_B, u_C, u_D, each of them corresponds to a block.

It can be easily verified that $\forall a \in A$, we have: $\forall d \in D, \lambda(a,d) = 1, \forall b \in B, \lambda(a,b) = 2, \forall c \in C, \lambda(a,c) = 3$. Hence the four super-nodes are connected by three edges of weight at most 3. However, in the contracted tree, u_A, u_B and u_C can not be leaf nodes: if u_A is a leaf node, then after removing the edge connected to u_A in the contracted tree from T, the nodes will be partitioned into two sets A and $B \cup C \cup D$ and we have $d(A) = 4$, which is a contradiction since the edge connecting u_A should be of weight at most 3; similarly, u_B and u_C can not be leaf nodes since $d(C) \geq d(B) \geq 5$. Since any tree must have at least two leaf nodes, we conclude that G does not have any Gomory-Hu tree. □

Throughout this paper, we will use Tarjan's $O(m)$ time algorithm [14] for computing the strongly connected components of a digraph as a subroutine.

3 Flow-Equivalent Tree for Directed Graph

We first show that a k-edge-connectivity tree can be efficiently constructed given the partitions $\Omega_1, \Omega_2, \ldots, \Omega_k, \Omega_{k+1}$. The following lemma is important for the proofs of Theorems 1 and 2.

Lemma 1 (Partitions to k-edge-connectivity tree). *For any integer $k \geq 1$, given partitions $\Omega_1, \Omega_2, \ldots, \Omega_{k+1}$ of graph $G(V, E)$, a k-edge-connectivity tree of G can be constructed in $O(kn)$ time.*

Proof. Let S be the set of seeds for blocks in Ω_{k+1}. Note that we only need to build a tree spanning all nodes in S since (1) $\forall u, v \in S$, we have $\lambda(u,v) \leq k$; (2) $\forall u \in V$, there exists $r \in S$ s.t. $\lambda(u,r) > k$. We initialize the flow-equivalent tree $T = (S, \emptyset)$ and add edges to the tree as follows.

For $i = k, k-1, \ldots, 1, 0$, for each $B \in \Omega_i$ with at least 2 child blocks, connect the seeds of the child blocks (which must be in S) by a single path in T. Set the weight of each edge in the path to be i. By the above construction, it is easy to check that after considering each i we have

- all edges added to the current flow-equivalent tree are of weight at least i.
- two seeds r_1 and r_2 are connected iff $\lambda(r_1, r_2) \geq i$ (can be easily proved by induction on i).

Hence, every time when we add an edge (u, v) to the flow-equivalent tree, u and v must not be connected and the weight of the edge is set to be $\lambda(u, v)$, which means that after the whole construction, T is tree spanning S and for any two seeds u and v, the minimum weight of edges in the unique path between u and v equals $\lambda(u, v)$.

Replacing each $r \in S$ in the tree by the $(k+1)$-ec block B such that $r(B) = r$ gives a k-edge-connectivity tree. Note that when considering Ω_i, we only need to scan each node once. Hence the total running time for constructing the flow-equivalent tree is $O(kn)$.   □

Using Lemma 1 with $k = \Delta$, we are able to prove Theorem 1. Recall that a Δ-edge-connectivity tree is a flow-equivalent tree.

Proof of **Theorem** 1: First, given a flow-equivalent tree $T(V, E_T)$ of $G(V, E)$, we can recover the edge-connectivities $\lambda(u, v)$ for all pairs of nodes as follows. Sort the edges in E_T by their weights in non-decreasing order in $O(n \log n)$ time. In every step we remove the edge e with minimum weight w_e from T, and set the edge connectivity $\lambda(u, v) = w_e$ for each pair of (u, v) that are disconnected in T by the removal of e. Hence the total running time can be bounded by $T(n) = \max_{i \in [n]}\{i(n-i) + T(i) + T(n-i) + O(n)\}$, which can be bounded by $O(n^2)$, using mathematical induction on $n \geq 1$.

By Lemma 1, to show that a flow-equivalent tree can be constructed given the edge-connectivities between all pairs of nodes, we only need to construct $\Omega_1, \Omega_2, \ldots, \Omega_\Delta$ of graph $G(V, E)$ in $O(n^2)$ time.

Given the edge-connectivities of all pairs of nodes, the above algorithm constructs each Ω_t in $O(\max\{1, |\Omega_t| - |\Omega_{t-1}|\}n)$ time for all $t = 1, 2, \ldots, \Delta$. Hence all partitions $\Omega_1, \Omega_2, \ldots, \Omega_\Delta$ of graph $G(V, E)$ can be constructed in $O(n^2)$ time,

---

Algorithm 1. all-partitions($G(V, E)$):

1: let $\Omega_0 = \{V\}$, set $r(V)$ to be an arbitrary node in V.
2: **for** each $t = 1, 2, \ldots, \Delta$ **do**
3:     $\Omega_t = \emptyset$, $i = 1$.
4:     **for** each $B \in \Omega_{t-1}$ **do**
5:         $B_i = \{v \in B | \lambda(r(B), v) \geq t\}$, $r(B_i) = r(B)$.
6:         $\Omega_t = \Omega_t \cup \{B_i\}$, $i = i + 1$.   ▷ $O(|B|)$ time
7:         **while** $\cup_{j=1}^{i-1} B_j \neq B$ **do**
8:             pick an arbitrary $u \in B \backslash \cup_{j=1}^{i-1} B_j$.
9:             $B_i = \{v \in B | \lambda(u, v) \geq t\}$, $r(B_i) = u$.
10:            $\Omega_t = \Omega_t \cup \{B_i\}$, $i = i + 1$.   ▷ $O(|B|)$ time

which by Lemma 1 means that a flow-equivalent tree can be constructed given the edge-connectivities between all pairs of nodes in $O(n^2)$ time.   □

4   Computing Low Edge Connectivities

For any integer $k \geq 1$, we describe in this section how to compute the partitions $\Omega_1, \Omega_2, \ldots, \Omega_k, \Omega_{k+1}$ of G in $O(kmn)$ time.

Note that given the i-ec partition Ω_i, the $(i + 1)$-ec partition Ω_{i+1} can be obtained by computing all child blocks of each i-ec block $B \in \Omega_i$. During each refinement step, we also need to assign seeds to the child blocks. The following algorithm applies the above steps to construct partitions $\Omega_1, \Omega_2, \ldots, \Omega_k, \Omega_{k+1}$, where function blocks$(B, r(B), r(B), i)$ returns all i-ec blocks of the $(i - 1)$-ec block B.

Algorithm 2. partitions$(G(V, E), k)$:

1: fix any node $s \in V$, let $\Omega_0 = \{V\}$, $r(V) = s$.
2: **for** each $i = 1, 2, \ldots, k, k + 1$ **do**
3:     $\Omega_i = \emptyset$.
4:     **for** each $B \in \Omega_{i-1}$ **do**
5:         $\Omega_i = \Omega_i \cup$ blocks$(B, r(B), r(B), i)$.   ▷ partition B into i-ec blocks
6: **return** $\Omega_1, \Omega_2, \ldots, \Omega_k, \Omega_{k+1}$.

4.1   Block Refinement

Suppose that before constructing Ω_i, for each pair of nodes (u, v): (1) we have already computed a current flow $\mathcal{F}(u, v)$, which is stored as a set of edge-disjoint paths from u to v; (2) $|\mathcal{F}(u, v)| = \min\{\lambda(u, v), i - 1\}$. Then for each $(i - 1)$-ec pair of nodes (u, v), we try to find an augmenting path from u to v (from v to u) in the residual graphs with flow $\mathcal{F}(u, v)$ (flow $\mathcal{F}(v, u)$). If we can increase the flow by 1 in both directions, then u and v are at least i-ec and we place them into the same i-ec block. Otherwise we find a min-cut $(W, V \backslash W)$ with $d(W) = i - 1$ that separates u and v in G and hence recursion can be applied.

Algorithm 3. blocks(B, r, u, i):

1: $S = \{u\}, r(S) = u, R = B \backslash S, \mathcal{B} = \emptyset$.   ▷ if $u \neq r$, then $\forall v \in B, \lambda(v, r) = i - 1$
2: **while** $R \neq \emptyset$ **do**
3:     pick an arbitrary node $v \in R$.
4:     **if** flow$(u, v, r, i) \neq$ NULL and flow$(v, u, r, i) \neq$ NULL **then**   ▷ $O(m)$ time
5:         $S = S \cup \{v\}, R = R \backslash \{v\}$.
6:     **else**   ▷ $\lambda(u, v) = i - 1$
7:         **if** flow$(u, v, r, i) ==$ NULL **then**
8:             let $\tilde{G}$ be the residual graph with flow $\mathcal{F}(u, v)$.
9:         **else**
10:            let $\tilde{G}$ be the residual graph with flow $\mathcal{F}(v, u)$.   ▷ $O(m)$ time
11:        let W be the set of nodes strongly connected to u in $\tilde{G}$.
12:        $\mathcal{B} = \mathcal{B} \cup$ blocks$(R \backslash W, r, v, i), R = R \cap W$.
13: **return** $\mathcal{B} \cup \{S\}$.

The above algorithm takes a subset $B \subseteq B'$ of an $(i-1)$-ec block B', together with the seed $r = r(B')$ and a starting node $u \in B$, computes all i-ec blocks that are subsets of B recursively. The function flow(u, v, r, i) computes a flow $|\mathcal{F}(u, v)| = i$ or returns NULL if $f(u, v) < i$.

Note that whenever we find a min-cut $(W, V \backslash W)$ with $d(W) = i - 1$ that separates u and v in G, then $\forall x \in B \backslash W$, we have $\lambda(u, x) = i - 1$, which means that $(W, V \backslash W)$ is a minimum u-x cut for all $u \in W$ and $x \in B \backslash W$. Hence we can recursively compute the i-ec blocks for $R \backslash W$ without splitting any i-ec block. Assume that flow(u, v, r, i) computes a flow $|\mathcal{F}(u, v)| = i$ or returns NULL if $f(u, v) < i$ (such an algorithm will be provided in Sect. 4.2), we have the following lemma immediately.

Lemma 2 (Block Refinement). *Given a subset $B \subseteq B'$ of an $(i - 1)$-edge-connected block B' and the seed $r = r(B)$, Algorithm 3 returns all i-ec blocks that are subsets of B.*

While we assume that before constructing Ω_i, we have already computed a current flow $|\mathcal{F}(u, v)| = \min\{\lambda(u, v), i - 1\}$ between any pair of nodes, it is easy to observe that the time and space complexity is too large: in the worst case we need to update $\Theta(n^2)$ current flows when constructing one partition and store $O(mn^2)$ edges, which may be $\omega(m^2)$ already. Hence, we need to represent all $\Theta(n^2)$ current flows using a sparse structure, i.e., $O(n)$ current flows, such that the current flow between any pair of nodes can be efficiently recovered.

4.2   Computing the Current Flow with Seed Replacement

Note that in order to test i-edge-connectivity in an $(i - 1)$-ec block B, we only need to do the test between every node $u \in B$ with the seed of B. Moreover, given any two nodes $u, v \in B$, if we have already computed $\mathcal{F}(u, r)$ and $\mathcal{F}(r, v)$ such that $|\mathcal{F}(u, r)| = i$ and $|\mathcal{F}(r, v)| = i$, then we can recover a flow $\mathcal{F}(u, v)$ with $|\mathcal{F}(u, v)| = i$ from u to v in $O(m)$ time using the following algorithm. We regard a path P as a sequence of edges and use $|P|$ to denote the number of edges in P.

Lemma 3 (Merge-Flows). *Given $u, v \in B \in \Omega_i$, $r = r(B)$, $|\mathcal{F}(u, r)| = i$ and $|\mathcal{F}(r, v)| = i$, Algorithm 4 computes a flow $\mathcal{F}(u, v)$ with $|\mathcal{F}(u, v)| = i$ from u to v in $O(m)$ time.*

Proof. Let $P_1, P_2, \ldots, P_i$ be i edge-disjoint paths from u to r and $Q_1, Q_2, \ldots, Q_i$ be i edge-disjoint paths from r to v. Note that there might exist edges that are used by both the P_j's and the Q_l's (such a *shared* edge will be given two labels by Algorithm 4). We argue that we can compute a matching between the P_j's and the Q_l's in $O(m)$ time such that each P_j is matched with $Q_{\mathrm{matchP}(j)}$ to form a new path H_j from u to v. Moreover, all H_j's are edge-disjoint. Then we have $\mathcal{F}(u, v) = \{H_1, H_2, \ldots, H_i\}$ as required.

In Algorithm 4, H_j is set to be $(p_{j,1}, p_{j,2}, \ldots, p_{j,\mathrm{top}(j)}, q_{l,x+1}, q_{l,x+2}, \ldots, q_{l,|Q_l|})$, where $l = \mathrm{matchP}(j)$, $p_{j,\mathrm{top}(j)} = q_{l,x}$ is a shared edge (or $\mathrm{top}(j) = |P_j|$ and $x = 0$). Note that by the end of the while loop, we have $\mathrm{matchP}(j) \neq 0$ for all

---

Algorithm 4. merge-flow(u, v, r, i)

1: let $\mathcal{F}(u,r) = \{P_1, P_2, \ldots, P_i\}$, $\mathcal{F}(r,v) = \{Q_1, Q_2, \ldots, Q_i\}$ and $M = 0$.
2: **for** each $j = 1, 2, \ldots, i$ **do**
3:     label edges in P_j as $(p_{j,1}, p_{j,2}, \ldots, p_{j,|P_j|})$.
4:     label edges in Q_j as $(q_{j,1}, q_{j,2}, \ldots, q_{j,|Q_j|})$.
5: $\forall j \leq i$, let top$(j) = 0$, head$(j) = 0$, matchP$(j) = 0$, matchQ$(j) = 0$.
6: **while** $M < i$ **do**
7:     pick an arbitrary $j \leq i$ s.t. matchP$(j) = 0$.
8:     **if** top$(j) == |P_j|$ **then**
9:         matchP$(j) = -1$, $M = M + 1$.   ▷ finished if r is reached
10:    **else**
11:        top$(j) =$ top$(j) + 1$.
12:        **if** $p_{j,\text{top}(j)}$ has another label $q_{l,x}$ **then**
13:            **if** matchQ$(l) == 0$ **then**   ▷ match Q_l with P_j, at position x
14:                matchQ$(l) = j$, head$(l) = x$.
15:                matchP$(j) = l$, $M = M + 1$.
16:            **else if** $x >$ head(l) **then**
17:                matchP(matchQ$(l)) = 0$, matchP$(j) = l$.
18:                matchQ$(l) = j$, head$(l) = x$.
19: set $\mathcal{F}(u,v) = \emptyset$.   ▷ add i edge-disjoint paths in $\mathcal{F}(u,v)$
20: **for** each $j = 1, 2, \ldots, i$ **do**
21:    **if** matchP$(j) > 0$ **then**
22:        $l =$ matchP(j), $x =$ head(l).
23:        $\mathcal{F}(u,v) = \mathcal{F}(u,v) \cup \{(p_{j,1}, p_{j,2}, \ldots, p_{j,\text{top}(j)}, q_{l,x+1}, q_{l,x+2}, \ldots, q_{l,|Q_l|})\}$.
24:    **else**   ▷ matchP$(j) = -1$
25:        pick any $l \leq i$ such that matchQ$(l) == 0$, set matchQ$(l) = j$.
26:        $\mathcal{F}(u,v) = \mathcal{F}(u,v) \cup \{(p_{j,1}, p_{j,2}, \ldots, p_{j,|P_j|}, q_{l,1}, q_{l,2}, \ldots, q_{l,|Q_l|})\}$.
27: **return** $\mathcal{F}(u,v)$.

$j \leq i$ and we have formed a partial matching between P_j's and Q_l's: the number of P_j's with matchP$(j) = -1$ equals the number of Q_l's with matchQ$(l) = 0$. We can form an arbitrary matching between P_j's with matchP$(j) = -1$ and Q_l's with matchQ$(l) = 0$.

At any moment during the execution of Algorithm 4, we use $E_p = \cup_{j \in [i]} \{p_{j,1}, p_{j,2}, \ldots, p_{j,\text{top}(j)}\}$ to denote the set of "p-edges" that are already scanned and $E_q = \cup_{l \in [i]} \{q_{l,\text{head}(l)+1}, q_{l,\text{head}(l)+2}, \ldots, q_{l,|Q_l|}\}$. We show that $E_p \cap E_q = \emptyset$ during the whole execution.

Showing that $E_p \cap E_q = \emptyset$ is trivial before the while loop since $E_p = \emptyset$. During each while loop (line 7-18) we increase top(j) by one, for some $j \leq i$, which include one more edge e in E_p. If e is not shared, then it is safe to include e in E_p. Otherwise (line 13-18), assume $e = p_{j,\text{top}(j)} = q_{l,x}$. The algorithm makes sure that head$(l) \geq x$ at the end of this iteration of while loop, which exclude $e = q_{l,x}$ from E_q and maintains $E_p \cap E_q = \emptyset$. Hence we conclude that $E_p \cap E_q = \emptyset$ at the end of the whole while loop. Since the H_j's only use edges in E_p once and use edges in E_q once, all paths H_j's are edge-disjoint.

It is easy to check that each while loop can be executed in $O(1)$ time and increases the size of E_p by exactly one, the first part (line 1-18) of Algorithm 4 executes in $O(m)$ time. Since the execution time of the second part (line 19-27) of Algorithm 4 can be bounded by $O(|E_p \cup E_q|) = O(m)$, we conclude that the i edge-disjoint paths $\mathcal{F}(u, v)$ can be computed in $O(m)$ time by Algorithm 4. □

Hence instead of computing and storing $\Theta(n^2)$ current flows, when constructing Ω_i, we only need to know the current flow between nodes in the same $(i-1)$-ec block. Moreover, to represent the current flow between two nodes in the same block, we only need to store the current flows between the seed of the block and all other nodes in the block, which reduces the total number of current flows we need to update from $\Theta(n^2)$ to $O(n)$.

Algorithm 5. flow(u, v, r, i)

1: **if** $u \neq r$ and $v \neq r$ **then**   ▷ $\min\{\lambda(u, r), \lambda(v, r)\} = i - 1$
2:      merge-flow$(u, v, r, i - 1)$.   ▷ $O(m)$ time
3: **if** v is reachable from u in the residual graph with flow $\mathcal{F}(u, v)$ **then**
4:      find an augmenting path from u to v in the residual graph.   ▷ $O(m)$ time
5:      merge the path with $i - 1$ paths in $\mathcal{F}(u, v)$.
6:      **return** $\mathcal{F}(u, v)$.
7: **else**
8:      **return** NULL.

Lemma 4. *For all $i \geq 1$, after constructing Ω_{i-1}, we have*

1. *$\mathcal{F}(u, v)$ contains $i - 1$ edge-disjoint paths from u to v if $u \in B \in \Omega_{i-1}, v = r(B)$ or $v \in B \in \Omega_{i-1}, u = r(B)$.*
2. *given $u, v \in B \in \Omega_{i-1}$ and $r = r(B)$, Algorithm 5 computes a flow $\mathcal{F}(u, v)$ with $|\mathcal{F}(u, v)| = i$ or returns NULL if $\lambda(u, v) = i - 1$ in $O(m)$ time.*

Proof. We prove the above statements by induction on $i \geq 1$.

The base case $(i = 1)$ is trivial since (1) $\Omega_0 = \{V\}$ and $|\mathcal{F}(u, v)| = 0$ for all $u, v \in V$; (2) computing a path from u to v in G (if possible) can be done in $O(m)$ time. Now assume the statement is true for i and consider $i + 1$. By induction hypothesis, $\forall u \in B \in \Omega_i$, we have $|\mathcal{F}(u, r(B))| = |\mathcal{F}(r(B), u)| = i$. Moreover, for any two nodes u and v, line 4 of Algorithm 3 computes in $O(m)$ time i edge-disjoint paths from u to v and i edge-disjoint paths from v to u.

By Algorithm 3, it is easy to observe that $\cup_{S \in \mathcal{B}} S = B$, where $\mathcal{B}$ is the set of i-ec blocks returned after executing blocks(B, r, u, i). Note that in every execution of Algorithm 3, we increase the size of $\mathcal{B}$ by exactly one. Let S be the last i-ec block that is included into $\mathcal{B}$ in some execution of Algorithm 3. Hence to prove statement-(1) for $i+1$, we only need to show that for all $v \in S$, we have $|\mathcal{F}(r(S), v)| = |\mathcal{F}(v, r(S))| = i$, which is obvious since every node v is included in S only if v passes of the test in line 4.

Assuming that statement-(1) is true for $i + 1$, proving statement-(2) is straightforward. By Lemma 3, line 2 of Algorithm 5 executes in $O(m)$ time and hence we can get a current flow $|\mathcal{F}(u, v)| = i - 1$ from u to v in $O(m)$ time. Then in $O(m)$ time, we can either increase the flow from u to v by one, or conclude that $\lambda(u, v) = i - 1$.   □

4.3   Complexity of the Construction

We have described how to compute all child blocks of any block B using recursive algorithm (Algorithm 3) which uses a subroutine Algorithm 5 to compute a flow from u to v in $O(m)$ time. We will analyze the total running time for the construction of partitions $\Omega_1, \Omega_2, \ldots, \Omega_k, \Omega_{k+1}$ in this section.

Notice that given an $(i-1)$-ec block B, Algorithm 3 (run as blocks($B, r(B)$, $r(B), i$)) computes all child blocks of B by using Algorithm 5 $O(|B|)$ times. Moreover, we have the following important observations:

- in every call of Algorithm 5 (run as flow(u, v, r, i)), exactly one of u, v has been assigned to be the seed for an i-ec block.
- before every recursive call of Algorithm 3 (line 11), we compute the strongly connected components of the residual graph with flow $\mathcal{F}(u, v)$ (or $\mathcal{F}(v, u)$), where $\lambda(u, v) = i - 1$ and exactly one of u, v has been assigned to be the seed for an i-ec block.
- in every recursive call of Algorithm 3 (line 12), node v will be assigned to be the seed of some child block of B.

Charging Argument. By the above observation, we can charge the running time $O(m)$ for computing a current flow from u to v using Algorithm 5 and the running time $O(m)$ for computing the strongly connected components to the non-seed node. Hence any node u will not be charged after being assigned to be the seed of some block by the above argument.

Lemma 5 (Running Time on Each Node).   *By the above charging argument, every node u will only be charged a total running time of $O(km)$.*

Proof. First observe that Algorithm 5 (run as flow(u, v, r, i) and flow(v, u, r, i)) is always called twice at a time (for computing flows $|\mathcal{F}(u, v)| = |\mathcal{F}(v, u)| = i$ from u to v and from v to u) and the computation cost will be charged to the same (non-seed) node. Suppose the node being charged is u, we say that u is charged *with requirement i* in the above case.

Note that for every node u, after being charged with requirement i, either u is placed in some i-ec block, or we conclude that $\lambda(u, v) = i - 1$. Note that a non-seed node in an i-ec block will not be charged with requirement i again. Moreover, if the requirement is not satisfied (flow(u, v, r, i) = NULL or flow(v, u, r, i) = NULL), then after being charged an extra $O(m)$ running time for computing the strongly connected components, u will be immediately assigned to be a seed.

Since all computation cost between u and v afterwards will be charged to the non-seed node v, we conclude that every node u will only be charged a total running time of $O(km)$.   □

Proof of **Theorem** 2: To prove Theorem 2, it suffices to show that partitions $\Omega_1, \Omega_2, \ldots, \Omega_k, \Omega_{k+1}$ can be constructed in $O(kmn)$ time, by Lemma 1.

By Lemma 2, we can use Algorithm 3 (block($B, r(B), r(B), i$)) to compute all the child blocks of any block B. Hence Algorithm 2 correctly computes partitions $\Omega_1, \Omega_2, \ldots, \Omega_k, \Omega_{k+1}$. Thus we only need to bound the running time.

By Lemma 5, the total running time charged on nodes is at most $O(kmn)$. Since the total running time not charged on any node, i.e., while loops and for loops, for the computation of partitions can be bounded by $O(\Delta n)$, we conclude that the computation of partitions $\Omega_1, \Omega_2, \ldots, \Omega_k, \Omega_{k+1}$ (and hence the k-edge-connectivity tree) of any graph G can be done in $O(kmn)$ time. $\square$

References

1. Benczúr, A.A.: Counterexamples for directed and node capacitated cut-trees. SIAM J. Comput. **24**(3), 505–510 (1995)
2. Cheng, C.-K., Hu, T.C.: Ancestor tree for arbitrary multi-terminal cut functions. In: Proceedings of the 1st Integer Programming and Combinatorial Optimization Conference, May 28–30 1990, Waterloo, Ontorio, Canada, pp. 115–127 (1990)
3. Cheung, H.Y., Lau, L.C., Leung, K.M.: Graph connectivities, network coding, and expander graphs. In: Ostrovsky, R. (eds.) IEEE 52nd Annual Symposium on Foundations of Computer Science, FOCS 2011, 22–25 October, Palm Springs, CA, USA, 2011, pp. 190–199. IEEE Computer Society (2011)
4. Dinits, E.A.: Algorithm of solution to problem of maximum flow in network with power estimates. Doklady Akademii Nauk SSSR **194**(4), 754 (1970)
5. Even, S., Tarjan, R.E.: Network flow and testing graph connectivity. SIAM J. Comput. **4**(4), 507–518 (1975)
6. Georgiadis, L., Italiano, G.F., Laura, L., Parotsidis, N.: 2-edge connectivity in directed graphs. In: Indyk, P. (eds.) Proceedings of the Twenty-Sixth Annual ACM-SIAM Symposium on Discrete Algorithms, SODA 2015, 4–6 January 2015, San Diego, CA, USA, pp. 1988–2005. SIAM (2015)
7. Gomory, R.E., Hu, T.C.: Multi-terminal network flows. J. Soc. Ind. Appl. Math. **9**(4), 551–570 (1961)
8. Gusfield, D.: Very simple methods for all pairs network flow analysis. SIAM J. Comput. **19**(1), 143–155 (1990)
9. Hariharan, R., Kavitha, T., Panigrahi, D.: Efficient algorithms for computing all low st edge connectivities and related problems. In: Proceedings of the Eighteenth Annual ACM-SIAM Symposium on Discrete Algorithms, pp. 127–136. Society for Industrial and Applied Mathematics (2007)
10. Hariharan, R., Kavitha, T., Panigrahi, D., Bhalgat, A.: An o (mn) gomory-hu tree construction algorithm for unweighted graphs. In: Proceedings of the Thirty-Ninth Annual ACM Symposium on Theory of Computing, pp. 605–614. ACM (2007)
11. Lee, Y.T., Sidford, A.: Path finding methods for linear programming: Solving linear programs in õ (vrank) iterations and faster algorithms for maximum flow. In: 2014 IEEE 55th Annual Symposium on Foundations of Computer Science (FOCS), pp. 424–433. IEEE (2014)
12. Menger, K.: Zur allgemeinen kurventheorie. Fundamenta Mathematicae **10**(1), 96–115 (1927)
13. Schnorr, C.-P.: Bottlenecks and edge connectivity in unsymmetrical networks. SIAM J. Comput. **8**(2), 265–274 (1979)
14. Tarjan, R.E.: Depth-first search and linear graph algorithms. SIAM J. Comput. **1**(2), 146–160 (1972)

Maximum Minimal Vertex Cover Parameterized by Vertex Cover

Meirav Zehavi[(⊠)]

Department of Computer Science, Technion IIT, 32000 Haifa, Israel
meizeh@cs.technion.ac.il

Abstract. The parameterized complexity of problems is often studied with respect to the size of their optimal solutions. However, for a maximization problem, the size of the optimal solution can be very large, rendering algorithms parameterized by it inefficient. Therefore, we suggest to study the parameterized complexity of maximization problems with respect to the size of the optimal solutions to their *minimization* versions. We examine this suggestion by considering the MAXIMUM MINIMAL VERTEX COVER (MMVC) problem, whose minimization version, VERTEX COVER, is one of the most studied problems in the field of Parameterized Complexity. Our main contribution is a parameterized approximation algorithm for MMVC, including its weighted variant. We also give conditional lower bounds for the running times of algorithms for MMVC and its weighted variant.

1 Introduction

The parameterized complexity of problems is often studied with respect to the size of their optimal solutions. However, for a maximization problem, the size of the optimal solution can be very large, rendering algorithms parameterized by it inefficient. Therefore, we suggest to study the parameterized complexity of maximization problems with respect to the size of the optimal solutions to their *minimization* versions. Given a maximization problem, the optimal solution to its minimization version might not only be significantly smaller, but it might also be possible to efficiently compute it by using some well-known parameterized algorithm—in such cases, one can know in advance if for a given instance of the maximization problem, the size of the optimal solution to the minimization version is a good choice as a parameter. Furthermore, assuming that an optimal solution to the minimization version can be efficiently computed, one may use it to solve the maximization problem; indeed, the optimal solution to the maximization problem and the optimal solution to its minimization version may share useful, important properties.

We examine this suggestion by studying the MAXIMUM MINIMAL VERTEX COVER (MMVC) problem. This is a natural choice—the minimization version of MMVC is the classic VERTEX COVER (VC) problem, one of the most studied problems in the field of Parameterized Complexity. Given a graph $G = (V, E)$ and a weight function $w : V \to \mathbb{R}^{\geq 1}$, WEIGHTED MMVC (WMMVC) seeks the

© Springer-Verlag Berlin Heidelberg 2015
G.F. Italiano et al. (Eds.): MFCS 2015, Part II, LNCS 9235, pp. 589–600, 2015.
DOI: 10.1007/978-3-662-48054-0_49

maximum weight of a set of vertices that is a *minimal* vertex cover of G. The MMVC problem is the special case of WMMVC where $w(v) = 1$ for all $v \in V$.

Notation: Let vc (vc_w) denote the size (weight) of a minimum vertex cover (minimum-weight vertex cover) of G, and let opt (opt_w) denote the size (weight) of a maximum minimal vertex cover (maximum-weight minimal vertex cover) of G. Clearly, $vc \leq \min\{vc_w, opt\} \leq \max\{vc_w, opt\} \leq opt_w$. Observe that the gap between $vc_w = \max\{vc, vc_w\}$ and $opt = \min\{opt, opt_w\}$ can be very large. For example, in the case of MMVC, if G is a star, then $vc_w = 1$ while $opt = |V| - 1$.

A problem is *fixed-parameter tractable (FPT)* with respect to a parameter k if it can be solved in time $\mathcal{O}^*(f(k))$ for some function f, where $\mathcal{O}^*$ hides factors polynomial in the input size. Given $v \in V$, $N(v)$ denotes the neighbor set of v. Let Δ (M) denote the maximum degree (weight) of a vertex in G. Given $U \subseteq V$, let $G[U]$ denote the subgraph of G induced by U. Finally, let γ denote the smallest constant such that it is known how to solve VC in time $\mathcal{O}^*(\gamma^{vc})$, using polynomial space. Currently, $\gamma < 1.274$ [12].

1.1   Related Work

MMVC: Boria et al. [4] show that MMVC is solvable in time $\mathcal{O}^*(3^m)$ and polynomial space, and that WMMVC is solvable in time and space $\mathcal{O}^*(2^{tw})$, where m is the size of a maximum matching of G, and tw is the treewidth of G. Since $\max\{m, tw\} \leq vc$ (see, e.g., [19]), this shows that WMMVC is FPT with respect to vc. Moreover, they prove that MMVC is solvable in time $\mathcal{O}^*(1.5874^{opt})$ and polynomial space, where the running time can be improved if one is interested in approximation.[1] Boria et al. [4] also prove that for any fixed $\epsilon > 0$, MMVC is inapproximable within ratios $O(|V|^{\epsilon - \frac{1}{2}})$ and $O(\Delta^{\epsilon - 1})$, unless P=NP. They complement this result by proving that MMVC is approximable within ratios $|V|^{-\frac{1}{2}}$ and $\frac{3}{2\Delta}$ in polynomial time. Recently, Bonnet and Paschos [3] and Bonnet et al. [2] obtained results related to the inapproximability of MMVC in subexponential time. Furthermore, Bonnet et al. [2] prove that for any $1 < r \leq |V|^{\frac{1}{2}}$, MMVC is approximable within ratio $\frac{1}{r}$ in time $\mathcal{O}^*(2^{\frac{|V|}{r^2}})$.

MMVC is the symmetric version of the well-known MINIMUM INDEPENDENT DOMINATING SET (MIDS) problem (also known as the MINIMUM MAXIMAL INDEPENDENT SET problem), where one seeks the minimum size of an independent set that is a dominating set in a given graph [4]. MIDS (and therefore also MMVC) has applications to wireless ad hoc networks (see, e.g., [24])

Vertex Cover: VC is one of the first problems shown to be FPT. In the past two decades, it enjoyed a race towards obtaining the fastest parameterized algorithm (see [1,7,8,10,12,17,18,28–30]). The best parameterized algorithm, due to Chen et al. [12], has running time $\mathcal{O}^*(1.274^{vc})$, using polynomial space. In a similar race [10,11,13,25,31,33], focusing on the case where $\Delta = 3$, the current winner is an algorithm by Issac et al. [25], whose running time is $\mathcal{O}^*(1.153^{vc})$.

---

[1] For example, they show that one can guarantee the approximation ratios 0.1 and 0.4 in times $\mathcal{O}^*(1.162^{opt})$ and $\mathcal{O}^*(1.552^{opt})$, respectively.

For WEIGHTED VC (WVC), parameterized algorithms were given in [21,22,29, 32]. The fastest ones (in [32]) use time $\mathcal{O}^*(1.381^{vc_w})$ and polynomial space, time and space $\mathcal{O}^*(1.347^{vc_w})$, and time $\mathcal{O}^*(1.443^{vc})$ and polynomial space.

Kernels for VC and WVC were given in [10,14,26], and results related to the parameterized approximability of VC were also examined in the literature (see, e.g., [5,6,20]). Finally, in the context of Parameterized Complexity, we would also like to note that vc is a parameter of interest; indeed, apart from VC, there are other problems whose parameterized complexity was studied with respect to this parameter (see, e.g., [9,23,27]).

1.2 Our Contribution

While it is easy to see that WMMVC is solvable in time $\mathcal{O}^*(2^{vc})$ and polynomial space (see Sect. 2), we observe that this result might be essentially tight. More precisely, we show that even if G is a bipartite graph and $w(v) \in \{1, 1 + \frac{1}{|V|}\}$ for all $v \in V$, an algorithm that solves WMMVC in time $\mathcal{O}^*((2 - \epsilon)^{vc_w})$, which upper bounds $\mathcal{O}^*((2 - \epsilon)^{vc})$, contradicts the SETH (Strong Exponential Time Hypothesis). We also show that even if G is a bipartite graph, an algorithm that solves MMVC in time $\mathcal{O}^*((2 - \epsilon)^{\frac{vc}{2}})$ contradicts the SETH.

Then, we turn to present our main contribution, ALG, which is a parameterized approximation algorithm for WMMVC with respect to the parameter vc. We prove the following theorem, where α is a user-controlled parameter that corresponds to a tradeoff between time and approximation ratio. Recall that M is the maximum weight of a vertex in G.

Theorem 1. *For any* $\alpha < \dfrac{1}{2 - \frac{1}{M+1}}$ *such that* $\dfrac{1}{x^x(1-x)^{1-x}} \geq 3^{\frac{1}{3}}$ *where* $x = 1 - \dfrac{1-\alpha}{M(2\alpha-1)+1-\alpha}$, ALG *runs in time* $\mathcal{O}^*((\dfrac{1}{x^x(1-x)^{1-x}})^{vc})$, *returning a minimal vertex cover of weight at least* $\alpha \cdot opt_w$. ALG *has a polynomial space complexity.*

In particular, the result holds for any $\frac{1}{2} < \dfrac{1}{2 - \frac{1}{7.35841 \cdot M+1}} \leq \alpha < \dfrac{1}{2 - \frac{1}{M+1}}$. For the smallest possible α, the time complexity is bounded by $\mathcal{O}^*(3^{\frac{vc}{3}}) < \mathcal{O}^*(1.44225^{vc})$, and for *any* fixed $\epsilon' > 0$ such that $\alpha \leq \frac{1}{2 - \frac{1}{M+1}} - \epsilon'$, there is a fixed $\epsilon > 0$ such that the time complexity is bounded by $\mathcal{O}^*((2 - \epsilon)^{vc})$. In other words, as α gradually increases, x gradually increases (from 0.11964 to $\frac{1}{2}$),[2] and consequently, the running time gradually increases (from $\mathcal{O}^*(3^{\frac{vc}{3}})$ to $\mathcal{O}^*(2^{vc})$). The approximation ratio(s) of ALG cannot be achieved by polynomial-time algorithms unless P=NP (see Sect. 1.1), and its running time(s) cannot be matched by exact parameterized algorithms unless the SETH fails.

ALG is a mix of two procedures that rely on the bounded search tree technique (see Sect. 3.1). The branching vectors of the first procedure are analyzed with

---

[2] Indeed, $x \approx 0.11964$ if $\frac{1}{x^x(1-x)^{1-x}} = 3^{\frac{1}{3}}$, and $x = 1 - \frac{1-\alpha}{M(2\alpha-1)+1-\alpha} = \frac{1}{2}$ if $\alpha = \frac{1}{2 - \frac{1}{M+1}}$.

respect to the size of *a minimum vertex cover of a minimum vertex cover* of G. Another interesting feature of this procedure is that once it reaches a leaf of the search tree, it does not immediately return a result, but to obtain the desired approximation ratio, it first performs a computation that is, in a sense, an exhaustive search. The design of our second procedure integrates rules that are part of the algorithm for VC by Peiselt [30]. Since ALG can be used to solve MMVC, in which case $M = 1$, we have the following corollary.

Corollary 1. *For any* $\alpha < \frac{2}{3}$ *such that* $\frac{1}{x^x(1-x)^{1-x}} \geq 3^{\frac{1}{3}}$ *where* $x = 2 - \frac{1}{\alpha}$,[3] ALG *runs in time* $\mathcal{O}^*((\frac{1}{x^x(1-x)^{1-x}})^{vc})$, *returning a minimal vertex cover of size at least* $\alpha \cdot opt$. ALG *has a polynomial space complexity.*

2   Upper and Lower Bounds

This section presents upper and conditional lower bounds related to the parameterized complexity of WMMVC and MMVC with respect to vc and vc_w. Recall that $vc \leq vc_w$, and that MMVC is a special case of WMMVC.

Observation 1. WMMVC *is solvable in time* $\mathcal{O}^*(2^{vc})$ *and polynomial space.*

Proof. The algorithm is as follows. First, compute a minimum vertex cover S of G in time $\mathcal{O}^*(\gamma^{vc})$ using polynomial space, and initialize A to S. Then, for every subset $S' \subseteq S$, if $B = S' \cup (\bigcup_{v \in S \setminus S'} N(v))$ is a minimal vertex cover of weight larger than the weight of A, update A to store B. Finally, return A.

Clearly, we return a minimal vertex cover. Let A^* be an optimal solution. Consider the iteration where $S' = A \cap S$. Then, since A^* is a vertex cover, $B \subseteq A^*$. Suppose, by way of contradiction, that there exists $v \in A^* \setminus B$. The vertex v does not belong to S (since $A \cap S \subseteq B$). Moreover, it should have a neighbor outside A^* (since A^* is a *minimal* vertex cover). Thus, since S is a vertex cover, v has a neighbor in $S \setminus A^*$. This implies that $v \in \bigcup_{u \in S \setminus S'} N(u)$, contradicting the assumption that $v \notin B$. Therefore, $B = A^*$. Thus, the algorithm is correct, and it clearly has the desired time and space complexities.   □

Now, we observe that even in a restricted setting, the algorithm above is essentially optimal under the SETH.

Lemma 1. *For any fixed* $\epsilon > 0$, WMMVC *in bipartite graphs, where* $w(v) \in \{1, 1 + \frac{1}{|V|}\}$ *for all* $v \in V$, *cannot be solved in time* $\mathcal{O}^*((2 - \epsilon)^{vc_w})$ *unless the SETH fails.*

Proof. Fix $\epsilon > 0$. Suppose, by way of contradiction, that there exists an algorithm, A, that solves WMMVC in the restricted setting in time $\mathcal{O}^*((2 - \epsilon)^{vc_w})$. We aim to show that this implies that there exists an algorithm that solves the HITTING SET (HS) problem in time $\mathcal{O}^*((2 - \epsilon)^n)$, which contradicts the SETH [15]. In HS, we are given an n-element set U, along with a family of subsets of

---

[3] In particular, the result holds for any $0.53183 \leq \alpha < \frac{2}{3}$.

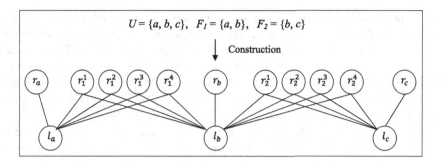

$U = \{a, b, c\}$, $F_1 = \{a, b\}$, $F_2 = \{b, c\}$

Construction

Fig. 1. The construction in the proof of Lemma 1.

U, $\mathcal{F} = \{F_1, F_2, \ldots, F_m\}$, and the goal is to find the minimum size of a subset $U' \subseteq U$ that is a hitting set (i.e., U' contains at least one element from every set in $\mathcal{F}$).

We next construct an instance $(G = (V, E), w : V \to \mathbb{R}^{\geq 1})$ of WMMVC in the restricted setting:

- $R_1 = \{r_u : u \in U\}$, and $R_2 = \{r_i^c : c \in \{1, \ldots, n+1\}, i \in \{1, \ldots, m\}\}$.
- $L = \{l_u : u \in U\}$, and $R = R_1 \cup R_2$.
- $V = L \cup R$.
- $E = \{\{l_u, r_u\} : u \in U\} \cup \{\{l_u, r_i^c\} : u \in F_i, i \in \{1, \ldots, m\}, c \in \{1, \ldots, n+1\}\}$.
- $\forall v \in L : w(v) = 1 + \frac{1}{|V|}$.
- $\forall v \in R : w(v) = 1$.

An illustrated example is given in Fig. 1. It is enough to show that (1) $vc_w \leq n+1$, and (2) the solution for $(U, \mathcal{F})$ is q *iff* the solution for (G, w) is $\frac{n-q}{|V|} + |R|$. Indeed, this implies that we can solve HS by constructing the above instance in polynomial time, running A in time $\mathcal{O}^*((2 - \epsilon)^n)$ (since $vc_w \leq n+1$), obtaining an answer of the form $\frac{n-q}{|V|} + |R|$, and returning q.

First, observe that L is a vertex cover of G (since every edge has exactly one endpoint in L), and that $|L| = n$. Therefore, $vc_w \leq w(L) = (1 + \frac{1}{|V|})|L| \leq n+1$.

Now, we turn to prove the second item. For the first direction, let q be the solution to $(U, \mathcal{F})$, and let U' be a corresponding hitting set of size q. Consider the vertex set $S = \{l_u : u \in U \setminus U'\} \cup \{r_u : u \in U'\} \cup R_2$. By the definition of w, the weight of S is $(1 + \frac{1}{|V|})|U \setminus U'| + |U'| + |R_2| = \frac{1}{|V|}|U \setminus U'| + |R_1| + |R_2| = \frac{n-q}{|V|} + |R|$. The set S is a vertex cover: since $R_2 \subseteq S$, every edge in G that does not have an endpoint in R_2 is of the form $\{l_u, r_u\}$, and for every $u \in U$ either $l_u \in S$ (if $u \notin U'$) or $r_u \in S$ (if $u \in U'$). Moreover, S is a *minimal* vertex cover. Indeed, we cannot remove any vertex $l_u \in S \cap L$ or $r_u \in S \cap R_1$ and still have a vertex cover, since then the edge of the form $\{l_u, r_u\}$ is not covered. Also, we cannot remove any vertex $r_i^c \in R_2$, since there is a vertex $l_u \notin S$ such that $\{l_u, r_i^c\} \in E$ (to see this, observe that because U' is a hitting set, there is a vertex $u \in U' \cap F_i$, which corresponds to the required vertex l_u).

For the second direction, let p be the solution to (G, w), and let S be a corresponding minimal vertex cover of weight p. Clearly, $p \geq w(R) = n + m(n+1)$, since R is a minimal vertex cover of G. Observe that for all $u \in U$, by the definition of G and since S is a minimal vertex cover, exactly one among the vertices l_u and r_u is in S. Suppose that there exists $r_i^c \in R_2 \setminus S$. Then, for all $u \in F_i$, we have that $l_u \in S$ (by the definition of G and since S is a vertex cover), which implies that for all $c \in \{1, \dots, n+1\}$, we have that $r_i^c \notin S$ (since S is a $minimal$ vertex cover). Thus, $p = w(S \cap (L \cup R_1)) + |S \cap R_2| \leq n(1 + \frac{1}{|V|}) + (m-1)(n+1) < n + m(n+1)$, which is a contradiction. Therefore, $R_2 \subseteq S$.

Denote $U' = \{u : r_u \in S \cap R_1\}$. By the above discussion and the definition of w, $|S| = \frac{|V \setminus R_2|}{2} + |R_2| = |R_1| + |R_2| = |R|$, and $p - |S| = \frac{1}{|V|}|S \cap L| = \frac{1}{|V|}(n - |S \cap R_1|)$. Denoting $|U'| = q$, we have that $p = \frac{n-q}{|V|} + |R|$. Thus, it remains to show that U' is a hitting set. Suppose, by way of contradiction, that U' is not a hitting set. Thus, there exists $F_i \in \mathcal{F}$ such that for all $u \in F_i$, we have that $u \notin U'$. By the definition of U', this implies that for all $u \in F_i$, we have that $l_u \in S$. Thus, $N(r_i^1) \subseteq S$, while $r_i^1 \in S$ (since we have shown that $R_2 \subseteq S$), which is a contradiction to the fact that S is a $minimal$ vertex cover. $\square$

Next, we also give a conditional lower bound for MMVC. Informally, to this end, we modify the above proof by adding a copy l_u' of every vertex $l_u \in L$, which is only attached to its "mirror vertex", r_u, in R_1. While previously the weights $1 + \frac{1}{|V|}$ encouraged us to choose many vertices from L, now the copies encourage us to choose these vertices.

Lemma 2. *For any fixed $\epsilon > 0$, MMVC in bipartite graphs cannot be solved in time $\mathcal{O}^*((2-\epsilon)^{\frac{vc}{2}})$ unless the SETH fails.*

3   A Parameterized Approximation Algorithm

In this section we develop ALG (see Theorem 1). When referring to α, suppose that it is chosen as required in Theorem 1, and define x accordingly. The algorithm is a mix of two procedures, ProcedureA and ProcedureB, that are based on the bounded search tree technique. These procedures are developed in the following subsections.[4] For these procedures, we will prove the following lemmas.

Lemma 3. *Let U be a minimum(-size) vertex cover of G, and let U' be a minimum(-size) vertex cover of $G[U]$. Moreover, suppose that $|U'| \leq \frac{vc}{2}$.*
    Then, ProcedureA$(G, w, \alpha, U, U', \emptyset, \emptyset)$ runs in time $\mathcal{O}^*((\frac{1}{x^x(1-x)^{1-x}})^{vc})$, uses polynomial space and returns a minimal vertex cover of weight at least $\alpha \cdot opt_w$.

Lemma 4. *Let U be a minimum(-size) vertex cover of G such that the size of a minimum(-size) vertex cover of $G[U]$ is larger than $\frac{vc}{2}$.*
    Then, ProcedureB$(G, w, U, \emptyset, \emptyset)$ runs in time $\mathcal{O}^*(3^{\frac{vc}{3}})$, uses polynomial space and returns a minimal vertex cover of weight at least $\dfrac{1}{2 - \frac{1}{M+1}} \cdot opt_w$.

---

[4] ProcedureA could also be developed without using recursion; however, relying on the bounded search tree technique simplifies the presentation.

Having these procedures, we give the pseudocode of ALG below. The algorithm computes a minimum vertex cover U' of a minimum vertex cover U, solving the given instance by calling either ProcedureA (if $|U'| \leq \frac{vc}{2}$) or ProcedureB (if $|U'| > \frac{vc}{2}$).

Algorithm 1. ALG$(G = (V, E), w : V \rightarrow \mathbb{R}^{\geq 1}, \alpha)$

1: Compute a minimum vertex cover U of G in time $\mathcal{O}^*(\gamma^{vc})$ and polynomial space.
2: Compute a minimum vertex cover U' of $G[U]$ in time $\mathcal{O}^*(\gamma^{vc})$ and polynomial space.

3: **if** $|U'| \leq \frac{vc}{2}$ **then**
4:     Return ProcedureA$(G, w, \alpha, U, U', \emptyset, \emptyset)$.
5: **else**
6:     Return ProcedureB$(G, w, U, \emptyset, \emptyset)$.
7: **end if**

Now, we turn to prove Theorem 1.

Proof. The correctness of the approximation ratio immediately follows from Lemmas 3 and 4. Observe that the computation of U' can indeed be performed in time $\mathcal{O}^*(\gamma^{vc})$ since $U' \subseteq U$, and therefore $|U'| \leq |U| \leq vc$. Moreover, since $\gamma < 1.274$, by Lemmas 3 and 4, the running time is bounded by $\mathcal{O}^*(\max\{(\frac{1}{x^x(1-x)^{1-x}})^{vc}, 3^{\frac{vc}{3}}\})$. Since α is chosen such that $(\frac{1}{x^x(1-x)^{1-x}})^{vc} \geq 3^{\frac{vc}{3}}$, the above running time is bounded by $\mathcal{O}^*((\frac{1}{x^x(1-x)^{1-x}})^{vc})$. $\qquad\square$

In the rest of this section, we give necessary information on the bounded search tree technique (Sect. 3.1), after which we develop ProcedureA (Sect. 3.2) and ProcedureB (Sect. 3.3).

3.1 The Bounded Search Tree Technique

Bounded search trees form a fundamental technique in the design of recursive parameterized algorithms (see [16]). Roughly speaking, in applying this technique, one defines a list of rules of the form Rule X. [condition] action, where X is the number of the rule in the list. At each recursive call (i.e., a node in the search tree), the algorithm performs the action of the first rule whose condition is satisfied. If by performing an action, the algorithm recursively calls itself at least twice, the rule is a *branching rule*, and otherwise it is a *reduction rule*. We only consider actions that increase neither the parameter nor the size of the instance, and decrease at least one of them. Observe that, at any given time, we only store the path from the current node to the root of the search tree (rather than the entire tree).

The running time of the algorithm can be bounded as follows. Suppose that the algorithm executes a branching rule where it recursively calls itself ℓ times, such that in the ith call, the current value of the parameter decreases by b_i. Then, $(b_1, b_2, \ldots, b_\ell)$ is called the *branching vector* of this rule. We say that β is the *root* of $(b_1, b_2, \ldots, b_\ell)$ if it is the (unique) positive real root of $x^{b^*} =$

$x^{b^*-b_1} + x^{b^*-b_2} + \ldots + x^{b^*-b_\ell}$, where $b^* = \max\{b_1, b_2, \ldots, b_\ell\}$. If $b > 0$ is the initial value of the parameter, and the algorithm (a) returns a result when (or before) the parameter is negative, (b) only executes branching rules whose roots are bounded by a constant c, and (c) only executes rules associated with actions performed in polynomial time, then its running time is bounded by $\mathcal{O}^*(c^b)$.

In some of the leaves of a search tree corresponding to our first procedure, we execute rules associated with actions that are not performed in polynomial time (as required in the condition (c) above). We show that for every such leaf ℓ in the search tree, we execute an action that can be performed in time $\mathcal{O}^*(g(\ell))$ for some function g. Then, letting L denote the set of leaves in the search tree whose actions are not performed in polynomial time, we have that the running time of the algorithm is bounded by $\mathcal{O}^*(c^b + \sum_{\ell \in L} g(\ell))$.

3.2 ProcedureA: The Proof of Lemma 3

The procedure ProcedureA is based on the bounded search tree technique. Each call is of the form ProcedureA$(G, w, \alpha, U, U', I, O)$, where G, w, α, U and U' always remain the parameters with whom the procedure was called by ALG, while I and O are disjoint subsets of U to which ProcedureA adds elements as the execution progresses (initially, $I = O = \emptyset$). Roughly speaking, the sets I and O indicate that currently we are only interested in examining minimal vertex covers that contain all of the vertices in I and none of the vertices in O. Formally, we prove the following result.

Lemma 5. ProcedureA *returns a minimal vertex cover S that satisfies the following condition:*

- *If there is a minimal vertex cover S^* of weight opt_w such that $I \subseteq S^*$ and $O \cap S^* = \emptyset$, then the weight of S is at least $\alpha \cdot opt_w$.*

Moreover, each leaf, associated with an instance $(G, w, \alpha, U, U', I', O')$, corresponds to a unique pair (I', O'), and its action can be performed in polynomial time if $I' \cup O' \neq U'$, and in time $\mathcal{O}^*(|\{\widetilde{U} \subseteq U : I' \subseteq \widetilde{U}, O' \cap \widetilde{U} = \emptyset, |\widetilde{U}| \geq (1-x)vc\}|)$ otherwise.

Let $vc' = |U'|$. To ensure that ProcedureA runs in time $\mathcal{O}^*((\frac{1}{x^x(1-x)^{1-x}})^{vc})$, we propose the following measure:

Measure: $vc' - |U' \cap (I \cup O)|$.

Next, we present each rule within a call ProcedureA$(G, w, \alpha, U, U', I, O)$. After presenting a rule, we argue its correctness (see Lemma 5). Since initially $I = O = \emptyset$, we thus have that ProcedureA guarantees the desired approximation ratio. For each branching rule, we also give the root of the corresponding branching vector (with respect to the measure above). We ensure that (1) the largest root we shall get is 2, (2) the procedure stops calling itself recursively, at the latest, once the measure drops to 0, and (3) actions not associated with leaves can be performed in polynomial time. Observe that initially the measure is vc'. Thus, as explained in Sect. 3.1, the running time of ProcedureA is bounded by

$\mathcal{O}^*(2^{vc'} + \sum\limits_{(I',O')\in\mathcal{P}} |\{\widetilde{U} \subseteq U : I' \subseteq \widetilde{U}, O' \cap \widetilde{U} = \emptyset, |\widetilde{U}| \geq (1-x)vc\}|)$, where $\mathcal{P}$ is the set of all partitions of U' into two sets. This running time is bounded by $\mathcal{O}^*(2^{vc'} + |\{\widetilde{U} \subseteq U : |\widetilde{U}| \geq (1-x)vc\}|) = \mathcal{O}^*(2^{vc'} + \max\limits_{i=(1-x)vc}^{vc} \binom{vc}{i})$. Since $x <$ $\frac{1}{2}$,[5] the running time is further bounded by $\mathcal{O}^*(2^{vc'} + (\frac{1}{x^x(1-x)^{1-x}})^{vc})$. Now, as $vc' \leq \frac{vc}{2}$ and $\frac{1}{x^x(1-x)^{1-x}} \geq 3^{\frac{1}{3}} > 2^{\frac{1}{2}}$, we have that $2^{vc'} \leq \mathcal{O}^*((\frac{1}{x^x(1-x)^{1-x}})^{vc})$. Thus, we have the desired bound $\mathcal{O}^*((\frac{1}{x^x(1-x)^{1-x}})^{vc})$ for the running time, concluding the correctness of Lemma 5, which, in turn, implies that Lemma 3 is correct.

Reduction Rule 1. [There is $v \in O$ such that $N(v) \cap O \neq \emptyset$] Return U.

In this case there is no vertex cover that does not contain any vertex from O, and therefore it is possible to return an arbitrary minimal vertex cover. The action can clearly be performed in polynomial time.

Reduction Rule 2. [There is $v \in X$ such that $N(v) \subseteq X$, where $X = I \cup (\bigcup_{u\in O} N(u))$] Return U.

Observe that any vertex cover that does not contain any vertex from O, must contain all the neighbors of the vertices in O. Thus, any vertex cover that contains all the vertices in I and none of the vertices in O, also contains the vertex v and all of its neighbors; therefore, it is not a *minimal* vertex cover. Thus, it is possible to return an arbitrary minimal vertex cover. The action can clearly be performed in polynomial time.

Reduction Rule 3. [$U' = I \cup O$] Perform the computation of Algorithm 2.

First, note that this rule ensures that at the latest, ProcedureA stops calling itself recursively once the measure drops to 0. Furthermore, by the pseudocode, the action can be performed in time $\mathcal{O}^*(|\mathcal{F}|)$ and polynomial space, which by the definition of $\mathcal{F}$, is the desired time. It remains to prove that Lemma 5 is correct.
    We begin by considering the set A. Since $U' = I \cup O$ is a vertex cover of $G[U]$ and the previous rules were not applied, we have that A is vertex cover of $G[U]$ such that $I \subseteq A$ and $O \cap A = \emptyset$. Thus, by its definition, A' is a *minimal* vertex cover of $G[U]$ such that $A' \subseteq A$. Thus, since U is a vertex cover, every edge either has both endpoints in U, in which case it has an endpoint in A', or it has exactly one endpoint in U and exactly one endpoint in $V \setminus U$, in which case it has an endpoint that is a vertex in A' or a neighbor in $V \setminus U$ of a vertex in $U \setminus A'$. Therefore, $\widetilde{A}$ is a vertex cover. Moreover, every vertex in A' has a neighbor in $U \setminus A'$ (by the minimality of A') and every vertex in $(\bigcup_{v\in U\setminus A'} N(v) \setminus U)$, by definition, has a neighbor in $U \setminus A'$. Thus, we overall have that $\widetilde{A}$ is a minimal vertex cover such that $\widetilde{A} \cap U \subseteq A$.

---

[5] This claim follows from the definition of x and since $\alpha < \dfrac{1}{2 - \frac{1}{M+1}}$.

Algorithm 2. Reduction Rule 3

Compute $\widetilde{A}$:

1: $A \Leftarrow I \cup (\bigcup_{v \in O} N(v) \cap U)$. While there is $v \in A$ s.t. $N(v) \cap U \subseteq A$, choose such a vertex (arbitrarily) and remove it from A. Let A' be the set obtained at the end of this process.

2: $\widetilde{A} \Leftarrow A' \cup (\bigcup_{v \in U \setminus A'} N(v) \setminus U)$.

Compute B:

3: Initialize $B = U$.

4: **for all** $F \in \mathcal{F}$, where $\mathcal{F} = \{\widetilde{U} \subseteq U : I \subseteq \widetilde{U}, O \cap \widetilde{U} = \emptyset, |\widetilde{U}| \geq (1-x)vc\}$ **do**

5:     $B_F \Leftarrow F \cup (\bigcup_{v \in U \setminus F} N(v))$.

6:     **if** B_F is a *minimal* vertex cover **then**

7:         **if** $w(B_F) > w(B)$ **then** $B \Leftarrow B_F$. **end if**

8:     **end if**

9: **end for**

10: Return the set of maximum weight among $\widetilde{A}$ and B.

Since $\widetilde{A}$ is a minimal vertex cover, by the pseudocode, we return a weight, W, of a minimal vertex cover. Assume that there is a minimal vertex cover S^* of weight opt_w such that $I \subseteq S^*$ and $O \cap S^* = \emptyset$. Now, to prove Lemma 5, it is sufficient to show that $W \geq \alpha \cdot opt_w$. Denote $F^* = S^* \cap U$. Since S^* is a minimal vertex cover, we have that $S^* = F^* \cup (\bigcup_{v \in U \setminus F^*} N(v))$. If S^* contains at least $(1-x)vc$ elements from U, there is an iteration where we examine $F = F^*$, in which case $B_{F^*} = S^*$, and therefore we return opt_w. Thus, we next suppose that $|S^* \cap U| < (1-x)vc$. Since B is initially U, to prove Lemma 5, it is now sufficient to show that $\max\{w(\widetilde{A}), w(U)\} \geq \alpha \cdot w(S^*)$.

Since S^* is a vertex cover such that $I \subseteq S^*$ and $O \cap S^* = \emptyset$, we have that $\widetilde{A} \cap U \subseteq A \subseteq F^*$. By the definition of $\widetilde{A}$ and since S^* is a *minimal* vertex cover, this implies that $S^* \setminus U \subseteq \widetilde{A} \setminus U$. Thus, overall we have that

$$
\frac{\max\{w(U), w(\widetilde{A})\}}{w(S^*)} = \frac{\max\{w(U), w(\widetilde{A})\}}{w(S^* \setminus U) + w(S^* \cap U)} \geq \frac{\max\{w(U), w(\widetilde{A})\}}{w(\widetilde{A} \setminus U) + w(S^* \cap U)}
$$

$$
= \frac{\max\{w(U), w(\widetilde{A})\}}{w(\widetilde{A}) + w(S^* \cap U) - w(\widetilde{A} \cap U)}
$$

$$
\geq \frac{w(U)}{w(U) + w(S^* \cap U)} = \frac{1}{2 - \frac{w(U \setminus S^*)}{w(U)}} \geq \frac{1}{2 - \frac{|U \setminus S^*|}{w(S^* \cap U) + |U \setminus S^*|}}
$$

$$
\geq \frac{1}{2 - \frac{x \cdot vc}{M \cdot (1-x) \cdot vc + x \cdot vc}} = \frac{1}{2 - \frac{x}{M - (M-1)x}}
$$

Since $x = 1 - \frac{1-\alpha}{M(2\alpha-1)+1-\alpha}$, we have that the expression above is equal to α.

Branching Rule 4. Let v be a vertex in $U' \setminus (I \cup O)$. Return the set of maximum weight among A and B, computed in the following branches.

1. $A \Leftarrow \text{ProcedureA}(G, w, \alpha, U, U', I \cup \{v\}, O)$.
2. $B \Leftarrow \text{ProcedureA}(G, w, \alpha, U, U', I, O \cup \{v\})$.

The correctness of Lemma 5 is preserved, since every vertex cover either contains v (an option examined in the first branch) or excludes v (an option examined in the second branch). Moreover, it is clear that the action can be performed in polynomial time and that the branching vector is (1,1), whose root is 2.

3.3   ProcedureB: The Proof of Lemma 4

This procedure is based on combining an appropriate application of the ideas used by the previous procedure (considering the fact that now the size of any vertex cover of $G[U]$ is larger than $\frac{vc}{2}$) with rules from the algorithm for VC by Peiselt [30]. The details can be found in the full version of this paper (see [34]).

References

1. Balasubramanian, R., Fellows, M., Raman, V.: An improved fixed-parameter algorithm for vertex cover. Inf. Process. Lett. **65**(3), 163–168 (1998)
2. Bonnet, E., Lampis, M., Paschos, V.T.: Time-approximation trade-offs for inapproximable problems. CoRR abs/1502.05828 (2015)
3. Bonnet, E., Paschos, V.T.: Sparsification and subexponential approximation. CoRR abs/1402.2843 (2014)
4. Boria, N., Della Croce, F., Paschos, V.T.: On the MAX MIN VERTEX COVER Problem. In: Kaklamanis, C., Pruhs, K. (eds.) WAOA 2013. LNCS, vol. 8447, pp. 37–48. Springer, Heidelberg (2014)
5. Bourgeois, N., Escoffier, B., Paschos, V.T.: Approximation of max independent set, min vertex cover and related problems by moderately exponential algorithms. Discrete Appl. Math. **159**(17), 1954–1970 (2011)
6. Brankovic, L., Fernau, H.: A novel parameterised approximation algorithm for minimum vertex cover. Theor. Comput. Sci. **511**, 85–108 (2013)
7. Buss, J., Goldsmith, J.: Nondeterminism within P. SIAM J. Comput. **22**(3), 560–572 (1993)
8. Chandran, L.S., Grandoni, F.: Refined memorization for vertex cover. Inf. Process. Lett. **93**(3), 123–131 (2005)
9. Chapelle, M., Liedloff, M., Todinca, I., Villanger, Y.: TREEWIDTH and PATHWIDTH parameterized by the vertex cover number. In: Dehne, F., Solis-Oba, R., Sack, J.-R. (eds.) WADS 2013. LNCS, vol. 8037, pp. 232–243. Springer, Heidelberg (2013)
10. Chen, J., Kanj, I.A., Jia, W.: Vertex cover: Further observations and further improvements. J. Algorithms **41**(2), 280–301 (2001)
11. Chen, J., Kanj, I.A., Xia, G.: Labeled search trees and amortized analysis: improved upper bounds for NP-hard problems. Algorithmica **43**(4), 245–273 (2005)
12. Chen, J., Kanj, I.A., Xia, G.: Improved upper bounds for vertex cover. Theor. Comput. Sci. **411**(40–42), 3736–3756 (2010)
13. Chen, J., Liu, L., Jia, W.: Improvement on vertex cover for low degree graphs. Networks **35**(4), 253–259 (2000)
14. Chlebík, M., Chlebíová, J.: Crown reductions for the minimum weighted vertex cover problem. Discrete Appl. Math. **156**(3), 292–312 (2008)

15. Cygan, M., Dell, H., Lokshtanov, D., Marx, D., Nederlof, J., Okamoto, Y., Paturi, R., Saurabh, S., Wahlström, M.: On problems as hard as CNF-SAT. In: CCC, pp. 74–84 (2012)

16. Downey, R., Fellows, M.: Fundamentals of parameterized complexity. Springer, Heidelberg (2013)

17. Downey, R.G., Fellows, M.R.: Fixed-parameter tractability and completeness II: on completeness for W[1]. Theor. Comput. Sci. 141(1–2), 109–131 (1995)

18. Downey, R.G., Fellows, M.R., Stege, U.: Parameterized complexity: a framework for systematically confronting computational intractability. DIMACS 49, 49–99 (1999)

19. Fellows, M.R., Jansen, B.M.P., Rosamond, F.A.: Towards fully multivariate algorithmics: Parameter ecology and the deconstruction of computational complexity. Eur. J. Comb. 34(3), 541–566 (2013)

20. Fellows, M.R., Kulik, A., Rosamond, F., Shachnai, H.: Parameterized approximation via fidelity preserving transformations. In: Czumaj, A., Mehlhorn, K., Pitts, A., Wattenhofer, R. (eds.) ICALP 2012, Part I. LNCS, vol. 7391, pp. 351–362. Springer, Heidelberg (2012)

21. Fomin, F.V., Gaspers, S., Saurabh, S.: Branching and treewidth based exact algorithms. In: Asano, T. (ed.) ISAAC 2006. LNCS, vol. 4288, pp. 16–25. Springer, Heidelberg (2006)

22. Fomin, F.V., Gaspers, S., Saurabh, S., Stepanov, A.A.: On two techniques of combining branching and treewidth. Algorithmica 54(2), 181–207 (2009)

23. Fomin, F.V., Liedloff, M., Montealegre, P., Todinca, I.: Algorithms parameterized by vertex cover and modular width, through potential maximal cliques. In: Ravi, R., Gørtz, I.L. (eds.) SWAT 2014. LNCS, vol. 8503, pp. 182–193. Springer, Heidelberg (2014)

24. Hurink, J., Nieberg, T.: Approximating minimum independent dominating sets in wireless networks. Inf. Process. Lett 109(2), 155–160 (2008)

25. Issac, D., Jaiswal, R.: An $O^*(1.0821^n)$-time algorithm for computing maximum independent set in graphs with bounded degree 3. CoRR abs/1308.1351 (2013)

26. Jansen, B.M.P., Bodlaender, H.L.: Vertex cover kernelization revisited - upper and lower bounds for a refined parameter. Theory Comput. Syst. 53(2), 263–299 (2013)

27. Chapelle, M., Liedloff, M., Todinca, I., Villanger, Y.: TREEWIDTH and PATHWIDTH parameterized by the vertex cover number. In: Dehne, F., Solis-Oba, R., Sack, J.-R. (eds.) WADS 2013. LNCS, vol. 8037, pp. 232–243. Springer, Heidelberg (2013)

28. Niedermeier, R., Rossmanith, P.: Upper bounds for vertex cover further improved. In: Meinel, C., Tison, S. (eds.) STACS 1999. LNCS, vol. 1563, pp. 561–570. Springer, Heidelberg (1999)

29. Niedermeier, R., Rossmanith, P.: On efficient fixed-parameter algorithms for weighted vertex cover. J. Algorithms 47(2), 63–77 (2003)

30. Peiselt, T.: An iterative compression algorithm for vertex cover. Ph.D. thesis Friedrich-Schiller-Universität Jena, Germany (2007)

31. Razgon, I.: Faster computation of maximum independent set and parameterized vertex cover for graphs with maximum degree 3. JDA 7(2), 191–212 (2009)

32. Shachnai, H., Zehavi, M.: A multivariate framework for weighted FPT algorithms. CoRR abs/1407.2033 (2014)

33. Xiao, M.: A note on vertex cover in graphs with maximum degree 3. In: Thai, M.T., Sahni, S. (eds.) COCOON 2010. LNCS, vol. 6196, pp. 150–159. Springer, Heidelberg (2010)

34. Zehavi, M.: Maximization problems parameterized using their minimization versions: the case of vertex cover. CoRR abs/1503.06438 (2015)

Fast Dynamic Weight Matchings in Convex Bipartite Graphs

Quan Zu$^{(\boxtimes)}$, Miaomiao Zhang, and Bin Yu

School of Software Engineering, Tongji University, Shanghai, China
{7quanzu,miaomiao,0yubin}@tongji.edu.cn

Abstract. This paper considers the problem of maintaining a maximum weight matching in a dynamic vertex weighted convex bipartite graph $G = (X, Y, E)$, in which the neighbors of each $x \in X$ form an interval of Y where Y is linearly ordered, and each vertex has an associated weight. The graph is subject to insertions and deletions of vertices and edges. Our algorithm supports the update operations in $O(\log^2 |V|)$ amortized time, obtains the matching status of a vertex (whether it is matched) in constant worst-case time, and finds the mate of a matched vertex (with which it is matched) in polylogarithmic worst-case time. Our solution is more efficient than the best known solution for the problem in the unweighted version.

1 Introduction

A graph $G = (V, E)$ is bipartite if V is partitioned into X and Y and $E \subseteq X \times Y$. A bipartite graph is a *convex* bipartite graph (CBG) if the neighbors of each $x \in X$ form an interval of Y, provided Y is totally ordered; i.e., for any $x \in X$ and y_1 and y_2 in Y where $y_1 < y_2$, if (x, y_1) and (x, y_2) belong to E, $(x, y) \in E$ for any $y \in Y$ where $y_1 < y < y_2$. A *vertex weighted* convex bipartite graph (WCBG) is a CBG in which each vertex has an associated weight and the *weight* of each edge is equal to the sum of the weights of its two endpoints.

A subset M of E is a *matching* if no two edges in M are adjacent. A *maximum cardinality matching* (MCM) is a matching of the maximum number of edges. The *weight* of a matching is the sum of the weights of its edges. A *maximum weight matching* (MWM) is a matching of the maximum weight. The MWM problem in a WCBG naturally corresponds to the classical maximum scheduling problem, in which the input is a collection of unit-time jobs and a collection of time slots. Each job has an associated release time, deadline and weight, and each time slot has an associated weight. The objective of the scheduling is to find a feasible maximum weight subset of the jobs and the time slots.

Static matching problems in CBGs have been extensively researched. Glover [5] proposed an algorithm to produce an MCM in a CBG G by iterating each $y \in Y$ in increasing order, and meanwhile matching y to the unmatched $x \in X$ where $(x, y) \in E$ and x has the earliest deadline, breaking ties according to a predefined order. Such a matching is called a *Glover matching*. Using a fast

© Springer-Verlag Berlin Heidelberg 2015
G.F. Italiano et al. (Eds.): MFCS 2015, Part II, LNCS 9235, pp. 601–612, 2015.
DOI: 10.1007/978-3-662-48054-0_50

priority queue [3], Glover's algorithm runs in $O(|Y| + |X| \log \log |X|)$ time. The best known MCM solution is Steiner and Yeomans's $O(|X|)$ time algorithm [9]. For the weighted version, Katriel [6] obtained an $O(|E| + |Y| \log |X|)$ time algorithm to find an MWM in a right weighted CBG (each $y \in Y$ associated with a weight). Recently, Plaxton [7] pointed out the MWM problem in a WCBG can be solved in $O(|V| \log^2 |V|)$ time, and obtained an $O(|V| \log |V|)$ time bound for the problem in a left weighted CBG (each $x \in X$ associated with a weight).

Dynamic matching problems in CBGs maintain a certain matching under update operations and query operations. The update operations support insertions and deletions of vertices and edges, and the query operations supprt answering whether a vertex is matched and finding the mate of a matched vertex. Brodal et al. [1] developed an algorithm to solve the dynamic MCM problem in a CBG, based on the binary computation tree data structure devised by Dekel and Sahni [2] for parallel computing. The algorithm performs the update operations in $O(\log^2 |V|)$ amortized time, reports the status of vertex in constant worst-case time, and reports the mate of a matched vertex in $O(\min\{k \log^2 |X| + \log |X|, |X| \log |X|\})$ worst-case time where $k < \min\{|X|, |Y|\}$, or amortized $O(\sqrt{|X|} \log^2 |X|)$ time. To solve the dynamic MWM problem in a left weighted CBG, Zu et al. [10] designed an algorithm to support the update operations in $O(\log^3 |V|)$ amortized time and the query operations in $O(k)$ worst-case time where k is not greater than the cardinality of the MWM.

In the dynamic MCM problem, maintaining the maximum cardinality of the matching is sufficient for a solution. However, in the weighted version of the problem, the weight of the matching has also to be considered, besides the cardinality aspect. For instance, when a left vertex is inserted into the graph, it is necessary to compute the *replaceable set*, i.e., the set of matched left vertices which can be replaced by the inserted vertex while keeping the cardinality of the matching. That makes the dynamic weighted problem different from and more difficult than the corresponding unweighted problem, and Brodal et al.'s solution cannot solve the dynamic MWM problem. In this paper, we consider the problem of maintaining an MWM in a dynamic WCBG. Despite the difficulty, our algorithm supports the update operations in the same time bound as their solution for the unweighted problem. Moreover, we find a polylogarithmic worst-case time solution for the query operations, which is a vast improvement on the previous results.

In a bipartite graph, all sets of left vertices participating in a matching form a matroid on X, and there is an *optimal* set with respect to the weight [4]. Symmetrically, it is the same for the right vertex. It is also known that the left and the right optimal matched sets admit the MWM [8]. Therefore, we observe that the key point to solve the problem is maintaining the optimal matched sets, in which the vertices participant in the MWM. The computation for the replaceable set plays an important role in the set maintenance. We prove that a replaceable set corresponds to an *M-tightest subgraph*, which admits a perfect matching, and the subgraph can be computed efficiently due to the convexity property of CBG. Moreover, based on the optimal matched sets, we propose a solution to perform the query operations very efficiently, also in light of the CBG's convexity property.

Our algorithm supports the update operations in $O(\log^2 |V|)$ amortized time, which achieves the same time bound as Brodal et al.'s solution for the problem in the unweighted version. More efficiently, in the query operations, our algorithm obtains the status of a vertex in constant worst-case time, finds the mate of a matched x vertex in $O(\log^2 |V|)$ worst-case time, and finds the mate of a matched y vertex in $O(\log^3 |V|)$ worst-case time.

2 Terminologies and Definitions

In this paper, a bipartite graph $G = (X, Y, E)$ is denoted by standard notations, such as $X(G)$ denoting the left vertex set X of G, $N_G(v)$ denoting the set of neighbors of the vertex v in G, $G[V']$ denoting the subgraph of G induced by the vertex subset $V' \subseteq V$ where $V = X \cup Y$, and $G[E']$ denoting the subgraph of G induced by the edge subset $E' \subseteq E$.

In a WCBG $G = (X, Y, E)$, each $y \in Y$ is represented by a pair $(y.id, y.w)$, where $y.id$ is a distinct integer and $y.w$ is a non-negative integer denoting the associated weight of y. Y is in id order of increasing $y.id$. For any subset $Y' \subseteq Y$, $Y'.min$ and $Y'.max$ respectively denote the minimum and the maximum y in Y' in id order. In G, each $x \in X$ is represented by a tuple $(x.id, x.s, x.e, x.w)$, where $x.id$ is a distinct integer, $x.s$ and $x.e$ respectively denote $N_G(x).min$ and $N_G(x).max$, $x.w$ is a non-negative integer denoting the associated weight of x. Since for any x and y, if $x.s \leq y \leq x.e$, $(x, y) \in E$, the edge set E can be represented implicitly in G. Then G can also be denoted by (X, Y).

Over the set Y, other than the id order, we define the *weight-id total order* by lexicographical order of $(y.w, -y.id)$. We also define several total orders over the set X that the *start-id total order* is decided by lexicographical order of $(x.s, x.id)$, the *end-id total order* by lexicographical order of $(x.e, x.s, x.id)$, and the *weight-id total order* by lexicographical order of $(x.w, -x.e, -x.s, -x.id)$. In this paper, we primarily use the weight-id order of x and adopt it for the x vertex comparisons unless stated otherwise.

We use M to denote a matching of G. Then $N_{G[M]}(v)$ denotes the vertex matched with v in M, and $N_{G[M]}(V')$ denotes the set of vertices matched with the vertices in V' in M. We say the vertices in $G[M]$ *participate* in M, and M *covers* these vertices.

For an $x \notin X$ of G, we use $G + x$ to denote the WCBG $(X + x, Y)$. Similarly, $G - x$, $G + y$, and $G - y$ are used. The dynamic MWM problem in a WCBG include the following update operations.

- *insert*($\hat{x}$): $G + \hat{x}$, i.e., $\hat{x} \notin X$ is inserted in X, and for each $y \in Y$ such that $\hat{x}.s \leq y \leq \hat{x}.e$, y is adjacent to $\hat{x}$ implicitly. If $\hat{x}.s$ (or $\hat{x}.e$) is not in Y, $\hat{x}.s$ (or $\hat{x}.e$) is inserted in advance.
- *insert*($\hat{y}$): $G + \hat{y}$, i.e., $\hat{y} \notin Y$ is inserted in Y, and for each $x \in X$ such that $x.s < \hat{y} < x.e$, x is adjacent to $\hat{y}$ implicitly.
- *delete*($\hat{x}$): $G - \hat{x}$, i.e., $\hat{x} \in X$ is deleted from X, and all edges incident with $\hat{x}$ are deleted implicitly.

- *delete*($\hat{y}$): $G - \hat{y}$, i.e., $\hat{y} \in Y$, which is not equal to $x.s$ or $x.e$ of some $x \in X$, is deleted from Y, and all edges incident with $\hat{y}$ are deleted implicitly.

We mainly discuss the concepts and the algorithm related to *insert*($\hat{x}$) operation. The solutions for the other update operations are similar and not included in this paper due to space limitations.

Arbitrary edge updates may break the convexity property and the update operations only support the restricted edge updates. The updates of edges are implicitly operated by the updates of vertices. Inserting the edge (x, y) is implemented by deleting x and then inserting x with the updated $x.s$ or $x.e$. Deleting an edge is similar. The query operations include:

- *status*(v): obtain the status of whether v is matched or not in the MWM.
- *pair*(v): given a matched vertex v, find its mate in the MWM that follows the Glover matching rule.

3   Optimal Matched Set, Replaceable Set, and Tight Subgraph

Given a matching M in a WCBG G, the set of vertices that participate in M is called a *matched set*, which is naturally partitioned into the *matched-x set* and the *matched-y set* by the type of vertex, i.e., x vertex or y vertex. We use Z and W to denote a matched-x set and a matched-y set, respectively. It is known that in G, all matched-x sets form a matroid on X, and there is an optimal one among the sets [4]. We define the *optimal matched-x set* (OMX in short) with respect to the weight-id total order in a WCBG.

Definition 1. *The matched-x set $Z = \{x_1, \ldots, x_k\}$, where $x_1 \geq \cdots \geq x_k$, is optimal if for any other matched-x set $Z' = \{x'_1, \ldots, x'_l\}$, where $x'_1 \geq \cdots \geq x'_l$, it satisfies that $k \geq l$ and $\forall i \in [1, l]$, $x_i \geq x'_i$.*

Symmetrically[1], the *optimal matched-y set* (OMY in short) with respect to the weight-id order is defined. It is easy to prove that the cardinalities of the OMX and the OMY are equal. The union of the OMX and the OMY is called the *optimal matched set*, OMS in short.

Let $\hat{Z}$ and $\hat{W}$ be the OMX and the OMY in G, respectively. By Lemma 2 in [8], rephrased in this paper's terms that there is a matching covering the OMX and the OMY, it implies $G[\hat{Z} \cup \hat{W}]$ admits perfect matchings, which clearly are MWMs. In the dynamic MWM problem in a WCBG, our algorithm maintains the OMX and the OMY in the update operations. Based on the two sets, following the Glover matching rule, the query operations are supported.

Given a matching M in a WCBG G, an *alternating path* of M is a path in which the edges alternate between $E - M$ and M. If both endpoints of an alternating path are not incident with some edge of M, it is an *augmenting path*.

---

[1] The matroid and optimal properties are of a bipartite graph without concerning the convexity property, and so it is symmetric for y vertex.

Let Z and W be the matched-x set and the matched-y set of M, respectively. For an unmatched $\hat{x} \in X$ in G, the set of $x \in Z$ which can be reached by $\hat{x}$ via an alternating path of M is called the *replaceable set* of $\hat{x}$ w.r.t. M, denoted by $RS(\hat{x}, M)$ or $RS(\hat{x})$ when M is clear. The set of $y \in Y - W$ which can be reached by $\hat{x}$ via an augmenting path of M is called the *complementary set* of $\hat{x}$ w.r.t. M, denoted by $CS(\hat{x}, M)$ or $CS(\hat{x})$ when M is clear. The following proposition is straightforward.

Proposition 1. *In G, letting Z and W be the matched-x set and the matched-y set of a matching M, for an $\hat{x} \notin Z$, $CS(\hat{x}, M) = N_G(RS(\hat{x}, M) \cup \hat{x}) - W$.*

The replaceable and the complementary sets of $\hat{x}$ are decided by Z and W.

Proposition 2. *Letting Z and W be the matched-x set and the matched-y set of a matching M in G, for any other matching M' covering Z and W, given an $\hat{x} \notin Z$, $RS(\hat{x}, M) = RS(\hat{x}, M')$ and $CS(\hat{x}, M) = CS(\hat{x}, M')$.*

Proof. $M \oplus M'$ are disjoint cycles. The x vertices reached by an alternating path of M starting from $\hat{x}$ must also be reached by an alternating path of M' starting from $\hat{x}$, That is, $RS(\hat{x}, M) = RS(\hat{x}, M')$. Then, by Proposition 1, $CS(\hat{x}, M) = CS(\hat{x}, M')$. $\square$

The following proposition can be directly implied from the definitions.

Proposition 3. *Given an M in G and an unmatched $\hat{x}$, $CS(\hat{x}, M) \neq \emptyset$ if and only if M is augmentable with $\hat{x}$.*

The following useful lemma summaries the results of Theorem 1 and 2 in [10].

Lemma 1. *[10] Suppose in G, there is a matching M covering the OMX $\hat{Z}$ and the OMY $\hat{W}$. In $G + \hat{x}$, where $\hat{x} \notin X$, if $CS(\hat{x}, M) \neq \emptyset$, $\hat{Z} + \hat{x}$ is the OMX, and $\hat{W} + y'$ is the OMY where $y' = \max\{y \in CS(\hat{x}, M)\}$; if $CS(\hat{x}, M) = \emptyset$, $\hat{Z} + \hat{x} - x'$ is the OMX where $x' = \min\{x \in RS(\hat{x}, M) + \hat{x}\}$, and $\hat{W}$ is still the OMY.*

In a WCBG G, *tight subgraph* is defined as follows.

Definition 2. *A subgraph is called tight if it has the following properties: it is a nontrivial induced subgraph which has a perfect matching, and for each x vertex x' of the subgraph, any neighbor that x' has in the supergraph is a y vertex of the subgraph. To put it in a more specific way, if X' is a nonempty subset of X where $|X'| = |N_G(X')|$ and $\forall X'' \subseteq X'$, $|X''| \leq |N_G(X'')|$, the induced subgraph $G[X' \cup N_G(X')]$ is a tight subgraph, which can be implied by Hall Theorem.*

Given a tight subgraph G', if there is an $\hat{x} \notin X(G')$ such that $N_G(\hat{x}) \neq \emptyset$ and $N_G(\hat{x}) \subseteq Y(G')$, G' is a *tight subgraph of $\hat{x}$ in G*. If $\hat{x} \notin X$, it can be checked that G' is still a tight subgraph of $\hat{x}$ in $G + \hat{x}$.

Let Z and W be the matched-x set and the matched-y set of a matching M in a WCBG G, and let $\widetilde{G} = G[Z \cup W]$. Given an $\hat{x} \notin Z$ with $N_G(\hat{x}) \cap W \neq \emptyset$, the *$M$-tightest subgraph of $\hat{x}$ in G* is defined in the following.

Definition 3. *The M-tightest subgraph of $\hat{x}$ in G is the tight subgraph G' of $\hat{x}$ in $\widetilde{G}$ with the minimal Y set; that is, if $\hat{G}$ is a tight subgraph of $\hat{x}$ in $\widetilde{G}$, and there is no other tight subgraph G' of $\hat{x}$ in $\widetilde{G}$ such that $Y(G') \subset Y(\hat{G})$, then $\hat{G}$ is the M-tightest subgraph of $\hat{x}$ in G.*

Note that $\hat{G}$ is an induced subgraph in $\widetilde{G}$ but not necessarily an induced subgraph in G, and $\hat{G}$ is tight in $\widetilde{G}$ but not necessarily tight in G. It can be proved that the M-tightest subgraph of $\hat{x}$ in G is unique. The following lemma is important.

Lemma 2. *Letting Z and W be the matched-x set and the matched-y set of a matching M in G, if there is an $\hat{x} \notin Z$ with $N_G(\hat{x}) \cap W \neq \emptyset$, the replaceable set of $\hat{x}$ $RS(\hat{x}, M) = X(\hat{G})$ where $\hat{G}$ is the M-tightest subgraph of $\hat{x}$ in G.*

Proof. Let $\widetilde{G} = G[Z \cup W]$. Any alternating path starting with the edge from $\hat{x}$ to a $y \in W$ cannot leave $\hat{G}$ without going to a $y \notin W$, which implies $RS(\hat{x}) \subseteq X(\hat{G})$. To prove $X(\hat{G}) \subseteq RS(\hat{x})$, initially, let $Y' = N_G(\hat{x}) \cap W$ and $X' = N_{G[M]}(Y')$. It is true that $X' \subseteq X(\hat{G})$, since otherwise $\exists x \in X(\hat{G})$ is unmatched, which is a contradiction. Then by recursively setting $Y' = N_{\widetilde{G}}(X')$ and $X' = N_{G[M]}(Y')$, X' can be extended until $|X'| = |N_{\widetilde{G}}(X')|$ and $X' \subseteq X(\hat{G})$. If $X' = X(\hat{G})$, the lemma is proved; otherwise, $G[X' \cup Y']$ is tight in $\widetilde{G}$ and $Y' \subset Y(\hat{G})$, which contradicts with the premise that $\hat{G}$ is the M-tightest subgraph of $\hat{x}$ in G. $\square$

Due to the convexity property of CBG, the neighbors of $X(\hat{G})$ form interval both in $\widetilde{G}$ and G. The following proposition is straightforward to prove.

Proposition 4. *Given the M-tightest subgraph $\hat{G}$ of $\hat{x}$ in G, letting $\hat{X} = X(\hat{G})$ and $\widetilde{G} = G[Z \cup W]$, $N_{\widetilde{G}}(\hat{X})$ forms an interval in W, and $N_G(\hat{X})$ forms an interval in Y.*

Hence, the convexity of the neighbors of the replaceable set leads to a fast computation for the complementary set, provided the M-tightest subgraph of $\hat{x}$ can be found efficiently.

Let Z and W be the matched-x set and the matched-y set of a matching M in G. Let $\widetilde{G} = G[Z \cup W]$, where $|M| = m$ and W is sorted in the ordering of increasing $y.id$. Assume $W[0].id = -\infty$ and $W[m + 1].id = +\infty$. We define:

$$\alpha(i) = |\{x \in Z : x.e < W[i + 1]\}|, \quad d_\alpha(i) = i - \alpha(i)$$
$$\beta(j) = |\{x \in Z : W[j - 1] < x.s\}|, \quad d_\beta(j) = m - j + 1 - \beta(j)$$

for $i \in [0, m]$ and $j \in [1, m + 1]$, where $\alpha(i)$ denotes the number of x that can only match with a $y \leq W[i]$, and $\beta(j)$ denotes the number of x that can only match with a $y \geq W[j]$. If there is an integer k such that $d_\alpha(k) = 0$, k is an α-*tight point*, which implies that $G' = G[Z[1, k] \cup W[1, k]]$ is a tight subgraph in $\widetilde{G}$ where Z is sorted in end-id order. Similarly, k is a β-*tight point* if $d_\beta(k) = 0$, implying $G'' = G[Z[k, m] \cup W[k, m]]$ is a tight subgraph in $\widetilde{G}$ where Z is in start-id order. Given a $y' \in Y$, let $k = max\{i \in [0, m] : W[i] \leq y'\}$. We define the α_{Post}-*tight point* of y' as the least α-tight point that is not less than k.

If $k = 0$, or k is an α-tight point and $Z[k].e < y'$ where Z is in end-id order, the α_{Pre}-*tight point* of y' is k; otherwise, it is the greatest α-tight point that is less than k. Given a $y' \in Y$, let $k' = min\{j \in [1, m+1] : W[j] \geq y'\}$. We define the β_{Pre}-*tight point* of y' as the greatest β-tight point that is not greater than k'. If $k' = m+1$, or k' is a β-tight point and $Z[k'].s > y'$ where Z is in start-id order, the β_{Post}-*tight point* of y' is k'; otherwise, it is the least β-tight point that is greater than k'. The fast computation for the M-tightest subgraph of $\hat{x}$ is based on the following lemma.

Lemma 3. *Given a matching M in G and an $\hat{x} \notin Z$, letting $\widetilde{G} = G[Z \cup W]$, if $\hat{x}.s \leq W.min$, or there is an $x' \in Z$ such that $x'.s \leq W.min$ and $x' \in RS(\hat{x})$, then $RS(\hat{x}) = Z[1, k]$ where k is the α_{Post}-tight point of $\hat{x}.e$ in $\widetilde{G}$ and Z is in end-id order. Symmetrically, if $\hat{x}.e \geq W.max$, or there is an $x' \in Z$ such that $x'.e \geq W.max$ and $x' \in RS(\hat{x})$, then $RS(\hat{x}) = Z[k, |Z|]$ where k is the β_{Pre}-tight point of $\hat{x}.s$ in $\widetilde{G}$ and Z is in start-id order.*

Proof. We will prove the α_{Post} situation and the proof for the β_{Pre} situation is symmetrical. Let $\widetilde{G} = G[Z \cup W]$. Suppose $\hat{x}.s \leq W.min$. By the definitions, if i is an α-tight point such that $W[i] \geq \hat{x}.e$ or $W[i+1] > \hat{x}.e$, $G_i = \widetilde{G}[Z[1, i] \cup W[1, i]]$ is a tight subgraph of $\hat{x}$ in $\widetilde{G}$ where Z is in end-id order. Then it is straightforward that G_k is the M-tightest subgraph of $\hat{x}$ in G where k is the α_{Post}-tight point of $\hat{x}$ in $\widetilde{G}$. By Lemma 2, $RS(\hat{x}) = Z[1, k]$.

In the other case, suppose there is an $x' \in Z$ such that $x'.s \leq W.min$ and $x' \in RS(\hat{x})$. Let $\hat{G}$ be the M-tightest subgraph of $\hat{x}$ in G. Since $x' \in RS(\hat{x})$ and $x'.s \leq W.min$, $W.min \in Y(\hat{G})$. By Proposition 4, $W[1, j] \subseteq Y(\hat{G})$ where $W[j] = N_{\widetilde{G}}(\hat{x}).max$. Then it is clear that $\hat{G} = \widetilde{G}[Z[1, k] \cup W[1, k]]$ where k is the α_{Post}-tight point of $\hat{x}$ in $\widetilde{G}$. By Lemma 2, $RS(\hat{x}) = Z[1, k]$. $\qquad\square$

4   Extended Binary Computation Tree

Our algorithm is based on the extension of the binary computation tree (BCT) data structure [2], which is a kind of augmented balanced binary search tree. Let $S = \{x.s : x \in X\}$ be sorted in id order and denoted as $\{s_1, \ldots, s_k\}$. Let $s_{k+1} = y'$ where $y'.id = +\infty$. In the BCT, the ith leaf node from left to right is associated with $G[X_i \cup Y_i]$ where $X_i = \{x \in X : x.s = s_i\}$, $Y_i = \{y \in Y : s_i \leq y < s_{i+1}\}$, and $1 \leq i \leq k$. Let P be an internal node, L its left child node, and R its right child node. P is associated with $G[X_P \cup Y_P]$ where $X_P = X_L \cup X_R$ and $Y_P = Y_L \cup Y_R$. Then the root node is associated with G.

In each node P, we maintain two collections of auxiliary sets of P, one for the dynamic updates, called U-model, and the other Q-model for the dynamic queries. We first discuss U-model, which includes:

- X_P and Y_P, which are the sets of x and y vertices associated with P, resp.;
- The free x set $F_P = \{x \in X_P : x.e > Y_P.max\}$;
- The matched-x set Z_P in $G_{\hat{P}}$ where $G_{\hat{P}} = G[(X_P - F_P) \cup Y_P]$; Z_P is partitioned into Z_P^l and Z_P^r, described in the next paragraph;

- The matched-y set W_P in $G_{\widehat{P}}$, which is partitioned into W_P^l and W_P^r; $W_P^l = \{y \in W_P : y < Y_P.mid\}$ where $Y_P.mid = Y_R.min$ if P is an internal node, and if P is a leaf, $Y_P.mid = Y_P.min$;
- The unmatched-x set $I_P = X_P - F_P - Z_P$;
- The unmatched-y set $J_P = Y_P - W_P$.

The key invariants maintained in U-model for each node P are listed as follows, where ϕ_1 implies that in the root $\mathcal{R}$ of the BCT, $Z_{\mathcal{R}}$ and $W_{\mathcal{R}}$ are the OMX and the OMY of G, respectively; ϕ_2 implies that Z_P^l is the matched-x set of the Glover matching in $G[Z_P \cup W_P^l]$.

ϕ_1. Z_P and W_P are the OMX and the OMY in $G_{\widehat{P}}$, resp.
ϕ_2. $\nexists x \in Z_P^r$ such that $Z_P^l + x - x'$ is a matched-x set in $G[Z_P \cup W_P^l]$ where $x' \in Z_P^l$ and $x < x'$ in end-id order.

In each node P, U-model are represented by two types of augmented balanced BSTs, which include:

- a corresponding tree for each of Z_P, Z_P^l, Z_P^r, F_P, I_P, and J_P.
- an *equal-start tree* corresponds to W_P^r, and two *equal-end trees* respectively correspond to W_P^l and W_P.

The first type of BST supports looking-up operations and update operations in a corresponding set. The looking-up operations locate the minimum or the maximum element in a part of the set. For instance, find $\min\{x \in Z_P^r : x.e \le y'\}$ in weight-id order. The update operations support element insertions or deletions in the set. These two kinds of operations can be implemented by standard techniques in logarithmic time. The equal-start tree supports the operations of finding α_{Pre} and α_{Post} tight points in $G[Z_P^r \cup W_P^r]$. Symmetrical to the equal-start tree, the equal-end trees support the operations of finding β_{Pre} and β_{Post} tight points in $G[Z_P^l \cup W_P^l]$ and $G[Z_P \cup W_P]$. By the implicit representation [1], the finding tight point operations can be implemented in logarithmic time. Due to the space limitation, the details of the representation and the operations of the two types of BST are not able to be included.

In each update operation, the algorithm travels in the BCT along the path from a certain leaf to the root, and meanwhile updates U-model for each nodes on the path. Then, $Z_{\mathcal{R}}$ and $W_{\mathcal{R}}$ in the root $\mathcal{R}$, i.e., the OMX and the OMY of G, are maintained. The algorithm will be shown in Sect. 5.

To support dynamic query operations, we use Q-model to maintain the Glover matching in $\widetilde{G} = G[Z_{\mathcal{R}} \cup W_{\mathcal{R}}]$. It is the fact that $Z_{\mathcal{R}}$ and $W_{\mathcal{R}}$ can change only by one element for each update [1]. After each change in $Z_{\mathcal{R}}$ or $W_{\mathcal{R}}$, the algorithm updates Q-model for each node on the path from a leaf to the root bottom-up in the BCT. In each node P, Q-model includes the following auxiliary sets.

- $\bar{X}_P$ and $\bar{Y}_P$: if P is a leaf, where $x.s = s_i$ in X_P, $\bar{X}_P = \{x \in Z_{\mathcal{R}} : x.s = s_i\}$ and $\bar{Y}_P = \{y \in W_{\mathcal{R}} : s_i \le y < s_{i+1}\}$; if P is an internal node, $\bar{X}_P = \bar{X}_L \cup \bar{X}_R$ and $\bar{Y}_P = \bar{Y}_L \cup \bar{Y}_R$. In the root $\mathcal{R}$, $\bar{X}_{\mathcal{R}} = Z_{\mathcal{R}}$ and $\bar{Y}_{\mathcal{R}} = W_{\mathcal{R}}$.

- $\bar{Z}_P$: the matched-x set in $G[\bar{X}_P \cup \bar{Y}_P]$, described in the next paragraph.
- T_P: the transferred x set. $T_P = \bar{X}_P - \bar{Z}_P$. $\forall x \in T_P$, $x.e > Y_P.max$.

We maintain the following two key invariants in Q-model for each node P. It is implied that $\bar{Z}_P$ is the matched-x set of the Glover matching in $G[\bar{X}_P \cup \bar{Y}_P]$.

ϕ'_1. $\bar{Z}_P$ is of maximum cardinality in $G[\bar{X}_P \cup \bar{Y}_P]$;

ϕ'_2. $\nexists x \in T_P$ such that $\bar{Z}_P + x - x'$ is a matched-x set in $G[\bar{X}_P \cup \bar{Y}_P]$ where $x' \in \bar{Z}_P$ and $x < x'$ in end-id order.

The weight of a vertex is omitted in Q-model and $\widetilde{G}$ always admits a perfect matching. Thus, it is a special dynamic MCM problem, and Brodal et al.'s algorithm can be simplified to update Q-model in $O(\log^2 |V|)$ time. Due to space limitations, the details of representation and update operations for Q-model are not able to be included in this paper.

5   Dynamic Updates

The dynamic update algorithm updates U-model. In each node P, it is maintained that Z_P and W_P are the OMX and the OMY in $G[(X_P - F_P) \cup Y_P]$. By Proposition 2, Z_P and W_P decide the replaceable set and the complementary set of $\hat{x}$, which are denoted as $RS(\hat{x}, P)$ and $CS(\hat{x}, P)$ in this section.

The algorithm traverses along the path from a leaf to the root in the BCT. The computation in a parent node depends on the computation of its child, whose result is transited between them by parameters rx, cy, lb, rb. In node P, if $\hat{x}$ augments Z_P, rx is null, denoted by the symbol Λ, and $cy = \max\{y \in CS(\hat{x}, P)\}$ in weight-id order; if $\hat{x}$ replaces x' in Z_P, $rx = x'$ and $cy = \Lambda$. It is possible that $rx = \hat{x}$, when $\hat{x}$ cannot augment Z_P and $\hat{x} < \min\{x \in RS(\hat{x}, P)\}$ in weight-id order. The value of lb is equal to $x.s$ of $\min\{x \in RS(\hat{x}, P) \cup \hat{x}\}$ in start-id order; and the value of rb is equal to $x.e$ of $\max\{x \in RS(\hat{x}, P) \cup \hat{x}\}$ in end-id order. That is, by Proposition 1, lb and rb denote the minimum and the maximum y in $CS(\hat{x}, P)$ in id order, respectively. Then, by Propositions 1 and 4, $CS(\hat{x}, P) = \{y \in J_P : lb \le y \le rb\}$.

5.1   Inserting x in a Leaf Node

Algorithm 1 shows the procedure of the insertion of $\hat{x}$ at the leaf P, where each $x \in X_P$ has $x.s = \hat{x}.s$. Thus, $lb = Y_P.min$. If $\hat{x}.e > Y_P.max$, $F_P = F_P + \hat{x}$ and $rb = Y_P.max$. Otherwise, by Lemma 3, $\alpha_{Post}(\hat{x})$ in $G[Z_P^r \cup W_P^r]$ decides $RS(\hat{x}, P)$. Then rb and $CS(\hat{x}, P)$ are computed. If $CS(\hat{x}, P) \ne \emptyset$, which means that $\hat{x}$ can augment Z_P, the maximum y in weight-id order in $CS(\hat{x}, P)$ is found to augment W_P. Otherwise, by Lemma 1, Z_P cannot be augmented by $\hat{x}$. In $RS(\hat{x}, P)$, the minimum element x' in weight-id order is replaced by $\hat{x}$ if $x' < \hat{x}$ in weight-id order. The subroutines in the algorithm can be implemented by the techniques discussed in the previous section in logarithmic time. Note that if $\hat{x}.s \notin \{x.s : x \in X - \hat{x}\}$, a new leaf node is augmented in the BCT and the balancing operation makes the time bound become amortized.

Algorithm 1. InsertXinLeaf($\hat{x}$)

1: $lb \leftarrow Y_P.min$, $X_P \leftarrow X_P + \hat{x}$
2: **if** $\hat{x}.e > Y_P.max$ **then**
3:    $rb \leftarrow Y_P.max$
4:    $rx \leftarrow \Lambda$, $cy \leftarrow \Lambda$, $F_P \leftarrow F_P + \hat{x}$
5: **else**
6:    $\alpha \leftarrow \alpha_{Post}(\hat{x}.e)$ computed in $G[Z_P^r \cup W_P^r]$
7:    $rb \leftarrow \max\{Z_P^r[\alpha].e, \hat{x}.e\}$ in id order
8:    $CS(\hat{x}, P) \leftarrow \{y \in J_P : y \leq rb\}$
9:    **if** $CS(\hat{x}, P) \neq \emptyset$ **then**
10:      $cy \leftarrow \max\{y \in CS(\hat{x}, P)\}$ in weight-id order
11:      $rx \leftarrow \Lambda$, update $Z_P^r, Z_P, W_P^r, W_P, J_P$ //augment with $\hat{x}$ and cy
12:    **else**
13:      $RS(\hat{x}, P) \leftarrow \{x \in Z_P^r : x.e \leq rb\}$
14:      $rx \leftarrow \min\{x \in RS(\hat{x}, P) + \hat{x}\}$ in weight-id order
15:      $cy \leftarrow \Lambda$, update Z_P^r, Z_P, I_P //replace rx with $\hat{x}$
16: **return** rx, cy, lb, rb

5.2 Inserting x in an Internal Node

Consider $\hat{x}$ is inserted in node P from its right child R. If $\hat{x}.e > Y_P.max$, $F_P = F_P + \hat{x}$ and $rb = Y_P.max$. By Lemma 3, $\beta_{Pre}(\hat{x}.s)$ in $G[Z_P \cup W_P]$ decides $RS(\hat{x}, P)$ and lb in P. Suppose $\hat{x}.e \leq Y_P.max$. There are three cases such that the operations in P are according to the parameters transmitted from R. If $rx = \hat{x}$, it implies that $CS(\hat{x}, R) = \emptyset$ and $\hat{x} < x'$ where $x' = \min\{x \in RS(\hat{x}, R)\}$ in weight-id order. The $CS(\hat{x}, R) = \emptyset$ implies $CS(\hat{x}, P) = \emptyset$ and $x' \leq x''$ where $x'' = \min\{x \in RS(\hat{x}, P)\}$ in weight-id order. Hence, $\hat{x}$ cannot augment Z_P and is added into I_P. The second case is $rx \neq \hat{x}$ and $rx \in Z_P$. It implies that $\hat{x}$ replaces rx in R and $rx = \min\{x \in RS(\hat{x}, P)\}$ in weight-id order. Thus $\hat{x}$ replacing rx in Z_P. The third case is $rx = \Lambda$ and $\delta = 0$ where $\delta = |\{y \in J_P : lb \leq y \leq rb\}| - |\{y \in J_R : lb \leq y \leq rb\}| - 1$. The former $rx = \Lambda$ implies that $\hat{x}$ augments Z_R, and the latter $\delta = 0$ implies that $CS(\hat{x}, P) = CS(\hat{x}, R)$ before $\hat{x}$ is inserted in R. Then $Z_P = Z_P + \hat{x}$ and $W_P = W_P + cy$.

As to the other two cases, the computation in P is required. If $rx = \Lambda$ and $\delta \neq 0$, it implies that $\exists y'$, $lb \leq y' \leq rb$, such that $y' \in J_R \cap W_P$. It can be implied that there is an alternating path from y' to x' where $x' \in F_L \cap Z_P^r$. That is, $x'.s < Y_P.mid$ and $x' \in RS(\hat{x}, P)$. By Lemma 3, $\alpha_{Post}(\hat{x}.e)$ in $G[Z_P^r \cup W_P^r]$ decides $RS(\hat{x}, P) \cap Z_P^r$ and rb. The last case is $rx \neq \Lambda$ and $rx \neq \hat{x}$ and $rx \notin Z_P$. Similar to the previous case, rb is decided by $\alpha_{Post}(\hat{x}.e)$ in $G[Z_P^r \cup W_P^r]$. Thus, the minimum element x_1 in start-id order in $RS(\hat{x}, P) \cap Z_P^r$ must have $x_1.s < Y_P.mid \leq x_1.e$ and $x_1 \in RS(\hat{x}, P)$. By Lemma 3, $\beta_2 = \beta_{Pre}(x_1.s)$ in $G' = G[Z_P^l \cup W_P^l]$ decides $RS(x_1)$ in G'. It can be proved that, by invariant ϕ_2, there is no $x \in Z_P^l$ such that $x.s \geq \beta_2$ and $x.e > rb$, which implies $RS(x_1)$ in G' is equal to $RS(\hat{x}, P) \cap Z_P^l$. Then lb can be computed from β_2 and x_1. Note that the updates for Z_P^l and Z_P^r maintain the invariant ϕ_2. The solution for inserting $\hat{x}$ from the left child is similar, and it can be checked that inserting $\hat{x}$ in an internal node takes logarithmic time. The update operation of $\hat{x}$ insertion, which includes logarithmic times inserting $\hat{x}$ in a node, takes $O(\log^2 |V|)$ amortized time.

6 Fast Dynamic Queries

Given an $\hat{x} \in X$ in G, the query of whether $\hat{x}$ is matched can be answered in constant time by maintaining an array of X to record the matching status of each x vertex, since each update operation changes the status of at most two $x \in X$. It is similar for a $y \in Y$. In this section, we use only the end-id order for the comparison between x vertices.

To support $pair(\hat{x})$ operation where $\hat{x}$ is matched, MateofX shown in Algorithm 2 finds the y vertex $\hat{y}$ that matches with $\hat{x}$ in the Glover matching in $G[Z_{\mathcal{R}} \cup W_{\mathcal{R}}]$, where $Z_{\mathcal{R}}$ and $W_{\mathcal{R}}$ are the OMX and the OMY in G, respectively. In MateofX, $\hat{x}$ is the queried x; P is the node in which $\hat{y} \in \bar{Y}_P$; Δ is the cardinality of the subset $\Omega = Z_{\mathcal{R}} - \bar{Z}_P$ in which $x.s < \bar{Y}_P.min$, $x \le \hat{x}$, and x is not matched with a $y < \bar{Y}_P.min$. The following straightforward lemma implies that $\forall x' \in \Omega$, if $x' \ne \hat{x}$, x' is matched before $\hat{x}$, i.e., if y' matches with x', $y' < \hat{y}$.

Lemma 4. *In a Glover matching, if x_1 matches with y_1 and x_2 matches with y_2, then it is true that: 1) if $x_1 < x_2$ and $x_1.s \le y_2 \le x_1.e$, $y_1 < y_2$; 2) if $x_1 < x_2$ and $x_1.s \le x_2.s$, $y_1 < y_2$.*

Algorithm 2. MateofX$(\hat{x}, P, \Delta)$

1: **if** P is a leaf **then**
2:     $\sigma \leftarrow |\{x \in \bar{Z}_P : x \le \hat{x}\}|$
3:     **return** $\bar{Y}_P[\Delta + \sigma]$
4: **else**
5:     $\sigma \leftarrow |\{x \in \bar{Z}_L : x \le \hat{x}\}|$
6:     $\delta \leftarrow |\{x \in T_L : x \le \hat{x}\}|$
7:     **if** $\Delta + \sigma > |\bar{Y}_L|$ **then**
8:         $\Delta \leftarrow \Delta + \sigma + \delta - |\bar{Y}_L|$
9:         **return** MateofX$(\hat{x}, R, \Delta)$
10:     **else**
11:         **if** $\hat{x}.s > \bar{Y}_L.max$ or $\hat{x} \in T_L$ **then**
12:             **return** MateofX$(\hat{x}, R, \delta)$
13:         **else**
14:             **return** MateofX$(\hat{x}, L, \Delta)$

Initially, MateofX is invoked with P equal to the root node and $\Delta = 0$. If P is a leaf, any $x' \in \bar{Z}_P$ such that $x' < \hat{x}$ is matched before $\hat{x}$ by Lemma 4-1, and other vertices in $\bar{X}_P$ is matched after $\hat{x}$ by Lemma 4-2. Suppose P is an internal node. Let $\Omega' = \{x \in \bar{Z}_L : x \le \hat{x}\}$ and $\sigma = |\Omega'|$. If $\Delta + \sigma > |\bar{Y}_L|$, assume $\hat{y} \in \bar{Y}_L$. Then there must be some $x' \in \Omega'$ that matches with a $y' \in \bar{Y}_R$. We have $x'.s > \hat{y}$ since if $x'.s \le \hat{y}$, it contradicts with Lemma 4-1. Similarly, for each x'' matches with a y where $\hat{y} < y < \hat{Y}_P.mid$, if $x''.s \le \hat{y}$, then $x'' > \hat{x}$ and there must be another $x' \in \Omega'$ such that $x'.s > \hat{y}$ and x' matches with a $y' \in \bar{Y}_R$. Then, letting $\hat{y} = \bar{Y}_L[i]$, we must have at least $|\bar{Y}_L| - i + 1$ number of $x' \in \bar{Z}_L$ such that $x'.s > \hat{y} = \bar{Y}_L[i]$, which is a contradiction. Thus, if $\Delta + \sigma > |\bar{Y}_L|$, $\hat{y} \in \bar{Y}_R$. The analysis for other situations is similar. By standard techniques, Δ, σ and other subroutines can be computed in worst-case logarithmic time. In one query for an x vertex, MateofX is invoked recursively $O(\log |Y|)$ times. Then $pair(x)$ query operation runs in $O(\log^2 |V|)$ worst-case time. Based on the $pair(x)$ operation, the $pair(y)$ query operation can run in $O(\log^3 |V|)$ worst-case time using the binary search technique. Our results are summarized in the following theorem.

Theorem 1. *The dynamic algorithm maintains a maximum weight matching in a weighted convex bipartite graph in $O(\log^2 |V|)$ amortized time, obtains the matching status of a vertex in constant worst-case time, finds the mate of a matched x vertex in $O(\log^2 |V|)$ worst-case time, and finds the mate of a matched y vertex in $O(\log^3 |V|)$ worst-case time.*

Acknowledgement. The authors would like to thank reviewers for invaluable comments which help to improve the presentation of this paper. This work was supported by NSF of China (Grant No. 61472279).

References

1. Brodal, G.S., Georgiadis, L., Hansen, K.A., Katriel, I.: Dynamic matchings in convex bipartite graphs. In: Kučera, L., Kučera, A. (eds.) MFCS 2007. LNCS, vol. 4708, pp. 406–417. Springer, Heidelberg (2007)
2. Dekel, E., Sahni, S.: A parallel matching algorithm for convex bipartite graphs and applications to scheduling. J. Parallel Distrib. Comput. 1(2), 185–205 (1984)
3. van Emde Boas, P.: Preserving order in a forest in less than logarithmic time and linear space. Inf. Process. Lett. 6(3), 80–82 (1977)
4. Gale, D.: Optimal assignments in an ordered set: an application of matroid theory. J. Comb. Theory 4(2), 176–180 (1968)
5. Glover, F.: Maximum matching in a convex bipartite graph. Naval Res. Logistics Q. 14(3), 313–316 (1967)
6. Katriel, I.: Matchings in node-weighted convex bipartite graphs. INFORMS J. Comput. 20(2), 205–211 (2008)
7. Plaxton, C.G.: Vertex-weighted matching in two-directional orthogonal ray graphs. In: Cai, L., Cheng, S.-W., Lam, T.-W. (eds.) Algorithms and Computation. LNCS, vol. 8283, pp. 524–534. Springer, Heidelberg (2013)
8. Spencer, T.H., Mayr, E.W.: Node weighted matching. In: Paredaens, J. (ed.) ICALP 1984. LNCS, vol. 172, pp. 454–464. Springer, Heidelberg (1984)
9. Steiner, G., Yeomans, J.S.: A linear time algorithm for maximum matchings in convex, bipartite graphs. Comput. Math. Appl. 31(12), 91–96 (1996)
10. Zu, Q., Zhang, M., Yu, B.: Dynamic matchings in left weighted convex bipartite graphs. In: Chen, J., Hopcroft, J.E., Wang, J. (eds.) FAW 2014. LNCS, vol. 8497, pp. 330–342. Springer, Heidelberg (2014)

Author Index

Printed in the United States
By Bookmasters

Printed in the United States
By Bookmasters